Cellular Diagnostics

The translation of this book is made possible by a grant from 'Wallace H. Coulter Foundation'.

Cellular Diagnostics

Basic Principles, Methods and Clinical Applications of Flow Cytometry

Editors

Ulrich Sack Leipzig
Attila Tárnok Leipzig
Gregor Rothe Bremen

184 figures, 29 in color, 117 tables, 2009

Basel · Freiburg · Paris · London · New York ·Bangalore · Bangkok · Shanghai · Singapore · Tokyo · Sydney

Prof. Dr. Ulrich Sack
Institute of Clinical Immunology and Transfusion Medicine
Medical Faculty, University of Leipzig
Johannisallee 30, 04103 Leipzig (Germany)
ulrich.sack@medizin.uni-leipzig.de

Prof. Dr. Attila Tárnok
Department of Pediatric Cardiology
Cardiac Center Leipzig GmbH, University of Leipzig
Strümpellstraße 39, 04289 Leipzig (Germany)
tarnok@medizin.uni-leipzig.de

Prof. Dr. Gregor Rothe
Laborzentrum Bremen
LADR Group
Friedrich-Karl-Straße 22
28205 Bremen (Germany)
gregor.rothe@laborzentrum-bremen.de

Cover Illustrations
Confocal laser scanning microscopy of activated mitochondria in HeLa cells that have been colored by mitotracker DeepRed (Molecular Probes). The figure was kindly provided by Marten Jakob, Kiel

Flow-cytometric analysis of a leukemic bone marrow sample with pseudo-coloring of the cells according to light scatter. The figure was kindly provided by Gregor Rothe, Bremen

Library of Congress Cataloging-in-Publication Data

Cellular Diagnostics. Basic Principles, Methods and Clinical Applications of Flow Cytometry / editors, Ulrich Sack, Attila Tárnok, Gregor Rothe.
XXX + 738 p.; 19.0 × 25.5 cm.
Includes bibliographical references and index.
ISBN 978-3-8055-8555-2 (hard cover: alk.paper)

www.karger.com
Printed in Germany on acid-free paper by Druckhaus "Thomas Müntzer", Bad Langensalza

ISBN 978-3-8055-8555-2

Contents

Background and Methodological Principles

Characterization and Phenotyping of Cells

Analysis of Cell Functions

Diagnostic Indications

Authors

Dr. Stefan Barlage
Laboratoriumsmedizin MVZ Leverkusen
51375 Leverkusen, Germany
barlage@labor-leverkusen.de

Dr. Nina Baudendistel
Research Center for Molecular Medicine
Medical and Health Science Center
University of Debrecen
Egyetem square 1, 4032 Debrecen, Hungary

Dr. József Bocsi
Department of Pediatric Cardiology,
Cardiac Center GmbH, University of Leipzig
Strümpellstraße 39, 04289 Leipzig, Germany
jobocsi@web.de

Dr. Andrea Bodnár
Cell Biology and Signaling Research Group
of the Hungarian Academy of Sciences
Medical and Health Science Center
University of Debrecen
Egyetem square 1, 4032 Debrecen, Hungary
bodnar@dote.hu

PD Dr. Gero Brockhoff
Clinic for Gynecology and Obstetrics
Caritas Hospital St. Josef
University of Regensburg
Franz-Josef-Strauss-Allee 11, 93053 Regensburg, Germany
gero.brockhoff@klinik.uni-regensburg.de

Prof. Dr. Dirk H. Busch
Institute for Microbiology, Immunology and Hygiene
Technical University Munich
Trogerstraße 30, 81675 Munich, Germany
dirk.busch@lrz.tum.de

Dr. Michael Cross
Department of Hematology/Oncology
IZKF – University of Leipzig
Inselstraße 22, 04103 Leipzig, Germany
crossm@medizin.uni-leipzig.de

Prof. Dr. Sándor Damjanovich
Cell Biology and Signaling Research Group
of the Hungarian Academy of Sciences
Medical and Health Science Center
University of Debrecen
Egyetem square 1, 4032 Debrecen, Hungary
dami@dote.hu

Prof. Dr. Zbigniew Darzynkiewicz
Brander Cancer Research Institute
New York Medical College
BSB, Room 438
Valhalla, NY 10595, USA
z_darzynkiewicz@nymc.edu

Viola Döbel
Department of Hematology/Oncology
IZKF – University of Leipzig
Inselstraße 22, 04103 Leipzig, Germany
doev@gmx.net

Dr. Alexandra Dorn-Beineke
Institute of Clinical Chemistry
Faculty for Clinical Medicine of the University of Heidelberg
Theodor-Kutzer-Ufer 1–3, 68167 Mannheim, Germany
alexandra.dorn beineke@ikc.ma.uni-heidelberg.de

Dr. Johannes Fischer
Institute for Transplantation Diagnostics and Cell Therapeutics
Heinrich-Heine University Hospital
Moorenstraße 5, 40225 Düsseldorf
Germany
jfischer@uni-duesseldorf.de

Prof. Dr. Stefan Frühauf
Zentrum für Tumordiagnostik und -therapie
Paracelsus-Klinik
Am Natruper Holz 69, 49076 Osnabrück, Germany
prof.stefan.fruehauf@pk-mx.de

Fee Gerling
Institute of Clinical Immunology and Transfusion Medicine
Medical Faculty, University of Leipzig
Johannisallee 30, 04103 Leipzig, Germany
feegerling@web.de

PD Dr. Andreas O.H. Gerstner
Department of Otorhinolaryngology/Head and Neck Surgery
University of Bonn
Sigmund-Freud-Straße 25, 53105 Bonn, Germany
andreas.gerstner@ukb.uni-bonn.de

PD Dr. Rudolf Gruber
synlab Weiden
Medizinisches Versorgungszentrum für Laboratoriumsmedizin und Mikrobiologie
Zur Kesselschmiede 4, 92637 Weiden, Germany
rudolf.gruber@synlab.de

PD Dr. Martin Grünewald
Medizinische Klinik I, Klinikum Heidenheim
Schloßhaustraße 100, 89522 Heidenheim, Germany
martin.gruenewald@kliniken-heidenheim.de

Dr. Conny Höflich
Institut für Medizinische Immunologie
Charite – Universitätsmedizin Berlin, Campus Mitte
Chariteplatz 1, 10117 Berlin, Germany
conny.hoeflich@charite.de

PD Dr. Roland Jacobs
Clinic for Immunology and Rheumatology
Carl-Neuberg-Straße 1, 30625 Hannover, Germany
jacobs.roland@mh-hannover.de,

Prof. Dr. Dieter Kabelitz
Institute of Immunology
Universitätsklinikum Schleswig-Holstein
Campus Kiel
Michaelisstraße 5, 24105 Kiel, Germany
kabelitz@immunologie.uni-kiel.de

Dr. Leonid Karawajew
HELIOS-Klinikum Berlin-Buch
Charité – Universitätsmedizin Berlin, Campus Berlin-Buch
Schwanebecker Chaussee 50, 13125 Berlin, Germany
leonid.karawajew@helios-kliniken.de

Prof. Dr. Florian Kern
BSMS, University of Sussex, Falmer
Brighton BN1 9PX, UK
F.Kern@bsms.ac.uk

Cornelia Keup
Institute of Clinical Chemistry
Faculty for Clinical Medicine of the University of Heidelberg
Theodor-Kutzer-Ufer 1–3, 68167 Mannheim, Germany
cornelia.keup@ikc.ma.uni-heidelberg.de

Dr. Hans-Dieter Kleine
Ansomed AG
Schillingallee 68, 18057 Rostock, Germany
hans-dieter.kleine@gst-rostock.de

Prof. Dr. Michael Köhler
Department of Transfusion Medicine, University Medical Center
Georg August University, Göttingen
37099 Göttingen, Germany
mkoehler@med.uni-goettingen.de

Prof. Dr. Tobias J. Legler
Department of Transfusion Medicine, University Medical Center
Georg August University, Göttingen
37099 Göttingen, Germany
tlegler@med.uni-goettingen.de

Dr. Wiebke Laffers
Department of Otorhinolaryngology/Head and Neck Surgery
University of Bonn
Sigmund-Freud-Straße 25, 53105 Bonn, Germany
wiebke.laffers@ukb.uni-bonn.de

Dr. Irina Lehmann
Department Umweltimmunologie
UFZ-Umweltforschungszentrum Leipzig-Halle GmbH
Permoserstraße 15, 04318 Leipzig, Germany
irina.lehmann@ufz.de

Dr. Jörg Lehmann
Fraunhofer-Institut für Zelltherapie und Immunologie IZI
Perlickstraße 1, 04103 Leipzig, Germany
joerg.lehmann@izi.fraunhofer.de

Dr. Dominik Lenz
Department of Pediatric Cardiology,
Cardiac Center GmbH, University of Leipzig
Strümpellstraße 39, 04289 Leipzig, Germany
d.lenz@web.de

Dr. Detlef Loppow
Laboratory Dr. Kramer & Collegues
LADR Group
Lauenburger Straße 67, 21502 Geesthacht, Germany
loppow@ladr.de

Prof. Dr. Wolf-Dieter Ludwig
HELIOS-Klinikum Berlin-Buch
Charité – Universitätsmedizin Berlin, Campus Berlin-Buch
Schwanebecker Chaussee 50, 13125 Berlin, Germany
wolf-dieter.ludwig@charite.de

PD Dr. Andreas Lun
Institut für Laboratoriumsmedizin und Pathobiochemie, Charité – Universitätsmedizin Berlin, Campus Virchow-Klinikum
Augustenburger Platz 1, 13353 Berlin, Germany
andreas.lun@charite.de;

Dr. Rainer Lynen
Department of Transfusion Medicine,
University Medical Center
Georg August University, Göttingen
37099 Göttingen, Germany
rlynen@med.uni-goettingen.de

Dr. Francis F. Mandy
International Centre for Infectious Diseases
Winnipeg, Manitoba, Canada
fmandy@rogers.com

Dr. Richard Mauerer
synlab Weiden
Medizinisches Versorgungszentrum für Laboratoriumsmedizin und Mikrobiologie
Zur Kesselschmiede 4, 92637 Weiden, Germany
rudolf.gruber@synlab.de

Dr. Christian Meisel
Institut für Medizinische Immunologie
Charite – Universitätsmedizin Berlin, Campus Mitte
Chariteplatz 1, 10117 Berlin, Germany
chr.meisel@charite.de

Anja Mittag
Department of Pediatric Cardiology,
Cardiac Center GmbH, University of Leipzig
Strümpellstraße 39, 04289 Leipzig, Germany
a.mittag@gmx.de

Dr. Béla Molnar
2nd Department of Medicine
Semmelweis University
Szentkiralyi u 46, 1085 Budapest VIII, Hungary
mb@bel2.sote.hu

Dr. Thomas Nebe
Institute of Clinical Chemistry
Faculty for Clinical Medicine of the University of Heidelberg
Theodor-Kutzer-Ufer 1–3, 68167 Mannheim, Germany
thomas.nebe@ikc.ma.uni-heidelberg.de

Dr. Richard Ratei
HELIOS-Klinikum Berlin-Buch
Charité – Universitätsmedizin Berlin, Campus Berlin-Buch
Schwanebecker Chaussee 50, 13125 Berlin, Germany
richard.ratei@helios-kliniken.de

Prof. Dr. J. Paul Robinson
Purdue University Cytometry Laboratories
Bindley Bioscience Center
1203 West State Street
Discovery Park, Purdue University
West Lafayette, IN 47907-2057, USA
jpr@flowcyt.cyto.purdue.edu

PD Dr. Joachim Roesler
Klinik und Poliklinik für Kinderheilkunde
Universitätsklinikum Carl Gustav Carus, Technische Universität Dresden
Fetscherstraße 74, 01307 Dresden, Germany
roeslerj@Rcs1.urz.tu-dresden.de

Prof. Dr. Gregor Rothe
Laborzentrum Bremen
LADR Group
Friedrich-Karl-Straße 22, 28205 Bremen, Germany
gregor.rothe@laborzentrum-bremen.de

PD Dr. Andreas Ruf
Zentrum für Labormedizin, Mikrobiologie und Transfusionsmedizin
Städtisches Klinikum Karlsruhe gGmbH
Moltkestraße 90, 76133 Karlsruhe, Germany
Andreas.Ruf@klinikum-karlsruhe.com

Prof. Dr. Ulrich Sack
Institute of Clinical Immunology and Transfusion Medicine
Medical Faculty, University of Leipzig
Johannisallee 30, 04103 Leipzig, Germany
Ulrich.Sack@medizin.uni-leipzig.de

Dr. Richard Schabath
HELIOS-Klinikum Berlin-Buch
Charité – Universitätsmedizin Berlin, Campus Berlin-Buch
Schwanebecker Chaussee 50, 13125 Berlin, Germany
richard.schabath@helios-kliniken.de

Dr. Alexander Scheffold
Miltenyi Biotec GmbH
Friedrich-Ebert-Straße 68, 51429 Bergisch Gladbach, Germany
alexandersc@miltenyibiotec.de

Dr. Matthias Schiemann
Institute for Microbiology, Immunology and Hygiene
Technical University Munich
Trogerstraße 30, 81675 Munich, Germany
matthias.schiemann@lrz.tum.de

Dr. Michael Schlesier
Division of Rheumatology and Clinical Immunology
University Clinic Freiburg
Hugstetter Straße 55, 79106 Freiburg, Germany
michael.schlesier@uniklinik-freiburg.de

Prof. Dr. Reinhold E. Schmidt
Clinic for Immunology and Rheumatology
Carl-Neuberg-Straße 1, 30625 Hannover, Germany
Schmidt.Reinhold.Ernst@mh-hannover.de

Dr. Ilka Schulze
Zentrum für Kinderheilkunde und Jugendmedizin
Universitätsklinikum Freiburg
Mathildenstraße 1, 79106 Freiburg, Germany
ilka.schulze@uniklinik-freiburg.de

Prof. Dr. Peter Sedlmayr
Center for Molecular Medicine
Institute for Cell Biology, Histology and Embryology
Medical University of Graz
Harrachgasse 21, 8010 Graz, Austria
peter.sedlmayr@meduni-graz.at

PD Dr. Rüdiger V. Sorg
Institute for Transplantation Diagnostics and Cell Therapeutics
Heinrich-Heine University Hospital
Moorenstraße 5, 40225 Düsseldorf, Germany
rsorg@itz.uni-duesseldorf.de

Prof. Dr. Janos Szöllősi
Department of Biophysics and Cell Biology
Medical and Health Science Center
University of Debrecen
Egyetem square 1, Life Sciences Building, 4032 Debrecen, Hungary
szollo@dote.hu

Prof. Dr. Attila Tárnok
Department of Pediatric Cardiology, Cardiac Center GmbH, University of Leipzig
Strümpellstraße 39, 04289 Leipzig, Germany
tarnok@medizin.uni-leipzig.de

Dr. Kathalin Tóth
Research Centre for Molecular Medicine
Medical and Health Science Center
University of Debrecen
Egyetem square 1, Life Sciences Building, 4032 Debrecen, Hungary

Prof. Dr. Guenter K. Valet
Steinseestraße 22, 81671 München, Germany
valet@classimed.de

Dr. György Vámosi
Cell Biology and Signaling Research Group of the Hungarian Academy of Sciences
Medical and Health Science Center
University of Debrecen
Egyetem square 1, 4032 Debrecen, Hungary
vamosig@dote.hu

Dr. György Vereb
Department of Biophysics and Cell Biology
Medical and Health Science Center
University of Debrecen
Egyetem square 1, 4032 Debrecen
Hungary
vereb@dote.hu

Prof. Dr. Hans-Dieter Volk
Institut für Medizinische Immunologie
Charite – Universitätsmedizin Berlin,
Campus Mitte
Chariteplatz 1, 10117 Berlin, Germany
hans-dieter.volk@charite.de

Prof. Dr. Volker Wahn
Klinik für Kinder und Jugendliche
Klinikum Uckermark
Auguststraße 23, 16303 Schwedt/Oder, Germany
v.wahn@klinikum-uckermark.de

PD Dr. Klaus Warnatz
Division of Rheumatology
and Clinical Immunology
University Clinic Freiburg
Hugstetter Straße 55, 79106 Freiburg, Germany
klaus.warnatz@uniklinik-freiburg.de

Dr. Mohammed Wattad
Abteilung Hämatologie/Onkologie
Evangelisches Krankenhaus Essen-Werden
gGmbH
Pattbergstraße 1–3, 45239 Essen, Germany
m.wattad.kmt@kliniken-essen-sued.de

Dr. Daniela Wesch
Institute of Immunology
Universitätsklinikum Schleswig-Holstein,
Campus Kiel
Michaelisstraße 5, 24105 Kiel, Germany
wesch@immunologie.uni-kiel.de

Dr. Michael Wötzel
Labor Dr. Reising-Ackermann & Partner
Strümpellstraße 40, 04289 Leipzig, Germany
m.woetzel@labor-leipzig.de

Abbreviations

AA	aplastic anemia
ABC	antibody-binding capacity
ABC transporters	ATP-binding cassette transporter
ACD	acid-citrate-dextrose
acLDL	acetylated low-density lipoprotein
AD	actinomycin D
ADA	adenosine deaminase
ADB	diacetoxdicyanobenzene
ADCC	antibody-dependent cellular cytotoxicity
7-ADD	7-aminoactinomycin
AID	activation-induced cytidine deaminase
AIDS	acquired immunodeficiency syndrome
AITP	autoimmune thrombocytopenia
ALCAM	activated leukocyte cell adhesion molecule
ALG	antilymphocyte globulin
ALL	acute lymphoblastic leukemia
AM	acetoxymethylester
AMCA-X	7-amino-4-methylcoumarin-3-acetic acid
AML	acute myelogenous leukemia
APC	allophycocyanine
APC	antigen-presenting cell
APL	acute promyelocytic leukemia
ART	antiretroviral therapy
AT	ataxia telangiectatica
ATG	antithymocyte globulin
BAFF-R	B-cell-activating factor receptor
BAL	bronchoalveolar lavage
BALF	bronchoalveolar lavage fluid
B-CLL	B-cell chronic lymphatic leukemia
BCRP	breast cancer resistance protein
BDV	borna disease virus
BFA	brefeldin A
BLNK	B-cell linker protein
BLS	bare lymphocyte syndrome
B-NHL	B-cell non-Hodgkin's lymphoma
BrdU	bromodeoxyuridine
BRET	bioluminescence resonance energy transfer

BSA	bovine serum albumin
BSA	bovine serum albumin
BSO	DL-buthionine-(S,R)-sulfoximine
BTK	Bruton's tyrosine kinase
BUDR	bromodeoxyuridine
$[Ca^{2+}{}_i]$	cytosolic free calcium concentration
CAC	circulating angiogenic cell
CALLA antigen	common acute lymphoblastic leukemia antigen
CARS	compensatory anti-inflammatory response syndrome
CCD	charge-coupled device
CCL	chemokine ligand
CCR	chemokine receptor
CD	cluster of differentiation
CEA	Commissariat à l'Energie Atomique
CFDA-SE	carboxyfluorescein diacetate succinimidyl ester
CFP	cyan fluorescent protein
CFSE	carboxyfluorescein succinimidyl ester
CFU-EC	endothelial cell colony-forming unit
CGD	chronic granulomatous disease
CID	combined immunodeficiency
CLL	chronic lymphatic leukaemia
CNS	central nervous system
CPD	citrate-phosphate-dextrose
CRTH2	chemoattractant receptor-homologous molecule expressed on Th2 cells
CsA	cyclosporine A
CSF	cerebrospinal fluid
CTL	cytotoxic T-lymphocytes
CV	coefficient of variation
CVID	common variable immunodeficiency
Cy	cyanine
DAF	decay-accelerating factor
DAG	diacylglycerol
DANS	1-dimethyl-amino-naphthalene-5-sulfonic acid
DAPI	4′,6-diamino-2-phenylindole
DC	desoxcytidine
DC	dendritic cell
DFP	diisopropyl fluorophosphates
DHR test	dihydrorhodamine 123 test
DIG	digoxigenin
DKFZ	Deutsches Krebsforschungszentrum
DMSO	dimethylsulfoxide
DNA	deoxyribonucleic acid
DNP	dinitrophenol
DPT	double-platform technology
DTCs	disseminated tumor cells
dUTP	deoxyuridine triphosphate
Dy dyes	Dyomics dyes
E/T ratio	effector to target ratio
EBV	Epstein-Barr virus
ECFC	endothelial colony-forming cell
EDTA	ethylenediamine tetraacetic acid

EdU	5-ethynyl-2′-deoxyuridine
EGF	epidermal growth factor
EGIL Classification	European Group for the Immunological Characterization of Leukemias Classification
ELISA	enzyme-linked immunosorbent assay
EMA	ethidium monoazide
EMEA	European Medicines Agency
emFRET	energy migration fluorescence resonance energy transfer
EMSA	electrophoretic mobility shift assay
EPC	endothelial progenitor cell
ER	endoplasmic reticulum
ESACP	European Society for Analytical Cellular Pathology
ESCCA	European Society for Clinical Cell Analysis
EWGCCA	European Working Group for Clinical Cell Analysis
F/P ratio	fluorochrome-to-protein ratio
FAB Classification	French-American-British Classification
FACS	fluorescence-activated cell sorting
FCI	flow-cytometric immunophenotyping
FCM	flow cytometry
FCS	fetal calf serum
FDA	Federal Drug Administration
FISH	fluorescence in-situ hybridization
FITC	Fluorescein isothiocyanate
fMLP	N-formyl-Met-Leu-Phe
FMO	fluorochrome minus one
FOXP3	forkhead box transcription factor
FP	fluorescent protein
FRET	florescence resonance energy transfer
FSC	forward scatter
G6DPH	glucose-6-phosphate dehydrogenase
GALT	gut-associated lymphatic tissue
GCP	good clinical practice
G-CSF	granulocyte colony-stimulating factor
GEF	guanine exchange factor
GEN	genisteine
GFP	green fluorescent protein
GLP	good laboratory practice
GM-CSF	granulocyte-macrophage colony-stimulating factor
GMP	good manufacturing practice
GPI	glycosylphosphatidylinositol
GSF	Gesellschaft für Strahlenforschung
GvHD	graft-versus-host disease
H-CAM	human cell adhesion molecule
HCMV	human cytomegalovirus
HEPES	4-(2-hydroxyethyl)-1-piperazineethaesulfonic acid
HG	heregulin
HIV	human immunodeficiency virus
HSA	human serum albumin
ICAM-1	intercellular adhesion molecule 1
ICOS	inducible costimulator
IFN-γ	interferon- γ

Ig	immunoglobulin
IL	interleukin
IL-1RA	IL-1 receptor antagonist
IP_3	inositol-3-phosphate
IPEX syndrome	immune dysregulation, polyendocrinopathy, enteropathy X-linked syndrome
ISAC	International Society for Analytical Cytology
ISCO	International Society of Cellular Oncology
ISCT	International Society for Cellular Therapy
ISDQP	International Society of Diagnostic Quantitative Pathology
ISHAGE	International Society of Hematotherapy and Graft Engineering
ISI	Institute for Scientific Information
iSP	induced sputum
ITAM	immunoreceptor tyrosin-based activating motif
ITIM	immunoreceptor tyrosin-based inhibitory motif
JAK3	Janus kinase 3
KIR	killer cell immunoglobulin receptor
KLR	killer cell lectin-like receptor
LAD	leukocyte adhesion deficiency
LAIP	leukemia-associated immunophenotypes
LAMP-1	lysosomal-associated protein 1
LBP	lipopolysaccharide-binding protein
LED	light-emitting diode
LGL	large granular lymphocyte
LPS	lipopolysaccharide
LRP	lung resistance-related protein
LSC	laser scanning cytometry
LT	leukotriene
mAb	monoclonal antibody
MALDI-TOF-MS	matrix-assisted laser desorption ionization time-of-flight mass spectrometry
MALT	mucosa- associated lymphatic tissue
MAPC	multipotent adult progenitor cell
MARS	mixed antagonistic response syndrome
MBL	mannan-binding lectin
mCherry	mCherry fluorescent protein
MDC	myeloid dendritic cell
MDR	multidrug resistance
MELC	multiepitope ligand cartography
MESF	molecules of equivalent soluble fluorochrome
MFI	mean fluorescence intensity
MHC	major histocompatibility complex
MIC Classification	Morphologic, Immunologic and Cytogenetic Working Classification
MIC	class-I-chain-related protein
MIP	macrophage inflammatory protein
MIRL	membrane inhibitor of reactive lysis
MM	multiple myeloma
MNC	mononuclear cell
MP	transmembrane potential
MPO	myeloperoxidase
MRD	minimal residual disease
mRFP	monomeric red fluorescent protein
MSC	multipotent mesenchymal stem cell

MTT	3-(4,5-dimethylthiazol-2-yl)2,5-diphenyltetrazolium bromide
MVP	human major vault protein
MW	molecular weight
NAIT	neonatal alloimmune thromboctopenia
NBS	Nijmegen breakage syndrome
NBT	nitroblue tetrazolium
N-CAM	neural cell adhesion molecule
NCCLS	National Committee for Clinical Laboratory Standards
NCR	natural cytotoxicity receptor
NEM	N-ethylmaleinimide
NF-AT	nuclear factor of activated T-cells
NF-κB	nuclear factor-κB
NHL	non-Hodgkin's lymphoma
NHS ester	N-hydroxysuccinimide ester
NK	natural killer cell
NKT cell	natural killer T-cell
NO	nitric oxide
NP	nitrophenyl
PAMP	pathogen-associated molecular pattern
PAS	periodic acid-Schiff
PB	peripheral blood
PBMC	peripheral blood mononuclear cell
PBS	phosphate-buffered saline
PCR	polymerase chain reaction
PCT	procalcitonin
PDB	phorbol-12,13-dibutyrate
pDC	plasmacytoid cell
PE	phycoerythrin
PECAM-1	platelet-endothelial cell adhesion molecule 1
PE-Cy5	phycoerythrin-cyanine 5
PerCP	peridinin chlorophyll protein
PETR	phycoerythrin-TexasRed
PFA	paraformaldehyde
Pgp	P-glycoprotein
pH_i	intracellular pH
PI	propidium iodide
PID	primary immunodeficiency disease
PMA	phorbol myristate acetate
PMN	polymorphonuclear neutrophil
PMT	photomultiplier tube
PNH	paroxysmal nocturnal hemoglobinuria
PNP	purine nucleoside phosphorylase
PR	peak ratio
Pre-B ALL	mature precursor B-cell acute lymphoblastic leukemia
PRR	pattern recognition receptor
PSA	prostate-specific antigen
PTP	posttransfusion purpura
REAL Classification	Revised European-American Lymphoma Classification
rFLIM	anisotropy fluorescence lifetime imaging microscopy
RFP	red fluorescent protein
RNA	ribonucleic acid

ROS	reactive oxygen species
RP	resolution parameter
RS-SCID	radiation-sensitive severe combined immunodeficiency
RT-PCR	reverse transcription polymerase chain reaction
S/N	signal-to-noise ratio
SA	streptavidin
SAC	Society for Analytical Cytology
SAP	lymphocyte activation molecule-associated protein
SBC	slide-based cytometry
SCID	severe combined immunodeficiency
SD	standard deviation
SFM	scanning fluorescence microscopy
SGD	neutrophil-specific granule deficiency
sIL-2R	soluble IL-2 receptor
SIRS	systemic inflammatory response syndrome
SNARF-1	seminaphtorhodafluor-1
SNP	single nucleotide polymorphism
SOP	standard operation procedure
SPF	S phase fraction
SPT	single-platform technology
SSC	side scatter
STAT	signal transducer and activator of transcription
TACI	transmembrane activator and CAML interactor
T-ALL	T-cell acute lymphoblastic leukaemia
TAP	transporter associated with antigen processing
TCR	T-cell receptor
Th cell	T-helper cell
Tie2/TEK	angiopoietin-1 receptor precursor or tunica intima endothelial cell kinase
TIRFM	total internal reflection fluorescence microscopy
TMD	transmembrane domain
TNF-α	tumor necrosis factor-α
TRAIL	TNF-related apoptosis-inducing ligand
T^{reg} cell	natural T-regulatory cell
UEA-1	Ulex europeus agglutinin 1
ULBP	UL-16-binding protein
UNG	uracil-DNA glycosylase
USSC	unrestricted somatic stem cells
UTP	uridine triphosphate
UV	ultraviolet
VE-cadherin	vascular endothelial cadherin
VEGFR	vascular endothelial growth factor receptor
vWF	von Willebrand factor
WAS	Wiskott-Aldrich syndrome
WASP	Wiskott-Aldrich syndrome protein
WHO	World Health Organization
XLA	X-chromosomal agammaglobulinemia
XLP	X-linked lymphoproliferative syndrome
X-SCID	X-linked severe combined immunodeficiency
YFP	yellow fluorescent protein
ZAP-70	æ-associated protein 70

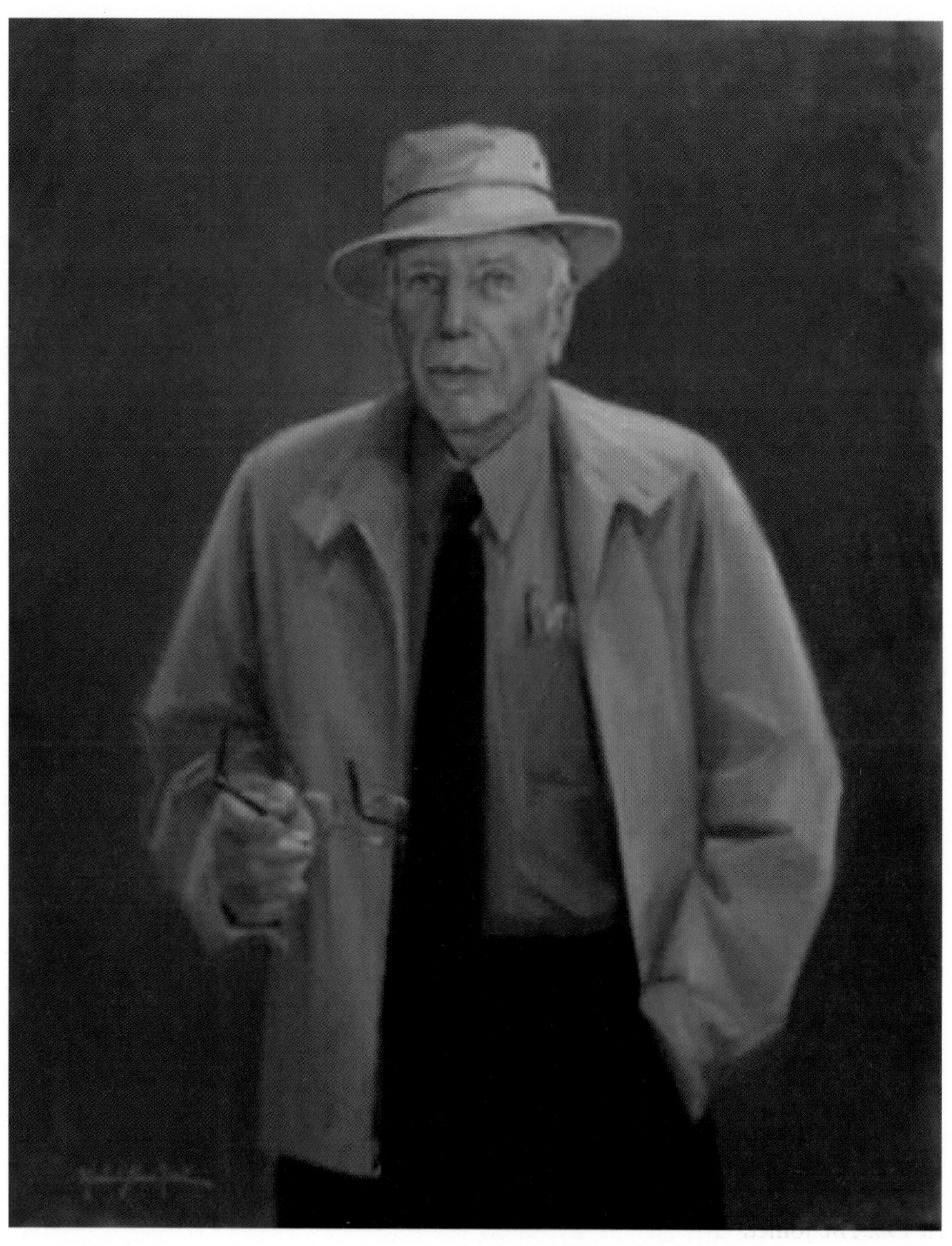

Michael Shane Neal, 2003, oil on canvas (40″ × 30″)

In tribute for his contributions to medicine, science and engineering

Wallace Henry Coulter – a Short Biography

Wallace Henry Coulter was an engineer, inventor, and entrepreneur. The two greatest passions of his life were applying engineering principles to scientific research and embracing the diversity of world cultures. The first passion led him to invent the Coulter Principle, the reference method for counting and sizing microscopic particles suspended in a fluid. This invention served as the cornerstone for automating the labor-intensive process of counting and testing blood cells. With this invention, Wallace Coulter helped create the diagnostics industry and founded the field of laboratory hematology. He was Chairman of Coulter® Corporation, and his early life experience inspired him to establish over twenty international subsidiaries. He recognized that it was imperative to employ locally based staff before this became a standard business practice for multinational companies.

Wallace, born in February, 1913, spent his early years in McGehee, AR, a small town near Little Rock, AR. Wallace had an inquisitive mind and was fascinated with numbers and gadgets. For his eleventh birthday, he asked for his first radio kit. He attended his first year of college at Westminster College in Fulton, MO. Following his interest in electronics, he transferred to the Georgia Institute of Technology for his second and third years of study. Due to the Great Depression of the 1930s, he was unable to complete his degree.

Wallace's Far East Adventures

In 1935, he joined General Electric X-Ray as a sales and service engineer in the Chicago area. This work familiarized Wallace with the testing procedures of hospital laboratories. When an opportunity to cover the Far East became available, he seized the chance to live and work abroad. The practice of employing expatriates by US companies was not commonplace before World War II. Wallace was based in three areas servicing the entire Far East; Manila, Shanghai and Singapore.

His first posting was Manila. Not only did he service the Philippines, but also traveled to the more remote regions of the territory like Borneo and Sumatra. It was here that he began his interest in exotic fruits. Manila was his first experience with the tropical climate, and he took an immediate liking to it. Many years later, it was a major factor in relocating his company to south Florida.

After Manila, Wallace was transferred to the Shanghai office where he was responsible for Hong Kong, Macao, and the major cities of China. This experience changed Wallace's life. He became fascinated with Chinese history, art, and culture. At that time, Shanghai was the center of business for Asia, and every global company had a presence there. This city, the Paris of the East, afforded him an education in international business.

Wallace was transferred to Singapore where he remained until the beginning of World War II, in late 1941. Wallace tried booking his departure on one of the many passenger ships leaving the country, but failed. With bombs falling on the city, he found a small cargo ship bound for India and left under cover of darkness in December 1941. After a few weeks in India, Wallace realized that returning to the USA through Europe was impossible. He chose a more circuitous route home, making his way through Africa and South America. It took him nearly 12 months, finally returning to the USA during Christmas 1942.

An Elegant Idea Becomes a Company

After the war, Wallace worked for several electronics companies, including Raytheon in Chicago, IL. He maintained a laboratory at home to work on promising ideas and projects. One such project was for the US Department of Naval Research. Wallace was trying to standardize the size of solid particles in the paint used on battleships in order to improve its adherence to the hull. He began tinkering in his laboratory in his spare time, experimenting with different applications of optics and electronics. Upon returning to the garage one cold, blustery evening, Wallace was faced with a challenge. The supply of paint for the experiment had frozen. Not wanting to go out in the cold, he asked himself, 'What substance has a viscosity similar to paint and is readily available?' Using his own blood, a needle and some cellophane, the principle of using electronic impedance to count and size microscopic particles suspended in a fluid was invented – the Coulter Principle.

He remembered observing hospital laboratory workers hunched over microscopes manually counting cells. Thus, this was the first application of his new invention. This instrument became known as the Coulter Counter (fig. 1) , the combination of engineering and biology.

Wallace's first attempts to patent his invention were turned away by more than one attorney who believed 'you cannot patent a hole'. Persistent as always, Wallace finally applied for his seminal patent in 1949 and it was issued in 1953. That same year, two prototypes were sent to the National Institutes of Health (NIH) for evalua-

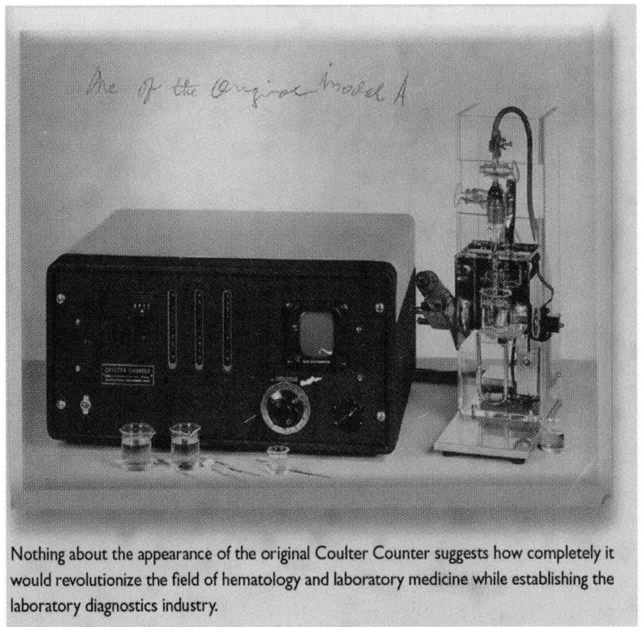

Nothing about the appearance of the original Coulter Counter suggests how completely it would revolutionize the field of hematology and laboratory medicine while establishing the laboratory diagnostics industry.

Fig. 1. One of the first Coulter Counters – Model A.

tion. Shortly after, the NIH published its findings in two key papers, citing improved accuracy and convenience of the Coulter method of counting blood cells. Wallace disclosed his invention in his one and only technical publication at the National Electronics Conference, 'High Speed Automatic Blood Cell Counter and Cell Size Analyzer'.

This simple device increased the sample size and therefore, the accuracy and precision, of the cell count from the manual method by counting in excess of 6,000 cells/s. It decreased the time it took to analyze from 30 min to less than 1 min and reduced the standard error by a factor of 10. The Coulter Principle is responsible for the current practice of laboratory hematology. The complete blood count (CBC) is one of the most commonly ordered diagnostic tests worldwide. Today, at least 90% of all CBCs are performed on instruments using or copying the Coulter Principle (fig. 2).

The Coulter Principle led to major breakthroughs in science, medicine, and industry. In fact, more than 50 years after its introduction, it still touches everyone's daily life in some manner from having a blood test, to painting your house, from drinking beer or a glass of wine to eating a bar of chocolate, swallowing a pill or applying cosmetics. It is critical to toners and ceramics as well as space exploration. The use of the Coulter Principle, or electronic sensing zone, modernized industry by establishing a method for quality control and standardization of particles used in each of these industries. The impact of the Coulter Principle to the medical, pharmaceutical, biotechnology, food, beverage and consumer industries is immeasurable!

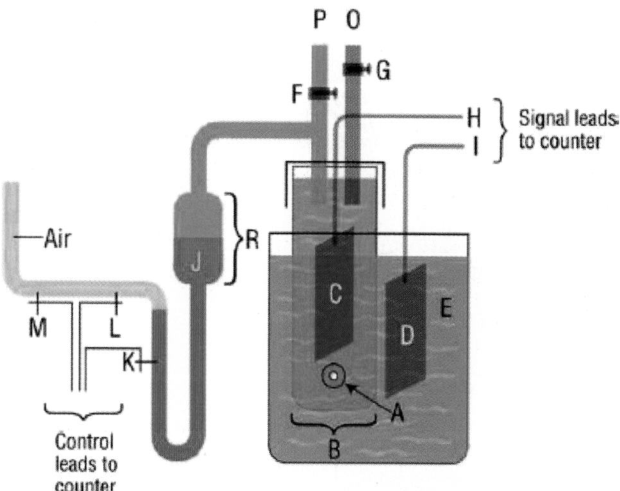

Figure 3. Functional schematic of the Coulter sample stand, co-invented by the Coulter brothers.[6] When stopcock **F** is opened, the mercury in the manometer reservoir **R** is drawn upward by a small vacuum pump connected to **P**, and the resulting pressure head causes movement of mercury column **J** after the stopcock is closed, drawing sample suspension **E** from the sample vessel through the hole in aperture wafer **A** into sample tube **B**. The aperture wafer and sample tube are made of dielectric materials having an electrical resistivity much greater than that of the suspending medium. Via connections **H** and **I**, electrodes **C** and **D** couple an electrical current through the aperture, and the resultant signal pulses to an amplifier and pulse counter (not shown). The volume of sample to be analyzed is determined by three control electrodes (**K**, **L**, **M**) penetrating the wall of the manometer tubing; when the flowing mercury causes electrical contact between **K** and **L**, the pulse counter starts, while mercury contact with **M** at a calibrated distance from **L** terminates it. Thus, cells are only counted in suspension flowing through the aperture at constant velocity, so permitting cell concentration to be determined as the count in the suspension volume equal to the volume of mercury between the electrodes **L** and **M**. The second stopcock **G** is only opened to fill or flush the sample tube with clean suspending media via **O**. The microscope for viewing the aperture is not shown. Adapted from Wallace's paper.

Fig. 2. The Coulter Principle – adapted with permission from Graham D: The Coulter Principal: Foundation of an Industry. Journal of the Association for Laboratory Automation (JALA) 2003;8(6):74, figure 3.

In 1958, Wallace and his brother, Joseph Coulter, Jr., founded Coulter® Electronics, Inc., to manufacture market and distribute Wallace's invention. Wallace and Joseph built the early models, loaded them in their cars, and personally sold each unit. Wallace went to Europe to market and sell the Coulter Counter; approaching the British Ministry of Health and other national health organizations. In 1959, to protect the patent rights in Europe, subsidiaries in the UK and France were established. Under his tenure as Chairman of the Corporation, the company developed into the industry leader in hematology, cytometry, and particle characterization.

Wallace continued to focus the resources of the company on advancing cellular analysis. Recognizing the potential of flow cytometry, Wallace invested in it in its earliest days, founding Particle Technologies, Inc. to commercialize the discoveries of the Los Alamos National Laboratory. In fact, in his book 'Practical Flow Cytometry, 3rd ed.' Howard Shapiro states: 'The simplest flow cytometers are the instruments which count cells based on the Coulter Principle'. Today, these technologies are used in the research and characterization of leukemia, cancer, stem cells, and infectious disease to name a few. Wallace also invested in the science of monoclonal antibodies. In fact, he envisioned their use as the 'magic bullet' to treat and cure cancer. The B-1 antibody (anti-CD20), marketed as Bexxar, was developed under his guidance, for treatment for non-Hodgkin's small cell lymphoma.

On the Personal Side

Wallace Coulter was a very private person who sought no public acclaim, yet his accomplishments are numerous. He received 82 patents, many of which were issued to him for discoveries made late in life. In 1960, Wallace Coulter was awarded the prestigious John Scott Award for Scientific Achievement. This award, established in 1816 for 'ingenious men and women', is given to inventors whose innovations have had a revolutionary effect on mankind. He joined Thomas Edison, Marie Curie, Jonas Salk and Guglielmo Marconi in receiving this award. He continued to receive many awards from business, industry, and academia. Although he was not a physician or hematologist, Wallace is the only person to receive the American Society of Hematology Distinguished Service Award for his enormous contribution to the field of hematology. In 1998, he was inducted into the National Academy of Engineering. In 2004, Wallace was posthumously inducted into the National Inventor's Hall of Fame.

Wallace was a humble and compassionate man who always encouraged his employees to dream and take risks. Wallace remained single; so, his company and its employees became his extended family. He never received a dividend rather investing his personal wealth into the company's research and development. He personally helped many employees, providing home and college loans and sponsorships. As an example, he funded an employee's bone marrow transplant while this technique was still very experimental and not covered by the company's health insurance. Upon the sale of Coulter Corporation in 1997, he ensured that his family of employees was 'taken care of' by setting aside a total fund of USD 100 million to be paid to each and every employee around the world based on their years of service.

Wallace Coulter passed away August 7, 1998. As a pioneer of the diagnostic industry he leaves behind a legacy of his achievements, including critical advancements in diagnosis and treatment of disease, a dynamic corporation that will continue to innovate in health care, as well as colleagues, associates, friends and family who were inspired by his influence. His fame and accomplishments continue to be recognized. He left his entire estate to form the foundation bearing his name, the Wallace H. Coulter Foundation. Its mission is dedicated to continuing his lifelong passions and pursuits of 'Science Serving Humanity.'

Wallace H. Coulter Foundation

Foreword

Progress in technology and development of new analytical methods provide the driving force for discovery in various branches of biology and medicine. During the past three decades we have witnessed a spectacular development of instrumentation and expansion of methods that utilize flow and laser-scanning cytometry. Application of these methods was particularly rewarding in clinical settings, where they are currently used as routine diagnosis and prognosis assays in numerous diseases, often providing invaluable information to the clinician and being life-saving to the patient.

Cellular Diagnostics, edited by Ulrich Sack, Attila Tárnok and Gregor Rothe, presents a very comprehensive review of the most useful cytometric methods that found clinical applications. The monograph consists of 37 chapters, many written by the renowned authors who contributed towards the development and application of the described method in clinical medicine. The first several chapters are introductory, describing the history of flow cytometry, principles of instrumentation, different fluorescence techniques, fluorescence measurements, cell sorting, data analysis, standards and controls. These chapters provide useful information, particularly for newcomers to the field, to develop knowledge on the background and capabilities of the technology, which is essential for its practical application.

The most frequent clinical uses of cytometry are for immunophenotyping, and this topic is extensively covered in numerous chapters of the monograph. In fact, this book contains an all-inclusive collection of chapters describing clinical applications of immunophenotyping in a variety of diseases. Thus, many chapters present the uses of cytometry in hematological disorders, HIV infection, organ or stem cell transplantation and sepsis. Applications of cytometry in oncology, particularly in leukemias and lymphomas, are covered in-depth as well. A large section of chapters of clinical relevance is devoted to cell function analysis. These chapters cover a variety of topics, including measurements of intracellular cytokines, metabolic parameters, oxidative stress, cell proliferation, apoptosis, differentiation markers, multidrug resistance, platelets function, etc.

The *Cellular Diagnostics* monograph, thus, provides a collection of valuable chapters that describe up-to-date developments in the most important areas of clinical cytometry. Certainly, the book will find wide readership among researchers who use cytometry in clinical settings as a routine tool for disease diagnosis and assessment of treatment efficiency and prognosis. It will also be useful for the laboratory personnel using the protocols for cell analysis and operating cytometry instrumentation. The researchers with a main interest in basic sciences, who however want to extend the relevance of their findings to the clinic, may also find this monograph worth close scrutiny. The text of this useful monograph, which was originally published in German, is now presented English, which extends its readership to a worldwide audience.

New York, NY, USA
Zbigniew Darzynkiewicz

Zbigniew Darzynkiewicz is director of the Brander Cancer Research Institute at the New York Medical College and professor in the Departments of Pathology and Medicine at the same medical school.

Preface

In the past 20 years, flow cytometry has developed from a 'science in itself' to an indispensable tool for both research and the diagnostic characterization of cells in health and disease. While several immunophenotyping techniques already are routine procedures in the laboratory, new methods for the functional characterization of cells, the analysis of rare cells, and the diagnosis of complex materials have only begun to gain wide recognition. Multiparameter approaches will further improve analysis.

The intention of this book is to provide a comprehensive and detailed compilation of all aspects of flow cytometry in clinical translational research and clinical practice. This is addressed in four sections. In the first section of the book, the background and common methodological principles of flow cytometry are introduced. The second section addresses the biology and immunophenotypic characterization of the various cell types of the immune and hematopoietic systems as well as of the cells involved in angiogenesis or tissue repair. This section also provides in depth information for advanced users of flow cytometry. The third section then addresses specific methods which allow the characterization of the functional state of cells, their immunological competence and their turnover within a given phenotype. Specific protocols are intended to support the adaptation of methods to various cell systems. Finally, the fourth section of the book addresses already firmly established diagnostic applications of flow cytometry and is intended to serve as a reference and to assist in the interpretation of results. Each chapter provides background information in conjunction analytical protocols for the various applications of flow cytometry.

The book is the updated English version of the 2006 handbook *Zelluläre Diagnostik* which was created by a working group of scientists active in the disciplines of laboratory medicine, immunology, hematology and transfusion medicine in Germany. The common goal of this group was to promote the use of flow cytometry by providing background information together with technical protocols for a broad range of research and clinical applications to colleagues and students as well as laboratory technicians. We are grateful to the Wallace H. Coulter Foundation who identified the potential of this concept of a combined textbook and manual not only to help new coworkers in our laboratories to become familiar with new applications, but also to promote cellular diagnostics in geographic regions where flow cytometry currently still finds only a limited use. The international orientation of the new edition of the book is reflected by additional information as well as new chapters on the history of flow cytometry and monitoring of HIV disease as well as by a selection of manuscripts focused on techniques or applications with a high potential to address current or emerging medical needs.

We are thankful to all contributing authors for the time they devoted to share their knowledge and experience. We are also thankful to the International Federation of Clinical Chemistry and Laboratory Medicine (IFCC) for adopting this project as part of its international educational efforts.

Ulrich Sack, Leipzig
Attila Tárnok, Leipzig
Gregor Rothe, Bremen

Sack U, Tárnok A, Rothe G (eds): Cellular Diagnostics. Basics, Methods and Clinical Applications of Flow Cytometry. Basel, Karger, 2009, pp 1–28

Cytometry – a Definitive History of the Early Days

J. Paul Robinson

Cytometry Laboratories, Bindley Bioscience Center, Purdue University, West Lafayette, IN, USA

Prelude

Cytometry is defined as the measurement of cells. Several chapters have been written about the discoveries that led to the establishment of the field of cytometry. The goal of this chapter is to lay out the fundamental discoveries that have been the foundation of the cytometry field that has contributed so much to the current state of knowledge in biological science. Without cytometry, many fundamental discoveries in immunology would yet remain to be made. Trying to imagine what the field of immunology would be like without flow cytometry is difficult at best.

A field of technology is often overlooked when it becomes one of the fundamental building blocks in other fields. Such is the case with cytometry. Immunology would arguably still be in the dark ages were it not for discoveries that built directly upon one another. The obvious example of course is the discovery of monoclonal antibodies (mAbs) by Köhler and Milstein [1] in 1975, exactly 10 years after the first cell sorter was built by Mack Fulwyler [2]. But it was Herzenberg, during a sabbatical in Milstein's lab in 1975, who recognized the power of mAbs, which would give him the tools to become one of the founding fathers of what became known as cytometry.

No introduction could be complete without mentioning the names of other founding fathers – such as Louis Kamentsky who, 4 years prior to Fulwyler, had himself designed a cell sorter. Despite the incredible power behind it, there was little understanding in the field of how powerful Kamentsky's ideas were, and his instrument was to all extents and purposes forgotten, to be rejuvenated and perhaps reinvented a few years later. Wolfgang Göhde, who worked primarily in Germany and did not have access to some of the reports and publications of his American colleagues, was simultaneously discovering the power of single-cell analysis with multiparameter technologies.

Table 1. Significant papers contributing to current cytometry capabilities (reproduced with permission from Purdue Cytometry DVD Volume 10)

Year	Development	Reference
1934	Photoelectric measurement of cells in a capillary	3
1941	Fluorescence antibody technique developed	4
	Nucleic acids shown necessary for protein synthesis	5
	Uterine cancer detection	6, 7
1944	DNA is carrier of genetic information – discovery of the 'transforming principle'	8
1947	Photoelectric particle counting	9, 10
1949	Particle counting by Coulter volume	11
1950	DNA and RNA shown to increase in actively growing cells	12
	Uterine cancer detection using fluorescence microscopy	13
1953	Hydrodynamic focusing for reproducible delivery of cells in a fluid	14
1955	Automated scanning instrument for screening cytological smears	15
1956	Particle counting by Coulter volume	16
1961	First use of fluorescence for quantitation	17
1963	First use of absorption for cancer cell detection	18
	Fluorescence for quantitation	19
1964	Electrostatic principle for ink jet	20
	Acridine orange differentiation of leukocytes	21, 22
	Need for automated imaging established	23
1965	Cell sorting	24
	Particle separator in principle capable of separating by volume, optical density, or fluorescence	25, 26
	Spectrophotometry of cells	27
1966	Fluorogenic esters in mammalian cells	28
1967	Fluorescence flow cytometry – 1st paper	29
1968	Automated imaging	30–32
	Fluorescence flow cytometry patent	33
	Scanning vs. flow for cancer cytology	34
1969	Fluorescence flow cytometry – 2nd paper	35
	Fluorescence flow cytometry – 3rd paper	36
	Fluorescence flow cytometry – 4th paper	37
	First paper describing light scatter	38
1972	Fluorescence-activated cell sorting	39
1973	Doublet discrimination patent	40
1974	Mathematical analysis of DNA distributions	41
1975	Monoclonal antibodies – invention	1
1977	Monoclonal antibodies – first use in flow cytometry	42
	Two-color fluorescence compensation	43
1978	Radiation collector methods – three patents	44
1979	Monoclonal antibodies – second use in flow cytometry	45, 46
	Flow imaging	47
	Radiant energy reradiating flow cell system – patent	48
1982	Slit-scanning flow cytometer	49
1983	DNA from paraffin tissue	50, 51

Table 1 continued on next page

Table 1. Continued

Year	Development	Reference
1984	Convention on nomenclature for DNA cytometry – guidelines for analysis	52
	Proposal for data file standard (FCS 1.0)	53
	Three-color immunofluorescence	54
1987	Time as a quality control parameter	55
	Dual-beam high-speed sorting	56
1988	4 pi light collection flow chamber – increased sensitivity	57
1990	Data file standard (FCS 2.0)	58
1991	Barcode reader – first automation of flow cytometry for clinical systems	59
1995	Five-color flow cytometry	60
1997	Eight-color, ten-parameter flow cytometry	61
2001	Eleven-color, 13-parameter flow cytometry	62
2004	Seventeen-color flow cytometry	63

This article is far more than a book chapter. It is really a great story of how our field emerged and developed over six decades (table 1). It is focused on the founders of our field, who for the most part have not been highly recognized by the world of science. We may be considered a niche technology as much as any other. But that is simply not the case. If you take away the discoveries of cytometry, you deplete the world of sorted chromosomes, cellular subset analysis – the bastion of immunology; you remove the rare-event sorting of stem cells, the power of cell cycle analysis, the simplicity of cell tracking, and the power of multiparameter clinical diagnostics of blood. The story of cytometry is a powerful one that appears to be always just beginning.

Introduction

Almost no one agrees on who invented technologies that emerged over the course of many years, and when it comes to documentation, the subsequent history is sometimes a series of 'unreliable memoirs' (with apologies to Clive James; 'Unreliable Memoirs', Picador Press, 1981). Each book chapter published includes some historical review of inventions; however, these are frequently recollections or personal experiences that can easily become somewhat biased. In the end, there may be no definitive history which is based on documented published works as well as on a detailed review of serious historical records such as patents, laboratory notes, and other data collected at the time of invention.

The goal of the present chapter is to determine a starting point for the beginning of cytometry and to follow the developing field up to the more recent past using the historical record as a guide. Usually, the best and most accurate documents are peer-reviewed published works. However, in the field of cytometry, two or three additional historical resources add great weight to our knowledge base. Some of these sources have never before been available. They are tremendously complete and in my opinion provide accurate documentation that establishes a number of dates, times, and discoveries for our field. One such valuable source has been Mack Fulwyler's original laboratory notebooks, from the late 1960s (Robert Auer of Beckman Coulter kindly provided access to Mack Fulwyler's original lab notebooks). Another is a set of videos made by the Smithsonian Institution in 1990–1991 (Smithsonian Tapes, RU 9554, 1991) that contain 14 h of personal interviews with many of the inventors of cytometry. In addition to that, the author has undertaken many more video interviews of those missed out by that original team. Two other valuable but not well-evaluated sets of records include patent documents (some in German) and internally published documents such as the Los Alamos reports [29], which, although public documents, rarely became available to anyone except those with direct contact with Los Alamos in the 1960s or even 1970s. The fact that some key discoveries were published in these types of journals almost certainly meant that they were not accessible or even known to the majority of scientists, particularly those outside the USA.

Two things in particular impact our understanding of how knowledge was distributed in the 1960s. First, there was no electronic source of communication like the internet of today. Second, it took months, sometimes a year or more, for a publication to be available in the public domain. These were the days when, even more than today, the scientific meeting was the primary mechanism for distribution of new knowledge. Thus, attending a congress was crucial to knowing what the latest discovery in any one field might be. This is one of the reasons that developments within the field of cytometry followed a series of scientific meetings and in fact are documented through those meetings very well. Further, those scientific meetings led to the formation of the Society for Analytical Cytology – an organization that has for 25 years carefully nurtured the field of cytometry.

Any history of a field risks being revisionist. While I have tried to focus on the facts with multiple sources of documentation, it is amazing how difficult it is to catch all the nuances of discovery. On many occasions, it appears that something was discovered and published or presented at a meeting, but the record of that discovery was not well documented or publicly distributed. In such cases, it remains each author's obligation to determine the accuracy or veracity of each piece of information. Where there is doubt or confusion, it should be so noted.

Three particular individuals in the field of cytometry have set the stage for recording the history of the field. First, Howard Shapiro documented in great detail the history of cytometry in his seminal work in the field, expanded significantly in the second, third, and fourth editions [64]. Anyone wishing to have a comprehensive, *down-to-the-wire* review of the history of cytometry should read Howard's book. Second, Guenter Valet documented and presented a history of the field from the European perspective, identifying and amplifying some of the work that had been unrecognized or not well identified earlier [65]. Third, Philip Dean dutifully recorded with great accuracy the details of the formation of SAC and ISAC and created an outstanding record which I have used as an important guide (*www.isac-net.org*). Finally, much of the work in this chapter is based directly on our own historical document [66], One of the goals of that publication was to bring to life the inventors themselves by showcasing them in dozens of video clips where they describe their own discoveries. Indeed many of the figures and photos used in this chapter were reproduced with permission from that electronic publication.

The Beginning

There are many arguments as to what constitutes the seminal discovery that established the fundamentals of cytometry. Of course, one could go back as far as Robert Hooke, who established the definition of the word 'cell' in 1665, or to Anton van Leeuwenhoek, who was the first to really document a great number of fundamental biological systems. It could be Paul Ehrlich, who manipulated the chemistry of the day to produce pathways for identifying cellular components and structural characteristics using stains. Or Robert Feulgen (1925), who first demonstrated that DNA was present in both animal and plant cell nuclei; he developed a stoichiometric procedure for staining DNA involving a derivatizing dye (fuchsin) and the formation of a Schiff base. It has been argued by some that Moldavan's publication [3] was the foundation for cell analysis. Others identify Gucker's instrument [9] since it was so close to the current particle detectors of today; it used a bright light source (a Ford car headlamp) and a photomultiplier tube (PMT), and was able to identify particles in air. Gucker's work was certainly more practical than Moldavan's as some scholars even doubt that Moldavan ever built the device described in his half-page article in *Science* [3]. Certainly many other investigators were focusing on understanding cellular processes. For example, Papanicolaou and Traut were defining the nature of abnormal cells in cervical smears [6]. Torbjorn Caspersson in 1941 demonstrated that 'nucleic acids, far from being waste products, were necessary prerequisites for the protein synthesis in the cell' [5] 'and that they actively participated in those processes' [67]. Not long after this initial discovery, Caspersson demonstrated in 1950 that both DNA and RNA increase in actively growing cells. His famous monograph '*Cell Growth and Cell Function*' [12] described nucleic acid and protein metabolism during normal and abnormal growth. These studies were made using a cadmium spark source for UV light and primitive electronic circuits for detection of signals. Again, crucial discoveries but not what we might call the '*eureka*' for cytometry.

Two other discoveries in the 1940s were important for cytometry. Coons and coworkers [4] developed the fluorescence antibody technique; they labeled anti-pneumococcal antibodies with anthracene, which allowed them to detect both the organism and the antibody in tissue using UV-excited blue fluorescence. Further, in 1950 Coons and Kaplan [68] conjugated fluorescein with isocyanate, which gave a better blue-green fluorescent signal somewhat further removed from tissue autofluorescence. It turned out that this discovery nearly 60 years ago set cytometry on the path which focused on the use of fluorescein isothiocyanate as the definitive fluorochrome to the present day.

The second key discovery that foreshadowed the cytometry era came in 1944 when Oswald T. Avery (1887–1955) demonstrated that DNA was the carrier of genetic information [8]. His discovery of the 'transforming principle' probably should have resulted in a Nobel Prize.

It was probably P.J. Crosland-Taylor in 1953 [14] who put together the final links in the cytometry chain. His publication of the sheath-flow principle used to this day in almost all flow cytometers transformed the field into one that was capable of rapid, accurate, and reproducible measurement of cell properties at high speed. At this time, the notion of analyzing single cells was just beginning.

The Beginning of Cytometry

So is there a key inventor, the one that made the real difference?

Probably it was Wallace H. Coulter who really made the difference. He submitted his patent application for what became known as the Coulter principle in 1949 and the patent was issued in 1953 [11]. The first commercial Coulter® Counter was marketed in 1956. Forty years later in 2006, Wallace Coulter was recognized in the American Institute for Medical and Biological Engineering (AIMBE) Hall of Fame for his invention. The introduction of the cell counter to clinical environments transformed the diagnostic toolsets available to physicians at the time. Is it accidental, or does it represent the enormous impact the Coulter Counter had that the same instrument with almost no fundamental changes in technology is still sold today? It is without doubt the most ubiquitous flow cytometer in the world. The invention of the Coulter Counter should be viewed with a great deal of admiration and recognition for three reasons. First, it satisfied an important clinical demand and has done so for over 50 years using almost the exact same technology with no significant variations. Second, it was fast and utilized a very small volume of blood, a criterion viewed with a high level of importance today. Third, it was statistically accurate and produced easily readable and instantaneous results. These three criteria can hardly be applied to many modern technologies, some of which pride themselves on the complexity of the analytical components.

The key value of the Coulter Counter was that it was able to give an immediate blood cell count. This was the first, most enduring, most successful, and most numerous flow cytometer. The success of the Coulter Counter generated competition.

Such competition has been the hallmark of innovation in the field of flow cytometry and has led to the evolution of today's very complex instruments.

Wallace Coulter – Inventor and Entrepreneur

The material on Wallace Coulter's personal life was provided by the Wallace H. Coulter Foundation and by Marshall (Don) Graham.

Wallace H. Coulter was an engineer, inventor, entrepreneur, and visionary. He was co-founder and Chairman of Coulter® Corporation, a worldwide medical diagnostics company headquartered in Miami, FL. The two great passions of his life were applying engineering principles to scientific research and embracing the diversity of world cultures. The first passion led him to invent the Coulter Principle, the reference method for counting and sizing microscopic particles suspended in a fluid.

This invention served as the cornerstone for automating the labor-intensive process of counting cells and performing blood cell analysis. With his vision and tenacity, Wallace Coulter was a founding father in the field of laboratory hematology, the science and study of blood. His global viewpoint and passion for world cultures inspired him to establish over twenty international subsidiaries. He recognized that it was imperative to employ locally based staff to service his customers well before this became standard business strategy.

He attended his first year of college at Westminster College in Fulton, MO; however, his interest in electronics led him to transfer to the Georgia Institute of Technology for his second and third years of study.

This was the early 1930s, and owing to the Great Depression he was unable to complete his formal education. Wallace's interest in electronics manifested itself in a variety of unconventional jobs. For example, he worked for WNDR in Memphis, TN, filling in as a radio announcer, maintaining the equipment, and conducting some of the earliest experiments on mobile communications.

In 1935, he joined General Electric X-Ray as a sales and service engineer in the Chicago area, servicing medical equipment. This work familiarized Wallace with the testing procedures in the hospital laboratory. When an opportunity to cover the Far East became available, he seized the chance to live and work abroad.

Wallace first went to the Philippines, where the local GE office was manned by technicians from many countries. After 6 months in Manila, Wallace was asked to make sales and service calls in the more remote regions of the territory, and he traveled to Hong Kong, Macao, and Canton, finally settling in Shanghai for 6 months.

Wallace transferred to Singapore, where he remained until the Japanese attacked the city in late 1941. It took him nearly 12 months to return finally to the USA at Christmas 1942.

After the war, Wallace worked for several electronics companies, including Raytheon and Mittleman Electronics in Chicago, IL. He maintained a laboratory at home to work on promising ideas and projects. One such project was for the Department of Naval Research, where Wallace was trying to standardize the size of solid particles in the paint used on US battleships in order to improve its adherence to the hull.

He began tinkering in his garage laboratory in his spare time, experimenting with different applications of optics and electronics. Upon returning to the garage one cold, blustery evening, Wallace was faced with a challenge. The supply of paint for the experiment related to battleship paint had frozen while he was out. Not wanting to go back out in the cold, he asked himself, 'What substance has a viscosity similar to paint and is readily available?' Using his own blood, a needle, and some cellophane, Wallace Coulter invented the principle of using electronic impedance to count and size microscopic particles suspended in a fluid – the Coulter Principle.

Remembering his visits to hospitals, where he observed lab workers hunched over microscopes manually counting blood cells smeared on glass, Wallace focused the first application on counting red blood cells. The instrument he constructed to accomplish this became known as the Coulter Counter. This simple device, using a blood sample 100 times the volume required by the usual microscope method for a blood test, counted in excess of 6000 cells/s. It decreased the time it took to analyze from 30 min to 15 s and in addition reduced the error by a factor of approximately 10.

Wallace's first attempts to patent his invention were turned away by more than one attorney who believed 'you cannot patent a hole.' Persistent as always, Wallace finally applied for his first patent in 1949 and it was issued on October 20, 1953. That same year, two prototypes were sent to the National Institutes of Health (NIH) for evaluation. Shortly afterwards, the NIH published its findings in two key papers, citing improved accuracy and convenience of the Coulter method of counting blood cells. That same year, Wallace publicly disclosed his invention at the National Electronics Conference in his one and only technical paper [16].

In 1958, Wallace and his brother, Joseph Coulter, Jr., founded Coulter Electronics to manufacture, market, and distribute their Coulter Counters. From the beginning, this was a family company, with father Joseph Coulter, Sr., serving as secretary treasurer. Wallace and Joseph, Jr., built the early models, loaded them in their cars, and personally sold each unit. In 1959, to protect the patent rights in Europe, subsidiaries in the UK and France were established. In 1961 the Coulter brothers relocated their growing company to the Miami area, where they remained for the rest of their lives.

Invention of the Electrostatic Cell Sorter

The link between Wallace Coulter and Mack Fulwyler has some interesting twists but was fundamentally driven by Fulwyler's interest in and understanding of the technology developed by Wallace Coulter. The story of how Mack Fulwyler came to invent the technology of electrostatic cell sorting is intriguing and even more impor-

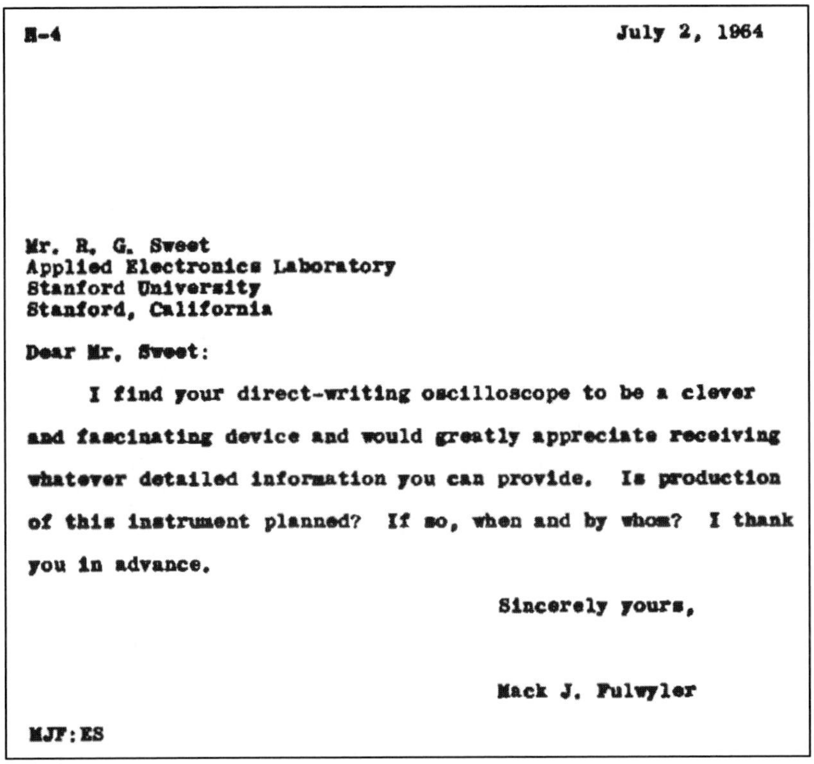

H-4 July 2, 1964

Mr. R. G. Sweet
Applied Electronics Laboratory
Stanford University
Stanford, California

Dear Mr. Sweet:

I find your direct-writing oscilloscope to be a clever and fascinating device and would greatly appreciate receiving whatever detailed information you can provide. Is production of this instrument planned? If so, when and by whom? I thank you in advance.

Sincerely yours,

Mack J. Fulwyler

MJF:ES

Fig. 1. The first documented letter from Mack Fulwyler to Richard Sweet asking for additional information on his 'direct-writing oscilloscope.' It was this letter written on July 2, 1964, that began the modern world of flow cytometry (Source: Robert Auer with permission)

tantly is well documented in both Mack's personal laboratory notebooks and a substantial record of letters and documents from the mid 1960s. Most of these documents had never been accessed until they were recently re-discovered in 2007 (as noted earlier, these materials were located by Bob Auer).

The interest in studying single red blood cells drove the next significant discovery which was intimately connected to the invention of the Coulter Counter. It is ironic that it was the incorrect use of the Coulter Counter by a pathologist that resulted in the development of the electrostatic cell sorter. In Mack Fulwyler's own words:

> 'We had a pathologist in our group who was using that device to analyze blood and he would adjust the aperture current and some other characteristics of the machine so that he could cause a small subpopulation of the red blood cell distribution to move away from the main distribution of red blood cells, and the pathologist thought that this represented immature RBC that had just been produced. Well, Marvin [Van Dilla] and I did not believe that that was the case, and so we set out to try to convince him that he was incorrectly using the device, and we were not successful in doing that. It occurred to me that if I could just pick out what was thought to be this abnormal population, just physically isolate those cells, run them back through the Coulter counter, get the same distribution, it would demonstrate that this was an artifact that he was misusing the machine. So my motivation in trying to come up with trying to sort cells was to disprove this fellow's interpretation of data' [69].

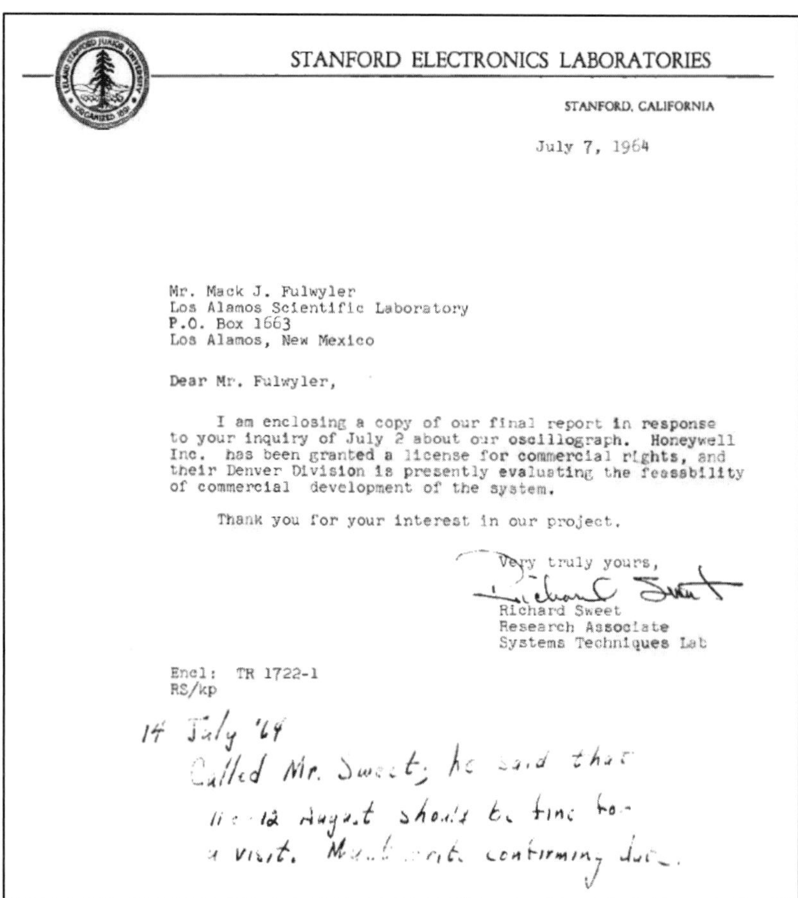

STANFORD ELECTRONICS LABORATORIES

STANFORD, CALIFORNIA

July 7, 1964

Mr. Mack J. Fulwyler
Los Alamos Scientific Laboratory
P.O. Box 1663
Los Alamos, New Mexico

Dear Mr. Fulwyler,

I am enclosing a copy of our final report in response to your inquiry of July 2 about our oscillograph. Honeywell Inc. has been granted a license for commercial rights, and their Denver Division is presently evaluating the feasbility of commercial development of the system.

Thank you for your interest in our project.

Very truly yours,

Richard Sweet
Research Associate
Systems Techniques Lab

Encl: TR 1722-1
RS/kp

14 July '64
Called Mr. Sweet; he said that 11-12 August should be fine for a visit. Mail [illegible] confirming date.

Fig. 2. Richard Sweet's reply to Fulwyler's letter of July 2, 1964. This was followed by Fulwyler's visit to Sweet in San Francisco on August 13, 1964.

That pathologist's name was Lushbaugh, and the story recounted by Fulwyler was that the laboratory had been using a Coulter Counter with an attached spectrum analyzer. Lushbaugh has been changing the settings of the pulse height analyzer and came up with some interesting distribution curves suggesting to him that he was able to identify possible subpopulations of red blood cells. Wright Langham (the lab director at the time) and Marvin Van Dilla were very concerned about the accuracy of the data, and Fulwyler also became involved. As an engineer, Fulwyler became concerned about how the Coulter technology was being used and sought to identify a mechanism to separate the cells physically based on measuring the same signals (impedance) being used in the Coulter Counter. To do this, he had to design and build a new technology.

Fulwyler had apparently heard a talk by Richard Sweet in which Sweet described the use of a piezoelectric crystal oscillator to create small droplets at high frequency.

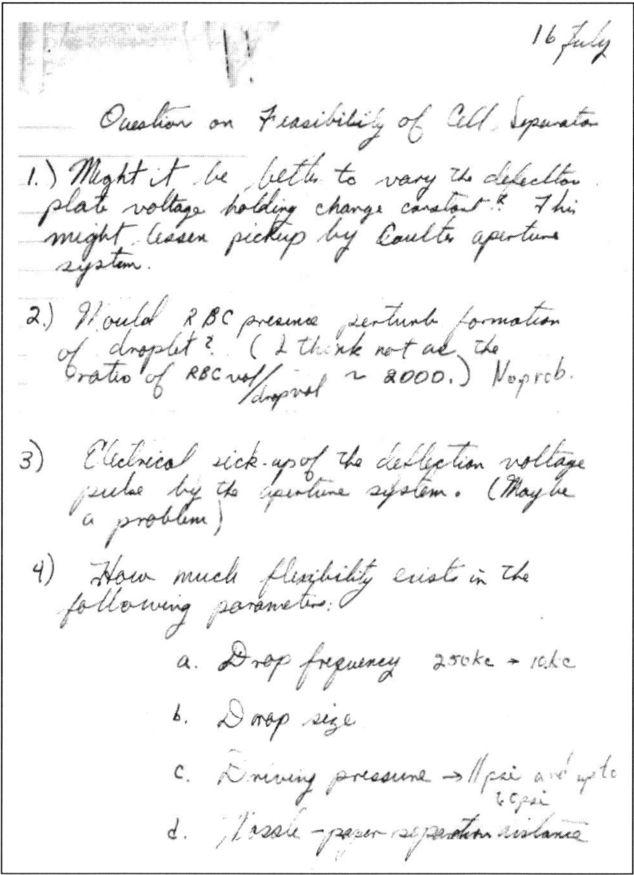

16 July

Question on Feasibility of Cell Separation

1.) Might it be better to vary the deflection plate voltage holding charge constant? This might lessen pickup by Coulter aperture system.

2.) Would RBC presence perturb formation of droplet? (I think not as the ratio of RBC vol/drop vol ~ 2000.) No prob.

3) Electrical pick-up of the deflection voltage pulse by the aperture system. (May be a problem)

4) How much flexibility exist in the following parameters:

a. Drop frequency 250kc → 10kc

b. Drop size

c. Driving pressure → 11 psi and up to 60 psi

d. Nozzle-paper separation distance

Fig. 3. Notes from Fulwyler's lab book showing his questions as to how well a design for a cell sorter might work.

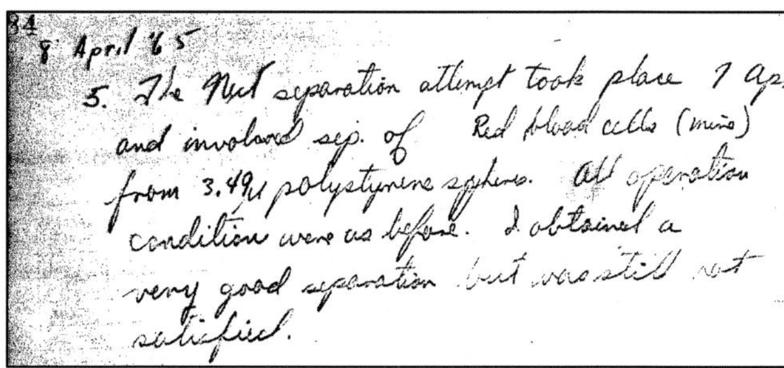

8 April 65

5. The next separation attempt took place 7 Ap. and involved sep. of Red blood cells (mine) from 3.49μ polystyrene spheres. All operation condition were as before. I obtained a very good separation but was still not satisfied.

Fig. 4. Notes From Mack Fulwyler's lab book dated 8 April, 1965, where he records the first successful separation of red blood cells (his) from polystyrene beads.

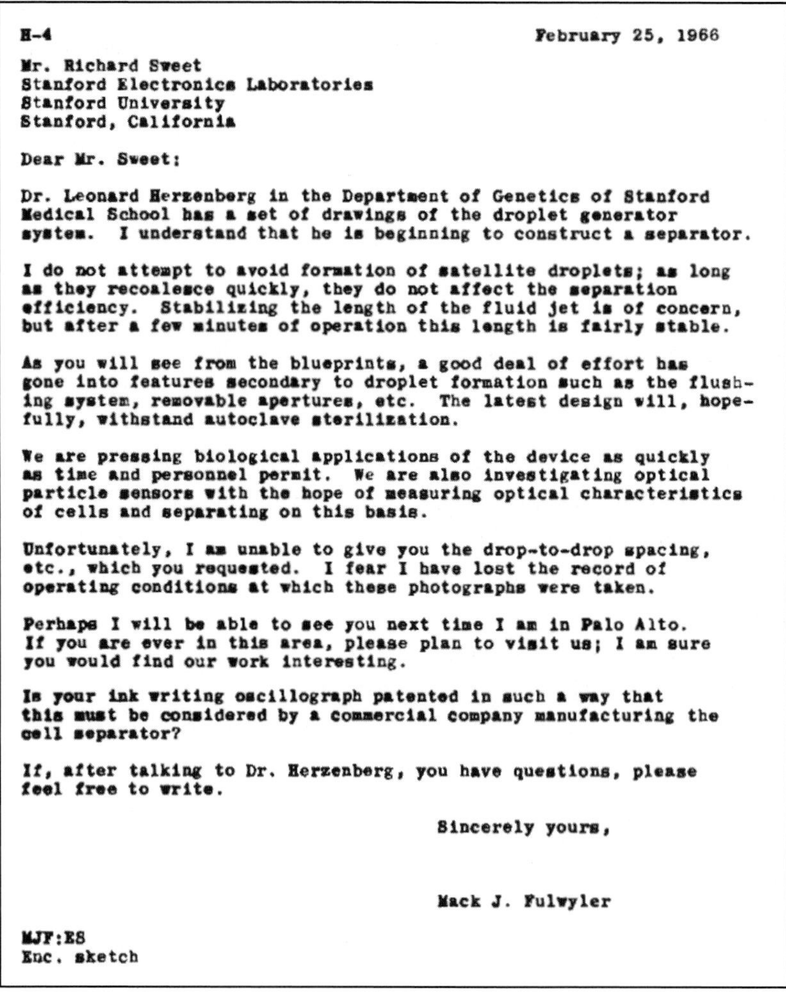

H-4 February 25, 1966

Mr. Richard Sweet
Stanford Electronics Laboratories
Stanford University
Stanford, California

Dear Mr. Sweet:

Dr. Leonard Herzenberg in the Department of Genetics of Stanford Medical School has a set of drawings of the droplet generator system. I understand that he is beginning to construct a separator.

I do not attempt to avoid formation of satellite droplets; as long as they recoalesce quickly, they do not affect the separation efficiency. Stabilizing the length of the fluid jet is of concern, but after a few minutes of operation this length is fairly stable.

As you will see from the blueprints, a good deal of effort has gone into features secondary to droplet formation such as the flushing system, removable apertures, etc. The latest design will, hopefully, withstand autoclave sterilization.

We are pressing biological applications of the device as quickly as time and personnel permit. We are also investigating optical particle sensors with the hope of measuring optical characteristics of cells and separating on this basis.

Unfortunately, I am unable to give you the drop-to-drop spacing, etc., which you requested. I fear I have lost the record of operating conditions at which these photographs were taken.

Perhaps I will be able to see you next time I am in Palo Alto. If you are ever in this area, please plan to visit us; I am sure you would find our work interesting.

Is your ink writing oscillograph patented in such a way that this must be considered by a commercial company manufacturing the cell separator?

If, after talking to Dr. Herzenberg, you have questions, please feel free to write.

Sincerely yours,

Mack J. Fulwyler

MJF:ES
Enc. sketch

Fig. 5. Letter from Fulwyler to Sweet dated February 25, 1966.

On July 2, 1964, Fulwyler wrote to Sweet asking for more information on the 'direct writing oscilloscope' (fig. 1).

Amazingly, Fulwyler received a written reply from Richard Sweet just 5 days after he wrote the original letter (fig. 2). Only a few days later after receiving Sweet's letter, Fulwyler noted down in his lab book some of the issues he felt would need to be solved in order to create the cell separation device (fig. 3). Within a month, Fulwyler had arranged to visit Sweet (which he did on August 13, 1964) where he obtained some parts that Sweet had used to build his 'direct writing oscilloscope' (see notebook page in fig. 3). Fulwyler designed, built, tested (for early results see fig. 4), and completed the first cell sorter within 8 months of conception of the idea. The key work was published in *Science* on November 12, 1965 [2].

It is apparent that communication between Mack Fulwyler and Len Herzenberg had occurred about this time. By the end of February 1966, Fulwyler noted in a letter (fig. 5) to Richard Sweet that he had sent a set of drawings to Herzenberg. Richard Sweet eventually went on to work in the Herzenberg lab. Mack Fulwyler did the same thing some years later.

The Other Side of Flow

Work developed by Louis Kamentsky in the early 1960s was a result of his knowledge of and keen interest in the work of Torbjorn Caspersson. Indeed, Mellors and Silver [70] developed a microfluorimetric scanner in 1951, and in 1964 Torbjorn Caspersson's son showed a comparison of DNA distributions in cervical cells, from normal, premalignant and carcinomatous cells, identifying that cytometric measurement was a useful tool for pathology [71]. This paper followed closely from Kamentsky's own publication of the spectroscopic variation between normal and cervical cancer cells using a new semi-automated instrument [18]. Single-cell analysis was becoming an important aspect of biological science.

At about the same time as Fulwyler was considering the red blood cell problem, Louis Kamentsky had initiated single-cell studies of cancer cells. Kamentsky was then at the IBM Watson Labs. He had been working for some time on a project very important to IBM – the concept of automated character recognition. Figure 6 is an example of the type of work on character recognition in which Kamentsky was involved – an area that strongly influenced his interest in identifying cancer cells. Fundamental problems in image analysis required huge computing capacity that simply was not available at the time. Kamentsky, however, was highly successful in solving the character recognition problems [72], leading him to consider other applications in the field of cell biology. Since the time of Papanicolaou and Traut in 1941 [6] a number of attempts had been made to develop automated image analysis systems. Kamentsky understood, more than most, the degree of difficulty in achieving these goals. Indeed, Kamentsky had access to the largest and most comprehensive computers in the world, and even these were insufficient to adequately solve the problems associated with image analysis of cancer cells. Louis Kamentsky has a great record of achievement in cytometry. Originating from the world of IBM and character recognition days, he came into the field influenced particularly by Myron Melamed. He developed the most advanced cell analysis tools in the early 1960s, publishing a seminal paper in 1963 in *Science* [18] in which he was able to differentiate cancer cells from normal cells using absorption profiles. He followed this with another key paper in 1965 [73] in which he was able to characterize the spectrophotometric properties of cells at a rate exceeding 500/s.

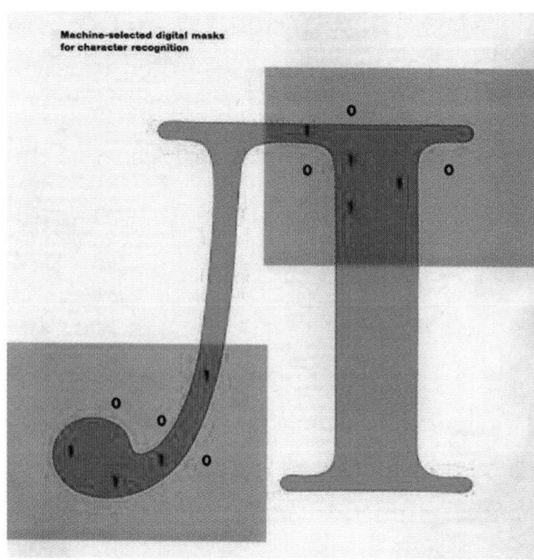

Fig. 6. Kamentsky was working on character recognition systems for IBM. It was this area of work that led him to consider using the same technology for analysis of images of cells (photos reproduced with permission – Elena Holden, Compucyte Corp.)

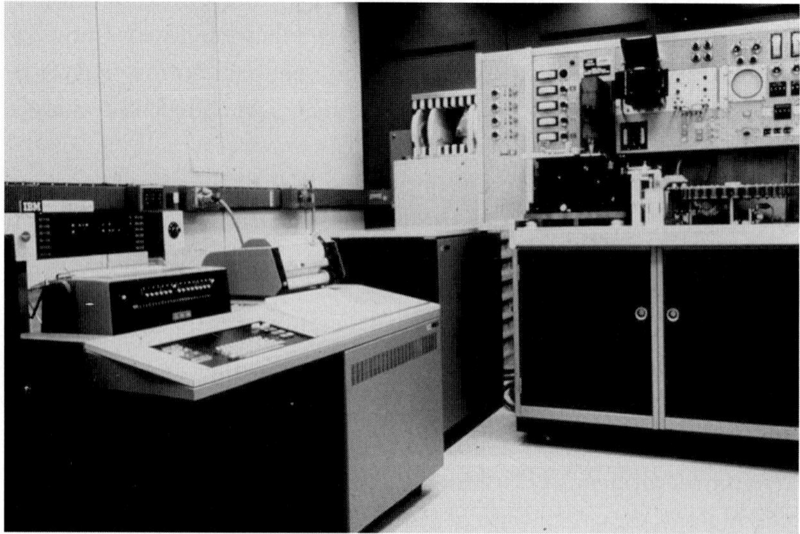

Fig. 7. Kamenstky built two of these instruments with four sensors, sorting, auto sampling, and multiparameter data reduction in 1966. One was supplied to Leonard Herzenberg at Stanford.

As Kamentsky recalls (Louis Kamentsky, personal communication), he felt that the power of the analytical tools he was developing could best be used in the biological sciences, and since his character recognition project was successfully completed, it was time to move on from IBM. Before long, he had identified the early work of Caspersson [e.g. 5] as critical to solving the analytical problems for cancer cell detection. He designed a spectrofluorimetric system that was capable of creating fingerprints of individual cells, with the goal of distinguishing cancer cells from normal ones. Kamentsky developed an incredibly complex multiparameter cell sorter using microfluidic sorting that was probably the earliest and most successful implementation of traditional cell sorting, albeit very slow and complex. With all the resources of IBM while he was there, Kamentsky built two enormous instruments for cell analysis (fig. 7). One of these instruments was shipped to Len Herzenberg at Stanford University in 1967. Louis Kamentsky is considered to be the first person to commercialize a flow cytometer with an inbuilt laser (Cytograph, 1970) (Howard Shapiro, personal communication). His company was eventually purchased by Ortho, a Johnson and Johnson company, about 1972, and much of his technology was fundamental in Ortho's huge success in flow cytometry instruments in the late 1970s and 1980s. Kamentsky worked for several years at expanding the cell analysis technologies before starting CompuCyte Corporation. Subsequently, Louis applied his tremendous knowledge of image analysis and automated classification techniques to the laser scanning cytometer built by the CompuCyte Corporation. From these efforts much of the fundamentals of the early really high-content screening devices were derived.

Background on Fulwyler's Invention

Mack Fulwyler built his cell sorter by taking the principles of Richard Sweet's ink-jet printer and combining it with Wallace Coulter's cell counter. It was an ingenious if not difficult technology to engineer. Fulwyler later recalled that the problem of the bimodal red blood cell populations was actually solved in one afternoon (Mack Fulwyler, personal communication). It was determined that this was an artifact of the way the cells were flowing through the Coulter aperture. Immediately after this, Mack and Marvin Van Dilla, his boss, had a conversation about what else could be run through the cell sorter and they decided that blood cells would be excellent targets. At the same time, Marvin Van Dilla had been intrigued by the work of Boris Rotman, who had published a seminal paper describing the use of fluorescence signals to define the presence of a bacterial enzyme, beta-D-galactosidase [17]. Van Dilla was interested in studying the cell cycle and identified fluorescence signals as a promising approach. Since Boris Rotman was moving from Stanford University to Brown University, he visited Van Dilla at Los Alamos and saw the cell sorter built by Fulwyler. By 1967, Fulwyler had already added a PMT with a fluorescence capability, and Rotman asked Fulwyler to build him a copy of his modified 1965 instrument. Rotman paid USD 5,000.– to have the instrument built and shipped to his lab at Brown University in 1967 (personal recollection based on the Smithsonian video collection from 1991. This instrument contained a Coulter volume sorting chamber (which Rotman says be never actually used), and a white-light source, a fluorescence PMT with a filter slot, and a sophisticated flow chamber to allow multiple samples to be mixed (fig. 8).

The first publicly documented report of fluorescence measurement may well be the report by Van Dilla et al. [29] on their work at the Los Alamos Scientific Laboratory between June 1966 and July 1967. The report was written in September 1967

Fig. 8. Fulwyler's 1967 instrument that was built for Boris Rotman in 1967 and shipped to Brown University. It was to be destroyed in 2005, but was rescued and is currently in the Bindley Bioscience Center at Purdue University. It is the only remaining system from the original invention.

and publicly distributed on January 23, 1968. It contained a detailed report on fluorescence cytometry showing a DNA histogram. Unfortunately, these reports from Los Alamos, although public, did not achieve much distribution, if at all, and were certainly not seen by those in Europe who were working on similar ideas.

The Parallel Discoveries in Europe

Caspersson had tremendous impact on most people in the field of cytometry. Wolfgang Göhde in Germany was no exception. Even in the early days, he was measuring dual-parameter fluorescence signals in his microscope-based flow systems. He has spent virtually his entire professional life developing and improving cytometric technologies, in addition to his work in radiobiology studying the effects of radiation. He had studied Caspersson's work and spent some time in his laboratory as well. It is clear that the microscope drove a great deal of the discovery in Europe. Göhde was working with the Zeiss UMSP-1 cytophotometer at the time, but it was very slow and difficult to use. Göhde modified the system 'by keeping the microscope's pinhole fixed while moving fluorescence stained cells in a closed cuvette within a narrowly focused epi-illuminated laminar fluid stream at ca 1 m/sec speed below the pinhole through the focal plane of the objective' [74].

Subsequently, Göhde went on to design in 1969 what became the first commercially available flow cytometer, the ICP-11 Impulscytophotometer sold by Phywe (Göttingen, Germany). The system had 2 PMTs which collected two wavelengths of emitted light and had a doublet discriminator using a pulse height divided by pulse area [75]. Not long after Phywe sold this technology and related patents to Ortho-Diagnostics in Raritan, NJ, this was combined with Lou Kamentsky's technology to create the basis for Ortho's powerful entry into cytometry. Independently, the Partec Company (Münster, Germany) continued to manufacture instruments to become an important player in Europe and more recently in Africa with their CD4 instruments.

This more recent activity (since 2000) focused Göhde on his efforts to develop low-cost CD4 tests for many countries in Africa. In 2000, Göhde heard Howard Shapiro challenge researchers in the field of cytometry to take a serious look at developing low-cost, effective tools for CD4 measurement in AIDS patients in Africa (this challenge was issued by Howard Shapiro on the Purdue Cytometry Discussion in 2000). Göhde took up the challenge with a passion. He visited Africa and realized how terrible the conditions were for those people who did not have access to CD4 counts. He began a program to bring low-cost CD4 counts to Africa and was probably the scientist most responsible for reducing the cost of CD4 testing in Africa. He continues this important work today, and it is possibly the most important mission in his life.

Leeonard Herzenberg and the Stanford Group

Leonard Herzenberg is one of cytometry's most important pioneers. As soon as Leonard heard about the technologies being developed to separate cells, he realized the enormous potential for cell biology. He knew that, if he could separate cells, he could better study their biochemistry and immunology. His primary interest in cell sorting was to try to test the clonal selection theory; he wanted to prove that cells that bound the particular antigens sorted by cell sorting techniques would then make antibodies against that particular antigen and not against other antigens. While not an engineer himself, he understood well before most biologists that the multidisciplinary approach was the only way to reach his objectives. Leonard's work led to the commercialization of cell sorters by Becton Dickinson (BD), and this has resulted in a significant number of quality instruments from several manufacturers available today.

The laboratory of Leonard Herzenberg at Stanford University has been the site of many developments and innovations since 1959. It has been the research home to many scientists and has hosted numerous visiting scientists. It is one of the most recognized places of cytometry. The story of how BD became involved is long and

Fig. 9. Leonard Herzenberg with the FACS I around 1972.

complex, but is documented in the Smithsonian interviews of Ramunas Kondratas (RU 9554, History of the Cell Sorter, Smithsonian Oral History Collection, 1991). Bernie Shoor, a long-time BD employee running a small satellite laboratory in California, heard about Leonard Herzenberg's development of high-speed cell sorters. Herzenberg had obtained an NIH grant to build two instruments, one for NIH and one for Stanford. Shoor worked out an agreement with Herzenberg to have BD actually build the machines, which were to cost around USD 180,000.– each at the time. In the deal, BD negotiated a license and built both machines, called FACS-1, which were delivered in early 1973 (fig. 9). BD came up with an interesting manufacturing plan whereby they used as many subsystems as possible. Each subsystem supplier would support its own technology – Spectra Physics for Lasers, Hewlett-Packard, Tektronics, and others. This allowed BD to rapidly build systems and supply them overseas as well as in the USA.

Shoor recalls receiving a phone call from Herzenberg one day in 1975 telling him to start making mAbs– Shoor's reaction was 'What are they?' (RU 9554, History of the Cell Sorter, Smithsonian Oral History Collection, 1991, p 14). Based on this conversation, BD very quickly established a laboratory to make mAbs run by one of Leonard's previous students, Chuck Metzler, who had gone to Yale University.

Leonard and his colleagues have been responsible for training and equipping an untold number of people in the field of cytometry. In recognition of his accomplishments, Leonard was awarded the Kyoto Prize in 2006 (*www.inamori-f.or.jp/laureates/k22_a_leonard/prf_e.html* referenced on May 30, 2008), truly an honor for the entire cytometry community. The recognition from the Kyoto prize read 'Dr. Leonard Arthur Herzenberg took the lead in developing a flow cytometer called the Fluorescence-Activated Cell Sorter (FACS) that automatically sorts viable cells by their properties. Combining fluorescent-labeled monoclonal antibodies as FACS reagents with this instrument, he made an enormous contribution towards the dramatic advancement of life sciences and clinical medicine'.

His many contributions include driving the commercialization of cell sorters into the most important field of all – immunology. Leonard spent time in César Milstein's laboratory right at the time that Milstein was creating the mAb [1]. César Milstein credits Herzenberg with coming up with the term 'hybridoma' while on sabbatical in his lab [76]. Because Leonard Herzenberg was such a driving force in the development of flow cytometry technology, and indeed was in Milstein's laboratory at the time of the initial mAb discoveries, it is not surprising that Milstein's lab was the first international laboratory to obtain a cell sorter, and it was the group of Milstein who was the first to publish a paper using mAbs in flow cytometry [77]. No doubt it was Leonard Herzenberg who showed Milstein the ropes of flow cytometry.

Particle Technologies and the Integration into Coulter Electronics

Mack Fulwyler left the Los Alamos Laboratory and completed his PhD in the department of biophysics at the University of Colorado in 1969 with a dissertation on 'Electronic Volume Analysis and Volume Fractionation as Applied to Mammalian Cells'. In 1971, Mack Fulwyler formed a small company in Los Alamos called Particle Technologies, Inc. – a company funded primarily by Wallace Coulter.

Robert Auer heard about a position at Particle Technologies (PTI), and he visited Mack J. Fulwyler, the president of PTI. From him Auer heard for the first time the story of flow cytometry and sorting, a technology based on Fulwyler's PhD thesis project (as told by Robert Auer in a video interview with J. Paul Robinson in 2006). Robert Auer learned that PTI was a subsidiary of a 'big' medical instrument company (Coulter Electronics, Inc.) and that PTI's charter was to spin the flow cytometry and microsphere technologies out of the H division of Los Alamos Scientific Laboratories in order to commercialize them. It was a dynamic, exciting group working in a new technology of incredible potential. In September of 1973, Robert Auer joined PTI in Los Alamos, NM.

When Auer arrived at PTI, there were four prototype instruments. The first was the SDS-1 (Super Duper Sorter), a sorter which measured two colors of fluorescence, forward light scatter (FSC), and electronic cell volume. It used a Spectra Physics 164, 5-W argon laser for excitation. All data acquisition and sort decisions were via a PDP-8 computer with graphic displays on a Tektronics 4010 display. Programs and data were stored on Tridata cartridge tape. All of the programs were written in PAL8. This instrument was large, occupying three six-foot high instrument racks. The second prototype instrument was a two-color fluorescence plus FSC analyzer with a Spectra-Physics 162, 150-mW argon laser for excitation that also used a PDP-8-based data acquisition system. The third prototype was the SPA-1, a single-color fluorescence-only analyzer with a Spectra-Physics 162, 15-mW argon laser with a hardwired pulse-height analyzer. Finally, John Glascow was in the process of building number four, the TPA-1. This was a single-color fluorescence plus FSC analyzer with a Spectra-Physics 162, 15-mW argon laser with a hardwired pulse height analyzer. The TPA-1 was to be transferred to the parent company (Coulter Electronics) for commercialization. Auer's initial task was to convert the TPA-1 into a sorter, the TPS-1. Soon after Auer arrived, John Glascow made the decision to leave the company, so both the TPA-1 and the TPS-1 projects became Auer's responsibility. In September and October of 1973, Auer worked on simplified flow chamber designs that would permit optical measurements with a 15-mW laser and droplet formation for sorting.

In November of 1973, Auer met for the first time Wallace H. Coulter, the principal shareholder of the parent company, Coulter Electronics. Wallace, Walt Hogg, the chief engineer, and Bob Klein, the director of product engineering, came to Los Alamos from Florida on their way to the Engineering Foundation meeting on Analytical Cytology in Asilomar. After several days of project reviews, Mack and Auer joined Wallace Coulter to travel to Asilomar. In Asilomar, Auer learned for the first time how broad the field of flow cytometry already was. On the way back to Los Alamos after the meeting, Auer paid a visit to Richard Sweet at Leonard Herzenberg's lab at Stanford. Sweet showed Auer his sorter that used in-the-jet sensing. This seemed like a simple solution to the flow chamber problem they had encountered but required larger lasers for high sensitivity.

Robert Auer was given the goal of having both the TPA-1 and TPS-1 prototypes at the 1974 FASEB meeting in Atlantic City, NJ. After a busy winter, that goal was realized. This was the meeting at which BD also showed their first FACS instrument. It was finally decided that Auer would undertake the re-engineering of the TPS-1 in Los Alamos and deliver a more refined prototype to Florida. This effort continued into the summer of 1975, when four redesigned TPS-1 prototypes were delivered to Florida. One of those prototypes went to Jerry Hudson's lab at the Miami VA Hospital and another went to the Wheeless/Horan lab at the University of Rochester. In addition in July 1975 a modified version of the SPS called the EPICS II was delivered to the NIH. As part of the TPS-1 project, Auer designed a scatter sensor that used a Fourier lens to reduce sensitivity to alignment and whose response was monotonic with particle size; this sensor became Auer's first cytometry-related patent.

New ideas and invention were an important part of the culture at PTI. This was led by Mack Fulwyler; he was always coming out of his office and saying, 'Think about this' as he described an idea. There were weekly meetings in Mack's office where, on a rotating basis, each staff member had to present a new idea and everyone else got to critique it. Some of these ideas turned into projects and patents; for example, an automated live cell / dead cell enumeration system based on trypan blue exclusion detected by FSC and extinction, and 'diagnosis on a microsphere'. This was Mack's idea for using microsphere substrates to hold antibodies and specific antigens as targets to diagnose diseases. The tests could be multiplexed using microsphere size, fluorescence color, and intensity as encoding. His patent on this was blocked in the USA by a similar patent from Technicon but did issue in the UK.

Based on input from potential customers, Mack Fulwyler continued to push to be able to build and sell the larger sorter systems. Early in 1975, PTI was awarded a contract by NIH to build and deliver a large multiparameter sorter to Chet Herman at the National Cancer Institute (NCI). Another engineer, Vaughn Rheems, was hired to work with Jim Corell to convert the SDS-1 prototype into a saleable system. This system, the EPICS II (Electronically Programmable Individual Cell Sorter), was shipped in July of 1975. At this time Mack Fulwyler decided to pursue other goals and left PTI. On November 14, 1975, it was announced to the employees that PTI would terminate operations on March 1, 1976.

Auer was asked to move to Florida and manage the TPS-1 project and did so in late January of 1976. The particle manufacturing technology was also moved. For the next year or so Mack worked at the Jovin lab in Germany in order to clear the non-competition clause in his contract.

The marketing group in Florida decided to skip the TPA-1 and go directly to the TPS-1. When Auer arrived in Florida, the TPS-1 was not yet in production. It took the better part of 1976 to get the first instruments out the door. At the same time the TPS-1 began to evolve, a second PMT was added to permit two-color fluorescence measurements. It became clear that the 15-mW laser was too low power for many applications when used with a sense-in-the-jet flow chamber, so an adaptation to a larger water-cooled laser was designed. A dedicated manufacturing group was formed that became the EPICS Division of the Coulter Corporation, and Auer was named the General Manager. Coulter Immunology was formed as a sister division to focus on mAb reagents.

In 1977 after spending more than one year at the Max Planck Institute with Tom Jovin, Mack Fulwyler returned to the USA to become the technical director for BD FACS Systems Division. He moved from that position in 1982 to become a professor at the University of California, San Francisco, and subsequently director of technical development for the Trancell Corporation. Mack Fulwyler passed away from cancer in 2001. A special memorial print was made to honor his memory and was featured on the cover of *Cytometry* Volume 67A, issue 2, 2005.

1, 2, 3, 4, . . . n Colors: Cytometry Gets Complex

It did not take long to see that two colors were better than one. In the early days of flow cytometry, there were few dyes specifically designed for use with proteins such as antibodies. In fact, it was just after the introduction of commercial instruments that the famous discovery of mAbs was made by Köhler and Milstein [1]. Milstein [78] remarked in his recent paper that the editors of *Nature* were not particularly impressed with the paper and would only allow it to be submitted as a letter to *Nature*. Parenthetically, in one of his papers reviewing the history of the invention of mAbs, Milstein notes that his institution failed to protect the intellectual property – a failure, he asserted, that allowed totally open access to mAb production and use, transforming the entire biomedical domain. Another interesting fact about Milstein was that he was the first person to publish flow cytometry fluorescence data using mAbs [42].

It was a fundamental move for the field of cytometry when Loken et al. [43] defined the process of spectral overlap using two-color analysis. This process was subsequently repeated as each color was added to bring the number of simultaneous colors up to 17 [63]. Naturally, most clinically related assays do not reach this level of complexity. Indeed, a great deal, if not the vast majority, of basic cytometry is performed using just a few different emission wavelengths. However, with the increasing ability to classify samples in a more defined way, 8–10 colors for clinical samples may well be effective.

Standards: Who Needs Them Anyway?

The creation of the Flow Cytometry Standard (FCS) file format was an important milestone in the history of cytometry. The impact was enormous, much more than the value of the standard itself. It was the opening of the field to those outside of the major companies who had almost complete control. At the time of the initial foray into file standards, there were virtually no options for data analysis except the tools provided by the instrument manufacturers and the 'do-it-yourself' writers. The first suggestion of the FCS standard transformed the field of cytometry. The FCS 1.0 standard was developed by Murphy and Chused [53] and discussed within the context of the Society for Analytical Cytology (SAC) meeting at Asilomar in the same year. Over the next several years it was continuously updated from FCS 1.0 to FCS 2.0, FCS 3.0 and more recently to FCS 4.0. At the same time the society changed to the International Society for Analytical Cytology (ISAC – 1993 Congress) and more recently to the International Society for Advancement of Cytometry (ISAC – 2007 International Congress). Continuous modification of standards provides a mechanism for increasing the stability and accuracy of the entire field. Cytometry is an excellent example of a field of technology that has taken full advantage of implementation of standards to the betterment of the entire field. Cytometry has been a leader in the implementation of standards that have allowed the field to expand from the two or three major companies in the 1970s to the dozens of small companies presently developing tools, devices, software, and reagents for the field.

Calibration: The Promise of Quantitative Fluorescence

Interestingly, one of the reasons PTI was formed by Mack Fulwyler was to create standard particles for calibration. As a physicist, Mack Fulwyler recognized the importance of calibration and absolute measurements. However, it took many years before the concept of creating really well defined particles for calibration became a part of cytometry. Even today, cytometry is more relative than absolute in its quantitative nature, a factor that must change before it becomes cemented into multiple fields. Over the years, several attempts have been made to use the true quantitative capacity of cytometry. For example, it is entirely possible to calibrate most assays so that the results expressed are converted to either an equivalent concentration of substrate or product in molar terms, or, in the case of surface antigens, the number of receptors per cell. The vast majority of reports in the literature, possibly as many as 99%, fail to provide more than a percentage increase over control and express results as 'relative fluorescence intensity', meaning a relationship between two different particles or cells. One of the individuals who did the foundation work in this area by creating defined relationships to facilitate calibrated measurement is Abe Schwartz, who over a number of years was probably the person most responsible for driving the calibrated measurement of cells [79–82].

The Emergence of Clinical Cytometry

DNA analysis was a natural application that related to clinical cytometry, and it was assumed from the very beginning that this would have great clinical significance. Van Dilla was studying the use of DNA dyes by 1968, and others were following closely, particularly in Europe. For example, Göhde et al. were driving this area very significantly [83–89] and established many of the early principles. Some years later Andreef [90], Crissman [91, 92], Vindelov [93], and Hedley [50] all developed fundamental components of diagnostic testing for DNA that held promise of use in clinical environments. These core technologies together with outstanding software packages developed to evaluate the cell cycle have given flow cytometry a decided edge over other technologies in the world of oncology.

The majority of clinical applications, however, are probably focused on lymphocyte phenotyping. The study of lymphocytes was heavily driven by the discovery of mAbs in 1975 [1].

The Reemergence of Cell Sorting

During the late 1980s a small group of individuals from Australia began to design computer systems that were more sophisticated than those provided by the major manufacturers. The idea was to add these computer systems for better management of data collection and analysis. Subsequently, a small company by the name of Cytomation emerged, and over a number of years it became the premier producer of high-speed cell sorters, significantly outperforming the traditional suppliers. High-speed cell sorting functioned as a powerful driver of cytometry for a number of years. Indeed, several companies emerged to produce instruments specifically designed for high-speed cell sorting (Cytomation, Cytopea). Interestingly, both companies have been absorbed by the major manufacturers in recent times. One reason may have been the emergence of yet another small company, very much in the mode of the original Cytomation. That company, iCyt Visionary Biosciences, is currently producing very high-speed, almost industrial strength sorters that have focused on advanced engineering, automation, and robustness, and particularly biosafety – all features that have been promised in the past, but not well implemented. It is interesting to speculate that another expansion period of high-speed sorting is about to enter the field of cytometry. A need for stem cells for transplantation is one of the clinical driving forces. Another is a desire to enhance the safety of cell sorters in routine research environments.

The Future: The Next Generation

Every generation talks of the future and what next-generation technologies will bring to the world. Upon review, promised 'next-gen' products are often true to their word – just promises. However, it is clear that there are well-defined periods of technology development over the past 40 years. The 1950s was the decade of cell counting, the 1960s that of the emergence of cell sorting, the 1970s that of fluorescence, the 1980s the second period of sorting, the 1990s the decade of multicolor fluorescence, and the turn of the century brought the re-emergence of high-speed cell sorting and low-cost analyzers. What is left? It can be only one thing: small point-of-test (POT) instruments (similar to the term point-of-care; POC) using microfluidic tools that are capable of simple rapid application across hundreds of areas. The irony of this prediction is that for the last 20 years we have been developing tremendously complex systems that are capable of performing every conceivable assay. The future will return to the simple, easily readable test, low-cost, and functionally distinct. Cytometry will become a tool for the masses, even those who have no idea what is inside the box or how it works. Companies whose focus has been on mass production of consumer products may well be the most capable of capitalizing on their technology; and low-cost, high distribution and compact electronics and optics will drive the 'next-gen' yet again. For example, companies that make high production runs of items like CD-ROMs and DVDs, which require accuracy and high degrees of reproducibility with low tolerance, will be best suited to capitalize on the future needs of cytometry. This will be the ultimate success for cytometry.

Postlude

My goal in writing this chapter was to document the earliest work in the field of cytometry in detailed fashion, not to cover every success in the field or even to venture into all the wonderful technology developments that currently give advanced cytometry the power it holds. Nor have I covered the aspects of imaging and the incredible story of discovery related to that. This is the task of another document. The recent decision of ISAC to change its name from 'analytical cytology' to 'advancement of cytometry' really carried with it a strong message that the field of cytometry is fundamentally about moving forward in all aspects of single-cell analysis. Clinical cytometry is a vital application of advances in cytometry. Developments and advances in the basic cytometry tools leads to advanced applications. Clinical cytometry may well be one of those applications with the greatest impact and the greatest opportunities in the future. For example, there is a clear need to introduce more sophisticated classification routines into the clinical field. Modeling must become more ingrained into the cytometry pathway. Advanced mathematics processing to produce faster and more meaningful results for the physician may be the 'holy grail' of clinical diagnostics. With automation comes better standardization, reproducibility, and accuracy. These are the hallmarks of quality cytometry.

Acknowledgments

Robert Auer for provision of the copies of Mack Fulwyler's original laboratory notebooks from 1964. Wolfgang Göhde, Lou Kamentsky, Robert Auer, Robert Hoffman, Howard Shapiro, Leonard Herzenberg, Marshall (Don) Graham, Michael Borowitz, and Gary Durack for kindly providing historical information.

Smithsonian Institution for recording and maintaining the historic video tapes of interviews with leading cytometry individuals in 1990. Boris Rotman and Brown University for donating the original Fulwyler sorter to Purdue University. Wallace H. Coulter Foundation for documentation, photographs and historical records relating to Wallace H. Coulter. Gretchen Lawler for review of this manuscript.

References

1 Köhler G, Milstein C: Continuous cultures of fused cells secreting antibody of predefined specificity. Nature 1975;256:495–497.
2 Fulwyler MJ: Electronic separation of biological cells by volume. Science 1965;150:910–911.
3 Moldavan A: Photo-electric technique for the counting of microscopical cells. Science 1934;80:188–189.
4 Coons AH, Creech HJ, Jones RN: Immunological properties of an antibody containing a fluorescent group. Proc Soc Exp Biol Med 1941;47:200–202.
5 Caspersson T: Studien über den Eiweißumsatz der Zelle. Naturwissenschaften 1941;29:3 33–43.
6 Papanicolaou GN, Traut HF: The diagnostic value of vaginal smears in carcinoma of the uterus. Am J Obst Gynec 1941;42:193–206.
7 Papanicolaou GN: Some improved methods for staining vaginal smears. J Lab Clin Med 1941;26:1200.

8 Avery OT, MacLeod CM, McCarty M: Studies on the chemical nature of the substance inducing transformation of pneumococcal types. J Exp Med 1944 79;2:137–158.

9 Gucker FT, Pickard HB, O'Konski CT: A photoelectric instrument for comparing the concentrations of very dilute aerosols, and measuring low light intensities. J Am Chem Soc 1947;69:429–438.

10 Gucker FT, O'Konski CT, Pickard HB, Pitts JN: A photoelectronic counter for colloidal particles. J Am Chem Soc 1947;69;2422–2431.

11 Coulter WH: Means for Counting Particles Suspended in a Fluid. U.S. Patent #2,656,508. Application August 27, 1949. Patented October 20, 1953.

12 Caspersson T: Cell Growth and Cell Function. A Cytochemical Study. Norton, New York, 1950.

13 Friedman HP Jr: The use of ultraviolet light and fluorescent dyes in the detection of uterine cancer by vaginal smear. Am J Obst Gynec 1950;59:852.

14 Crosland-Taylor PJ: A device for counting small particles suspended in a fluid through a tube. Nature 1953;171:37–38.

15 Tolles WE: The cytoanalyzer: an example of physics in medical research. Trans N Y Acad Sci 1955;17:250–256.

16 Coulter WH: High speed automatic blood cell counter and cell size analyzer. Proc Natl Electronics Conf 1956;12:1034–1042.

17 Rotman B: Measurement of activity of single molecules of β-d-galactosidase. Proc Natl Acad Sci USA 1961;47:1981–1991.

18 Kamentsky LA, Derman H, Melamed MR: Ultraviolet absorption in epidermoid cancer cells. Science 1963;142:1580–1583.

19 Rotman B, Zderic JA, M Edelstein: Fluorogenic substrates for β-d-galactosideases and phosphatases derived from fluorescein (3,6-dihydroxyfluoran) and its monomethyl ether. Proc Natl Acad Sci USA 1963;50:1–6.

20 Sweet RG: High frequency recording with electrostatically deflected ink jets. Rev Sci Instrum 1965;36:131–136.

21 Hallerman L, Thom R, Gerhartz H: Elektronische Differentialzählung von Granulocyten und Lymphocyten nach intravitaler Fluorochromierung mit Acridinorange. Verh Dtsch Ges Inn Med 1964;70:217.

22 Kosenow W: Die Fluorochromierung mit Acridinorange, eine Methode zur Lebendbeobachtung gefärbter Blutzellen. Acta Haematol 1952;7:217.

23 Ingram M, Preston K Jr: The importance of automatic pattern recognition techniques in early detection of altered blood cell production. Ann N Y Acad Sci 1964;113:1066.

24 Fulwyler MJ: Particle Separator. U.S. Patent #3,380,584. Application June 4, 1965. Patented April 20, 1968.

25 Fulwyler MJ: An Electronic Particle Separator with Potential Biological Application. Los Alamos Scientific Laboratory Annual Report of the Biological and Medical Research Group (H–4) of the Health Division, July 1964 through June 1965, Written July 1965.

26 Fulwyler MF: Electronic separation of biological cells by volume. Science 1965;150:910–911.

27 Kamentsky LA, Melamed MR, Derman H: Spectrophotometer: new instrument for ultrarapid cell analysis. Science 1965;150:630–631.

28 Rotman B, Papermaster BW: Membrane properties of living mammalian cells as studied by enzymatic hydrolysis of fluorogenic esters. Proc Natl Acad Sci USA 1966;55:134–141.

29 Van Dilla MA, Mullaney PF, Coulter JR: The Fluorescent Cell Photometer: A New Method for the Rapid Measurement of Biological Cells Stained with Fluorescent Dyes. Los Alamos Scientific Laboratory Annual Report of the Biological and Medical Research Group (H–4) of the Health Division, July 1966 through June 1967, Written September 1967.

30 Ingram M, Norgren PE, Preston K Jr: Automatic differentiation of white blood cells; in Ramsey DM (ed): Image Processing in Biological Science. Los Angeles, University of California Press, 1968, pp 97–117.

31 Wald N, Preston K Jr: Automatic screening of metaphase spreads for chromosome analysis; in Ramsey DM (ed): Image Processing in Biological Science. Los Angeles, University of California Press, 1968, pp 9–34..

32 Ingram M, Preston K Jr:. Automatic analysis of blood cells. Sci Am 1970;223:72.

33 Göhde W: Automatisches Meß- und Zählgerät für die Teilchen einer Dispersion. German Patent #DE1815352, priority date Dec. 18, 1968.

34 Wied GL, Bahr GF (eds): Automated Cell Identification and Cell Sorting (Proceedings of 1968 Conference in Chicago). New York, Academic Press, 1970.

35 Dittrich W, Göhde W: Impulsfluorimetrie bei Einzelzellen in Suspensionen. Z Naturforsch 1969;24b:360–361.

36 Hulett HR, Bonner WA, Barrett J, Herzenberg LA: Cell sorting: automated separation of mammalian cells as a function of intracellular fluorescence. Science 1969;166:747–749.

37 Van Dilla MA, Trujillo TT, Mullaney PF, Coulter JR: Cell microfluorometry: a method for rapid fluorescence measurement. Science 1969;163:1213–1214.

38 Mullaney PF, Van Dilla MA, Coulter JR, Dean PN: Cell sizing: a light scattering photometer for rapid volume determination. Rev Sci Instrum 1969;40:1029–1032.

39 Bonner WA, Hulett HR, Sweet RG, Herzenberg LA: Fluorescence activated cell sorting. Rev Sci Instrum 1972;43:404–409.

40 Göhde W: Process for Automatic Counting and Measurement of Particles. US patent #4,021,117, priority date Jun 23, 1973 (DE).

41 Dean PN, Jett JT: Mathematical analysis of DNA distributions derived from flow microfluorometry. J Cell Biol 1974;60:523.

42 Williams AF, Galfrè G, Milstein C: Analysis of cell surfaces by xenogeneic myeloma-hybrid antibodies: differentiation antigens of rat lymphocytes. Cell 1977;12:663–673.

43 Loken MR, Parks DR, Herzenberg LA: Two-color immunofluorescence using a fluorescence-activated cell sorter. J Histochem Cytochem 1977;25:899–907.

44 Hogg WR, Brunsting A: Ellipsoid-Conic Radiation Collector Method. US Patent #4,189,236 Ellipsoid Radiation Collector Apparatus and Method. US Patent #4,188,543. Mirror Image Ellipsoid Radiation Collector and Method. US Patent #4,188,542.

45 Kung PC, Goldstein G, Reinherz E, Schlossman SF: Monoclonal antibodies defining distinctive human T cell surface antigens. Science 1979;206:347–349.

46 Reinherz E, Schlossman SF: The differentiation and function of human T lymphocytes. Cell 1980;19:821.

47 Kachel V, Benker G, Lichtnau K, Valet G, Glossner E: Fast imaging in flow: a means of combining flow-cytometry and image analysis. J Histochem Cytochem 1979; 27:335–341.

48 Brunsting A, Hogg WR: Radiant Energy Reradiating Flow Cell System and Method. US Patent #4,523,841.

49 Weller LA, Wheeless LL: EMOSS: an epiillumination microscope objective slit-scan flow system. Cytometry 1982;3:15–18.

50 Hedley DW, Friedlander ML, Taylor IW, Rugg CA, Musgrove EA: Method for analysis of cellular DNA content of paraffin embedded pathological material using flow cytometry. J Histochem Cytochem 1983;31:1333–1335.

51 Hedley DW, Friedlander ML, Taylor IW, Rugg CA, Musgrove EA: DNA flow cytometry of paraffin-embedded tissue. Cytometry 1984;5:660.

52 Hiddemann W, Schumann J, Andreeff M, Barlogie B, Herman CJ, Leif RC, Mayall BH, Murphy RF, Sandberg AA: Convention on nomenclature for DNA cytometry. Cytometry 1984;5:445–446.

53 Murphy RF, Chused TM: A proposal for a flow cytometric data file standard. Cytometry 1984;5:553–555.

54 Parks DR, Hardy RR, Herzenberg LA: Three-color immunofluorescence analysis of mouse B-lymphocyte subpopulations. Cytometry 1984;5:159–168.

55 Watson JV: Time, a quality-control parameter in flow cytometry. Cytometry 1987;8:646–649.

56 Gray JW, Dean PN, Fuscoe JC, Peters DC, Trask BJ, van den Engh GJ, Van Dilla MA: High-speed chromosome sorting. Science 1987;238:323–329.

57 Watson JV: Flow cytometry chamber with 4π light collection suitable for epifluorescence microscopes. Cytometry 1989;10:681–688.

58 Data File Standards Committee of the Society for Analytical Cytology: Data File Standard for Flow Cytometry. Cytometry 1990;11:323–332.

59 Robinson JP, Maguire D, King G, Kelley S, Durack G: Integration of a barcode reader with a commercial flow cytometer. Cytometry 1992;13:193–197.

60 Roederer M, Bigos M, Nozaki T, Stovel RT, Parks DR, Herzenberg LA: Heterogeneous subsets revealed calcium flux in peripheral T cell by five-color flow cytometry using log-ratio circuitry. Cytometry 1995;21:187–196.

61 Roederer M, De Rosa S, Gerstein,R, Anderson M, Bigos M, Stovel R, Nozaki T, Parks DR, Herzenberg L, Herzenberg L: 8 color, 10-parameter flow cytometry to elucidate complex leukocyte heterogeneity. Cytometry 1997;29:328–339.

62 De Rosa SC, Herzenberg LA, Roederer M: 11-color, 13-parameter flow cytometry: identification of human naive T cells by phenotype, function, and T-cell receptor diversity. Nat Med 2001;7:245-248.

63 Perfetto SP, Chattopadhyay PK, Roederer M: Seventeen-colour flow cytometry: unravelling the immune system. Nat Rev Immunol 2004;4:648–655.

64 Shapiro HM: Practical Flow Cytometry, 3rd ed. New York, Wiley Liss, 1995.

65 Valet G: Past and present concepts in flow cytometry: a European perspective. J Biol Regul Homeost Agents 2003;17:213–222.

66 Robinson JP: Cytometry – 60 Years of Innovation. West Lafayette, Purdue University Cytometry Laboratories, 2007.

67 Caspersson TO: History of the development of cytophotometry from 1935 to the present. Anal Quant Cytol Histol 1987;9:2–6.
68 Coons AH, Kaplan MH: Localization of antigen in tissue cells. II. Improvements in a method for the detection of antigen by means of fluorescent antibody. J Exp Med 1950;91:1–13.
69 Robinson JP: Mack Fulwyler in his own words. Cytometry Part A 2005;67A:61–67.
70 Mellors RC, Silver R: A microfluorometric scanner for the differential detection of cells: application to exfoliative cytology. Science 1951;114:356–360.
71 Caspersson O: Quantitative cytoochemical studies on normal, malignant, premalignant and atypical cell populations from the human uterine cervix. Acta Cytol 1964;8:45–60.
72 Kamentsky LA, Liu CN: Computer-automated design of multifont print recognition logic. IBM J Res Dev 1963;7:2–13.
73 Kamentsky LA, Melamed MR, Derman H: Spectrophotometer: new instrument for ultrarapid cell analysis. Science 1965;150:630–631.
74 Valet G: Past and present concepts in flow cytometry: a European perspective. J Biol Regul Homeost Agents 2003;17:213–222.
75 Göhde W, Schumann J, Fruh J: Coincidence eliminating device for pulse-cytophotometry; in Göhde W, Schumann J, Büchner T (eds): Pulse Cytophotometry II. Ghent, European Press, 1976, pp 71–78.
76 Milstein C: The hybridoma revolution: an offshoot of basic research. Bioessays 1999;21:966–973.
77 Williams AF, Galfre G, Milstein C: Analysis of cell surfaces by xenogeneic myeloma-hybrid antibodies: differentiation antigens of rat lymphocytes. Cell 1977;12:663–673.
78 Milstein C: With the benefit of hindsight. Immunol Today 2000;21:359–364.
79 Schwartz A, Sugg H, Ritter TW, Fernandez-Repollet E: Direct determination of cell diameter, surface area, and volume with an electronic volume sensing flow cytometer. Cytometry 1983;3:456–458.
80 Schwartz A, Fernández-Repollet E: Development of clinical standards for flow cytometry, Ann N Y Acad Sci 1993;677:28–39.
81 Schwartz A, Marti GE, Poon R, Gratama, JW, Fernandez-Repollet E: Standardizing flow cytometry: a classification system of fluorescence standards used for flow cytometry. Cytometry 1998;33:106–114.
82 Schwartz A, Fernandez-Repollet E: Quantitative flow cytometry. Clin Lab Med 2001;21:743–761.
83 Göhde W, Dittrich W: Simultane Impulsfluorimetrie des DNS – und Proteingehaltes von Tumorzellen. Z Anal Chem 1970;252:328–330.
84 Büchner T, Dittrich W, Göhde W: Impulsecytophotometrie für Blut- und Stammzellen. Verh Dtsch Ges Inn Med 1971;77:416–418.
85 Büchner T, Dittrich W, Göhde W: Impulsecytophotometrie in der Hämatologischen Cytologie. Klin Wochensch 1971;49:1090–1092.
86 Büchner T, Göhde W, Dittrich W, Barlogie B: Proliferation kinetics of leukemias before and during therapy as based on impulse cytophotometry (in German). Verh Dtsch Ges Inn Med 1972;78:159–162.
87 Büchner T, Hiddemann W, Schneider R, Kamanabroo D, Göhde W Cell-kinetic effects of leukemia therapy in chemical medicine as based on the DNA-histogram with impulse cytophotometry (in German). Med Welt 1973;24:1616–1617.
88 Büchner T, Asseburg U, Kamanabroo D, Hiddemann W, Hiddemann RM, Barlogie B, Göhde W: Clinical studies on combined chemotherapy of leukemia with partial cell synchronization (in German). Verh Dtsch Ges Inn Med 1974;80:1674–1678.
89 Büchner T, Barlogie B, Asseburg U, Hiddemann W, Kamanabroo D, Göhde W: Accumulation of S-phase cells in the bone marrow of patients with acute leukemia by cytosine arabinoside. Blut 1974;28:299–300.
90 Andreeff M: Impulscytophotometrie. Berlin, Springer, 1975.
91 Crissman HA, Steinkamp JA: Rapid, simultaneous measurement of DNA, protein, and cell volume in single cells from large mammalian cell populations. J Cell Biol 1973;59:766–771.
92 Crissman HA, Tobey RA: Cell-cycle analysis in 20 minutes. Science 1974;184:1297–1298.
93 Vindelov LL: Microfluorometric analysis of nuclear DNA in cells from tumors and cell suspensions. Virchows Archiv B 1977;24:227–227.

Sack U, Tárnok A, Rothe G (eds): Cellular Diagnostics. Basics, Methods and Clinical Applications of Flow Cytometry. Basel, Karger, 2009, pp 29–52

Concept Developments in Flow Cytometry

Günter Valet

Max Planck Institute for Biochemistry, Martinsried, Germany

Introduction

Cytometry fundamentally influences the development of modern biosciences because even our highly detailed knowledge on the sequenced genomes does not currently explain how the coded biomolecules assemble in order to ultimately develop the architecture and function of living cells. The multitude of theoretical possibilities for the formation of biostructures and the highly compartmentalized functionality of complex metabolic pathways limit the precise understanding of the molecular networks in living cells by deductive hypothesis and mathematical modeling. In this context, cytometry opens the possibility of performing high-speed simultaneous measurements of several molecular parameters of single cells in heterogeneous cell mixtures consisting of many different cell types. This can be done under fairly physiological conditions. Comparing cell measurements from healthy and diseased individuals permits detection of disease-associated changes in molecular expression or functionality in the form of differential molecular cell phenotypes. Measurement parameters can be selected according to system analysis concepts rather than to prove or disprove a particular molecular pathway hypothesis. For example, data analysis can be focused on the capacity of cell parameters to predict the therapy-dependent outcome of disease. Such parameters and parameter patterns can be used for inductive hypothesis development in an effort to unify the observations in particular molecular pathways and to understand the molecular causes of disease.

Given the multitude of published articles in the field, this review predominantly centers on work that initiated essential further developments in flow cytometry.

Cytophotometry

The stimulus for the development of cytophotometric instrumentation originated from endeavors to distinguish cells from healthy and diseased organisms based on their molecular properties [1]. Cells in such studies were typically attached to microscopic glass slides. Light absorption at a given wavelength was measured by stepwise overlap-free motion of the microscope stage in front of the opening of a small diaphragm in the focus plane of the microscope. The integrated optical density of a given area of interest such as the cell nucleus could be determined in this way. The wavelength was between 250 and 260 nm to determine the cellular DNA content from the photometric absorption of the DNA bases. The use of the stoichiometric Feulgen stain [2] with a maximum light absorption between 550 and 570 nm [3] permitted measurements without ultraviolet (UV) optics as required otherwise. The cell protein content was determined at 280 nm by the light absorption of aromatic amino acids like tyrosine, tryptophan and phenylalanine. The Zeiss-UMSP1 universal microspectrophotometer (Oberkochem, Germany) was frequently used for such investigations. The comparatively slow speed of measurement of between 5 and 10 min per cell nucleus or cytoplasm area, however, precluded the investigation of larger cells numbers.

Flow Technology

Blood cell counters worked much faster, counting between 500 and 5,000 cells/s. Light scatter [4, 5] or electrical voltage pulse counting during the transit of electrically nonconductive blood cells in physiological saline suspension through a cylindrical capillary orifice of typically between 70 and 100 μm as in the Coulter [6, 7] were used to enumerate erythrocyte, leukocyte and thrombocyte concentrations in blood. As the electrical pulse height is proportional to cell volume, cell volume distribution curves can be recorded as well. Furthermore, the absolute cell volume can be calculated from capillary geometry, conductivity of the suspension medium and cell form during capillary transit. The first cell sorter, for example, used electrical cell sizing to sort erythrocytes according to cell volume [8].

Impulse Cytophotometry

A significant improvement compared with slide-oriented cytophotometry was achieved by aspirating cells into a laminar and hydrodynamically focused [9] narrow beam through the capillary orifice of a measurement chamber at a speed of around 1 m/s. Following specific staining with a fluorescent dye, the cell beam can be axially Koehler epi-illuminated through a 40× or 63× high numerical aperture objective by UV light from a high-pressure mercury arc lamp (HBO-100, Osram, Munich, Germany) [10]. The emerging fluorescence is observed in the focus plane of the microscope through the same objective as for epi-illumination through a confocal pinhole to lower the amount of captured fluorescence noise around the cell beam. The fluorescence pulses from the stained cells are amplified by a photomultiplier tube, followed by analogue-to-digital conversion and classified according to their maximum amplitude or pulse area in a multichannel analyzer.

The first commercial flow cytometer, the ICP-11 impulse cytophotometer (Phywe Company, 1969, Göttingen, Germany) [10], typically measured the fluorescence of several thousand cells per second. Two photomultipliers permitted simultaneous collection of fluorescence from two types of molecules of similar fluorescence excitation but different emission characteristics. The separation of the emitted fluorescence light was achieved by a wavelength-sensitive dichroic mirror and long- and short-pass filters that collected the light in defined wavelength windows in each of the two light paths. Cell doublets and cell aggregates were eliminated by setting thresholds for the pulse height over pulse area ratio [11, 12], a procedure that is widely used today in flow cytometers or cell sorters.

The ICP-11 instrument [10, 13] was available at a time when doubts still existed as to whether fluorescence cytometry would be able to reach the same sensitivity as absorption cytometry using specific cytochemical cell staining [14–16]. Progress in absorption cytometry led to the development of the Hemalog-D [17, 18] blood cell analyzers (Technicon, Tarrytown, NY, USA). Although its technology is still used in today's instruments, most researchers considered fluorescence the more promising approach.

As a consequence, the Bio/physics CytoFluorograph (Bio/Physics Systems Inc, Mahopac, NY, USA) with a 488-nm argon ion laser for fluorescence excitation was developed (1970) from the initial Cytograph absorption cytometer with its 633-nm HeNe laser light source for light absorption measurements in cells.

The Phywe ICP-11 and ICP-22 as well as the Biophysics developments were purchased by Ortho Diagnostics (Johnson and Johnson Unternehmensbereich, Westwood, MA, USA) in 1976. The CytoFluorograph development was continued while the Phywe development disappeared from the market. The Coulter Company (Miami, FL, USA) followed with the EPICS cell sorter in 1977, and Becton Dickinson (Mountain View, CA, USA), with the FACSIV cell sorter [19] in 1978.

The Partec Company (Münster, Germany) was at the origin of the early Phywe ICP-11/ICP-22 development and continued the development of flow cytometers on its own after the takeover of Phywe by Ortho Diagnostics. A closed piezocrystal cell sorter [20–22] became available around 1986, proving particularly useful for sorting large particles (pancreatic islands [23]) and infectious material. Partec also developed various new types of instruments like the PASI, PASII, PASIII flow cytometers with different possibilities for illumination, measurement and piezo cell-sorting chambers as well as a number of analysis parameters.

Nomenclature

The variety of initial nomenclature and publication languages for the new technology explains the difficulty of gathering an adequate impression of the early developments, especially in Europe. Terms like 'Impulsfluorimetrie' [13, 24], 'Impulsfluorometrie' [25], 'Impulszytophotometrie' [26, 27, 29, 30], 'Impulsmicrophotometrie' [28], 'pulse cytophotometry' [30], 'micro-flow fluorometry' [31], and 'microflow fluorometry' [32] were used in Europe whereas 'flow cytometry' [33], 'flow microfluorimetry' [34] or 'flow microfluorometry' [35] were preferred in the USA. The term 'flow cytofluorometry' [36] was used both in Europe and the USA. A further term, 'flow DNA analysis' [37], was also coined.

The term 'flow cytometry' was accepted in 1976 by an international consensus at the 5th American Engineering Foundation Conference in Pensacola, FL, USA. Due to the nomenclature change and the publication habits, especially of the German proponents of flow cytometry, a significant part of the groundbreaking work in the technical, and especially in the clinical areas is not adequately appreciated. In view of the various terms, the ISI Web of Knowledge during the first decade of flow cytometry (1969–1978) mentioned 60 articles for the various European terms versus 44 articles for the terms flow cytometry, flow microfluorimetry, flow cytoflorimetry and flow cytofluorometry, predominantly used by US authors. This comparison does not take into account the significant number of European articles in catalogued books and periodicals that were not listed by the Institute for Scientific Information (ISI).

Clinical Impulse Cytophotometry

A first meeting of the growing group of clinical scientists interested in 'impulse cytophotometry' was organized by Michael Andreeff in 1972 in Heidelberg, Germany, with follow-up meetings in Nijmegen, The Netherlands (1973), Münster, Germany (1975), Vienna, Austria (1977), Voss, Norway (1979) and Rome, Italy (1980). The organizers were Clemens Haanen, Wolfgang Göhde, Dieter Lutz, Ole Laerum and Francesco Mauro. The scientific contributions were edited and published by C.A.M. Haanen, H.F.P. Hillen, J.M.C. Wessels (*Pulse Cytophotometry I*), W. Göhde, J. Schumann, T. Büchner (*Pulse Cytophotometry II*) and D. Lutz (*Pulse Cytophotometry III*) and printed by European Press, Ghent, Belgium in 1975, 1976 and 1978, respectively. *Flow Cytometry IV* was edited by O.D. Laerum, T. Lindmo, E. Thorud and printed by Universitetsforlaget, Bergen, Norway in 1980. A reprint of this book appeared in *Acta Pathologica et Microbiologica Immunologica Scandinavica A, Supplement* (1981;274:1–535). The abstracts of the Rome Congress were edited by F. Mauro and G. Mazzini for *Basic and Applied Histochemistry* (1980;24:229–398). A separate congress proceedings monograph with the publications of these presentations is not available.

The significance of the early flow cytometry congresses held between 1972 and 1980 is shown by the fact that 20 of the cited publications in this review mark conceptual focus points for later developments. Seven additional articles in *The* Journal of *Histochemistry and Cytochemistry* represent European contributions at the American Engineering Foundation Congresses, which were organized during the same period and alternated between the USA and Europe (*http://www.isac-net.org/content/category/4/132/42*).

The Society for Analytical Cytology (SAC) was founded in 1978 during the 6th American Engineering Foundation Conference on Automated Cytology in Schloss Elmau near Mittenwald (Germany). This resulted in the initially controversial tendency to abandon the organization of cytometry meetings in Europe and to work within SAC. SAC congresses were subsequently organized in the USA and Europe, and the journal *Cytometry* was founded in 1980.

The European scene was reorganized between 1985 and 2000 in the form of national cytometric societies in Italy (GIC, SICICS), France (AFC), Portugal/Spain (SIC), Germany (DGfZ), Denmark, UK (section of the Royal Microscopic Society, RMS), Belgium (BVC/ABC), Switzerland (SCS), Sweden, Poland (PCS) and Austria (OEGfZ) totaling more than 2,000 members to date. In addition to SAC, the significance of national and regional organizations is evident, as shown by the foundation of the European Society for Analytical Cellular Pathology (ESACP) in 1986 with its journal *Analytical Cellular Pathology* (ACP) published since 1989, and the foundation of the European Working Group for Clinical Cell Analysis (EWGCCA) in 1996, which cooperates with the *Journal of Biological Regulators and Homeostatic Agents* (JBRHA) since 2002. Quality assurance for clinical immunophenotyping and other clinical cytometry applications at the European level started with the collaboration between EWGCCA and Eurostandards/UK-NEQAS (Sheffield, UK) and resulted in the foundation of the European Society for Clinical Cell Analysis (ESCCA) in 2007 and participation in the journal *Cytometry Part B* (*Clinical Cytometry*). ESACP and the Society for Diagnostic Quantitative Pathology (ISDQP) merged in 2003 as the International Society for Cellular Oncology (ISCO) with the journal ACP reappearing as *Journal of Cellular Oncology* (JCO). SAC emphasized their international ambition by changing their name to International Society for Analytical Cytology (ISAC) in 1990 and, under the same acronym, to International Society for Advancement of Cytometry in 2008.

The changes in organizational frameworks mirror the remarkable developments in flow and image cytometry.

Further Instrument Developments

During the initial expansion phase of the ICP-11 flow cytometers, several research groups developed their own instrumentation because some of the measurement tasks were only partially possible with the ICP-11 and its successor ICP-22. The HBO-100 mercury arc lamp was suitable for measurements in the UV fluorescence excitation range, but no light scatter or electrical cell volume measurements were possible. It was also difficult to measure the low-intensity immunofluorescences of fluorescein isothiocyanate (FITC)-labeled antibodies on cells with the limited excitation intensity of the mercury arc lamp between 470 and 490 nm. The CytoFluorograph (Bio/Physics 1970, Ortho Diagnostics 1976) as well as the Coulter EPICS and the Becton Dickinson FACS instruments were typically equipped with argon ion laser fluorescence excitation at 488 nm and captured small-angle (1–3 degrees) forward-scatter and orthogonal (90 degrees) side scatter light.

The necessity to determine more than two fluorescence parameters led to the development of the first double-laser [38] flow cytometer and cell sorter at the Deutsches Krebsforschungszentrum (DKFZ, Heidelberg, Germany). An instrument with a particularly tightly bundled laser beam [39, 40] for fast and precise length measurement of cells and cell aggregates was developed by the Gesellschaft für Strahlenforschung (GSF), Hanover, Germany, and later commercialized by Kratel Instrumente (Stuttgart-Leonberg, Germany).

The Metricell [41] and Fluvo-Metricell instruments [42] were developed at the Max Planck Institute for Biochemistry to determine relative analyte concentrations in cells and average surface densities of molecules on the cell surface. A measurement chamber with a hydrodynamically focused electrical sizing orifice for the fast determination of cell volumes was initially combined with the optical bench of an ICP-11 instrument. A multichannel analyzer and a computer served for histogram display and storage of listmode data [43]. The Fluvo-Metricell was subsequently redesigned with a new optical setup, a Z-80 microprocessor computer, and manufactured by HEKA Elektronik (Forst/Weinstraße, Germany) between 1985 and 1990.

The generally observed right skew of erythrocyte volume distribution curves by electrical sizing with Coulter measurement capillaries initially presented a significant obstacle for the precise characterization of normal or abnormal erythrocyte populations, in particular the analysis of mixed, discrete erythrocyte populations of various size as sometimes found in newborn or x-irradiated mammalian organisms, after a hemorrhage or induced erythropoiesis in the post-partum phase [44]. Several cell volume peaks were observed under these conditions for certain time periods.

Significant work on the right skew issue led to the development of the piezo-crystal-driven [45] droplet cell sorter [8]. The exact cause of the artifact was finally elucidated by high-speed photography of native and fixed erythrocytes during their passage through the orifice using microsecond laser pulses [46] or an argon arc flashlight [47]. These investigations revealed that the right skew is caused by cells flowing close to the edge of the capillary entrance through zones of increased electrical field strength [9, 46, 48], leading to M-shaped electronic pulses. The electronic rejection of M-shaped pulses [49], the use of short measurement capillaries, or the widening of the orifice outlet side [50] of the capillary only diminishes the right skew, while hydrodynamic focusing of the cells as a narrow cell beam through the orifice reliably avoids it altogether [9, 51–53]. AEG Telefunken (Ulm, Germany) used hydrodynamically focused electrical sizing [51] in its AEG-Telefunken particle analyzer (1972). The AEG development was acquired by the Coulter Company (Miami, FL, USA) and subsequently taken off the market.

The AEG engineers observed a transcellular ion flux through erythrocytes or nucleated cells during elevated electrical-field strength in the measurement capillary. The ion flux is caused by a temporary dielectric breakdown of the cell membrane [54–56]. These observations are used for controlled molecule transport through cell membranes by electroporation [57], a method frequently applied for transfection in molecular biology.

High-speed photography enables fast imaging in cell flow [58], an idea taken up recently as high-throughput technology by the ImageStream100 instrument (Amnis, Seattle, WA, USA) [59], taking advantage of far more sensitive imaging technologies. The narrow focusing of the fluorescence exciting laser beam to a focus of around 0.5 μm permits the determination of single-cell [60] or chromosome [61] slit-scan profiles in flow.

The requirement for fast signal processing, histogram display and signal ratios led to the use of software-driven microprocessors [62–64] instead of a hardware circuit or computer control. The early modular instrumentation was partly developed by Dr. O. Ahrens Meßtechnik (Bargteheide, Germany), a company which continues to develop DNA image analysis systems.

The interest in time- and temperature-controlled flow-cytometric cell function experiments led to the development of a particularly adapted laser flow cytometer and cell sorter in Cambridge (UK) [65]. Besides several lasers, particular measurement chambers for the sensitive fluorescence light collection from large spatial angles were developed [66, 67].

The measurement of microorganisms like bacteria or yeast cells [68] initiated the development of a particularly sensitive epi-illumination system [69] with an HBO-100 mercury arc lamp for fluorescence excitation in Oslo (Norway). The instrument was successively produced under the names MPV flow cytometer by Leitz (Wetzlar, Germany), Argus100 by Skatron (Tranby, Norway) and Bryte HS by BioRad-Laboratories (Hercules, CA, USA).

Bruker-Odam (Wissembourg, France) produced the ATC3000 flow cytometer and cell sorter between 1990 and 1993. The instrument was developed under the patronage of the French Commissariat à l'Energie Atomique (CEA) with a hydrodynamically focused electrical cell-sizing chamber. Fast graphics with the possibility to evaluate a high number of simultaneous histogram windows of polygonal or elliptic shape represented the particular feature of this instrument.

Ethidium bromide (2,7-diamino-10-ethyl-9-phenylphenantridiniumbromide) [13, 24], acriflavin-auramine [70], propidium iodide [71], mithramycin [72], ethidium bromide-mithramycin [73], chromomycin A3 [74], acridine orange [75, 76], DAPI (4′,6-diamidino-2-phenyindol) [77], Hoechst 33342 and Hoechst 33258 (2,6-bis-benzimidazole derivatives) [78] for DNA measurements or DANS (1-dimethylaminonaphthalene-5-sulfonic acid) [24] and FITC [79] as fluorescent protein stains were introduced for flow-cytometric measurements. DNA against protein was the first two-parameter fluorescence combination in flow cytometry [24], including mathematical histogram analysis [80, 81].

Cell nuclei were prepared from biological tissues by digestion with 0.5% acid pepsin solution at pH 1.8 [29, 82] or with pronase [83]. Alternatively, cellular RNA was removed by tissue treatment with 0.1–1% RNase [13, 29, 30, 84]. High and low salt concentrations at pH 10 and pH 5.8 in the presence of RNase and detergents [85], or trypsin in combination with detergent [86] proved useful for the measurement of narrow DNA distribution curves of cell nuclei. The successful enzymatic preparation of cell nuclei from paraffin block [83] material permitted to access archive material from pathological institutes, a methodology that became quite popular later on [87].

The low coefficients of variation (CV $=$ 100 $\times$ standard deviation/mean) (see also 'Quality Control and Standardization', pp. 159) of the DNA distribution of cell nuclei as obtained by measurements with Phywe or Partec instruments, and the use of the intensive ethidium bromide-mithramycin fluorescence staining [73, 88] finally opened the way for the analytical and preparative x- and y-sperm cell separation [89, 90] and flow-cytometric chromosome analysis [91, 92] with mercury arc lamp flow cytometers.

The clinical interest in cellular DNA measurements was often related to the detection of DNA aneuploidy as an indicator of malignancy. The degree of aneuploidy was expressed by the DNA index. Furthermore, the fraction of cells in the S-phase of the cell cycle was determined when possible [29, 30, 82, 93–95]. The characterization of precancerous lesions [96], stomach cancers [27, 97], leukemias and lymphomas [93, 98–100] or abnormal granulopoiesis or erythropoiesis [101] and the measurement of synovial [102], skin [32] or bladder cancer [103, 104] cells account for the immediate and widespread clinical interest in the new technology.

The use of DNA aneuploidy and S-phase determinations in daily clinical routine remainedlimited, however [105–109] despite the concerted efforts of many scientists, reflected in more than 1,000 clinically oriented scientific publications since 1969.

Chromosome Analysis, FISH

The use of intensive and narrowly focused lasers allowed the analytical and preparative separation of chromosomes [35, 110–113] to establish specific DNA libraries as one of the preconditions for the subsequent human genome project. Fluorescence in situ hybridization (FISH) was important for the visualization of specific DNA strands in chromosomes [114–117].

DNA Cell Cycle Analysis, Micronuclei, Hematopoietic Stem Cells

An essential part of the early experimental work was devoted to DNA cell cycle analysis under different growth conditions of cells, especially during and after the action of cytostatic drugs like Velbe [24], daunomycin [26], bleomycin [118], combinations of Adriamycin and bleomycin [79] or after ionizing irradiation [79, 119–121]. Cell cycle duration [122], synchronization within the cell cycle by x-irradiation and daunomycin [123], mechanisms of contact inhibition [124, 125] and lectin (concanavalin A)-induced cell agglutination [126] concerned other areas of the initial interest.

The flow-cytometric bromodeoxyuridine (BrdU)/Hoechst 33258 fluorescence quenching technique [127] offered a fast and excellent nonradioactive alternative for the study of cell regulation, similar to the use of fluorescence-labeled anti-BrdU antibodies [128–130]. The monoclonal Ki-67 antibody for the analysis of cell proliferation in healthy and diseased tissues was frequently used [131].

The flow-cytometric determination of micronuclei in peripheral blood or in tissue culture opened new possibilities for the evaluation of the mutagenic potential of substances, cytostatic drugs or ionizing radiation [132].

Hematopoietic stem cells were characterized and enriched based on their light scatter properties [133] and antibody-binding characteristics [134] following centrifugal elutriation and cell sorting.

Predictive Cytology and Cytopathology by DNA Image Cytometry

While flow-cytometric DNA analysis is typically faster and superior in precision to DNA image cytometry on Feulgen stained-cell nuclei, the combination of the morphological analysis of normal, dysplastic or tumor cells with DNA measurement for aneuploidy provides sensitive information. Several investigations show that DNA aneuploidy in dysplastic lesions of the lung [135], larynx [136] and cervix uteri [137] predisposes a high percentage of cases to future malignant tumors. DNA image cytometry also permits to reliably detect few DNA aneuploid cells as an early sign of tumor relapse [138].

DNA image cytometry is therefore increasingly used as a reference method for the detection of malignant cells in cytological slides [139, 140], with a better discrimination rate for malignant cells by DNA measurement as compared to the exclusive use of morphological criteria. DNA image cytometry achieves >95% correct single-case predictions for the development of subsequent malignancies in oral leukoplakia [141, 142].

Quantitative DNA image cytometry paves the way for cytology and cytopathology in predicting malignant disease, with the apparent exception of DNA euploid malignant tumors, which cannot be detected this way. This suggests that the early recognition of DNA aneuploid cells by DNA image techniques may finally be of more clinical value to the individual patient than the preferred flow-cytometric investigations.

Immunophenotyping

The discrimination of lymphocytes, monocytes and granulocytes by their different forward and sideward light scatter characteristics [143, 144] is an essential prerequisite for the generalized use of flow-cytometric immunophenotyping in clinical medicine and research. The technique combines the differential binding of fluorescently labeled antibodies [145] and fluorescence compensation to remove overlap in the individual light collection paths of the flow cytometer [146], and may include cell sorting [145, 147].

Fluorescence Anisotropy, FRET, Polarized Light and Raman Scatter

Fluorescence anisotropy [148] and fluorescence resonance energy transfer (FRET) [38, 88, 149–152] in single cells can be determined by flow cytometry to investigate membrane fluidity and the spatial proximity of biomolecules, including their degree of interaction. The degree of polarization of scattered light [153] from leukocytes permits stain-free discrimination of lymphocytes, monocytes, granulocytes, and basophil and eosinophil granulocytes [154] in diluted blood. Confocal Raman microscopy permits label-free visualization of the functionality of proteins, lipids or other molecules in viable cells [155].

Measurement of Cell Functions

The interest in cell functions as fast-reacting parameters for cell biochemical, cell physiological or clinical alterations of cells led to the development of fluorescence indicator molecules for various specific cell functions. The indicators frequently pervade the cell membrane in a diffusion-controlled way as electrically uncharged and nonfluorescent precursor molecules. The fluorescent indicator molecules of positive or negative electric charge are released following intracellular enzymatic cleavage or activation by other mechanisms. Positively charged molecules tend to auto-accumulate inside of cells or organelles due to the electrically negative transmembrane and mitochondrial membrane potentials while negatively charged molecules are easier to expel by the cells through active excretion or diffusion. The tendency of positively charged molecules to auto-accumulate significantly increases their detection sensitivity.

The flow-cytometric determination of esterases with fluorescein diacetate [17, 156, 158, 159], of phosphatases and β-d-glucuronidase [160], the simultaneous determination of esterases, phosphatases (umbelliferone phosphate) [161, 162], and peptidases and transpeptidases [163, 164] in flow, represented initial challenges for enzyme activity assessment in single cells. Simultaneous DNA staining of viable cells with Hoechst 33342 and in dead cells with propidium iodide discriminated both types of cells in the same assay [165] while phagocytosis of fixed and FITC-labeled bacteria [166] or of monodisperse fluorescence particles [167] or viable bacteria [168] provided insight into cell function changes during phagocytosis.

Interest in the intracellular pH value as an indicator of the metabolic state of viable cells in experimental and clinical settings led to the use of fluorescent pH indicator substances like fluorescein, which maintains a pH-dependent fluorescence excitation spectrum at a constant emission range. This required the sequential excitation of cells in the flow cytometer with a single laser in two runs of a cell batch at two different wavelengths. The observed fluorescence excitation ratio from the double wavelength excitation represents a measure of the average intracellular pH of all cells [169], but not that of a single cell. The technically more demanding setup of two sequential laser beams in the same instrument, with 4-methylumbelliferone [170]) as pH indicator, provided single-cell pH values. The earlier alternative idea of using substances with pH dependent fluorescence emission spectra such as 1,4-dicyano-hydroquinone [171]) required only a single light source and two fluorescence emission channels to determine intracellular pH values from fluorescence emission ratios. The use of fluorescence emission ratio dyes made intracellular pH measurements accessible to most standard flow cytometers without hardware modification. The use of fluorescence emission ratios determined from various emission channels has become common practice [172] for many flow- and image-cytometric applications.

Further challenges concerned the determination of cell stimulation from changes of intracellular Ca^{2+} levels (Indo 1 [173]), stopped-flow calcium kinetics [174, 175], the assessment of oxidative burst activities with dihydrorhodamine 123 [176] or hydroethidine [177]), of intracellular free glutathione with *o*-phthaldialdehyde [178] or monobromobimane with N-ethylmaleimide protein thiol group blocking [179] as indicators of the reductive cell potential, as well as of the negative surface charge density with FITC-fluoresceinated polycations like polylysin or polyornithine as a measure of the electrophoretic mobility of cells [180] (see 'Determination of Cell Physiological Parameters: pH, Ca^{2+}, Glutathione, Transmembrane Potential', pp. 325).

The accurate determination of protease activities in viable cells was of interest for granulocyte function studies in intensive-care patients as early indicators of imminent sepsis or polytraumatic shock [181]. Efforts to increase sensitivity and specificity of the initially used fluorogenic endopeptidase substrate (Z-Arg_2-4-trifluoromethyl-coumarinyl-7-amide [182] led to the development of the significantly more sensitive and specific rhodamine110 proteinase substrates for serin-, cystein- [183–185] and aminopeptidases [186].

Of further interest were not only cell function assays for the assessment of the effect of cytostatic drugs on patient cells [187] but also the simultaneous determination of cell concentration and function of blood lymphocytes, monocytes, granulocytes, erythrocytes and thrombocytes [188, 189] in peripheral blood samples. The long-lasting lack of an absolute cell counter device in commercial flow cytometers led to the addition of a known amount of fluorescent monodisperse microparticles as internal counting and fluorescence standard for flow-cytometric measurements [189], a method that is widely used in the clinical environment under the name of 'single-platform absolute cell counting'. In flow cytometers with an integrated absolute counting feature [190], monodisperse fluorescent particles remain useful to monitor the long-term performance of fluorescence and light scatter measurements.

Apoptosis

In situ nick translation with DNA polymerase I in the presence of either fluorescein-12-dUTP (uridine triphosphate) or digoxigenin-labeled 11-dUTP (deoxyUTP) opened the way for the flow-cytometric assessment of DNA fragmentation during apoptosis [191].

Microbiology and Biotechnology

The interest in flow-cytometric determination of microorganisms led to the detection of DNA, RNA and protein in yeast cells [192, 193] and bacteria [68] as an early effort toward biotechnology as well as food quality testing and antibiotics efficiency assessment.

Data Analysis

Data analysis in flow cytometry is of major importance for the evaluation of the histograms or of multiparametric listmode data. The initial interest concerned the exact determination of cell fractions in various phases of the cell cycle (cell cycle analysis)

[80, 194] from one-parameter DNA histograms and the mathematical analysis of two-parameter flow-cytometric DNA/protein measurements [81]. Furthermore, one- [195, 196], two- [180] or three-parameter [197] linear or logarithmic Gaussian distributions were adapted to single or multiparametric flow cytometer measurements in order to simplify the results as well as the development of scientific hypotheses concerning the biological regulation of various cell populations by the organism.

Besides the above primary data evaluation, significant efforts have been invested in extracting further information or knowledge from the measurements, usually by parametric mathematical or statistical methods like cluster [198, 199] or principal-component analysis [200], multivariate statistics [201–203], knowledge-based [204, 205] or hierarchical classification [206–208], fuzzy logic [201, 209], neuronal networks [201, 210–212] or self-organizing matrices [212].

These methods depend in part on mathematical assumptions for observed value distributions, for example, that they are Gaussian. Frequently, multipoint clusters in multidimensional space are evaluated in which the coherence of the data points of the individual experiment or patient is lost. Data complexity in this instance is tentatively reduced by making *models* to evaluate the conceptually most promising parameters or those being most significantly different between experimental series or patient groups by trial and error. The consequence for medicine is that therapy-dependent extrapolations for predicting the course of a disease are expressed as *prognoses,* indicating the *statistical* future of patient groups (Kaplan-Meier statistics) but not, as would be desirable, the *individualized future* of a given patient before the start of a calculated therapy.

Individualized patient prognosis is obtained by the linked evaluation of the multiparameter values (data pattern) of each patient by algorithmic (nonparametric) data pattern analysis (DIAGNOS1 [213], CLASSIF1 [214–216]). In case of large numbers of available parameters per patient, they are split into portions of, for example, 50 parameters. Then, at the first level, the five most discriminatory parameters between the investigated patient groups are determined for each parameter portion. At the second evaluation level, the most discriminatory parameters of the first level are subsequently merged with 50 new parameter portions. The five most discriminatory parameters of each portion of the second level are merged with the third level and so on until the most discriminatory data pattern of all parameters is obtained at the last level.

This hierarchical information concentration of data pattern analysis can be imagined as a multilevel sieve cascade with increasingly coarser mesh size (data sieving). The calculations can be automated and parallel computed in an unattended way with an essentially unlimited number of parameters. Data sieving does not depend upon data models. The resulting knowledge extraction is of importance for standardized flow-cytometric diagnostics as well as for therapy-dependent predictions of disease course in individual patients, i.e. predictive medicine, personalized or individualized medicine.

Predictive Medicine by Cytomics

The concept of predictive medicine by cytomics is a consequence of the initial observation that the results of cell function measurements provided a >80% correct extrapolation for the later occurrence of sepsis, posttraumatic shock, an intermediate state or normal recovery in surgical intensive-care patients three days in advance [181].

The multimolecular cytometric analysis of the heterogeneity of cells and cell systems (*cytomes, system cytometry*) [217], in conjunction with exhaustive bioinformatics knowledge extraction (*cytomics*), constitutes the basis of this concept [218–223]. Likewise, the analysis of data from flow-cytometric multiplex particle arrays, multiparameter cell-oriented DNA or proteomics arrays, and of clinical and clinical chemistry or image analysis data is equally possible. This permits generalized access to therapy-dependent predictions for the further course of disease in individual patients, often with >95% accuracy. These predictions may, among others, reduce tissue damage or tissue loss, thus supporting the efforts of preventive medicine. Improved adaptation of therapy to the individual patient also permits the pretherapeutic exclusion of therapy-resistant patients. Additionally, there is the potential to reduce adverse drug effects.

Reverse engineering of the predictive data patterns by cell systems biology [224] has the potential to analytically explore molecular disease courses in patients instead of exploring them in disease models that are not necessarily representative of the patient's situation. This may lead to the discovery of new drug targets for the pharmaceutical industry [225]. Such goals can furthermore be realized within the framework of the proposed human cytome project [226].

Outlook

Rapid technological progress is currently leading to a merger between flow and image cytometry, integrating them as cell-oriented bioinformatics into cytomics. Examples of this development are the fast acquisition of fluorescence images in a flow cytometer [59], the laser scanning cytometer [227] (see 'Technical and Methodological Basics of Slide-Based Cytometry', pp. 89), and the 4Pi-microscope [228]. Chip and particle arrays [229], instrument miniaturization by nanotechnologies and progress in bioinformatics generate an increasing interest in the molecular analysis of single cells and their molecular environment.

The interaction with systems biology [230, 231] opens new potentials as biomedical cell systems biology [232, 233] for modeling disease pathways and optimizing therapy.

The fascination with the biocomplexity of intact cells as the organism's elementary function units and their intriguing molecular analysis will remain a primary driving force behind this cross-disciplinary approach. Concerning the medical field, implementing the acquired knowledge for the patients' benefit will be a major goal.

Summary

The development of flow cytometry has provided an important driving force for the advancement of molecular single-cell research in biomedical and clinical domains. Continually evolving research concepts for instrument development combined with the analysis of multiparameter results from immunological data, cell function tests, DNA measurements, and microbiological and biotechnological applications have profoundly shaped this area of research. They remain essential prerequisites for the solution of future scientific challenges.

Flow cytometry addresses the high complexity of the assembled molecular architecture of single cells and cell systems (cytomes). Molecular cell phenotypes develop during the life of an organism and represent the cumulative result of genotype and exposure influences. This makes them particularly interesting for biomedical research. The merging of flow and image cytometry with multiparametric bioinformatics into cytomics initiates a new approach to predictive medicine. Some of its potentials include therapy-dependent prediction of further disease course for individual patients (i.e. personalized, individualized medicine, improved detection of molecular disease pathways, and the identification of new targets in pharmaceutical research and drug discovery. These developments are accompanied by rapid progress in molecular fluorescence technologies and microscopy instrumentation. New insights into the molecular expression of the genomic information derived from the observed heterogeneity of cells and cell systems may contribute to the organization of large scale international research projects such as the human cytome project.

Acknowledgments

The support of W Göhde, M Stöhr, A Böcking, and A Tárnok in the supply of information concerning the widely distributed and often electronically uncatalogued publications of the initial development phase as well as valuable discussions and comments are gratefully acknowledged.

References

1 Caspersson TO: Cell Growth and Cell Function: A Cytochemical Study. New York, Norton, 1950, pp 1–185.
2 Chieco P, Derenzini M: The Feulgen reaction 75 years on. Histochem Cell Biol 1999;111:345–358.
3 Kasten FH: The chemistry of Schiff's reagent. Int Rev Cytol 1960;10:1–100.
4 Moldavan A: Photoelectric technique for the counting of microscopic cells. Science 1934;80:188–189.

5 Mullaney PF, Van Dilla MA, Coulter JR, Dean PN: Cell sizing – a light scattering photometer for rapid volume determination. Rev Sci Instrum 1969;40:1029–1032.
6 Coulter WH: Means for counting particles suspended in a fluid. US patent 2656508 priority Aug 27, 1949.
7 Coulter WH: High speed automatic blood cell counter and cell size analyzer. Proc Natl Electronics Conf 1956; 12:1034–1040.
8 Fulwyler MJ: Electronic separation of biological cells by volume. Science 1965;150:910–911.
9 Crosland-Taylor PJ: A device for counting small particles suspended in a fluid through a tube. Nature 1953;171:37–38.
10 Göhde W: Automatisches Meß- und Zählgerät für die Teilchen einer Dispersion. Patent DE1815352, priority date Dec 18, 1968.
11 Göhde W: Process for automatic counting and measurement of particles. US patent 4021117, priority date Jun 23, 1973(DE).
12 Göhde W, Schumann J, Fruh J: Coincidence eliminating device for pulse-cytophotometry; in Göhde W, Schumann J, Büchner Th (eds): Pulse Cytophotometry II. Ghent, European Press, 1976, pp 71–78.
13 Dittrich W, Göhde W: Impulsfluorimetrie bei Einzelzellen in Suspensionen. Z Naturforsch 1969;24b:221–228.
14 Kamentsky LA, Derman H, Melamed MR: Spectrophotometer: New instrument for ultrarapid cell analysis. Science 1965;150:630–631.
15 Kamentsky LA: Discussion remarks; in Evans DMD (ed): Cytology Automation. Proc 2nd Tenovus Symp, Cardiff Oct 24–25, 1968. Edinburgh, Livingstone, 1970, pp 55, 141, 196.
16 Kamentsky LA: Rapid cell spectrophotometry for cell identification and sorting; in Evans DMD (ed): Cytology Automation. Proc 2nd Tenovus Symposium, Cardiff, 1968. Edinburgh, Livingstone, 1970, pp 177–185.
17 Ornstein L, Ansley HR: Spectral matching of classical cytochemistry to automated cytology. J Histochem Cytochem 1974;22:453–469.
18 Mansberg HP, Saunders AM, Groner W: The Hemalog D white cell differential system. J Histochem Cytochem 1974;22:711–724.
19 Fulwyler MJ, McDonald CW, Haynes JL: The Becton Dickinson FACS IV; in Melamed MR, Mullaney PF, Mendelsohn ML (eds): Flow Cytometry and Sorting. New York, Wiley, 1979, pp 653–667.
20 Dühnen J, Stegemann C, Wiezorek C, Mertens H: A new fluid switching flow sorter. Histochemistry 1983;77:117–121.
21 Göhde H, Schumann J: Method and apparatus for sorting particles. US patent 4756427, priority date Sep 11, 1984 (CH).
22 Göhde W: Verfahren und Vorrichtung zur Sortierung von mikroskopischen Partikeln. Patent EP0177718, priority date Aug 08, 1985.
23 Derek WR, Gray W, Göhde W, Carter N, Heiden T, Morris PJ: Separation of pancreatic islets by fluorescence-activated sorting. Diabetes 1989;38:133–135.
24 Göhde W, Dittrich W: Simultane Impulsfluorimetrie des DNS- und Proteingehaltes von Tumorzellen. Z Anal Chem 1970;252:328–330.
25 Göhde W, Dittrich W: Impulsfluorometrie – ein neuartiges Durchflußverfahren zur ultraschnellen Mengenbestimmung von Zellinhaltsstoffen. Acta Histochem 1971; 41(suppl 10):429–437.
26 Göhde W, Dittrich W: Cytostatische Wirkung von Daunomycin im Impulscytophotometrie Test. Arzneimittelforschung1971;21:1656–1658.
27 Schwabe M, Wiendl HJ: Die Bedeutung der Impulszytophotometrie für die Früherkennung benigner und maligner Erkrankungen der Magenschleimhaut. Münch Med Wochenschr 1974;116:1005–1008.
28 Göhde W: Automation of cytofluorometry by use of the Impulsmicrophotometer; in Thaer AA, Sernetz M (eds): Fluorescence Techniques in Cell Biology. Berlin, Springer, 1973, pp 79–88.
29 Reiffenstuhl G, Severin E, Dittrich W, Göhde W: Die Impulscytophotometrie des Vaginal- und Cervicalsmears. Arch Gynäkol 1971;211:595–616.
30 Schumann JF, Ehring W, Göhde W, Dittrich W: Impulscytophotometrie der DNS in Hauttumoren. Arch Klin Exp Dermatol 1971;239:377–389.
31 Farsund T: Preparation of bladder mucosa cells for micro-flowfluorometry. Virchows Arch B 1974;16:35–42.
32 Clausen OPF, Lindmo T, Sandnes K, Thorud E: Separation of mouse epidermal basal and differentiating cells for microflow fluorometric measurements – Methodological study. Virchows Arch B 1976;20:261–275.
33 Barrett DL, King EB, Jensen RH, Merrill JT: Cytomorphology of gynecologic specimens analyzed and sorted by 2-parameter flow cytometry. Acta Cytol 1976;20:585–586.
34 Yataganas X, Mitomo Y, Traganos F, Strife A, Clarkson B: Evaluation of a Feulgen-type reaction in suspension using flow microfluorimetry and a cell separation technique. Acta Cytol 1975;19:71–78.

35 Gray JW, Carrano AV, Moore DH, Steinmetz LL, Minkler J, Mayall BH, Mendelsohn ML: High speed quantitative karyotyping by flow microfluorometry. Clin Chem 1975;21:1258–1262.

36 Stephens SO: Analysis of effect of fluorescence intensity on distributions obtained by flow cytofluorometry. Exp Cell Res 1974;89:228–230.

37 Auer G, Tribukait B: Comparative single cell and flow DNA analysis in aspiration biopsies from breast carcinomas. Acta Pathol Microbiol Scand A 1980;88:355–358.

38 Stöhr M: Double beam application in flow techniques and recent results; in Göhde W, Schumann J, Büchner T (eds): Pulse Cytophotometry II. Ghent, European Press, 1976, pp 39–45.

39 Eisert WG, Ostertag R, Niemann EG: Simple flow microphotometer for rapid cell-population analysis. Rev Sci Instrum 1975;46:1021–1024.

40 Eisert WG, Nezel M: Internal calibration to absolute values in flowthrough particle-size analysis. Rev Sci Instrum 1978;49:1617–1621.

41 Kachel V: Basic principles of electrical sizing of cells and particles and their realization in the new instrument Metricell. J Histochem Cytochem 1976;24:211–230.

42 Kachel V, Glossner E, Kordwig E, Ruhenstroth-Bauer G: Fluvo-Metricell, a combined cell volume and cell fluorescence analyzer. J Histochem Cytochem 1977;25:804–812.

43 Benker G, Kachel V, Valet G: A computer-controlled data management system for multiparameter flow cytometric analysis; in Laerum OD, Lindmo T, Thorud E (eds): Flow Cytometry IV. Bergen, Universitetsforlaget, 1980, pp 116–119.

44 Valet G, Hofmann H, Ruhenstroth-Bauer G: The computer analysis of volume distribution curves: demonstration of two erythrocyte populations of different size in the young guinea pig, and analysis of the mechanism of immune lysis of cells by antibody and complement. J Histochem Cytochem 1976;24:231–246.

45 Sweet RG: High frequency recording with electrostatically deflected ink jets. Rev Sci Instrum 1965;36:131–136.

46 Thom R, Hampe A, Sauerbrey G: Die elektronische Volumen-Bestimmung von Blutkörperchen und ihre Fehlerquellen. Z Gesamte Exp Med 1969;151:331–349.

47 Kachel V, Metzger H, Ruhenstroth-Bauer G: Der Einfluß der Partikeldurchtrittsbahn auf die Volumenverteilungskurven nach dem Coulter-Verfahren. Z Gesamte Exp Med 1970;153:331–347.

48 Thom R, Kachel V: Fortschritte für die elektronische Größenbestimmung von Blutkörperchen. Blut 1970;21:48–50.

49 Kachel V: Eine elektronische Methode zur Verbesserung der Volumenauflösung des Coulter-Partikelvolumenverfahrens. Blut 1973;27:270–274.

50 Göhde W: Kapillare zur Zählung und Messung kleiner Teilchen. Patent DE1973692, priority date: Aug 25, 1967.

51 Thom R: Anordnung zur Gewinnung von Größen, die den Mengen von in der Untersuchungsflüssigkeit enthaltenen Teilchen verschiedenen Volumens entsprechen. Patent DE1806512, priority date Nov 02, 1968.

52 Thom R: Vergleichende Untersuchung zur elektronischen Zellvolumen-Analyse. Telefunken Hochfrequenztechnik N1/EP/V 1698, Ulm, 1972, pp 1–59.

53 Spielman L, Goren SL: Improving resolution in Coulter counting by hydrodynamic focusing. J Colloid Interface Sci 1968;26:175–182.

54 Schulz J, Nitsche HJ: Nachweis des transzellulären Ionenflusses bei der Volumenbestimmung von nativen Humanerythrozyten; in Thom R (Hrsg): Vergleichende Untersuchung zur elektronischen Zellvolumen-Analyse. Telefunken Hochfrequenztechnik N1/EP/V 1698, Ulm, 1972, pp 60–61.

55 Zimmermann U, Schulz J, Pilwat G: Transcellular ion flow in *Escherichia-coli* B and electrical sizing of bacteria. Biophys J 1973;13:1005–1013.

56 Zimmermann U, Pilwat G, Riemann F: Dielectric breakdown of cell membranes. Biophys J 1974;14:881–899.

57 Riemann F, Zimmermann U, Pilwat G: Release and uptake of hemoglobin and ions in red blood-cells induced by dielectric breakdown. Biochim Biophys Acta 1975;394:449–462.

58 Kachel V, Benker G, Lichtnau K, Valet G, Glossner E: Fast imaging in flow: a means of combining flow-cytometry and image analysis. J Histochem Cytochem 1979;27:335–341.

59 Basiji DA, Ortyn WE: Imaging and analyzing parameters of small moving objects such as cells. US patent 6249341 priority date Jan. 24, 2000.

60 Wheeless LL, Patten SF: Slit scan cytofluorometry. Acta Cytol 1973;17:333–339.

61 Gray JW, Peters D, Merrill JT, Martin R, Van Dilla MA: Slit-scan flow cytometry of mammalian chromosomes. Cytochemistry 1979;27:441–444.

62 Ahrens O, Albrecht U, Rajewsky MF: Microprocessor-based data acquisition-system for flow cytometers; in Laerum OD, Lindmo T, Thorud E (eds): Flow Cytometry IV. Bergen, Universitetsforlaget, 1980, pp 112–115.

63 Kachel V, Meier H, Stuhlmüller P, Ahrens O: Ultra fast digital calculation of ratios of flow cytometric values; in Laerum OD, Lindmo T, Thorud E (eds): Flow Cytometry IV. Bergen, Universitetsforlaget, 1980, pp 109–111.

64 Kachel V Benker G, Weiss W, Glossner E, Valet G, Ahrens O: Problems of fast imaging in flow; in Laerum OD, Lindmo T, Thorud E (eds): Flow Cytometry IV: Bergen, Universitetsforlaget, 1980, pp 49–55.
65 Watson JV: A twin laser multi-parameter analyzing flow cytometer (abstract 42). Br J Cancer 1980;42:184.
66 Watson JV: A method for improving light collection by 600% from square cross-section flow-cytometry chambers. Br J Cancer 1985;51:433–435.
67 Watson JV: Flow-cytometry chamber with 4-Pi light collection suitable for epifluorescence microscopes. Cytometry 1989;10:681–688.
68 Steen HB, Boye E: Bacterial-growth studied by flow-cytometry. Cytometry 1980;1:32–36.
69 Steen HB, Lindmo T: Flow cytometry – high resolution instrument for everyone. Science 1979;204:403–404.
70 Van Dilla MA, Trujillo TT, Mullaney PF, Coulter JR: Cell microfluorometry: a method for rapid fluorescence measurements. Science 1969;163:1213–1213.
71 Crissman HA, Steinkamp JA: Rapid simultaneous measurement of DNA, protein and cell volume in single cells from large mammalian cell populations. J Cell Biol 1973;59:766–771.
72 Crissman HA, Tobey RA: Cell-cycle analysis in 20 minutes. Science 1974;184:1297–1297.
73 Zante J, Schumann J, Barlogie B, Göhde W, Büchner T: New preparation and staining procedures for specific and rapid analysis of DNA distributions; in Göhde W, Schumann J, Büchner T (eds): Pulse-Cytophotometry II. Ghent, European Press, 1976, pp 97–106.
74 Jensen RH: Chromomycin A3 as a fluorescent probe for flow cytometry of human gynecologic samples. J Histochem Cytochem 1977;25:573–579.
75 Darzynkiewicz Z, Traganos F, Sharpless T, Melamed MR: Thermal denaturation of DNA in situ as studied by acridine orange staining and automated cytofluorometry. Exp Cell Res 1975;90:411–428.
76 Darzynkiewicz Z, Traganos F, Sharpless T, Melamed MR: Conformation of RNA in situ as studied by acridine-orange staining and automated cytofluorometry. Exp Cell Res 1975;95:143–153.
77 Göhde W, Schumann J, Zante J: The use of DAPI in pulse cytophotometry; in Lutz D (ed): Pulse Cytophotometry III. Ghent, European Press, 1978, pp 229–232.
78 Lämmler G, Schütze HR: Vital-Fluorochromierung tierischer Zellkerne mit einem neuen Fluorochrom. Naturwissenschaften 1969;56:286–286.
79 Göhde W, Schumann J, Büchner T, Barlogie B: Influence of irradiation and cytostatic drugs on proliferation patterns of tumor cells; in Haanen CAM, Hillen HFP, Wessels JMC (eds): Pulse Cytophotometry I. Ghent, European Press, 1975, pp 138–152.
80 Baisch H, Göhde W, Linden WA: Analysis of PCP-data to determine the fraction of cells in the various phases of cell cycle. Radiat Environ Biophys 1975;12:31–39.
81 Baisch H, Göhde W: Mathematical analysis of DNA-protein two-parameter pulse-cytophotometric data; in Göhde W, Schumann J, Büchner Th (eds): Pulse Cytophotometry II. Ghent, European Press, 1976, pp 71–78.
82 Berkhan E: DNS Messung von Zellen aus Vaginalabstrichen. Ärztl Lab 1972;18:77–79.
83 Schonefeld J: DNS Verteilungsmuster in Hauttumoren – Impulszytophotometrische Messungen an fixierten Geweben; Dissertation Münster, 1974.
84 Göhde W, Dittrich W: Die cytostatische Wirkung von Daunomycin im Impulscytophotometrie-Test. Arzneimittelforschung 1971;21:1656–1658.
85 Vindelov LL: Flow microfluorometric analysis of nuclear-DNA in cells from solid tumors and cell-suspensions – New method for the rapid isolation and staining of nuclei. Virchows Arch B 1977;24:227–242.
86 Vindelow LL, Christensen IJ, Nissen NI: A detergent-trypsin method for the preparation of nuclei for flow cytometric DNA analysis. Cytometry 1983;3:323–327.
87 Hedley DW, Friedlander ML, Taylor IW: Application of DNA flow-cytometry to paraffin-embedded archival material for the study of aneuploidy and its clinical significance. Cytometry 1985;6:327–333.
88 Zante J, Schumann J, Göhde W, Hacker U: DNA-fluorometry of mammalian sperm. Histochemistry 1977;54:1–7.
89 Meistrich ML, Göhde W, White RA: Resolution of x and y spermatids by pulse cytophotometry. Nature 1978;274:821–823.
90 Meistrich ML, Göhde W, White RA, Longtin JL: Cytogenetic studies of spermatids of mice carrying Cattanach's translocation by flow cytometry. Chromosoma 1979;74:141–151.
91 Otto FJ, Oldiges H: Flow cytogenetic studies in chromosomes and whole cells for the detection of clastogenic effects. Cytometry 1980;1:13–17.
92 Otto F, Oldiges H, Göhde W, Dertinger H: Flow cytometric analysis of mutagene induced chromosomal damage; in Laerum OD, Lindmo T, Thorud E(eds): Flow Cytometry IV. Bergen, Universitetsforlaget, 1980, pp 284–286.
93 Büchner T, Dittrich W, Göhde W: Die Impulscytophotometrie in der hämatologischen Cytologie. Klin Wochenschr 1971;40:1090–1092.

94 Büchner T, Göhde W, Schneider R, Hiddemann W, Kamanabroo D: Zellsynchronisation und cytocide Effekte durch Chemotherapie der Leukämie in der Klinik anhand der Impulscytophotometrie; in Andreeff M (ed): Impulscytophotometrie. Berlin, Springer-Verlag, 1975, pp 77–86.

95 Barlogie B, Göhde W, Johnston DA, Smallwood L, Schumann J, Drewinko B, Freireich EJ: Determination of ploidy and proliferative characteristics of human solid tumors by pulse cytophotometry. Cancer Res 1978;38:3333–3339.

96 Fey F, Gibel W, Schramm T, Teichmann B, Ziebarth D: Untersuchung über den Wert der Impulscytophotometrie bei der Erkennung präkanzeröser Veränderungen. Arch Geschwulstforsch 1972;39:1–7.

97 Gibel W, Weiss H, Schramm T, Gütz HJ, Wolff G: Untersuchungen über die Möglichkeiten einer Magenkrebsfrüh- und Differentialdiagnostik mittels Impulscytophotometrie. Arch Geschwulstforsch 1972;40:263–267.

98 Büchner T, Dittrich W, Göhde W: Impulscytophotometrie von Blut- und Knochenmarkszellen. Verh Dtsch Ges Inn Med 1971;77:416–418.

99 Müller D, Reichert E, Lang HD, Simon A, Benöhr HC: Die Möglichkeiten der Impulszytophotometrie für die Bestimmung der Zellproliferation bei Hämoblastosen; in Groß R, van de Loo J (eds): Leukämie. Berlin, Springer, 1972, pp 221–228.

100 Büchner Th: Impulszytophotometrie in der Hämatologie. Blut 1974;28:1–7.

101 Müller D, Reichert E, Lang HD, Orywall D: Zellcyclusänderungen und ineffektive Zellneubildung bei gestörter Granulo- und Erythropoese. Klin Wochenschr 1974;52:384–393.

102 Klein G, Altmann H: Impulscytophotometrische Untersuchungen zur Proliferationskinetik von Synovialzellen. Wien Klin Wochenschr 1973;85:774–778.

103 Pedersen T, Larsen JK, Krarup T: Characteriztaion of bladder tumors by flow cytometry on bladder washings. Eur Urol 1978;4:351–355.

104 Tribukait B, Esposti PL: Quantitative flow-microfluorometric analysis of DNA in cells from neoplasms of urinary-bladder – Correlation of aneuploidy with histological grading and cytological findings. Urol Res 1978;6:201–205.

105 Wheeless LL, Badalamant RA, deVere White RW, Fradet Y, Tribukait B: Consensus review of the clinical utility of DNA cytometry in bladder cancer. Cytometry 1993;14:478–481.

106 Hedley DW, Clark GM, Cornelisse CJ, Killander D, Kute T, Merkel D: Consensus review of the clinical utility of DNA cytometry in carcinoma of the breast. Cytometry 1993;14:482–485.

107 Bauer KD, Bagwell BC, Giaretti W, Melamed M, Zarbo RJ, Witzig TE, Rabinovitch PS: Consensus review of the clinical utility of DNA flow cytometry in colorectal cancer. Cytometry 1993;14:486–491.

108 Duque RE, Andreeff M, Braylan RC, Diamond LW, Peiper SC: Consensus review of the clinical utility of DNA flow cytometry in neoplastic hematopathology. Cytometry 1993;14:492–496.

109 Shankey TV, Kallionemi OP, Koslowski JM, Lieber ML, Mayall BH, Miller G, Smith GJ: Consensus review of the clinical utility of DNA content cytometry in prostate cancer. Cytometry 1993;14:497–500.

110 Gray JW, Carrano AV, Steinmetz LL, Van Dilla MA, Moore DH, Mayall BH, Mendelsohn ML: Chromosome measurement and sorting by flow systems. Proc Natl Acad Sci U S A 1975;72:1231–1234.

111 Stubblefield E, Cram S, Deaven L: Flow microfluorometric analysis of isolated Chinese hamster chromosomes. Exp Cell Res 1975;94:464–468.

112 Gray JW, Langlois RG, Carrano AV, Burkhart-Schultz K, Van Dilla MA: High resolution chromosome analysis: One and two parameter flow cytometry. Chromosoma 1979;73:9–27.

113 Carrano AV, Gray JW, Langlois RG, Burkhart-Schultz KJ, Van Dilla MA: Measurement and purification of human chromosomes by flow cytometry and sorting. Proc Natl Acad Sci U S A 1979;76:1382–1384.

114 Bauman JGJ, Wiegant T, Borst P, Van Duijn P: A new method for fluorescent microscopical localization of specific DNA-sequences by in-situ hybridization of fluorochrome-labeled DNA. Exp Cell Res 1980;128:485–490.

115 Landegent JE, Dewal NJI, Baan RA, Hoeijmakers HJ, Van Der Ploeg: 2-Acetylaminofluorene-modified probes for the indirect hybridocytochemical detection of specific nucleic-acid sequences. Exp Cell Res 1984, 153:61–72.

116 Hopman AHN, Wiegant J, Van Duijn P: Mercurated nucleic-acid probes, a new principle for non-radioactive insitu hybridization. Exp Cell Res 1987;169:357–368.

117 Pinkel D, Landegent J, Collins C, Fuscoe J, Seagraves R, Lucas J, Gray J: Fluorescence in situ hybridization with human chromosome-specific libraries: Detection of trisomy 21 and translocations of chromosome 4. Proc Natl Acad Sci U S A 1988;85:9138–9142.

118 Schumann J, Göhde W: Die zellkinetische Wirkung von Bleomycin auf das Ehrlich-Karzinom der Maus in vivo. Strahlentherapie 1974;147:298–307.

119 Dittrich W, Göhde W: Phase progression in two dose response of Ehrlich ascites tumour cells. Atomkernenergie 1970;15–36:174–176.

120 Kal HB: Distribution of cell volume and DNA content of rhabdomyosarcoma cells growing in vitro and in vivo after irradiation. Eur J Cancer 1973;9:77–79.

121 Otto F, Göhde W: Effects of fast neutrons and x-ray irradiation on cell kinetics; in Göhde W, Schumann J, Büchner Th (eds): Pulse Cytophotometry II. Ghent, European Press, 1976, pp 244–249.

122 Reddy SB, Erbe W, Linden WA, Landen H, Baigent C: Die Dauer der Phasen im Zellzyklus von L-929 Zellen. Vergleich von Impulscytophotometrischen und Autoradiographischen Messungen. Biophysik 1973;10:45–50.

123 Linden WA, Zywietz F, Landen H, Wendt C: Synchronisation des Teilungszyklus von L-Zellen in der G2-Phase durch fraktionierte Röntgenbestrahlung und Daunomycin. Strahlentherapie 1973;146:216–225.

124 Smets LA: Contact inhibition of transformed cells incompletely restored by dibutyryl cyclic AMP. Nat New Biol 1972;239:123–124.

125 Smets LA: Activation of nuclear chromatin and the release from contact-inhibition of 3T3 cells. Exp Cell Res 1973;79:239–343.

126 Smets LA: Agglutination with ConA dependent on cell cycle. Nat New Biol 1973;245:113–115.

127 Böhmer RM: Flow cytometric cell-cycle analysis using the quenching of 33258 Hoechst fluorescence by bromodeoxyuridine incorporation. Cell Tissue Kinet 1979;12:101–110.

128 Gratzner HG, Leif RC, Ingram DJ, Castro A: The use of antibody specific for bromodeoxyuridine for the immunofluorescent determination of DNA replication in single cells. Exp Cell Res 1975;95:88–94.

129 Gratzner HG, Pollack A, Lingram DJ, Leif RC: Deoxyribonucleic acid replication in single cells and chromosomes by immunologic techniques. J Histochem Cytochem 1976;24:34–39.

130 Gratzner HG, Leif RC: An immunofluorescent method for monitoring DNA synthesis by flow cytometry. Cytometry 1981;1:385–389.

131 Gerdes J, Lemke H, Baisch H, Wacker HH, Schwab U, Stein H: Cell-cycle analysis of a cell proliferation-associated human nuclear antigen defined by the monoclonal-antibody KI-67. J Immunol 1984;133:1710–1715.

132 Nüsse M, Kramer J: Flow cytometric analysis of micronuclei found in cells after irradiation. Cytometry 1984;5:20–25.

133 Van den Engh G, Visser J: Light scattering properties of pluripotent and committed haematopoietic stem cells. Acta Haematol 1979;62:289–298.

134 Van den Engh G, Visser J, Bol S, Trask B: Concentration of hematopoietic stem-cells using a light-activated cell sorter. Blood Cells 1980;6:609–623.

135 Auffermann W, Böcking A: Early detection of precancerous lesions in dysplasias of the lung by rapid DNA image cytometry. Anal Quant Cytol Histol 1985;7:218–226.

136 Böcking A, Auffermann W, Vogel H, Schlöndorff G, Goebbels R: Diagnosis and grading of malignancy in squamous epithelial lesions of the larynx with DNA cytophotometry. Cancer 1985;56:1600–1604.

137 Böcking A, Hilgarth M, Auffermann W, Hack-Werdier C, Fischer-Becker D, von Kalkreuth G: DNA-cytometric diagnosis of prospective malignancy in borderline lesions of the uterine cervix. Acta Cytol 1986;30:608–615.

138 Sun D, Biesterfeld S, Adler CP, Böcking A: Prediction of recurrence in giant cell bone tumors by DNA cytometry. Anal Quant Cytol Histol 1992;14:341–346.

139 Böcking A, Motherby H: Abklärung zervikaler Dysplasien mittels DNA-Bild-Zytometrie. Pathologe 1999;20:25–33.

140 Grote HJ, Nguyen HVQ, Leick AG, Böcking A: Identification of progressive cervical epitelial cell abnormalities using DNA-image cytometry. Cancer Cytopathol 2004;102:373–379.

141 Remmerbach TW, Weidenbach H, Pomjanski N, Knops K, Mathes S, Hemprich A, Böcking A: Cytologic and DNA-cytometric early diagnosis of oral cancer. Anal Cell Pathol 2001;22:211–221.

142 Remmerbach TW, Weidenbach H, Hemprich A, Böcking A: Earliest detection of oral cancer using non-invasive brush biopsy including DNA-image-cytometry: Report on four cases. Anal Cell Pathol 2003;25:159–166.

143 Salzman G, Crowell JM, Martin JC, Trujillo TT, Romero A, Mullaney PF, Labauve OM: Cell classification by laser light-scattering – Identification and separartion of unstained leukocytes. Acta Cytol 1975;19:374–377.

144 Loken MR, Sweet RG, Herzenberg LA: Cell discrimination by multiangle light scattering. J Histochem Cytochem 1976;24:284–291.

145 Bonner WA, Sweet RG, Hulett HR, Herzenberg LA: Fluorescence-activated cell sorting. Rev Sci Instrum 1972;43:404–409.

146 Loken MR, Parks DR, Herzenberg LA: Two-color immunofluorescence using a fluorescence-activated cell sorter. J. Histochem Cytochem 1977;25:899–907.

147 Julius MH, Masuda T, Herzenberg LA: Demonstration that antigen binding cells are the precursors of antibody producing cells after purification using a fluorescence activated cell sorter. Proc Natl Acad Sci U S A 1972;69:1934–1938.

148 Jovin TM, Arndt-Jovin DJ: The measurement of structural changes in cells using fluorescence emission anisotropy in flow systems; in Göhde W, Schumann J, Büchner T (eds): Pulse Cytophotometry II. Ghent, European Press, 1976, pp 33–38.

149 Jovin TM: Fluorescence polarization and energy transfer: theory and application; in Melamed MR, Mullaney PF, Mendelsohn ML (eds): Flow cytometry and sorting. New York, Wiley, 1979, pp 137–165.

150 Chan SS, Arndt-Jovin DJ, Jovin TM: Proximity of lectin receptors on the cell-surface measured by fluorescence energy-transfer in a flow system. J Histochem Cytochem 1979;27:56–64.

151 Szollosi J, Tron L, Damjanovich S, Helliwell SH, Arndt-Jovin D, Jovin TM: Fluorescence energy-transfer measurements on cell-surfaces – A critical comparison of steady-state fluorimetric and flow cytometric methods. Cytometry 1984;5:210–216.

152 Tron L, Szollosi J, Damjanovich S, Helliwell SH, Arndt-Jovin DJ, Jovin TM: Flow cytometric measurement of fluorescence resonance energy-transfer on cell surfaces – quantitative evaluation of the transfer efficiency on a cell-by-cell basis. Biophys J 1984;45:939–946.

153 De Grooth BG, Terstappen LWMM, Puppels GJ, Greve J: Light scattering polarization measurements as a new parameter in flow-cytometry. Cytometry 1987;8:539–544.

154 Terstappen LWMM, De Grooth BG, Visscher K, Van Kouterik FA, Greve J: 4-Parameter white blood-cell differential counting based on light-scattering measurements. Cytometry 1988;9:39–43.

155 Puppels GJ, Demul FFM, Otto C, Greve J, Robert-Nicoud M, Arndt-Jovin DJ, Jovin TM: Studying single living cells and chromosomes by confocal Raman microsprectroscopy. Nature 1990;347:301–303.

156 Hulett HR, Bonner WA, Barrett J, Herzenberg LA: Cell sorting – automated separation of mammalian cells as a function of intracellular fluorescence. Science 1969;166:747–749.

157 Sengbusch GV, Couwenbergs C, Kühner J, Müller U: Fluorogenic substrates in single living cells. Histochem J 1976;8:341–350.

158 Wilder ME, Cram LS: Differential fluorochromasia of human lymphocytes as measured by flow cytometry. J Histochem Cytochem 1977;25:888–891.

159 Watson JV: Enzyme kinetic-studies in cell populations using fluorogenic substrates and flow cytometric techniques. Cytometry 1980;1:143–151.

160 Dolbeare FA, Phares WF: Napthol AS-BI(7-bromo-3-hydoxy-2-naphtho-*o*-anisidine) phosphatase and naphtol AS-BI-beta-d-glucoronidase in chinese-hamster ovary cells – biochemical and flow cytometric studies. J Histochem Cytochem 1979;27:120–124.

161 Watson JV: Enzyme kinetic-studies in cell-populations using fluorogenic substrates and flow cytometric techniques. Cytometry 1980;1:143–151.

162 Malin-Berdel J, Valet G: Flow cytometric determination of esterase and phosphatas activities and kinetics in hematopoietic cells with fluorogenic substrates. Cytometry 1980;1:222–228.

163 Dolbeare FA, Smith RE: Flow cytometric measurement of peptidases with use of 5-nitrosalicylaldehyde and 4-methoxy-beta-napthylamine derivatives. Clin Chem 1977;23:1485–1491.

164 Vanderlaan M, Cutter C, Dolbeare F: Flow microfluorometric identification of liver cells with elevated gamma-glutamyltranspeptidase activity after carcinogen exposure. J Histochem Cytochem 1979;27:114–119.

165 Stöhr M, Vogt-Schaden M: A dual staining technique for simultaneous flow cytometric DNA analysis of living and dead cells; in Laerum OD, Lindmo T, Thorud E (eds): Flow Cytometry IV. Bergen, Universitetsforlaget, 1980, pp 96–99.

166 Bassoe CF, Solsvik J, Laerum OD: Quantitation of single cell phagocytic capacity by flow cytometry; in Laerum OD, Lindmo T, Thorud E (eds): Flow Cytometry IV. Bergen, Universitetsforlaget, 1980, pp 170–174.

167 Raffael A, Valet G: Distinction of macrophage subpopulations: measurement of functional cell parameters by flow-cytometry; in Norman SJ, Sorkin E (eds): Macrophages and Natural Killer Cells. New York, Plenum Press, 1982, pp 453–459.

168 Rothe G, Valet G: Phagocytosis, intracellular pH and cell volume in the multifunctional analysis of granulocytes by flow-cytometry. Cytometry 1988;9:316–324.

169 Visser JWM, Jongeling AAM, Tanke HJ: Intracellular pH determination by fluorescence measurements. J Histochem Cytochem 1979;27:32–35.

170 Gerson DF, Kiefer H, Eufe W: Intracellular pH of mitogen-stimulated lymphocytes. Science 1982;1009–1010.

171 Valet G, Raffael A, Moroder L, Wünsch E, Ruhenstroth-Bauer G: Fast intracellular pH determination in single cells by flow-cytometry. Naturwissenschaften 1981;68:265–266.

172 Valet G, Ruhenstroth-Bauer G, Wünsch E, Moroder L: Process for determining the pH value in the interior of a cell. US patent 4677060 priority date Feb 6, 1981 (DE).

173 Valet G, Raffael A: Determination of intracellular calcium in vital cell by flow cytometry. Naturwissenschaften 1985;72:600–602.

174 Dubben H, Meyerhoff W, Tarnok A: Selection strategy for neuropeptide receptor expressing cells. Cell Prolif 1991;1:81.

175 Tarnok A: Improved kinetic analysis of cytosolic free calcium in pressure sensitive cells by fixed time flow-cytometry. Cytometry 1996;23:82–89.

176 Rothe G, Oser A, Valet G: Dihydrorhodamine 123: a new flow cytometric indicator for respiratory burst activity. Naturwissenschaften 1988;75:354–355.

177 Rothe G, Valet G: Flow cytometric analysis of respiratory burst activity in phagocytes with hydroethidine and 2′7′-dichlorofluorescein. J Leukoc Biol 1990;47:440–448.

178 Treumer J, Valet G: Flow-cytometric determination of glutathione alterations in vital cells by *o*-phthaldialdehyde (OPT) staining. Exp Cell Res 1986:163:518–524.

179 Poot M, Verkerk A, Koster JF, Jongkind JF: De novo synthesis of glutathione in human-fibroblasts during in vitro aging and some metabolic diseases as measured by a flow cytometric method. Biochim Biophys Acta 1986;883:580–584.

180 Valet G, Bamberger S, Ruhenstroth-Bauer G: Flow cytometric determination of surface charge density of the erythrocyte membrane using fluorescinated polycations. J Histochem Cytochem 1979;27:342–349.

181 Rothe G, Kellermann W, Valet G: Flow cytometric parameters of neutrophil function as early indicators of sepsis- or trauma-related pulmonary or cardiovascular organ failure. J Lab Clin Med 1990;115:52–61.

182 Rothe G, Assfalg-Machleidt I, Machleidt W, Valet G: Independent regulation of endopeptidase activity and respiratory burst activity of neutrophils analyzed by flow cytometry; in Burger G, Oberholzer H, Voijs GP (eds): Advances in Analytical Cellular Pathology. Amsterdam, Excerpta Medica, 1990, pp 119–120.

183 Assfalg-Machleidt I, Rothe G, Klingel S, Banati R, Mangel WF, Valet G, Machleidt W: Membrane permeable fluorogenic rhodamine substrates for selective determination of cathepsin L. Biol Chem Hoppe Seyler 1992;373:433–440.

184 Rothe G, Klingel S, Assfalg-Machleidt I, Machleidt W, Zirkelbach C, Banati RB, Mangel WF, Valet G: Flow cytometric analysis of protease activities in vital cells. Biol Chem Hoppe Seyler 1992;373:547–554.

185 Klingel S, Rothe G, Kellermann W, Valet G: Flow cytometric determination of cysteine and serine proteinase activities in living cells with rhodamine110 substrates. Methods Cell Biol 1994;41:449–459.

186 Ganesh S, Klingel S, Kahle H, Valet G: Flow cytometric determination of aminopeptidase activities in viable cells using fluorogenic rhodamine110 substrates. Cytometry 1995;20:334–340.

187 Valet G, Warnecke HH, Kahle H: New possibilities of cytostatic drug testing on patient tumor cells by flow cytometry. Blut 1984;49:37–43.

188 Valet G: A new method for fast blood cell counting and partial differentiation by flow cytometry. Blut 1984;49:83–90.

189 Valet G: Method for the simultaneous quantitative determination of cells and reagent therefor. US patent 4751188, priority date: Oct 15, 1982 (DE).

190 Cassens U, Greve B, Tapernon K, Nave B, Severin E, Sibrowski W, Göhde W: A novel true volumetric method for the determination of residual leucocytes in blood components. Vox Sang 2002;82:198–206.

191 Gold R, Schmied M, Rothe G, Zischler H, Breitschopf H, Wekerle H, Lassmann H: Detection of DNA fragmentation in apoptosis – application of in-situ nick translation to cell-culture systems and tissue-sections. J Histochem Cytochem 1993;41:1023–1030.

192 Hutter KJ, Göhde W, Emeis CC: Untersuchungen über die DNS-, RNS- und Proteinsynthese ausgewählter Mikroorganismenpopulationen mit Hilfe der Zytophophotometrie und der Impulscytophotometrie. Food Chem Mikrobiol Techn 1975;4:29–32.

193 Hutter KJ, Stöhr M, Eipel H: Simultaneous DNA and protein measurements of microorganisms; in Laerum OD, Lindmo T, Thorud E (eds): Flow cytometry IV: Bergen, Universitetsforlaget, 1980, pp 100–102.

194 Baisch H, Beck HP, Christensen IJ, Hartmann NR, Fried J, Dean PN, Gray JW, Jett JH, Johnston DA, White RA, Nicolini C, Zietz S, Watson JV: Comparison of evaluation methods for DNA histograms measured by flow-cytometry; in Laerum OD, Lindmo T, Thorud E (eds): Flow Cytometry IV. Bergen, Universitetsforlaget, 1980, pp 152–155.

195 Ruhenstroth-Bauer G, Valet G, Kachel V, Boss N: Die elektrische Volumenmessung von Blutzellen bei der Erythropoese, bei Rauchern, Herzinfarkt- und Leukämiepatienten, sowie von Leberzellkernen. Naturwissenschaften 1974;61:260–266.

196 Valet G, Hofmann H, Ruhenstroth-Bauer G: The computer analysis of volume distribution curves: demonstration of two erythrocyte populations of different size in the young guinea pig and analysis of the mechanism of immune lysis of cells by antibody and complement. J Histochem Cytochem 1976;24:231–246.

197 Valet G: Graphical representation of three-parameter flow cytometer histograms by a newly developed FORTRAN IV computer program; in Laerum OD, Lindmo T, Thorud E (eds): Flow Cytometry IV. Bergen, Universitetsforlaget, 1980, pp 125–129.

198 Eisen MB, Spellman PT, Brown PO, Botstein D: Cluster analysis and display of genome-wide expression patterns. Proc Natl Acad Sci USA 1998;95:1486–1488.

199 Wilkins MF, Hardy SA, Boddy L, Morris CW: Comparison of five clustering algorithms to classify phytoplankton from flow cytometric data. Cytometry 2001;44:210–217.

200 Leary JF: Strategies for rare cell detection and isolation; in Z Darzynkiewicz, JP Robinson, HA Crissman (eds): Methods in Cell Biology: Flow Cytometry. San Diego, Academic Press, 1994, ed 2, pt B, vol 42, pp 331–358,

201 Molnar B, Szentirmay Z, Bodo M, Sugar J, Feher J: Application of multivariate, fuzzy set and neural network analysis in quantitative cytological examination. Anal Cell Pathol 1993;5:161–175.

202 Davey HM, Jones A, Shaw AD, Kell DB: Variable selection and multivariate methods for the identification of microorganisms by flow cytometry. Cytometry 1999;35:162–168.

203 Hokanson JA, Rosenblatt JI, Leary JF: Some theoretical and practical considerations for multivariate statistical cell classification useful in autologous stem cell transplantation and tumor cell purging. Cytometry 1999;36:60–70.

204 Thews O, Thews A, Huber C, Vaupel P: Computer-assisted interpretation of flow cytometry data in hematology. Cytometry 1996;23:140–149.

205 Diamond LW, Nguyen DT, Andreeff M, Maiese RL, Braylan RC: A knowledge based system for the interpretation of flow cytometric data in leukemia and lymphoma. Cytometry 1994;17:266–273.

206 Beckman RJ, Salzman GC, Stewart CC: Classification and regression trees for bone marrow immunophenotyping. Cytometry 1995;20:210–217.

207 Dybowski R, Grant VA, Riley PA, Phillips I: I. Rapid compound pattern classification by recursive partitioning of feature space. An application in flow cytometry. Pattern Recog Lett 1995;16:703–709.

208 Decaestecker C, Remmelink M, Salmon I, Camby I, Goldschmidt D, Petein M, Van Ham P, Pasteels JL, Kiss R: Methodological aspects of using decision trees to characterise leiomyomatous tumors. Cytometry 1996;24:83–92

209 Peltri G, Bitterlich N: Increased predictive value of parameters by fuzzy logic-based multiparameter analysis. Cytometry B Clinical Cytometry 2003;53B:75–77.

210 Zheng Q, Milthorpe BK, Jones AS: Direct neural network application for automated cell recognition. Cytometry A 2003;57A:1–9.

211 Frankel DS, Frankel SL, Binder BJ, Vogt RF: Application of neural networks to flow cytometry data analysis and real-time cell classification. Cytometry 1996;23:290–302.

212 Boddy L, Wilkins MF, Morris CW: Pattern recognition in flow cytometry. Cytometry 2001;44:195–209.

213 Valet G: Automated diagnosis of malignant and other abnormal cells by flow cytometry using the newly developed DIAGNOS1 program system; in Burger G, Ploem B, Goerttler K (eds): International Symposium Histometry. London, Academic Press, 1987, pp 58–67.

214 Valet G, Valet M, Tschöpe D, Gabriel H, Rothe G, Kellermann W, Kahle H: White cell and thrombocyte disorders: Standardized, self-learning flow cytometric list mode classification with the CLASSIF1 program system. Ann NY Acad Sci 1993;677:183–191.

215 Valet G: Human cytome project: A new potential for drug discovery; in Las omicas genomica, proteomica, citomica y metabolomica: modernas tecnologias para desarrollo de farmacos. Madrid, Real Academia Nacional de Farmacia, 2005, pp 207–228.

216 Valet G, Höffkes HG: Automated classification of patients with chronic lymphatic leukemia and immunocytoma from flow cytometric three colour immunophenotypes. Cytometry Comm Clin Cytometry 1997;30:275–288.

217 Valet G, Cytometry, a biomedical key discipline; in Robinson JP (ed): Purdue Cytometry CD-ROM, vol 4. Purdue University, West Lafayette, 1997. *www.cyto.purdue.edu/cdroms/flow/vol4/8_websit/valet/keyvirt1.htm.*

218 Valet G: Human disease; in Robinson JP (ed): Purdue Cytometry CD-ROM, vol 2. West Lafayette, Purdue University, 1996. *www.cyto.purdue.edu/cdroms/flow/vol2/14/valet/disease.htm.*

219 Valet G. Cytometry and human disease; in Robinson JP (ed): Purdue Cytometry CD-ROM, vol 5. West Lafayette, Purdue University, 2000. *www.cyto.purdue.edu/cdroms/flow/vol5/websites/cytrelay/disease.htm.*

220 Valet G: Predictive medicine by cytomics: potential and challenges. J Biol Regul Homeost Agents 2002;16:164–167.

221 Valet G, Tarnok A: Cytomics in predictive medicine. Cytometry B Clin Cytometry 2003;53B:1–3.

222 Valet G, Repp R, Link H, Ehninger G, Gramatzki M, SHG-AML study group. Pretherapeutic identification of high-risk acute myeloid leukemia (AML) patients from immunophenotype, cytogenetic and clinical parameters. Cytometry 2003;53B:4–10.

223 Valet G: Concepts and history of flow cytometry and cytomics at Max-Planck-Institut für Biochemie, Martinsried (1960–2006); in ed JP Robinson, 60 Years Innovation in Cytometry, Purdue University Cytometry Laboratories, ISBN 978–1–890473–10–5 (2007)
224 Valet G, Murphy RF, Robinson JP, Tárnok A, Kriete A: Cytomics – from cell states to predictive medicine; in Kriete A, Eils R (eds): Computational Systems Biology. Elsevier, Amsterdam 2006, p 363–381.
225 Valet G: Cytomics as a new potential for drug discovery. DDT 11:785–791(2006).
226 Valet G, Leary JF, Tarnok A: Cytomics – new technologies: towards a human cytome project. Cytometry A 2004;59A:167–171.
227 Martinreay DG, Kamentsky LA, Weinberg DS, Hollister KA, Cibas ES: Evaluation of a new slide-based laser-scanning cytometer for DNA analysis of tumors – comparison with flow-cytometry and image-analysis. Am J Clin Pathol 1994;102:432–438.
228 Hell S, Stelzer EHK: Properties of a 4PI confocal fluorescence microscope. J Opt Soc Am A – Opt Imag Sci Vis 1992;9:2159–2166.
229 Khan SS, Smith MS, Reda D, Suffredini AF, McCoy JP: Multiplex bead array assays for detection of soluble cytokines: comparison of sensitivity and quantitative values among kits from multiple manufacturers. Cytometry 2004;61B:35–39.
230 Hood L, Heath JR, Phelps ME, Lin B: Systems biology and new technologies enable predictive and preventive medicine. Science 2004;306:640–643.
231 Liebman M: Systems biology: top-down or bottom-up. Bio-IT World 2004: *www.bioitworld.com/archive/031704/horizons_horizons_comm.html.*
232 Valet G: Cytomics: an entry to biomedical cell systems biology. Cytometry A 2005;63A:67–68.
233 Valet G: Human cytome project, cytomics and systems biology: the incentive for new horizons in cytometry. Cytometry A 2005;64A:1–2.

Sack U, Tárnok A, Rothe G (eds): Cellular Diagnostics. Basics, Methods and Clinical Applications of Flow Cytometry. Basel, Karger, 2009, pp 53–88

Technical Background and Methodological Principles of Flow Cytometry

Gregor Rothe

Laborzentrum Bremen, LADR Group, Bremen, Germany

Introduction

Flow cytometers are highly flexible technical platforms which allow the quantitative analysis and molecular characterization of intact cells. The hydrodynamic transport of cells and their optical analysis after specific staining form the central components of cell analysis. In more detail, the following basic principles apply:

- A suspension of cells is injected into a sheath flow (table 1) and passes as a focused sequence of single cells across an orthogonal light source (fig. 1). A laser serves as a monochromatic light source. Alternatively, mercury arc lamps are also used.
- At the intersection of sample flow and the exciting beam of light, the cellular scattering of light and the emission of fluorescent probes are detected. This allows the simultaneous analysis of molecular and physical characteristics of cells such as size and granularity (fig. 2). Depending on staining, the brightness of fluorescence signals correlates to the amount of bound antibodies, nuclear DNA content or biochemical characteristics of cells.
- All light signals encoding physical and molecular characteristics of cells, the 'parameters', are quantified and recorded in a database for each single cell. Data analysis is performed in a separate second step. Thus different questions may be addressed during data analysis regarding various populations of cells independent of analysis settings during initial data acquisition.
- Data acquisition for a large number of cells allows statistical analysis. At the same time, the stability of the instrument setup allows calibration of fluorescent signals and thus quantitative analysis of molecular characteristics of cells.

Table 1. Technical terms in flow cytometry

Term	Explanation
Channels	Historically, the use of classes of digital resolution as a scale for the quantification of a parameter. Today data are acquired with high digital resolution and data are presented as the underlying signal intensities
Compensation	Correction of spectral overlap between fluorochromes
Contour plot	A two-dimensional display of parameters where contour lines connect fields with the same number of events similar to the indication of height on a map
Coulter volume	The precise analysis of cell size based on the detection of changes in electrical resistance when a cell passes through an orifice
Dot plot	A two-dimensional display of parameters where each cell or event is shown as a dot at the x- and y-coordinates of the corresponding signal intensities
Emission	Light with a higher wavenlength which is emitted by a fluorochrome following absorption of exciting light
Excitation	Light which is absorbed by a fluorochrome and results in the emission of fluorescence
FL-1, FL-2	Detectors for fluorescent light as numbered in the sequence of lasers and with increasing wavelength
Forward scatter	Light deflected at a small angle to the exciting laser beam. It correlates to the size of cells even though it depends on shape and does not continuously increase with size
Gate	A region or logical combination of regions which selects a subset of events for further selective analysis
Histogram	A graph showing the intensity distribution of one parameter on the x-axis and the number of events on the y-axis
List mode data	A data file which contains all parameters for each single cell in the sequence of data acquisition
log amplifier	An analog circuit in older instruments which logarithmically transforms signals before analog-to-digital conversion. log amplifiers are problematic for the quantification of fluorescence as they largely deviate from a mathematically correct transformation. Modern instruments therefore use software log transformation following high-resolution digital conversion of raw signals
MESF	Molecules of equivalent soluble fluorochrome: A measure for signal intensity specifically calibrated for each fluorochrome
Photomultiplier	A device which converts light into electrical signals. The sensitivity of photomultipliers is regulated by voltage
Pulse height	The highest light signal during the flow of a cell across a laser beam. Pulse height is falsely lower than the total cellular fluorescence for large cells, e.g. tumor cells or cultured cells, which are broader than the focus of the laser beam. Pulse area is a correct measure under these circumstances
Region	A region defines a subpopulation of events in a one- (see histogram) or two-dimensional (see dot plot) graphical presentation of cells for statistical analysis or further selective analysis by gating (see gate)
Sheath fluid	A fluid mantle which focuses and dilutes sample flow in the analysis cuvette. Buffered saline solutions are typically used as a sheath fluid
Side scatter	Light which is detected at a right angle to the exciting laser beam and that correlates to the internal complexity or granularity of cells
Threshold	A value defined for the leading parameter which discriminates cells of interest from noise and initiates data acquisition. All signals from cells which do not reach the threshold are not stored during data acquisition

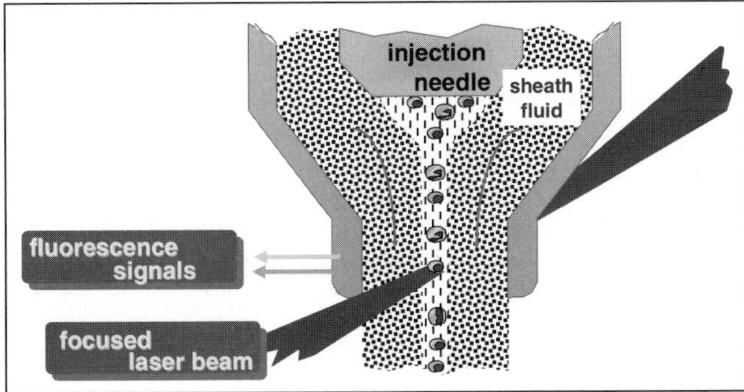

Fig. 1. Principle of cell analysis in flow. Cells in suspension are passing by a point of optical detection. A sheath flow is focusing and 'diluting' the sample flow and single cells are crossing the point of detection in sequence (modified from R. Murphy, Carnegie Mellon University, *www.isac-net.org*).

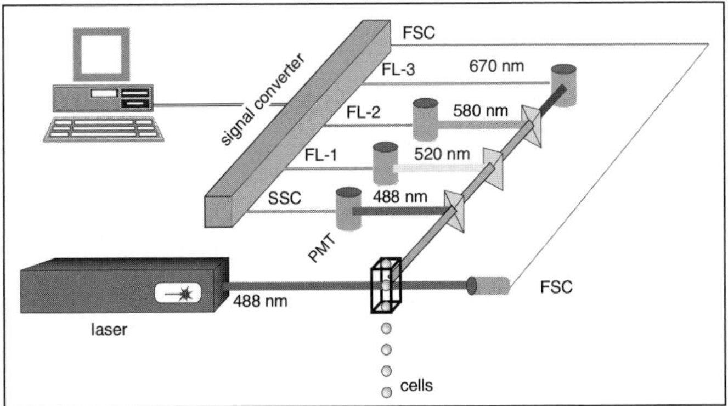

Fig. 2. General structure of a flow cytometer: Light is scattered at the intersection of a laser beam with the sample flow and fluorescence signals are elicited. FSC is detected as a correlate of cell size. Dichroic mirrors and filters are spectrally separating SSC as a measure of granularity and fluorescence signals (FL1, FL2, FL3). PMTs are detecting the light signals and data of each cell are stored in digital form.

The analysis of DNA content, e.g. of tumor cells (see 'DNA and Proliferation Analysis in Flow Cytometry', pp. 390), or the determination of absolute counts of cells identified using specific antibodies, e.g. CD4+ T cells or CD34+ hematopoietic stem cells and progenitor cells, are typical applications of flow cytometry. Patterns of antigen coexpression are analyzed, which indicate, e.g. the activation of cells of the immune system or aberrant maturation in leukemia. Furthermore, quantification of fluorescence signals allows the functional characterization of cells. The analysis of biochemical signals such as ion concentrations or enzymatic reactions in response to stimulation (see 'Determination of Cell Physiological Parameters: pH, Ca^{2+}, Glutathione, Transmembrane Potential', pp. 325), the quantification of receptor expression density (see 'Quality Control and Standardization', pp. 159), or the analysis of the interaction of receptors in complexes or with their ligands (see 'Fluorescence Resonance Energy Transfer', pp. 141) are different examples of functional assays.

Detailed descriptions of the history of flow cytometry and its clinical applications can be found in 'Cytometry – A Definitive History of the Early Days' (pp. 1), and in 'Concept Developments in Flow Cytometry' (pp. 29), of this book as well as in a review by Janossy [1]. It is the goal of this chapter to illustrate some common principles of flow-cytometric methods used in diagnostic or clinical research applications. Furthermore, technical platforms are described. For a detailed introduction to the technology of flow cytometry, readers are referred to the handbook by Shapiro [2].

Technical Background

Principles of Analysis

Flow-cytometric analysis is based on the measurement of light scatter and fluorescence parameters when a linear stream of cells is passing a laser beam at a right angle. Cells from 0.2 to 20 μm in size are hydrodynamically focused by a sheath fluid and pass the focus of the laser beam at a rate of 200–2,000 events/s. For analysis of rare cells or for high sample throughput, some instruments will also allow data acquisition at a rate up to 50,000 cells/s. Typically, the sample stream is slow at 10–60 μl/min. Some instruments will also operate at 300 μl/min and higher.

Optics

As physical measures of cells, light scattering at a low angle to the laser beam is detected as forward scatter (FSC) and light scattering at 90° as side scatter (SSC) (fig. 3). FSC will typically resolve cells with a size of 0.5–1 μm and larger and correlates to the size of cells even though it depends on shape and does not continuously increase with size. SSC depends on the multiple refraction of light through intracellular compartments and thus correlates to the granularity and internal structure of cells. Pretreatment of cells, e.g. by erythrocyte lysis or fixation, will affect scatter characteristics. In order to optimize cell separation in instruments with different types of optics, specific lysis and fixative solutions are used (fig. 4). Electrical resistance also changes when cells pass a flow channel, and these signals provide a more precise correlate of cell size. Electrical resistance volumetry, which was initially introduced by Coulter, is currently also available in some instruments from Partec for instance.

Cells emit fluorescent light in all directions. Fluorescence signals are detected at an angle of 90° to excitation as scattered light is lowest at this angle to the laser

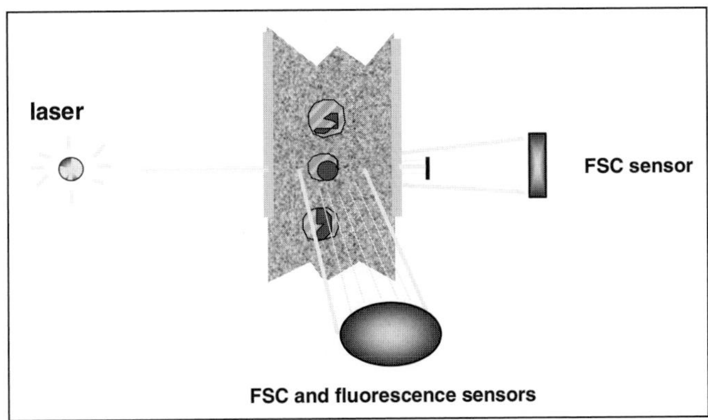

Fig. 3. Geometry of the detection of light scatter and fluorescence signals. FSC, which is detected at a low angle to the laser beam, represents an imprecise correlate for the size of cells. SSC and fluorescence signals are detected at a right angle (modified from R. Murphy, Carnegie Mellon University, *www.isac-net.org*).

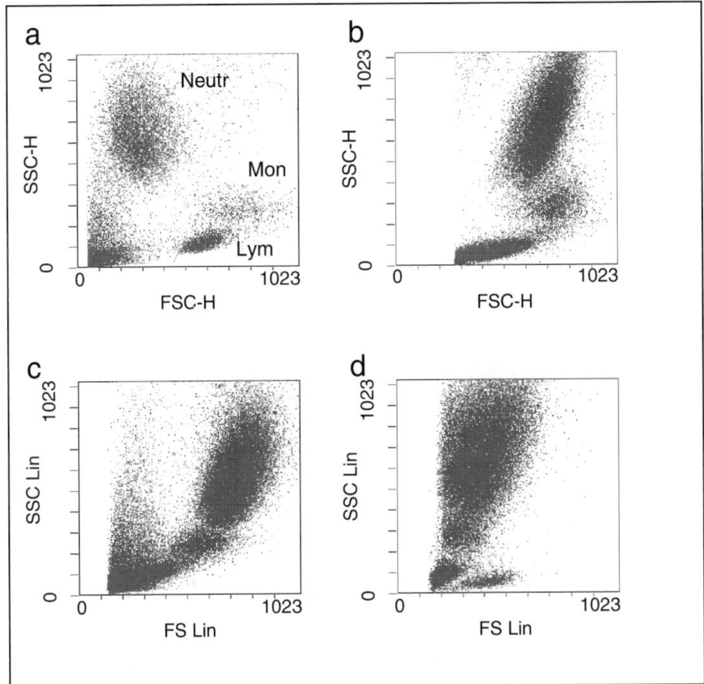

Fig. 4. Impact of optical layout and preanalytical treatment on light scatter. Leukocyte light scatter as analyzed using a Becton Dickinson FACSCalibur (**a, b**) and a Beckman Coulter FC500 (**c, d**) following erythrocyte lysis with a reagent optimized for the optics of Beckman Coulter (**a, c**) and Becton Dickinson (**b, d**), respectively. Lymphocytes (Lym), monocytes (Mon) and neutrophils show increasing SSC or SS. The relative extent of FCS in contrast differs dependent on the angle of detection as well as lysis and fixation.

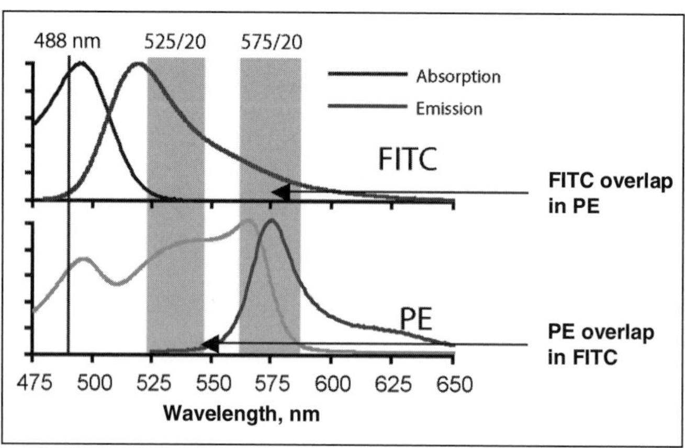

Fig. 5. Spectral separation of fluorescence signals using band-pass filters. The green-yellow fluorescence of FITC as well as the orange fluorescence of R-phycoerythrin (PE) are excited using the blue light (488 nm) of an argon laser. Band-pass filters of 20 nm width with center wavelengths of 525 and 575 nm allow the detection of light within the main emission spectrum of FITC and PE, respectively. The spectral overlap of the dyes needs to be corrected by compensation (by courtesy of M. Schiemann and D. Busch).

beam. Dichroic mirrors and band-pass filters are used for the spectral separation of light emission from different fluorochromes (fig. 5) (see 'Selection and Combination of Fluorescent Dyes', pp. 107). Photomultiplier tubes (PMT) detect light and convert it into electric signals. By convention, these detectors and their signals, which are called 'parameters', are numbered according to the laser sequence and the increasing wavelength of detected light (FL1, FL2, FL3 ...). As an example in a typical configuration with a 488-nm argon ion laser exciting with blue-green light, the FL1 detector will collect green-yellow light, the FL2 detector will collect orange light whereas red light is collected by the FL3 detector. Electrical signals are finally converted from analog to digital format and stored as a list of data sets of all signals for consecutive cells in a list mode data file.

Data Acquisition

For the discrimination of cellular signals from continuous 'background noise', a leading parameter is defined as the 'trigger'. The acquisition of a list mode data set will start as soon as the leading signal reaches a predefined 'threshold' value. As an example, for FSC as the trigger, data sets will be acquired for all cells with a size above the threshold whether they are fluorescent or not. In the case of samples with a lot of cellular 'debris', e.g. lysed whole blood, a DNA dye or an antibody which binds to all cells of interest can be used as the trigger in order to acquire data selectively for cells of interest without generating a large data set with noise signals or irrelevant events.

Pulse processing of signals represents a further critical issue. In immunological assays, the highest light signal, i.e. 'pulse height' (e.g. FL1-Height or FL1-H), is often stored as a correlate of cellular fluorescence intensity. However, this signal is only proportional to cell fluorescence if the cell is completely excited when it passes the laser beam. Thus, for cells which are larger than the width of the laser beam, acquisition of the integral curve under the light pulse, i.e. 'pulse area' (e.g. FL1-Area or FL1-A) is needed. The simultaneous determination of pulse width (e.g. FL1-Width or FL1-W) and pulse area furthermore allows doublet discrimination between signals caused by the passage of one large cell and those caused by the simultaneous passage of two smaller cells (coincidence of cells).

Finally, there are different methods for the generation of logarithmic data and compensation of spectral overlap between different fluorochromes which would affect data quality. Representation of data on a logarithmic scale usually is needed as e.g. staining with monoclonal antibodies frequently leads to a 100- or 1,000-fold difference of signal intensities between strongly positive and negative cells. Weaker signals in such cases are only visible on a logarithmic scale. Typically, logarithmic scales are used which categorize a four-decade range of signal intensities with a resolution of 1,024 or 4,096 channels, e.g. digital classes of cells with similar signal intensity. Electronic logarithmic amplification is a simple form of logarithmic conversion. However, this method leads to significant deviations from logarithmic order in the lower and upper range of intensities. Therefore, modern flow cytometers address this problem using high-resolution linear analog-to-digital conversion which allows computational conversion into logarithmic format (tables 2–4).

Data acquisition in linear format has advantages if only DNA is detected as a parameter. Traditionally, light scatter parameters, i.e. FSC and SSC, are mostly acquired on a linear scale. As a drawback, light scatter in linear format does not allow the discrimination of large or highly granulated cells such as plasma cells and eosinophil granulocytes in peripheral blood, cell doublets or cells enlarged following stimulation or due to proliferation.

Compensation

The compensation of fluorescence signals for spectral overlap represents a similar problem. As the underlying problem, fluorochromes typically overlap in their emission spectra and mixtures with different proportions of each fluorochrome are detected on each PMT (fig. 5) (see 'Selection and Combination of Fluorescent Dyes', pp. 107). An uncorrected, or uncompensated, measurement in dual color analysis will thus result in signals on both fluorescence detectors even for cells which only carry either of the fluorochromes (fig. 6; see fig. 11 for explanations). Similar to logarithmic conversion (see section on 'Data Acquisition' above), this problem can also be addressed using analog circuits for the percent correction of signals before digitization and storage. This approach suffices for many immunological applica-

Table 2. Technical specifications of analytical flow cytometers from Beckmann Coulter according to the manufacturer's information

Parameter	BC Cytomics FC 500	BC EPIC XL/XL-MCL	BC CyAn ADP 9 colors
Laser	20-mW, 488-nm argon laser and optionally 25-mW red solid-state diode 635 nm	20-mW, 488-nm argon laser	25-mW, 405-nm diode laser, 20-mW, 488 nm solid-state laser and 25-mW 635-nm diode laser elliptical beam-shaping optics
Excitation	n/a for 1 laser configuration; collinear for 2 laser configurations	n/a; no laser alignment needed	spatially separated through pinholes, fixed time delay for signal processing
Detectors	FSC (two selectable angles), SSC, 5 fluorescences	FSC, SSC, 4 fluorescences	FSC, SSC, 9 fluorescences
Standard filters	FITC 525BP; PE 575BP, ECD 620BP; PC5 675BP, PC7755, dichroic LP: 488, 550, 600, 710, dichroic SP: 615, 645 LP: 500, SP: 620	FITC 525 BP; PE 575 BP, ECD 620 BP; PC5 675 BP, dichroic LP: 488, 550, 600, blocking filter: 488	FITC: 530/40BP – (545DLP) – PE: 575/25BP – (595DLP) – PE-TxR/PI: 613/20BP – (640DLP) – PerCP/7AAD: 680/30BP – (730DLP) – PE-Cy7: 750LP; APC: 665/20BP – (730DLP) – APC-Cy7: 750LP PB: 450/50BP – (485DLP) – CY: 530/40BP standard 12-mm filters exchangeable by user
Standard fluorochromes	FITC, PE, PE-Texas Red (ECD), PE-Cy5 (PC5) (1 laser configuration) or APC (2 laser configuration), PE-Cy7 (PC7)	FITC, PE, PE-Texas Red (ECD), PE-Cy5 (PC5)	FITC, PE, PE-TxR/PI, PerCP/PerCP-Cy5.5, PE-Cy7; APC, APC-Cy7; Pacific Blue, Cascade Yellow
Fluorescence sensitivity	PE < 300 MESF, FITC < 600 MESF	FITC and PE < 1,000 MESF)	PE < 50 MESF, FITC < 00 MESF
Scatter resolution	FSC 0.5 μm, SSC < 0.5 μm	FSC 0.5 μm, SSC < 0.5 μm	FSC 1 μm, SSC 0.5 μm

Table 2. Continued on next page

Table 2. Continued

Parameter	BC Cytomics FC 500	BC EPIC XL/XL-MCL	BC CyAn ADP 9 colors
Signal processing	digital, 20 bit	digital, 10 bit	digital, full resolution in all decades by synthetic pulse transformation 16 bit
Compensation	inverted matrix, on line and off line	inverted matrix, online	Inverted matrix, on- and offline
Maximum acquisition rate	3,300 events/s fully compensated	3,300 events/s fully compensated	50,000 events/s fully compensated
Data resolution	1,048,576 channels	1,024 channels	65,536 channels
Data format	FCS 3.0, downward compatible for data analysis	FCS 2.0	FCS 2.0, 3.1
Computer/ operating system	PC/Microsoft Windows 2000	PC/Microsoft Windows 98 and/or Microsoft DOS 6.22	PC/Microsoft Windows XP Pro
Sample loading	carousel loader for 32 tubes and single tube position with integrated vortex (MCL) or Multi Plate Loader for tubes and microtiter plates (MPL)	single tube loading optionally with carousel loader for 32 tubes with integrated vortex (MCL)	single-tube loading
Media supply	integrated media supply for sheath and cleansing fluid; external waste container	integrated media supply for sheath and cleansing fluid; external waste container	external media supply for 20 liters sheath fluid, 20 liters waste and 5 liters cleansing fluid
Flow rate	10-, 30- or 60-μl sample/min	10-, 30- or 60-μl sample/min	0- to 300-μl sample/min variable, 3 user-defined presets
Sample carryover	<1.0%	<1.0%	<1%
Power supply	220/240 V, 16 A	220/240 V, 16 A	230 V, 16 A

Table 3. Technical specifications of analytical flow cytometers from Becton Dickinson according to manufacturer's information

Parameter	BD FACSCanto	BD LSR II	BD FACSCalibur
Laser	20-mW, 488-nm solid-state laser and 17-mW, 633-nm He–Ne laser, glass fiber coupling	25-mW, 355-nm solid-state laser, 25-mW, 405-nm diode laser, 20-mW, 488-nm solid-state laser and 17-mW, 633-nm He–Ne laser	15-mW, 488-nm argon laser and 10-mW, 633-nm diode laser
Excitation	spatially separate, fixed time delay for signal processing	spatially separate, fixed time delay for signal processing	spatially separated, fixed time delay for signal processing
Detectors	FSC, SSC, 6 fluorescences	FSC, SSC, up to 18 fluorescences	FSC, SSC, 4 fluorescences
Standard filters	FITC: 502LP, 530/30BP; PE 556LP, 585/42BP; PerCP/PerCP-Cy5.5 655LP, 670LP; PE-Cy7 735LP, 780/60BP; APC-Cy7 735LP, 780/60BP; APC no LP, 660/20BP filters exchangeable by user	contact BD Biosciences filters exchangeable by user	FITC 530/30 BP; PE 585/42 BP, PerCP/PerCP-Cy5.5 670 LP; APC 661/16 BP
Standard fluorochromes	FITC, PE, PerCP/PerCP-Cy5.5, PE-Cy7; APC, APC-Cy7	contact BD Biosciences	FITC, PE, PerCP/PerCP-Cy5.5; APC
Fluorescence sensitivity	PE < 50 MESF, FITC < 100 MESF	<200 MESF	<200 MESF
Scatter resolution	FSC 1 μm, SSC 0.5 μm	allows the separation of fixed platelets from background noise	optimized for the resolution of lymphocytes, neutrophils and monocytes
Signal processing	digital, 18 bit	digital, 18 bit	analog
Compensation	inverted matrix, on line and off line	inverted matrix, on line and off line	on line

Table 3. Continued on next page

Table 3. Continued

Parameter	BD FACSCanto	BD LSR II	BD FACSCalibur
Maximum acquisition rate	10,000 evens/s fully compensated	more than 20,000 events/s	3,500 events/s
Data resolution	262,144 channels	262,144 channels	1,024 channels
Data format	FCS 2.0 and 3.0	FCS 2.0 and 3.0	FCS 2.0
Computer/ operating system	PC/Microsoft Windows XP Pro	PC/Microsoft Windows XP Pro	Mac/OSX
Sample loading	single tube position optional carousel for 40 tubes or HTS option for microtiter plates	single tube position HTS option for microtiter plates	single tube position optional carousel for 40 tubes or HTS option for microtiter plates
Media supply	external media supply for sheath, waste and cleansing fluids	external media supply with optionally selectable container sizes	integrated or optionally external media supply for sheath fluid and waste with a five-fold longer interruption-free operation time
Flow rate	10-, 60- or 120-μl sample/min	12-, 35- or 60-μl sample/min, with additional fine adjustment in the range of 0.5–2×	12-, 35- or 60-μl sample/min
Sample carryover	<0.1%	<1.0%	<1.0%
Power supply	230 V, 16 A	230 V, 16 A	230 V, 16 A

Table 4. Technical specifications of analytical flow cytometers from Partec according to manufacturer's information

Parameter	Partec CyFlow® SL	Partec CyFlow® space	Partec CyFlow® ML
Laser	20-mW, 488-nm solid-state laser or 100-mW, 532-nm solid-state laser	selection of up to 3 lasers: 20-/50-/200-/500-mW, 488-nm solid-state laser, 25-mW, 638-nm laser diode, 50-/100-mW, 405-nm diode laser or 10-mW, 375-nm diode laser	selection of up to 4 lasers: 20/50/200/500 mW, 488 nm, 25 mW, 638 nm, 50/100 mW, 405 nm, 10 mW, 375 nm, 50 mW, 562 nm, 100 mW, 532 nm, HBO lamp for high-resolution UV excitation
Excitation	direct, laser spot 60 × 15μm	direct, optical paths and laser spots 60 × 15 μm spatially separate, fixed time delay for signal processing	direct, optical paths and laser spots 60 × 15 μm spatially separated, adjustable time delay for signal processing
Detectors	FSC, SSC, 3 fluorescences	FSC, SSC, 6 fluorescences	FSCI, FSCII, SSC, 13 fluorescences
Standard filters	freely exchangeable standard filters FL1: 537/35 FL2: 590/50 FL3: 630LP	filter set dependent on laser configuration, filter inserts freely exchangeable	filter set dependent on laser configuration, filter inserts freely exchangeable
Standard fluorochromes	FITC, Alexa-488, Syto9, PE, PerCP, PE-Cy5, PI, PE-Cy7,	FITC, Alexa-488/405/633, Syto9, PE, PerCP, PE-Cy5, PI, PE-Cy7, APC, APC-Cy7, DAPI, Hoechst among others dependent on laser configuration	FITC, Alexa-488/405/633, Syto9, PE, PerCP, PE-Cy5, PI, PE-Cy7, APC, APC-Cy7, DAPI, Hoechst among others dependent on laser configuration
Fluorescence sensitivity	<100 MESF (FITC), <50 MESF (PE)	<100 MESF (FITC), <50 MESF (PE)	<100 MESF (FITC), <50 MESF (PE)
Scatter resolution	FSC 0.7 μm, SSC 0.2 μm	FSC 0.7 μm, side scatter SSC 0.2 μm	FSC 0.7 μm, SSC 0.2 μm

Table 4. Continued on next page

Table 4. Continued

Parameter	Partec CyFlow® SL	Partec CyFlow® space	Partec CyFlow® ML
Signal processing	digital, 16 bits	digital, 16 bits	Digital, 16 bits
Compensation	on-line and off-line n-color software compensation	on-line and off-line n-color software compensation	online and off-line n-color software compensation
Maximum acquisition rate	15,000 events/s, 8 plots, fully compensated	15,000 events/s, 8 plots, fully compensated	15,000 events/s, 8 plots, fully compensated
Data resolution	16 bits > 65,000 channels	16 bits > 65,000 channels	16 bits > 65,000 channels
Data format	FCS 2.0	FCS 2.0	FCS 2.0
Computer/ operating system	PC/Microsoft Windows XP Pro	PC/Microsoft Windows XP Pro	PC/Microsoft Windows XP Pro
Sample loading	automated single tube loading, optionally automated loader for carousel or microtiter plates	automated single tube loading, optionally automated loader for carousel or microtiter plates	automated single-tube loading, optionally automated loader for carousel or microtiter plates
Media supply	5-liter container for sheath fluid and waste	5-liter container for sheath fluid and waste	5-liter container for sheath fluid and waste
Flow rate	continuously adjustable sample flow from 0 to 1,200 μl/min	continuously adjustable sample flow from 0 to 1,200 μl/min	continuously adjustable sample flow from 0 to 1,200 μl/min
Sample carry over	<0.5%	<0.5%	<0.5%
Power supply	100/240 V AC, 60VA, 50/60 Hz or 12 VDC/5A	100/240 V AC, 60VA, 50/60 Hz	100/240 V AC, 60 VA, 50/60 Hz

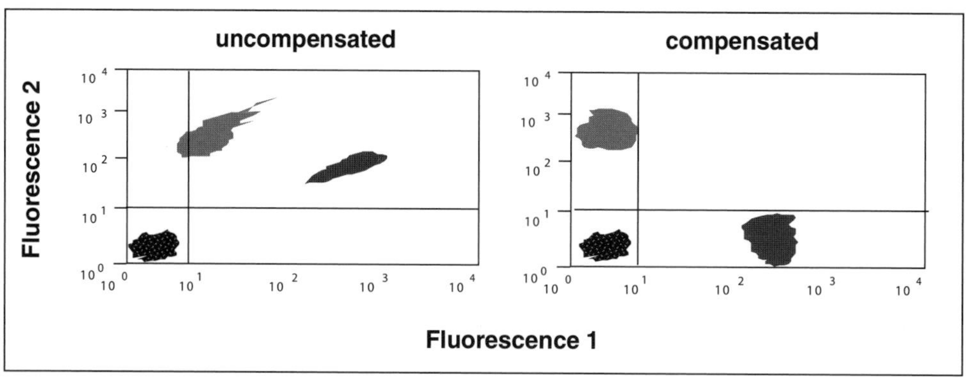

Fig. 6. Principle of compensation. The left panel shows the detection of spectrally overlapping fluorochromes due to incomplete optical separation (see fig. 5). A computational correction for the percentage of spectral overlap leads to the 'compensated' presentation of data in the rigth panel.

tions. As a problem, any errors during compensation will lead to an irreversible corruption of data. High-resolution digitization and list mode storage of native fluorescence signals are a newer alternative approach. In this case, software compensation can be adjusted during re-analysis. Automated software compensation, i.e. the automated generation of a multiparametric compensation matrix from list mode data files, has recently allowed to establish complex multicolor assays with five or more different fluorochromes as routine methods.

Multiple Excitation

Multilaser excitation has recently been introduced into routine assays in order to allow simultaneous detection of four, five or more fluorochromes. As described in detail in 'Selection and Combination of Fluorescent Dyes' (pp. 107), flow cytometers with multiple lasers allow the use of a larger range of fluorochromes as excitation of fluorescence can be performed at different wavelengths. Such instruments are available with two different optical layouts. In a simpler design, two or more laser beams intersect with the sample stream at a single point (collinear laser beams; tables 2–4). As a consequence, the selected fluorochromes must have distinct emission spectra. In an alternative design, two or more laser beams intersect with the sample stream at subsequent points (spatially separate laser beams). This design allows simultaneous use of fluorochromes with similar emission as long as they differ in excitation by the different laser beams. Some instruments use high-pressure mercury arc lamps as the exciting light source. The emission wavelength is selected using band-pass filters for this polychromatic light source. Multiple excitation is collinear unless combination with a laser is performed in a spatially separated form.

Data Formats: Flow Cytometry Standard Data Format

The Flow Cytometry Standard (FCS) data format has been created several years ago as an instrument-independent standard for the storage of measurement data in list mode data files. In its current FCS 3.0 version, the header of the list mode file contains the description of instrument settings during measurement as well as user-defined input such as the sample description, staining and parameter description and a description of the experiment in plain text format. This header is especially useful to retrieve instrument settings for subsequent experiments. The main part of the list mode data file contains data sets composed of light scatter and fluorescence signals for each cell and the standardized format allows data analysis with any FCS-compatible software. Time can be stored as an initial parameter and allows the kinetic analysis of measurements, e.g. in stimulation experiments.

Technical Layout of Widely Used Instruments

Flow cytometers which are used for analytical rather than cell-sorting applications and especially for diagnostic applications are characterized by a closed design with a preadjusted optical bench and the analysis of cells in a closed flow cell. Mainly due to the traditionally higher availability of fluorochrome-conjugated antibodies, an argon ion laser with 488 nm emission is typically used as the main light source, frequently in combination with further lasers. Open technical layouts which allow modifications of the optical pathway, e.g. an exchange of optical filters are described in 'Flow-Cytometric Cell Sorting' (pp. 178). A detailed description of the spectral properties of different lasers as well as of compatible fluorochromes can be found in 'Selection and Combination of Fluorescent Dyes' (pp. 107).

The Cytomics FC500 from Beckman Coulter is an example of an instrument which allows detection of up to five fluorescence signals either with single argon ion laser excitation or additional excitation with a collinear red laser. In the instrument, dichroic mirrors and band-pass filters are in close proximity to the flow cell (fig. 7). Furthermore, similar to the older Epics XL models from the same company, digital logarithmic conversion, integrated automated sample loading and a high degree of automation of sample processing and quality control as well as connectivity to laboratory information systems are characteristics of the instrument.

The FACSCanto from Becton Dickinson is an example of an instrument with spatially resolved dual laser excitation and the acquisition of up to six fluorescence parameters. Fiber-optic coupling of lasers and detection optics are characteristics of the more complex optical design (fig. 8). The instrument has digital processing of primary analysis data and automated sample loading is optional. The FACSCalibur is an older instrument from the same company which can still be used for many applications; it also has spatially separate dual laser excitation but only analog loga-

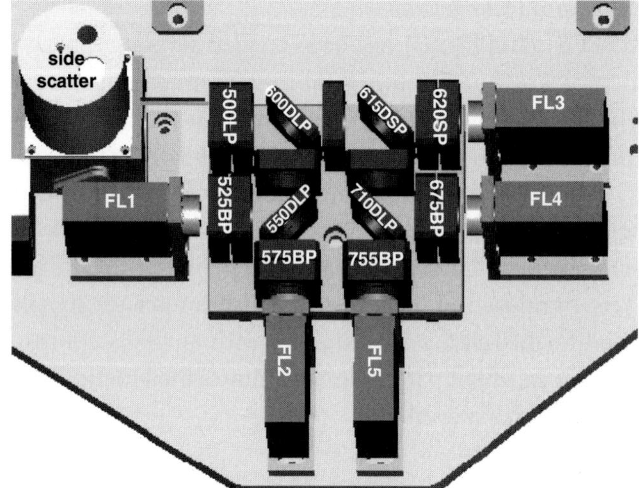

Fig. 7. Optical bench of the Beckman Coulter FC 500.

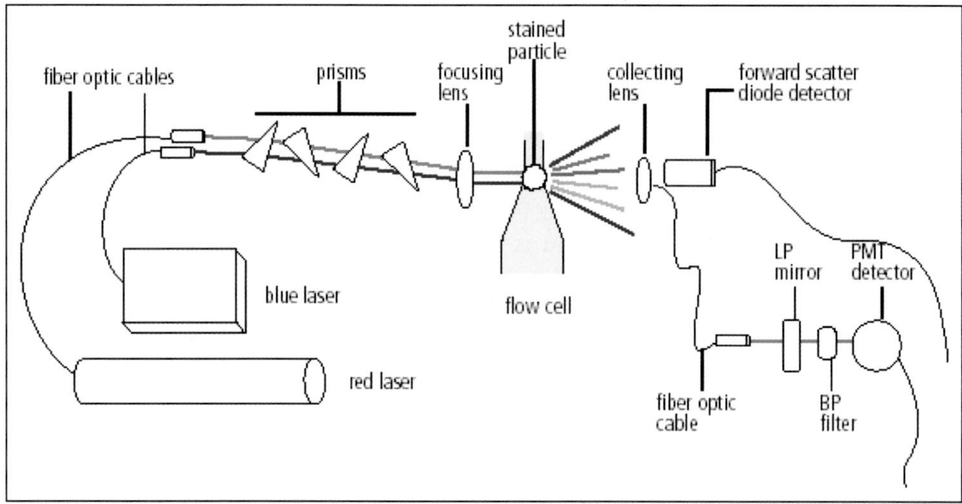

Fig. 8. Fiber optics of the Becton Dickinson FACSCanto.

rithmic conversion and compensation. Furthermore, the LSR II from Becton Dickinson deserves mentioning as an analytical research instrument which allows the use of up to four lasers and detection of up to eighteen fluorescence parameters.

The CyAn from Beckman Coulter, previously available from Cytometry and Dako, integrates spatially separate up to triple laser excitation and up to nine fluorescence parameters in a very compact design (fig. 9). Special features are the digital parallel processing of primary analysis data at up to 50,000 events per second and a high maximal flow rate of the sample fluid.

The precise control of the sample flow rate is a special feature of flow cytometers from Partec (fig. 10). It allows absolute cell counts (see pp. 84) without the need for

Fig. 9. Spatial configuration of the Dako CyAn.

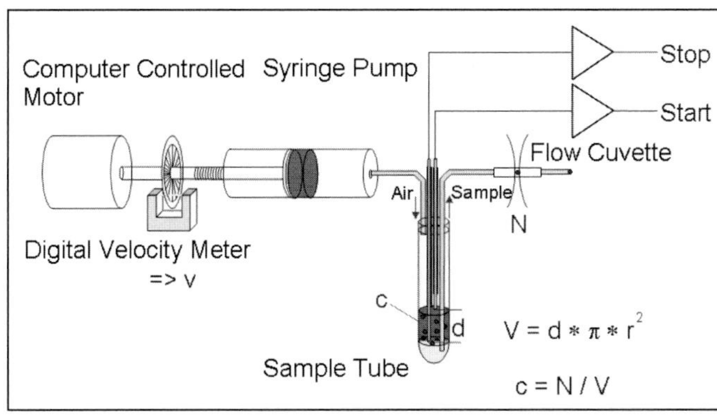

Fig. 10. Control and calibration of the sample volume flow in instruments from Partec.

microparticles as volume standards. Instruments from Partec, futhermore, are relatively compact and are flexible regarding the choice of excitation wavelengths and detection optics.

The technical properties of the described instruments are shown in more detail in tables 2–4. It is important to understand in this context that hardly any of the applications further described in this book requires a special type of instrument. Thus, descriptions of certain instruments are only examples. Furthermore, regarding the selection of instruments, data acquisition and analysis software, the suitability of automated sample preparation as well as connectivity to other IT components such as laboratory information systems are often more relevant. It is, however, difficult to describe such features objectively in a table.

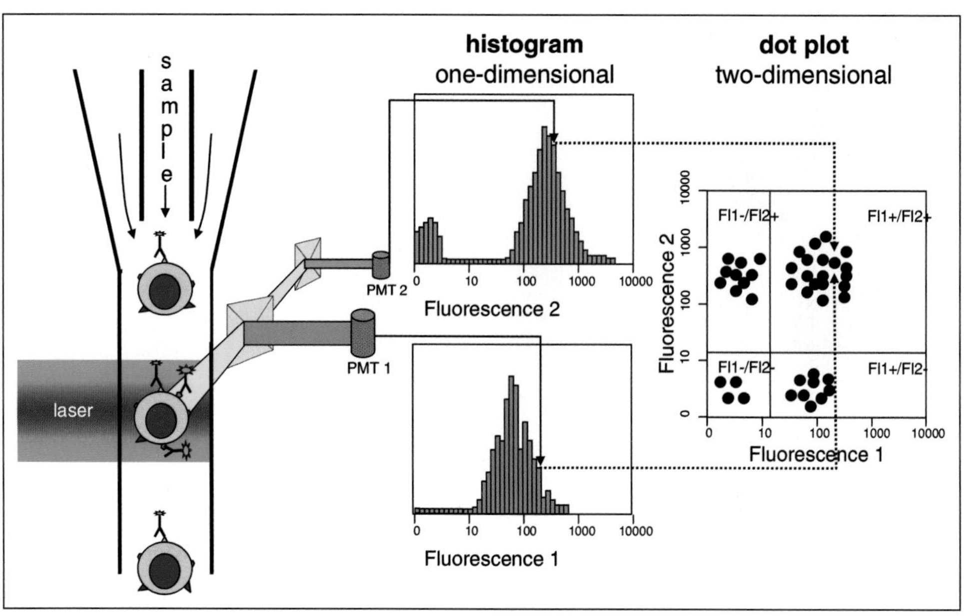

Fig. 11. Correlated presentation of analysis parameters. Multiple fluorescence signals of each cell are detected by flow cytometry. Histograms are showing the frequency of different fluorescence intensities. Dot plots allow the correlated presentation of fluorescence signals.

Principles of Data Analysis

As already described above, in flow cytometry the standardized storage of all parameters together with the acquisition protocol in FSC 3.0 format allows separation of data analysis from data acquisition. Therefore, data analysis, therefore, also can be performed with independent software and computers. Moreover, modern flow cytometers with digital compensation allow storage of the original uncompensated acquisition data. Thus suboptimal compensation can be adjusted during analysis. Loss of data during acquisition due to an inappropriate trigger parameter or to parameter settings in which cells are above or below the dynamic range of the detectors cannot be corrected during analysis.

The primary goal of data analysis is to identify cells with similar properties, i.e. populations, in the data set. These populations are then described regarding their number in relation to all cells or as an absolute count, the pattern of expression and the analytical parameters, including heterogeneity. For this purpose, multidimensional acquisition parameters are presented in easily understandable one- or two-dimensional graphs (fig. 11).

Histograms

A histogram shows the intensity distribution for a single parameter and represents the simplest form of data presentation. The x-axis of a histogram shows the intensity

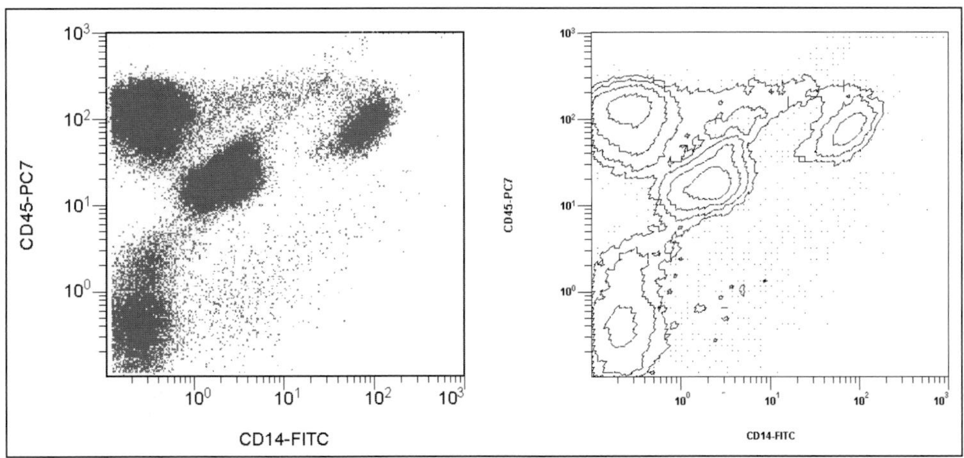

Fig. 12. Two-dimensional data presentation. In a dot plot (left panel) each cell is shown as a single dot. In a contour plot (right panel) lines are connecting fields with the same number of cells similar to contour lines on a map.

of a parameter, e.g. of fluorescence, whereas the y-axis presents the number of cells with a given signal intensity. The typical resolution of 1,024 levels of intensity or 'channels' on the x-axis results in a serrated curve if only a few hundred cells are shown or the distribution is heterogeneous. Smoothing algorithms which group channels or use statistical modeling for a more linear curve are thus often applied optionally or as a predefined setting.

Histograms are not only used for graphical presentation. The definition of regions allows the definition of populations within histograms, enumeration of cells and statistical analysis of their signal intensities.

Overlay histograms show the distribution of intensities either of the same selection of cells from two measurements or of two different populations as defined by gating (see below) in one graph. Different colors or line types are used to distinguish the different curves. Determination of a positive cell fraction by comparison with a specific antibody-stained sample and an isotype-stained control is a typical application for an overlay histogram. In this case, a region is defined which allows quantification of cells whose intensity is just superior to that of the majority of the control population.

Dot Plots and Contour Plots

For the presentation of two parameters in one graph, dot plots and contour plots are alternatives. In a dot plot (fig. 12) each single cell is plotted as a dot on the x- and y-axis according to the signal intensity of two parameters. This allows one to establish whether two parameters are correlated to each other. In this case, cells are distributed on a diagonal for the two parameters. Otherwise there are two different populations of cells that show signals for one parameter or the other.

With dot plots, predefined settings are frequently used as well for a clearer presentation. Only a representative fraction of cells, e.g. 1,000 cells or 10%, may be selected in order to allow the analysis of heterogeneities despite a high number of cells which would otherwise lead to broad blackening. Such a selection, however, at the same time may suppress the presentation of small populations of cells.

The selection of a baseline offset is an important detail for the interpretation of dot plots. The high quality of signal detection typically leads to the fact that large populations of cells are negative for a parameter and that all events are plotted directly onto the x- or y-axis. This leads to a strongly erroneous impression about the size of positive and negative populations in dot plot graphs. Selection of baseline offset adds electronic noise to all parameters; as a consequence, negative events also become visible on the logarithmic intensity scales. While minimally decreasing the quality of analysis data, baseline offset thus improves visibility on dot plots. Some instrument manufacturers incorporate baseline offset as a standard while others store native data and implement baseline offset as a selectable option during data analysis.

Quadrant analysis is a simple method for the quantitative analysis of cells within dot plots. Similar to the definition of a region in a histogram, thresholds are defined on the x- and y-axis immediately above the negative population and thus result in a cross. The percentage of cells which are double negative, double positive or positive for either of the two parameters is then calculated. Drawing a polygonal region represents a more flexible method for the determination of the amount of cells in a population or for the calculation of their signal intensities. Regions are typically automatically numbered in the sequence of their definition either with numbers R1, R2, R3) or alphabetically (A, B, C). Renaming regions according to the specific populations, e.g. 'CD4+ T-cells', facilitates the interpretation of region statistics.

The contour plot represents an alternative way of presenting two-dimensional data. In a two-dimensional signal intensity scale in these plots, the number of cells at a specific coordinate is shown using contour lines similar to the presentation of altitude on a map. Contour plots thus are very useful for analysing the heterogeneity of large populations. Small populations are not visible with a linear graduation of the contour lines. Alternatively, a logarithmic graduation can be selected and events below the lowest contour line can be shown as dots. The density plot is a variant of the contour plot in which the area between neighboring contour lines always contains the same number of cells. Finally, smoothing algorithms are also used for contour plots and the more appealing presentation of data bears the risk that information on subpopulations is suppressed.

Regions and Gates

Histograms and dot plots allow the retrieval of information on the distribution of signal intensity for one parameter or the association of two parameters. Additional steps are needed in order to identify e.g. the pattern of expression of a third fluorescence for cells that are positive for fluorescence 1 and 2 (fig. 11).

In a first step for dot plot analysis, pseudocoloring of dots can be selected. In this case the definition of regions will assign a selectable color to all cells within the regions for each of the different dot plots. Figure 13 shows the definition of three regions with the typical light scatter characteristics of lymphocytes, monocytes and neutrophil granulocytes in an FSC vs. SSC dot plot. As a consequence, the cells are pseudocolored according to their light scatter characteristics in further dot plots showing the expression of fluorescence parameters.

In a second step, populations of cells can be defined using more than two parameters by combining regions as gates. According to the rules of Boolean algebra, regions can be combined for this purpose using 'and', 'or' and 'not'. Such gates can then be used for pseudocoloring of events in dot plots. Alternatively, dot plots and histograms which will only show events within a certain gate can also be generated. Finally, the combination of gates allows the definition of populations according to all parameters in the data set. The definition of hematopoietic stem and progenitor cells as described in Gratama et al. [3] is an example for the hierarchical combination of gates.

When applying gates for pseudocoloring of cells in dot plots, alternative concepts exist in different software packages for the selection of colors in case that two populations overlay at a certain coordinate of fluorescence intensities. One philosophy relies on a hierarchy of gates and in the case of such a conflict, the hierarchically higher gate determines the color of a dot plot. This improves the visualization of small populations or the population of interest. As an alternative, the color of a dot can be determined based on the population which predominates at a certain coordinate of fluorescence intensities.

The conflict can be addressed in an alternative way when a single cell belongs to more than one region as defined in multiple dot plots. In this case as well either a hierarchy of gates is defined or color blending can be selected, and the cells belonging to two regions will be presented in a complementary color. This option, however, is difficult to apply for more than three regions.

Automation of Data Analysis

Predefined regions and gates only represent a limited approach when trying to apply a similar analysis to different cell samples and their corresponding data sets. One of the problems encountered in this context is the relatively higher heterogeneity of light scatter data due to differences in preanalytical sample storage as well as lower stability of instruments regarding the determination of light scatter. Furthermore, differences in the cellular composition of samples, activation, or abnormal cells can lead to errors in the detection of cells using predefined gates. Different strategies are available to overcome these problems.

Backgating

Backgating relies on the identification of populations based on their most stable characteristics in a first step. In most cases, this is a characteristic antigenic pheno-

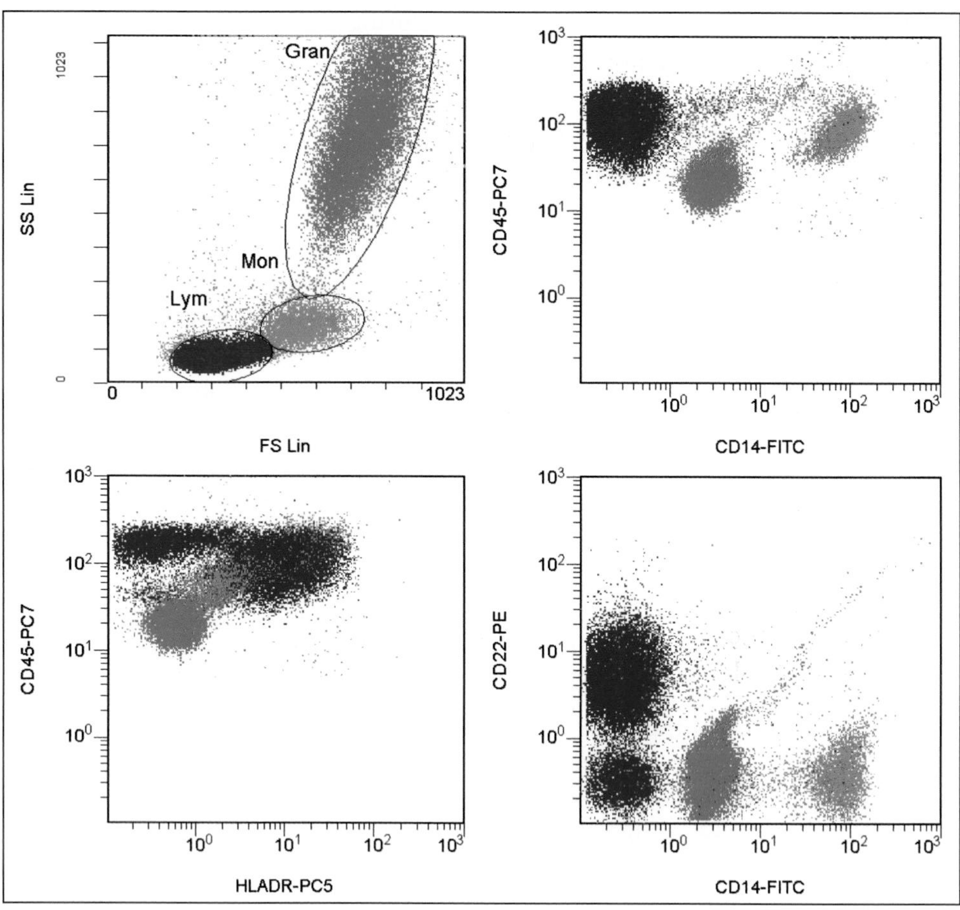

Fig. 13. Pseudocolor coding of populations in dot plots. In the left upper dot plot, a region assigns blue to all events with the light scatter characteristics of lymphocytes. Monocytes are defined as red and neutrophils as green. Pseudocolor coding facilitates the interpretation of antigen expression as analyzed by four-color immunophenotyping as an example.

type. In a second step, the corresponding light scatter characteristics are established for this population, and the newly defined light scatter gate is used for all further analyses. The definition of a light scatter gate for lymphocytes (fig. 13) based on the high expression the pan-leukocyte antigen CD45 and the lack of expression of the monocyte antigen CD14 is a classical example for backgating. Following the definition of the corresponding light scatter region, the purity (percentage of cells within the light scatter region with the CD14– CD45++ phenotype) and the integrity (percentage of cells with the CD14– CD45++ immunophenotype in the region) of the gate can be determined as indicators of the quality of the gate as a descriptor of lymphocytes in subsequent measurements with further antigens.

Autogating and Clustering

Autogating represents an automated procedure for the improvement of predefined regions. The type of regions, e.g. rectangular, circular or elliptical, are predefined, but the precise location and/or size are identified by algorithms which search the center of populations within the predefined regions.

Cluster analysis reaches a step further. Here populations are identified in multidimensional space by software algorithms without any previous definition of regions of interest or expected populations.

Descriptive Statistics

Independent of the definition of populations with the use of either regions in histograms, regions in dot plots, quadrants or complex gates, a number of similar options are typically available in software for the statistical description of populations. These statistics are used for data evaluation as well as for quality control (see 'Quality Control and Standardization', pp. 159). The most important statistical measures are:

- *Mean value*: this number represents the arithmetic mean of a parameter for the whole population. It is important to know that some software packages give the option to calculate the mean value of the 'channel', i.e. the number of the digital class according to the resolution of intensity, e.g. from 1 to 1,024. In this case, the intensity is only calculated for the 'mean channel' of the population in a second step. This method is adequate in case of an assumed logarithmically normal distribution. However, this option is often preferred only because it generates smaller coefficients of variation (CV) of the mean value.
- median: this is the value which separates a cell population into 50% of cells with higher values from the 50% with lower values for a given parameter. The median thus divides a histogram into two parts with an equal area under the curve.
- Mode: this is the value which is most frequently observed in a population and represents the peak of a histogram.

In addition, the statistical parameters typically include the minimum and maximum values of the population as well as the 5% and 95% ordinal values.

Generation of Derived and Corrected Values

The direct arithmetic computation of derived or corrected parameters for each cell in the list mode data set is a further important method in data analysis. Software compensation of spectral overlap in stored data is an example for such a manipulation. Furthermore, complex formulas can be applied to generate new parameters from one or several parameters. The computation of ratios between parameters when analyzing Ca^{2+} concentrations, pH or membrane potential with spectrally sensitive fluorochromes are interesting applications of this technology (see 'Determination of Cell Physiological Parameters: pH, Ca^{2+}, Glutathione, Transmembrane Potential', pp. 325).

Fluorescent Detection of Biomarkers

Various methods for the staining of cells are described in detail in the different chapters on specific applications. The following sections describe some common aspects of fluorochromes (see 'Selection and Combination of Fluorescent Dyes', pp. 107) and antibodies.

Principles of Fluorochromes

The capability of flow cytometers to quantify cell-associated fluorescence independent of localization and dependent on light source over a wide spectrum allows the use of various classes of fluorochromes for the molecular description of cells.

Fluorescent Probes
Biomolecules such as monoclonal or polyclonal antibodies for the immunological detection of protein antigens, oligonucleotides for in-situ hybridization reactions or receptor ligands can be directly or indirectly coupled to fluorochromes in order to quantify their cellular binding using fluorescence. Derivatization of antibodies with FITC or the integration of Cy5-conjugated nucleotides into nucleic acid probes are examples for direct methods. The biotinylization of antibodies followed by cellular staining with fluorochrome-conjugated streptavidin is an example of an indirect method.

Fluorochromes with Stoichiometric Binding
The total DNA or RNA of cells as well as their hydrophobic lipid compartments can be quantified using the stochiometric binding of fluorochromes. Dyes such as propidium iodide also increase in quantum efficiency following binding to DNA, which allows quantification of DNA with a good signal-to-noise ratio (see 'DNA and Proliferation Analysis in Flow Cytometry', pp. 390).

Fluorescent Substrates
Low-molecular-weight fluorescent compounds, which are metabolized similarly to physiological substrates, can be used for the analysis of biochemical pathways in cells. As an example, the incorporation of desoxy-UTP during DNA synthesis allows the analysis of cell proliferation (see 'DNA and Proliferation Analysis in Flow Cytometry', pp. 390).

Fluorogenic Substrates
Low-molecular-weight nonfluorescent compounds which are intracellularly metabolized to fluorescent products can be used either for the quantitative labeling of biomolecules or the determination of enzyme activity. The derivatization of glutathione after formation of a ring structure with *o*-phthadialdehyde and the detection of the

formation of reactive oxygen species based on the intracellular oxidation of dihydrorhodamine 123 to rhodamine 123 (see 'Determination of Cell Physiological Parameters: pH, Ca^{2+}, Glutathione, Transmembrane Potential', pp. 325, and 'Oxidative Burst, Phagocytosis and Expression of Adhesion Molecules', pp. 343) are examples of this approach.

Potential-Sensitive Dyes
Fluorescent anionic and cationic fluorochromes can be used for detecting the plasma membrane or the mitochondrial membrane potential. The spectral change in the fluorescence of JC-1 following accumulation in the presence of a high membrane potential and the selective and potential-dependent accumulation of rhodamine 123 at mitochondrial membranes are examples (see 'Determination of Cell Physiological Parameters: pH, Ca^{2+}, Glutathione, Transmembrane Potential', pp. 325).

Microenvironment-Sensitive Fluorophores
Dyes which change their fluorescent properties dependent on ion concentration, the fluidity of their microenvironment or their interaction with other fluorochromes can be used for the characterization of cellular compartments. The spectral change of emission in case of the pH indicator carboxy-SNARF-1, excimer formation with a shift to longer wavelengths in case of the membrane dye pyrene decanoic acid and fluorescence resonance energy transfer in case of spatial association of dye-conjugated probes are examples (see 'Determination of Cell Physiological Parameters: pH, Ca^{2+}, Glutathione, Transmembrane Potential', pp. 325, and 'Fluorescence Resonance Energy Transfer (FRET), pp. 141).

A number of reviews [2] (also see *http://probes.invitrogen.com/handbook/*) provide a detailed description of the diversity of general methods for the fluorescent characterization of cells. When applying different staining methods for specific purposes, it is important to know that the specificity of fluorochromes or fluorescent probes can change in their interaction with cells. Furthermore, staining methods can interact in combination even if there is no spectral overlap. Therefore, careful adaptation of methods to different cell types is needed. This is especially when staining is performed for in vitro stimulation experiments and an undisturbed cellular response is required [4].

CD Antigens
Similarly, a large number of monoclonal antibodies directed against leukocyte antigens are available and these antibodies can generally be used in multiple combinations for immunophenotyping. Antibodies with specificity for the same antigen are grouped into a 'cluster of differentiation' and designated with the same CD number. Further information on the currently classified CD antigens is available in different on-line directories (e.g., *www.hcdm.org, http://mpr.nci.nih.gov/prow, www.immundefekt.de/cd.shtml*). Some antigens relevant to clinical diagnostics are described in table 5. Detailed information can be found in the corresponding chapters of the sec-

Table 5. Selection of diagnostically relevant antigens

Antibody cluster	Pattern of expression	Function
CD1a	cortical thymocytes, dendritic cells, Langerhans cells	MHC-class-I-related molecule, function in the presentation of nonpeptidic and lipid antigens
CD2	T-cells, NK cells	cell activation
CD3	T-cells	T-cell-receptor associated, signal transduction
CD4	T-cells, monocytes, myeloid progenitor cells	co-receptor for MHC class II molecules, human immunodeficiency virus receptor
CD5	T-cells, B-cell subpopulation	cellular activation
CD7	hematopoietic stem cells, early myeloid cells, T-/NK cell subpopulations	cellular activation
CD8	T-cell subset, NK cells	co-receptor for MHC class I molecules
CD10	B-cell precursors	CALLA, endopeptidase
CD11a	pan-leukocyte antigen	LFA-1a, ICAM-1/2/3 receptor in combination with CD18
CD11b	myeloid cells, NK cells, T-/B-cell subsets	Mac-1, subunit of complement receptor 3 (CR3)
CD11c	monocytes, macrophages, neutrophils, NK-cells, T-/B-cell subsets	Subunit of complement receptor 4 (CR4)
CD13	myelomonocytic cells	aminopeptidase N
CD14	monocytes, weakly on neutrophils	lipopolysaccharide receptor, glycosylphosphatidylinositol-anchor-protein
CD15	neutrophils, monocytes	Lewis-x antigen
CD16a	NK cells, macrophages	Fcγ receptor IIIa
CD16b	Neutrophils	Fcγ receptor IIIa, glycosylphosphatidylinositol-anchor-protein
CD18	pan-leukocyte antigen	β2-integrin
CD19	pan-B-cell antigen	B-cell activation and differentiation
CD20	mature B-cells	B-cell activation and differentiation
CD22	B-cells, basophils (dependent on clone)	B-cell activation
CD23	activated B-cells, macrophages, eosinophils	low affinity IgE receptor FcγRII
CD25	activated T-/B-cells, activated monocytes	IL-2 receptor α-chain
CD33	myeloid cells	lectin activity
CD34	hematopoietic stem and progenitor cells	cell adhesion, control of differentiation?

Table 5. Continued on next page

Table 5. Continued

Antibody cluster	Pattern of expression	Function
CD36	platelets, monocytes, macrophages	glycoprotein IV, collagen/thrombospondin receptor, scavenger receptor for oxidized lipoproteins
CD38	hematopoietic precursor cells, activated T-cells, plasma cells	leukocyte activation, -proliferation
CD41	thrombocytes, megakaryocytes	fibrinogen receptor in combination with CD61
CD42a–d	thrombocytes, megakaryocytes	von Willebrand receptor complex, CD42a GpIX, CD42b GpIbα, CD42c GpIbβ, CD42d GpV
CD45	pan-leukocyte antigen	tyrosine phosphatase, differential expression of the splicing-variants CD45RA and CD45RO on T-cell subsets
CD52	lymphocytes, monocytes	CAMPATH-1
CD55	leukocytes, thrombocytes, erythrocytes	decay-accelerating factor, glycosylphosphatidylinositol-anchor-protein
CD56	NK cells, T-cell subset	NCAM
CD57	NK cells, subpopulation of T-cells, monocytes	HNK-1
CD61	thrombocytes, megakaryocytes	fibrinogen receptor in combination with CD41, vitronectin receptor in combination with CD51
CD64	monocytes/makrophages, activated neutrophils	Fcγ receptor I
CD66b	neutrophils	glycosylphosphatidylinositol anchor protein
CD69	activated T-/B-cells, NK cells, neutrophils, eosinophils, thrombocytes	cellular activation antigen
CD71	proliferating cells	transferrin receptor
CD79a	B-cells	cytoplasmic component of the B-cell antigen receptor
CD103	intestinal epithelial lymphocytes	HML-1, tissue specific lymphocyte retention
CD117	myeloid blasts, mast cells	c-kit, stem cell factor-receptor
CD138	plasma cells, pre-B cells	syndecan-1
CD235a	erythrocytes and erythroid precursor cells	glycophorin A
Myeloperoxidase	myeloid cells	lysosomal enzyme
TdT	lymphoblasts, myeloblast-subset	nuclear terminal desoxynucleotidyl-transferase
HLA-DR	precursor cells, B-cells, monocytes, activated T-cells	HLA-class II receptor

Table 6. Antigen coexpression for peripheral blood leukocytes and normoblasts

Marker	T-cells	B-cells	NK-cells	Mono-cytes	Imma-ture neutro-phils	Neutro-phils	Eosino-phils	Baso-phils	Normo-blasts
FS	+	+	+	++	+++	+++	+	+	+
SS	+	+	+	++	+++	+++	++++	+	+
CD1a	–	–	–	–	–	–	–	–	–
CD2	++	–	(+)	–	–	–	–	–	–
CD3	++	–	–	–	–	–	–	–	–
CD4	(++)	–	–	+	+	–	+	–	–
CD5	++	(+)	–	–	–	–	–	–	–
CD7	(++)	–	(++)	–	–	–	–	–	–
CD8	++	–	(+)	–	–	–	–	–	–
CD10	–	–	–	–	+	+	–	–	–
CD13	–	–	–	++	+	++	+	+	–
CD14	–	(+)	–	++	–	(+)	–	–	–
CD15	–	–	–	+	+	++	–	–	–
CD16	(+)	–	(+)	(+)	–	++	–	–	–
CD19	–	++	–	(+)	–	–	–	–	–
CD20	–	++	–	–	–	–	–	–	–
CD22*	–	++	–	–	–	–	–	+	–
CD25	(+)	(+)	–	(+)	–	–	–	–	–
CD33	–	–	–	++	+	+	+	+	–
CD45	+++	+++	+++	++	+	+	++	+	–
CD56	–	–	(+)	–	–	–	–	–	–
CD57	(+)	–	++	–	–	–	–	–	–
CD65	–	–	–	+	+	++	+++	–	–
CD79a	–	+	–	–	–	–	–	–	–
CD103	–	–	–	–	–	–	–	–	–
CD117	–	–	–	–	–	–	–	–	–
CD235a	–	–	–	–	–	–	–	–	+
HLA-DR	(+)	+	(+)	++	–	–	–	–	––
TdT	–	–	–	–	–	–	–	–	–
MPO	–	–	–	+	+	+	–	–	–

+, ++, +++ = Relative expression densities; (+) or (++) = expression only on subpopulations or under abnormal conditions; * = dependent on clone.

tions on 'Characterization and Phenotyping of Cells' and 'Diagnostic Indications' in this book.

Selection and Combination of Antibodies

A unique identification of cells in a given type of sample is of primary importance in immunophenotyping. Many antigens are expressed on multiple populations of blood cells. Therefore, typically only selected combinations of antibodies will allow a unique description of cells (table 6). Furthermore, a meaningful selection of fluorochrome conjugates with respect to their different signal intensity is needed (see 'Selection and Combination of Fluorescent Dyes', pp. 107). In addition, the spectral neighborhood of fluorochromes binding to the same cell is important too as this plays a role in potential compensation artifacts [5].

Basic Methods of Cell Preparation

Isolation and Purification of Cells

In general, all single-cell suspensions, e.g. blood, bone marrow, cerebrospinal fluid or other body fluids can be analyzed by flow cytometry. Adherent cells or solid tissue samples need to be isolated as a single-cell suspension in a first step. Compared with enzymatic methods such as trypsinization, mechanical methods of cell separation have the advantage that destruction of epitopes and disturbances of cell function are avoided even if only a smaller fraction of intact cells is left for analysis. While slide-based cytometry such as laser scanning cytometry may allow the analysis of adherent cells or cells in tissue without separation (see 'Technical and Methodological Basics of Slide-Based Cytometry', pp. 89), experience with this technology is still limited, however.

The repertoire of methods for mechanical or physical separation of cells includes tissue chopping, repeated aspiration through a small-gauge needle, the application of cold for dissociation of adherent cells and density gradient centrifugation. If remaining cell aggregates cannot be excluded following the dissociation of tissue, filtration through a Nylon mesh with a pore width of 50 μm is performed as the last step in order to avoid clogging of the flow cytometer.

Enrichment of Cells

In many cases, enrichment of cells also has advantages. Thus leukocytes can generally be analyzed in the presence of red blood cells if parameters with a good discrimination capacity such as the pan-leukocyte antigen CD45 or staining of DNA are used. However, the high excess of red blood cells leads to long analysis times before a large number of leukocytes can be analyzed.

Alternatively, leukocytes can be enriched in blood by depletion of red blood cells. As a classical method, isolation of mononuclear cells, i.e. of lymphocytes and monocytes, by density gradient centrifugation ($d = 1.077$) was applied for this purpose. However, this method is only rarely used for analytical purposes today because it requires time-consuming washing steps and as selective cell loss is difficult to control. Preparation of cell suspensions for cell culture, cryopreservation and immunomagnetic enrichment is a typical further application of this technology.

Today, whole-blood lysis is the method of choice for the depletion of erythrocytes from blood. In this method, the lower osmotic resistance of erythrocytes is used for selective lysis. Different methods exist, either in combination with fixation or without fixation, such as ammonium chloride lysis. Methods with fixation improve cell stability pending flow-cytometric analysis. As a drawback, selective cell loss occurs when cells are washed by centrifugation following lysis in the presence of fixatives. This problem can be addressed either by lyse-no-wash methods or the use of lysing reagents without fixatives. In the latter case, light scatter properties of cells are only stable for up to 2 h. Storage can be extended to several days if samples are finally stabilized by formaldehyde fixation after washing.

Immunomagnetic enrichment methods are applied for the analysis of rare cells (see 'Enrichment of Disseminated Tumor Cells', pp. 174).

Preanalytical Aspects and Sample Stabilization

The choice of antiocoagulants for the analysis of blood depends on the analytical methodology. Thus, EDTA is preferred in case of absolute counting methods or simultaneous analysis with a hematology analyzer or by blood smear review. Heparin has the advantage of high sample stability. Citrate reduces the alteration of thrombocyte function.

Similarly, the maximum sample storage time and the optimal temperature for storage also depend on the method of analysis. They are described in the different chapters of the book. Cooling of samples is not generally recommended as alterations of temperature lead to cellular activation [6]. (see 'Determination of Cell Physiological Parameters: pH, Ca^{2+}, Glutathione, Transmembrane Potential', pp. 325)

Some types of sample are stable only for a short period of time. Sample stabilization for immunophenotyping, e.g. using commercial preparations of formaldehyde donors, polyethylene glycol and EDTA, allows immunophenotyping for up to 7 days. However, potential alterations of monoclonal antibody epitopes need to be carefully controlled when using such methods.

Cell death is associated with selective cell loss and nonspecific antibody binding. Thus, especially in the case of unknown preanalytical conditions, analysis of cell viability is recommended. DNA dyes such as propidium iodide which are excluded by the membrane of intact cells are used in flow cytometry for this purpose (see 'Determination of Cell Physiological Parameters: pH, Ca^{2+}, Glutathione, Transmembrane Potential', pp. 325).

Antibody Staining

In unseparated whole blood, the unspecific antibody binding to Fc receptors is low only due to the high concentration of autologous immunoglobulins. Using a native sample for antibody staining also has the advantage that artifacts introduced by washing or isolating cells are reduced. However, as a drawback, higher amounts of antibodies are needed than for staining cells enriched by lysis or density gradient centrifugation. Saturating antibody concentrations must be determined and controlled for reproducible staining. For antibody titration, a sample with a relatively high number of cells expressing the target antigen or an activated sample in case of activation-dependent upregulation should be used.

Compared to indirect staining, the use of directly conjugated antibodies leads to lower nonspecific signals. For signal amplification, a biotin/streptavidin system is much more effective than a secondary antibody.

Washing by centrifugation after incubation with the antibody reduces background signals due to the elimination of unbound antibodies. At the same time, cell loss occurs, which may be selective. Thus, in the case of highly expressed antigens and especially for cell counting, methods with only few processing steps are used, e.g. lyse-no-wash methods, where erythrocyte lysis and dilution by addition of the lysis buffer to whole blood incubated with the antibodies is followed by measurement without further centrifugation. As a drawback of these methods, the cell concentration in the diluted sample is low and long data acquisition times are needed. In the case of weakly expressed antigens, e.g. for the analysis of aberrantly expressed antigens in leukemias, a reduction of background signals by centrifugation is needed (see 'Flow-Cytometric Immunophenotyping of Acute Leukemias', pp. 612, and 'Flow Cytometry in the Diagnosis of Non-Hodgkin's Lymphomas', pp. 642).

Buffers used for incubation and washing need to be selected according to type of sample material and processing. Thus protein-free buffers can be used for the dilution and washing of whole blood; however, in order to reduce nonspecific antibody binding and cell loss due to adherence. supplementation with protein is needed if already washed or especially fixed cells are further processed.

Stringent centrifugation, i.e. centrifugation at a high speed or for a long time, reduces cell loss during washing. However, with increasing intensity of centrifugation, disturbances of cell function and even cell death increase. Therefore, validation for the specific type of sample is needed.

Intracellular antigen staining can be performed by cross-linking fixation in a first step followed by incubation with antibodies in the presence of a reversible detergent in a second step. Commercial preparations differ in their ability to open different intracellular compartments. Furthermore, antigenic epitopes recognized by specific antibodies can be destroyed. Thus careful adaptation of methods and antibodies is needed [7].

The staining specificity, i.e. the fraction of the signal which results from an interaction between an antibody and an antigen, but is neither caused by the fluorochrome nor by binding to Fc receptors needs to be controlled for each antibody staining. In the case of indirect staining, isotype controls, i.e. antibodies of the same isotype (e.g. IgG1, IgG2a, IgG2b) and therefore the same affinity for Fc receptors but with irrelevant specificity are used at the same concentration as the specific antibody. In the case of directly fluorochrome-conjugated antibodies, however, the fluorochrome-to-protein ratio (F/P-ratio) and the manufacturer-specific type of purification of the conjugates also determine the nonspecific part of the signal.

Furthermore, nonspecific staining is typically lower as only purified antibodies are conjugated. The use of isotype controls is thus only of limited value and usually impracticable in direct staining methods. In case of heterogeneous cell populations or multiple stainings of the same sample, the background signals of the cells which are negative for a given antibody are informative as well regarding background staining intensity. Accreditation therefore typically requires objective methods for the determination of positive and negative cell fractions, and isotype controls are not requested as the method of choice.

Methodological Developments in Cytometry

Flow cytometry is moving from an open research method to a validated diagnostic method. This process requires improvement of the precision of analysis as well as biological calibration and quality-controlled results. An improvement of absolute counting, antigen quantification and the development of standards and control materials are central to this process. Finally, integration of data reporting into the clinical flow of information is needed, which improves the availability of technical results and at the same time provides transparency regarding the interpretative part of the report and the underlying data.

Absolute Counting

The clinical need for a precise determination of absolute concentrations of CD4+ T-lymphocytes in peripheral blood in HIV infection has stimulated the development of methods for absolute counting although flow cytometry traditionally did not include volume calibration due to the required staining and washing steps before measurement. In the meantime, the quantification of CD34-expressing hematopoietic stem cells and progenitor cells has established itself as a further important clinical application of absolute counting.

In the classical dual-platform method for counting CD4+ T-cells, the flow cytometer was used for determining the relative fraction of CD4+ T-cells among lymphocytes while the absolute concentration of lymphocytes in blood was derived from a hematology analyzer [8]. Alternatively, the population of all leukocytes was also used as the common denominator for both platforms. These methods are flawed by large analytical errors due to differences in cell identification, especially in lymphopenic samples, in both types of platform. Thus, normoblasts and basophil granulocytes interfere on both platforms depending on methodology, and disturbances of automated leukocyte differentiation in the hematology analyzer lead to broadly variable results. Furthermore, selective cell loss may occur and is difficult to control with the classical methods of sample processing for flow cytometry including centrifugation.

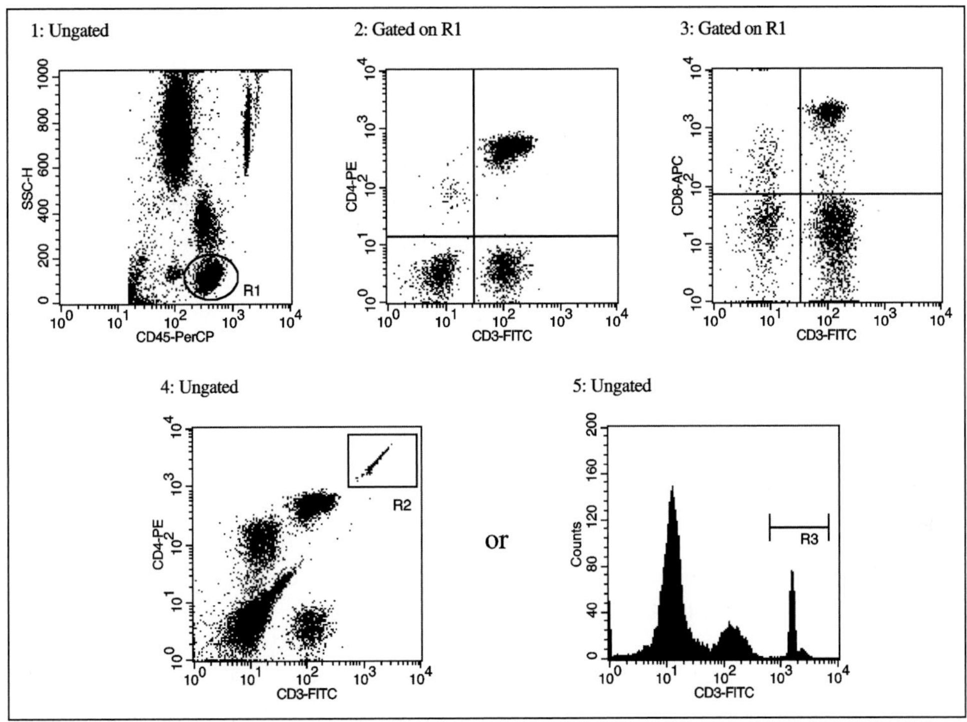

Fig. 14. Principle of gating as performed for single-platform counting. Lysed whole blood is supplemented with counting beads and analyzed according to a Center for Disease Control guideline [9]. Dot plot 1 shows the counting beads as events with high fluorescence intensity and SSC. Region 1 (R1) is drawn around lymphocytes with dense expression of CD45. Only events in region 1 are shown in dot plots 2 and 3. A quadrant analysis is performed in order to quantify T-helper cells with expression of CD3 and CD4 and cytotoxic T-cells with expression of CD3 and CD8. Dot plot 4 and histogram 5 are ungated and show regions for the counting of beads which are used as volume calibrators for the calculation of absolute counts.

In single-platform methods for direct absolute cytometric counting, a fixed relation between the initial sample volume and the volume before analysis is maintained by lyse-no-wash methods (see above), which do not require centrifugation. Volume is then determined during analysis using microbeads initially added to the blood sample as counting standards or a volume-calibrated flow cytometer. Figure 14 illustrates the use of counting beads according to the current guidelines of the American Centers for Disease Control and Prevention for CD4+ T-cell counting [9]. Rare cells such as CD34-expressing cells or antigen-specific T-cells are also counted using single-platform methods. It is important to know in this context that counting beads have been developed for use with whole blood. In sample materials with low protein content such as apheresis products, counting beads can clump and precipitate. Therefore, sample processing in a buffer with protein supplementation is needed in order to avoid falsely high results.

Antigen Quantification

In many cases, the expression density of antigens has immediate diagnostic relevance. Some examples are the expression of CD38 on CD8+ T-cells in HIV infection, of HLA-DR on monocytes in sepsis and of activation antigens on platelets in prothrombotic diseases. Furthermore, antigen expression density can support the identification of cell types, especially pathological cells displaying a characteristic abnormal expression such as CD10 overexpression on B cells in acute lymphoblastic leukemia or CD34 overexpression in variants of acute myeloblastic leukemia. Finally, even the determination of the fraction of cells that are positive for an antigen, e.g. for CD38 or ZAP-70 in B-cell chronic lymphatic leukemia, is a problem of antigen quantification as the detection sensitivity depends on the selected fluorochrome and antibody preparation as well as on the instrument and instrument settings.

Several approaches have been attempted towards a calibrated or absolute determination of antigens [10]:

- The flow cytometer can be calibrated using particles with calibrated expression of the same fluorochrome as that used for immunophenotyping in several bead populations with different expression densities.
- Microbeads with a defined number of binding sites for mouse immunoglobulin can be used for calibration after loading with the antibody that is also used for cell staining
- Cells with a stably expressed antigen can be used as a reference population. Either stabilized control cell suspensions or intrinsic populations with stable expression, e.g. of CD4 on T-helper lymphocytes, can be used for this purpose.

Methods for antigen quantification are still problematic as the various available methods do not achieve the same results even when highly and homogeneously expressed antigens are analyzed [11]. Furthermore, variability of some calibration reagents from one lot to another is problematic. The lack of saturation of binding of many commercially available antibodies also constitutes a problem. This is especially true of conjugates with bright fluorochromes such as R-phycoerythrin or allophycocyanin. The fluorochrome-to-protein ratio, the accessibility of epitopes to specific clones or differences in their immunoreactivity, bivalent or monovalent binding of antibodies as well as the linearity and stability of flow cytometers will further affect the results depending on the methods. Especially the compensation of spectral overlap in multicolor experiments remains an unresolved problem.

Standard Operation Procedures and Control Materials

A lack of standardized methods frequently prevents multicenter comparison of results and thus the routine diagnostic use of flow cytometry. Validated test systems and to some extent control materials have become available only for some clearly defined parameters such as the enumeration of CD4-expressing T-cells, CD34-expressing hematopoietic stem cells and progenitor cells or the counting of residual cells in blood products. For some of these applications, quality control modules have become available as well as part of the data acquisition software.

For more complex methods such as the immunophenotyping of leukemias and lymphomas or the analysis of platelets, standardization still is poor, however. These methods are currently on the level of a literature- and experience-based consensus discussion while cross-validation of alternative methodological approaches is lacking.

Integration into Clinical Reporting and Interpretation
Information logistics also represents a major problem in the clinical application of flow cytometry. Thus an automated and reliable import of sample identification, e.g. using a barcode, as well as of order information and of patient master data is required for an efficient processing of samples. The export of numerical analysis results into laboratory information systems and hospital information systems typically does not present a problem. Standards for the generation and documentation of graphical results such as informative dot plots in leukemia and lymphoma analyses are still missing. Integration of such information into clinical reports would be an important step towards increased transparency of the interpretative parts of the report and thus also towards improved clinical acceptance.

Conclusion

Flow cytometry is a general technique for the simultaneous analysis of size, granularity and multiple fluorescence properties of single cells. Specific staining with fluorescent antibodies or other fluorescent staining methods allows flexible assessment of immunological, biochemical or functional properties of single cells in heterogeneous cell suspensions. Complex methods for data analysis allow assignment of the measured signals to cell populations with similar properties. While analysis of fluorescence generally general provides relative data, new methodological approaches using microparticles allow absolute cell counts as well as quantitative analysis of antigen expression.

References

1 Janossy G: Clinical flow cytometry, a hypothesis-driven discipline of modern cytomics. Cytometry A 2004;58:87–97.
2 Shapiro HM: Practical Flow Cytometry, ed 4. New York, Wiley-Liss, 2003.
3 Gratama JW, Orfao A, Barnett D, Brando B, Huber A, Janossy G, Johnsen HE, Keeney M, Marti GE, Preijers F, Rothe G, Serke S, Sutherland DR, Van der Schoot CE, Schmitz G, Papa S: Flow cytometric enumeration of CD34+ hematopoietic stem and progenitor cells. Cytometry 1998;34:128–142.
4 Rothe G, Klouche M: Phagocyte function. Methods Cell Biol 2004;75:679–708.

5 Maecker HT, Frey T, Nomura LE, Trotter J: Selecting fluorochrome conjugates for maximum sensitivity. Cytometry A 2004;62:169–173.
6 Elghetany MT, Davis BH: Impact of preanalytical variables on granulocytic surface antigen expression: a review. Cytometry B Clin Cytom 2005;65:1–5.
7 Kappelmayer J, Gratama JW, Karaszi E, Menendez P, Ciudad J, Rivas R, Orfao A. Flow cytometric detection of intracellular myeloperoxidase, CD3 and CD79a. Interaction between monoclonal antibody clones, fluorochromes and sample preparation protocols. J Immunol Methods 2000;242:53–65.
8 Brando B, Barnett D, Janossy G, Mandy F, Autran B, Rothe G, Scarpati B, D'Avanzo G, D'Hautcourt JL, Lenkei R, Schmitz G, Kunkl A, Chianese R, Papa S, Gratama JW: Cytofluorometric methods for assessing absolute numbers of cell subsets in blood. European Working Group on Clinical Cell Analysis. Cytometry 2000;42:327–346.
9 Mandy FF, Nicholson JK, McDougal JS: Guidelines for performing single-platform absolute CD4+ T-cell determinations with CD45 gating for persons infected with human immunodeficiency virus. Centers for Disease Control and Prevention. MMWR Recomm Rep 2003;52:1–13.
10 Gratama JW, D'Hautcourt JL, Mandy F, Rothe G, Barnett D, Janossy G, Papa S, Schmitz G, Lenkei R: Flow cytometric quantitation of immunofluorescence intensity: problems and perspectives. European Working Group on Clinical Cell Analysis. Cytometry 1998;33:166–178.
11 Lenkei R, Gratama JW, Rothe G, Schmitz G, D'Hautcourt JL, Arekrans A, Mandy F, Marti G: Performance of calibration standards for antigen quantitation with flow cytometry. Cytometry 1998;33:188–196.

Sack U, Tárnok A, Rothe G (eds): Cellular Diagnostics. Basics, Methods and Clinical Applications of Flow Cytometry. Basel, Karger, 2009, pp 89–106

Technical and Methodological Basics of Slide-Based Cytometry

Anja Mittag[a] · Jozsef Bocsi[b] · Wiebke Laffers[c] · Andreas O. H. Gerstner[c]

[a] Translational Center for Regenerative Medicine,
[b] Department of Pediatric Cardiology, Heart Center, University of Leipzig,
[c] Department of Otorhinolaryngology/Head and Neck Surgery, University of Bonn, Germany

Introduction/Background

Multiparametric analyses of clinical samples at the cellular level are of increasing relevance to both clinical routine and basic research [e.g. 1–4]. Flow cytometry (FCM) is the standard technique used to cover a plethora of applications in these fields (see 'Technical Background and Methodological Principles of Flow Cytometry', pp. 53). However, while FCM is unsurpassed in the routine analysis of clinical blood specimens, the analysis of solid tumors poses unique challenges for which this technology is less suited. The most important issues in tumor tissue analysis are: i) the requirement to assay interacting cells in their spatial and topological context and ii) the limited amount of sample material normally available for the detailed functional and/or phenotypic analysis of specific cell subsets [5, 6]. The same issues are also relevant to a wide range of related applications.

Slide-based cytometry (SBC) has been developed in order to obtain quantitative and objective data from tissue specimens [7]. It is a high-content screening method characterized by high reproducibility as well as the capacity for high-throughput analyses, and can be controlled by standardization methods similar to those used in FCM (see 'Quality Control and Standardization', pp. 159). To date, SBC techniques have mainly been used for research purposes rather than for routine clinical applications. In recent years, however, SBC has begun to enter the field of clinical diagnostics, e.g. analyses of fine-needle biopsies and swabs [e.g. 8–12] or of a variety of tissue specimens [e.g. 13–16]. The high-throughput capacity also makes these slide-based systems suitable for single-cell analyses in drug-screening exercises [17–19].

Several different SBC instruments based on various technical systems are currently available. All are designed to analyze specimens fixed on microscope slides or culture plates, whereby the analysis is not limited to one field of view as in common microscopy, but can encompass a substantially wider area of interest ranging from defined fields to an entire slide. Cross-platform comparison is difficult since almost all instruments utilize unique data formats. Furthermore, each instrument is controlled by specific proprietary software. FCS files, developed for FCM, are not currently used as a standard for SBC, so that data exchange is hardly feasible. The formats of the microscopic images generated also differ widely and are unique to a single instrument in some cases. Some of the systems do offer the possibility to export data, and should therefore allow the conversion of data into alternative formats. Unfortunately, however, there is as yet no commercially available common software for this purpose so one has to rely on 'home-made' solutions.

Cytometry demands the quantitative analysis of a signal emanating from an intact single cell – a requirement that is not necessarily satisfied by all of the available 'SBC' systems. For example, systems based on confocal microscopy, which analyze a very limited depth of focus (normally in the submicrometer range), cannot routinely provide cytometric quantitative data that represent the whole cell.

The scientific field of SBC analyses has been highly innovative in recent years and this is likely to continue in the future. Although major conceptual changes in the instrumentation are unlikely, the range of different instruments is likely to be extended and new approaches to data analysis developed to answer the demand from an ever-growing market of applications.

While we cannot hope to provide an exhaustive account of all systems, concepts and approaches relevant to the use of SBC, it is our intention here to review the technology and a selection of applications that have so far been established for the quantitative analysis of clinical samples.

Slide-Based Cytometry

In general, SBC instruments have a conventional inverted or upright fluorescence microscope as core. As a minimal requirement, the objective stage is motorized in the x- and y-directions, with some instruments including movement in the z-dimension to allow autofocus and to enable the creation of z-stacks, e.g. for confocal imaging. Most systems offer a port for direct visualization of the specimen by eye or via a digital camera. As in conventional fluorescence microscopy, light sources include xenon or mercury arc lamps, lasers and diodes, with most instruments offering a combination of lasers and diodes with different excitation wavelengths.

Fluorescence detection in SBC obeys the same physical principles as in FCM, so that staining procedures are similar for both types of cytometric analyses. The fundamental difference is the physical state of the cells to be analyzed. While in FCM cells are dispersed in suspension, transported by a sheath fluid to the detection spot, excited by lasers and analyzed by photomultipliers (PMTs), in SBC, cells are placed, e.g., on slides, culture plates, and moved along the focal plane by the motorized microscope stage. Since there is no fluidic system that can be blocked by particles too large to pass the nozzle, the only limit in sample size for SBC is set by the dimensions of the slide or plate. This makes it feasible to analyze not only multicellular objects, but also entire tissue sections, small organisms, cell cultures or bacterial colonies. The thickness of a sample can range from very thinly sliced tissue to several cell layers. Another important consequence of the SBC principle, especially for small-volume clinical samples, is that the sample is not lost by the analysis. While in FCM the sample is lost in the waste container after the analysis (unless cells are sorted), the SBC sample remains on the slide and can be reanalyzed or stored for documentation. As long as the cells stay in their initial location on the slide, they can be relocated and visualized during or after analysis. This is particularly important in medical issues since it is better to base therapeutic decisions on actual cells, rather than on dots representing events of unknown origin and/or limited discriminatory value, especially in oncology. To this end, cells can be stained (or restained following cytometric analysis) by conventional dyes such as hematoxylin and eosin [20]. In addition, cells can be reanalyzed one by one for additional sets of markers. Examples include supravital staining before and after the induction of apoptosis [21, 22] and the iterative staining of CD markers [23]. This means that cells can effectively be 'asked' a second time or even more often. In principle, the experimental setup can be a never-ending circle of staining, followed by analysis followed by staining and so on [24, 25]. Although this would also be feasible using FCM, it is impossible to combine the results of the first analysis with those of the second one on an individual single-cell basis since cells would never run through the instrument in the same order. In SBC, individual cells can be addressed by their x-y-coordinates on the slide, which allows the sequential assembly of a multilayer data cube for every single cell.

There are many different instruments and a corresponding number of different principles of data acquisition and analysis available on the market. All of them share a basis in fluorescence microscopy technology. The main differences between them concern the way in which images are generated: some instruments use high-resolution digital cameras, while others rely on PMT-generated images. These two main principles are described below on the basis of one typical instrument example each.

PMTs, also used in FCM, provide information about fluorescence intensities which is the result of signal amplification, pulse processing, and analog-to-digital conversion. The signal can be analyzed immediately and/or stored. However, there is presently, to the best of our knowledge, only one PMT-based SBC instrument available on the market.

In comparison, data extraction from CCD-camera-generated images often requires extensive computation by image analysis, although image acquisition is faster

Table 1. Overview of the presented examples of commercial slide-based systems[a]

	Instrument name						
	iCys® research imaging cytometer	AxioVision SFM	Scan^R	Opera™ LX	TissueFAXS®	ArrayScan® VTI HCS Reader	Toponome® Imaging Cycler® MM3
Manufacturer	CompuCyte Corp. (Cambridge, MA, USA) *www.compucyte.com*	Carl Zeiss MicroImaging GmbH (Hallbergmoos, Germany) *www.zeiss.com*	Olympus (Hamburg, Germany) *www.microscopy.olympus.eu/microscopes/Life_Science_Microscopes_scan_R.htm*	PerkinElmer (Waltham, MA, USA) *www.perkinelmer.com*	TissueGnostics GmbH (Vienna, Austria) *www.tissuegnostics.com*	Thermo Scientific; Cellomics, Inc. (Pittsburgh, PA, USA) *www.cellomics.com*	meltec GmbH & Co. KG (Magdeburg, Germany) *www.meltec.de*
Excitation							
Laser(s)	1–3 lasers (488, 633 and 405 nm)	–	–	up to 4 (405, 488, 532 and 635 nm)	–	–	–
Lamp(s)	–	mercury arc lamp (HBO-100) or X-Cite 120 XL FL light source	MT20 illumination system	UV, Xenon lamp	halogen 12 V, 100 W	UV, mercury-xenon lamp	xenon lamp
Emission	PMTs	Carl Zeiss AxioCam MRm 12 bit CCD camera	high-sensitivity cooled CCD camera	12-bit cooled CCD cameras	monochrome and color CCD camera	12-bit cooled CCD camera	14-bit cooled CCD camera
Detectors/filters	2–4 interchangeable filter blocks	6- or 10-position reflector turret, motorized	8-position filter wheel	up to 4 parallel detection channels	up to 10 fluorescence filter cubes	10-position excitation filter wheel	4-filter sets
Microscope	Olympus IX-71 microscope	Carl Zeiss Axio Imager.M1 microscope	Olympus IX-81 microscope	confocal microplate imaging reader	microscope from Zeiss, Leica or Nikon (e.g. Carl Zeiss Axiolmager)	Carl Zeiss 200M Axiovert microscope	Leica DMI 6000
System	nonconfocal	widefield	widefield	confocal (widefield with UV excitation)	widefield	widefield	confocal
Objectives	10×, 20×, 40×	10×, 20×, 40×	10×, 20×, 40×	20×, 40×	10×, 20×, 40×, 63×	5×, 10×, 20×	20×, 40×, 63×
Autofocus	yes	Yes	yes	yes	Yes	yes	yes
Motorized stage	x- and y-direction	x-, y- and z-direction	x-, y- and z-direction	x-, y- and z-direction	x-, y- and z-direction	x-, y- and z-direction	x-, y- and z-direction
Analysis software	iCys® Cytometric Analysis Software and iBrowser® Data Analysis Software	AxioVision SFM package	Scan^R analysis software	Acapella™ software	TissueQuest and HistoQuest	Cellomics® BioApplications	MELK process control software
Specimen carriers	slides, multiwell plates, Petri dishes	Slides	multiwell plates, slides	multiwell plates	slides, multiwell paltes (only in TissueFAXSi™)	slides, multiwell plates	slides

[a]A short overview is given here but it has to be pointed out that in most cases additional options (e.g. filters, objectives) are available.

than in PMT-based systems. Inexpensive personal computers and reasonably priced, low-noise, high-resolution CCD cameras are now available for such applications.

The following section aims to give an overview of present instruments. It covers the instruments currently most popular, but may not be comprehensive (table 1).

Laser Scanning Cytometry (Photomultiplier-Generated Images)

iCys®

The laser scanning technology relies on a fluorescence microscope (inverted iCys or upright in their preceding model LSC™ (Both Compucyte Corp.)). Up to three lasers serve as light source (fig. 1). The standard configuration is a 488 nm argon laser although other lasers can be added. Generally lasers with 405 nm and 633 nm excitation wavelength line are added since the combination of the three lasers covers broadest range of applications [e.g. 26–29] for the spectral characteristics of the fluorochromes.

Image Generation
The laser light is coaxially guided to the mechanically most sensitive part of the instrument, an oscillating scanning mirror, which converts the laser spot into a line. This line is directed to the slide, resulting in a scanning line in the y-direction. The length and width of this line depend on the chosen objective as well as the movements of the motorized stage and hence the movement of the slide in the x-direction. This results in the scanning of a rectangular segment of the slide, the dimensions of which depend on the objective used. If two or all three lasers are used, this segment is first scanned with the 488-nm laser followed by the other lasers, running in combination. The respective inactive laser(s) are blocked by mechanical bars moved into the laser path. After processing the scanned segment images, the next field is scanned until the entire defined scan area of the slide has been analyzed. This means that the entire scan area is eventually covered by a mosaic of scanned segments.

The fluorescence signals from the sample are collected by the microscope objective and directed backwards to the scanning mirror. This beam hits a dichroic mirror that reflects the excitation wavelengths while allowing the emission light to pass to a cascade of a maximum of four filter sets. In order of ascending wavelengths, each of these filter sets reflects a defined spectral range to a PMT tube to generate a digital output. In addition to the fluorescence, light scattered by the sample can also be detected. Unlike FCM, however, where forward and sideward scatter can be detected, SBC generally only allows the detection of forward scatter. The sideward scatter is in the plane of the slide and therefore not detectable. The forward scatter is not fluorescence specific and is detected by a photodiode. Scatter detection is possible in two

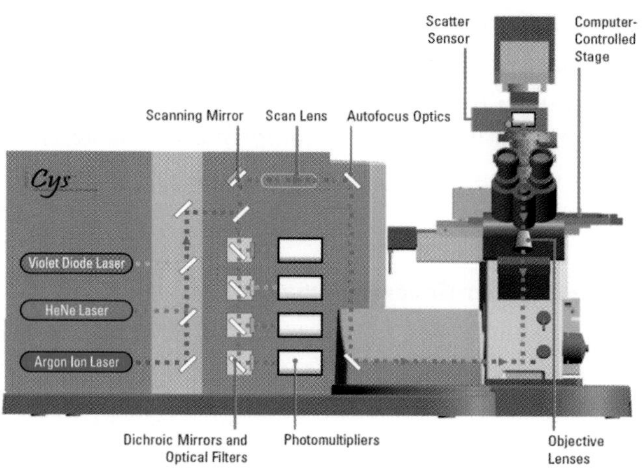

Fig. 1. The way of light in LSC. The scheme shows an iCys® Reaserch imaging cytometer (CompuCyte Corp., Cambridge, MA, USA). Light emitted by lasers is directed via different mirrors and lenses through the inverted microscope to the sample on the stage. The emitted fluorescence of the stained sample is going back the same way and is directed to four filter cubes (consisting of a dichroic and an optical band-pass filter each) in ascending wavelength order. Thus, the fluorescence emission of the sample is divided into different spectral ranges and is led to a corresponding PMT which generates digital scan images. Additionally, the scattered laser light on the sample is detected with a sensor above the stage. These generated images (by the PMTs or Scatter sensor) serve as basis for data analysis (by courtesy of CompuCyte Corp., with permission).

ways: i) shaded relief, to generate relief details of the sample in an otherwise flat image, or ii) light loss.

For each scanned segment of the slide, pixel maps (black and white images) are generated for each color-channel measured by the respective PMTs, and for the scatter channel of the photodiode. The fluorescence intensity of a pixel is expressed in gray scale values from 0 to 16,384 (14 bit). Each pixel map corresponds to the respective 'color channel', defined by the specification of the PMT's filter set. These pixel maps are analyzed during scanning and can additionally be stored and analyzed later offline. iCys® analysis in principle resembles conventional image analysis. In older instruments such as the LSC, the pixel maps are analyzed immediately before the next segment is scanned, but the images of these segments cannot be saved.

The hardware of the iCys® is limited to four PMTs. By three-laser (405, 488 and 633 nm) excitation and the original filter settings, a maximum of five fluorochromes is distinguishable. However, the corresponding software offers the possibility of merging subsequent measurements of the same sample by overlaying scanned images. Therefore, this method can be used to increase the information derived from a single cell by modifying the sample or instrument settings between subsequent measurements. It has been used, for example, for restaining [23] or increasing the detectable fluorochromes in one sample to eight by changing the emission filters [2].

Data Analysis

The user has to define what a 'cell', or more generally an 'object of interest', exactly is. The user first has to select a single channel or a combination of several channels that best represent the specific characteristics of the cell. This channel does not necessarily have to be an original pixel map generated by the PMTs. Also virtual channels generated by subtraction, multiplication or inversion of the pixel maps from the different real channels may be used. Additionally, the minimal area of an event has to be defined in order to exclude data smaller than the objects of interest. The software draws a trigger contour around the detected 'cell'. The user finally defines the distance (in pixels) from the trigger contour where the final analysis contour is set (e.g. 'add 4 pixels'). Another possibility to include events for analysis is to use a fluorescence signal or, respectively, a sample-independent trigger, called 'phantom contouring'. These phantoms are random or lattice-like circular contours of a user-defined size that are applied over the channel image. In the absence of an appropriate trigger signal, phantom contouring allows a quantitative analysis of tissue and gives an overview e.g. of labeled proteins in a tissue map. It is possible to use independent triggers within a single analysis, the triggers being set as different thresholds of one defined channel (multiple thresholding [11]) or different channels with independent triggers.

Since the fluorescence background across a slide can vary, the local background can be calculated and subtracted for every single cell. The user has to define the distance from the trigger contour and the width of the two background contours (e.g. 'between 8 and 10 pixels'). Alternatively, a fixed manual background can be set.

For each event, a number of general as well as channel-specific parameters are calculated: these include (but are not limited to) area, perimeter, position (in terms of x-y-coordinates), integral fluorescence intensity, intensity of the brightest pixel ('max pixel'), and homogeneity of the fluorescence. In addition to the primary triggering contour, a secondary triggering can be set to events within the object of interest, as exemplified by FISH signals [30].

Data analysis is comparable to FCM. The information obtained concerning the analyzed events in each channel can be displayed as histograms or scattergrams. Measured and calculated parameters (see above) are shown as mean values in region statistics for every histogram/scattergram and every gated population. Single-cell parameters can be obtained by exporting the measurement as *.txt file. Compensation of data is possible by creating virtual channels in which spillover signals are subtracted.

Despite the possibility of reanalyzing the sample on the basis of the saved scanned images, it is recommended to monitor the scanning and calculation algorithm carefully at the beginning of the scanning process. The software allows visualization of the steps online, so that corrections can be implemented. Saturated or invisible fluorescence signals cannot be optimized by different PMT settings in offline analysis.

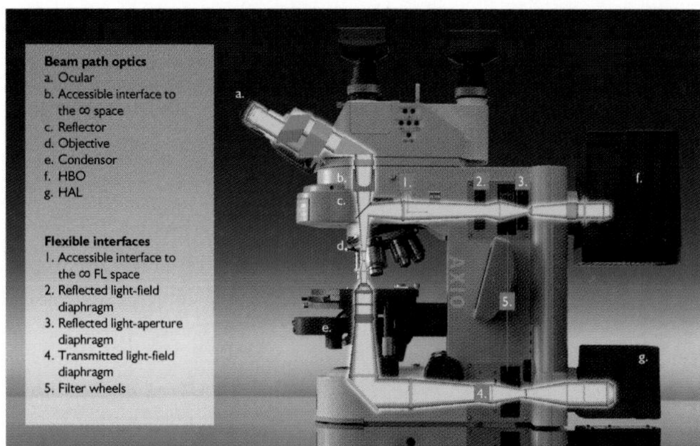

Fig. 2. Light path in fluorescence microscopy. The figure shows the AxioImager of the AxioVision SFM System (Carl Zeiss MicroImaging GmbH, Hallbergmoos, Germany). The stained sample on the stage is excited by light, emitted by a mercury arc lamp (HBO) or can be illuminated with a halogen lamp. Examination of the slide via oculars is possible or images are taken by a digital camera. These images are then analyzed by the appropriate software (by courtesy of Carl Zeiss MicroImaging GmbH, with permission).

Cytometric Analyses Using Fluorescence Microscopy (Camera-Based Images)

Scanning Fluorescence Microscopy™

Principle

The SFM™ technology is a slide-based approach based on camera-captured images and is provided by Carl Zeiss MicroImaging GmbH (Hallbergmoos, Germany). The AxioVision SFM has a technically robust platform based on a conventional fluorescence microscope. It features a motorized conventional filter wheel and an improved digital camera in combination with motorized stage and objective revolver (fig. 2). The whole system is computer controlled and overcomes the need for manual qualitative analysis and morphological documentation.

A fully automatic version (MIRAX SCAN, 3DHistech Budapest, Hungary) of this system is also available, which can automatically digitize up to 300 slides in one run. The MIRAX SCAN loads and unloads slides, detects the sample on the slide and scans it in constant high quality.

The data of the different 'colors' defined by the filter combinations are merged to generate a multiparametric image set, resulting in a multichannel, manual or autofocused scan of slides. The object stained with a fluorescent dye is excited by a mercury arc lamp. By controlling the power of the mercury arc lamp, it is possible to derive fluorescence data that are linear and stoichiometric. The emitted fluorescence light is detected by a charge-coupled device (CCD) digital camera with increased sensitivity for weak signals and a broad detection spectrum (400–1,000 nm). After digitizing the fluorescence signals and creating multichannel images, the resulting digital slides can be analyzed quantitatively.

The primary images corresponding to every field of view can be stored and recalled. Hence, the SFM can also be used as a virtual microscope. The manipulation of fluorescence filters, magnification, and movements in x-y-direction are possible offline. This means that analysis can also be performed offline and cell galleries can be generated. The standard cytometric analysis techniques (gating, scattergrams, frequency histograms, cell galleries) are all possible.

Virtual Microscopy

Each of the microscopic fields of view is recorded in every selected fluorescence channel in a digital format. If a virtual slide is created, the slide is divided into several parts. A digital image is generated for each of these fields of view. These single images are compiled by the software to create a continuous (mosaic) image of the whole slide. Based on this digital slide, virtual microscopy software simulates the basic functions of the microscope. SFM technology has been shown to be both linear and stoichiometric by the analysis of fluorescent calibration beads [31], confirming the feasibility of cytometric measurements [32]. Inhomogeneities of the fluorescence in a field of view caused by inhomogeneous excitation by the mercury arc lamp can be compensated mathematically by the software.

The advantage of the SFM method over the earlier fluorescent microscopy scanning techniques is the application of multichannel digital slides. This means that images of the same field of view are generated using different ranges of the spectrum (using different band-pass filters) representing different fluorochromes, parameters or channels.

The technology of virtual microscopy has the advantage of separated slide digitization and permits postscanning evaluation of data away from the machine.

Data Analysis

Positive events are defined on the basis of their fluorescence signals in every chosen channel by thresholding. Thresholds are defined by histograms of pixel intensities of the respective channel on the virtual slide. The pixel islets on the image (above the background) are then defined as events. In combining single images into a multichannel feature, the defined threshold values are also combined, the maximum size being that taken for analysis. For every detected event, standard cytometric characteristics for every channel (e.g. maximal or sum of intensity, diameter, perimeter, and area) are calculated and can be displayed as histograms and scattergrams.

Other Slide-Based Cytometry Systems Based on Camera-Generated Images

Widefield Microscopy

ArrayScan® V^{TI} HCS Reader

The ArrayScan HCS Reader (Thermo Fisher Scientific – Cellomics, Pittsburgh, PA, USA) is based on a Zeiss fluorescence microscope and equipped with a broad white-light source and a 12-bit cooled CCD camera. It includes the ApoTome optical-sectioning device developed by Carl Zeiss. The ApoTome technology allows confocal-style three-dimensional imaging with a conventional fluorescence microscope. ArrayScan® is developed for high-content screening by using an automated imaging system, robotic handling and scan automation. The system can be upgraded with a live-cell module enabling long-term experiments and quantitative kinetic imaging in a controlled, incubator-like environment. The appropriate image analysis module (BioApplications) converts images automatically into numeric data. Results can be obtained for individual cells or for populations and the associated images are shown.

Scan^{R}™

The Scan^{R} platform is a cell analysis system distributed by Olympus (Hamburg, Germany), which can be upgraded to allow stand-alone high-throughput screening using a plate-loading robot, auto-focus, multiparametric imaging and analysis software. The system runs a Xenon™ burner, controlled by a proprietary illumination system, an automated filter wheel with 8 positions, and is driven by software that synchronizes all steps in order to minimize bleaching and scanning time. Images are generated in a multidimensional space with x, y, z-coordinate, time, and wavelength. It can be equipped with an environmental control chamber to modulate and control humidity, temperature, and CO_2 concentration, which allows long-term life-cell imaging.

TissueFAXS® and TissueFAXSi®

An alternative approach, based on high-end microscopes (Zeiss, Leica, Nikon) is offered by the upright (TissueFAXS) or inverted (TissueFAXSi®) systems distributed by TissueGnostics GmbH (Vienna, Austria). These microscopes are equipped with a fully motorized stage and the proprietary software TissueQuest®. The platform can be equipped with up to seven objective lenses (ranging from 1× to 100× oil immersion) and up to ten filter cubes in a motorized filter revolver. Images are generated by a high-end monochrome camera and data are analyzed offline. The software offers the unique feature of bidirectional gating: in addition to relocating a dot in the analysis and evaluating the cell behind it (from dot to cell), the software also allows identification of the dot generated by a cell (from cell to dot).

Triggering can include nuclear bodies as well as ring structures, while the data processing includes a watershed function to help distinguish closely neighboring cells. The analysis of all kinds of specimens, as long as they are fixed on a slide, is feasible using this platform.

Confocal Microscopy

Opera™ LX

The Opera™ LX family of SBC instruments, originally developed by Evotec Technologies GmbH (Hamburg, Germany) and now distributed by PerkinElmer (Waltham, MA, USA), uses laser light sources for illumination (405, 488, 532 and 635 nm) and a CCD camera for imaging. The fully equipped Opera QEHS™ functions confocally with four lasers. An additional UV excitation by a Xenon™ lamp enables a nonconfocal mode. Four parallel CCD detection channels allow up to fourteen emission bands. Images are analyzed by the proprietary Acapella™ software, which offers an open architecture for development of individual cell recognition algorithms. There is a variety of objectives available (20× to 60×, water and oil immersion). The hardware configuration allows automated handling of different formats, including microtiter and nanotiter plates. The system can be upgraded with an environmental control unit for temperature, CO_2, and humidity, enabling live-cell screening.

Toponome® Imaging Cycler®

Although unsuitable for high-throughput analysis, one further system of potential interest should be presented here. The Toponome® Imaging Cycler® was developed by meltec GmbH (Magdeburg, Germany) and offers the possibility of expanding the research field of toponomics [33]. The microscope frame is based on a conventional inverted fluorescence microscope with a motorized stage and z-drive, allowing 3D image exposure. In addition, it consists of a multiwell ligand container that stores different ligands for staining. The system is completed by a pipetting unit and a cooled CCD camera. The platform engages robotic staining, imaging, and bleaching procedures of repetitive cycles (more than 100) in situ: the intermittent bleaching after each analysis being used to erode the staining of the previous cycle. This provides the unique opportunity of unraveling complex networks of proteins and cell subsets with a theoretically unlimited number of members. The iterative restaining protocol which this technology employs is termed MELK (Multi-Epitope-Ligand-Kartographie) [34]. It should be pointed out that this very complex data acquisition procedure is not capable of high-throughput analyses of large numbers of cells. Nevertheless, it offers very interesting possibilities to understand intracellular connections and interactions in high-content analyses.

This technical platform has generated a previously unimaginable density of data. It does not yield single-cell-based data immediately, but assembles all data from the same slide instead of relying on serial sections. Complex 3D images of cells and tissues can be reconstructed by using image stack algorithms and single color, multiligand libraries. The colocalization of several proteins in one sample can be analyzed by overlaying the fluorescence images. Using extended ligand libraries, even complex protein networks and complete signaling pathways can be displayed and analyzed with respect to their action at their appropriate location (toponome) instead of analysis in cell lysates.

Principal Considerations for Slide-Based Cytometry

The general principles of fluorescence staining can be applied to SBC in essentially the same way as to any other fluorescence technique, the size and form of samples being limited only by the physical dimensions of the solid carrier. Robotic automation allows the sequential feeding of stacks of several microwell slides (as in the iCys®); alternatively, several slides are initially loaded into a carrier.

The fewer mechanical moving parts the instrument has, the more stable the system can be expected to run. For example, a laser source permits the generation of highly stable excitation power, but the light has to be transformed to an evenly illuminated area, requiring correction of the emission signals according to the angle at which the laser hits the slide. This means that the stable emission of a laser is not easily transformed into stable illumination of an area.

When nonlaser light sources are used to illuminate an area, even illumination can be achieved either by electronic correction of the brightness (as is the case in the SFM) or by engaging stable high-power light sources, such as light-emitting diodes (LEDs) (e.g. in TissueFAXS™).

Regardless of differences in the illumination systems, the optical quality of the microscope is clearly a crucial determinant of the achievable quality of data. Practically, data quality is also highly dependent on the quality of sample processing. Important steps in this cascade are sample acquisition, deposition on the slide, fixation and staining procedures. Although any suboptimal condition in any of these steps has the potential to influence the quality of the entire procedure, optimal conditions in only one step do not guarantee satisfactory results.

While placing the cells on slide, every effort should be taken to guarantee an even dispersion of the cells. This cannot be achieved in cells growing on slides, in chamber slides or culture plates. An even dispersion of cells has several crucial advantages: cells can easily be identified by various algorithms and, more or less arbitrary, watershed functions can be used in order to separate closely situated cells.

The preparation of the slide, the staining protocol, and the analysis algorithm are fundamentally influenced by the biological context. For example, if a subset of cells is specified by a certain marker, e.g. CD3, all CD3+ events can be recognized as 'cells'. Therefore it would suffice to stain for CD3 in order to mark the cells of interest; the other detection channels can be used to detect further cell-specific parameters. If the aim is to include all cells in an analysis, then a global marker should be used. This can be a contour generated by the scattered light, such as interference contrast or forward scatter. With a scatter signal as trigger, it is important not to disrupt the membranes of the cells during preparation since this would lead to altered scattering characteristics and an increased amount of debris and artifacts, which in the worst case are impossible to discriminate from intact cells. In addition, air drying as well as centrifugation with high speed can lead to flattened cell bodies which lack any significant scatter. Cell density should not be too high since overlapping cell bodies cannot be separated. Alternatively, nuclear DNA can be used as a global marker for cell recognition, with erythrocytes and thrombocytes being the only cellular constituents in the blood that would not be detected by this marker. Nuclear DNA is a very stable and reliable marker: any object with a sufficient amount of DNA is a 'cell', and anything without DNA can be excluded.

If the nuclear DNA serves just as a stable marker for cell recognition, a broad range of DNA dyes is available in a wide variety of fluorescence spectra. However, if the DNA has to be analyzed stoichiometrically (e.g. in cell cycle analyses), the dyes that can be used are limited. For example, propidium iodide has been shown to yield reliable data in clinical samples, similar to DAPI and the Hoechst dyes. The concentration of the staining solutions has to be even across the slide. This is especially important for DNA dyes in stoichiometric analyses or activation markers.

Sample Preparations for Slide-Based Cytometry

Independent of the system used for analysis, high-quality sample preparation is the key to obtaining good results.

Cytometric analyses require a parameter for triggering, i.e. the information of a fluorescence or absorbed or scattered light to recognize an event by the software and apply it for analysis. Unlike in FCM, where usually forward and side scatter are used for triggering (see 'Technical Background and Methodological Principles of Flow Cytometry', pp. 53), in SBC a specific fluorescence is mostly taken as trigger signal. While the scattered light can be sufficient for triggering in cell suspensions and cell cultures, it is almost impossible in tissue analyses. Since the detection of an event and therefore the quality of the analysis depend on the triggering signal, it should be bright and well distinguishable from the background noise.

Since SBC offers the possibility of analyzing a variety of different types of samples, e.g. cell suspension, cells in culture, smears and tissue sections, many different protocols and staining procedures can be relevant. Unfortunately, there is no space to cover all of them here. For this reason, only a few suggestions will be considered concerning what is important and what one has to take care of in sample preparation for SBC.

First of all, staining of cells is very similar to that for FCM. The main difference in sample preparation is to put the respective sample on the slide.

Cell Suspensions/Smears

For multiparametric measurements or analyses, including manipulation (e.g. activation or restaining), SBC is very helpful even for cell suspensions. However, the cells have to be immobilized on the slide to perform such analyses. There are two different ways to stain them for this purpose: i) staining of cells in suspension before they are immobilized on the slide or ii) fixation of the cells on the slide, followed by staining directly on the surface.

The first method is very similar to commonly used staining procedures for FCM (for details, see 'Selection and Combination of Fluorescent Dyes', pp. 107). One should bear in mind that many fluorochromes are sensitive to drying. In such cases, it is recommended not to fix the cells on the slide by drying or freezing, but by mixing with a microscopy medium. Viscous mounting media have the advantage that the cells become immobilized on the slide following mixing, due to the hardening of the medium.

The second method should be used if the cells have to be permeabilized prior to staining. The probability of cell loss after removing the cover glass for restaining is lower if the cells are fixed onto the slide with e.g. acetone than if a mounting medium is used alone for immobilization. Note that some fixing procedures can have an influence on the epitopes on the cell surface and thus on the susceptibility to staining.

Cell Cultures

Analyses of cells in culture offer the possibility to follow the culture or even single cells over a long period of time although not all systems are capable of measuring living cells. The best way of staining living cells in culture is to add the dyes to the culture medium.

Even if an analysis of living cells is not possible, SBC offers the opportunity to characterize cells and their interactions in their 'natural' environment. These interactions are not visible after detachment. Following fixation, cells remain in the state before death and can be stained and analyzed in the culture plate.

Tissues

Paraffin sections as well as frozen sections or native tissue can be analyzed. Optimal section thickness depends on the type of the tissue as well as the aim of the study. To be cytometric, at least one cell layer should be analyzed. Thicker sections are used in 3D analyses, e.g. to study interactions of cells. One has to remember that the thicker the section, the longer the staining time should be. Also the background noise is increased in thick sections. Therefore, it is recommended to use a minimal section thickness, optimal for the respective analyses.

It is optimal to stain cohesive tissue in the free-floating state. The working surface for the antibodies is higher in this case and good staining results can be obtained. Other, softer or discontiguous tissue should be stained directly on the slide.

As for common histological or pathology tissue preparation procedures, paraffin sections have to be deparaffinized before staining. Normally, common preparation steps can be used. There are many different treatments for antigen retrievals after formaldehyde fixation and embedding [35, 36] to achieve better straining or a better discrimination of the fluorescence from the background, respectively.

The sample should be planar on the slide to avoid changes of fluorescence intensities in nonconfocal tissue analyses. If no damage of the tissue is anticipated, the tissue should be weighed down on the slide before embedding with a mounting medium.

General Comments

It is important to note that SBC is based on microscopic techniques. For this reason, the sample should be embedded in a fluorescence mounting medium or a mixture of PBS, glycerol and antifading agents (e.g. DABCO) following preparation to protect fluorescence and slow down photobleaching. The pH can also influence fluorescence intensities of fluorochromes. FITC, for example, is sensitive to acid conditions [37]. These problems can be avoided by using buffered fluorescence mounting media.

During and after staining of the samples, attention should be paid to avoiding long-term exposure to light. All samples should be stored in the dark.

A wide range of fluorochromes is available for cell staining. Generally, dyes used for FCM can also be applied in SBC analyses (see 'Selection and Combination of Fluorescent Dyes', pp. 107). However, it should be considered that SBC is a microscope-based technology and therefore the stability and in particular the sensitivity to light exposure of the fluorochromes play an important role. Very photosensitive dyes such as phycoerythrin (PE) or peridinin chlorophyll protein (PerCP) are used only rarely or not at all. Although they are very bright, they can be hardly seen in microscopy due to photobleaching. The longer exposure time associated with some SBC approaches compared to FCM accentuates this problem. On the other hand, in iCys® analyses, where the sample is scanned and not illuminated as in camera-based systems, PE gives a very good bright signal, optimally distinguishable from the background. New dyes with increased photostability (e.g. Alexa dyes or Quantum dots) have recently become available and are especially suited to SBC applications. As always, one should consider that weakly expressed antigens should be labeled with bright fluorochromes whereas highly expressed markers can be well discriminated from the negative population with dimmer fluorochromes. The fluorescence of weakly expressed antigens can also be increased by amplification steps. Amplification is accomplished by indirect staining. Directly labeled antibodies (e.g. CD45-APC) offer only a certain number of fluorochromes, coupled to an antigen. This can be amplified by indirect staining. In this case, an initial staining step with a biotinylated antigen is performed, e.g. CD45, and followed by the second step using streptavidin linked to the chosen fluorochrome (e.g. streptavidin-APC). More fluorochrome molecules can be coupled to a single antigen with this method than would be possible with a directly conjugated antibody. Thus, the discrimination between fluorescence-positive and negative cells can be improved. This can also be achieved by the use of primary and secondary antibodies. In multicolor experiments, biotin and streptavidin should be used only once. Otherwise it cannot be ruled out that the streptavidin, despite extensive washing procedures and use of saturating concentrations, binds to the 'wrong' biotin and increases the background noise or in the worst case gives false-positive results. In this case, primary and secondary antibodies are the better choice.

For multicolor experiments, fluorochrome combinations should be combined in a way that a clear distinction is possible. If band-pass filters are used for data acquisition, information of spillover signals must be obtained for a correct interpretation of the data. If possible, compensation of the data allows the subtraction of unwanted spillover signals (see 'Technical Background and Methodological Principles of Flow Cytometry', pp. 53). Acquisition of a whole spectrum range (without physical band-pass filters) will overcome this problem in most cases. Every fluorochrome offers a characteristic spectrum and can be separated from other fluorescence spectra. Nevertheless, in multicolor experiments, compensation of different curves is helpful, as is the use of virtual band-pass filters (e.g. in TissueFAXS™).

The aim of the analysis as well as the point of action of the fluorochrome plays a role in choosing the optimal dye for staining. For cell physiology experiments (see 'Determination of Cell Physiological Parameters: pH, Ca^{2+}, Glutathione, Trans-Membrane Potential', pp. 325), it is appropriate to use fluorochromes other than those normally employed for surface staining or in fluorescence live-cell imaging. Moreover, the instrument itself should be carefully selected. Not all instruments are capable of measuring time-related changes in samples, for example, the time needed for image acquisition can be too long to measure calcium uptake.

In most cases, a DNA staining is recommended. Not only in tissue analyses is it useful to discriminate cells from debris, but in almost all SBC applications. In addition, it also gives information about the cell cycle of the cells. DNA fluorescence signals are suitable for triggering. Concentration of DNA dyes (and fluorochromes in general) is very important. With too high concentrations, the background noise is increased (see 'Quality Control and Standardization', pp. 159), i.e. the discrimination of positive and negative events is hindered. On the other hand, a sufficient concentration must be used to stain all cells in an adequate way. Insufficient staining has a distinct effect on the fluorescence intensity of the stained DNA and therefore the analysis of the cell cycle distribution. This will lead to very high coefficients of variation or, in the worst case, to nonanalyzable measurements. Additional DNA dye can be added to the embedding medium to avoid heterogeneous staining or loss of DNA fluorescence due to washing out effects.

In combining a DNA dye with other staining, one has to take care of the usually very broad emission range of a DNA dye. This sometimes hampers the options for combinations with other dyes.

Statement

The authors want to point out that this list of SBC systems is not exhaustive. They assume no responsibility for completeness, possible modifications and updates of the presented systems. The descriptions of the listed systems are intended to provide an overview of the instrumentation only and are not comprehensive. For further detailed information about these imaging systems they refer the readers to the respective homepages of the manufacturers or contributors.

References

1 Steiner GE, Ecker RC, Kramer G, Stockhuber F, Marberger MJ: Automated data acquisition by confocal laser scanning microscopy and image analysis of triple stained immunofluorescent leukocytes in tissue. J Immunol Methods 2000;237:39–50.

2 Mittag A, Lenz D, Gerstner AO, Sack U, Steinbrecher M, Koksch M, Raffael A, Bocsi J, Tárnok A: Polychromatic (eight-color) slide-based cytometry for the phenotyping of leukocyte, NK, and NKT subsets. Cytometry A 2005;65:103–115.

3 Tárnok A, Gerstner AO: Clinical applications of laser scanning cytometry. Cytometry 2002;50:133–143.

4 Basso G, Buldini B, De Zen L, Orfao A: New methodologic approaches for immunophenotyping acute leukemias. Haematologica 2001;86:675–692.

5 Laffers W, Schlenkhoff C, Pieper K, Mittag A, Tarnok A, Gerstner AOH: Concepts for absolute immunophenosubtyping by slide-based cytometry. Transfus Med Hemother 2007;34:188–195.

6 Gerstner AO, Machlitt J, Laffers W, Tárnok A, Bootz F: Analysis of minimal sample volumes from head and neck cancer by laser scanning cytometry. Onkologie 2002;25:40–46.

7 Kamentsky LA, Kamentsky LD: Microscope-based multiparameter laser scanning cytometer yielding data comparable to flow cytometry data. Cytometry 1991;12:381–387.

8 Schwock J, Ho JC, Luther E, Hedley DW, Geddie WR: Measurement of signaling pathway activities in solid tumor fine-needle biopsies by slide-based cytometry. Diagn Mol Pathol 2007;16:130–140.

9 Gerstner AOH, Thiele A, Tárnok A, Tannapfel A, Weber A, Bootz F: Prediction of upper aerodigestive tract cancer by slide-based cytometry. Cytometry A 2006;69:582–587.

10 Gerstner AOH, Thiele A, Tárnok A, Machlitt J, Oeken J, Tannapfel A, Weber A, Bootz F: Preoperative detection of laryngeal cancer in mucosal swabs by slide-based cytometry. Eur J Cancer 2005;41:445–452.

11 Gerstner AOH, Trumpfheller C, Racz P, Osmancik P, Tenner-Racz K, Tárnok A: Quantitative histology by multicolor slide-based cytometry. Cytometry A 2004;61:210–219.

12 Gerstner AOH, Müller AK, Machlitt J, Tárnok A, Tannapfel A, Weber A, Bootz F: Slide-based cytometry for predicting malignancy in solid salivary gland tumors by fine needle aspirate biopsies. Cytometry B Clin Cytom 2003;53:20–25.

13 Mosch B, Morawski M, Mittag A, Lenz D, Tarnok A, Arendt T: Aneuploidy and DNA replication in the normal human brain and Alzheimer's disease. J Neurosci 2007;27:6859–6867.
14 Persohn E, Seewald W, Bauer J, Schreiber J: Cell proliferation measurement in cecum and colon of rats using scanned images and fully automated image analysis: validation of method. Exp Toxicol Pathol 2007;58:411–418.
15 Kayser K, Radziszowski D, Bzdyl P, Sommer R, Kayser G: Digitized pathology: theory and experiences in automated tissue-based virtual diagnosis. Rom J Morphol Embryol 2006;47:21–28.
16 Haider AS, Grabarek J, Eng B, Pedraza P, Ferreri NR, Balazs EA, Darzynkiewicz Z: In vitro model of 'wound healing' analyzed by laser scanning cytometry: accelerated healing of epithelial cell monolayers in the presence of hyaluronate. Cytometry A 2003;53:1–8.
17 Esposito A, Dohm CP, Bähr M, Wouters FS: Unsupervised fluorescence lifetime imaging microscopy for high content and high throughput screening. Mol Cell Proteomics 2007;6:1446–1454.
18 Galanzha EI, Tuchin VV, Zharov VP: Advances in small animal mesentery models for in vivo flow cytometry, dynamic microscopy, and drug screening. World J Gastroenterol 2007;13:192–218.
19 Lövborg H, Gullbo J, Larsson R: Screening for apoptosis – classical and emerging techniques. Anticancer Drugs 2005;16:593–599.
20 Gerstner AOH, Gutsche M, Bücheler M, Machlitt J, Emmrich F, Tannapfel A, Tárnok A, Bootz F: Eosinophilia in nasal polyposis: its objective quantification and clinical relevance. Clin Exp Allergy 2004;34:65–70.
21 Bedner E, Li X, Kunicki J, Darzynkiewicz Z: Translocation of bax to mitochondria during apoptosis measured by laser scanning cytometry. Cytometry 2000;41:83–88.
22 Li X, Darzynkiewicz Z: The Schrodinger's cat quandary in cell biology: integration of live cell functional assays with measurements of fixed cells in analysis of apoptosis. Exp Cell Res 1999;249:404–412.
23 Laffers W, Mittag A, Lenz D, Tárnok A, Gerstner AO: Iterative restaining as a pivotal tool for n-color immunophenotyping by slide-based cytometry. Cytometry A 2006;69:127–130.
24 Schubert W: A three-symbol code for organized proteomes based on cyclical imaging of protein locations. Cytometry A 2007;71:352–360.
25 Mittag A, Lenz D, Gerstner AO, Tárnok A: Hyperchromatic cytometry principles for cytomics using slide based cytometry. Cytometry A 2006;69:691–703.
26 Ozawa K, Hudson CC, Wille KR, Karaki S, Oakley RH: Development and validation of algorithms for measuring G-protein coupled receptor activation in cells using the LSC-based imaging cytometer platform. Cytometry A 2005;65:69–76.
27 Pachmann K, Camara O, Kavallaris A, Krauspe S, Malarski N, Gajda M, Kroll T, Jörke C, Hammer U, Altendorf-Hofmann A, Rabenstein C, Pachmann U, Runnebaum I, Höffken K: Monitoring the response of circulating epithelial tumor cells to adjuvant chemotherapy in breast cancer allows detection of patients at risk of early relapse. J Clin Oncol 2008;26:1208–1215.
28 Darzynkiewicz Z, Galkowski D, Zhao H: Analysis of apoptosis by cytometry using TUNEL assay. Methods 2008;44:250–254.
29 Harnett MM: Laser scanning cytometry: understanding the immune system in situ. Nat Rev Immunol 2007;7:897–904.
30 Kamentsky LA, Kamentsky LD, Fletcher JA, Kurose A, Sasaki K: Methods for automatic multiparameter analysis of fluorescence in situ hybridized specimens with a laser scanning cytometer. Cytometry 1997;27:117–125.
31 Varga VS, Bocsi J, Sipos F, Csendes G, Tulassay Z, Molnar B: Scanning fluorescent microscopy is an alternative for quantitative fluorescent cell analysis. Cytometry A 2004;60:53–62.
32 Bocsi J, Varga VS, Molnar B, Sipos F, Tulassay Z, Tarnok A: Scanning fluorescent microscopy analysis is applicable for absolute and relative cell frequency determinations. Cytometry A 2004;61:1–8.
33 Schubert W: Topological proteomics, toponomics, MELK-technology. Adv Biochem Eng Biotechnol 2003;83:189–209.
34 Schubert W: Exploring molecular networks directly in the cell. Cytometry A 2006;69:109–112.
35 Kanai K, Nunoya T, Shibuya K, Nakamura T, Tajima M: Variations in effectiveness of antigen retrieval pretreatments for diagnostic immunohistochemistry. Res Vet Sci 1998;64:57–61.
36 Leong AS, Sormunen RT: Microwave procedures for electron microscopy and resin-embedded sections. Micron 1998;29:397–409.
37 Huth U, Wieschollek A, Garini Y, Schubert R, Peschka-Süss R: Fourier transformed spectral bio-imaging for studying the intracellular fate of liposomes. Cytometry A 2004;57:10–21.

Sack U, Tárnok A, Rothe G (eds): Cellular Diagnostics. Basics, Methods and Clinical Applications of Flow Cytometry.
Basel, Karger, 2009, pp 107–140

Selection and Combination of Fluorescent Dyes

Matthias Schiemann · Dirk Hans Busch

Institute for Medical Medicine, Immunology and Hygiene, Technical University Munich, Germany

Introduction

The selection and combination of the most suitable fluorochromes for each application are prerequisites for an optimal quality of flow-cytometry-based cell analysis and purification. Fluorochrome-labeled antibodies are generally used to identify defined extra- and/or intracellular antigens (immunofluorescence). In addition, fluorescence-marked substrates, receptors and/or ligands, e.g. major histocompatibility complex (MHC) multimers, have been developed for flow cytometry applications to further improve the characterization of cell populations.

Simultaneous detection of several different parameters on the single-cell level is unquestionably a scientific advance (multiparameter flow cytometry). However, it also confronts researchers with the important question of finding the optimal combination of antibodies and/or fluorochromes for their specific applications. The first steps when selecting combinations of fluorescent dyes usually focus on the available technical equipment, which may limit the repertoire of important parameters. The most relevant technical issues are the lasers (number, wavelength, power), optical layout, e.g. the number of photomultipliers (PMTs) that are colocalized with each laser, and the specification of dichroic and other filters.

Biological parameters can also guide decisions for setting up multiparameter flow cytometry applications, as different antigens are expressed and detectable with very different qualities (e.g. low, intermediate or high expression). In addition, the availability of fluorochrome conjugates (e.g. commercial source versus the necessity for home-made conjugations) can often become a crucial aspect. If the sample contains significant numbers of dead cells, which is often not predictable, specific labeling of dead cells might be required.

All this basic information is necessary for the selection of optimal combinations of antibodies and fluorochromes.

This chapter covers basic aspects of fluorochrome selection and combination. We provide an overview of the most common fluorescent dyes and their specific applications; in some instances, we also discuss potential problems or precautions for individual fluorochromes.

Due to the constantly increasing number of available reagents, we can give only a rough overview of the most common dyes and techniques. The interested reader will find links to additional literature and relevant web sites at the end of the chapter. Dyes that are specifically used for cell cycle and proliferation analysis (Hoechst, DAPI, chromomycin A_3, Indo-1, rhodamin-123, CFSE and others) will be discussed in more detail in the chapters 'DNA and Proliferation Measurements in Flow Cytometry' (pp. 390), 'Oxidative Burst, Phagocytosis and Expression of Adhesion Molecules' (pp. 343) or 'Determination of Cell Physiological Parameters: pH, Ca^{2+}, Glutathione, Trans-Membrane Potential' (pp. 325). Most staining examples included in this chapter were generated on a CyAn ADP analyzer (Beckman Coulter, FL, USA).

Basics

Fluorescence

Some molecules (mostly polyaromatic or heterocyclic carbon hydrogens) have the property of absorbing light of a distinct wavelength and to emit light with lower energy (higher wavelength). This process is illustrated in figure 1A as an electron status diagram, which has originally been described by Jablonski.

During the excitation phase ①, electrons move to a higher energy level; when dropping to the original status, released energy is emitted as light (emission ③). The transient status ② between excitation and emission exists only for a very short time ($1-10 \times 10^{-9}$ s). The dimensionless *fluorescent quantum yield* Φ describes the relation between the light delivered with fluorescence (③) and the amount of absorbed light (①). In order to estimate the brightness of individual fluorochromes, the *coefficient of extinction* ε and the *fluorescence quantum yield* Φ have to be considered. The spectrum of the emission is independent of the wavelength used for excitation. However, the intensity of the emitted light is proportional to the amplitude of the exciting light wave (fig. 1B). The difference between the maxima for excitement and fluorescence is called the Stokes shift S.

Tandem Conjugates/Energy Transfer

Some combinations of fluorochromes provide a situation where the emission wavelength of a first dye corresponds to the absorption wavelength of a second fluorochrome. If such dyes can be brought to sufficient physical proximity, energy transfer

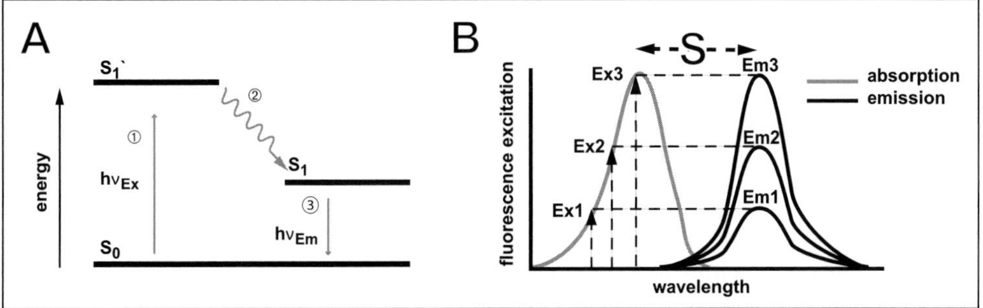

Fig. 1. A The Jablonski diagram illustrates the ground status S_0, in addition to the first (S1) and second (S_1') excitation level (horizontal lines). **B** Excitation of a fluorochrome at three different wavelengths (Ex1, Ex2, Ex3) induces different intensities of emission (Em1, Em2, Em3).

occurs. Ideally, the emission light of the first fluorochrome is completely absorbed and transferred to the second dye. In such a case, larger Stokes shifts can be observed than with a single fluorochrome. A major advantage of this technology is that a panel of fluorescence conjugates with the same excitation wavelength can be generated and discriminated by different emissions derived from the tandem dye. The development of tandem conjugates had a significant impact on the rapid development of multiparameter flow cytometry applications over the last years.

Antibodies

Vaccination with foreign substances usually results in immune reactions with antigen-specific antibody formation. Since a complex antigen usually possesses several immunogenic determinants (epitopes), immune serum can contain a multiplicity of different antigen-specific antibodies (polyclonal antibodies). By further enrichment of antibodies for defined specificity, e.g. by affinity purification, unwanted cross-reactivities can be further reduced.

For flow cytometry applications monoclonal antibodies (mAbs) are preferred as their quality, specificity and sensitivity can be better defined. For the generation of mAbs, antibody-producing B-lymphocytes are fused to a tumor cell line [1]. Such hybridomas can be expanded and cultured very easily, thereby maintaining their ability to produce the B-cell-associated antibody. An individual B-cell hybridoma clone secretes only one ('monoclonal') antibody into the culture supernatant, from which the mAbs can be further purified in virtually unlimited quantities.

Direct and Indirect Immunofluorescence

'Direct immunofluorescence' describes the detection of defined antigens by a probe which is directly labeled with the detecting fluorescence. *'Indirect immunofluorescence'* describes a procedure in which the detecting fluorescence is linked to a secondary reagent for example an isotype-specific antibody which is specific for the primary antibody used to detect the antigen. Alternatively, the primary detection reagent can be labeled with a hapten (like DIG or NP) which can subsequently be detected by a fluorescence-conjugated antihapten antibody (e.g. anti-DIG or anti-NP). The advantage of indirect immunofluorescence is that it often results in a reinforcement of signal intensity compared to staining with direct conjugates. The disadvantage is that an increase in nonspecific background staining can take place. Alternatively to hapten, biotinylated antibodies are frequently used and visualized in the second staining step with streptavidin-conjugated fluorochromes (e.g. SA-PE). The biotin-avidin system is characterized by very high binding affinity and permits a flexible choice between different fluorochromes with minimal unspecific background staining.

Excitation of Fluorescent Dyes

Which of the fluorescent dyes can be used for an anticipated flow cytometry experiment is primarily dependent on the available device and laser equipment. Tables 1–3 give a short overview (which is certainly incomplete!) of the most commonly used laser types for flow cytometry applications.

Standard flow cytometers already come with a setup laser layout. The following table (table 4) describes the lasers mounted in most common flow cytometers and their specific performance parameters. The specifications are based on the manufacturer's information for the current standard configurations.

Laser Power

Similar to the wavelength, the laser's general power output also has a substantial impact on the quality of fluorescence detection. This is important to know as a weaker laser may not allow detection of a fluorescent dye, which should generally be excitable by a defined wavelength (fig. 2).

Most fluorochromes can readily be detected with the most common flow cytometers without further modifications of the laser's performance. However, for optimization of specific applications, improved signals can be received by changing the laser's output. Thus, the term 'optimal laser performance' cannot be standardized and will always depend on the specific requirements of the experiment. If certain fluorescences should become 'too bright' due to high laser performance, selective signal weakening before detection (PMT) might become necessary for the combination with other fluorescent dyes. On the other hand, certain fluorochromes can also suffer from exposure to too high laser powers and lose their emission signal ('bleaching'); as an example, PerCP can be damaged at a laser power (488 nm) above >15 mW.

Table 1. Water-cooled lasers

Wavelength nm	Innova 70C-4	Innova 70CSpectrum	Innova 90C-A4	Innova 302C-UV Krypton	Innova 305C-UV	Innova 306C-UV	Innova Enterprise II
647		0.25 W					
568.3		0.15 W					
530.9		0.13 W					
528.7	0.30 W		0.30 W		0.35 W	0.42 W	
520.8		0.13 W		0.03 W			
514.5	1.70 W	0.25 W	1.70 W		2.00 W	2.40 W	
488	1.30 W	0.25 W	1.30 W		1.50 W	1.80 W	0.18 W
457.9	0.30 W	0.03 W	0.30 W		0.35 W	0.42 W	
413.1				0.30 W			
406.7				0.20 W			
406.7–415.4				0.60 W			
351.1–363.8			0.20 W		0.40 W	0.50 W	0.05 W
350.7–356.4				0.50 W			
350.7–363.8		0.05 W					

Table 2. Air-cooled lasers

Wavelength nm	JDS Uniphase	Melles Griot 543	Melles Griot 532	Midwest ILT5500	Spectra Physics 163	Spectra Physics 163
457–514		0.195 W	0.05 W	0.50 W		0.075 W
488	0.03 W	0.15 W			0.025 W	0.075 W
458		0.039 W				0.015 W

Table 3. Solid-state and other lasers

Wavelength nm	Coherent Sapphire 488-20	Coherent OPSL 488-200	ICyt Lyt200S	Lasiris DLS	DLC Red Diode	Power Technology LD1510	Point Source
635				0.035 W	0.027 W		
405						0.05 W	
488	0.020 W	0.2 W	0.2 W				
407							0.017 W

Optical Layout of a Flow Cytometer

Depending on the area of research and specific requirements, various flow cytometers for standard applications are currently available (for more detailed information also see 'Technical Background and Methodological Principles of Flow Cytometry', pp. 53). Most of these 'standard' machines come with a defined opti-

Table 4. Standard configurations in common flow cytometers

Wavelength, nm	Laser	P, mW	Amount PMT
CyAn ADP LX (Beckman Coulter)			
405	Compass 405-25 CW	25	2
488	Sapphire 488-20	20	5
635	Diode	25	2
CyAn ADP MLE (Beckman Coulter)			
351	Enterprise II 621	$\leq$50	2
488	Enterprise 621	$\leq$150	5
635	Diode	25	2
CyFlow ML (Partec)			
360–370	Hg lamp	100	3
405	Compass 405-25	25	2
488	Sapphire 488-200	$\leq$200	2
635	Diode	25	2
CyFlow Space (Partec)			
488	Sapphire 488-200	$\leq$200	5
635	Diode	25	1
Cytomics FC 500 (Beckmann Coulter)			
488	JDS Uniphase Argon	25	4
635	Diode	25	1
Epics XL (Beckmann Coulter)			
488	JDS Uniphase Argon	15	4
FACSCalibur (BD Biosciences)			
488	Spectra-Physics	15	3
635	Diode	10	1
FACSCanto (BD Biosciences)			
488	Sapphire 488-20	20	4
633	JDS Uniphase He-Ne	17	2
LSRII (BD Biosciences)			
488	Sapphire 488–20	25	4
633	JDS Uniphase He-Ne	17	2
405	Vioflame	25	2
355	Lightwave OPSL	20	2

cal layout which supports the detection of many common fluorescent dyes. However, in some cases optimization of signal detection by changes in filters or mirrors could substantially improve performance for individual applications. Helpful information for the optimization of optical layouts is provided online by manufacturers of optical filters (e.g. see under *www.chroma.com*).

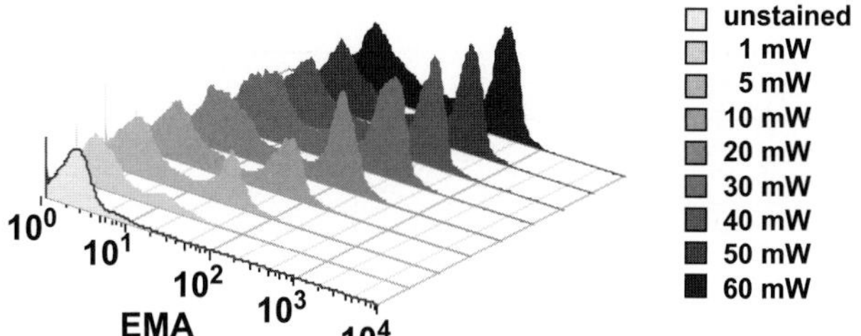

Fig. 2. Fluorescence intensity (EMA detected by MLUV) as a function of laser power output.

Spillover

Due to the individual emission spectra, the detection of different fluorochromes is often affected by partial overlap of the spectra. This phenomenon is called 'spillover'. For data analysis and correct annotation of the staining signals, a correction (compensation) has to be performed, which can take place on the hardware level or via software compensation.

Compensation

For analysis or phenotyping of lymphocytes at least two – usually even much more – fluorochrome-conjugated markers are used. For clinical diagnostics, especially those fluorochrome-combinations are interesting that can be simultaneously excited at 488 nm (ionized argons laser). As most fluorochromes show spectral overlap to each other, compensation is required in order to clearly distinguish marker-positive and marker-negative populations.

For the compensation procedure, unwanted emission detected in the 'false' fluorescence channel, which is proportionate to the voltage level, is subtracted (e.g. FITC spillover into the PE channel). In order to do this, single-color staining controls need to be provided for each fluorochrome conjugate [2, 3].

A good guide on compensation procedures can be found under *www.drmr.com.* For standardization, fluorochrome-labeled microparticles (e.g. available from Bangs) [4] are commonly used (table 5).

Fluorescent Dyes Suitable for Immune Staining

Fluorochromes Excitable at 488 nm (table 6)

Fluorescein-5-isothiocyanate
Fluorescein-5-isothiocyanate (FITC) is a xanthene dye with intensive yellow-green fluorescence; this isothiocyanate derivative is by far the most common fluorochrome

Table 5. Selections of dyes with standard filters[a]

Application	Excitation					
	UV 351	UV 351–364	Violet 405	Blue 488	Yellow 595	Red 635/647
Immunophenotype						
Alexa 350	450/40	450/40				
AMCA 350	450/40	450/40				
Cascade Blue			450/65			
Cascade Yellow			550/30			
Alexa 405			450/50			
FITC				530/30		
PE				585/42		
PerCP				682/33		
PETR				630/30		
PE-Cy5				682/33		
PE-Cy5.5				720/45[b]		
PerCP-Cy5.5				720/45[b]		
PE-Cy7				780/60		
APC						670/14
Alexa 660						695/40
APC-Cy5.5						705/50[c]
APC-Cy7						750LP[d]
Live/Dead (DNA/RNA)						
Ethidium monoazide bromide				600/20		
PI	630/30			630/22		
7-AAD				630/30		
To-Pro®-3						660/20
LDS-751 (RNA)			475LP			
DNA cell cycles						
DAPI	450/40					
Hoechst 33258	450/40					
Chromosomal DNA						
Chromomycin A_3		560/30				
Hoechst 33258	450/40					
Cell function						
Fluo-3				530/30		
Indo-1 (bound)	405/22					
Indo-1 (free)	519/40					
Rhodamine 123 (TRITC)				530/30		
CFSE				530/30		

[a] Filters (nm) are given x/y, where x is the wavelength of the center of the band pass and y is the width of the band (LP = long-pass filter).

[b] In the absence of PE-Cy5, 705/50 is better.

[c] In combination with Cy5, 720/45 is better.

[d] 785/50 can also be used.

Table 6. Fluorochromes excitable at 488 nm

Fluorochrome	Molecular weight, Da	Excitation nm	Emission nm	Coefficient of extinction M^{-1} cm^{-1}
FITC	389	488	519	73,000
PE	240,000	488	578	1,960,000
PETR	240,625	488	615	n.s.
PerCP	35,000	488	677	n.s.
PE-Cy5	240,792	488	680	250,000
PE-Cy5.5	241,128	488	694	250,000
PerCP-Cy5.5	36,128	488	695	250,000
PE-Cy7	240,818	488	767	250,000

n.s. = No specification.

currently used in flow cytometry, and many FITC-conjugated antibodies and proteins are commercially available. FITC can be detected after excitation with a 488 nm laser. FITC has a high quantum yield (efficiency of energy transfer from absorption to fluorescence emission). Approximately half of the absorbed photons are emitted as fluorescence light (fig. 3).

R-Phycoerythrin

Phycoerythrin (PE) is a photosynthesis pigment discovered in two red algae *Porphyridium cruentum* and *Porphyridium sordidum;* it transfers light energy to chlorophyll during photosynthesis. It is commonly used as a fluorochrome in flow cytometry, which is also reflected in the large product list of PE conjugates provided by manufacturers. In addition, similar to allophycocyanin (APC), PE is well-suited for the production of tandem conjugates.

One molecule of PE consists of 34 individual phycoerythrobilin subunits. Because of this high number of fluorochrome subunits per molecule, PE is characterized by a high 'brightness' which makes it ideal for most flow cytometry applications, including the detection of antigens with low expression levels. For some applications, in particular for intracellular staining [5], the large molecule size of PE has to be considered (fig. 4).

Peridinin Chlorophyll Protein

Peridinin chlorophyll protein (PerCP) is a light-collecting protein from the fire algae *Amphidinium carterae* with an unusual absorption band at 450 nm. PerCP forms a trimer, each subunit consisting of 8 peridinin and 2 chlorophyll molecules (fig. 5).

PerCP is a fluorochrome patented by BD; it is not recommended for 'jet in air' flow cytometry. PerCP is unique regarding its Stokes shift (difference between excitation and emission wavelength), but its applications are limited by its low stability. Furthermore, PerCP is very sensitive to 'photobleaching', which explains the limited tolerance to stronger laser performance. Samples stained with PerCP need to be pro-

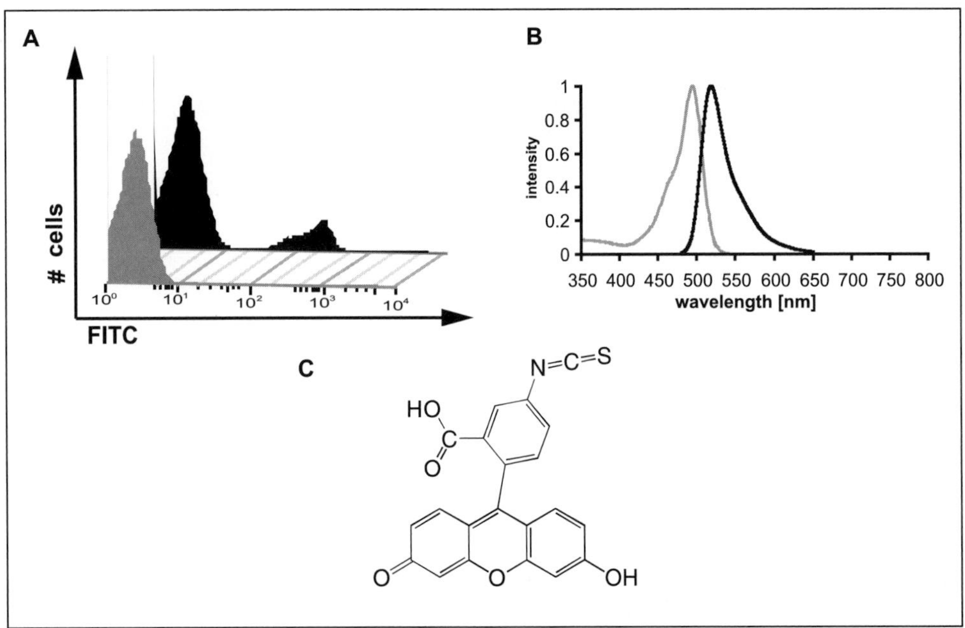

Fig. 3. A Spleen cells (mouse) stained with CD8α FITC (clone 53–6.7, BD) (black); gray = unstained control. **B** Absorption and emission spectrum. **C** FITC.

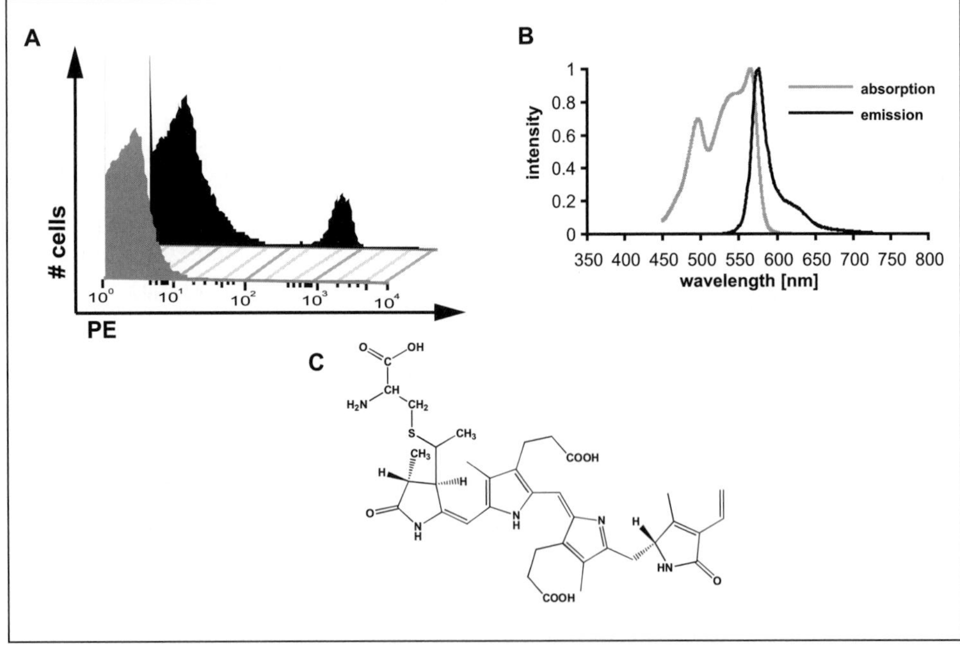

Fig. 4. A Spleen cells (mouse) were stained with CD8α PE (clone 53–6.7, BD) (black); gray = unstained control. **B** Absorption and emission spectrum. **C** R-PE.

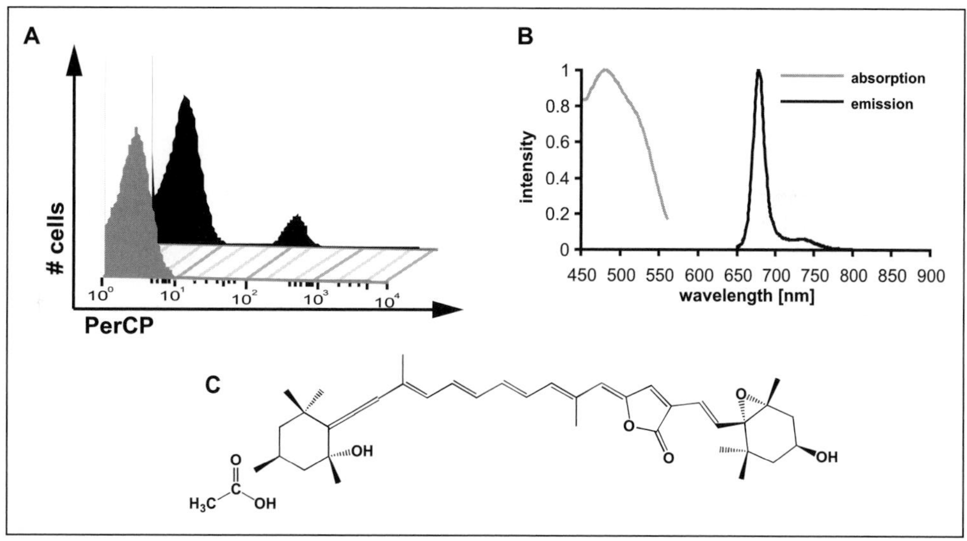

Fig. 5. **A** Spleen cells (mouse) stained with CD8α PerCP (clone 53–6.7, BD) (black); gray = unstained control. **B** Absorption and emission spectrum. **C** PerCP.

cessed under strict protection against light and data need to be acquired rapidly. PerCP can be replaced by the tandem conjugate PE-TexasRed® (PETR) for most multicolor experiments. PETR is characterized by an emission spectrum with only small spectral overlap with PE and is not excited by 635-nm laser light (only minimal compensation necessary) (fig. 5).

Phycoerythrin-TexasRed® (PETR, RED613, ECD®)

PETR is a tandem conjugate combining PE and TexasRed®. TexasRed is a sulfonyl chloride derivative of sulforhodamine 101.

As a first component, the PE subunit is excited via 488-nm laser light. The emitted light of approximately 570 nm subsequently activates the TexasRed part via energy transfer, which results in the final emission of 610 nm (fig. 6).

Phycoerythrin-Cy5 (Cy-Chrome™, TriChrome™, PC5™, TriColor™)

PE-Cy5 is another commonly used tandem conjugate; it is characterized by very high energy transfer efficiency (more than 95%). Because of its broad absorption width it is not recommended for use on flow cytometers that support simultaneous excitation of 488-nm laser light together with a red laser (633 or 647 nm) line (fig. 7).

Phycoerythrin-Cy5.5

PE-Cy5.5 is a tandem conjugate, which combines PE with a cyanine fluorochrome (Cy5.5) (fig. 8).

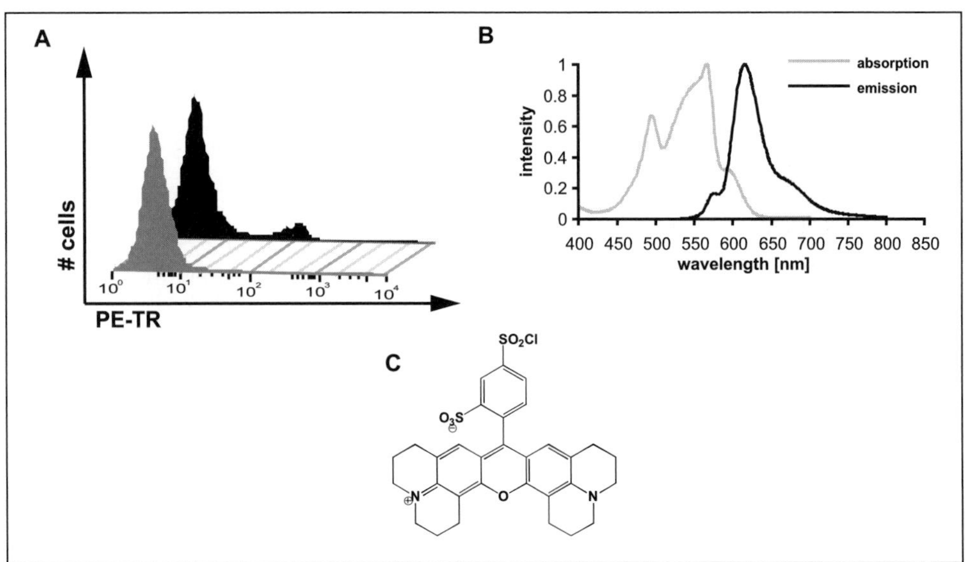

Fig. 6. A Spleen cells (mouse) stained with CD8α PETR (clone 5H10, Invitrogen); gray = unstained control. **B** Absorption and emission spectrum. **C** TR.

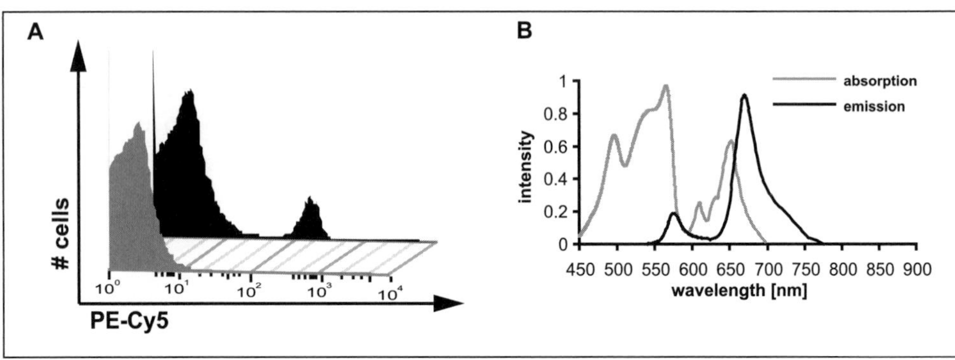

Fig. 7. A Spleen cells (mouse) stained with CD8α PE-Cy5.5 (clone 5H10, Invitrogen) (black); gray = unstained control. **B** Absorption and emission spectrum.

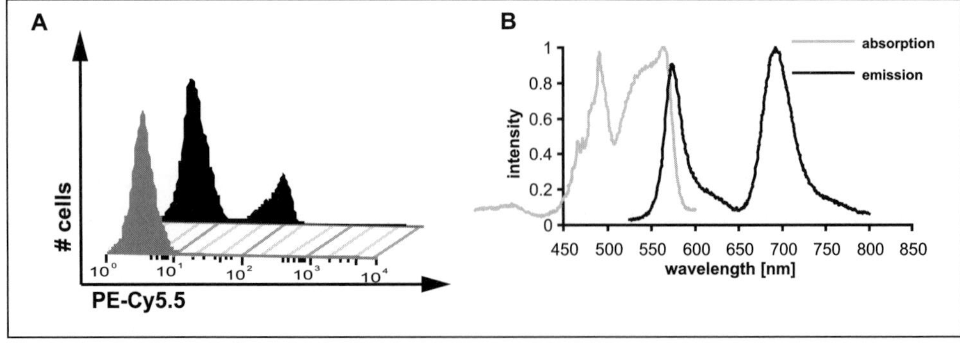

Fig. 8. A Spleen cells (mouse) stained with CD8α PE-Cy5.5 (clone 5H10, Invitrogen) (black); gray = unstained control. **B** Absorption and emission spectrum.

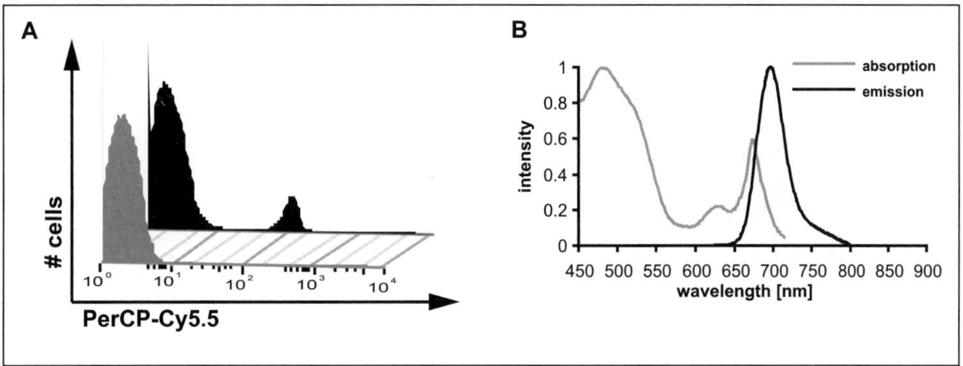

Fig. 9. A Spleen cells (mouse) stained with CD8α PerCP-Cy5.5 (clone 53–6.7, BD) (black); gray = unstained control. **B** Absorption and emission spectrum.

Peridinin Chlorophyll Protein -Cy5.5

PerCP-Cy5.5 represents a tandem conjugate of PerCP and a cyanine fluorochrome (Cy5.5). PerCP-Cy5.5 is also recommended for 'jet in air' flow cytometry (in contrast to PerCP) (fig. 9).

Phycoerythrin-Cy7

PE-Cy7 is a tandem conjugate consisting of PE and a cyanine fluorochrome (Cy7) (6). PE-Cy7 can be used for cell staining in combination with APC, with usually only limited need for compensation (fig. 10).

Fluorochromes Excitable at 635/647 nm (Table 7)

Allophycocyanin

APC is a component of a photochemical pigmentation from blue-green seaweed. APC has 6 phycocyanobilin chromophores per molecule and is structurally similar to phycoerythrobilin in R-PE (fig. 11).

Allophycocyanin-Cy5.5

Allophycocyanin-Cy5.5 is a tandem conjugate consisting of APC and a cyanine fluorochrome (Cy5.5). Excited through red laser light, APC-Cy5.5 can be used as a third color in this range; it can be well separated from APC and APC-Cy7.

Allophycocyanin -Cy7 (PharRed)

APC-Cy7 is a tandem conjugate combining APC and a cyanine fluorochrome (Cy7). Photobleaching needs to be considered. It is recommended to use a 750-nm pass filter together with a sensitive detector for red light (e.g. Hamamatsu R3896 PMT). Some APC-Cy7 conjugates change their emission spectrum when in contact with formaldehyde. Therefore, some manufacturers recommend acquisition of data for fixed samples within 4 h (fig. 12).

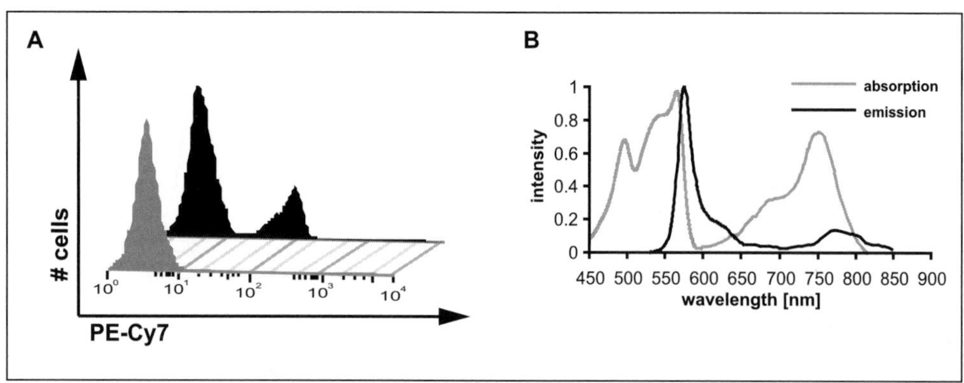

Fig. 10. A Spleen cells (mouse) stained with CD8α PE-Cy7 (clone 53–6.7, BD) (black); gray = unstained control. **B** Absorption and emission spectrum.

Table 7. Fluorochromes excitable at 635/647 nm

Fluorochrome	Molecular weight, Da	Excitation nm	Emission nm	Coefficient of extinction M^{-1} cm^{-1}
APC	105,000	635/647	660	700,000
APC-Cy5.5	106,128	635/647	694	250,000
APC-Cy7	105,818	635/647	767	250,000

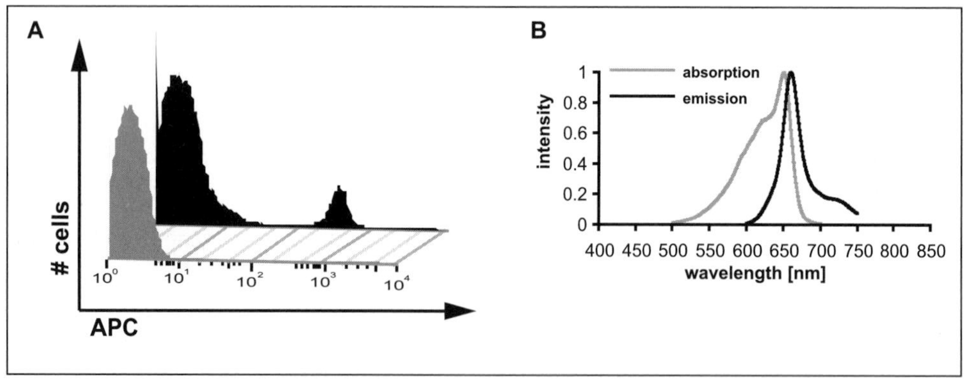

Fig. 11. A Spleen cells (mouse) stained with CD8α APC (clone 53–6.7, BD) (black); gray = unstained control. **B** Absorption and emission spectrum.

UV-light-Excitable Dyes (Table 8)

7-Amino-4-Methylcoumarin-3-Acetic Acid

The major characteristic of 7-amino-4-methylcoumarin-3-acetic acid (AMCA-X) is that a so-called 'spacer' is added between the fluorochrome and the marked molecule. The spacer can reduce the quenching that typically follows the conjugation (fig. 13).

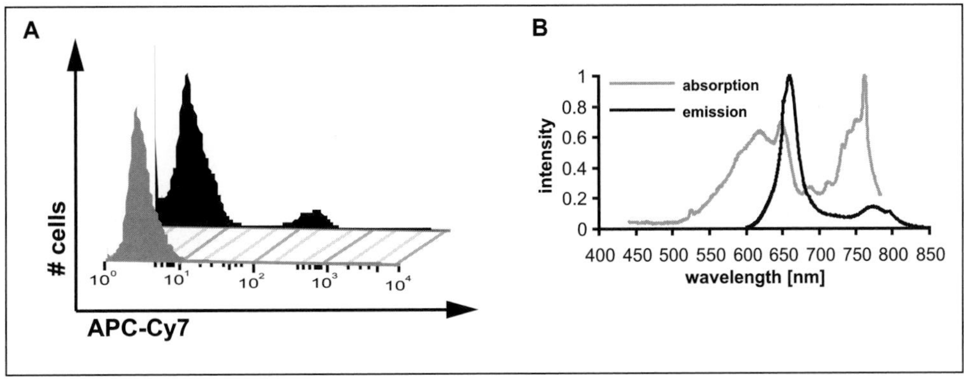

Fig. 12. A Spleen cells (mouse) stained with CD8α APC-Cy7 (clone 53–6.7, BD) (black); gray = unstained control. **B** Absorption and emission spectrum.

Table 8. Fluorochromes excitable at UV light

Fluorochrome	Molecular weight, Da	Excitation nm	Emission nm	Coefficient of extinction M^{-1} cm^{-1}
AMCA-X	443	353	442	19,000
Marina Blue	367	362	459	19,000
Cascade Blue	607	399	423	30,000
Cascade Yellow	563	409	558	24,000
Pacific Blue	339	416	451	37,000

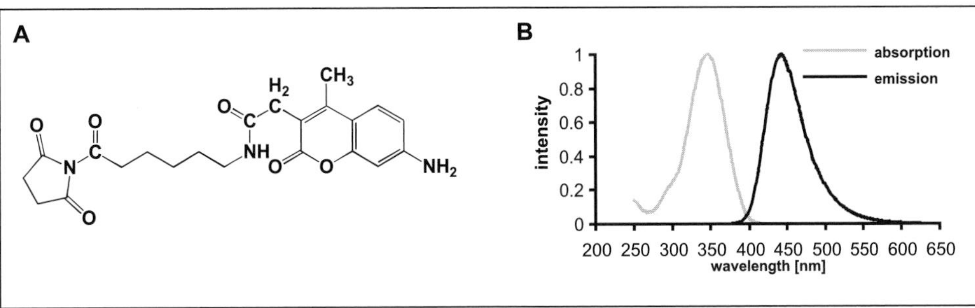

Fig. 13. A AMCA. **B** Absorption and emission spectrum.

Marina Blue® and Pacific Blue®

Marina Blue and Pacific Blue are patented dyes distributed by Molecular Probes and both are based on the 6,8-difluoro-7-hydroxycoumarine fluorochrome. Both fluoresce most strongly at pH 7 (fig. 14).

Cascade Blue®

Cascade Blue is a derivative of a patented sulfonated pyrene. Because of its strong fluorescence, it is often used in multicolor flow cytometry (fig. 15).

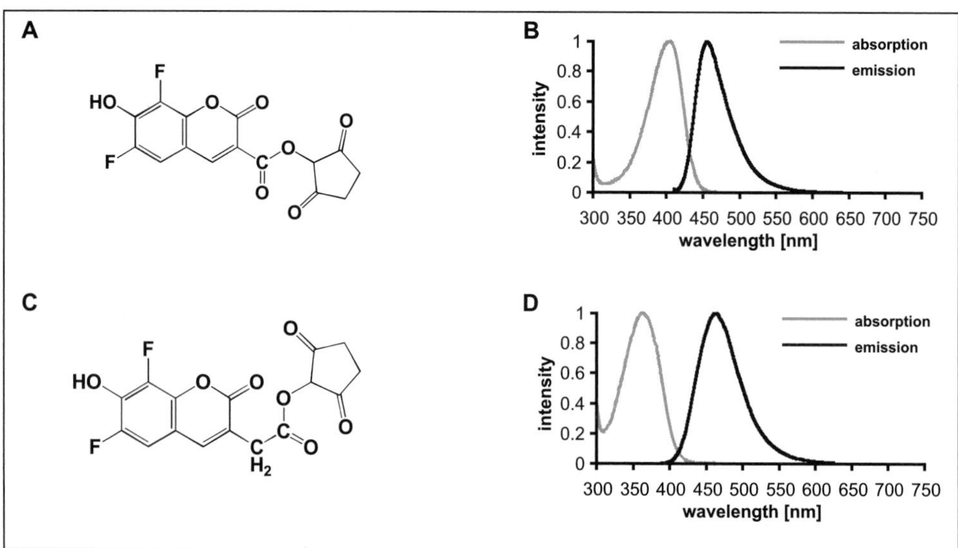

Fig. 14. A Pacific Blue. **B** Absorption and emission spectrum. **C** Marina Blue. **D** Absorption and emission spectrum.

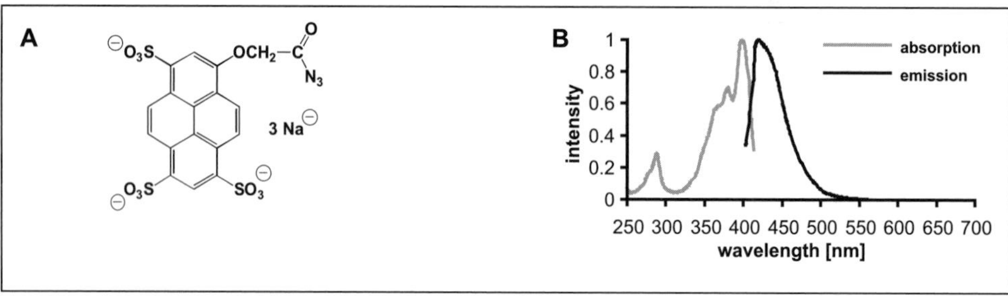

Fig. 15. A Cascade Blue. **B** Absorption and emission spectrum.

Cascade Yellow®

Cascade Yellow is a sulfonated pyridyl oxazol dye with an unusually large Stokes shift, which allows detection outside the autofluorescence seen in most cases. Furthermore, Cascade Yellow is a perfect second dye excited by violet light (fig. 16).

Other Dyes

Cy™ Dyes

Cy fluorochromes are distributed by Amersham Biosciences and belong to the group of cyanine dyes (table 9). Cyanine dyes show high emission intensity, durability in a wide pH range from 3 to 10, weak unspecific binding, photostability, solubility in water and a high extinction coefficient (fig. 17). However, these dyes tend to form

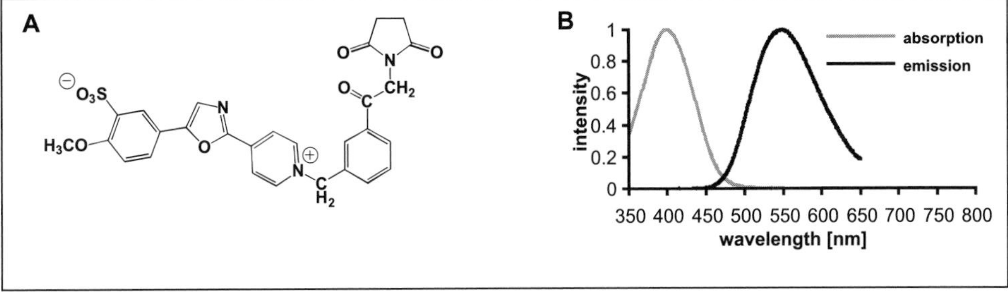

Fig. 16. A Cascade Yellow. **B** Absorption and emission spectrum.

Table 9. Cy™ dyes

Fluorochrome	Molecular weight, Da	Absorption maxima, nm	Emission maxima, nm	Coefficient of extinction $M^{-1}cm^{-1}$
Cy™2Dye	714	489	506	~150,000
Cy™5Dye	792	649	670	250,000
Cy™5.5Dye	1128	675	694	190,000
Cy™7Dye	818	743	767	~200,000

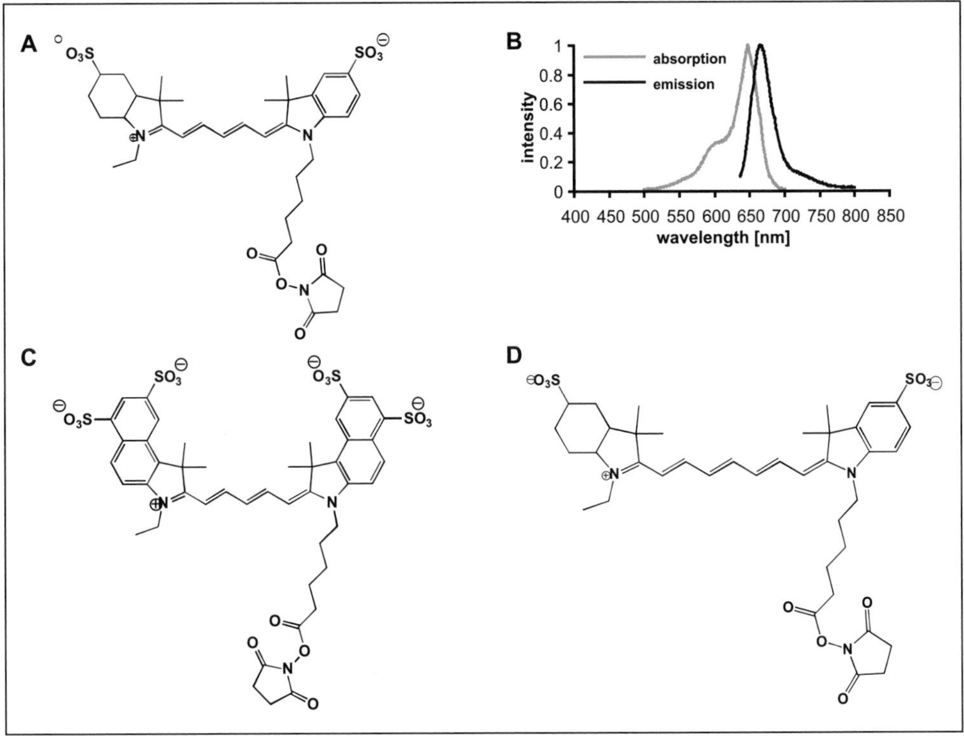

Fig. 17. A Cy5. **B** Absorption and emission spectrum Cy5. **C** Cy5.5. **D** Cy7.

Table 10. Alexa Fluor dyes

Fluorochrome	Molecular weight, Da	Absorption maxima, nm	Emission maxima, nm	Coefficient of extinction M^{-1} cm^{-1}
Alexa Fluor 350	410	346	442	19,000
Alexa Fluor 405	1,028	401	421	34,500
Alexa Fluor 430	702	434	531	16,000
Alexa Fluor 488	643	495	519	71,000
Alexa Fluor 500	700	502	525	71,000
Alexa Fluor 514	714	517	542	80,000
Alexa Fluor 532	721	532	554	81,000
Alexa Fluor 546	1,079	556	573	104,000
Alexa Fluor 555	~1,250	555	565	150,000
Alexa Fluor 568	792	578	603	91,300
Alexa Fluor 594	820	590	617	73,000
Alexa Fluor 610	1,172	612	628	138,000
Alexa Fluor 633	~1,200	632	647	100,000
Alexa Fluor 647	~1,300	650	665	239,000
Alexa Fluor 660	~1,100	663	690	132,000
Alexa Fluor 680	~1,150	679	702	184,000
Alexa Fluor 700	~1,400	702	723	192,000
Alexa Fluor 750	~1,300	749	775	240,000

aggregates when dissolved or while binding to proteins. In part through 'internal quenching', this phenomenon causes a drastic loss of detectable fluorescence (>50%). Therefore, most cyanine conjugates are used in a fluorochrome to protein ratio of ≤4. Cyanine dyes offer a wide range of applications. Because of the convenient procedures for antibody conjugation, they are good alternatives to the following dyes:

- Cy2 alternative to FITC,
- Cy5 alternative to APC.

These dyes are also suitable for FRET analysis (see following chapter).

Alexa Fluor™ Dyes

Alexa Fluor dyes (table 10) show high photostability, low background staining, high water solubility, very high FRET efficiency and good fluorescence over a pH range of 4–10. Furthermore, they provide 'brighter' fluorescence compared to other dyes (fig. 18). Labeling kits are available from Molecular Probes e.g. there are amine-reactive succinimidyl-ester and maleimide conjugation kits. Also available from Molecular Probes are Alexa Fluor bioconjugates (R-PE and APC tandem conjugates, Zenon labeling kits, and several others).

Alexa Fluor dyes also can be worthwhile alternatives to other dyes such as:

- Alexa Fluor 488, 500 or 514 as alternatives to FITC,
- Alexa Fluor 568 or 594 as alternatives to TexasRed,
- Alexa Fluor 633, 635 or 647 as alternatives to APC or Cy5.

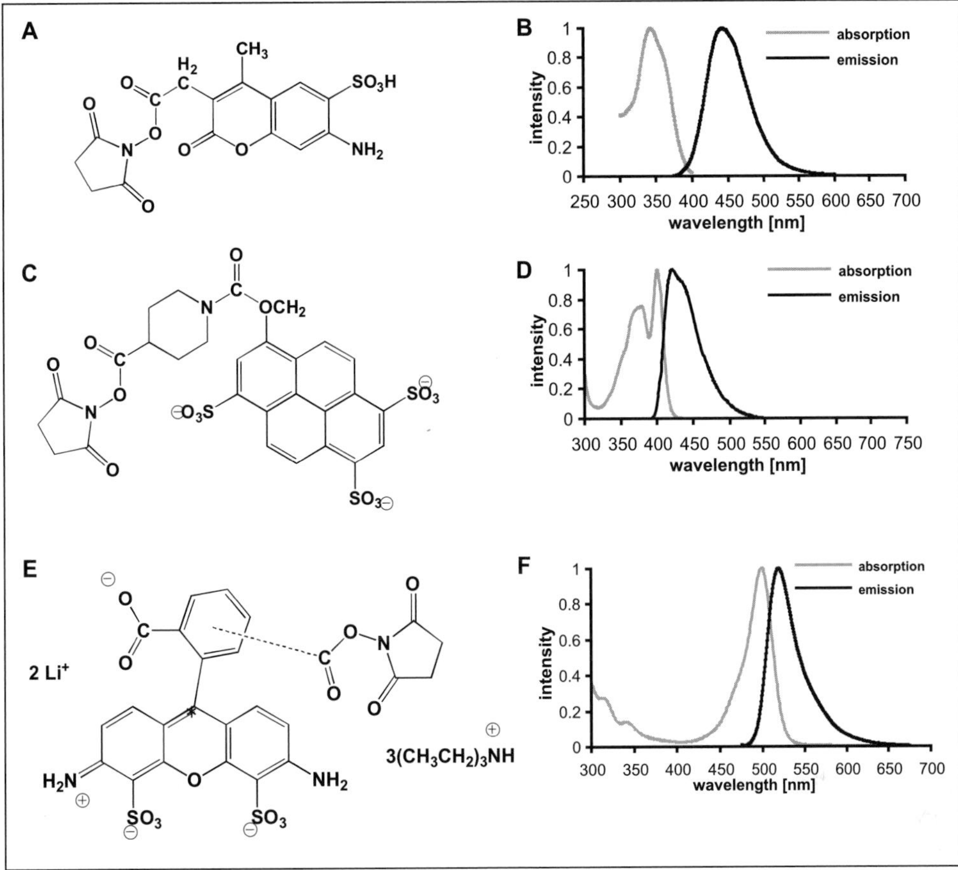

Fig. 18. A Alexa Fluor 350. **B** Absorption and emission spectrum Alexa Fluor 350. **C** Alexa Fluor 405. **D** Absorption and emission spectrum Alexa Fluor 405. **E** Alexa Fluor 488. **F** Absorption and emission spectrum Alexa Fluor 488.

Table 11. Oyster dyes

Fluorochrome	Molecular weight, Da	Excitation nm	Emission nm	Coefficient of extinction (M^{-1} cm^{-1})
Oyster-556	850	556	570	155,000
Oyster-645	1,000	645	666	250,000
Oyster-665	900	656	674	220,000

Oyster® Dyes

Oyster dyes belong to a new class of cyanine dyes patented by Denovo Biolabels (table 11). Oyster dyes combine excellent quantum efficiency and high photostability with low absorption and a low tendency to form aggregates. The astonishing brightness of these dyes originates from the high dye to protein ratio which reaches values of 10 and more. Oyster dyes are commercially available as activated succinimidylester for antibody conjugation (fig. 19).

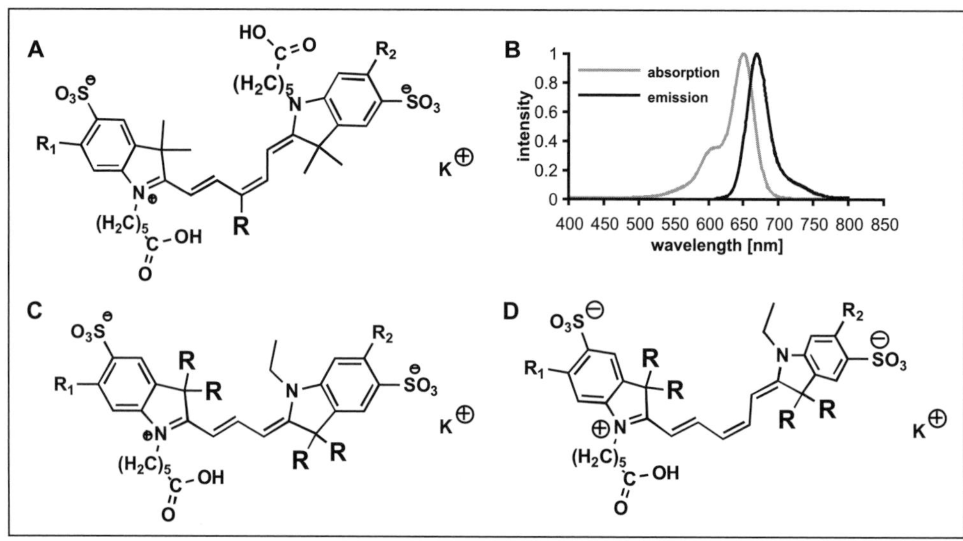

Fig. 19. A Oyster 645. **B** Absorption and emission spectrum Oyster 645. **C** Oyster 556. **D** Oyster 665.

Table 12. ATTO dyes

Fluorochrome	Molecular weight, Da	Excitation nm	Emission nm	Coefficient of extinction (M^{-1} cm^{-1})
ATTO 425	498	436	484	45,000
ATTO 495	549	495	527	80,000
ATTO 520	564	525	545	110,000
ATTO 532	743	532	553	115,000
ATTO 550	791	554	576	120,000
ATTO 565	708	563	592	120,000
ATTO 590	788	594	624	120,000
ATTO 610	588	615	634	150,000
ATTO 620	709	619	643	120,000
ATTO 635	628	635	659	120,000
ATTO 647	690	645	669	120,000
ATTO 655	625	663	684	125,000
ATTO 680	623	680	700	125,000

ATTO Dyes

ATTO dyes are patented ATTO-TEC products (table 12, fig. 20). These dyes are fluorescent chromophores which have been developed for labeling of many different bio-molecules. Their major characteristics are: strong light absorption, high quantum efficiency, high photostability, good solubility in water, and little triplet allocation. Several of the available ATTO chromophores have been specifically developed for excitation and emission in the red light spectrum. By using them, trouble with autofluorescence of biological samples can be efficiently avoided. For conjugation,

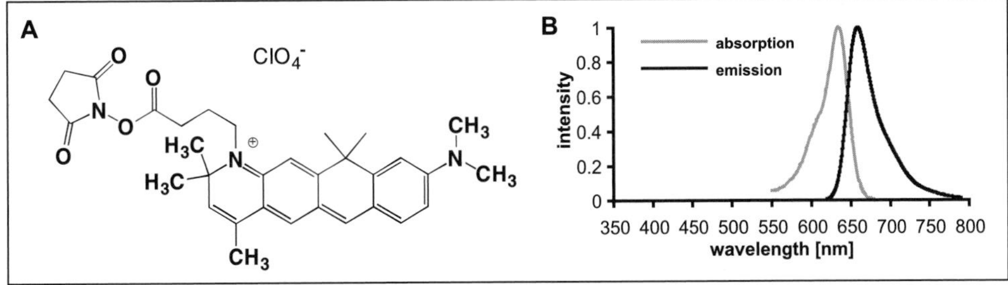

Fig. 20. A ATTO 635. **B** Absorption and emission spectrum ATTO 635.

Table 13. Dy dyes

Fluorochrome	Molecular weight, Da	Excitation nm	Emission nm	Coefficient of extinction (M^{-1} cm^{-1})
Dy 555	536.18	555	580	100,000
Dy 615	578.73	621	641	200,000
Dy 631	736.88	637	658	185,000
Dy 636	760.91	645	671	190,000
Dy 647	664.78	652	673	250,000
Dy 651	788.96	653	678	160,000
Dy 676	807.95	674	699	145,000
Dy 681	736.88	691	708	125,000
Dy 701	770.90	706	731	115,000
Dy 731	762.92	736	759	225,000
Dy 751	814.99	751	779	220,000
Dy 781	762.92	783	800	100,000

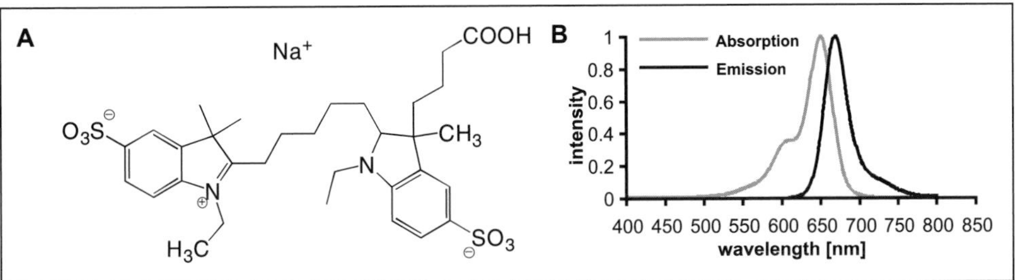

Fig. 21. A Dy 647. **B** Absorption and emission spectrum Dy 647.

ATTO dyes are available with a free carboxyl group or as an NHS ester. ATTO dyes can also be used for FRET analysis.

Dy Dyes

Dyomics has patented the 'Dy' dyes (table 13, fig. 21). All dyes from this company are offered as NHS esters and therefore can easily be covalently conjugated to antibodies. Dy dyes are characterized by high light absorption, high quantum

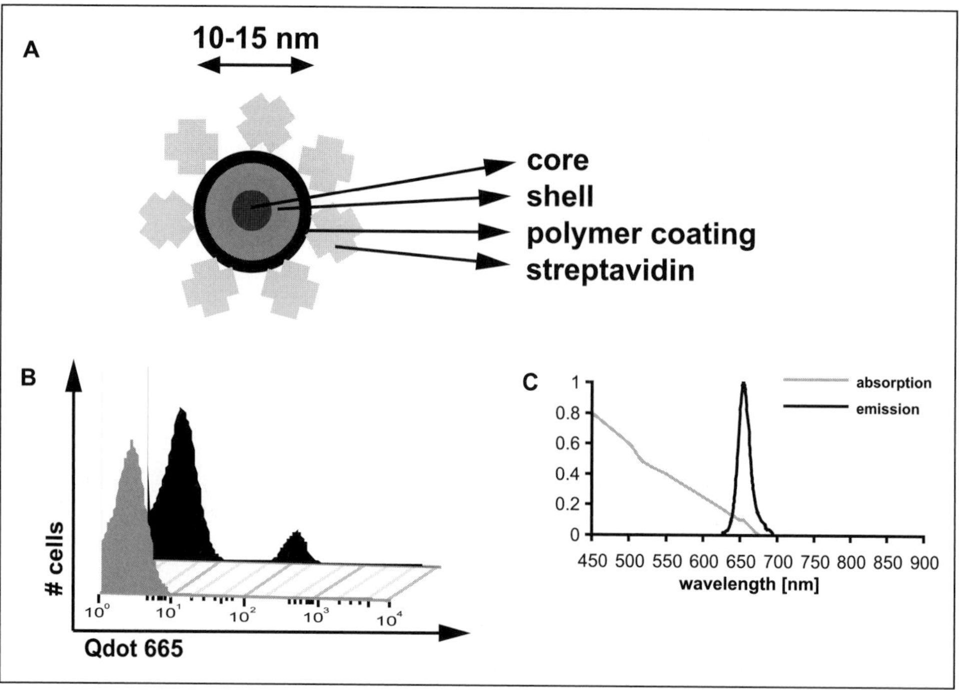

Fig. 22. A Structure of nanoparticles. **B** Spleen cells (mouse) primary stained with CD8α bio (clone 53–6.7, BD), secondarily stained with SA QDot 665 – excited at 351 nm (60 mW) (black); gray = unstained control. C Absorption and emission spectrum QDot 665.

efficiency, high photostability, good solubility in water and a low tendency to form aggregates. Dy dyes can also be used for FRET analysis or as substitutes for other dyes such as:

- Dy-636 as alternative to Cy5,
- Dy-647 as alternative to Cy5,
- Dy-676 as alternative to Cy5.5,
- Dy-751 as alternative to Cy7.

Nano 'Dyes' (QDots™)

QDots consist of semiconductor material (CdSe) of nanometer size with a special coating. The material shows optical characteristics that depend on the size of the nanobody and on the wavelength of the excitation light source which creates a symmetrical emission spectrum, differentiating these fluorochromes from all previously mentioned dyes (fig. 22). Normally, the Stokes shift of fluorescent dyes is relatively low. However, QDots have an absorption which dramatically increases with decreasing wavelength of the light source. An additional coating allows direct conjugation with biological molecules (e.g. through streptavidin).

Other Dyes
Other dyes have been established for cell cycle analysis and functional measurements in flow cytometry (Hoechst, DAPI, Chromomycin A_3, Fluo-3, Indo-1, Rhodamine 123, CFSE and more), which cannot be covered in detail in this chapter.

Live/Dead Dyes

Live/dead discrimination is essential, especially in multiparameter (>4 colors) flow cytometry. Different staining techniques are available, and the optimal choice mainly depends on the specific application. An important aspect is whether the samples have to be fixed during or after staining. Most of the dyes used for live/dead discrimination are small fat-soluble molecules carrying positive charges at physiological pH. In dead and apoptotic cells, the disturbed membrane integrity and increased fat solubility make it easier for the dye to pass the two-layer cell wall (bilayer phase) and enter the cell. Positively charged dyes bind to negatively charged components of the cell, as for example glucosaminoglycans and nucleic acids (table 14). Live cells have an electrical gradient which opposes the entry and binding of these dyes.

Without Cell Fixation

Propidium Iodide
3.8-diamino-5-[3-(diethylmethylammonio)propyl]-6-phenylphenanthridinium diiodide (fig. 23).

Stock solution: usually 2 mg/ml in PBS.

Working solution: 4 µg/ml.

Note: propidium iodide (PI) is mutagenic; always wear nitrile gloves when working with PI, and carefully read the product information provided by the manufacturer.

PI is usually added shortly before sample acquisition. PI has a wide emission spectrum supporting its detection mostly in the PE, PETR and PE-Cy5 channels. If available, PI is also readily excited by a UV laser. In this case, PI can be handled as a 'real' color independent of other laser excitations. The specific advantage of this set-up is that live/dead discrimination can be performed without any disturbing signals in the PE and PETR/PE-Cy5 channels.

To-Pro®-3 Iodide
Quinolinium-4-[3-(3-methyl-2(3H)-benzothiazolylidene)-1-propenyl]-1-[3-(trimethylammonio)propyl]-diiodide (fig. 24).

Table 14. DNA Dyes

DNA dye	Molecular weight, Da	Excitation nm	Emission nm	Coefficient of extinction (M^{-1} cm^{-1})
PI	668.4	535	617	5,400
EMA	420.3	462	625	4,000
7-AAD	1270.45	546	647	25,000
To-Pro-3	671.42	642	661	102,000

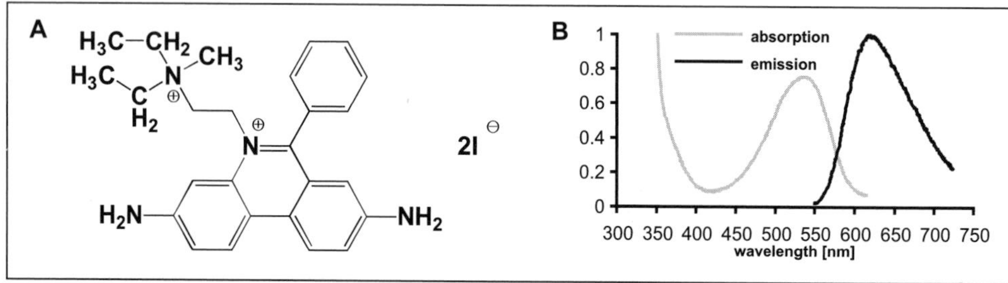

Fig. 23. A Propidium iodide. **B** Absorption and emission spectrum.

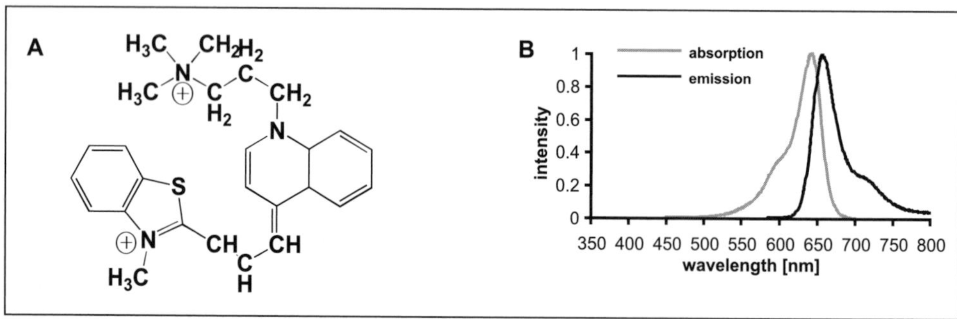

Fig. 24. A To-Pro-3. **B** Absorption and emission spectrum.

Stock solution: usually 1 mmol/l in dimethyl sulfoxide (DMSO).

Working solution: 10 μmol/l.

Note: According to present knowledge, To-Pro3 itself is not poisonous; however, toxicity issues of the solvent (DMSO) must be considered.

This dye is well detected in the APC channel since it corresponds to the APC emission spectrum. To-Pro3 can be used together with annexin (V) on FITC for analysis of apoptosis; in this case, no compensation is needed between To-Pro3 and annexin (V).

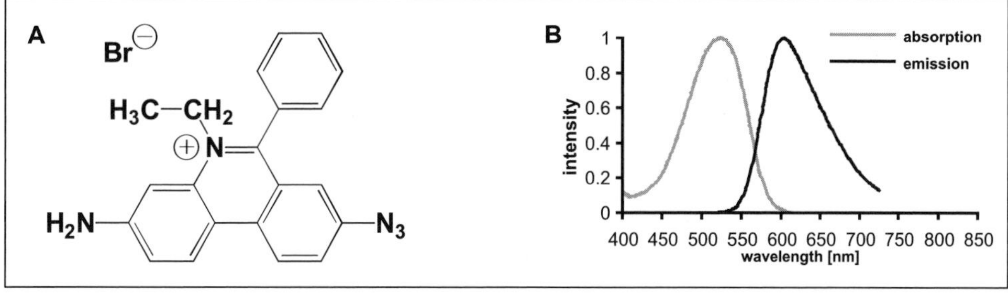

Fig. 25. A Ethidium monoazide. **B** Absorption and emission spectrum.

With Cell Fixation

Ethidium Monoazide Bromide

Phenanthridium-3-amino-8-azido-5-ethyl-6-phenyl bromide (fig. 25).

Stock solution: usually 2 mg/ml in ethanol.

Working solution: 2 μg/ml.

Note: Ethidium monoazide (EMA) might be carcinogenic, wear nitrile gloves, and follow the specific recommendations provided by the manufacturer!

The unbound fluorochrome emits only very weakly. After binding to DNA, however, the intensity of emission light increases 15 times. Like PI, EMA has a wide emission spectrum and is detected in the PE and PETR/PE-Cy5 channels of a common flow cytometer. EMA is unique by its irreversible binding to DNA through 'photocross-linking'. It is the ideal dye for live/dead discrimination if fixation of the sample is required (e.g. intracellular staining). The cells are usually stained before fixation for 20 min on ice under light exposure (normal 'white' light bulb, 60 mW, approximately 20 cm distance from the sample). Subsequently, the cells are washed and can then be fixed and further stained. EMA staining is extremely stable for several days.

7-Aminoactinomycin D

Stock solution: 1 mg/ml in (50 μl absolute methanol/950 μl PBS).

Working solution: 10 μg/ml.

Note: 7-Aminoactinomycin (7-AAD) is potentially mutagenic and possibly carcinogenic. Wear gloves, and follow the recommendations provided by the manufacturer.

Usually, this dye is detected in the PETR/PE-Cy5 channels. 7-AAD can be well compensated against other dyes (fig. 26).

7-AAD is often used in combination with annexin (V) on FITC for analyzing apoptosis as no compensation between To-Pro3 and annexin (V) is needed.

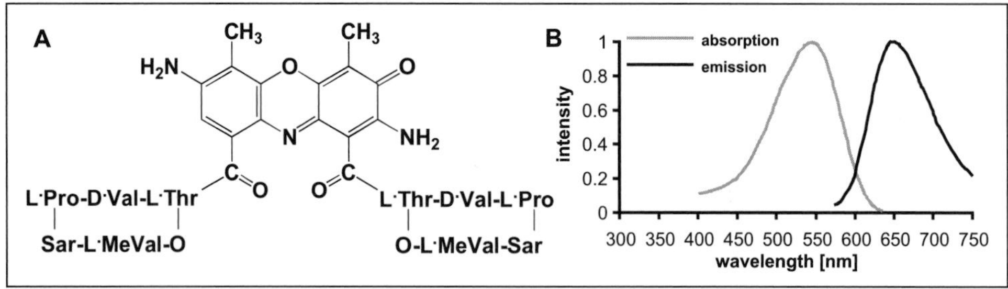

Fig. 26. A 7-Amino-actinomycin. **B** Absorption and emission spectrum.

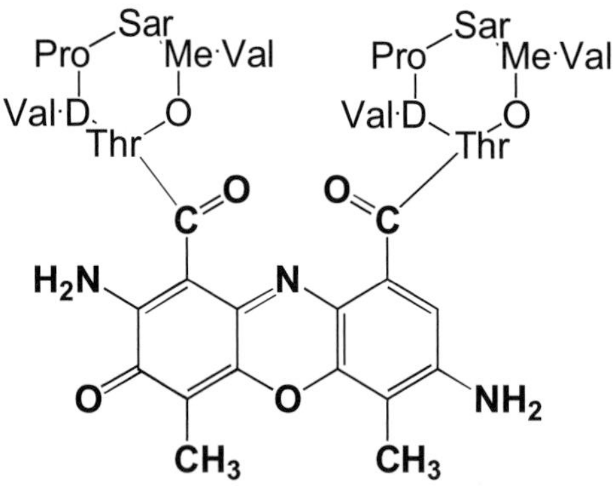

Fig. 27. Actinomycin D.

Actinomycin D
2-Amino-4.6-dimethyl-3-oxo-N,N′-bis[7.11.14-trimethyl-2.5.9.12.15-pentaoxo-3.10-di(propan-2-yl)-8-oxa-1.4.11.14-tetrazabicyclo[14.3.0]nonadecan-6-yl]phen-oxazine-1.9-dicarboxamide (fig. 27).

Samples stained with 7-AAD can be fixed with PFA, but only if PFA contains actinomycin D (AD) at a concentration of 5 µl/ml. AD stably occupies 7-AAD binding sites and prevents a virtual increase of dead cells during cell storage of fixed samples.

Stock solution: 1 mg/ml in 50 ml absolute ethanol/950 ml PBS.

Fixing solution: 5 µg/ml PFA, 5 µl/ml AD.

For this live/dead discrimination method, cells are first stained with 7-AAD for 20 min and washed as usual; at the end, cells are resuspended in the fixing solution. Live/dead staining stays stable for up to 3 days.

In-House Conjugation of Reagents

The most common methods to conjugate biological molecules with fluorescent dyes use primary amines, which are commonly found on peptides, oligonucleotides or proteins. The underlying chemical reaction is summarized in figure 28.

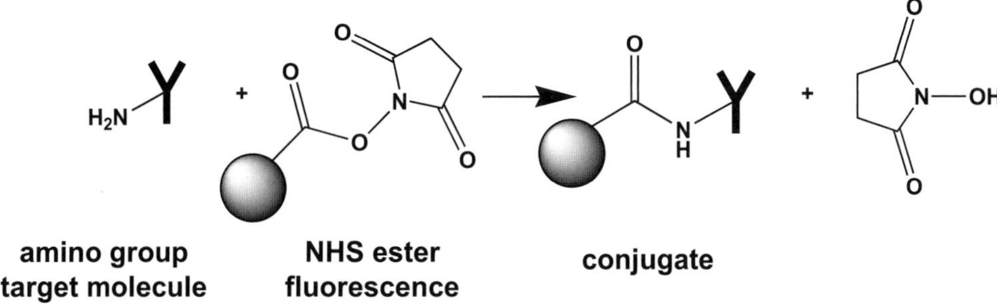

Fig. 28. N-Hydroxysuccinimide (NHS) ester for the conjugation of amine groups.

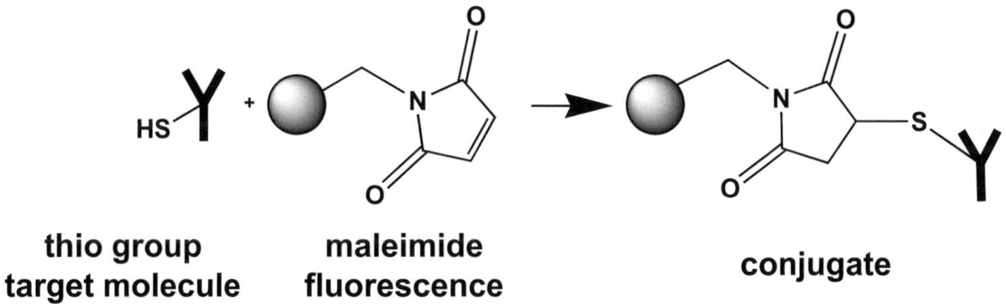

Fig. 29. Maleimide for the conjugation of thioles.

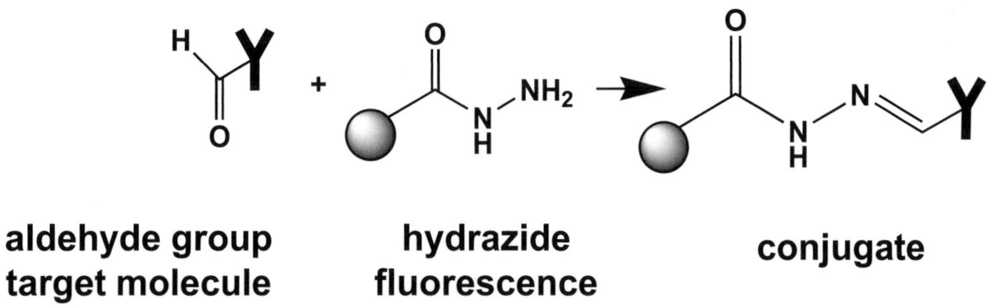

Fig. 30. Hydrazide for the conjugation of aldehydes.

Thioles can be used for the conjugation instead of primary amines. The principal mechanism of this reaction is summarized in figure 29.

If neither primary amines nor thioles are present on the target molecule, aldehydes can be targeted for conjugation (fig. 30).

Usually, it is relatively easy to find setups for staining with up to 4 combined fluorochromes. Frequently used antibodies are commercially available in several conjugations; in addition, secondary staining procedures can be used for individual antibodies to achieve the detection fluorescence. However, with increasing numbers

Table 15. Selection of commercially available conjugation kits

Company	Dye	Internet address
Ambion	Cy3, Cy5, DIG	*www.ambion.com*
Amersham	CyDye	*www.amersham.com*
Atto-Tec GmbH	ATTO dyes	*www.atto-tec.com*
BD Bioscience, Clontech	biotin	*www.clontech.com*
BioCat	Dy dyes, biotin	*www.biocat.com*
Dyomics	Dy dyes, EVO	*www.dyomics.com*
emp Biotech GmbH	Cy-, Dy- and ATTO dyes	*www.empbiotech.com*
MoBiTec	FITC, DIG, DNP, Cy3, Cy5, Pacific Blue, Alexa	*www.mobitec.com*
Invitrogen	diverse conjugations kits	*www.invitrogen.com*
Prozyme	APC	*www.prozyme.com*
Qbiogen	biotin, FITC	*www.qbiogene.com*
Roche	FITC, DIG, biotin	*www.roche.com*
Sigma-Aldrich	psoralen derivative	*www.sigma-aldrich.com*
Stratagene	diverse dyes	*www.stratagene.com*

of combined parameters it will become more and more difficult to find commercially available fluorescence conjugates for the entire panel.

Several companies offer conjugation kits which strongly facilitate work. A short selection is provided in table 15.

Principles for the Selection of Fluorescent Dyes

The properties of fluorescent dyes used for flow-cytometric analysis differ from those used e.g. for fluorescence microscopy. This is mainly due to the very short excitation and detection times available for each cell passing through the flow chamber.

In principle, an ideal fluorescent dye for flow-cytometric applications must fulfill the following requirements:

The dye should

- be excitable by lasers commonly mounted on flow cytometers,
- emit light within a very short time frame upon excitation,
- be well detectable by the optical layout of common flow cytometers,
- create little overlap with the emission spectra of other dyes,
- be biologically inert,
- bind stably to the carrier (e.g. antibody),
- produce minimal unspecific background staining,
- give ‘bright’ detection signals,
- be stable and – depending on the application – enable fixation,
- be commercially available.

Preliminary Tests with Fluorescent Dyes

Before using new fluorescent dyes, preliminary tests should be performed with cell material comparable to the cell material that will later be used in the anticipated experiment or application. These first tests should also include titration of the reagent conjugated with the fluorescent dye. On the one hand, titration experiments are essential to determine the optimal conditions to achieve the highest staining quality (too high concentrations of the staining probe can substantially increase background staining); on the other hand, it may be an important factor for cost reduction (many reagents can be used with similar or even better staining performance at concentrations significantly below the manufacturer's recommendation). New titration experiments should also be performed whenever a new batch/lot of a staining probe is implemented in an established multicolor panel. In addition, for quality control it should be remembered that some fluorescent probes have only limited stability and can lose activity over time (especially tandem conjugates).

Choosing Antibody/Dye Combinations

At least when more than two fluorescent dye conjugates are needed in one staining, the question comes up 'Which is the best combination?' Often many combinations are 'theoretically' possible, making the decision difficult. Based on the topics discussed in the previous sections of this chapter, the following questions need to be answered:

1. What excitation lines are available on my instrument?
   The table of the available lasers with their wavelengths might be helpful for this selection.
   → The answer should tell the user which fluorescent dyes he can use.
2. What laser powers does my instrument provide?
   → Some dyes are sensitive to high laser power (e.g. PerCP), others require high laser power to be efficiently excited (UV dyes).
3 Does my instrument offer inter-laser compensation?
   → Many dyes emit in similar wavelength bands. Without this tool, direct compensation might not be possible.
4. What kind of detection functions is available?
   → The optical layout gives information on the available filter configuration and determines if the PMTs are suitable to detect the required wavelength ranges. If necessary, the optical layout needs to be changed (filter adoption).

After having screened the fluorescent dyes detectable with the available hardware, further selections have to be done.

In the previous sections we summarized specific characteristics of commonly used dyes. All these dyes can substantially differ in fluorescence intensity (brightness). The commercial availability will also influence the choice of combination of fluorescence-labeled staining probes.

As a general guidance, the following example ranks the fluorescence intensities on a molar basis for some dyes measured with a CyAn MLE (Beckman Coulter, 40 mW MLUV, 180 mW, 488 nm and red diode 15 mW):

PE, PE-Cy5, APC > PE-Cy7, FITC, PE-Cy5.5, PETR, APC-Alexa750 ≫ Alexa350

The signal to noise (S/N) ratio can differ very much between antibodies. In addition, the S/N ratio is also dependent on the utilized flow cytometer.

The capability to detect and differentiate the specifically stained population by a staining probe is determined by the expression intensity of the target antigen and the selected dye. Furthermore, the number of fluorochromes per antibody (F/P ratio) can influence the brightness (see detailed descriptions for each dye). In many cases, optimization requires home-made conjugation of antibodies, which has been facilitated by the availability of easy-to-use conjugation kits.

In principle, most fluorescence dyes can be used for a highly expressed antigen. However, the detection of weakly expressed antigens, which requires a high S/N ratio (e.g. PE or APC) to separate stained from unstained cells, is more challenging.

Amplification Techniques

As the 'per cell expression' of an antigen of interest is often low, the specific fluorescence can hardly be distinguished from background signal. Since antibodies can only be coupled with a limited number of dye molecules, the total signal intensity from directly fluorochrome-conjugated reagents might be too low in some cases to detect antigen-expressing populations. The following techniques can be used to amplify the detection of weakly expressed molecules:

- indirect immunofluorescence techniques (e.g. biotin, DIG, dinitrophenol),
- coupling of the staining probe to large fluorescence structures (e.g. dye-loaded liposomes [7]),
- nanoparticles.

Indirect Immunofluorescence

Among other purposes, indirect immunofluorescence techniques can be used for signal amplification (see 'Direct and Indirect Immunofluorescence', p. 109). The use of isotype-specific antibodies for indirect immunofluorescence can already result in some signal amplification compared to directly conjugated primary staining reagents; however, this approach is often limited in multicolor flow cytometry because of the high chance that at least some primary antibodies will be of the same isotype. Therefore, antibody-conjugates which can be amplified independent of the isotype are used more and more frequently. We have already discussed the option to biotinylate the primary antibody in order to use streptavidin-fluorescence conjugates for visualization; especially with PE- or APC conjugates, this strategy can go along with signal amplification. Alternatively, primary staining probes can be conjugated with digoxigenin (DIG) or dinitrophenol (DNP), and subsequently detected with fluorescence-conjugated secondary reagents (monoclonal anti-DIG or anti-DNP antibodies).

Furthermore, amplification techniques have been developed which utilize anti-fluorescence antibodies (e.g. anti-FITC) coupled to fluorescence with an absorption and emission spectrum similar to the first fluorochrome (e.g. Alexa 488, Alexa Fluor 633, Oyster 645, ATTO 635, Dy 647).

Liposomes

Liposomes consist of amphipathic phospholipids which spontaneously form vesicles surrounded by a double-layer membrane (liposomes). A large number of fluorescence molecules can be trapped inside the hydrophobic lipid layer as well as inside liposomes. Antibodies can be conjugated with the surface of liposomes to generate antigen-targeted staining probes with very high fluorochrome concentrations. Compared to fluorescence-labeled antibodies, liposomes can provide up to 1,000-fold higher fluorescence signals. Less than 100 expressed molecules per cell can be detected with this highly sensitive reagent [7].

Background Signal

Many reagent-specific aspects can cause 'background staining'. Therefore, in the following we will only discuss some more general aspects regarding hardware, autofluorescence and Fc receptor binding.

Amplifier Noise
The sensitivity of PMTs is optimal for certain wavelengths. If the amplification is increased, the noise gets amplified as well.

Autofluorescence
Some molecules, for example riboflavin, cause cell autofluorescence. With increasing cell size (e.g. activated cells), this autofluorescence can further increase just by an enlarged cytosolic compartment. Monocytes, granulocytes and dendritic cells are

known to show relatively high autofluorescence as compared to lymphocytes. After fixation of cells (especially with paraformaldehyde; PFA), their autofluorescence can get further amplified (this phenomenon is even utilized to selectively identify eosinophilic granulocytes).

Autofluorescence can be reduced by a variety of techniques, for example through additional staining with trypan blue [8] or special detection techniques [9].

Unspecific Binding

Antigen-unspecific binding of antibodies can occur for example by Fc receptor binding. This can be prevented by preincubation with Fc-receptor-blocking antibodies (anti-CD16/CD32; 'Fc block') or by preincubation with autologous serum. Binding of antibodies among each other can also cause unspecific binding. Dyes containing carbocyanine (Cy3, Cy5 and Cy5.5), TexasRed and other tandem dyes (e.g. PE-Cy5) sometimes tend to show unspecific binding with a preference for certain cell populations. It could be shown especially for Cy5 that background binding to low-affinity Fc receptors causes staining.

Absorption

If protein-free buffers are used to wash cells during the staining procedure, especially antibodies with lower affinity tend to increase in background staining. This can be prevented by using protein-containing wash buffers (e.g. 0.5% BSA).

Noise Dependent on Dyes

Especially when using home-made conjugates, too high fluorochrome to protein ratios (normally around 3–6) can be the reason for unspecific binding. Fluorochromes remaining unbound can cause unspecific labeling as well (see Cy5-conjugates and PerCP). In this case, further purification steps should eliminate unbound fluorochromes.

Vitality of Cells

Dead cells can take up unspecific antibodies because of the loss of the membrane barrier. Therefore, parallel live/dead discrimination should be performed whenever possible. In addition, working at temperatures above 4 °C can cause rapid internalization of fluorescence-labeled molecules (e.g. MHC multimer staining [10]).

Fluorochrome Proximity

Especially with multiparameter staining, the proximity of antigens can generate unintentional energy transfer phenomena (e.g. PE emission to Cy5). Similarly, steric hindrance of antigens can cause staining interferences. For example, simultaneous detection of antigen-specific T cells with MHC multimers together with staining for CD8α or for TCR-Vβ segments can result in decreased MHC multimer staining.

Contamination

Careful cleaning of flow cytometers is a prerequisite for good performance. Cell deposits, cell fragments and remaining dyes (e.g. PI or Indo) within the tubings or inside the flow cell can create enormous spillover problems. Regular cleaning (sodium hypochlorite/water) can prevent such phenomena.

Stability of Reagents

If reagents are used beyond the expiration date given by the manufacturer, further use of the conjugates needs quality controls (e.g. titration experiments). Some reagents do not even reach this anticipated 'life span', which is mostly due to improper handling (for example through bleaching or exposure to high or low temperatures). Many dyes are degraded by bacterial contamination (if possible, addition of sodium azide can reduce this problem). Some dyes are intrinsically unstable (e.g. some tandem conjugates) and the binding between fluorochromes falls apart during storage.

pH

The fluorescence intensity of some dyes is strongly pH dependent (e.g. FITC).

Biological Durability

Most of the fluorescent dyes used for conjugation of antibodies nowadays should be biologically inert, meaning they should not bind to cellular components, they should not influence the biological activity and they should not be toxic.

Summary

Flow cytometry has developed very rapidly over the last 50 years. Starting from the first attempts to detect blood cells in liquid media, many markers and fluorescent dyes can now be detected simultaneously on individual cells.

This chapter provides an overview on the most important fluorescent dyes used in multi-parameter flow cytometry. Characteristics of different dyes, advantages for specific applications, and aspects regarding the selection of optimal fluorochrome combinations are discussed.

Interesting Web Sites and Literature

www.biochem.mpg.de/valet/cytorel.html.

www.cyto.purdue.edu.

The German Society of Immunology provides additional information under *www.immunologie.de.*

Recommended Literature

Darzynkiewicz Z, Crissman HA, Robinson JP: Cytometry, ed 3, part B. Methods in Cell Biology. San Diego, Academic Press, 2000, vol 64.

Stewart CC, Nicholson JKA (eds): Immunophenotyping. New York, Wiley-Liss, 2000.

Radbruch A (ed): Flow Cytometry and Cell Sorting (Springer Lab Manual). Heidelberg, Springer, 2000.

Ormerod MG: Flow Cytometry: A Practical Approach, ed 3. Oxford, Oxford University Press, 2000.

Robinson JP, Darzynkiewicz Z (eds): Current Protocols in Cytometry. New York, Wiley, 1997.

Shapiro HM: Practical Flow Cytometry, ed 4. New York, Wiley Liss, 2003.

George McNamara, Congressman Julian Dixon Cellular Imaging Facility (Image Core) of The Saban Research Institute of Childrens Hospital Los Angeles (CHLA), Los Angeles. ,http://home.earthlink.net/~mpmicro/mpmicro.doc

Longobardi Givan A: Flow Cytometry: First Principles. New York, Wiley-Liss, 2001.

References

1 Kohler G, Milstein C: Continuous cultures of fused cells secreting antibody of predefined specificity. Nature 1975;256:495–497.

2 Roederer M: Compensation is not dependent on signal intensity or on number of parameters. Cytometry 2001;46:357–359.

3 Roederer M: Spectral compensation for flow cytometry: visualization artifacts, limitations, and caveats. Cytometry 2001;45:194–205.

4 Zhang YZ, Kemper C, Bakke A, Haugland RP: Novel flow cytometry compensation standards: internally stained fluorescent microspheres with matched emission spectra and long-term stability. Cytometry 1998;33:244–248.

5 Schiemann M, Busch V, Linkemann K, Huster KM, Busch DH: Differences in maintenance of CD8+ and CD4+ bacteria-specific effector-memory T cell populations. Eur J Immunol 2003;33:2875–2785.

6 Roederer M, Kantor AB, Parks DR, Herzenberg LA: Cy7PE and Cy7APC: bright new probes for immunofluorescence. Cytometry 1996;24:191–197.

7 Scheffold A, Assenmacher M, Reiners-Schramm L, Lauster R, Radbruch A: High-sensitivity immunofluorescence for detection of the pro- and anti-inflammatory cytokines gamma interferon and interleukin-10 on the surface of cytokine-secreting cells. Nat Med 2000;6:107–110.

8 Mosiman VL, Patterson BK, Canterero L, Goolsby CL: Reducing cellular autofluorescence in flow cytometry: an in situ method. Cytometry 1997;30:151–156.

9 Alberti S, Parks DR, Herzenberg LA: A single laser method for subtraction of cell autofluorescence in flow cytometry. Cytometry 1987;8:114–119.

10 Knabel M, Franz TJ, Schiemann M, Wulf A, Villmow B, Schmidt B, et al: Reversible MHC multimer staining for functional isolation of T-cell populations and effective adoptive transfer. Nat Med 2002;8:631–637.

Sack U, Tárnok A, Rothe G (eds): Cellular Diagnostics. Basics, Methods and Clinical Applications of Flow Cytometry. Basel, Karger, 2009, pp 141–158

Fluorescence Resonance Energy Transfer (FRET)

György Vámosi[a,c] · György Vereb[b,c] · Andrea Bodnár[a,c] · Katalin Tóth[c] · Nina Baudendistel[c,d] · Sándor Damjanovich[a,c] · János Szöllősi[a,b,c]

[a] Cell Biology and Signaling Research Group of the Hungarian Academy of Sciences,
[b] Department of Biophysics and Cell Biology,
[c] Research Center for Molecular Medicine recognized as Center of Excellence by the European Commission Medical and Health Science Center, University of Debrecen, Hungary
[d] Division Biophysics of Macromolecules, German Cancer Research Center, Heidelberg, Germany

Introduction

Fluorescence techniques are widely used to quantify molecular parameters of various biochemical and biological processes in vivo because of their inherent sensitivity, specificity and temporal resolution. FRET is a phenomenon in fluorescence spectroscopy during which energy is transferred from an excited donor molecule to an acceptor molecule under favorable spectral, proximity and orientation conditions. For reviews see [1–10].

FRET processes during fluorescence measurements in flow and image cytometry can either compromise results or open new applications for these techniques. Some applications of FRET are summarized in table 1. When we apply multiple fluorescent probes simultaneously, FRET processes should be controlled if we want to quantify the cell surface expression of various antigens at the same time. In contrast to this adverse effect, FRET can also be used to improve the spectral characteristics of fluorescent dyes and dye combinations such as tandem dyes in flow and image cytometry, and FRET primers in DNA sequencing and the polymerase chain reaction. The incentive to apply this technique is that it enables the use of single wavelength excitation while providing various dye combinations with a wide range of Stokes shifts that permit simultaneous detection of three or four fluorescent dyes.

Combination of FRET with monoclonal antibodies facilitates structural analysis of proteins in solutions and in intact cells. Applying donor- or acceptor-labeled anti-

Table 1. Some biological applications of FRET

Measurement of inter- and intramolecular distances
Association, interaction of proteins in cells
Nucleic acid structure
Protein structure
Conformational changes of macromolecules
DNA sequencing
Quantitative PCR

bodies, lipids and various types of fluorescent proteins (such as green fluorescent protein (GFP) and its spectral variants), the FRET technique can be used to determine inter- and intramolecular distances of cell surface components in biological membranes and other cellular compartments [1–3, 6–8, 11, 12]. FRET allows the determination of molecular distances in live cells, thus providing information which would be impossible to obtain with other classical approaches, e.g. with electron-microscopic methods.

In this chapter we describe the basic concepts of FRET, delineate measurement and evaluation techniques, characterize available parameters and instruments, discuss limitations of FRET and show a few applications.

Basic Principles of FRET

FRET is a physical process in which energy is transferred from an excited donor molecule to an acceptor molecule by means of dipole-dipole coupling. One of the most important factors influencing the strength of coupling is the distance between the donor and acceptor dyes. FRET can occur with measurable efficiency over the 1- to 10-nm distance range, which coincides with the dimensions of biological macromolecules [13]. FRET is nonradiative; i.e. the donor does not actually emit and the acceptor does not absorb a photon.

In order to explain the mechanism of FRET, let us consider a system with two different fluorophores where the molecule with higher energy absorption is defined as the donor (D) and the one with lower energy absorption as the acceptor (A). If the donor gets into the excited state, it will lose energy by internal conversion without emission until it reaches the lowest vibrational level of the first excited state (Kasha's rule). If the emission spectrum of the donor overlaps with the absorption spectrum of the acceptor, the following resonance can occur through coupling between the emission and absorption transition dipoles of the donor and the acceptor, respectively:

$$D^* + A \Leftrightarrow D + A^* \tag{1}$$

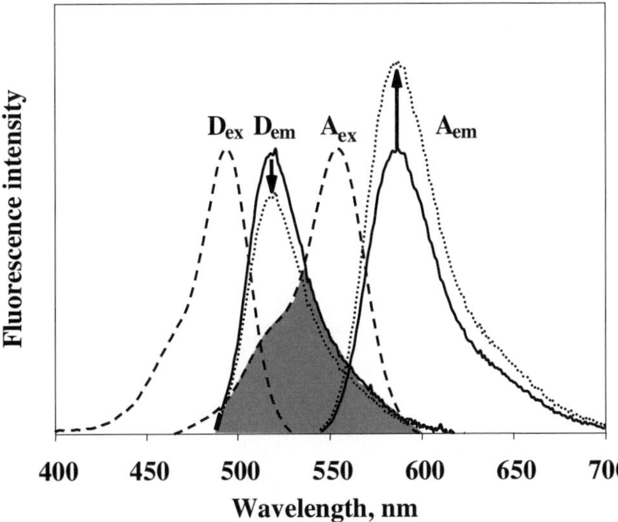

Fig. 1. Fluorescence excitation and emission spectra of an ideal donor-acceptor pair. The shaded area represents the overlap integral (J). The spectra are normalized. ↓ indicates quenching of the donor; ↑ shows the sensitized emission of the acceptor. The amounts of quenching and sensitized emission are exaggerated for illustrative purposes.

where D and A denote the donor and the acceptor molecules in ground state, while D^* and A^* denote the first excited states of the fluorophores. As a result of FRET, the donor molecules become quenched, while the acceptor molecules become excited and, under favorable conditions, can emit fluorescent light. This latter process is called 'sensitized emission' (fig. 1).

The efficiency of FRET (E) is a quantitative measure of the number of quanta that are transferred from the donor to the acceptor and can be expressed as:

$$E = \frac{\text{the number of quanta transferred from the donor to the acceptor}}{\text{the number of quanta absorbed by the donor}} \quad (2)$$

According to Förster's theory, the rate (k_T) and efficiency (E) of energy transfer is:

$$k_T = \text{const } k_F J n^{-4} R^{-6} \kappa^2 \quad (3)$$

$$E = \frac{k_T}{k_T + k_F + k_D} \quad (4)$$

where k_F is the rate constant of fluorescence emission of the donor and k_D is the sum of the rate constants of all other de-excitation processes of the donor. R is the distance between the donor and acceptor molecules, and κ^2 is an orientation factor, which is a function of the relative orientation of the donor's emission dipole and the acceptor's absorption dipole in space. Other parameters are n, the refractive index of the medium, and J, the spectral overlap integral, which is proportional to the overlap in the emission spectrum of the donor and the absorption spectrum of the acceptor:

$$J = \frac{\int F_D(\lambda)\, \varepsilon_A(\lambda)\, \lambda^4\, d\lambda}{\int F_D(\lambda)\, d\lambda} \tag{5}$$

where $F_D(\lambda)$ is the fluorescence intensity of the donor at wavelength λ and $\varepsilon_A(\lambda)$ is the molar extinction coefficient of the acceptor.

From theoretical considerations, κ^2 is in the range between 0 and 4. Uncertainties in the value of κ^2 cause the greatest error in distance determination by FRET. Fortunately, R depends on $(\kappa^2)^{1/6}$ so it is relatively insensitive to changes of κ^2 in a wide range (e.g. between 0.3 and 3). Direct measurement of its value is impracticable. Fluorescence anisotropy measurements on donor and acceptor molecules can limit the range of possible values of κ^2, but can rarely completely eliminate the uncertainty of R arising from this factor [14]. If the donor or the acceptor or both have a dynamically randomized orientation due to rapid rotational diffusion, κ^2 can be approximated by 2/3. This condition is often satisfied by fluorophores attached to biomolecules at the cell surface [14].

It can be shown that:

$$E = \frac{R^{-6}}{R^{-6} + R_0^{-6}} \tag{6}$$

From equations (3), (4) and (6) it follows that:

$$k_t = \frac{1}{\tau}\left(\frac{R_0}{R}\right)^6 \tag{7}$$

where τ is the donor's lifetime in the absence of the acceptor, and R_0 is the Förster radius of the donor-acceptor pair, i.e. the characteristic distance between the two dyes where FRET efficiency is 50%.

$$R_0 = \text{const}\,(J\kappa^2 Q_D n^{-4})^{1/6} \tag{8}$$

In this equation, Q_D is the fluorescence quantum yield of the donor in the absence of the acceptor. To observe effective transfer in the 1- to 10-nm range, the fluorescence emission spectrum of the donor and the absorption spectrum of the acceptor should overlap adequately, and both the quantum yield of the donor (Q_D) and the absorption coefficient of the acceptor (ε_A) should be sufficiently high ($Q_D \geq 0.1$ and $\varepsilon_A \geq 1{,}000\ M^{-1}cm^{-1}$).

A list of R_0 values for more than 70 donor-acceptor pairs is provided in Wu and Brand [15]. The largest R_0 value reported for a donor-acceptor pair is 8.0 nm for malachite green and rhodamine B [16]. If acceptor-labeled molecules form clusters, this can increase the apparent value of R_0. Along this line, Mathis [17] reported an exceptionally large R_0 of 9.0 nm, using europium cryptate as donor and allophycocyanine as acceptor. When applying phycobillin proteins, however, we should keep in mind that these molecules have bulky dimensions, which can interfere with the original goal, i.e. with accurate distance measurements.

How to Measure FRET Efficiency

FRET efficiency, as follows from the above formulas, can be determined in a number of different ways. Since energy is transferred from the excited donor to the acceptor, the lifetime (τ), quantum efficiency (Q) and fluorescence intensity (F) of the donor decrease in the presence of the acceptor (equation 9). At the same time, the fluorescence intensity of the acceptor increases (sensitization) in the presence of the donor (equation 10).

$$1 - E = \frac{\tau_D^A}{\tau_D} = \frac{F_D^A}{F_D} = \frac{Q_D^A}{Q_D} \tag{9}$$

$$\frac{F_A^D}{F_A} = 1 + \left(\frac{\varepsilon_D C_D}{\varepsilon_A C_A}\right) E \tag{10}$$

In the above equations, the lower indices refer to the donor (D) or acceptor (A) while the upper indices indicate the presence of the donor (D) or the acceptor (A) in the system. C_D and C_A are the molar concentrations; ε_D and ε_A are the molar absorption coefficients of the donor and the acceptor, respectively.

FRET can be determined by measuring the fluorescence characteristics of either the donor or the acceptor. The simplest way to measure energy transfer is to measure donor quenching in the presence of the acceptor. The fractional decrease in the donor fluorescence with the acceptor present is equal to the efficiency of FRET. Donor quenching cannot be used to determine FRET efficiency on a cell-by-cell basis since the expression levels of various proteins vary from cell to cell and the donor intensity cannot be measured in the absence and in the presence of the acceptor on the same cell. However, a mean E value over the whole cell population can be obtained. For more accurate and sensitive measurements of E on a cell-by-cell basis, the sensitized emission of the acceptor can be utilized since with multiple excitations the direct and sensitized emission of the acceptor can be determined on the same cell.

Technical details of flow-cytometric energy transfer measurements using fluorescein and rhodamine as a donor-acceptor pair can be found in the original papers and in several recent reviews [5, 6, 18–20]. Nowadays, the original fluorescein-rhodamine dye pair is usually replaced by donor-acceptor pairs having longer excitation and emission wavelengths. The reason is that in the red and far red spectral ranges the cellular autofluorescence is lower than in the green regime, thereby the signal-to-noise ratio is superior. In addition, new dyes such as the Alexa, ATTO or Cy series have higher quantum yields and/or absorption coefficients as well as a better photostability. In the following description we give the details of measuring FRET using green excitation (532 nm) for the donor (e.g. Alexa 546, Alexa 568, Cy3 or ATTO 532) and red excitation (632 nm) for the acceptor (e.g. Alexa 647, Cy5 or ATTO 647) (see 'Selection and Combination of Fluorescent Dyes', pp. 107). These excitation wavelengths are available in several commercial flow cytometers such as Becton Dickinson FACSDIVA or FACSArray flow cytometers (see 'Technical Background and Methodological Principles of Flow Cytometry', pp. 52). Spectral ranges of fluorescence are detected around the emission maximum of the donor (~ 580 nm) and the acceptor (> 650 nm). The calculation of FRET efficiency is based on measuring three background-corrected fluorescence signals I_1, I_2 and I_3. I_D and I_A stand for the unquenched intensity of the donor (excited at 532 nm, emission detected around 580 nm) and of the nonenhanced intensity of the acceptor (excited at 632 nm, detected at > 650 nm), respectively. Since the emission spectra of the donor and the acceptor usually overlap, and both molecules can be excited at 532 nm, correction factors S_1 and S_2 are introduced:

$S_1 = I_2 / I_1$ determined using donor-only labeled cells

$S_2 = I_2 / I_3$ determined using acceptor-only labeled cells

In case we use a donor dye that can be excited to some extent at 632 nm, we also introduce S_3:

$S_3 = I_3 / I_1$ determined using donor-only labeled cells.

The three detected intensities can be expressed as:

$$I_1(532 \rightarrow 580) = I_D(1 - E) \tag{11}$$

$$I_2(532 \rightarrow 650) = I_D(1 - E)\, S_1 + I_A S_2 + I_D E\alpha \tag{12}$$

$$I_3(632 \rightarrow 650) = I_D(1 - E)\, S_3 + I_A + \frac{S_3}{S_1}\, I_D E\alpha \tag{13}$$

If $E > 0$, then $I_1 < I_D$ because of donor quenching. I_2 consists of three additive terms: i) the overlapping fraction of the quenched donor intensity; ii) the direct contribution of the acceptor and iii) sensitized emission of the acceptor due to FRET. I_3 is the sum of i) a fraction of the quenched donor intensity; ii) the acceptor intensity and iii) the sensitized emission of the acceptor corrected for the lower molar extinction coefficient of donor at 532 nm than at 632 nm. The proportionality factor α is the ratio of I_2 for a given number of excited acceptor molecules and I_1 for the same number of excited donor molecules and can be determined as:

$$\alpha = \frac{I_2^A}{I_1^D} \frac{\varepsilon^D(532)}{\varepsilon^A(532)} \frac{L^D}{L^A} \frac{N^D}{N^A} \tag{14}$$

Where I_1^D and I_2^A are the I_1 and I_2 intensities measured with a donor- or an acceptor-labeled sample, respectively, ε^D (532) and ε^A (532) are the extinction coefficients of the donor and the acceptor at 532 nm, L^D and L^A are the dye-to-protein labeling ratios of the donor- and acceptor-labeled antibodies, and N^D and N^A are the number of binding sites for donor- and acceptor-labeled antibodies (if the same antibody is used for both dyes, this factor cancels out). α is constant for each experimental setup and dye pair, and has to be determined for every defined case. From equations 11–13, E can be expressed as follows:

$$E = 1 - \frac{1}{1 + \frac{1}{\alpha}\left(\frac{I_2 - S_2 I_3}{(1 - S_2 S_3/S_1) I_1} - S_1\right)} \tag{15}$$

Based on equation 15, E can be calculated on a cell-by-cell basis, resulting in a frequency distribution histogram providing information on the heterogeneity of the cell population with a high statistical accuracy. Equations 11–15 are equally valid for other donor-acceptor pairs as well if the appropriate spectral characteristics are considered.

Usually, it is assumed that cellular autofluorescence is negligible compared to specific fluorescence. If this is not the case, appropriate corrections must be made. If the signal-to-autofluorescence ratio is above 4, the average autofluorescence intensities of the cell population can be subtracted. If the autofluorescence is more significant, such that its value is comparable with the specific signal, a more elaborate correction method is also possible. The basis for this correction is that in most cell types, autofluorescence has a well-defined spectral shape, i.e. there is a good correlation between the autofluorescence detected in different spectral regions. In such situations, a fourth independent autofluorescence signal is detected in a region where the specific fluorescence signal is negligible, and the autofluorescence content of each signal can then be corrected on a cell-by-cell basis. However, it should be remembered that high autofluorescence is likely to decrease the precision of FRET measurements. A detailed description of an improved flow-cytometric energy transfer method has been published [21–23]. The main advantage of the approach is the reduction of autofluorescence-related errors, which is achieved by application of long-wavelength dyes and cell-by-cell correction of autofluorescence. The improved flow-cytometric energy transfer method allows energy transfer efficiencies to be determined in cellular systems even when the expression of cell surface molecules is low. In addition, the lower variance of energy transfer efficiency distributions enables a much more accurate discrimination of subpopulations having distinct FRET efficiencies.

In the past few years, molecular biological techniques have become widespread in cell-biological applications. Among these techniques, labeling with fluorescent proteins GFP and its spectral variants, cyan fluorescent protein (CFP), yellow fluorescent protein (YFP), monomeric red fluorescent protein (mRFP), mCherry has obviated some of the unwanted side effects due to the use of antibodies such as induced aggregation and internalization. In FRET applications, the commonly used donor-acceptor pairs are the CFP-YFP [24] and the GFP-mRFP [12] or GFP-mCherry pairs. A problem compared to antibody labeling is the determination of the α-factor, i.e. the equivalent amount of fluorescence arising from the same number of donor and acceptor molecules, since the expressions of the donor-labeled and acceptor-labeled proteins are independent of each other. In order to be able to calculate α, we need samples that express the same amount of donor and acceptor. The solution to this problem is the use of a fusion protein that contains both the donor and acceptor protein expressed from the same promoter in a single polypeptide [12, 24]. In this case, FRET may appear between the two proteins. The equations to be used are the following for the GFP-mRFP pair [12]:

$$I_1(488, 500-540) = I_D(1-E) \tag{16}$$

$$I_2(488, 597-639) = I_D(1-E)\,S_1 + I_A S_2 + I_D E\alpha \tag{17}$$

$$I_3\,(532, 597-639) = I_A \tag{18}$$

$$\alpha = \frac{I_3 S_2 (1-E)}{I_1}\frac{\varepsilon^D(488)}{\varepsilon^A(488)} \tag{19}$$

$$E = 1 - \frac{1}{1 + \frac{1}{\alpha}\left(\frac{I_2 - S_2 I_3}{I_1} - S_1\right)} \tag{20}$$

Equations 16–18 describe the background-corrected fluorescence intensities as measured by a FACSDIVA flow cytometer equipped with 488 and 532 nm excitation wavelengths and appropriate emission filters for the detection of GFP and mRFP fluorescence. To solve these equations, we first need the value of α, for which we use the sample expressing the above-mentioned fusion of GFP and mRFP. The expression for α (equation 19) contains E, and the expression for E (equation 20) contains α; therefore we have to solve these equations by successive approximation. In the first approximation, we assume $E = 0$ in equation 19 and get the initial guess for α. With this α we get the first approximation of E from equation 20; this value can be plugged into equation 19 to get the second approximation of E. With this iterative method, we reach convergence in about 3–5 steps. Once we have determined the value of α with the GFP-mRFP fusion protein, we can simply use equation 20 to calculate FRET efficiency between further donor- and acceptor-labeled epitopes.

Brief Protocol for FRET Measurements using Flow Cytometry

Although experimental protocols will vary considerably depending on the precise cellular system being analyzed, the following method can be used for the analysis of cell surface associations between different cell surface proteins.

1) Prepare a minimum set of samples: unlabeled cells, cells labeled with donor (e.g. Cy3) only, cells labeled with acceptor (e.g. Cy5) only, and cells labeled with both donor and acceptor. Use antibodies, preferably Fab fragments, conjugated directly with fluorescent dyes. If the expression levels of proteins under investigation are comparable, prepare an additional sample for which proteins originally labeled with the donor-conjugated antibody will be labeled with acceptor-conjugated antibody and vice versa. If the expression level of the two proteins under investigation is different, then the protein with the higher expression level should be chosen as acceptor.
2) Analyze the samples on a flow cytometer having at least two excitation wavelengths, one for the donor (e.g. 488 or 532 nm) and one for the acceptor (e.g. 514, 532 or 632 nm). Choose the appropriate filters according to your donor and acceptor dye. Collect at least three fluorescence intensities (for cell-by-cell autofluorescence correction, collect a 4th fluorescence intensity). Whenever possible, use linear rather than logarithmic gain. Do not use hardware compensation.
3) Store the data in list mode. Analyze the unlabeled and singly labeled cells to determine the spectral overlap S factors. For determining the α-factor, use two samples: a singly donor-labeled and a singly acceptor-labeled sample, preferably using the same kind of antibody. For fluorescent protein labeling (such as GFP and mRFP), use a sample expressing the fusion protein of the donor and the acceptor in a single polypeptide.
4) Finally analyze the double-labeled (FRET) sample, and calculate the FRET efficiency distribution histogram.

These calculations can be carried out with the help of a custom-made program (REFLEX) written in Delphi, where one can define histograms and dot plots of fluorescence intensities, select populations, and store the mean values of fluorescence intensities or any derived parameters in an Excel spreadsheet [22]. The program is

Table 2. FRET donor-acceptor pairs with possible excitation wavelengths

Donor	Excitation wavelength nm	Acceptor	Excitation wavelength nm
Fluorescein, Alexa Fluor 488	488	tetramethylrhodamine, Cy3 Alexa Fluor 546 Alexa Fluor 568	514, 532, 546
Cy3	488	Alexa Fluor 633 Alexa Fluor 647 Cy5	633
Alexa Fluor 546 Alexa Fluor 555 Alexa Fluor 568 Cy3	532, 543	Alexa Fluor 633 Alexa Fluor 647 Cy5	633
Alexa Fluor 568	568	Alexa Fluor 633 Alexa Fluor 647 Cy5	633
Cyan fluorescent protein	442, 405	yellow fluorescent protein	488, 514
Green fluorescent protein	488	monomeric red fluorescent protein mCherry	532, 546, 568

available on the web as a freeware with help instructions at the following site: *www.freewebs.com/cytoflex.*

Table 2 lists some possible FRET donor-acceptor pairs and their excitation wavelengths.

Limitations of FRET Measurements

Although FRET can provide very useful information about molecular associations, it has its limitations. The most serious limitation is the difficulty to determine absolute distances because E depends not only on the distance, but also on the relative orientation (κ^2) of the dyes. However, FRET is quite reliable at determining relative distances, namely, whether two points are getting closer or farther upon a stimulus. Even when measuring relative distances, care must be taken to ensure that the orientation factor (κ^2) does not change between the systems to be compared. If the fluorescent dye is attached to an antibody or Fab fragment via a carbon linker having 6–12 carbon atoms, the linker often allows relatively free rotation of the dye, which minimizes the uncertainty of κ^2.

Another problem is that due to the very sharp distance dependence (6th power of R) E vanishes quickly as the distance gets larger. Energy transfer tends to be all or none: if the donor and the acceptor are within $1.63 \times R_0$ distance, there is energy transfer, if they are farther apart, energy is transferred with very little efficiency. Due to this sharp decrease, the absence of FRET is not a direct proof of the absence of molecular proximity between the epitopes investigated, as it can be caused by steric hindrance even for neighboring molecules or protein domains. On the other hand, the presence of FRET to any appreciable extent above the error of determination is strong evidence of molecular interactions.

Indirect immunofluorescent labeling may be applied to FRET measurements if suitable directly labeled monoclonal antibodies are not available, or as an approach to enhance the specific fluorescence signal. In such cases it has to be taken into account that the size of the antibody complexes affects the FRET efficiency values. Application of a larger antibody complex causes a decrease in E since the actual distance between the donor and acceptor fluorophores increases (fig. 2) [21]. This explains the decreased FRET efficiency when fluorescent secondary Fab fragments rather than primary fluorescent antibodies are used on both the donor and the acceptor side. It also explains a further decrease in FRET efficiency when whole fluorescent secondary antibodies are used. Such findings underline the fact that FRET values cannot be directly compared if they are obtained using different labeling strategies [21]. In order to increase the signal, we can use phycoerythrin(PE)- or allophycocyanine(APC)-labeled antibodies since PE and APC have exceptional brightness; however, these molecules are even larger than whole antibodies. Due to the steric limitations, the measurable FRET efficiency values can be low [25]. Although it should be mentioned that even these low FRET efficiency values might have biological meanings since the accuracy of the measurements are greatly improved due to the high level of specific signals. Appropriate positive and negative controls can help decide whether the studied molecules are associated or not on the basis of the measured FRET efficiency values [25].

When studying cells labeled with donor- and acceptor-conjugated monoclonal antibodies, averaging is performed at different levels. The first averaging follows from the random conjugation of the fluorescent label. An additional averaging is brought about by the distribution of separation distances between the epitopes labeled with monoclonal antibodies. This multiple averaging, an inevitable consequence of the nonuniform stoichiometry, explains why FRET measurements are usually carried out with different goals on purified molecular systems, on the one side, and on the surface of the cytoplasmic membrane, on the other side. In the former case, FRET efficiency values can be converted into absolute distances whereas in the latter case they are rather only indicators of molecular proximity.

Calculation of distance relationships from energy transfer efficiencies is easy in the case of a single-donor-single-acceptor system (i.e. each donor has a single acceptor in its vicinity) if the localization and relative orientation of the fluorophores are known. If FRET measurements are performed on the cell surface, many molecules might not be labeled at all, many could be alone without a FRET pair, while others

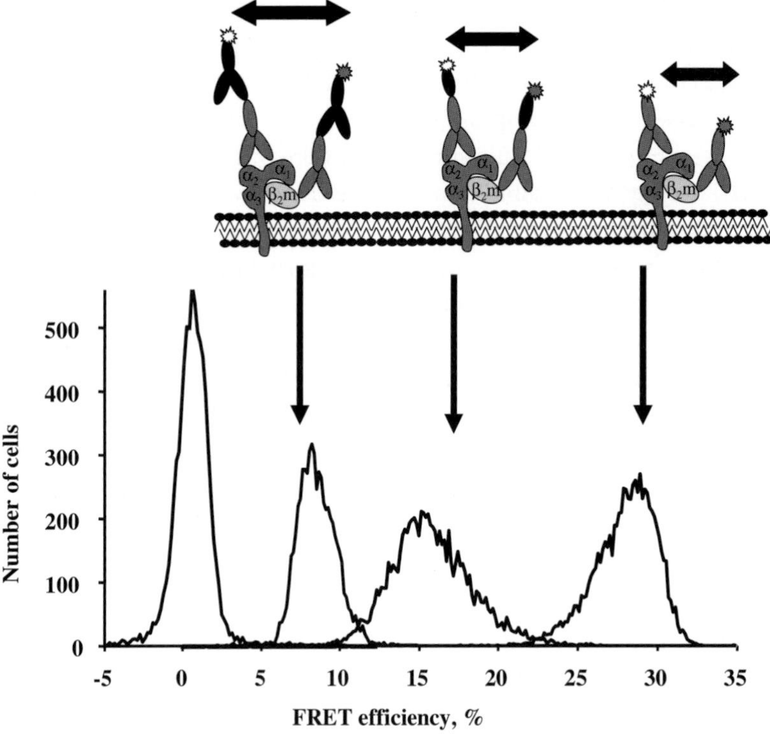

Fig. 2. The labeling strategy affects the efficiency of FRET. The top half of the figure shows the labeling strategy. The heavy chain and the light chain (β_2m) of class I HLA molecules were labeled directly with fluorescently tagged primary monoclonal antibodies, indirectly with unlabeled monoclonal antibodies and subsequently with fluorescent Fab fragments of goat-antimouse IgG or whole rat-antimouse IgG molecules. In the case of indirect staining, more than one polyclonal antibody or Fab fragment can bind to a single primary antibody. For the sake of simplicity, only one secondary antibody or Fab fragment is shown in the scheme. In these measurements, Cy3 dye was used as donor and Cy5 as acceptor. The bottom half of the figure shows histograms displaying FRET efficiency distributions measured between the light and the heavy chains of class I HLA molecules using direct and indirect labeling. The first distribution curve on the left represents the FRET distribution of cells in the absence of energy transfer when cells were labeled only with a donor-conjugated primary monoclonal antibody. Note that upon increasing the complexity of the labeling scheme by introducing secondary antibodies, the FRET efficiency values decrease, shifting the histograms to the left.

may be in smaller groups of hetero-oligomers, creating higher rates of FRET than expected from stand-alone pairs. If there are a large number of acceptor-labeled epitopes in the vicinity of the donor, E increases simply because the net rate of FRET is increased due to alternative transfer from the donor to neighboring acceptors. This can be misinterpreted as a short distance between the donor and the acceptor. If reversing labeling by donor and acceptor still results in the same large E, the proximity can be considered to be verified. However, if reversing the labeling results in a donor-to-acceptor ratio >1 and this makes FRET disappear, chances are that we have previously seen a multiple acceptor situation, or even random co-localizations due to the high number of acceptors. In addition, if cell membrane components are investigated, a two-dimensional restriction applies to the labeled molecules. Analytical solutions for randomly distributed donor and acceptor molecules and numerical solutions for nonrandom distributions have been elaborated by different groups [26–28]. In order to differentiate between random and nonrandom distributions, FRET efficiencies must be determined at different acceptor concentrations, e.g. by varying the extent of labeling of the acceptor antibody and mixing labeled and unlabeled acceptor antibodies at different ratios. In any case, the donor/acceptor ratio should be in the range of 0.1–10 [29] if we want to obtain reliable FRET measurements. Outside this range, errors arising from noise and data irreproducibility are too high, rendering accurate FRET efficiency calculations impossible. Berney and Danuser [29] also suggested that in order to obtain stable FRET measurements, transfer has to be observed in the FRET channel, i.e. by excitation of the donor and measurement of the acceptor emission. Methods estimating FRET efficiency from the donor signal alone in the presence and absence of acceptor are less robust. In spite of the increased cross-talk, donor-acceptor dye pairs should be chosen with maximal spectral overlap to increase R_0, and thereby E, since spectral overlap can be corrected using samples labeled with donor or acceptor only [29].

Applications

FRET is widely utilized for a variety of applications. In one series of studies, FRET was used to obtain structural information that is otherwise difficult to obtain. The major advantage of FRET for structural studies is that due to the specificity of labeling, the experimental object can be investigated in situ and/or in vivo with little or no interference from the rest of the system. Even complex and heterogeneous systems can be studied this way. Many molecular interactions have been subjected to FRET analysis, including – just to mention a few examples – the homo- and hetero-associations of MHC class I and class II [30], the interleukin-2 receptor α-subunit and ICAM-1 [20, 31], the TCR/CD3 complex [32], tetraspan molecules (CD53, CD81, CD82) and CD20 with MHC class I and class II [33], the three subunits of the multi-subunit IL-2 receptor [34], the IL-12/IL-15 receptor complex [11, 35, 36], the TNF receptor [37], Fas (CD95) [38], Kv1.3 K^+ channels [39] and the AP1 complex [12]. FRET-based co-localization has been extended to three proteins using three-color labeling and a combination of acceptor and donor photobleaching in confocal microscopy. In this procedure, the acceptor of the first FRET process serves as a donor in the second FRET process; this way, the heteroassociation of β2-integrin with erbB2 and erbB2 with CD44 was detected on breast tumor cells [40]. Excellent reviews are available on the applicability of FRET to biological systems as well as descriptions and comparisons of various approaches. Only a few are quoted here [1–10, 29, 41–44].

These reviews deal with associations of various membrane proteins, structures of receptors, conformational changes in transmembrane proteins evoked by ligands and membrane potential changes.

In another series of studies FRET was used as a tool to provide high sensitivity in various biological assays. The biotechnological applications of FRET are summarized in reviews [5, 42]. These reviews thoroughly discuss the working principles of FRET-based enzyme assays, immunoassays, designing tandem dyes and FRET primers for DNA analysis [5, 42].

Luminescent lanthanide complexes produce emissions with the narrowest-known width at half maximum; however, their significant use in cytometry required an increase in luminescence intensity. To this end two methods have been introduced: adding a second lanthanide ion in micellar or dry form. Both methods involve the resonance energy transfer enhanced luminescence (RETEL) effect as the mechanism for the luminescence enhancement, which increases luminescence and is compatible with standard slide microscopy [45, 46].

A detailed list of possible FRET applications is beyond the scope of this chapter. The readers are referred to the reviews given above to find useful examples of FRET studies. Here we still would like to mention a couple of examples for demonstrating new and interesting applications of FRET.

A steadily increasing new field in FRET studies is based on the application of GFP as an intrinsic reporter. These newly developed fluorescent probes provide high sensitivity and great versatility while minimally perturbing the cell under investigation. Genetically encoded reporter constructs that are derived from GFP and RFP are leading to a revolution in the real-time visualization and tracking of various cellular events [9, 47–50]. Recent advances include the continued development of 'passive' markers for the measurement of biomolecule expression and localization in live cells, and 'active' indicators for monitoring more complex cellular processes such as small-molecule-messenger dynamics, changes in Ca^{2+} concentration, enzyme activation and protein-protein interactions [51–53]. The simultaneous measurement of the activities of two caspases was carried out using a fusion protein of CFP, YFP and mRFP with appropriate cleavage sites between the fluorescent proteins [54]. Photophysical properties of GFP and RFP and their application in quantitative microscopy are excellently summarized in a recent review [55]. Using this approach, the spatial and temporal interaction of Bcl-2 and Bax proteins [56] and Jun and Fos transcription factors [12] was studied at the single-cell level by monitoring FRET efficiency.

In a recent report, a mutant protein is described, which is capable of unique irreversible photoconversion from the nonfluorescent to a stable bright-red fluorescent form '('kindling'). This 'kindling fluorescent protein' can be used for precise in vivo photolabeling to track the movements of cells, organelles, and proteins and also as an acceptor in future FRET experiments [57, 58].

The combination of FRET with total internal reflection fluorescence microscopy (TIRFM) conditions is an interesting application. The developed imaging system enables screening of large numbers of cells under TIRFM illumination combined with FRET imaging, thereby providing the means to record, e.g. FRET efficiency of a membrane-associated protein labeled with a donor-acceptor pair. With this system, high-throughput analysis of stoichiometric FRET constructs can be performed on live cells [59, 60].

A new extension of the application of FPs in FRET studies has been achieved by combining GFP with luciferase. In bioluminescence resonance energy transfer (BRET), one protein is fused to Renilla luciferase and the other to a mutant of GFP [61, 62]. The luciferase can be activated by the addition of its substrate, and if the proteins in question interact, resonance energy transfer occurs between the excited luciferase and the mutant GFP. BRET can be detected by monitoring the fluorescence signal emitted by the mutant GFP. By choosing the proper luciferase/GFP mutant combinations, BRET can be used to measure protein-protein interactions in vitro and in vivo. BRET is perfectly suited for cell-based proteomics applications, including receptor research and mapping to signal transduction pathways [61, 63, 64]. The BRET method was used to assay interactions between proteins encoded by the circadian clock genes *kaiA* and *kaiB* in cyanobacterium [61].

Another new FRET modality is homotransfer or energy migration FRET (emFRET) [65]. This approach exploits fluorescence polarization measurements in flow cytometry, wide-field or confocal laser-scanning microscopy, or in the form of anisotropy fluorescence lifetime imaging microscopy (rFLIM). These methods permit the assessment of rotational motion, association and proximity of cellular proteins in vivo. They are particularly applicable to probes generated by fusions of visible fluorescent proteins and are capable of monitoring homoassociations of various signaling proteins [65–68].

Since Förster [69] first described the FRET phenomenon, the number of its applications has increased enormously in various fields of research and biotechnology. Technical improvements in spectrofluorimeters, flow cytometers and microscopes, and introduction of new fluorescent probes with better photophysical properties have opened up a new area for the innovative and successful application of the FRET method.

Summary

With the onset of modern proteomics, hundreds of pairs of cellular proteins that are capable of interacting with each other in vitro have been identified. However, the extent to which these theoretically possible interactions appear in live cells is not clear. Combination of fluorescence spectroscopy with flow and image cytometry provides a basis for the rapid and continuous development of new technologies capable of studying molecular interactions in vitro and in vivo. The FRET method is a most versatile spectroscopic modality. FRET has been applied widely for many purposes. In one group of studies, FRET was used as a tool to enhance sensitivity. FRET methods can be incorporated into chromatographic assays, electrophoresis, microscopy, and flow cytometry. FRET can also be used for improving spectral characteristics of fluorescent dyes. In another group of studies, FRET was used to obtain structural information. The major advantage of applying FRET for structural studies is that the specifically labeled object can be investigated in situ and/or in vivo with little interference regardless of the complexity and heterogeneity of the system. This chapter describes the basic principles of the FRET process. In addition, it characterizes parameters for flow-cytometric FRET measurements, limitations of FRET and discusses a few FRET applications.

References

1 Selvin PR: The renaissance of fluorescence resonance energy transfer. Nat Struct Biol 2000;7:730–734.
2 Jares-Erijman EA, Jovin TM: Fret imaging. Nat Biotechnol 2003;21:1387–1395.
3 Sekar RB, Periasamy A: Fluorescence resonance energy transfer (FRET) microscopy imaging of live cell protein localizations. J Cell Biol 2003;160:629–633.
4 Scholes GD: Long-range resonance energy transfer in molecular systems. Annu Rev Phys Chem 2003;54:57–87.
5 Szöllősi J, Damjanovich S, Matyus L: Application of fluorescence resonance energy transfer in the clinical laboratory: Routine and research. Cytometry 1998;34:159–179.
6 Szöllősi J, Alexander DR: The application of fluorescence resonance energy transfer to the investigation of phosphatases. Methods Enzymol 2003;366:203–224.
7 Szöllősi J, Nagy P, Sebestyen Z, Damjanovicha S, Park JW, Matyus L: Applications of fluorescence resonance energy transfer for mapping biological membranes. J Biotechnol 2002;82:251–266.
8 Piston DW, Kremers GJ: Fluorescent protein FRET: the good, the bad and the ugly. Trends Biochem Sci 2007;32:407–414.
9 Dobbie IM, Lowndes NF, Sullivan KF: Autofluorescent proteins. Methods Cell Biol 2008;85:1–22.
10 Szöllősi J, Damjanovich S, Nagy P, Vereb G, Matyus L: Principles of resonance energy transfer; in Robinson P (ed): Current Protocols in Cytometry. New York, Wiley, 2006, pp 1.12.11–11.12.16.
11 Vámosi G, Bodnár A, Vereb G, Jenei A, Goldman CK, Langowski J, Tóth K, Matyus L, Szöllősi J, Waldmann TA, Damjanovich S: IL-2 and IL-15 receptor alpha-subunits are coexpressed in a supramolecular receptor cluster in lipid rafts of T cells. Proc Natl Acad Sci U S A 2004;101:11082–11087.
12 Vámosi G, Baudendistel N, von der Lieth CW, Szaloki N, Mocsar G, Muller G, Brazda P, Waldeck W, Damjanovich S, Langowski J, Tóth K: Conformation of the c-fos/c-jun complex in vivo: a combined FRET, FCCS, and MD-modeling study. Biophys J 2008;94:2859–2868.
13 Stryer L: Fluorescence energy transfer as a spectroscopic ruler. Annu Rev Biochem 1978;47:819–846.
14 Dale RE, Eisinger J, Blumberg WE: The orientational freedom of molecular probes. The orientation factor in intramolecular energy transfer. Biophys J 1979;26:161–193.
15 Wu P, Brand L: Resonance energy transfer: Methods and applications. Anal Biochem 1994;218:1–13.
16 Yamazaki I, Tami N, Yamazaki T: Electronic excitation transfer in organized molecular assemblies. J Phys Chem 1990;94:516–525.
17 Mathis G: Rare earth cryptates and homogeneous fluoroimmunoassays with human sera. Clin Chem 1993;39:1953–1959.
18 Tron L, Szöllősi J, Damjanovich S, Helliwell SH, Arndt-Jovin DJ, Jovin TM: Flow cytometric measurement of fluorescence resonance energy transfer on cell surfaces. Quantitative evaluation of the transfer efficiency on a cell-by-cell basis. Biophys J 1984;45:939–946.
19 Szöllősi J, Tron L, Damjanovich S, Helliwell SH, Arndt-Jovin D, Jovin TM: Fluorescence energy transfer measurements on cell surfaces: A critical comparison of steady-state fluorimetric and flow cytometric methods. Cytometry 1984;5:210–216.

20 Szöllősi J, Damjanovich S, Goldman CK, Fulwyler MJ, Aszalos AA, Goldstein G, Rao P, Talle MA, Waldmann TA: Flow cytometric resonance energy transfer measurements support the association of a 95-kDa peptide termedTt27 with the 55-kDa Tac peptide. Proc Natl Acad Sci U S A 1987;84:7246–7250.

21 Sebestyen Z, Nagy P, Horvath G, Vámosi G, Debets R, Gratama JW, Alexander DR, Szöllősi J: Long wavelength fluorophores and cell-by-cell correction for autofluorescence significantly improves the accuracy of flow cytometric energy transfer measurements on a dual-laser benchtop flow cytometer. Cytometry 2002;48:124–135.

22 Szentesi G, Horvath G, Bori I, Vámosi G, Szöllősi J, Gaspar R, Damjanovich S, Jenei A, Matyus L: Computer program for determining fluorescence resonance energy transfer efficiency from flow cytometric data on a cell-by-cell basis. Comput Methods Programs Biomed 2004;75:201–211.

23 Nagy P, Vereb G, Damjanovich S, Matyus L, Szöllősi J: Measuring FRET in flow and image cytometry; in Robinson P (ed): Current Protocols in Cytometry. New York, Wiley, 2006, pp 12.18.11–12.18.11.

24 Nagy P, Bene L, Hyun WC, Vereb G, Braun M, Antz C, Paysan J, Damjanovich S, Park JW, Szöllősi J: Novel calibration method for flow cytometric fluorescence resonance energy transfer measurements between visible fluorescent proteins. Cytometry A 2005;67:86–96.

25 Batard P, Szöllősi J, Luescher I, Cerottini JC, MacDonald R, Romero P: Use of phycoerythrin and allophycocyanin for fluorescence resonance energy transfer analyzed by flow cytometry: Advantages and limitations. Cytometry 2002;48:97–105.

26 Wolber PK, Hudson BS: An analytic solution to the Förster energy transfer problem in two dimensions. Biophys J 1979;28:197–210.

27 Dewey TG, Hammes GG: Calculation on fluorescence resonance energy transfer on surfaces. Biophys J 1980;32:1023–1035.

28 Snyder B, Freire E: Fluorescence energy transfer in two dimensions. A numeric solution for random and nonrandom distributions. Biophys J 1982;40:137–148.

29 Berney C, Danuser G: FRET or no FRET: a quantitative comparison. Biophys J 2003;84:3992–4010.

30 Szöllősi J, Damjanovich S, Balazs M, Nagy P, Tron L, Fulwyler MJ, Brodsky FM: Physical association between MHC class I and class II molecules detected on the cell surface by flow cytometric energy transfer. J Immunol 1989;143:208–213.

31 Burton J, Goldman CK, Rao P, Moos M, Waldmann TA: Association of intercellular adhesion molecule 1 with the multichain high-affinity interleukin 2 receptor. Proc Natl Acad Sci U S A 1990;87:7329–7333.

32 de la Hera A, Muller U, Olsson C, Isaaz S, Tunnacliffe A: Structure of the T cell antigen receptor (TCR): Two CD3 epsilon subunits in a functional TCR/CD3 complex. J Exp Med 1991;173:7–17.

33 Szöllősi J, Horejsi V, Bene L, Angelisova P, Damjanovich S: Supramolecular complexes of MHC class I, MHC class II, CD20, and tetraspan molecules (CD53, CD81, and CD82) at the surface of a B cell line JY. J Immunol 1996;157:2939–2946.

34 Damjanovich S, Bene L, Matko J, Alileche A, Goldman CK, Sharrow S, Waldmann TA: Preassembly of interleukin 2 (IL-2) receptor subunits on resting KIT 225 K6 T cells and their modulation by IL-2, IL-7, and IL-15: A fluorescence resonance energy transfer study. Proc Natl Acad Sci U S A 1997;94:13134–13139.

35 Bodnár A, Nizsaloczki E, Mocsar G, Szaloki N, Waldmann TA, Damjanovich S, Vámosi G: A biophysical approach to IL-2 and IL-15 receptor function: Localization, conformation and interactions. Immunol Lett 2008;116:117–125.

36 Bene L, Kanyari Z, Bodnár A, Kappelmayer J, Waldmann TA, Vámosi G, Damjanovich L: Colorectal carcinoma rearranges cell surface protein topology and density in CD4+ T cells. Biochem Biophys Res Commun 2007;361:202–207.

37 Chan FK, Chun HJ, Zheng L, Siegel RM, Bui KL, Lenardo MJ: A domain in TNF receptors that mediates ligand-independent receptor assembly and signaling. Science 2000;288:2351–2354.

38 Siegel RM, Frederiksen JK, Zacharias DA, Chan FK, Johnson M, Lynch D, Tsien RY, Lenardo MJ: Fas preassociation required for apoptosis signaling and dominant inhibition by pathogenic mutations. Science 2000;288:2354–2357.

39 Panyi G, Bagdany M, Bodnár A, Vámosi G, Szentesi G, Jenei A, Matyus L, Varga S, Waldmann TA, Gaspar R, Damjanovich S: Colocalization and nonrandom distribution of Kv1.3 potassium channels and CD3 molecules in the plasma membrane of human T lymphocytes. Proc Natl Acad Sci U S A 2003;100:2592–2597.

40 Fazekas Z, Petras M, Fabian A, Palyi-Krekk Z, Nagy P, Damjanovich S, Vereb G, Szöllősi J: Two-sided fluorescence resonance energy transfer for assessing molecular interactions of up to three distinct species in confocal microscopy. Cytometry A 2008;73:209–219.

41 Bastiaens PI, Squire A: Fluorescence lifetime imaging microscopy: Spatial resolution of biochemical processes in the cell. Trends Cell Biol 1999;9:48–52.

42 Clegg RM: Fluorescence resonance energy transfer. Curr Opin Biotechnol 1995;6:103–110.

43 Gordon GW, Berry G, Liang XH, Levine B, Herman B: Quantitative fluorescence resonance energy transfer measurements using fluorescence microscopy. Biophys J 1998;74:2702–2713.

44 Vereb G, Szöllősi J, Matko J, Nagy P, Farkas T, Vigh L, Matyus L, Waldmann TA, Damjanovich S: Dynamic, yet structured: The cell membrane three decades after the Singer-Nicolson model. Proc Natl Acad Sci U S A 2003;100:8053–8058.

45 Leif RC, Vallarino LM, Becker MC, Yang S: Increasing the luminescence of lanthanide complexes. Cytometry A 2006;69:767–778.

46 Leif RC, Vallarino LM, Becker MC, Yang S: Increasing lanthanide luminescence by use of the RETEL effect. Cytometry A 2006;69:940–946.

47 Zhang J, Campbell RE, Ting AY, Tsien RY: Creating new fluorescent probes for cell biology. Nat Rev Mol Cell Biol 2002;3:906–918.

48 Aoki K, Nakamura T, Matsuda M: Spatio-temporal regulation of Rac1 and Cdc42 activity during nerve growth factor-induced neurite outgrowth in PC12 cells. J Biol Chem 2004;279:713–719.

49 Scarlata S, Dowal L: The use of green fluorescent proteins to view association between phospholipase C beta and G protein subunits in cells. Methods Mol Biol 2004;237:223–232.

50 Shaner NC, Patterson GH, Davidson MW: Advances in fluorescent protein technology. J Cell Sci 2007;120:4247–4260.

51 Lippincott-Schwartz J, Altan-Bonnet N, Patterson GH: Photobleaching and photoactivation: following protein dynamics in living cells. Nat Cell Biol 2003;suppl:S7–14.

52 Adachi T, Tsubata T: FRET-based Ca^{2+} measurement in B lymphocyte by flow cytometry and confocal microscopy. Biochem Biophys Res Commun 2008;367:377–382.

53 Lippincott-Schwartz J, Patterson GH: Development and use of fluorescent protein markers in living cells. Science 2003;300:87–91.

54 Wu X, Simone J, Hewgill D, Siegel R, Lipsky PE, He L: Measurement of two caspase activities simultaneously in living cells by a novel dual FRET fluorescent indicator probe. Cytometry A 2006;69:477–486.

55 Subramaniam V, Hanley QS, Clayton AH, Jovin TM: Photophysics of green and red fluorescent proteins: Implications for quantitative microscopy. Methods Enzymol 2003;360:178–201.

56 Mahajan NP, Linder K, Berry G, Gordon GW, Heim R, Herman B: Bcl-2 and Bax interactions in mitochondria probed with green fluorescent protein and fluorescence resonance energy transfer. Nat Biotechnol 1998;16:547–552.

57 Chudakov DM, Belousov VV, Zaraisky AG, Novoselov VV, Staroverov DB, Zorov DB, Lukyanov S, Lukyanov KA: Kindling fluorescent proteins for precise in vivo photolabeling. Nat Biotechnol 2003;21:191–194.

58 Chudakov DM, Feofanov AV, Mudrik NN, Lukyanov S, Lukyanov KA: Chromophore environment provides clue to 'kindling fluorescent protein' riddle. J Biol Chem 2003;278:7215–7219.

59 Nagy P, Szöllősi J: Seeing through protein complexes by high-throughput FRET. Cytometry A 2008;73A:388–389.

60 Paar C, Paster W, Stockinger H, Schutz GJ, Sonnleitner M, Sonnleitner A: High throughput fret screening of the plasma membrane based on TIRFM. Cytometry A 2008;73A:442–450.

61 Xu Y, Piston DW, Johnson CH: A bioluminescence resonance energy transfer (BRET) system: application to interacting circadian clock proteins. Proc Natl Acad Sci U S A 1999;96:151–156.

62 Bacart J, Corbel C, Jockers R, Bach S, Couturier C: The BRET technology and its application to screening assays. Biotechnol J 2008;3:311–324.

63 Issad T, Boute N, Boubekeur S, Lacasa D, Pernet K: Looking for an insulin pill? Use the BRET methodology! Diabetes Metab 2003;29:111–117.

64 Devost D, Zingg HH: Identification of dimeric and oligomeric complexes of the human oxytocin receptor by co-immunoprecipitation and bioluminescence resonance energy transfer. J Mol Endocrinol 2003;31:461–471.

65 Tramier M, Coppey-Moisan M: Fluorescence anisotropy imaging microscopy for homo-FRET in living cells. Methods Cell Biol 2008;85:395–414.

66 Clayton AH, Hanley QS, Arndt-Jovin DJ, Subramaniam V, Jovin TM: Dynamic fluorescence anisotropy imaging microscopy in the frequency domain (rFLIM). Biophys J 2002;83:1631–1649.

67 Lidke DS, Nagy P, Barisas BG, Heintzmann R, Post JN, Lidke KA, Clayton AH, Arndt-Jovin DJ, Jovin TM: Imaging molecular interactions in cells by dynamic and static fluorescence anisotropy (rFLIM and emFRET). Biochem Soc Trans 2003;31:1020–1027.

68 Tramier M, Piolot T, Gautier I, Mignotte V, Coppey J, Kemnitz K, Durieux C, Coppey-Moisan M: Homo-FRET versus hetero-FRET to probe homodimers in living cells. Methods Enzymol 2003;360:580–597.

69 Förster T: Energiewanderung und Fluoreszenz. Naturwissenschaften 1946;6.

Sack U, Tárnok A, Rothe G (eds): Cellular Diagnostics. Basics, Methods and Clinical Applications of Flow Cytometry. Basel, Karger, 2009, pp 159–177

Quality Control and Standardization

Anja Mittag[a] · Dominik Lenz[b] · Attila Tarnok[b]

[a] Translational Center for Regenerative Medicine (TRM), University of Leipzig,
[b] Department of Pediatric Cardiology, Heart Center Leipzig, Germany

Introduction

Background

Almost 150 years have elapsed since Rudolf Virchow's first publication on cellular pathology in 1859 [1], and cytologic analyses now play an eminent role in clinical routine diagnostics. In many different diseases, cytometric tests, which allow fast and quantitative examination of single cells in mixtures, have occupied a central position for the last 20 years. The frequency of cells and cell subtypes can be determined based on the expression pattern of different cell surface markers. Furthermore, a stoichiometric (quantitative) analysis of defined cell contents is possible if they were tagged with fluorochromes (see 'Selection and Combination of Fluorescent Dyes', pp. 107). Many chapters of this book illustrate the importance of fluorescence dyes for the identification of different disease patterns.

Despite the importance of standardization, there are no generally accepted cross-national rules. Quality management in cytometric laboratories can be integrated in a laboratory- or companywide quality policy. All general aspects of accreditation and regulation by European or international agencies such as the Global Consensus Standardization for Health Technologies of the National Committee for Clinical Laboratory Standards (NCCLS) or The European Medicines Agency (EMEA) cannot be discussed here. We focus on general aspects allowing high-quality cytometry. Readers interested in information on accreditation, i.e. the formal authentication of the professional competence of laboratories, are referred to the respective national administrative offices. To give a short overview, there are standards and statutory guidelines for the following fields of application (they may show variations from one country to another, but in most cases they are similar):

- medical laboratory diagnostics (in Germany: ISO/EN/DIN 15189);

- laboratory diagnostics in other controlled areas (in particular pharmaceutical products): Good Laboratory Practice (GLP) (rarely cytometric);
- quality control in pharmaceutical production, in particular biological: Good Manufacturing Practices (GMP) (product-dependent, specified by authorities), and
- monitoring of clinical studies: Good Clinical Practice (GCP).

This chapter is mainly based on the global consensus protocol of the NCCLS [2]. For further details on standardization, we refer the readers to this protocol as well as other reference works [3, 4].

Cytometry

For cytometric analyses of a great number of single cells in heterogeneous systems (peripheral blood leukocytes, tissue cells), the overall fluorescence intensity of every fluorochrome used is determined. The brightness (integral fluorescence intensity) of the cells is proportional to the number of fluorochrome molecules per cell. If cells are tagged with antibodies linked with a certain fluorochrome, the brightness, i.e. fluorescence intensity, of a cell correlates with the number of antibodies bound. Since this number depends on the number of antigen molecules of a cell identified by the antibodies used, the fluorescence intensity of the cell directly shows the number of molecules per cell. If the tagged antigen is a marker of cell activation, then brightness is a direct measure of cell activation. An instrument can be checked by quality control procedures, and this is another important point. If standardization is not possible despite appropriate preparative procedures and instrument settings, service should be contacted.

In principle, cytometric analyses of biological material are possible with two types of instruments:

- flow cytometer/cytometry (FCM; see 'Technical Background and Methodological Principles of Flow Cytometry', pp. 53) and
- slide-based cytometer/cytometry (SBC; see 'Technical and Methodological Basics of Slide-Based Cytometry', pp. 89).

Both analytical methods have in common that, for example, surface antigens on the cells of interest are tagged with fluorochrome-labeled antibodies. Fluorochromes are excited by light from lasers or arc lamps. Fluorochromes excited this way emit fluorescence light at certain wavelengths which is transformed into electric signals by photomultipliers (or is detected by a digital camera). The amplitude of this electric signal correlates with the actual brightness of the object.

The differences between FCM and SBC are mainly the analyzable samples. Unlike FCM, which can only analyze cells in suspension, measurements by SBC instruments are based on cells fixed on a slide. Hence, FCM is best suited for the analysis of blood cells or other body fluids whereas SBC is appropriate for cell cultures or solid tissues. While in FCM the measured samples are lost, in SBC analyses the sample remains on the slide and can be used for further preparations and/or repeat analyses. Basically, both systems determine the overall fluorescence per cell or cell compartment so that standardization and calibration procedures are comparable.

Principle of Cytometric Methods

With cytometers, several fluorochromes can be measured simultaneously. Thereby, it is possible to label a multitude of cell types or to detect different characteristics on a single-cell basis. However, the number and type of lasers in the instrument are limiting factors in setting up a multiparametric measurement. The most frequently used lasers are argon lasers (Ar) with an excitation wavelength of 488 nm and helium-neon lasers (He-Ne) with an excitation wavelength of 633 nm. In recent years, so-called diode lasers with comparable wavelengths have increasingly been used. Of course, only fluorochromes excitable with lasers available in the instrument are suitable for cytometric measurements.

Most of the commercially available fluorochromes can be excited at 488 nm, e.g. fluorescein isothiocyanate (FITC), phycoerythrin (PE), peridinin chlorophyll protein (PerCP) or some tandem conjugates (Pe-Cy5, PE-Cy5.5, PE-Cy7, etc.). There are also many fluorochromes for 633-nm lasers, e.g., APC and Cy5, as well as tandem conjugates (APC-Cy5.5, APC-Cy7). Furthermore, there is a plethora of fluorescence dyes for staining of DNA, RNA and proteins [review in 5]. Each fluorochrome possesses its own emission spectrum. To detect only one specific fluorochrome out of many, the emitted fluorescence passes different optical filters (bandpass and dichroic filters). These filters are connected upstream of the photomultiplier and allow the passage of only a defined wavelength range. Most cytometric instruments offer at least three of these filters, arranged according to afferent wavelengths.

Cytometric measurements enable the detection of every fluorescence-labeled parameter. Most commonly used cytometry applications are:

- cell surface antigen expression;
- cell physiology (intracellular free Ca^{2+}, intracellular pH, oxidative burst);
- apoptosis (annexin V binding, mitochondrial membrane potential, DNA fragmentation);
- intracellular expression of cytokines, and
- DNA content (fluorescence in-situ hybridization, intranuclear antigens).

Flow Cytometry

Since its introduction, the flow cytometer (ICP-11, Phywe GmbH) is the most frequently used instrument for multiparametric analyses of cells in suspension [6]. FCM allows detection and differentiation of cell types, even without fluorescence labeling, using two morphological parameters: forward (FSC) and side scatter (SSC) signals. Additional tagging of cells with fluorochromes enables further differentiation.

High velocity as well as the possibility of standardization makes FCM a versatile tool in clinical routine diagnostics. Main applications are analyses of tumor cell cycles and of immune deficiencies. However, examination of cell morphology by FCM analysis is restricted to FSC and SCC signals. Following the requests of pathologists to examine cell morphology, electrostatic cell sorters were installed in FCM instruments in the 1970s [7] and new mechanical cell sorters were developed simultaneously. Sorting of cells (on the basis of specific parameters, detected, e.g. by fluorescence) allows their purification. However, sorting is expensive and, in most cases, not applicable in the clinical routine. The demand to combine multiparametric FCM and morphological documentation has therefore grown in the past.

Slide-Based Cytometry

Even though the concept of SBC was already presented in the 1980s, the first type of such instruments, the laser scanning cytometer (LSC), became commercially available only in 1996 [8]. It combines the requested multiparametric cytometry with morphological analysis and documentation [8–10]. The cytometer evolved from a conventional epifluorescence microscope. Samples are fixed on a slide and excited by two (Ar and He–Ne) or up to three lasers (additional violet laser: 405 nm). UV or violet excitation wavelength is suitable for DNA dyes such as DAPI or Hoechst.

In recent years, various SBC instruments were developed that are now commercially available, e.g. the scanning fluorescence microscope (SFM) [11] or the laser scanning cytometer [12] (also see 'Technical and Methodological Basics of Slide-Based Cytometry', pp. 89).

For every measured event (cell), the information of its exact x-y position on the slide is recorded. Thus it is possible to relocate a measured event any time after finishing the measurement. At the same time, both light-microscopic image and fluorescence can be visualized. Hence, this feature enables one to check:

- whether measured events are single cells, doublets or artifacts and
- whether cell morphology is correctly documented.

Standardization

Calibration of the Instrument

Standardization, control and calibration allow quantification of measured fluorescence and guarantee that results will vary only within certain boundaries. Cytometers are complex instruments, composed of a multitude of electronic, optical, and mechanical devices. Even two instruments constructed in the same way are not identical and will provide marginally different results from the same sample [13, 14]. Additionally, over the years, aging effects and a loss of transparency in optics occur. For these reasons, the instruments have to be regularly calibrated and standardized. Calibrated results obtained from standardized instruments can be compared in an objective and quantitative way with measurements from other laboratories.

Most results of cytometric analyses are expressed as 'percentage positive' or specified more qualitatively as 'dim' or 'bright'. These terms are relative: what one laboratory describes as 'negative', 'dim' or 'bright' can be totally different from the description by other laboratories. Such relative terms are appropriate when observing cells on fluorescence microscopes. Cytometers are able to generate objective criteria to evaluate results such as fluorescence intensities.

For quantitative and repeated long-term analyses of antigen expressions per cell, instrument setup and sample preparation (pre-analytic steps) must be standardized, whichs is particularly important for long-term monitoring of antigens in patients or comparison of experiments with different laboratories.

The fluorescence brightness of a cell is a direct measure of the number of bound dye molecules. Changes in brightness therefore indicate alterations in the cell's physiological status (e.g. activation or differentiation). However, the exact brightness of a cell also depends on instrument settings such as voltage of the photomultipliers, optical filters. Hence, setup changes exert a direct influence on the measured brightness. On the other hand, even an identical setup can generate substantially different results with instruments of identical design. Such differences are product dependent or are due to a loss of sensitivity caused by aging and long-term usage. This is the reason why calibration of instruments is of fundamental relevance for plausible results. Daily calibration is not always necessary, but after calibration a daily quality control is essential.

After changing instruments or procedures as well as after opening new batches of antibodies, or beads, one has to verify that nothing has changed and identical results can be obtained. This can be done easily by parallel measurements of the old and the new batch. If necessary, correction factors should be calculated.

Internal Calibration

Basically, two types of standardization can be distinguished: an internal and external standardization. In internal standardization, calibration particles (microbeads: synthetic microparticles of the size of a cell) with a defined number of antimouse immunoglobulin antibodies on the surface (table 1) are used. These particles are treated the same way as cells and stained with the same mouse antibodies. Under saturating conditions, the brightness of the particles correlates directly with the number of fluorescence molecules per cell [15, 16]. Antigens on cells constitute another possible internal standard. For this purpose, cells with a known and relatively stable number of surface antigens are required. Reference antigens such as CD45, CD3, CD4 and other typical surface molecules of human lymphocytes are suited for this type of standardization [15]. Activation antigens are not suitable for standardization due to the fact that their expression depends on the biological activity of the cell. Labeling of cells with appropriate anti-CD antibodies under saturating conditions, i.e. all antigens of a sample are bound to the respective antibody, enables a correlation between the brightness of cells and the number of surface antigens as well as the generation of calibration curves [16]. For internal standardization with cells, fresh material from healthy volunteers and commercial cell products are suitable (table 1).

Pro: The same antibodies (and fluorochromes) as in cell staining can be used for internal calibration. This means that the same staining conditions (temperature, pH, light exposure) apply for calibration and for the sample. The results are thus directly comparable. Internal calibration is ideal to adjust compensation and determine the optimal measuring range.

Contra: With beads, due to the antimouse antibodies used, only one antibody (and hence only one fluorochrome) can be tested at a same time. Further antibodies have to be tested separately, which increases the calibration costs of multicolor experiments.

External Calibration

The second type of standardization is the so-called external calibration. Even though external calibration is not as precise as internal calibration, it is still an easy and reproducible method of standardization. In the simplest case, standardization beads with a defined amount of fluorochromes are used. Such beads can be purchased from several companies (table 1). Beads with different fluorescent dye concentrations, i.e. different fluorescence intensities, are most appropriate. There are beads made up of particles with usually six or eight different brightness levels (rainbow beads, Spherotech, Libertyville, Ill., USA) (fig. 1). Prior to analysis, beads are diluted tenfold in PBS (e.g. Sigma-Aldrich, St. Louis, Mo., USA) and approximately 10,000 events are measured with exactly the same settings as used for biological material. Further analysis is shown in figure 1. For all fluorescence channels, the mean fluorescence intensities (MFI) of each brightness level of the particles (eight in the example) are determined. These fluorescence intensities are plotted vs. the number of molecules of equivalent soluble fluorochrome (MESF), provided by the manufacturer for every type of particle (fig. 1C). A calibration curve can then be calculated (see 'Relevant Equations' below). Using this calibration curve, MFI values of the cells can be converted to MESF values.

For further analyses, data are displayed as scatter plot (two-parameter histogram or dot plot) with FSC vs. SSC (fig. 1A). A gate is set around the main population of measured events and only these gated events are further analyzed. Histograms for each fluorescence channel are created to show the respective fluorescence intensities

Table 1. Examples of calibration particles and their applicability for calibration of cytometers, staining, compensation as well as determination of MESF and cell counting

Name	Calibration of				
	instrument	staining	Compensation	molecule number (MESF)	cell count
Synthetic					
Rainbow[a, c]	+++	–	–	(++)	–
Flow-Check[b, h]	+++	–	+	–	–
Flow-Set[b]	+++	–	–	–	–
CaliBRITE[c]	++	–	++	–	–
Calibration Beads[f]	++	–	++	–	–
QuantiBrite[c]	+	–	+	++	–
Quantum MESF[*, d]	++	–	++	++	–
Quantum Simply Cellular[*, d]	++	++	++	++	–
QIFI[e*]	++	++	++	++	–
TruCount[c]	++	–	–	–	++
Flow-Count[b]	++	–	–	–	++
CountCheck[f]	++	–	–	–	++
Biological					
CYTO-Comp[*, b]	+	+++	+++	+++	–
Immuno-Trol[*, b]	+	+++**	+++	+++	–
Leukocytes (normal donor)*	+	+++**	+++	+++	–
DNA (nuclei normal blood)*	+++	+++	++	+	–
GCRBC[g, i]	+++	++	+	+	–

GCRBC = Glutaraldehyde-fixed chicken red blood cells.
* Staining analogous to cells.
** Standardization of staining, including lysis of erythrocytes.
[a] Spherotech Inc.
[b] Beckman Coulter.
[c] BD Biosciences.
[d] Bangs Laboratories Inc.
[e] DAKOCytomation.
[f] Partec.
[g] Innovative Research.
[h] Polysciences Inc.
[i] Rockland Immunochemicals Inc.

of the beads. Excluded events are mainly aggregates or debris. Additional analysis steps are shown in figure 1. MFI values of every 'color', and brightness level are determined and plotted vs. MESF per particle (fig. 1C). The resulting calibration curve allows an estimation of the MESF values based on the measured MFIs.

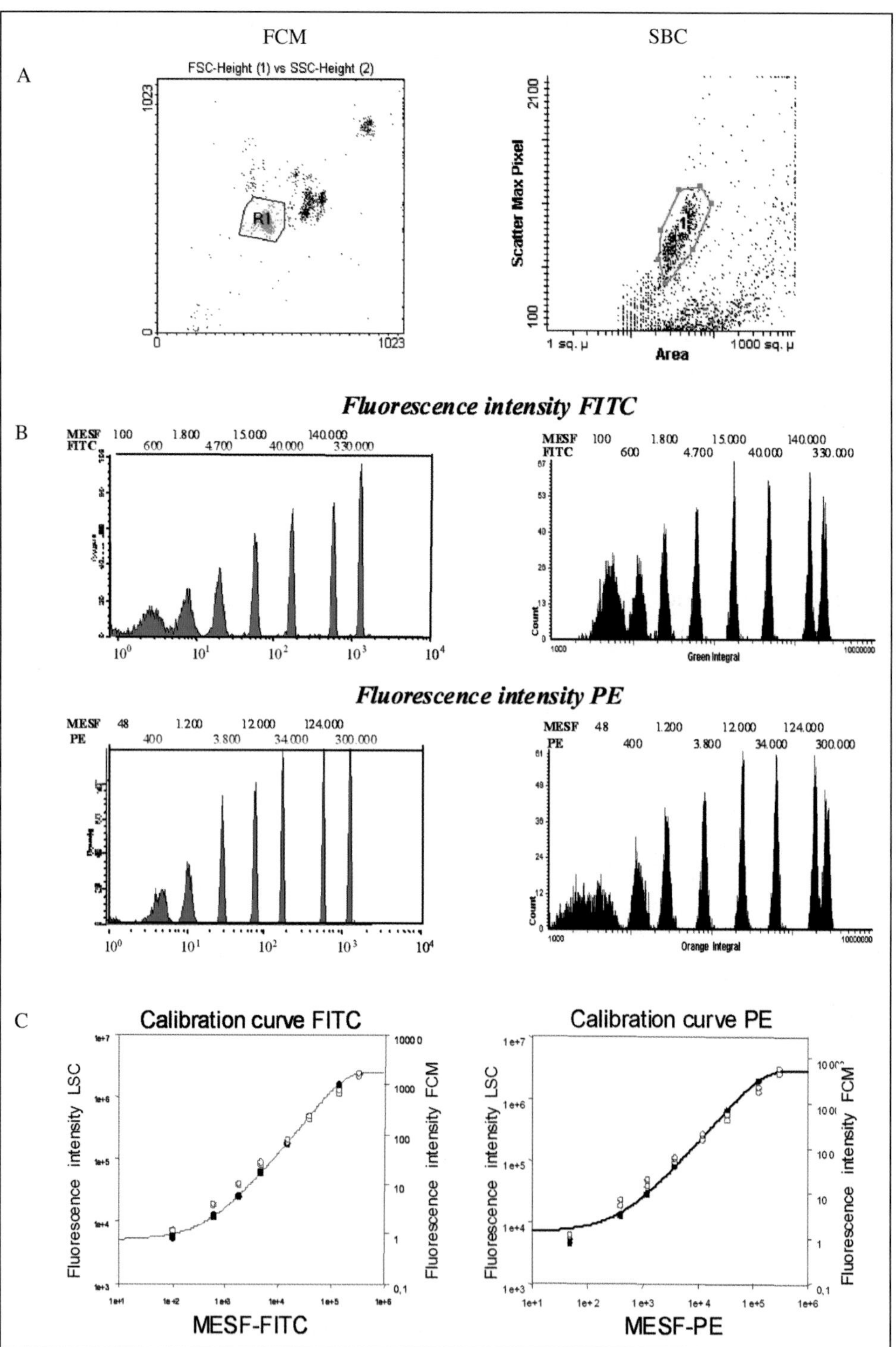
FCM
SBC
A
FSC-Height (1) vs SSC-Height (2)
R1
Scatter Max Pixel
Area
1 sq. µ
1000 sq. µ
B
Fluorescence intensity FITC
MESF FITC
100
600
1.800
4.700
15.000
40.000
140.000
330.000
Count
Green Integral
Fluorescence intensity PE
MESF PE
48
400
1.200
3.800
12.000
34.000
124.000
300.000
Orange Integral
C
Calibration curve FITC
Calibration curve PE
Fluorescence intensity LSC
Fluorescence intensity FCM
MESF-FITC
MESF-PE

After calibrating a cytometer instrument, settings should not be changed for the following measurements. Otherwise, the existing calibration curve cannot be used for determination of antigen numbers.

Pro: This method is well suited for testing the sensitivity of the instrument and calibrating the measuring range. An example for a long-running control of instrument settings is shown in figure 2. Single-fluorochrome dyed beads are adequate for setting up compensation.

Contra: There is no control whether the cell preparation worked faultlessly. Furthermore, beads possess other fluorescence and scatter characteristics than cells. Beads containing a mixture of fluorescent dyes cannot be used for adjustment of compensation.

Additional Parameters to Be Calibrated

Optical Alignment

The purpose of optical alignment in FCM analyses is to center the sample stream in the laser beam. Generally, this is done by custom services and is necessary in 'open' systems. To this end, particles with very uniform scatter and fluorescence are used.

Light Scatter

Due to its dependency on the measured light scatter angle and geometric properties of the optical system, the scatter signal is very difficult to standardize. Although it is difficult to standardize the light scatter of two different types of instruments, it is possible with instruments of identical construction as well as for monitoring the relative performance of an instrument. Light scatter can be standardized by measuring the relative and absolute scatter intensities of two particles of different sizes. The elliptical shape of fixed chicken red blood cells generates a characteristic scattergram, with two peaks in FSC. Calibration with standard particles of known size allows estimation of cell size.

Fig. 1. Calibration of the scale of a cytometer on the basis of MFI values of calibration beads with different brightness (Spherotech Rainbow^{TM}). Calibration data of FCM are on the left side, SBC data on the right. **A** Dot plot of beads (left: FSC vs. SSC; right: FSC vs. area). Only data of gate R1 are used for further analysis. **B** Histogram of fluorescence intensity distribution. The upper two histograms show fluorescence distribution in the FITC channel, the lower two, PE channel distribution. Digits above the histograms indicate the molecule numbers (MESF) provided by the manufacturer. **C** Calibration curves for FITC and PE. Molecule numbers are plotted vs. fluorescence integral measured by FCM (FACSCalibur, BD Biosciences) and SBC (LSC, CompuCyte Corp.).

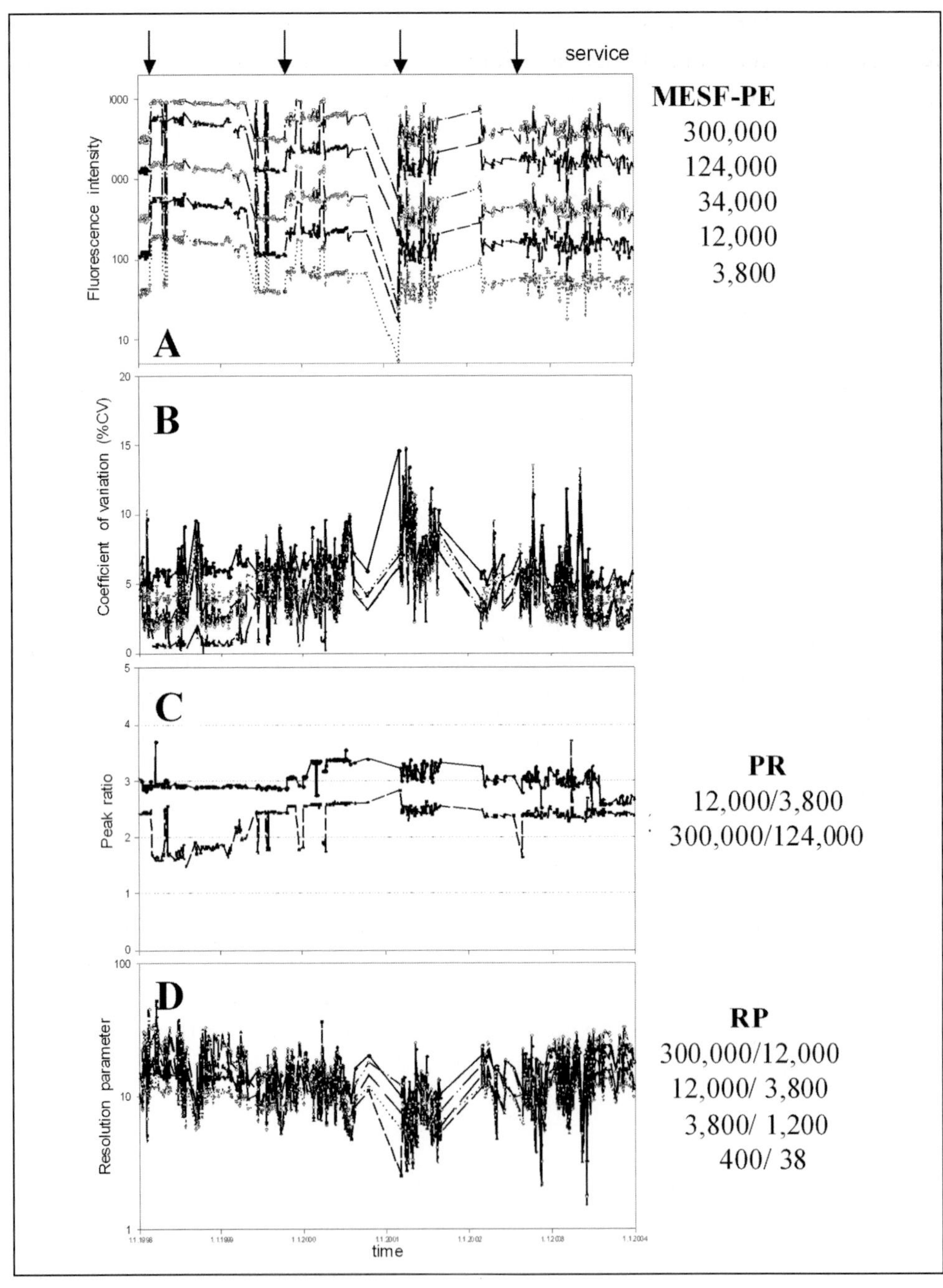

Fig. 2. Monitoring of the quality of an FCM instrument over a period of 4 years. Results of the six brightest Spherotech calibration beads. The arrows indicate the time points of service. All these data were obtained with identical instrument settings. **A** Fluorescence intensity. **B** CV values. **C** PR. **D** RP.

Fluorescence and Scatter Resolution
Resolution, i.e. the ability to distinguish two cell populations of different brightness, is commonly determined with uniform particles by measuring the coefficient of variation (CV; see 'Relevant Equations' below). In fluorescence measurements, CV is a good way to specify the resolution because fluorescence signals are generally proportional to the number of fluorescence dye molecules bound to a particle.

Measuring Range
The measuring range in cytometric analyses can be reported in relative (arbitrary) or absolute units. Determining the relative range is quite simple: the sum of the MFI signals of two adjacent particles should be twice the amount of a single particle. The absolute range calibrates parameters in units like MESF or antibody-binding capacity (ABC). For testing the relative linearity of measurements, small beads or other stained particles are typically used. In a relative scale, only relational data can be described, i.e. bright and dim populations. To convert this relative scale into an absolute scale, a bead mixture of many different fluorescence values with known relative intensities is used. Calibration of these mixed beads in MESF units enables definition of the absolute log range [17, 18] (fig. 1).

For adjusting the instrument, beads should be applied. The brightness of positive cells will fall within the central part of the logarithmic scale (2nd and 3rd log). Sensitivity, linearity, and resolution are optimal in this range whereas in the first and last decade of a logarithmic scale, resolution and linearity are reduced.

Peak Ratio
The standard definition of the lowest resolvable signal is $S/N = 1$, i.e. in this case, the signals of positive cells (S) and background noise (N) are equal. In cytometry, the CV value is commonly used instead of the peak ratio. The simple equation $CV = N/S$ can be used to convert both values reciprocally. The background noise can also be expressed in MESF terms. The MESF value belonging to $S/N = 1$ (or other ratios) is a measure of sensitivity. More helpful equations for the calculation of measuring and staining quality can be found in 'Relevant Equations' below.

Cell Count
In clinical diagnostics, determination of the absolute cell count (cells per volume) of specific blood leukocytes is often required (e.g. T-helper cell count in AIDS diagnostics). To perform an absolute cell count, the instrument should previously be calibrated. Commercially available beads with a defined number of particles per volume (table 1) are suitable for this purpose. Since most FCMs have a stable flow rate (volume per time unit), cell count per volume can also be calculated by the measuring time and the total number of measured cells. Generally, for an absolute cell count, the cell sample (blood) has to be stained directly, without additional washing steps, due to the fact that washing and centrifugation lead to an unknown cell loss.

Standardization of Sample Preparation

Other factors besides those mentioned so far can influence the results. A well-adjusted instrument and accurate quality control cannot correct errors caused by improper preparations. Furthermore, apparently good measurements cannot guarantee correct results in case of a wrong data analysis. The hardware of a cytometer is only part of the measuring unit which has to work faultlessly. Moreover, photobleaching, duration of storage as well as storage conditions, measuring protocol, type and concentration of antibodies used may also have an influence of the results. Hence, it is not sufficient to calibrate the sole instrument: the whole preparative procedure must be standardized as well. This includes, among others, the staining method, the utilization of the same antibodies as well as storage conditions and time elapsed until measurements [19]. Strict compliance with an established standard protocol is essential for immune monitoring (e.g. monitoring of transplantation and therapy) and predictive medicine [20].

Titration of Antibodies and Dyes

New monoclonal antibodies should be titrated to determine saturating concentrations (fig. 3). For this purpose, a defined number of cells are stained with different antibody concentrations. The respective MFI is measured. By increasing the concentration, the fluorescence intensity of labeled cells should rise to a plateau (saturated conditions). It is important to include MFI of nonspecific bindings (negative cells) because MFI also rise at increasing antibody concentrations (fig. 3). Nonspecific binding can be checked within the sample by unstained cells. Optimal concentration is where the peak ratio is maximal. The specific fluorescence intensity is the difference between the brightness of stained cells and the background. Final antibody concentration should be slightly above saturated conditions to adjust variations in cell counts in different samples.

Negative Control

For every cytometric measurement, it is important to measure the brightness of cells that are non-specifically bound by an antibody. This can easily be done by measuring unlabeled cells treated the same way as stained cells. However, it is also feasible to use control antibodies that only bind nonspecifically (fig. 3, upper picture) though it has become less important to use such isotypic controls in recent years. Fluorescence intensities of cells stained with control antibodies provide information on background staining. Control antibodies should be chosen in such a way that they are on par with the respective specific antibodies, i.e. same species and same class of immunoglobulins. Specific and unspecific staining can be distinguished based on the position of the cells stained with control antibodies.

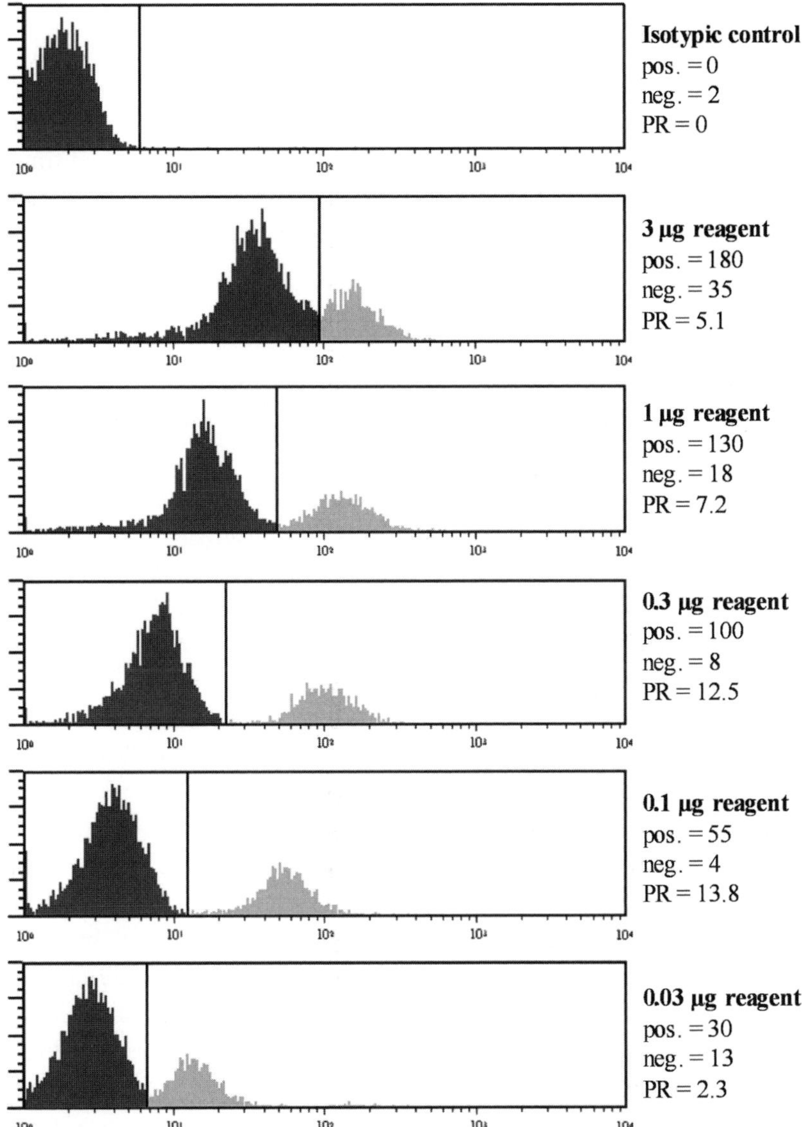

Fig. 3. Example of antibody titration for determining the antibody concentration for optimal discrimination of stained and unstained cells. Results are shown for decreasing antibody concentrations. Only part of the cell population should be stained. Lines delimit positive and negative cells. The optimal concentration of this antibody is in the range of 0.1–0.3 μg as shown by the highest PR values.

DNA Measurements

Guaranteed linearity and resolution of fluorescence are the preconditions for DNA measurements. Sample preparation and data analysis have to be carefully checked as well. For standardization, either fluorescent beads or stained cells or nuclei are suitable. Internal controls are necessary for the identification of abnormal DNA content. Chicken, trout, or human leukocytes are commonly used for this purpose.

Fluorescence Media

As SBC measurements are based on analyses of cells fixed on slides, the mounting medium is frequently used to cover the cells. In most cases, this medium is an anti-fading mounting medium. The main property of such mounting media is to keep fluorochromes (and thus also their fluorescence) stable over a longer period of time. The added anti-fading and buffering agents also stabilize the pH. However, not all fluorochromes react positively to such fluorescence media. Hence an appropriate medium should be selected.

Calibration Particles

There are two different types of calibration particles: synthetic or biological.

Synthetic particles are made of polymers (beads, plastic beads, latex beads, microbeads). Their sizes range from the submicron range to 100 μm in diameter. This usually covers the range of cell sizes analyzed cytometrically. Most of these synthesized particles are produced by polymerization. Very uniform beads can also be manufactured by using the principle of droplet generation, also applied in flow-cytometric cell sorting [21]. Colored or fluorescent beads are obtained by staining such beads with one or several fluorochromes. Nonfluorescent beads as well as most fluorescent beads are stable over a long period of time. Two methods are used to stain particles: solvent (or 'hard') dying and surface staining. In solvent staining, water-insoluble dyes and particles are mixed in an organic solvent. The particles take up the dye and are then suspended in an aqueous solution. The dye is embedded within the beads, becoming basically the hard-dyed synthetic material. In some cases, hard-dyed particles can be synthesized directly from fluorescence monomers [22]. As most of the dyes or fluorochromes used for cell staining are water soluble, they cannot always be used for solvent staining. If solvent staining of water-soluble fluorochromes is possible, their spectral characteristics can differ significantly from those of fluorochromes in aqueous solutions or on the surface of cells. Surface staining allows utilization of many conventional fluorochromes, in particular those used for labeling of antibodies. In this case, a chemical group on the surface of the particle (e.g. amino groups) is covalently bound to a reactive group on the fluorochrome (e.g. carboxyl groups).

Biological particles can be stained with the same fluorochromes as those used for cell staining.

For standardization of cytometers, both synthetic and biological particles may be used. The emission characteristics of particles (beads) can be adapted to the characteristics of the fluorochromes used or can also be effective in a wider spectral range. Spectrally adapted beads allow standardization and calibration of instruments with different optical filters [23, 24].

Calibration via Particle Fluorescence

To compare a known concentration of particles directly with a fluorochrome solution, the MESF value is necessary. MESF is the standardized measure of particle fluorescence. The signal of a fluorescent particle corresponds to a known number of molecules in solution (fig. 1).

Beads and particles with known dye concentrations (molecules/ml) can be analyzed with respect to MESF if a fluorochrome solution is used as reference. If excitation and emission wavelength are close to each other, bandpass filters should be used to minimize light-scattering artifacts from the excitation light source.

First of all, the signal intensity of a reference fluorochrome at a given concentration is determined. Then, the particle suspension to be calibrated is measured with the same instrument settings. The fluorescence of this particle suspension can be expressed as MESF by comparing it with the reference fluorochrome solution. If necessary, particle concentration (particle/ml) has to be corrected (doublets, artifacts). The MESF value per particle is calculated as the ratio of the fluorochrome concentration equivalent of a bead suspension and the particle concentration. At low MESF values, the measurement or the instrument might become imprecise or less sensitive. In case of a precise calibration of the logarithmic scale of fluorescence channels, the upper range of the fluorescence channel is calibrated with brighter particles. The same process can be applied to linear amplification. Dimmer particles can now be investigated on the calibrated instrument.

Details for the Use of Particles for Standardization

Two important factors have to be kept in mind if manufactured particles are measured by FCM instead of cells. First, beads are not cells and might scatter light in a different way than cells. Second, the fluorescence of a bead may be similar to that of a cell stained with the same fluorochrome, but is almost never identical.

Different light-scattering characteristics depend on differences in the refraction index of beads and cells. Beads with high water content have a lower index of refraction (e.g. Sephadex, chromatography beads) and hence their scatter signal is similar to that of cells. Fluorochromes used for solvent staining of beads are rarely the same as those used to stain cells. Even if the fluorescence emission spectrum of a hard-dyed bead is equal to that of a fluorescent cell, the excitation spectrum is rarely identical, unless beads and cells are stained with the same fluorochrome (because it is not exposed to the solvent).

Fluorochromes used for antibody labeling are also available on surface-stained beads. The fluorescence characteristics of such beads correspond to those of antibody-labeled cells. Surface-stained beads have the same excitation spectrum as cells stained with the same fluorescence dyes.

Calibration for Slide-Based Cytometers

Basically, SBC instruments are calibrated in the same way as FCM instruments. The main difference is that calibration particles have to be immobilized in the slide. Slides are commercially available but can also be prepared in the lab.

Relevant Equations

Based on data obtained by calibration, the following device-related parameters can be calculated:

- mean or median fluorescence intensity of the calibration particle;
- CV: standard deviation (SD) divided by the mean brightness (MFI) of a given cell population in percent

$$\%CV = 100\frac{\mathrm{SD}}{\mathrm{Mean}} \tag{1}$$

- peak ratio (PR): median of the brightness of a positive population divided by the fluorescence integral of a negative population

$$\mathrm{PR} = \mathrm{Median}^{+}/\mathrm{Median}^{-} \tag{2}$$

- resolution parameter (RP): mean brightness (MFI) of a positive population minus the fluorescence integral of a negative population divided by the sum of the respective standard deviations

$$\frac{\mathrm{RP} = \mathrm{Mean}^{+} - \mathrm{Mean}^{-}}{\mathrm{SD}^{+} + \mathrm{SD}^{-}} \tag{3}$$

CV, PR, and RP are measures of instrument sensitivity. The smaller the CV or the higher the PR or RP, the better the discrimination between stained and unstained cells [24] (see examples in figs. 2, 3). Deviant values indicate the need to clean and adjust the instrument or dysfunctions of its individual components and assemblies.

- Four-parametric sigmoid function to determine the number of molecules per fluorescence

$$y = y_0 + \frac{a}{1 + e^{-\left(\frac{x - x_0}{b}\right)}} \rightarrow x : \mathrm{mean}, \quad y : \mathrm{MESF} \tag{4}$$

This function is helpful for converting MFI to MESF. Data of the bead's MFIs are plotted vs. the respective MESF values provided by the company. Then a fitting function is calculated with the equation specified above using appropriate software (e.g. Sigma Plot or SPSS, both SPSS Inc., Chicago, Ill., USA). Fluorescence intensities of the cells can thus be converted to MESF (example: fig. 1C).

Examples

A typical calibration of FCM and SBC instruments is shown in figure 1B. In these measurements, Rainbow™ beads consisting of eight populations (fig. 1B) of different brightness are used. Such beads are suitable for calibrating the green (FITC), orange (PE), and red (PerCP, PE-Cy5) channels. For calibrating the APC channel (He–Ne laser excitation), other particles are available.

The best resolution can be achieved in the green and orange channels at 488 nm excitation. As shown in figure 1B, C, even very dim signals (<500 molecules/cell) can be separated from the background in the green channel. This difference in brightness is not distinguishable by eye in a microscope. FITC, PE, and APC are thus best suited to detect low-expressed antigens.

Each MESF value can be assigned to a bead population according to the manufacturer's specifications. On the basis of these data, a calibration curve can be plotted with which the respective molecule number can be assigned to any fluorescence (fig. 1, bottom). To plot this calibration curve, either the manufacturer's software or a four-parameter sigmoid regression curve (SPSS) can be used.

Figure 2 shows continuous monitoring of the measuring quality of an FCM instrument over a period of 4 years. Fluorescence intensity (fig. 2A) as well as % CV (fig. 2B), PR value (fig. 2C), and RP value (fig. 2D) are helpful to decide whether maintenance is necessary.

Troubleshooting

Flow Cytometer

Beads are not measurable or only in very low numbers:
- increase the concentration or flow rate in the instrument settings;
- check threshold: beads are often smaller than leukocytes; if necessary, reduce threshold value;
- a tube or nozzle might be clogged; perform cleaning procedures according to the manufacturer's specifications.

Slide-Based Cytometer

No scatter signal is detectable:
- this problem can often be solved by adjusting the obscuration bar; the quantity of laser light falling on the photodiode is thus altered; when running stem cell measurements, keep in mind that these cells emit a lower scatter signal than, e.g. leukocytes.

Scatter signal is too bright:

– if the background of the scatter signal is too high, imprecise data are generated. In this case, the focus should be checked and the obscuration bar adjusted if necessary; details are listed in the instrument manual.

Summary

Standardization, calibration, and control are essential for quality assurance. The purpose of standardization is to develop a standard protocol for sample preparation and measurement that allows comparison of data over a long period of time and between different instruments and laboratories. Calibration enables one to verify whether the instrument fulfills the requirements. Such requirements may be therapeutic questions that can be answered by cytometric measurements. The process of quality assurance quantifies the variance from the desired value. Results can thus be compared objectively with those of other laboratories. Standardization is the basis of cytometry and a prerequisite for obtaining reliable data.

In this chapter, different ways to standardize fluorescence are described and discussed. For further details on standardization, we refer the reader to the global consensus protocol of the NCCLS [2] as well as to other reference works [3, 4].

References

1 Virchow R: Die Cellularpathologie in ihrer Begründung auf physiologische und pathologische Gewebelehre. Berlin, Hirschwald, 1859.

2 Marti GE, Vogt RF, Gaigalas AK, Hixson CS, Hoffman RA, Lenkei R, Magru-der LE, Purvis NB, Schwartz A, Shapiro HM, Waggoner A: Fluorescence Calibration and Quantitative Measurement of Fluorescence Intensity; Approved Guideline. NCCLS document I/LA24-A, 2004.

3 Robinson JP, Darzynkiewicz Z, Hoffman R, Nolan JP, Orfao A, Rabinovitch P, Watkins S (eds): Current Protocols in Cytometry. New York, Wiley, 2007.

4 Shapiro HM: Practical Flow Cytometry. New York, Wiley, 2003.

5 Haugland RP: The Handbook – A Guide to Fluorescent Probes and Labeling Technology, ed 10. Eugene, Molecular Probes, 2005.

6 Dittrich W, Göhde W: Impulszytophotometrie bei Einzelzellen in Suspension. Z Naturforsch 1969;B24:221–228.

7 Crissman HA, Mullaney PF, Steinkamp JA: Methods and applications of flow systems for analysis and sorting of mammalian cells. Methods Cell Biol 1975;9:179–246.

8 Kamentsky LA, Kamentsky LD: Microscope-based multiparameter laser scanning cytometer yielding data comparable to flow cytometry data. Cytometry 1991;12:381–387.

9 Tárnok A, Gerstner AO: Clinical applications of laser scanning cytometry. Cytometry 2002;50:133–143.

10 Kamentsky LA, Burger DE, Gershman RJ, Lamentsky LD, Luther E: Slide-based laser scanning cytometry. Acta Cytol 1997;41:123–143.

11 Varga VS, Bocsi J, Sipos F, Csendes G, Tulassay Z, Molnar B: Scanning fluorescent microscopy is an alternative for quantitative fluorescent cell analysis. Cytometry A 2004;60:53–62.

12 Bajaj S, Welsh JB, Leif RC, Price JH: Ultra-rare-event detection performance of a custom scanning cytometer on a model preparation of fetal nRBCs. Cytometry 2000;39:285–294.

13 McLaughlin BE, Baumgarth N, Bigos M, Roederer M, De Rosa SC, Altman JD, Nixon DF, Ottinger J, Oxford C, Evans TG, Asmuth DM. Nine-color flow cytometry for accurate measurement of T cell subsets and cytokine responses. I. Panel design by an empiric approach. Cytometry A 2008;73A:400–410.

14 McLaughlin BE, Baumgarth N, Bigos M, Roederer M, De Rosa SC, Altman JD, Nixon DF, Ottinger J, Li J, Beckett L, Shacklett BL, Evans TG, Asmuth DM: Nine-color flow cytometry for accurate measurement of T cell subsets and cytokine responses. II. Panel performance across different instrument platforms. Cytometry A 2008;73A:411–420.

15 Bikoue A, George F, Poncelet P, Mutin M, Janossy G, Sampol J: Quantitative analysis of leukocyte membrane antigen expression: normal adult values. Cytometry 1996;26:137–147.

16 Bikoue A, Janossy G, Barnett D: Stabilised cellular immuno-fluorescence assay: CD45 expression as a calibration standard for human leukocytes. J Immunol Methods 2002;266:19–32.

17 Schwartz A, Fernandez-Repollet E: Development of clinical standards for flow cytometry. Ann NY Acad Sci 1993;677:28–39.

18 Schwartz A, Fernandez-Repollet E, Vogt R, Gratama J: Standardizing flow cytometry: construction of a standardized fluorescence calibration blot using matching special calibrators. Cytometry 1996;26:22–31.

19 Prince HE, Arens L: Effect of storage on lymphocyte surface markers in whole blood units. Transplantation 1986;41:235–238.

20 Valet GK, Tárnok A: Cytomics in predictive medicine. Cytometry B Clin Cytom 2003;53B:1–3.

21 Fulwyler MJ: Standards for flow cytometry; in Melamed MR, Mullaney PF, Mendelsohn ML (eds): Flow Cytometry and Sorting. New York, Wiley, 1973, pp 351–358.

22 Rembaum A: Microspheres as immunoreagents for cell identification; in Melamed MR, Mullaney PF, Mendelsohn ML (eds): Flow Cytometry and Sorting. New York, Wiley, 1979, p 335.

23 Poon R, Fernandez-Repollet E, Ottinger J, Schwartz A: Characterization of available standards for flow cytometry. Flow Cytometry Standards FCS Forum 6 Num, San Juan, 1994, vol 2, pp 1–4.

24 Caldwell CW, Maggi J, Henry LB, Taylor HM: Fluorescence intensity as a quality control parameter in clinical flow cytometry. Am J Clin Pathol 1987;88:447–456.

Sack U, Tárnok A, Rothe G (eds): Cellular Diagnostics. Basics, Methods and Clinical Applications of Flow Cytometry. Basel, Karger, 2009, pp 178–189

Flow-Cytometric Cell Sorting

Michael Cross · Viola Döbel

Division of Hematology/Oncology, IZKF, Leipzig, Germany

Introduction

Many of those who have used fluorescence markers to analyze their cells would like to go on to purify subpopulations with particular characteristics for further study. Compared to the accurate detection of a marked cell, however, the successful recovery of that cell involves a higher level of technology, some delicate and expensive instrumentation and a great deal more training. For these reasons it is unlikely that an experimenter be let loose directly on a high-speed cell sorter, but rather that he or she will be expected to interact with a trained and experienced operator who will at least oversee the procedure. However, even for the novice who has the option of simply using an all-inclusive cell-sorting service, it is invaluable to understand the basic principles behind the technology, the different modes of operation, settings, capabilities and limitations of the machine, in order to appreciate: i) which sort procedure can best meet the demands of each particular experiment; ii) how to prepare your samples to ensure the best possible results and iii) what you can realistically expect to get back in terms of yield and purity.

Experience has shown that it is advisable to discuss the design of each experiment with the operator beforehand, in order to balance the experimental requirements with the capabilities of the machine. This chapter is intended to arm you with the information necessary to make these discussions as informative, constructive and effective as possible.

The Principle behind Cell Sorting

In analytical flow cytometry, it really does not matter which cell generates which signal since everything that flows past the laser beam is on its way to the waste anyway. In a cell sorter, on the other hand, a cell with particular properties needs not only to

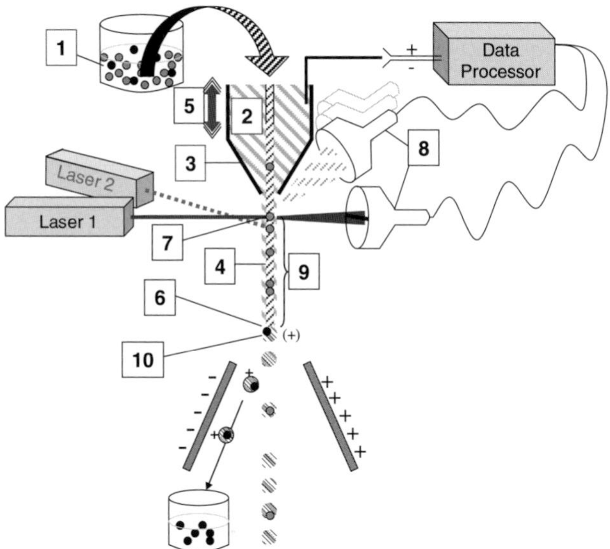

Fig. 1. The principle of preparative flow cytometry. The cell suspension (1), surrounded by sheath fluid (2) is passed under pressure through a nozzle (3) in order to create a stream in air (4). Oscillation of the nozzle assembly (5) breaks the stream up into single drops at the drop-off point (6). When a desired cell passes by the laser intercept (7) it is recognized by the optical system (8), which then pauses for a length of time equal to the drop delay (9) before depositing a charge on the drop expected to contain the target cell (10). The charged drop is then deflected out of the main stream in the field between two charged plates and collected.

be recognized but thereafter to be rescued away from all other cells. In short, this is achieved by passing the cell suspension under pressure through a nozzle in order to generate a 'stream in air' (fig. 1). The cells enter the nozzle as a tributary into a carefully controlled flow of 'sheath fluid' which (as long as the flow remains nonturbulent) uniformly envelops the inner core of cell suspension. Regulating the flow of sheath fluid allows precise hydrodynamic focusing of the inner core, so that the cells emerge from the nozzle in a tight line. At the same time the nozzle assembly is subject to continuous, high-frequency oscillation in the vertical plane, causing the continuous stream-in-air to break up at some point below the laser intercept. This has the aim of isolating single cells into single droplets.

The laser illumination of the cells and the collection of the various scatter and fluorescence signals involve exactly the same principles as used in analytical machines. Many models incorporate quartz windows similar to those in an analyzer, so that optical information is gathered before the cells enter the nozzle. In others the laser intercept is below the nozzle at the 'stream-in-air' stage. This compromises the quality of the optical signals somewhat, but simplifies the coordination required to correctly predict which cell ends up in which drop (see below).

In either case, the scatter and fluorescence signals corresponding to a cell passing the laser intercept are not simply stored for later analysis, but are used immediately to decide whether the sum of the optical properties associated with a particular event represent a 'desired' cell or not. Should this be the case, then the drop into which this cell segregates is given an electrostatic charge, allowing it to be deflected out of the droplet stream in an electrical field between two charged plates. Since the drop can be given either a positive or a negative charge (+, –) it is possible to sort two independent target cell types simultaneously (left and right) out of the main stream. Some machines manage even more, by delivering charges of different strength to the drops concerned (2+, +, –, 2–) resulting in different degrees of deflection.

This means that successful sorting depends ultimately on the accurate delivery of a precise charge to a cell-containing droplet that does not even exist at the point in time at which the cell is recognized. In effect, the machine must recognize a desired cell as it passes the laser beam(s) and predict the precise instant at which this cell will reach the point at which the stream breaks up into drops (the 'drop off point'). At this moment, the whole stream (including the drop which is forming but still attached) is given a charge. This charge is removed from the system through an earth contact almost immediately, but not before the drop containing the cell of interest has parted company with the stream. The end result is therefore a charge remaining on one drop only, which is consequently deflected out of the stream by the charged plates. If this sounds complicated, then consider that the whole process needs to be repeated many thousand times per second during a typical sort. It is this that makes the preparative technology so much more challenging than the analytical.

Quality versus Quantity

The resolution and precision of a sort depends on a number of parameters, including:

- The establishment of a stable flow in which the cells are neither too close together nor too far apart.
- The optimization of the signal strength and signal to noise ratio through alignment and focusing both of the lasers, and of the detectors.
- The separation of the stream into clean, single droplets.
- The precise coordination between signal recognition, separation of the corresponding cell at the drop off point and the transient charging of the stream.
- The accurate deflection of the charged drop into a collecting vessel.

There are of course a number of limitations which have to be taken into account. Although all users would like to receive a sorted population which i) contains all the cells of interest, ii) contains nothing else and iii) is preferably ready within the next 10 min, this is normally unrealistic. In practice, high purity can only be bought at the expense of yield and/or speed. To explain the reasons for this we must distinguish between a series of *parameters* which need to be optimized to make the sort process as efficient as possible, and different *strategies* which can be employed to optimize the balance between speed, yield and purity for each particular application.

Parameters

The first question asked by most new users is how many cells can be 'sorted' per second. The theoretical maximum naturally depends on the frequency at which the single drops can be produced (the 'drop drive frequency'), which in turn depends on the diameter of the nozzle opening and the pressure with which the cell suspension and the sheath fluid enter the nozzle.

Nozzle Size

Standard practice is to use a nozzle with a 70 μm aperture and to adjust the fluid pressures to generate a 30 μm core of cell suspension surrounded by 20 μm of sheath fluid. This is appropriate for most common cell types and is optimal for the majority of hematopoietic cells. Although other nozzle sizes are available for special applications, a different nozzle size will require different fluid pressures to generate a clean stream of the required dimensions and a different oscillation frequency ('drop drive frequency') to break the stream up into clean drops at the required point. As a general rule, the larger the nozzle the lower both the drop drive frequency and the sheath pressure and therefore the lower the number of cells which can be put through the machine per second. While it may therefore be necessary to use a nozzle size of 300 μm for very large cells, this will involve a marked reduction in the drop drive frequency and a corresponding increase in the time required to process a given number of cells.

Sheath Pressure

The maximum attainable pressure in the fluidics system depends on the machine – the older models run a pressure of around 15 psi, while the newer 'turbo' machines use up to 60 psi or even more. Since a higher pressure can drive a higher flow rate, this allows the machine to generate more drops per second and increases the theoretical maximum sort speed. However, maximum is not necessarily optimum, firstly since high pressure and high speeds make it more difficult to maintain stability, and secondly because a very high pressure (or rather the large pressure drop as the cells leave the nozzle) can damage the cells.

Drop Drive Frequency

Using a system pressure of 30 psi and a standard 70 μm nozzle, it is possible (depending on the machine concerned) to generate a stable stream of 60,000 drops/s. However, this is not to say that one can push 60,000 cells/s through the machine and hope to get a usable result, the reason being that the cells in suspension are not distributed evenly but rather tend to arrive at the laser intercept like city buses – long periods of nothing being punctuated by groups of more than one. This means that some of the desired cells end up so close to undesired ones that they cannot be reliably separated. Furthermore, even a modern sorter equipped with high-speed processors needs a little time to assess the optical data and make up its mind whether or not a particular event is a good cell or a bad one, so that two signals that happen to arrive within a few microseconds of each other are effectively excluded. 'Conflicts' of this nature begin to make a noticeable impact from a throughput frequency of around 10,000 cells/s. For this reason, a good rule of thumb is to adjust the flow rate to aim for an average of one cell every three drops, so that a drop drive frequency of 60,000 Hz would correspond to a sort frequency of 10,000–20,000 cells/s. Incidentally, each adjustment of the drop drive frequency necessitates re-tuning the amplitude of the oscillation. Too high an amplitude leads to the production of satellite drops (these being small droplets splitting off from the main ones) which can interfere dramatically with the charging and deflection process. Too low an amplitude can result in an extension of the drop off point (see below).

Sample Pressure and Particle Rate

The frequency with which cells flow past the laser intercept (the particle rate) depends on the one hand on their concentration in the sample suspension and on the other hand on the sample pressure. It is therefore possible to adjust the sample pressure in order to achieve the required particle rate. There are limits, however. Firstly, it is important that the sample pressure does not fall below the sheath fluid pressure since this can result in sheath fluid flowing back into the sample tube. Secondly, increasing the sample pressure necessarily increases the diameter of the core of cell suspension within the stream. This allows cells to wander out of line as they pass the laser and results in a less uniform illumination, which in turn reduces the quality of the optical signals. For these reasons, it is better to start with a cell suspension of the correct concentration than to try to compensate too much by playing with the sample pressure. As a guideline, if your intention is to sort relatively few cells as cleanly as possible, then aim for a cell density of around 1×10^6/ml. Should you wish on the other hand to process as many cells as possible in a given time (naturally at the expense of the purity and yield) then it is possible for some applications to go up to 10^8/ml – but please only after consulting your machine operator.

Drop-Off Point

The 'drop-off point' is the position at which the stream in air breaks up into a series of single drops. The stream is illuminated in this region by a strobe light coupled directly to the drop-drive frequency allowing the position and form of the first drops to be monitored by a video camera. As long as the machine is running stably, the drop-off point will not wander and the average drop visible in the monitor will remain large and clearly defined. Any instability, be it through progressive changes in temperature or fluid pressures or through the sudden formation of air bubbles or blockages in the system, can change the position of the drop-off point and/or the form of the drops and result in the incomplete, overlapping and inaccurate delivery of charges to the drops. During setup, it is normal to tune the drop drive frequency to achieve the shortest distance from nozzle to the drop-off point, as this should be the most stable state.

Side Streams

Statically charged drops are diverted out of the main stream between two charged plates. The path of the side stream thus formed depends on the size of the drops, the size of the charge deposited on them and the size of the charge on the plates. The combination of evenly sized drops and accurate charging will produce a tight and accurate side stream. However, if the drops are heterogeneous, or if the drop-charging is brought out of phase by poor machine settings or instability, then the side stream becomes broader or separates into a fan of substreams which, even if they do carry the desired cells, are difficult to catch. In practice, the side streams are set up via a test sort procedure, in which the machine 'imagines' a fluorescence signal corresponding to a desired cell in every third drop and tries to hit precisely the right drops with a charge. At this high frequency of positive events, the side stream becomes clearly visible. One can then adjust the phasing of the charge deposition to obtain a single, tightly focused side stream.

Drop Delay

Assuming that the drop-off point is stable and the charges are being placed accurately onto single drops to generate a clean side stream, it only remains to be determined in which of the drops our cell of interest is to be found. In other words, how long does it take our cell to flow from the point of analysis (laser intercept) to the drop-off point? This is the 'drop delay' which has to be taken into account in order to hit the right drop with the right charge.

It is normal to begin by setting an approximate value for the drop delay using a rule-of-thumb procedure based on measurement of the distance between the laser intercept and the drop-off point as a function of the distance between single drops. Fine tuning then follows empirically, by 'trial-and-error' sorting of fluorescent beads. Originally, this involved running a series of sorts under incremental increases of the drop delay, catching the products on a microscope slide and counting the beads under fluorescence to determine which setting produced the best results. These days the job of has become much easier and involves direct visualization of highly fluorescent beads. In short, a position below the separation of main and side streams is illuminated by an additional laser and monitored by video, allowing the path of the fluorescent beads to be seen in real time. One can then simply adjust the drop delay until the maximum number of desired particles is deflected into the side stream.

Monitoring the Machine

The drop delay setting is only good as long as the drop-off point remains stable. For this reason it is normal to keep an eye on both the drop-off point and the side streams during the entire sort. Should problems occur, it is usually best to remove the valuable cell suspension being sorted immediately, and to replace it only when the problem has been solved and the machine is running stably. Note that the alarms with which some machines are equipped can draw your attention to a change in the drop-off point but do nothing to solve the problem. Your sort should therefore be accompanied at all times.

Sort Strategies

Sorted Drops

The aim of the machine settings described above is to make the recognition and sorting of desired cells as efficient as possible. In an ideal world, every positive cell would be recognized and the corresponding drop always precisely charged. Reality looks a little different. Even during absolutely stable operation, some of the cells fall so close to the division between two drops that it is not possible to decide with any certainty in which drop they will actually land. With every decrease in stability, the fate of any particular cell becomes progressively harder to predict and uncertainties increasingly frequent.

To counteract this problem, one has the option of sorting up to three consecutive drops for each event, to increase the probability of catching the target cells. To gain advantage from this expansion of the 'sort window' one must be prepared to pay the price of a slower sort rate since the larger window results in an increased frequency of conflicts (two signals occurring in the same window), which are usually sent to the waste. A standard compromise is to choose a sort window of 1.5 sorted drops. This, of course, is not to say that the machine tries to sort half-drops, but rather that 50% of the time the one 'most likely' drop will be charged, and 50% of the time this drop will be charged together with either the drop preceding it or the drop following it. Which of these is chosen depends on the calculated position of the target cell in the most likely drop. If the cell is expected to be in the upper quarter of one drop, then this drop will be taken together with the next drop. If the cell is predicted to be

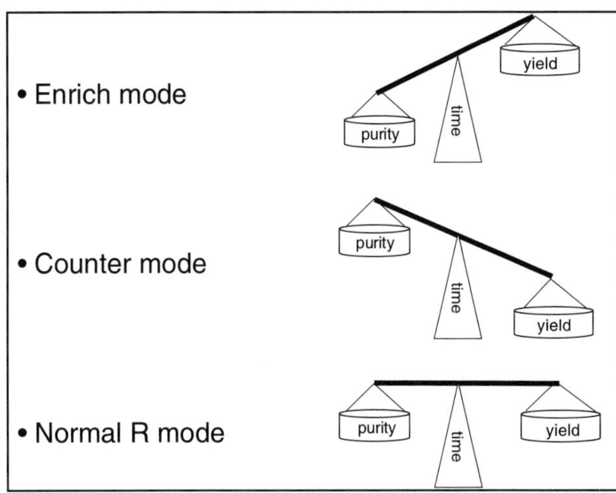

Fig. 2. Sort modes. Some desired cells occur close to undesired cells, leading to uncertainty regarding their separation (conflict). In 'enrich mode' all desired cells are collected, regardless of what may be copurified (highest yield, lowest purity). In 'counter mode' all conflicts are left to flow through to waste (highest purity, lowest yield). In 'normal R mode' only those desired cells which are very close to undesired ones are allowed to flow to waste, while all others are collected (compromise between purity and yield). Since the frequency of conflicts depends on the particle rate (cells/s), both purity and yield can be increased by sorting more slowly.

in the lower quarter of a drop, then this drop will be taken together with the preceding drop.

Sort Modes

It remains unavoidable, at least with the current technology, that some desired cells flow through the system in very close proximity to undesired ones, and the frequency of such conflicts increases with the particle rate. The fate of these events can be determined by the operator to match the requirements and priorities of a particular experiment (fig. 2).

Should it be important to recover as many positive cells as possible, the sort can be run in 'enrich mode'. In this case, every positive cell is diverted into the side stream, regardless of what else might be in the vicinity, so that the increase in yield is paid for by a decrease in purity.

On the other hand, should purity be the highest priority, one can choose to run a sort in 'counter mode', in which every positive cell which happens to be anywhere near an unwanted event is simply ignored. Only those positive cells that can be separated with a high degree of certainty are collected. This means that counter mode should use slow particle rates to ensure good separation, otherwise a great many 'good' cells end up going to waste.

A standard compromise is to use the 'normal R mode'. In this mode, a certain risk of contamination with accompanying cells is accepted. Only when cells occur so

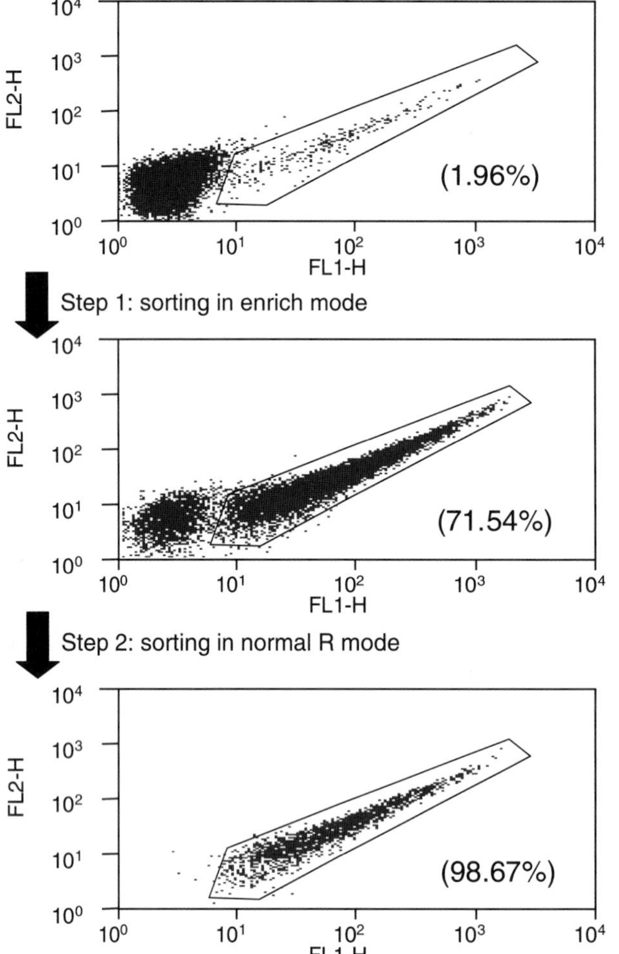

Fig. 3. Sequential sorting. Following transient transfection, 2% of this cell population expressed the GFP reporter gene (shown here in FL1 without compensation). To recover these cells at both high yield and high purity they were first concentrated to 70% purity by sorting in enrich mode to minimize losses, and then purified to >98% purity by sorting in normal R mode.

close together that even a machine that is set up optimally and running stably cannot be expected to separate them are they sent to the waste. As a very rough guideline, a routine sort in normal R mode using a sort window of 1.5 sorted drops should be capable of taking desired cells at a starting frequency of 10% and delivering them at 98% purity. A 5% starting population can be enriched to 95%, and even a 1–2% population comes out at around 80% pure. As long as only one population is to be sorted in any one run (so that only one of the two available side streams is occupied) one has the option of using the 'abort save' mode to direct those conflicting cells which do not qualify for collection into the other side stream. This population will then contain many impurities, but can normally be further processed by re-sorting into a highly pure population. Indeed, the combination of sequential sorts is one way of attaining both high purity and high yield, both of which in this case are paid for by increasing the sort duration. An example of this is shown in figure 3.

In this case, a population of GFP-expressing cells has been sorted from <2 to >70% frequency in enrich mode to minimize cell loss, and then purified to >98% using a standard normal R sort. Alternatively, it often makes sense to enrich target cells from large populations using magnetic-bead-coupled antibodies. Magnetic affinity purification is well suited to the positive or negative enrichment of cell populations which can be distinguished by antibodies to surface antigens. Thereafter, preparative flow cytometry enables high-definition sorting on the basis of combinations and intensities of multiple signals.

Special Applications

Sterile Sorting

Users who are used to the stringent conditions of modern cell culture and who wish to culture their cells after sorting are often surprised to find that the cell sorter is not usually situated in a sterile room. The fact is that as long as the room and the machine are kept clean, the sort chamber regularly sterile-wiped (for instance with 70% ethanol) and the fluidics system sterile-flushed, the frequency of contamination is very low. Many sorters are run continuously on a sterile basis. However, if your machine is used with nonsterile samples and nonsterile sheath fluid, it is important to warn your operator well beforehand that you wish to conduct a sterile sort, since the sterilization of the machine and the preparation of the sterilized sheath fluid (normally autoclaved PBS) requires some time.

ACDU – Single Cell Sorting

If the machine is equipped with a single-cell sorting option (special software, robotic platform, diverse electronic modifications), you will have the option to sort your cells not only into standard tubes, but in defined numbers and formats onto microscope slides or into multiwell plates. This function uses a platform driven by stepper motors to bring each 'target' (e.g. defined wells of a 96-well plate) into position under the side stream. After the required number of cells has been directed towards the target, the platform is moved to bring the next target into the firing line. This operation is always performed in counter mode, whereby in this case the machine ignores not only those positive cells which are too close for comfort to negative cells, but also all events which pass the laser while the platform is moving. Special software enables you to define how many cells (from 1 upwards) should be placed in which position. Using the 'indexing' option it is also possible to save the fluorescence information corresponding to each sorted cell together with the position into which that cell is sorted. Should single cells go on to display interesting properties, one can then take a detailed, retrospective look at their original fluorescence characteristics.

Probe Preparation

After you have agreed with your operator the relevant points of experimental design and sort strategy, it only remains to prepare the probe. Here, there are a few important things to take into account.

Antibody Staining

The optical efficiency of a preparative sorter can be slightly inferior in some respects to that of an analytical machine. Nonetheless, it should be possible to sort almost anything that you can analyze reliably. An important exception are antibodies labeled with phycoerythrin, which can be bleached so rapidly by the strong laser used in the sorter that the signal is subsequently undetectable. Use a different fluorochrome if you can.

The Cell Suspension

The enrich mode can be run with cell concentrations up to 10^8/ml, assuming the priority being to process a lot of cells in a reasonably short time, rather than to produce a sorted population of high purity.

For normal R mode, the sample should be prepared at between 10^6 and 10^7/ml, depending on the desired purity of the product and, of course, the number of cells to be processed in a given time.

Sorting in counter mode is best performed on a sample concentration below 10^6/ml to minimize sort conflicts.

It is recommended to supply the cell suspension to be sorted on ice, and in phosphate-buffered saline (PBS) containing 0.5% bovine serum albumin (BSA) (to prevent cell clumping). The capture tube should also contain a little PBS. It is not recommended to use medium as a sample buffer. Not least because non-heat-inactivated serum can trigger a complement reaction to antibody-labeled cells, which effectively annihilates your fluorescence signal.

Should your cells not tolerate PBS it is very important to notify your FACS operator well beforehand, since it will then be necessary to choose and prepare an alternative sheath buffer as well as an alternative suspension buffer. Hanks balanced salt solution is a good fallback.

The prepared cell suspension will normally be held on ice before the sort, then at 4 °C during the sort. The capture tube is similarly cooled in order to minimize the temperature shock to your cells, although some transient increase during the journey through the tubing, through the nozzle and past the lasers is unavoidable.

It is very important that the cells do not clump. This is not simply because cell clumps are excluded from consideration due to their unusual scatter and fluorescence properties, but more importantly because they can result in blockage in the sample tube or in the nozzle. Even if the operator manages to remove the blockage, this usually results in loss of valuable sample. Should the blockage prove to be persistent it can take a long time to clean and reset the machine. For this reason, dirty and clumped cell suspensions interfere not only with your sort but can delay those that follow you, and you can expect your operator to be correspondingly grateful. A simple way to reduce the risk of cell clumps interfering with the sort is to pass your cell suspension through a cell sieve beforehand.

Dead Cells

As long as the dead cells don't clump (see above) they do not in principle represent a danger to your sort since they can be easily excluded on the basis of their typical scatter properties and/or by propidium iodide staining. However, a high proportion of dead cells increases the number of sort conflicts for any given particle rate. Should your cell preparation contain more than 10–20% dead cells, then it is a good idea to remove them beforehand, for instance by density gradient centrifugation.

Collection Tubes

Please do not forget that your precious cells will not want to land in a dry capture tube. Prepare enough tubes containing 0.5–1 ml PBS beforehand. Although many cell types can be sorted directly into growth medium, some react badly to the dilution of medium with PBS (most of which comes over as sheath fluid) as happens in the capture tube. To be on the safe side, it is usually better to sort cells in PBS (sample buffer) through PBS (sheath fluid) and into PBS (capture tube) and then to recover your sorted cells by centrifugation and resuspend them in the medium of your choice.

Sack U, Tarnok A, Rothe G (eds): Cellular Diagnostics. Basics, Methods and Clinical Applications of Flow Cytometry. Basel, Karger, 2009, pp 190–199

Characterization of T-Lymphocytes

Richard Mauerer · Rudolf Gruber

Synlab Laboratory, Weiden, Germany

Introduction

Staining of different molecules such as proteins and glycoproteins on the surface or in the cytoplasm of peripheral blood cells with specific fluorescence-labeled monoclonal antibodies (mAbs) allows extensive characterization of these cells. mAbs which recognize identical structures are assigned to clusters, called 'clusters of differentiation' (CD), in international workshops [1, 2]. This has resulted in a generally accepted CD nomenclature, which makes the exchange of data and scientific knowledge in this field much easier. Since the 8th CD Workshop in 2004, 339 CD numbers have been established [3, 4]. Fluorescence-labeled mAbs in combination with flow cytometry have clear advantages compared to polyclonal antisera or enzyme-marked antibodies. Nevertheless, in principle, cell subpopulations can also be determined with enzyme-marked antibodies in immune-histological staining followed for instance by microscopic analysis.

Peripheral blood leukocytes can be readily identified as granulocytes, monocytes and lymphocytes based on morphological differences using routine stainings and a light optical microscope. With few exceptions, lymphocytes cannot be differentiated further this way. The development of mAbs together with flow cytometry has enabled classification of lymphocytes into T-, B- and natural killer (NK) cells and definition of subpopulations of these main subgroups for broad use in both scientific and diagnostic applications. Both quantitatively and qualitatively, T-cells, notably CD4+ and CD8+ T-lymphocytes, are the most important subgroup of lymphocytes. CD markers are used for the definition of T-cells, analysis of T-cell development during hematopoiesis, definition of functional subgroups such as naive and memory T-cells, T-helper (Th) 1/2/17 cell populations and α/β- versus γ/δ-T-cell-receptor-positive lymphocytes. Different T-cell receptor (TCR)-specific mAbs are available for a more detailed classification and quantification of restricted TCR repertoires. The activation state of the cells can also be analyzed by means of different surface markers and cytokine-specific mAbs. Variations within T-cell subpopulations can be found physiologically (e.g. circadian rhythm, stress), but are mainly of interest for the assessment of pathological changes in diseases. Different distributions of T-cell subpopulations in various tissues (gut-associated lymphatic tissue, GALT, or mucosa-associated lymphatic tissue, MALT, and spleen) and interspecies differences are interesting especially for research on the immune system. There are many well-established indications for the determination of T-cell subpopulations in routine diagnostics, notably of primary or acquired immunodeficiency syndromes, including HIV infection, and the application of flow cytometry in the diagnoses of leukemias and lymphomas, which will be discussed in detail in the second part of this book.

Markers for the Definition of T-Cells

T-lymphocytes can be defined unambiguously and quantified by using mAbs against the TCR. The TCR is expressed on all normal peripheral T-cells and almost exclusively on T-cells. Nevertheless, for practical reasons, CD3 has become the standard marker to define T-cells in almost all applications. On the one hand, there are two different TCRs (α/β-and γ/δ-TCR) due to which γ/δ-TCR was discovered relatively late and thus no mAb has been available for its detection for a long time. On the other hand, both TCRs cannot be detected with a single mAb. Furthermore, easily applicable highly affine mAbs against CD3 have long been available for routine use. Indeed, CD3 was also described to be expressed on a small subgroup of cells which show characteristic features of NK cells (NKT cells, see below). T-cells also share further surface antigens and functional activities with NK cells. T-cell antigens such as CD2, CD7, CD28 are also found on NK cells, mostly limited to minor subpopulations. Moreover, typical NK cell molecules such as CD16, CD56, CD57, CD94, KIRs and KARs can also appear on subgroups of T-cells. Coexpression of CD3 and CD56 defines a rare subgroup in peripheral blood, with 1–5%, appearing not to be major histocompatibility complex (MHC) restricted [5]. Later, this subgroup was analyzed in more detail and called 'NKT'. These cells are characterized by the simultaneous expression of T-cell markers and NK cell markers. They bear the NK cell receptor NK1.1 (CD161 or NKR-P1) and a TCR, but the TCR repertoire is extremely limited. In mice, only the Vα14-Jα281 TCR is expressed, whereas the homologous TCR Vα24-JαQ (AV24-AJ18) is expressed in humans [6]. Another lymphocyte subpopulation, called 'large granular lymphocytes' (LGLs), represents a small subgroup of cells which can also show characteristic features of T-cells as well as NK cells. In healthy individuals, about 5–15% of peripheral lymphocytes are LGLs. LGLs are defined primarily by morphologic criteria. LGL leukemias derive from CD3+ T-cells in 85% and from CD3– NK cells in 15% [7].

For the correct identification of a certain lymphocyte subpopulation, the absence of an antigen that is typically expressed on another cell population can also be of great value. This plays a role especially in the classification of lymphomas because lymphoma cells often show a pathological expression pattern. Thus the CD19 expression is an important indication that this clone is usually not a T-, but a B-cell, even if T-cell or NK cell markers are coexpressed.

T-Cell Development

Like all hematopoietic cells, T-cells originate from pluripotent stem cells which migrate from the primary lymphatic organs, the fetal liver and bone marrow into the thymus where they develop to functional T-cells. During development from progenitor cells to mature T-cells, several stages of differentiation can be defined on the basis of intracellular and cell surface markers [8]. This development is a continuous process, i.e. coexpression of markers of different stages can be found physiologically on a single cell. In the thymus, the developmental potential of the multipotent progenitor cells gets lost with the beginning of TCR rearrangements. In adult bone marrow, most T-lymphocytes are mature and express CD3 and a TCR on their surface. By contrast, in the fetal bone marrow, T-cell progenitor cells characterized by coexpression of CD34 and CD7 can be found. CD7 is the earliest antigen expressed on the cell surface of a cell which develops into a T-cell. These progenitor T-cells leave the bone marrow and settle in the thymus where they develop to mature T-cells. The CD10 molecule disappears during development before the TCR complex is expressed on the surface. Most T-cells in the thymus are 'double positive', i.e. they express CD4 as well as CD8 on their surface. This is in contrast to mature peripheral blood T-cells, where double-positive T-cells account for less than 2%. Mature T-cells express either CD4 or CD8. Some important molecules on the surface of mature T-cells and their sequential expression during T-cell differentiation are shown schematically in figure 1.

Functional T-Cell Subpopulations

T-cells can be classified into functional subgroups by the expression of different intracellular markers and surface molecules but also by measurement of characteristic cytokine patterns secreted by them. The two major subgroups are defined by CD4 and CD8. CD4 is the ligand of MHC-II, whereas CD8 binds to MHC-I. CD8+ cells have been named 'cytotoxic and suppressor cells'. The idea that CD8+ expression defines a major group of suppressor cells is a historical concept that is not supported by current data. The most important 'suppressor activity' is now thought to be car-

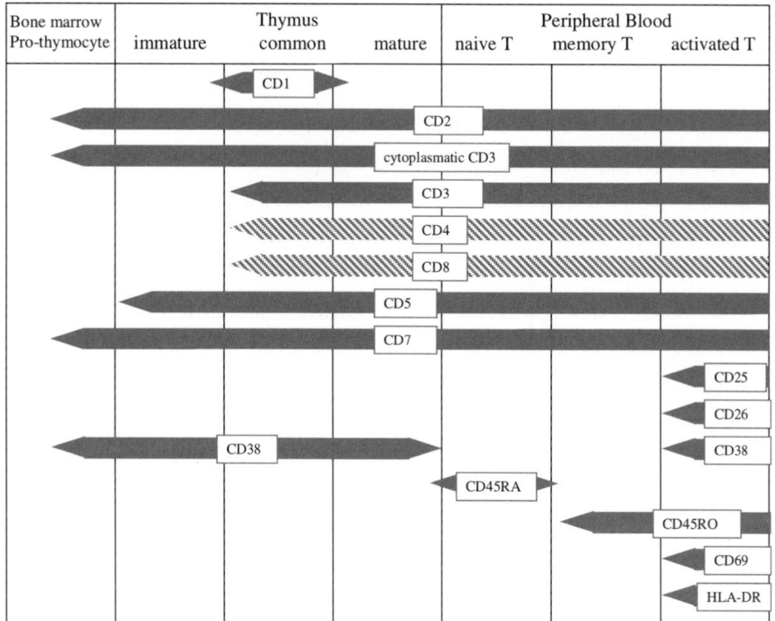

Fig. 1. Sequential expression of CD markers during T-cell differentiation.

ried out by the CD4+ CD25+ cells, called 'regulatory T-cells' or T_{reg} [9]. The concept of antigen-specific CD8+ suppressor cells is not clearly defined, and the phrase 'suppressor cells' should not be used any more as a synonym for CD8+ cells as a whole.

Regulatory T-Cells

T-cells coexpressing the IL-2α chain receptor (CD25) are known to increase during T-cell activation and therefore CD25 is used as a flow-cytometric marker of activated T-cells. In addition, a truncated form of the IL-2 receptor shed by activated T-cells and thus called 'soluble IL-2 receptor' (sIL-2R) can be measured in serum as a marker of an active cellular-mediated immune process. But in 1995 a Japanese group identified a subset of CD25-expressing CD4+ T-cells [10] which exert an immunoregulatory function. More precisely, these CD4+ CD25+ T-cells, later called 'natural T-regulatory cells' (natural T_{reg}), were shown to maintain immunologic self-tolerance. Subsequent work identified the forkhead box transcription factor FOXP3 as the key regulator in the development of T_{reg}. However, as FOXP3 is an intracellular marker, flow-cytometric assessment of human T_{reg} cells is fraught with some practical difficulties in distinguishing CD25-expressing activated and regulatory T-cells. To avoid the disadvantages of intracellular staining, measurement of CD127 (IL-7 receptor α-chain) surface expression was suggested in order to be able to distinguish between these populations. It was shown that high coexpression of CD127 (CD127hi) on CD25+ T-cells rather indicates activated T cells whereas low expression (CD127lo) argues for regulatory T-cells. This is true for both the effector/memory (CD45RO+RA-) and the naive T-cell compartment (CD45RA+RO-) in peripheral blood and lymph nodes [11, 12]. However, although several additional surface markers (GITR, CTLA-4, LAG-3) have been proposed for the correct identification of T_{reg} by flow cytometry, intracellular staining of FOXP3 still remains the 'gold standard'. Mutations in FOXP3 cause a lethal autoimmune syndrome in scurfy mice and humans with IPEX (immune dysregulation, polyendocrinopathy, enteropathy, X-linked) syndrome [13, 14]. CD4+/CD25+/FOXP3+ T-cells are now widely accepted as the most important and biologically most relevant regulatory T-cell population [15]. A number of additional regulatory T-cell subpopulations have been described. These include among others CD4+ FOXP3– IL-10-producing T-cells (also called Tr1), CD8+ and 'double-negative' (CD4– CD8–) regulatory T-cells, but their relevance and role in physiological and pathological processes have not been fully established so far.

T-Helper Subpopulations

CD4+ T-cells are also called Th cells. Both CD4+ and CD8+ cells can be further subdivided into naive and memory T-cells. Different restricted forms of CD45 were set up for their definition, namely CD45RA for naive and CD45RO for memory cells. Furthermore, high expression of CD62L and low expression of CD11a can only be found on naive T-cells.

T-helper 1 (Th1), Th2 and Th17 cells are functionally important subgroups of CD4+ helper cells. Comparable functional differences and different cytokine patterns are also seen within CD8+ cells, but they are less clearly distinctive. Th1 and Th2 cells were described first and best in mice. It was shown that these two subsets produce a reciprocal pattern of immunity by secreting different cytokines, i.e. Th1 cells produce interferon-γ (IFN-γ) and mediate cellular immunity, whereas Th2 produce interleukin 4 (IL-4), IL-5 and IL-13 and promote humoral immunity and allergic response. These cytokines also support the development of their own subset while suppressing the other subset [16]. Another subgroup of Th cells producing IL-17 and therefore called 'Th17' has recently been described. IL-17 seems to play an important role in the regulation of inflammatory responses, especially in autoimmune disorders such as rheumatoid arthritis or systemic lupus erythematosus [17]. For the identification of these different subpopulations by flow cytometry, intracellular staining of the respective characterizing cytokines is still the 'gold standard' [18, 19]. Different methods for cell fixation and membrane permeabilization have been set up and can be used with most flow cytometers (see 'Flow Cytometric Analysis of Intracellularly Strained Cytokines, Phosphoproteins and Microbial Agents', pp. 459). However, detection of intracellular antigens remains sophisticated, and a definition of these subpopulations by surface markers would greatly facilitate integration into a routine panel. Unfortunately, the markers described up to now have not yet proven to be specific enough. Some chemokine receptors have been shown to be preferentially expressed on either Th1 (CXCR3, CXCR6 and CCR5) or Th2 cells (CCR3, CCR4, CCR8). In addition, the receptor for prostaglandin D2, CRTH2 (chemoattractant receptor-homologous molecule expressed on Th2 cells) seems to be the best surface marker for the selective staining of Th2 cells so far [20, 21]. It has been shown very recently that the simultaneous expression of CCR6 and CCR4 identifies IL-17-producing memory T-cells [22]. However, the value of these markers in routine use remains to be established.

NKT Cells

NKT are a rare subpopulation of lymphocytes in peripheral blood. NKT cells coexpress both T- and NK lineage markers, including CD56 or NK1.1 (CD161) in addition to semi-invariant CD1d-restricted αβ-TCRs [23]. NKT cells can be further subdivided into type I NKT cells, also called 'invariant NKT' (iNKT) cells, and type II, non-iNKT, cells. Whereas iNKTs exclusively express TCR Vα24-Jα18 and Vβ11 chains in humans, non-iNKTs bear a very limited but still more diverse TCR repertoire [24]. NKTs recognize lipid and glycolipid antigens presented by antigen-presenting CD1d molecules rather than peptide antigens presented by the MHC, and have been implicated in a broad variety of immunological processes such as immune responses against infectious agents, tumors and tissue grafts. Furthermore, a lack or dysfunction of NKTs has been shown to promote the development of autoimmune diseases. A multicolor flow cytometry assay combining staining for the TCR α-chain segment Vα24, in conjunction withVβ11 or an αGalCer-loaded CD1d tetramer has recently been proposed for the proper identification of iNKT cells in peripheral blood [25].

γ/δ-T-Cells

Whereas the T-lymphocyte subpopulations mentioned before carry α/β-TCRs, another small subgroup (1–5% in peripheral blood) expresses γ/δ-TCRs instead. In humans, 85–98% of peripheral blood T-cells carry the 'classical' α/β-TCR. Flow cytometry offers a quick and relatively favorable possibility to analyze the TCR-Vβ repertoire using a panel of different mAbs against TCRs (TCR-typing panel), which detects more than 80% of the whole repertoire (see also 'T-Cell Receptor Repertoire Analysis by Flow Cytometry', pp. 200). Nearly 100% can be defined by the polymerase chain reaction (PCR) using a panel of TCR-specific primers whereas only few mAbs specifically recognize certain α-chains. This is primarily due to the huge variety of possible TCR-α V and J recombinations and the polymorphous nature of the α-gene [26].

In contrast, TCR-γ/δ is clearly less polymorphic. About 75% of all blood γ/δ-T-cells express the Vγ9δ2 TCR and nearly 100% of the subgroups can be grasped with mAbs for flow cytometry [27]. In addition to these phenotypic differences there are also some important functional differences. Whereas α/β-T-cells are part of the adaptive immune system, γ/δ-T-lymphocytes have an exceptional position among immunocompetent cells as they combine both properties of the innate and adaptive immune system [28]. Neither the antigenic molecules that activate γ/δ-T-cells nor the exact mechanism of ligand recognition are understood so far. Several lines of evidence suggest an implication of these cells in the regulation of immune responses, including responses to pathogens or allergens, and in wound healing [28]. Despite some interesting findings, the use of quantifying or typing of γ/δ-T-cells in routine applications has to be established in further studies.

Activation Marker

To obtain information on the activation state of T-cells, surface markers as well as intracellular or secreted cytokines can be analyzed. For the qualitative and quantitative detection of different cytokines, it is necessary to fix and permeabilize the cells (see 'Flow Cytometric Analysis of Intracellularly Stained Cytokines, Phosphoproteins and Microbial Antigens', pp. 459). In addition, even secreted cytokines can now be measured by flow cytometry in small sample volumes with comparable or even higher sensitivity than with conventional immunoassays. Nevertheless, the determination of cell surface activation markers has long become a classical application of flow cytometry. Besides the upregulation of constitutively expressed markers, de novo expression of antigens which appear on the cell surface only upon activation can be analyzed by specific mAbs. CD38, CD40L, CD69, CD71, HLA-DR and CD25 are important activation markers for T-cells. Expression of CD38 on CD8+ cells for example could be identified as a prognostically relevant parameter in patients with HIV infection (see 'Immunophenotyping CD4 T-Cells to Monitor HIV Disease', pp. 505). Expression of costimulatory markers such as CD27 and CD28 or of cytotoxic markers such as CD57 is also found on CD8+ cells, and indicates immune activation. Also the loss of certain antigens, for example shedding of the adhesion molecule L-selectin (CD62L) upon activation, can be measured by flow cytometry.

Table 1. Factors interfering with determination of lymphocyte subpopulations

Preanalytical factors (blood withdrawal, storage, transport)	
Patient position	As for all cellular elements approximately 5–10% lower values in supine position (absolute cell counts only)
Time	Circadian rhythm
Storage and transport	Prolonged storage and transport influence the results (e.g. storage of blood at 4 °C causes selective changes in certain lymphocyte subpopulations)
Individual factors	Lifestyle and nutrition (e.g. smoking, alcohol abuse)
Permanent factors	Ethnical background
	Sex
	Hereditary factors (e.g. CD45RA heterozygosity; loss of certain CD 4 epitopes
Long-time effects	Age-dependence of certain lymphocyte subpopulations (especially in children, see 'Lymphocyte Subsets in the Peripheral Blood of Healthy Children', pp. 261)
	Gravidity (many laboratory parameters vary physiologically in pregnancy)
Short-time effects	Immobilization
	Nutrition (fasting)
	Different concomitant diseases
	Different therapies (e.g. chemotherapy, immunostimulation, immunosuppression)
	Diagnostic procedures

Determination of T-Cell Subpopulations in Clinical Applications

Peripheral blood lymphocytes with an absolute number of 10×10^9 cells represent only 2% of the whole lymphocyte pool of the human body. Therefore it is evident that when determining peripheral blood lymphocyte subpopulations only a minor fraction of the lymphocyte subpopulation of interest can be analyzed. Peripheral blood lymphocytes are exchanged approximately 50 times per day, which means that daily 500×10^9 lymphocytes pass the blood circulation. Many physiological factors influence lymphocyte distribution. Thus for example a circadian rhythm exists for different lymphocyte subpopulations. Also stress, physical effort, age, ethnic background, drugs and other factors can influence lymphocyte distribution in peripheral blood (table 1). In lymphatic organs and lymphatic tissues (GALT, MALT, spleen) the distribution of lymphocyte subpopulations differs from that in peripheral blood. For patient care, the analysis of lymphocyte subpopulations is relevant for example in bronchoalveolar lavage fluid (BALF) (see ' Flow-Cytometric Analysis of Cells from the Respiratory Tract: Bronchoalveolar Lavage and Induced Sputum', pp. 582) and in cerebrospinal fluid (CSF). Comparison of lymphocyte subpopulations in humans with those of other species can contribute a lot to the understanding of the im-

Table 2. Indications for determination of T-cell subpopulations

Indication		Parameter	Value
Infectiology	HIV staging	Absolute count of CD4+ lymphocytes CD4/CD8 ratio Activation markers (e.g. CD38)	Reference method; Basis for the WHO classification of HIV/AIDS
Immunology	Primary immuno-deficiencies	Quantification of T-/B-cells B-cell differentiation	Essential analysis
	Sarcoidosis	CD4/CD8 ratio in BAL > 3	Important part of diagnosis
	Hypersensitivity pneumonitis	CD4/CD8 ratio in BAL < 1.5	Important part of diagnosis
Hematology/ Oncology	Diagnosis of leukemias/ lymphomas	Immunophenotyping	Important analysis for the classification of lymphomas according to WHO classification

mune system and the role of T-cells. Even primates exhibit relevant differences to the human immune system. Thus a higher number of CD8+ T-cells in comparison to CD4+ cells in chimpanzees is probably associated with a more favorable course of HIV infection [29, 30]. Pathological changes of lymphocyte subpopulations with different immunological, but also nonimmunological diseases are of special interest. The most important indications for flow-cytometric analysis of lymphocytes arise from the determination of subpopulations in several diseases, for diagnosis, prognosis and therapeutic monitoring. Primary and secondary immunodeficiency syndromes and leukemias are major, undisputed targets for flow-cytometric analysis of lymphocytes (table 2). In addition, 'immunomonitoring' is often carried out in other conditions, e.g. in oncological patients and in 'anti-aging' medicine. This includes analyses of major lymphocyte subpopulations in peripheral blood to obtain an overview of cellular immune system function. However, it is extremely doubtful whether a relevant statement on a patient's immune system can be made outside of a clinical study; anyhow, such analyses are expensive.

In conclusion, the qualitative and quantitative analysis of T-cells by flow cytometry has substantially increased our knowledge of the function of the immune system in health and disease. Furthermore quantification of T-cell subsets is still one of the most important applications in the routine use of flow cytometry, e.g. in monitoring HIV patients.

References

1 Knowles R: Immunochemical analysis of the T-cell-specific antigens; in Reiherz E, Haynes B, Nadler L, Berstein I (eds): Leukocyte Typing II: Human T Lymphocytes. New York, Springer-Verlag, 1986, pp 259–288.

2 Moebius U. Cluster Report: CD8; in Knapp W, Dörken B, Wilks W (eds): Leukocyte Typing IV: White Cell Differentiation Antigens.. New York, Oxford University Press, 1989, pp 342–343.
3 Mason D, Andre P, Bensussan A, Buckley C, Civin C, Clark E, et al: CD antigens 2001. Eur J Immunol 2001;31:2841–2847.
4 Mason D, Andre P, Bensussan A, Buckley C, Civin C, Clark E, et al: Reference: CD Antigens 2002. J Immunol 2002;168:2083–2086.
5 Schmidt-Wolf GD, Negrin RS, Schmidt-Wolf IG: Activated T cells and cytokine-induced CD3+ CD56+ killer cells. Ann Hematol 1997;74:51–56.
6 Elewaut D, Kronenberg:M. Molecular biology of NK T cell specificity and development. Semin Immunol 2000;12:561–568.
7 Lamy T, Loughran TP, Jr: Current concepts: large granular lymphocyte leukemia. Blood Rev 1999;13:230–240.
8 Macey M: Leukocyte immunobiology; in McCarthy D, Macey M (eds): Cytometric Analysis of Cell Phenotype and Function. Cambridge, Cambridge University Press, 2001, pp 118–137.
9 Schwartz RH: T cell anergy. Annu Rev Immunol 2003;21:305–334.
10 Sakaguchi S, Sakaguchi N, Asano M, Itoh M, Toda M: Immunologic self-tolerance maintained by activated T cells expressing IL-2 receptor alpha-chains (CD25). Breakdown of a single mechanism of self-tolerance causes various autoimmune diseases. J Immunol 1995;155:1151–1164.
11 Seddiki N, Santner-Nanan B, Martinson J, Zaunders J, Sasson S, Landay A, et al: Expression of interleukin (IL)-2 and IL-7 receptors discriminates between human regulatory and activated T cells. J Exp Med 2006;203:1693–1700.
12 Sigal LH: Basic science for the clinician. 45 CD4+ T-cell subsets of probable clinical consequence. J Clin Rheumatol 2007;13:229–233.
13 Brunkow ME, Jeffery EW, Hjerrild KA, Paeper B, Clark LB, Yasayko SA, et al: Disruption of a new forkhead/ winged-helix protein, scurfin, results in the fatal lymphoproliferative disorder of the scurfy mouse. Nature Genet 2001;27:68–73.
14 Bennett CL, Christie J, Ramsdell F, Brunkow ME, Ferguson PJ, Whitesell L, et al: The immune dysregulation, polyendocrinopathy, enteropathy, X-linked syndrome (IPEX) is caused by mutations of FOXP3. Nature Genet 2001;27:20–21.
15 Shevach EM: From vanilla to 28 flavors: multiple varieties of T regulatory cells. Immunity 2006;25:195–201.
16 Weaver CT, Hatton RD, Mangan PR, Harrington LE. IL-17 family cytokines and the expanding diversity of effector T cell lineages. Annu Rev Immunol 2007;25:821–852.
17 Dong C: Diversification of T-helper-cell lineages: finding the family root of IL-17-producing cells. Nat Rev Immunol 2006;6:329–333.
18 Mills KH, McGuirk P: Antigen-specific regulatory T cells – their induction and role in infection. Semin Immunol 2004;16:107–117.
19 Romagnani S: T-cell subsets (Th1 versus Th2). Ann Allergy Asthma Immunol 2000;85:9–18; quiz , 21.
20 Sallusto F, Geginat J, Lanzavecchia A: Central memory and effector memory T cell subsets: function, generation, and maintenance. Annu Rev Immunol 2004;22:745–763.
21 Sallusto F, Mackay CR, Lanzavecchia A: The role of chemokine receptors in primary, effector, and memory immune responses. Annu Rev Immunol 2000;18:593–620.
22 Acosta-Rodriguez EV, Rivino L, Geginat J, Jarrossay D, Gattorno M, Lanzavecchia A, et al: Surface phenotype and antigenic specificity of human interleukin 17-producing T helper memory cells. Nature Immunol 2007;8:639–646.
23 Bendelac A, Savage PB, Teyton L: The biology of NKT cells. Annu Rev Immunol 2007;25:297–336.
24 Van Kaer L: NKT cells: T lymphocytes with innate effector functions. Curr Opin Immunol 2007;19:354–364.
25 Gonzalez VD, Bjorkstrom NK, Malmberg KJ, Moll M, Kuylenstierna C, Michaelsson J, et al: Application of nine-color flow cytometry for detailed studies of the phenotypic complexity and functional heterogeneity of human lymphocyte subsets. J Immunol Meth 2008;330:64–74.
26 Hodges E, Krishna MT, Pickard C, Smith JL: Diagnostic role of tests for T cell receptor (TCR) genes. J Clin Pathol 2003;56:1–11.
27 Carding SR, Egan PJ: γδ T cells: functional plasticity and heterogeneity. Nat Rev Immunol 2002;2:336–345.
28 Born WK, Jin N, Aydintug MK, Wands JM, French JD, Roark CL, O'Brien RL: γδ T lymphocytes – selectable cells within the innate system? J Clin Imunol 2007;27:133–144.
29 Famularo G, Moretti S, Marcellini S, Nucera E, De Simone C: CD8 lymphocytes in HIV infection: helpful and harmful. J Clin Lab Immunol 1997;49:15–32.
30 Folks TM, Chused TM, Portnoy D, Edison L, Leiserson W, Sell KW: Increased number of Leu-2-bearing non-T cells with natural killer activity in chimpanzees. Cell Immunol 1986;97:164–172.

Sack U, Tárnok A, Rothe G (eds): Cellular Diagnostics. Basics, Methods and Clinical Applications of Flow Cytometry. Basel, Karger, 2009, pp 200–210

T-Cell Receptor Repertoire Analysis by Flow Cytometry

Dieter Kabelitz · Daniela Wesch

Institute of Immunology, University Hospital Schleswig-Holstein Campus Kiel, Kiel, Germany

Introduction

T-lymphocytes (or T-cells) recognize antigen via a clonally distributed T-cell receptor (TCR) molecule which is noncovalently associated with the signal-transducing CD3 molecular complex. Most mature T-cells in the peripheral blood and lymphoid organs carry a heterodimeric TCR which is composed of disulfide-linked α and β chains. Such αβ T-cells recognize antigens as short peptides that are presented on the surface of antigen-presenting cells (e.g. dendritic cells) in the context of major histocompatibility complex (MHC) class I (for CD8+ T-cells) or class II (for CD4+ T-cells) molecules [1]. In addition to αβ T-cells, there is a second population of CD3+ T-cells which express an alternative TCR composed of γ and δ chains. Although there is considerable variability between individuals, γδ T-cells usually account for approximately 5% of CD3+ T-cells in the peripheral blood of healthy adults. However, γδ T-cells are more frequent in other anatomical localizations such as the intestine, where they constitute approximately 20–30% of intraepithelial lymphocytes [2]. In striking contrast to conventional αβ T-cells, γδ T-cells usually do not recognize peptides in an MHC-restricted fashion but rather unconventional ligands in an MHC-nonrestricted way. Most interestingly, the dominant subset of γδ T-cells in the peripheral blood, which specifically expresses Vγ9 paired with Vδ2, recognizes phosphorylated metabolites of the bacterial isoprenoid biosynthesis pathway, collectively termed 'phosphoantigens' [3]. Many pathogenic bacteria and some parasites, that are potent activators of human γδ T-cells, produce and secrete such ligands. So far, the most potent identified microbial ligand is hydroxyl-methyl-butenyl pyrophosphate, which is recognized at picomolar concentrations by human Vγ9Vδ2 T-cells [4]. Interestingly, the bacteria-reactive Vγ9Vδ2 T-cells are the major γδ T-cell subpopulation in the peripheral blood of healthy adults where they account for 50 to >95% of all γδ T-cells [5].

Molecular methods of TCR repertoire analysis are based on the polymerase chain reaction (PCR). The simultaneous amplification of all Vβ genes with primer pairs for all Vβ families is widely used; subsequent sequencing of the CDR3 region allows the identification of clonal expansions within the analyzed T-cell population [6]. More recently, CDR3 length analysis ('CDR3 spectratyping') has been introduced and provides a useful analysis of (oligo)clonality [7, 8]. Alternatively, the quantitative distribution of expressed TCR V gene families within a given T-cell population can be conveniently analyzed by flow cytometry. A large number of monoclonal antibodies (mAbs) directed against many of the expressed human Vα, Vβ, Vγ and Vδ genes has been developed over the last 15 years. Multicolor analysis allows the simultaneous detection of expressed TCR V genes on various subsets of T-cells such a CD4+, CD8+ [9]. More importantly, such T-cell populations expressing particular V genes can be conveniently isolated by magnetic or flow-cytometric cell sorting for further functional analysis. It is also possible to combine the flow-cytometric TCR repertoire analysis with molecular methods such as CDR3 spectratyping [10]. Many mAbs directed against TCR Vβ gene families are commercially available (e.g. from Pierce Biotechnology, BD Biosciences, Beckman Coulter), whereas only a few Vα-specific mAbs are available. For the time being, the Vα repertoire can thus not be analyzed comprehensively by flow cytometry.

Generally speaking, the flow-cytometric TCR repertoire analysis is indicated whenever the (oligo)clonality or polyclonality of a T-cell population is to be determined, e.g. in the peripheral blood or in cell cultures. In a clinical setting, such an analysis is of interest for the diagnosis of hematological diseases (leukemias, reactive T-cell expansion during e.g. EBV-induced infectious mononucleosis [11–14]) and in infectious diseases such as HIV infection, in order to characterize the relationship between disturbance of the TCR repertoire and the stage of infection [15–19]. Moreover, this method is very useful in the research laboratory of cellular immunologists, for instance for the identification of transgenic T-cells with clonotypic anti-TCR Vβ mAbs [20], or to monitor the frequency of superantigen-reactive T-cells which are known to express selective TCR Vβ families [21].

The flow-cytometric determination of the TCR repertoire can point to the clonality of a given T-cell population. However, one should keep in mind that, although expressing the same TCR Vβ family, T-cells might still differ in their hypervariable CDR3 region and thus represent different T-cell clones – despite their identical reactivity pattern with anti-Vβ mAbs. In a diagnostic setting, therefore, the flow-cytometric TCR repertoire analysis is generally followed by a molecular analysis of clonality (CDR3 sequencing, Vβ-Cβ RT-PCR heteroduplex analysis).

Protocol for Flow-Cytometric Analysis of the Human TCR Repertoire

Principle

Usually, the TCR repertoire is analyzed by cell surface staining with anti-TCR mAbs because all mature T-cells express a CD3-associated TCR molecule on their surface. At immature stages of T-cell differentiation (e.g. in leukemias), the TCR protein

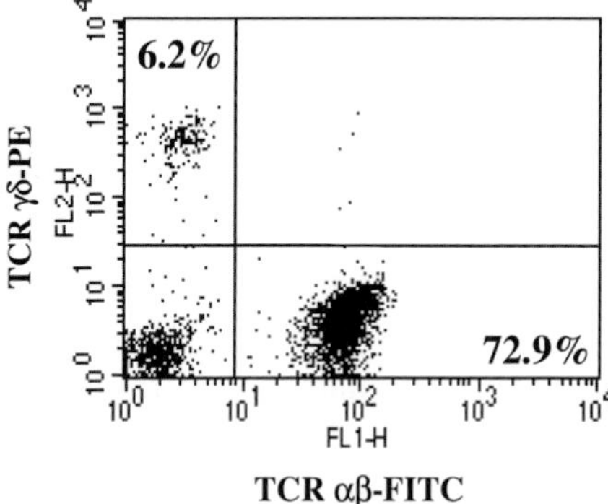

Fig. 1. Distribution of αβ and γδ T-cells in peripheral blood. Ficoll-Hypaque-separated mononuclear cells from a healthy adult donor were stained with FITC-conjugated pan-αβ mAbs BMA031 and PE-conjugated pan-γδ mAbs TCRδ1. The fluorescence intensity is recorded at a 4-log scale.

might be expressed intracellularly [22]; in this case, the staining protocol must be adapted to the analysis of intracellular antigens with permeabilization and fixation steps (see 'Flow-Cytometric Analysis of Intracellularly Stained Cytokines, Phosphoproteins, and Microbial Antigens', pp. 459). In the following section, we will restrict ourselves to protocols for cell surface analysis of TCR expression.

As a first step of TCR repertoire analysis, the relative proportion of αβ and γδ T-cells should be determined. Usually, there are approximately 90–95% αβ T-cells and 5–10% γδ T-cells among CD3+ T-cells in the peripheral blood of adult donors [2]. This can be analyzed by two- or three-color analysis with directly labeled anti-CD3, pan-αβ TCR (e.g. BMA031, anti-Cβ), and pan-γδ TCR (e.g. TCRδ1, anti-Cδ) mAbs. An example of a two-color analysis of αβ versus γδ T-cells within peripheral blood mononuclear cells stained with FITC-conjugated BMA031 (anti-Cβ; pan αβ marker) and PE-conjugated TCRδ1 (anti-Cδ; pan γδ marker) is shown in figure 1. Since αβ and γδ T-cells are mutually exclusive T-cell populations, indirect conclusions can also be drawn if mAbs are available against only one of the two TCR isoforms. In other words, CD3+ Cβ– (e.g. CD3+ BMA031–) cells are most likely γδ T-cells, whereas CD3+ Cδ– (e.g. CD3+ TCRδ1–) cells are most likely αβ T-cells. It is useful to combine the analysis of particular anti-V TCR mAbs with the staining of the relevant constant region (Cβ or Cδ) in order to be able to calculate the relative proportion of a given TCR V family within the total population of αβ or γδ T-cells (see below).

Depending on individual needs, the TCR repertoire analysis can be easily combined with the analysis of additional surface markers such as CD4, CD8, chemokine receptors. Commercial anti-TCR mAbs are usually available as PE or FITC conjugates, in some instances also biotinylated. The reader is referred to the chapter by Schiemann and Busch ('Selection and Combination of Fluorescent Dyes', pp. 107) for the use of other fluorochromes for multicolor analyses.

Although TCR repertoire analysis is frequently done with isolated mononuclear cells, it is also possible to perform whole blood analysis [23] (see 'Technological and Methodological Basics of Flow Cytometry', pp. 53, for more information on whole blood analysis).

Material

We perform the staining of cell populations in 96-well V-bottom microtiter plates (e.g. Nerbe Plus # 101 11 0000). It is mandatory to use a cooled centrifuge for washing steps. The washing buffer is calcium- and magnesium-free PBS with 1% BSA and 0.1% sodium azide. The fixing buffer is PBS with 1% paraformaldehyde. After the staining and fixation steps, cells are resuspended in 100 μl and transferred to tubes (# 2054, BD Biosciences) which fit into 96-well round-bottom plates.

For the detection of biotinylated primary mAbs, we use PE-labeled streptavidin (SA-PE, e.g. from BD Biosciences).

The following companies offer anti-TCR mAbs and anti-CD3 mAbs for the flow-cytometric analysis of the human TCR repertoire:

- Pierce Biotechnology (*www.piercenet.com*)
- Beckman Coulter (*www.beckmancoulter.com/eCatalog/Catalog/Flow_Cytometry/Discipline/T_Cell_Receptors*)
- AbD Serotec (*www.ab-direct.com*)
- BD Biociences (*www.bdbiosciences.com*)
- Invitrogen (*www.invitrogen.com*)
- Interchim (*www.interchim.com/interchim/inter_intro.htm*)

Table 1 summarizes the available mAbs with specificity for expressed constant or variable segments of the human TCR.

We recommend determination of the optimal concentration of the commercial mAbs by titration. It is our general experience that most of these mAbs can be diluted at least 1:2 or 1:5 without any loss of sensitivity. This will significantly help to reduce the costs of flow-cytometric TCR repertoire analysis in the research laboratory.

Working Steps

1) The cells to be stained (e.g. Ficoll-Hypaque-separated peripheral blood mononuclear cells or cells taken from the in vitro culture) are adjusted to 5×10^5/ml; after careful resuspension, 100 μl are transferred with a 12-channel multipipette to wells of V-bottom 96-well microtiter plates. The plates are centrifuged (cooled centrifuge) for 5 min at 600–800 rpm. Subsequently, the supernatant is removed by flicking the plate above the sink.
2) The cells are washed twice with the washing buffer; to this end, cell pellets are resuspended in 100 μl of the washing buffer per well with the multipipette, the plates are centrifuged as above, and the supernatant is removed as above.
3) Following the washing steps, the cells are resuspended in 5–10 μl of the desired fluorochrome-labeled anti-TCR mAbs (use an Eppendorf pipette). The plates are subsequently incubated for 20 min on ice under a cover to protect them from light.
4) Two washing steps follow as above.
5) After the last washing step, cell pellets are each resuspended in 100 μl fixation buffer and transferred to tubes that fit into 96-well round bottom plates. The

Table 1. Available mAbs with specificity for constant or variable regions of the human TCR chains

Specificity	mAb/clone	Company*	Catalogue number**
CD3	HIT3a	BD Biosciences	555337 (0), 555336 (NA/LE), 555339 (FITC), 555340 (PE), 555342 (APC), 555341 (PE-Cy5)
CD3	SK7	BD Biosciences	347340 (0), 339186 (AmCyan)
CD3	CRIS-7	Invitrogen	AHS0312 (0), AHS0318 (FITC), AHS0317 (PE), AHS0319 (biotin)
Cβ (pan-αβ)	BMA031	Pierce	TCR1043 (0)
	BMA031	Beckman Coulter	IM1466 (0),
	BMA031	AbD Serotec	MCA990FT (FITC)
Pan-αβ	T10B9.1A-31	BD Biosciences	555546 (0), 555547 (FITC), 555548 (PE)
Pan-αβ	WT31	BD Biosciences	347770 (0), 33140 (FITC)
Cδ (pan-γδ)	TCRδ1	Pierce	TCR1061 (0), TCR2061 (FITC)
Pan-γδ	11F2	BD Biosciences	347900 (0), 347903 (FITC), 333141 (PE)
Pan-γδ	B1	BD Biosciences	555715 (0), 559878 (FITC), 555717 (PE), 555718 (APC), 555716 (biotin)
Pan γδ	IMMU 510	Beckman Coulter	Im1349 (0), IM1571U (FITC), IM1418U (PE)
Vα2	F1	Pierce	TCR1663 (o), TCR2663 (FITC)
Vα12.1	6D6.6	Pierce	TCR1764 (0), TCR2764 (FITC)
Vα24	C15	Beckman Coulter	IM1588 (0), IM1589 (FITC), IM2283 (PE), IM2027 (biotin)
Vβ1	BL37.2 (rat)	Beckman Coulter	IM2219 (0), IM2406 (FITC), IM2355 (PE)
Vβ2	MPB2D5	Beckman Coulter	IM2006 (0), IM2407 (FITC), IM2213 (PE), IM2081 (biotin)
Vβ3	CH92	Beckman Coulter	IM2372 (FITC)
Vβ3.1	8F10	Pierce	TCR1740 (0), TCR2740 (FITC)
Vβ5.1	IMMU157	Beckman Coulter	IM1551 (0), IM1552 (FITC), IM2285 (PE)
Vβ5.1	LC4	Pierce	TCR2660 (FITC)
Vβ5.2	36213	Beckman Coulter	IM1482 (FITC), IM2286 (PE)
Vβ5.3	3D11	Beckman Coulter	IM2002 (PE)
Vβ5.3	W112	Pierce	TCR2645 (FITC)
Vβ5.2, 5.3	1C1	Pierce	TCR1642 (0), TCR2642 (FITC)
Vβ5.2, 5.3	MH3-2	BD Biosciences	555600 (0), 555602 (FITC), 555603 (PE)
Vβ6.7	OT145	Pierce	TCR1657 (0), TCR2657 (FITC)
Vβ7	ZOE	Beckman Coulter	IM2037 (0), IM2408 (FITC), IM2287 (PE)
Vβ7.1	3G5	Pierce	TCR1665 (0)
Vβ8	56C5.2	Beckman Coulter	IM1148 (0), IM1233 (FITC), IM2289 (PE)
Vβ8	JR2	BD Biosciences	555604 (0), 555606 (FITC), 555607 (PE)
Vβ8 (a)	16G8	Pierce	TCR1648 (0), TCR2648 (FITC)
Vβ8 (b)	MX6	Pierce	TCR1750 (0), TCR2750 (FITC)
Vβ9	FIN9	Beckman Coulter	IM2003 (PE)
Vβ11	C21	Beckman Coulter	IM1582 (0), IM1586 (FITC), IM2290 (PE), IM2018 (biotin)
Vβ12	VER2.32.1	Beckman Coulter	IM1583 (0), IM1587 (FITC), IM2291 (PE)
Vβ12	S511	Pierce	TCR2654 (FITC)
Vβ13.1, 13.3	BAM13	Pierce	TCR2601 (FITC; part of anti-Vβ panel)
Vβ13.1	IMMU 222	Beckman Coulter	IM1553 (0), IM1554 (FITC), IM2292 (PE)

Table 1 continued on next page

Table 1. Continued

Specificity	mAb/clone	Company*	Catalogue number**
Vβ13.6	JU74.3	Beckman Coulter	IM1330 (FITC)
Vβ14	CAS1.1.3	Beckman Coulter	IM1558 (FITC), IM2047 (PE)
Vβ16	TAMAYA1.2	Beckman Coulter	IM1559 (0), IM1560 (FITC)
Vβ17	E17.5F3.15.13	Beckman Coulter	IM1189 (0), IM1234 (FITC), IM2048 (PE)
Vβ17	C1	Pierce	TCR1668 (0)
Vβ18	BA62.6	Beckman Coulter	IM1261 (0), IM2049 (PE)
Vβ20	ELL1.4	Beckman Coulter	IM1562 (FITC), IM2295 (PE)
Vβ21.3	IG125	Beckman Coulter	IM1483 (FITC)
Vβ22	IMMU 546	Beckman Coulter	IM1365 (0), IM1484 (FITC), IM2051 (PE)
Vβ23	AF23	Beckman Coulter	IM2004 (PE)
Vβ23	AHUT7	BD Biosciences	555892 (FITC)
Vδ1	δTCS1 (TS-1)	Pierce	TCR1055 (0), TCR2055 (FITC)
Vδ1	TS8.2	Pierce	TCR1730 (0), TCR2730 (FITC)
Vδ1	R9.12	Beckman Coulter	IM1761 (0)
Vδ2	IMMU389	Beckman Coulter	IM1464 (FITC)
Vδ2	15D	Pierce	TCR1732 (0), TCR2732 (FITC)
Vδ2	B6	BD Biosciences	555738 (FITC), 555739 (PE)
Vγ2,3,4	23D12	own production	[24]
Vγ3,5	56.3	own production	[25]
Vγ9	IMMU 360	Beckman Coulter	IM1366 (0) IM1463 (FITC)
Vγ9	B3	BD Biosciences	555732 (FITC), 555733 (PE)
Vγ9	7A5	Pierce	TCR1720 (0), TCR2720 (FITC) [26]

* Please note: most of the anti-TCR mAbs provided by Beckman Coulter are also available from Interchim.

** The mAbs are available as FITC-conjugates (FITC), PE-conjugates (PE), unconjugated (0), or biotinylated (biotin) as indicated. Certain mAbs (especially anti-CD3) are also available as conjugates with a range of other fluorochromes.

samples are then ready for flow-cytometric analysis. Once fixed, the samples (covered with Parafilm) can be kept up to 4 weeks before analysis.

Specific Remarks

1) The described protocol makes use of directly fluorochrome-conjugated mAbs. Different fluorochromes (e.g. PE, FITC, APC, TC, Cy5.5, Cy7) may be combined with different mAbs (e.g. CD4, CD8, Vβ) in one single staining procedure; in the case of biotinylated or nonconjugated primary mAbs, a second staining step with e.g. SA-PE (for the detection of biotinylated primary mAbs) or fluorochrome-conjugated $F(ab)_2$ goat-anti-mouse IgG (for the detection of nonconjugated primary mAbs) must follow step 4 above.
2) It is mandatory to include appropriate controls, i.e. fluorochrome-labeled Ig of the same species and the same isotype and subclass as the primary anti-TCR mAbs.

Flow Cytometry

The flow-cytometric analysis of TCR expression follows standard procedures as described in other chapters of this book. We routinely use an FACScalibur (BD Biosciences) with the Cell Quest software. A first gate is set on the lymphocyte population on the basis of forward scatter (FSC) and side scatter (SSC); the different size of resting lymphocytes (peripheral blood mononuclear cells) and activated lymphocytes (e.g. taken from the cell culture) needs to be taken into account. Subsequently, the fluorescence is recorded. Since the proportion of cells expressing a particular TCR V gene family is usually quite small, we recommend recording at least 10,000 cells per sample.

Data Analysis

To obtain a quantitative estimation of the TCR repertoire, the distribution of individual V gene expression must be correlated to the relevant TCR constant chain. Therefore, we recommend combination of the staining of Vβ or Vα families with Cβ (e.g. BMA031) and the staining of Vγ or Vδ families with Cδ (e.g. TCRδ1). The relative proportion of a given subfamily within all αβ or γδ T-cells can then be determined. In addition, the relative distribution of αβ versus γδ T-cells within the total population of CD3+ T-cells should be determined. Of course, it is possible to calculate absolute numbers of T-cells expressing particular TCR V gene families by including standard beads for quantification (e.g. BD Truecount Tubes # 340334) as described in other chapters of this book.

Representative Result

Typically, individual Vβ families account for 2–5% of the αβ TCR repertoire [9]. Within the peripheral blood γδ T-cells of healthy adult individuals, Vγ9Vδ2 T-cells are the dominant population and can account for up to 95% of the γδ T-cells [24]. Figure 2 shows a representative result of the flow-cytometric TCR repertoire analysis with anti-Vβ, anti-Vγ and anti-Vδ mAbs of peripheral blood mononuclear cells in a healthy adult donor. The left dot plots illustrate αβ subfamilies, the right dot plots, γδ subfamilies. The relative proportion of the various TCR V gene subfamilies within the αβ or γδ T-cells can be easily calculated by setting the proportion of αβ T-cells (62.5%; fig. 2, upper left dot plot) or γδ T-cells (7.2%; fig. 2, upper right dot plot) to 100%. Whereas the three analyzed Vβ families (Vβ6, 8, 19) each identified approximately 5% of αβ T-cells (62.5% total αβ T-cells; 2.3–3.3% Vβ+ Cβ+ T-cells), both Vδ2 and Vγ9 identified approximately 90% of the γδ T-cells in this donor (7.2% total γδ T-cells; 6.0% Vγ9+ Cδ+, 6.7% Vδ2+ Cδ+ T-cells). In contrast, Vδ1 accounted for only 0.2% of the γδ T-cells in this donor (table 2).

General Remarks

Technically, the flow-cytometric TCR repertoire analysis is not more demanding than other multicolor flow-cytometric analyses. As already mentioned, it is no problem to combine this method with the analysis of additional surface markers.

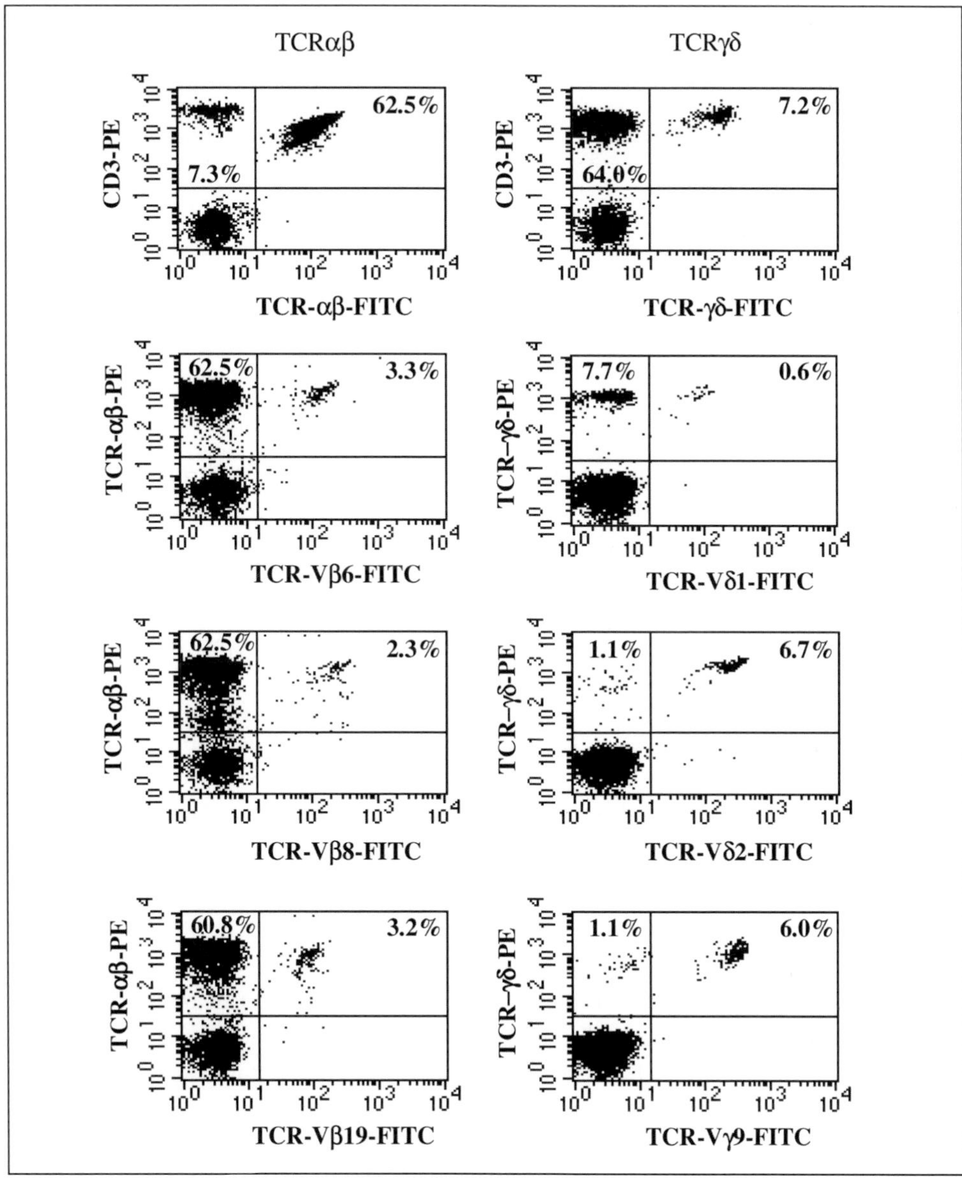

Fig. 2. Representative analysis of Vβ and Vγ/Vδ usage in peripheral blood αβ and γδ T-cells. Ficoll-Hypaque-separated mononuclear cells from a healthy adult donor were stained with FITC-labeled pan-αβ mAbs BMA031 or the indicated anti-Vβ mAbs plus PE-labeled anti-CD3 or pan αβ mAbs (left). Alternatively, mononuclear cells were stained with FITC-labeled pan-γδ mAbs TCRδ1 of FITC-labeled anti-Vδ/anti-Vγ mAbs plus PE-labeled anti-CD3 or pan γδ mAbs as indicated in the right part.

Troubleshooting

Some TCR Vβ families are used only at low frequency. In addition, there is considerable heterogeneity among individuals in the usage of different Vβ families. Therefore, it is sometimes difficult to demonstrate the presence of certain TCR V gene

Table 2. Calculation of the relative frequency of specific $V\beta$, $V\delta$ or $V\gamma$ subsets within the $\alpha\beta$ or $\gamma\delta$ T-cells based on the flow-cytometric result presented in figure 2

Staining	Cells in the upper right quadrant, %	$\alpha\beta$ T-cells or $\gamma\delta$ T-cells expressing the indicated V gene family, %
$V\beta6$//TCR$\alpha\beta$	3.3	5.0
$V\beta8$/TCR$\alpha\beta$	23	3.5
$V\beta19$/TCR$\alpha\beta$	3.2	5.0
$V\delta1$/TCR$\gamma\delta$	0.6	7.2
$V\delta2$/TCR$\gamma\delta$	6.7	85.9
$V\gamma9$/TCR$\gamma\delta$	6.0	84.5

families in certain healthy subjects. In these instances, positive controls are very useful. Established T-cell clones or leukemic T-cell populations with well-characterized TCR repertoires are very helpful positive control samples.

Possible Misinterpretations

1) Theoretically, the process of allelic exclusion should result in the expression of only one TCR in a given T-cell, leading to reactivity with only one and not several anti-TCR V gene mAbs. However, this allelic exclusion is not complete. In fact, it has been shown that a sizeable proportion of T-cells (both $\alpha\beta$ and $\gamma\delta$) can actually express two different TCR molecules on their surface [27, 28]. As a consequence, it is very well possible that a given T-cell population might be stained with more than one anti-V gene family mAb.
2) In rare instances, a transrearrangement between TCR genes can occur, giving rise to T-cells that express a given TCR V gene together with the 'wrong' TCR constant region. For instance, a $V\gamma$ gene instead of a $V\beta$ can be rearranged to the $C\beta$ gene, giving rise to T-cells that are normal $\alpha\beta$ T-cells (i.e. they express $C\alpha$ and $C\beta$), but the β chain carries a $V\gamma$ gene instead of the $V\beta$ gene [29]. Such cells occur at very low frequency in healthy individuals and can be identified by flow cytometry, e.g. as $V\gamma9+$ $C\beta+$ double-positive T-cells [30].

Summary

T-cells recognize antigen via a CD3-associated TCR molecule. Most mature T-cells express a TCR composed of a heterodimeric $\alpha\beta$ protein complex, whereas a minor subset (approximately 5%) expresses the alternative $\gamma\delta$ TCR. In both cases, the TCR protein chains are composed of variable (V) and constant (C) regions. Functional human $V\beta$ and $V\alpha$ are grouped into 24 and 29 families, respectively. In contrast, there are only 6 expressed human $V\gamma$ genes and a similarly small number of Vδ genes. Most Vβ but only a few Vα families can be detected by mAbs. Several mAbs specifically detect human Vγ and Vδ genes. The TCR repertoire of αβ and γδ T-cells can be analyzed with available anti-V gene antibodies in combination with antibodies directed against the constant region (e.g. anti-Cβ, anti-Cδ) of the αβ- or γδ TCR. In pathological situations of lymphocyte expansions, such an analysis can point to a possible mono- or oligoclonal proliferation of a particular T-cell subset. Therefore, the flow-cytometric TCR repertoire analysis is indicated in cases of hematological and infectious (e.g. HIV) diseases.

References

1 Davis MM, Boniface JJ, Reich Z, Lyons D, Hampl J, Arden B, Chien Y: Ligand recognition by αβ T cell receptor. Annu Rev Immunol 1998;16:523–544.

2 Hayday AC: γδ T cells: a right time and a right place for a conserved third way of protection. Annu Rev Immunol 2000;11:637–685.

3 Constant P, Davodeau F, Peyrat MA, Poquet Y, Puzo G, Bonneville M, Fournié JJ: Stimulation of human γδ T-cells by nonpeptidic mycobacterial ligands Science 1994; 264:267–270.

4 Kabelitz D: Small molecules for the activation of human γδ T cell responses against infection. Recent Patents Anti-Infect Drug Disc 2008;3:1–9.

5 Wesch D, Hinz T, Kabelitz D: Analysis of the TCR Vγ repertoire in healthy donors and HIV-1-infected individuals. Int Immunol 1998;10:1067–1075.

6 Marguerie C, Lunardi C, So A: PCR-based analysis of the TCR repertoire in human autoimmune diseases. Immunol Today 1992;13:336–338.

7 Manfras B, Rudert WA, Trucco M, Boehm B: Analysis of the α/β T-cell receptor repertoire by competitive and quantitative family-specific PCR with exogenous standards and high resolution fluorescence based CDR3 imaging. J Immunol Meth 1997;210:235–249.

8 Berovici N, Duffour MT, Agrarwal S, Salcedo M, Abastada JP: New Methods for assessing T-cell responses. Clin Diagn Lab Immunol 2000;7:859–864.

9 Van den Beemd R, Boor PPC, van Lochem EG, Hop WCJ, Langerak AW, Wolvers-Tettero ILM, Hooijkaas H, van Dongen JJM: Flow cytometric analysis of the Vβ repertoire in healthy controls. Cytometry 2000;40:336–345.

10 Pilch H, Höhn H, Freitag K, Neukirch C, Necker A, Haddad P, Tanner B, Knapstein PG, Maeurer MJ: Improved assessment of T-cell receptor (TCR) Vβ repertoire in clinical specimens: combination of TCR-CDR3 spectratyping with flow cytometry-based TCR Vβ frequency analysis. Clin Diagn Lab Immunol 2002;9:257–266.

11 Langerak AW, van den Beemd R, Wolvers-Tettero ILM, Boor PPC, van Lochem EG, Hooijkaas H, van Dongen JM: Molecular and flow cytometric analysis of the Vβ repertoire for clonality assessment in mature TCRαβ T-cell proliferations. Blood 2001;98:165–173.

12 Lima M, Almeida J, Santos AH, dos Anjos Teixeira M, del Carmen Alguero M, Queirós ML, Balanzategui A, Justica B, Gonzalez M, San Miguel JF, Orfao A: Immunophenotypic analysis of the TCR-Vβ repertoire in 98 persistent expansions of $CD3^+$/TCR-$\alpha\beta^+$ large granular lymphocytes. Utility in assessing clonality and insights into the pathogenesis of the disease. Am J Pathol 2001;159:1861–1868.

13 Beck RC, Stahl S, O'Keefe CL, Maciejewski JP, Theil KS, His ED: Detection of mature T-cell leukemias by flow cytometry using anti-T-cell receptor B antibodies. Am J Clin Pathol 2003;120:785–794.

14 Lima M, dos Anjos Teixeira M, Queirós ML, Santos AH, Goncalves C, Correia J, Farinha F, Mendonca F, Soares JM, Almeida J, Orfao A, Justica B: Immunophenotype and TCR-Vβ repertoire of peripheral blood T-cells in acute infectious mononucleosis. Blood Cells Mol Dis 2003;30:1–12.

15 Cossarizza A: T-cell repertoire and HIV infection: facts and perspectives. AIDS 1997; 11:1075–1088.

16 Gea-Banacloche JG, Weiskopf EE, Hallahan C, Bernaldo de Quirós JCL, Flanigan M, Mican JM, Falloon J, Baseler M, Stevens R, Lane HC, Connors M: Progression of human immunodeficiency virus disease is associated with increasing disruptions within the $CD4^+$ T cell receptor repertoire. J Clin Invest 1998;177:579–585.

17 Soudeyns H, Champagne P, Holloway CL, Silvestri GU, Ringuette N, Samson J, Lapointe N, Sékaly RP: Transient T cell receptor β-chain variable region-specific expansions of $CD4^+$ and $CD8^+$ T cells during the early phase of pediatric human immunodeficiency virus infection: characterization of expanded cell populations by T cell receptor phenotyping. J Infect Dis 2000;181:107–12.

18 Bodman-Smith MD, Williams I, Johnstone R, Boylston A, Lydyard PM, Zumla A: T cell receptor usage in patients with non-progressing HIV infection. Clin Exp Immunol 2002;130:115–120.

19 Kabelitz D: Disturbance of the T-cell receptor repertoire in HIV infection; in Hengge UR, Kabelitz D (eds): Immunotreatment and Gene Therapy of HIV Infection. Bremen, UNI-MED, 2004, pp 14–24.

20 Alferink J, Schittek B, Schönrich G, Hämmerling GJ, Arnold B: Long life span of tolerant T cells and the role of antigen in maintenance of peripheral tolerance. Int Immunol 1995;7:331–336.

21 Jenkinson EJ, Kingston R, Smith CA, Williams GT, Owen JJ: Antigen-induced apoptosis in developing T cells: a mechanism for negative selection of the T cell recepotr repertoire. Eur J Immunol 1989;19:2175–2177.

22 Inukai T, Saito M, Mori T, Nishino K, Abe T, Kinoshita A, Suzuki T, Kurosawa Y, Okazaki T, Sugita K, et al: Analysis of cytoplasmic and surface antigens in childhood T-cell acute lymphoblastic leukaemias: clinical relevance of cytoplasmic TCR β chain expression. Br J Haematol 1994;87:273–281.

23 MacIsaac C, Curtis N, Cade J, Visvanathan K: Rapid analysis of the Vβ repertoire of CD4 and CD8 T lymphocytes in whole blood. J Immunol Meth 2003;283:9–15.

24 Kabelitz D, Ackermann T, Hinz T, Davodeau F, Band H, Bonneville M, Janssen O, Arden B, Schondelmaier S: New monoclonal antibody (23D12) recognizing three different Vγ elements of the human γδ T cell receptor: $23D12^+$ cells comprise a major subpopulation of γδ T cells in postnatal thymus. J Immunol 1994;152:3128–3136.

25 Hinz T, Wesch D, Halary F, Marx S, Choudhary A, Arden B, Janssen O, Bonneville M, Kabelitz D: Identification of the complete expressed human T-cell receptor Vγ repertoire by flow cytometry. Int Immunol 1997;9:1065–1072.

26 Janssen O, Wesselborg S, Heckl-Östreicher B, Bender A, Schondelmaier S, Pechhold K, Moldenhauer G, Kabelitz D: T-cell receptor/CD3-signalling induces death by apoptosis in human T-cell receptor γδ-positive T-cells. J Immunol 1991; 146: 35–39.

27 Hinz T, Marx S, Nerl C, Kabelitz D: Clonal expansion of γδ T cells expressing two distinct T cell receptors. Br J Haematol 1996;94:62–64.

28 Hinz T, Weidmann E, Kabelitz D: Dual TCR-expressing T lymphocytes in health and disease. Int Arch Allergy Immunol 2001;125:16–20.

29 Allam A, Kabelitz D: TCR trans-rearrangements: biological significance in antigen recognition vs the role as lymphoma biomarker. J Immunol. 2006;176:5707–5712.

30 Hinz T, Allam A, Wesch D, Schindler D, Kabelitz D: Cell surface expression of transrearranged Vγ-Cβ T cell receptor chains in healthy donors and in ataxia telangiectasia patients. Br J Haematol 2000;109:201–210.

Sack U, Tárnok A, Rothe G (eds): Cellular Diagnostics. Basics, Methods and Clinical Applications of Flow Cytometry. Basel, Karger, 2009, pp 211–229

Characterization of B-Lymphocytes

Michael Schlesier · Klaus Warnatz

Department of Rheumatology and Clinical Immunology, Medical Clinic, University Hospital of Freiburg, Germany

Introduction

B-lymphocytes are the main mediators of the specific humoral immune response. They initially develop from hematopoietic stem cells of the fetal liver and are subsequently produced in the bone marrow. During this antigen-independent differentiation, a step-by-step rearrangement of the immunoglobulin (Ig) heavy chain (μ-chain) and light chain locus (κ- or λ-chain) occurs, ending with the surface expression of a clonotypic antigen receptor in the form of surface-bound IgM and IgD. These antigen receptors are associated with the signal transduction molecules Igα (CD79a) and Igβ (CD79b). Mature B-cells circulate in the blood as the main B-cell population until their homing-receptor-mediated migration to the secondary lymphoid organs via high endothelial venules. Regulated by chemokines, they then populate B-cell follicles. If B-cells are activated by their antigen receptors while passing through the T-cell areas and costimulated by T-helper cells, they will form germinal centers in the B-cell follicles and display high proliferation and somatic hypermutation of their V-regional genes. These centroblasts mature into centrocytes whose mutated antigen receptor is tested on follicular dendritic cells laden with immune complexes. Centrocytes not sufficiently stimulated in this phase undergo apoptosis and die. Only those centrocytes with high affinity for the antigen are further expanded. Contact and signals from antigen-specific T-helper cells trigger the IgG, IgA or IgE class switch of the constant heavy-chain region, thus promoting differentiation towards either long-lived memory cell or plasmablast. Memory cells recirculate in the blood and lymphoid organs while plasmablasts primarily migrate to the bone marrow where they undergo final differentiation into long-lived, antigen-producing plasma cells [for reviews see 1–5].

Human B-cells are produced continuously throughout life, although their quantity decreases with progressing age [4, 6, 7]. The various stages of B-cell develop-

Table 1. B-cell differentiation markers in bone marrow[a]

Marker	Pro-B	Pre-B-I	Pre-B-II		Immature B	Mature B
			large	small		
Lineage-specific						
CD22	+	+	+	+	+	++
CyCD79a	–/+	+	+	+	+	+
CD19	–	+	+	+	+	+
Differentiation						
CD20	–	–	–	(+)	+	+
CD10	–/(+)	+	+	+	+	–
CD34	+	+	–	–	–	–
CD24	–	++	++	++	++	+
CD38	++	++	++	++	++	+
Cy-TdT	–/+	+	+/–	–	–	–
Cy-VpreB (CD179a)	–	+	+	–	–	–
Cy-μ	–	–	+	+	+	+
Sm-IgM	–	–	–	–	+	+
Sm-IgD	–	–	–	–	–/+	+
Recombination						
RAG	–	+	–	+	–	–
Ig heavy chain	germline	DJ	VDJ	VDJ	VDJ	VDJ
Ig light chain	germline	germline	germline	VJ	VJ	VJ

[a] Modified according to Ghia et al. [7] and Noordzij et al. [9].
– = Negative; –/+ = negative and positive subpopulations; (+) = weakly positive; + = positive; ++ = strongly positive.

ment can be determined based on differentiation-dependent molecules, pre-B-cell receptor expression, proliferation and recombination of the Ig heavy- and light-chain genes. In the literature, there is no uniform nomenclature denoting these stages of maturation [2, 3, 7–9]. This article uses the nomenclature of Ghia et al. [7] and Noordzij et al. [9], where the pre-B-I state is defined by complete DJ rearrangement of the heavy-chain locus and intracelluar detection of the surrogate light-chain VpreB (table 1).

The various stages of B-cell differentiation in peripheral blood and bone marrow can be accurately characterized by expression of surface proteins and intracelluar proteins (tables 1, 2). On this basis, fluorochrome-conjugated antibodies and flow cytometry represent an excellent tool to identify defects and malignancies of the B-cell system [8–22].

This report describes phenotyping of B-cells in peripheral blood and bone marrow. B-cell subpopulations from lymph nodes and tonsils can be determined in a similar fashion [23, 24]. Furthermore, this article is restricted to B-cell analysis in

Table 2. Markers for B-cell subpopulations in peripheral blood

Marker	Transitional B-cell	Naïve B-cell	Marginal zone like B-cell	Class-switched memory B-cell	Plasmablast
CD19	+	+	+	+	(+)
CD20	+	+	+	+	–
CD21	(+)	+	+	+	–
CD24	++	+	++	++	–
CD27	–	–	+	+	++
CD38	++	+	–/+	–/+	++
Sm-IgD	++	++	+	–	–
Sm-IgM	++	+	++	–	–/(+)

immunodeficiencies, but similar techniques, which are discussed in another chapter of this book, can be employed for leukemia and lymphoma diagnostics.

Protocol

Principle

All measurements described here employ standard methods of four-color flow cytometry with plain surface staining and combined staining of surface molecules and intracellular proteins. All protocols are based on analyses using a FACSCalibur with a 488-nm laser, a 635-nm diode laser and CellQuest software. For other flow cytometers, appropriate modifications might be required. The basics of machine operation and settings, compensations, two-dimensional dot plot analyses and setting of regions or gates are assumed to be known. Mononuclear cells (MNCs) harvested from EDTA blood by Ficoll gradient density centrifugation are used for staining. Since anti-immunoglobulin antibodies are employed, whole blood [13] and washed blood [25] stainings are not recommended as residual serum immunoglobulin interferes with the staining. The protocol describes stainings with antibody combinations proven to be reliable markers for characterizing the various stages of differentiation in blood and bone marrow. These combinations can easily be extended or modified to meet the anticipated requirements.

Material

Sample Material
8 ml EDTA-anticoagulated blood or bone marrow aspirate (not older than 24 h).

Reagents
- Ficoll separation solution (1.077 g/ml; Biochrom, Berlin, Germany);
- RPMI-1640 (Biochrom);
- FCS (PAN Biotech, Aidenbach, Germany);
- FACSFlow (BD Biosciences, Heidelberg, Germany);
- Optilyse B (Beckman-Coulter, Krefeld, Germany);
- IntraPrep (Beckman-Coulter), and
- Deionized water.

Equipment
- FACSCalibur with a 488-nm laser and a 635-nm diode laser (BD Biosciences);
- Standard pipettes;
- Centrifuge;
- Counting chamber;
- Vortex;
- 50-ml Falcon polypropylene tubes with lid (BD Biosciences);
- 15-ml Falcon polypropylene tubes with lid (BD Biosciences);
- 12 × 75-mm Falcon polystyrene tubes (BD Biosciences), and
- 0.5-ml sample tubes with lid (Biozym, Hessisch Oldendorf, Germany).

Antibodies
All antibodies used are listed in table 3.

Procedures

Preparation of Mononuclear Cell Populations

Using EDTA-anticoagulated blood or bone marrow aspirate has worked reliably in the authors' laboratory. Material should be processed as soon as possible and not later than 24 h after obtaining the sample. As compared to heparin blood, staining with EDTA blood is less susceptible to unspecificity, especially when using older samples.

RPMI-1640 1:2 prediluted blood (approximately 8 ml) is placed in a 50-ml tube and carefully underlayed with 15 ml Ficoll separation solution. After 20 min centrifugation at 900 × *g* at room temperature, the mononuclear interphase is carefully pipetted into a 50-ml tube and diluted 1:2 with RPMI, followed by centrifugation for 10 min at 900 × *g*. Supernatant is then discarded, the cellular sediment resuspended in 5–10 ml RPMI-1640 with 10% FCS and centrifuged for 10 min at 200 × *g*. After decanting of supernatant and resuspension of sediment in RPMI-1640 with 10% FCS, cell concentration is determined and adjusted to 5 to 10 × 10^6 MNCs/ml. This cell suspension can either be processed immediately or be refrigerated until staining for 24 h without affecting the outcome. PBS with 2% FCS and 0.1% sodium azide can

Table 3. List of antibodies

Antibody	Clone	Fluorochrome	Order No.	Provider
CD3	SK7	APC	345767	BD
CD3	SK7	PerCP	345766	BD
CD4	13B8.2	FITC	IM0448	BC
CD8	B9.11	PE	IM0452	BC
CD10	ALB1	FITC	IM2720	BC
CD16	3G8	PE	IM1238	BC
CD16	3G8	PerCP-Cy5.5	338440	BD
CD19	J4,119	FITC	IM1284	BC
CD19	J4,119	PC7	IM3628	BC
CD19	SJ25C1	APC	345791	BD
CD20	L27	PE	345793	BD
CD21	BL13	FITC	IM0473	BC
CD21	B-ly4	PE	555422	BD
CD22	SJ10.1H11	PE	IM1835	BC
CD24	ALB9	PE	IM1428	BC
CD27	M-T271	FITC	F7178	Dako
CD27	M-T271	PE	R7179	Dako
CD33	P67.6	PerCP-Cy5.5	333146	BD
CD34	8G12	FITC	345801	BD
CD36	FA6,152	FITC	IM0766	BC
CD38	HIT2	FITC	555459	BD
CD45	2D1	PerCP	345809	BD
CD56	NKH-1	PE	IM2073	BC
CD79a (Igα)	HM47	PE	IM2221	BC
CD179a (VpreB)	HSL96	PE	IM3647	BC
Anti-TdT	HT-6	FITC	F7139	Dako
Anti-IgM	rabbit F(ab′)2	FITC	F0058	Dako
Anti-IgM	goat F(ab′)2	PE	2020–09	SB
Anti-IgM	goat F(ab′)2	Cy5	109–176–129	JIR
Anti-IgD	rabbit F(ab′)2	FITC	F0059	Dako
Anti-IgD	goat F(ab′)2	PE	2032–09	SB
Anti-IgA	goat F(ab′)2	PE	2052–09	SB
Anti-IgG	rabbit F(ab′)2	FITC	F0056	Dako
Anti-κ	G20–193	FITC	555791	BD
Anti-λ	JDC-12	PE	555797	BD
Isotype control	mouse IgG1	FITC	IM0639	BC
Isotype control	mouse IgG1	PE	IM0670	BC
Isotype control	rabbit F(ab′)2	FITC	X0929	Dako
Isotype control	goat F(ab′)2	PE	0110–09	SB
Isotype control	goat F(ab′)2	Cy5	005–170–006	JIR

BD = BD-Biosciences (Heidelberg, Germany); BC = Beckman-Coulter (Krefeld, Germany); Dako = DakoCytomation (Hamburg, Germany); SB = SouthernBiotech (Biozol, Eching, Germany); JIR = Jackson ImmunoResearch Laboratories (Dianova, Hamburg, Germany).

Table 4. Antibody mixtures for the B-cell panel[a]

	FITC 100 μl	PE 100 μl	PC7 25 μl	FL4 25 μl
B1	CD27 (1:5)	goat F(ab')2	CD19	goat F(ab')2 Cy5 (1:40)
B2	mouse IgG1 (1:2)	anti-IgD (1:40)	CD19	CD45 APC
B3	CD27 (1:5)	anti-IgD (1:40)	CD19	anti-IgM Cy5 (1:40)
B4	anti-κ (1:2)	anti-λ (1:2)	CD19	anti-IgM Cy5 (1:40)
B5	anti-IgG (1:20)	anti-IgA (1:40)	CD19	anti-IgM Cy5 (1:40)
B6	CD38	CD24	CD19	anti-IgM Cy5 (1:40)
B7	CD38	CD27 (1:5)	CD19	anti-IgM Cy5 (1:40)
B8	CD38	CD21	CD19	anti-IgM Cy5 (1:40)

[a] 10 μl/50 μl cell suspension.

Table 5. Antibody mixtures for bone marrow panel part 1[a]

	FITC 100 μl	PE 100 μl	PerCP 100 μl	APC 20 μl
BM1	CD19	CD16 (1:2) + CD56 (1:2)	CD45 (1:2)	CD3 (1:2)
BM2	CD4	CD8	CD45 (1:2)	CD3 (1:2)
BM3	rabbit F(ab')2 (1:5)	goat F(ab')2	MIX[b]	CD19
BM4	CD10	CD20 (1:8)	MIX	CD19
BM5	CD34	CD20 (1:8)	MIX	CD19
BM6	anti-IgD (1:10)	anti-IgM (1:40)	MIX	CD19
BM7	CD36	CD22	MIX	CD19

[a] 20 μl/50 μl cell suspension (modified according to [9]).
[b] MIX = 33.3 μl CD3-PerCP + 33.3 μl CD33-PerCP-Cy5.5 + 33.3 μl CD16-PerCP-C5.5.

be used instead of RPMI-1640 with 10% FCS. For this cell suspension, however, we have no experience regarding results after overnight storage.

Preparation of Antibody Mixtures

Stainings are done much easier and faster if antibodies are premixed for the four different fluorochromes. For this, pretitrated antibodies (combinations, predilutions and volumes specified in tables 4 and 5) are all placed in 0.5-ml sample tubes, mixed und labeled (panel number, date). The lid should also be labeled to avoid mix-ups. The indicated antibody predilutions done with FACSFlow are merely reference points and should be verified for each charge. These antibody mixtures can be kept in the refrigerator for several months. For tests that are combined with intracellular stainings, pipetting of each individual antibody is recommended, except for the mix in FL3 (table 6).

Surface Staining

The following protocol is an adaptation of whole blood staining according to the Opti-Lyse B-method (Beckman-Coulter, Krefeld). Alternatively, steps 3–4 can be left out.

Table 6. Bone marrow panel part 2

	FITC	PE	PerCP	APC
BM8[a]	–	–	MIX 5 µl	CD19 (1:5) 5 µl
	Cy-isotype[b] (1:5) 10 µl	Cy-mouse-IgG1 (1:2) 10 µl	–	–
BM9	–	Anti-IgM (1:40) 5 µl	MIX 5 µl	CD19 (1:5) 5 µl
	Cy-anti-IgM (1:10) 10 µl	–	–	–
BM10	–	CD20 (1:4) 5 µl	MIX 5 µl	CD19 (1:5) 5 µl
	Cy-anti-IgM (1:10) 10 µl	–	–	–
BM11	–	–	MIX 5 µl	CD19 (1:5) 5 µl
	Cy-anti-IgM (1:10) 10 µl	Cy-CD179a (1:4) 10 µl	–	–
BM12	–	–	MIX 5 µl	CD19 (1:5) 5 µl
	Cy-anti-TdT 10 µl	Cy-CD79a 10 µl	–	–
BM13	CD34 10 µl	–	MIX 5 µl	CD19 (1:5) 5 µl
	–	Cy-CD79a 10 µl	–	–

[a] Upper line: surface staining; lower line: intracellular staining after permeabilization (modified according to [9]).
[b] Rabbit F(ab′)2.

During each step, all staining solutions must be protected from sunlight and kept in the dark during incubation.

Step 1: specified volumes of antibody mixture (10–20 µl) as listed in tables 4 and 5 are placed into prelabeled FACS tubes (12 × 75 mm).

Step 2: 50 µl cell suspension each (250,000–500,000 MNCs) are added to the antibodies, vortexed and incubated for 15–30 min at 0–4 °C.

Step 3: after thorough mixing, 25 µl of Optilyse B are added to each sample, mixed again thoroughly and incubated for 15 min at room temperature.

Step 4: samples are mixed, 500 µl deionized water are added, followed by further incubation for 15 min.

Step 5: After addition of 1 ml FACSFlow each, samples are centrifuged for 5 min at 300 × *g*.

Step 6: cell-free supernatant is then carefully decanted or suctioned off (possible cell loss!) and cells are resuspended in 500 µl FACSFlow.

Samples are stored in the refrigerator until flow-cytometric analysis which should be carried out within a few hours.

Intracellular Staining

Combined surface and intracellular staining (table 6) is carried out according to the IntraPrep protocol of the manufacturer. To begin with, surface staining is performed as delineated in steps 1 and 2 (see above). In this case, it is better to pipette the antibodies individually at 5 µl each.

Step 3: cell fixation and erylysis by adding IntraPrep reagent 1 (100 µl) during a 15-min incubation period.

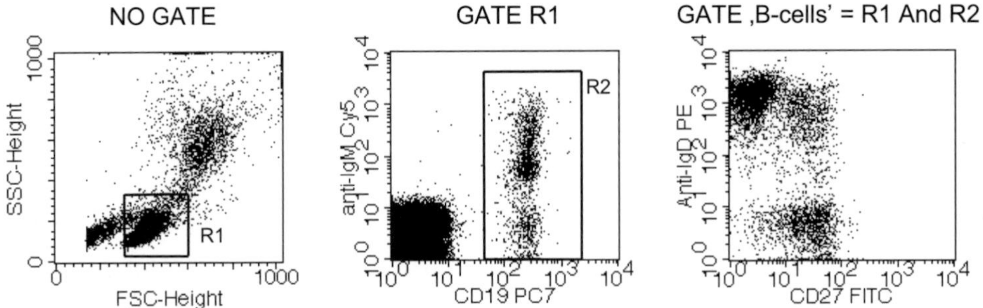

Fig. 1. B-cell measurement. Lymphocyte region R1 and CD19+ region R2 are combined to form a B-cell gate for acquiring at least 5,000 events.

Step 4: samples are washed with 4 ml FACSFlow or PBS (5 min, 300 × *g*).

Step 5: after supernatant is suctioned off, the cell pellet is carefully resuspended in 100 μl IntraPrep reagent 2 (saponin). 5 min later, antibodies for intracellular staining (10 μl of each antibody) are added, mixed gently and incubated for another 15 min.

Step 6: after addition of 4 ml FACSFlow each, samples are centrifuged for 5 min at 300 × *g*.

Step 7: Cell-free supernatant is then carefully suctioned off (possible cell loss!) and cells are resuspended in 500 μl FACSFlow.

Samples are stored in the refrigerator until flow-cytometric analysis which should be carried out within a few hours.

Measurements

A prerequisite for B-cell measurements is a correct instrument setting for the photomultiplier tube (PMT) and for fluorochrome compensations. Appropriately titrated antibodies show uniform compensation characteristics within the panel. As a rule, instrument settings are very stable over time so that a control with beads in all colors is sufficient rather than preparing each run with single stainings.

Measuring B-cell Subpopulations in Peripheral Blood (B-cell Panel, Table 4)

Results are displayed in a two-dimentional acquisition window with two axes for forward light scatter (FSC) and sideward light scatter (SSC). A light scatter analysis gate R1 is set to define the lymphocyte population. Another window displaying an FL3 and FL4 axis contains a region R2 that includes all CD19+ B-cells. Alternatively, a histographic display can be chosen. Regions R1 and R2 are combined into one 'B-cell' gate. A third window depicts FL1 versus FL2 and is confined to presentation of B-cells (fig. 1). For the 'B-cell' gate, events to be saved to a data file are set to 5,000 and a time limit of 5 min (for 500 μl sample volume). It is recommended to record all cells, or at least all lymphocytes, in order to re-examine the specificity of unexpected results with other cells. Choose instrument settings that have already been tested in trial runs. The eight B-cell panel stainings are then done consecutively. The first two stainings contain isotope controls, which allow for an initial verification of the correct settings. 1,000–5,000 B-cells are taken up during the analysis (5 min maximum). When dealing with a very low B-cell count in peripheral blood, the number and/or the volume of cells and measuring time might have to be increased accordingly.

Measuring B-Progenitors in Bone Marrow (Bone Marrow Panel, Tables 5, 6)
The two parts of the bone marrow panel comprise plain surface staining on one hand and combined surface and intracellular staining on the other. Assays contain the line-specific, APC-marked antibody (CD19 or CD3) and the PerCP-marked antibodies in FL3. These analyses must be done with the appropriate compensation settings. The assays for intracellular staining (BM8–BM13) show differences in SSC and have to be adjusted accordingly. For data acquisition, a scattered-light window R1 for lymphocytes and a line-specific FL4 histogram region R2 are defined and combined into one acquisition gate as described above. If possible, at least 5,000 CD19+ events should be acquired due to the complex analysis of B-cell precursors.

Data Analysis

Peripheral B-Cell Subpopulations
CD19 in FL3 is used as the line-specific marker (fig. 2) for determining the B-cell subpopulations from the eight stainings of the B-cell panel (B1–B8, table 4). For the B1 staining, create a two-dimensional data display showing CD19 versus SSC and define an analysis gate R2 that must also include cells with a weak CD19 expression (plasmablasts). In a second window with an FSC and SSC axis define a lymphocyte region R1 that should exclude cell debris and apoptotic cells on the left side, since they could cause unspecific staining and cannot be eliminated with propidium iodide (four colors allocated, fixed cells). On the right side, however, and in the transient area extending into the monocytes, draw a generous region in order to include the larger plasmablasts. A further FSC-SSC window limited to region R2 (gating) displays all CD19+ events as a control for the lymphocyte window. By combining regions R1 and R2, define a 'B-cell' gate for subsequent analysis of all stainings for FL1, FL2 and FL4. In order to ensure a uniform staining quality, these three windows should be used for all eight stainings (the staining example in figure 2 only depicts B1).

Using the 'B-cell' gate, B1–B2 are analyzed with quadrant or region statistics as shown in figure 2. B1 and B2 have isotype controls that can detect unspecific staining and incorrect compensation. Thus, stainings FL2 and FL4 in B1 test for unspecific anti-Ig antiserum binding and correct compensation of the PC7 conjugate

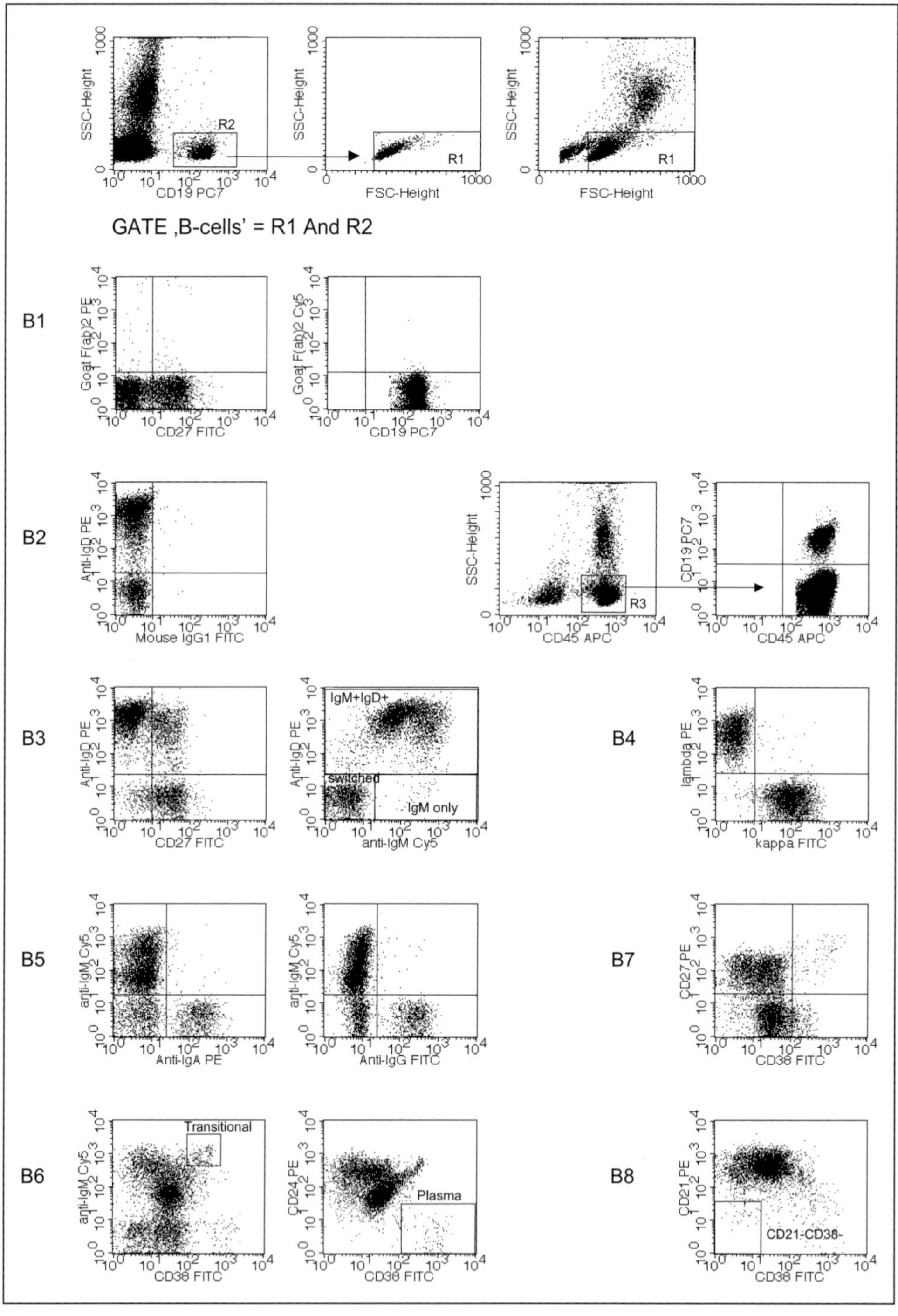

Fig. 2. Analysis of the peripheral B-cell subpopulations. Lymphocyte region R1 and CD19+ region R2 are combined to form a B-cell gate for assessing stainings B1–B8. Analysis is done by using either quadrant or region statistics.

against FL2. The isotype control used in B2 for FL1 together with anti-IgD-PE is important for the correct placement of the quadrant in B3 (CD27/anti-IgD-PE), which can be problematic for some patient samples. As customary for whole blood samples, B2 furthermore contains a CD45 label to define a lymphocyte area free of erythrocytes and basophils in a SSC/CD45 display that is used to calculate the absolute B-cell count.

Owing to the differential expression of IgM, IgD and CD27, B3 can distinguish between naïve-, memory- and class-switched cells. CD27+ memory B-cells can be further separated into IgD+ IgM+ and IgD− [10, 15].

B4 serves to exclude monoclonal B-cell populations while B5 analyzes IgM class-switched B-cells for IgA or IgG surface expression. Where required, B5 can also be construed as an intracellular staining. The following three assays use CD38 as an excellent marker for analyzing various B-cell differentiation stages. Thus, CD38 is highly expressed on all B-cell precursors (see bone marrow stainings and table 2). Naïve B-cells still show an intermediate CD38 surface expression, while memory B-cells show intermediate to negative CD38 expression. During the germinal center reaction, B-cells regain their CD38 expression and plasmablasts are once again strongly positive. In B6, transitional B-cells are defined by high IgM and high CD38 expression, plasmablasts by absent CD24 but high CD38 expression. Here, CD20 can be substituted for CD24. B7 serves as supplement and plausibility control for transitional cells (CD38++ CD27−), plasmablasts (CD38++ CD27+) and memory B-cells (CD38−/+ CD27+). Staining assay B8 is used for differentiating various CD21− B-cell populations.

Report

The main B-cell subpopulations used for analysis are listed in table 7. It is practical to specify the portion of total B-cells as percent of lymphocytes and subpopulations as percent of B-cells. If required, these specifications can then be used to calculate lymphocyte-related or absolute values.

Table 7 shows normal values from an adult collective. Reference values for human peripheral B-cell subpopulations were published by Deneys et al. [13]. In our opinion, however, most norm values are not very helpful since they are based on one-parameter data that disregard differential expression on subpopulations (e.g. CD10, CD24, CD38, IgD).

B-Cell Precursors in Bone Marrow

In many respects, analysis of B-cell precursors in bone marrow is considerably more complicated than for peripheral B-cell populations, since bone marrow aspirate contains a mixture of mature and immature B-cells, and memory B-cells can migrate there as well [26].

With progressing age, the relative amount of mature B-cells can increase in the bone marrow [6, 7]. Additionally, the bone marrow aspirate shows variable contam-

Table 7. Reference values of B-cell subpopulations in peripheral blood

B-cell population		Reference range[a]
CD19+ in CD45+ lymphocytes	B-cells	4.9–18.4%
IgM+ IgD+ in CD19+ B cells	naïve and marginal zone like B-cells	67.9–91.7%
IgM– IgD– in CD19+ B cells	class-switched B- cells	7.6–31.4%
IgD+ CD27– in CD19+ B cells	naïve B-cells	42.6–82.3%
IgD+ CD27+ in CD19+ B cells	marginal zone like B-cells	7.4–32.5%
IgD– CD27+ in CD19+ B cells	class-switched memory B-cells and plasmablasts	6.5–29.1%
IgA+ in CD19+ B-cells	IgA+ B-cells	2.7–13.8%
IgG+ in CD19+ B-cells	IgG+ B-cells	3.6–13.4%
κ/λ	light chain restriction	1.2–2.0
$CD21^{low}$ CD38– in CD19+ B-cells	$CD21^{low}$ subpopulation	0.9–7.6%
M++ D++ CD24++ CD38++ in CD19+ B-cells	transitional B-cells	0.6–3.4%
CD24– CD38++ CD27+ in CD19+ B-cells	plasmablasts	0.4–3.6%

[a] 5th to 95th percentile, based on 54 healthy donors (age range 19–61 years).

ination through blood. Some differentiation markers change gradually (CD20, CD179a, IgM, IgD), or they are also differentially expressed on mature B-cells (CD20, CD22, CD38, IgM, IgD) (tables 1, 2). Therefore, the stages of B-cell differentiation can usually not be defined by one staining alone, but must be deduced from combinations and plausibility controls of several stainings. For this reason, percent values for B-cell precursors cannot be as exact as those for peripheral subpopulations. Patients with B-cell maturation malfunction, however, display such pronounced changes in B-cell precursor composition that they can be reliably diagnosed. This situation will improve with the introduction of the five-color flow cytometer that allows incorporation of immaturity markers like CD10 or CD38 in all stainings. The protocol for analyzing B-cell precursors described here is based on the work of the J.J.M. van Dongen group (Rotterdam) [9, 20–22].

As for the peripheral B-cell system, an analysis gate is defined for evaluation of all bone marrow stainings, combining a generous FSC-SSC window R1 for the lymphocyte population with a line-specific region. Back-gating of line-specific events on a light scatter gate in turn enables confirmation of the location of the lymphocyte gate. The first two stainings (BM1 and BM2) are used to quantify the main lymphocyte populations (CD3+, CD4+, CD8+ T-cells, CD16/56+ NK cells and CD19+ B-cells) in the bone marrow. For this, an SSC-CD45 region is defined to include all lymphocytes and exclude all myeloid and erythropoietic cells. Here, it is to be noted that immature B-cells show significantly lower CD45 expression than mature B-cells (fig. 3A).

For BM3–BM6, define a line-specific analysis window with a two-dimensional display of CD19 versus a combination of the exclusion markers CD3, CD33 and CD16 (fig. 3B). CD20 versus CD10 (BM4) shows the stages of differentiation from

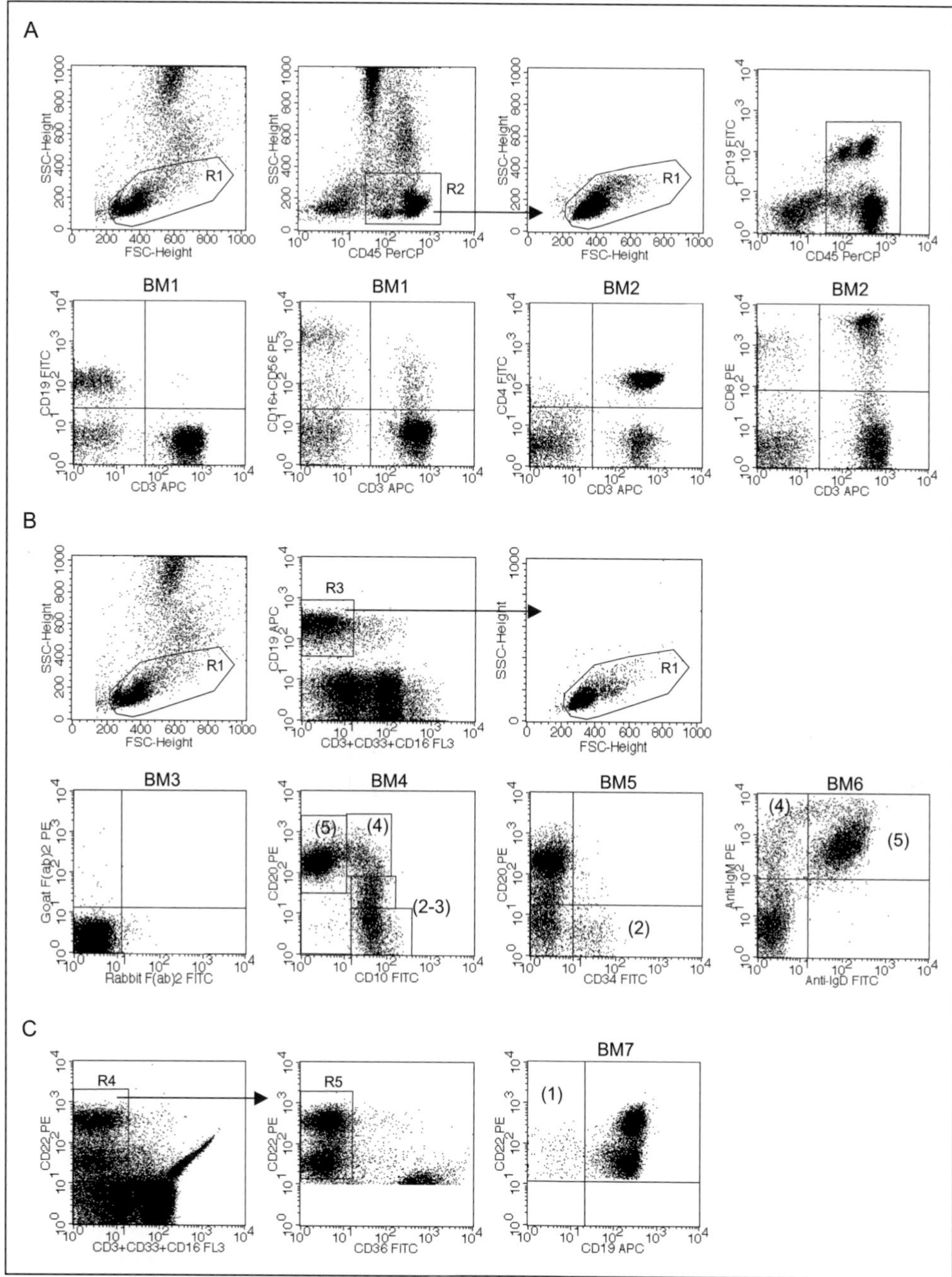

Fig. 3. Analysis of B-cell precursors in bone marrow with surface stainings. **A** A lymphocyte scattered-light region R1 and a lymphocyte region R2 defined by CD45/SSC are combined to analyze stainings BM1 and BM2. **B** A lymphocyte region R1 and a CD19+ region R3 are combined to analyze stainings BM3–BM6. **C** For BM7, lymphocyte region R1 is combined with two consecutively defined CD22 regions R4 and R5 used for purification, and pro-B-cells are quantified as the CD22+ CD19– population. The different B-cell populations are coded with numbers in the text and in the figures: (1) pro-B-cells, (2) pre-B-I-cells, (3) pre-B-II-cells, (4) immature B-cells, (5) mature B-cells.

Pre-B-I (2) to mature B-cell (5), whereby all pre-stages are CD10+ and express increasing amounts of CD20. This depiction allows one to define immature (4) and mature (5) B-cells by regions. In most bone marrow samples, mature and immature B-cells will exhibit high CD20 expression while pre-B-cells are negative or weakly positive. The combination CD20/CD34 (BM5) presents Pre-B-I cells (2) as CD20– CD34+. Staining BM6 in turn characterizes immature (4) and mature (5) B-cells through differential surface expression of IgM and IgD. Since immature B-cells exhibit a continually increasing IgM and IgD expression, they cannot be defined quite accurately. Just an increased level of marginal zone like B-cells or plasma cells can greatly interfere with correct data interpretation.

Pro-B-cells are defined as CD19– CD22+ (table 1) [9]. Compared to mature B-cells, all immature precursors display markedly reduced CD22 expression. Consequently, adequate purity cannot be achieved by exclusion markers such as CD3, CD33 and CD16. For this reason, staining BM7 contains a further exclusion marker CD36 for cleansing the CD22 gate, which develops as a result of combining both analysis regions from displays CD22/MIX (R4) and CD22/CD36 (R5) (fig. 3C). By using this CD22 gate, pro-B-cells can be analyzed in the CD22/CD19 dot plot.

Stainings BM8–BM13 evaluate intracellular proteins. As described above, a CD19 gate is defined by combining a lymphocyte region R1 from an FSC-SSC display with a CD19 region R2 from the CD19-MIX display. Staining BM9 analyses pre-B-I (2), pre-B-II (3), immature (4) and mature (5) B-cells through differential expression of the intracellular μ-chain and surface IgM. Here, the pre-B-I-region (2) can be contaminated by class-switched mature B-cells; however, they can be separated out in the following stainings BM10 and BM11 (5a). For this, Pre-B-I cells (2) are defined as CD20– Cy-VpreB+ Cy-IgM– and Pre-B-II cells (3) as CD20–/^{low} Cy-VpreB+/– Cy-IgM+ (table 2, fig. 4A). Surrogate light chain VpreB surface expression is too weak for pre-B-II cell phenotyping.

Most pro-B-cells already express the Igα-chain CD79a [9] intracellularly. For BM12 and BM13, a purified CD79a gate is used to analyze pro-B-cells (fig. 4B). In display Cy-TdT/CD19, TdT– and TdT+ pro-B-cells (1) as well as pre-B-I cells (2) can be differentiated, the last of which can be contaminated by pre-B-II cells still expressing TdT. Staining BM13 characterizes pro-B-cells (1) and pre-B-I cells (2) by their CD34 expression and differential CD19 expression. CD19– CD34– events can be classified as unspecific.

If required, these bone marrow analyses can be complemented with stainings from peripheral B-cell phenotyping (see there), thus permitting quantification of, e.g., memory cells and plasmablasts as well as exclusion of monoclonality.

Report

For final evaluation, mean values have to be calculated for the various stages of differentiation, taking into account plausibility factors and exclusion of nonspecific populations. The fractions of pre-B-I, pre-B-II and immature B-cell subpopulations should

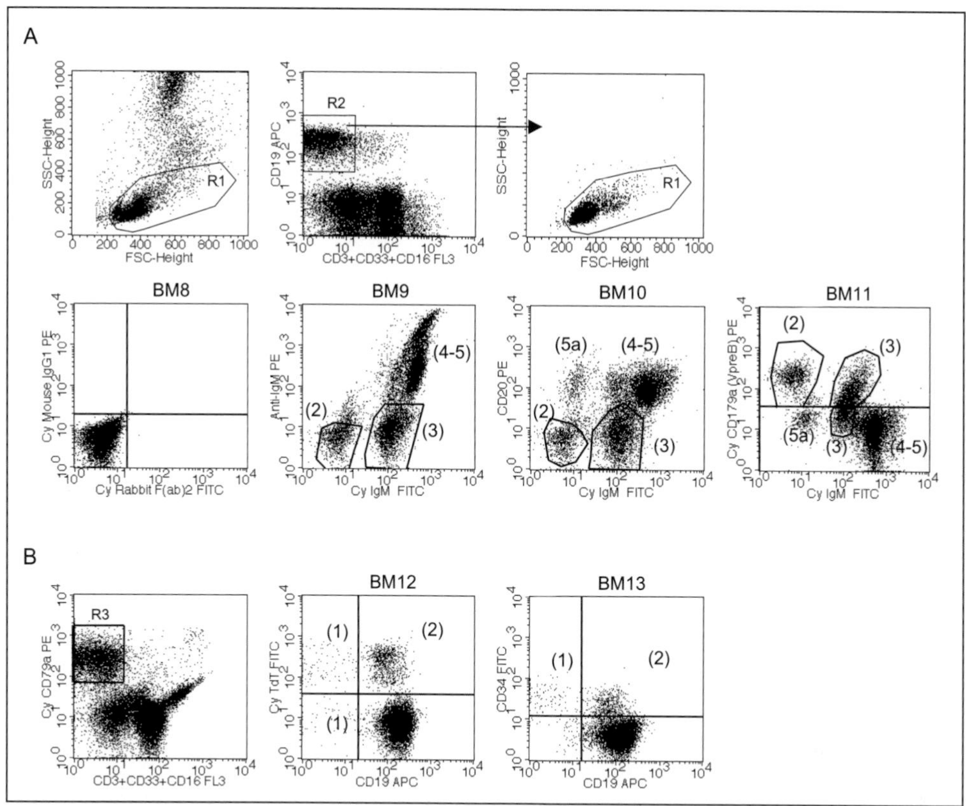

Fig. 4. Analysis of B-cell precursors in bone marrow with intracellular stainings. **A** For analysis of BM8–BM11, a lymphocyte scattered-light region R1 is combined with a purified CD19+ region R2. **B** For analysis of pro-B-cells in BM12 and BM13, a scattered-light region R1 and a purified CD79a+ region R3 are combined. Designation of B-cell populations corresponds to figure 3; class-switched B-cells are marked with '5a'

be specified as percent of B-cell precursors. For this, CD19+ CD10+ precursors can be used as a reference base. It is better, however, to include the pro-B-cells in the B-cell precursors based on CD22+ or CD79a+ cells. Table 8 gives an example for calculating subpopulations. A typical result is shown in table 9. At this time, we cannot provide norm values for healthy adults. More than 15 bone marrow samples collected for clarification or elimination of hematological disorders from patients aged between 25 and 75 showed a similar precursor composition as described for healthy children [9].

Comment

In our laboratory, B-cell stainings on Ficoll-separated MNCs from EDTA-anticoagulated material have been successfully tried and tested. Some laboratories perform B-

Table 8. Analysis of B-cell precursors in bone marrow

B-cell population	In lymphocytes, %	In CD19+, %	In CD19+ CD10+, %	In CD22+/ CD79a+, %	In all precursors, %
Conversion factors			UF1 = 2.48[a]		UF2 = 2.44[b] UF3 = 0.97[c]
CD19+	22.4				
CD19+ CD10+		40.4			
Pre-B-I		BM4: 5.6 BM9: 6.0 BM10: 5.7 BM11: 6.8 BM12: 7.7 Mean: 6.4	16		16
Pre-B-II		BM9: 20.2 BM10: 22.1 BM11: 21.8 Mean: 21.4	53		51
Immature B		BM4: 11.9 BM6: 8.8 Mean: 10.4	26		25
Pro-B-cells				BM7: 0.9 BM12: 1.4 BM13:1.1 Mean: 1.1	3
Sum		38.2	95		95

[a] Conversion factor UF1 converts the percentage of precursors based on all CD19+ B-cells to values based on 100% CD19+ precursors (example: UF1 = 100/40.4 = 2.48).
[b] Conversion factor UF2 corrects the percentage of pro-B-cells to values based on all precursors composed of pro-B-cells and CD19+ B-cell precursors (UF2 = 100/(1.1+(100−1.1)/UF1) = 2.44).
[c] Conversion factor UF3 finally serves to correct percentages for CD19+ precursors to values based on all B-cell precursors (UF3 = (100-%pro-B-cells)/100 = 0.97).

cells stainings with washed whole blood. This method certainly represents a valid alternative, but it hardly saves time due to the required intensive washing steps. Owing to the large amount of cells in these samples, there is also the risk that residual serum immunoglobulin inhibits the staining with antibodies against IgM, IgG, IgA κ or λ.

Although we use isolated MNCs, we successfully employ a lysis method (OptilyseB) generally used for whole blood stainings. This technique prevents contamination of the lymphocyte gate by erythrocytes and antibody-mediated cell aggregates (escapees) [27]. Furthermore, the cells are fixed and do not have to be processed immediately, but can be stored in the refrigerator for several hours if necessary.

For staining of the B-cell panel (table 4), we use a PC7-conjugated CD19 antibody which has an advantage over a PerCP or PerCP-Cy5.5 conjugate insofar as it

Table 9. Typical report for lymphocyte phenotype in bone marrow

Lymphocyte subpopulations	%
CD34+	4.4
CD3+ T-cells	60.6
CD3+ CD4+ T-cells	30.9
CD3+ CD8+ T-cells	29.1
CD3− CD16+/CD56+ NK cells	14.9
CD19+ B-cells	22.4
CD10+ precursor in CD19+ (40.4% of CD19+)	9.0
B-cell precursor subpopulations	
Pro-B-cells	3
Pre-B-I cells	16
Pre-B-II cells	51
Immature B-cells	25

exhibits a brighter fluorescence and no compensation problems with respect to the IgM staining in FL-4 (Cy5).

Quality Control

Prior to every analysis, the flow cytometer has to be tested for correct function with a mixture of beads in all four colors. In our experience, established instrument settings can generally be used again without correction of compensations (e.g. with single stainings). Special care should be taken to maintain correct compensation of CD19-PC7 against PE conjugates, especially after a batch change. Conspicuous deviations for stainings with isotype controls might be an indicator for the need to reset the instrument. If possible, a healthy control should be run with B-cell stainings of peripheral blood to eliminate staining errors in strongly deviating subpopulations. This is particularly important after production of fresh antibody mixtures.

Troubleshooting

Increased fluorescence of a negative subpopulation probably stems from a too high concentration of antibodies. This may especially be true for anti-Ig antisera. Poor separation of negative and positive populations on the other hand can be caused by overly diluted antibodies. These probable causes should be remembered particularly when using antibodies other than described in this protocol. Moreover, monoclonal antibodies or anti-Ig antisera from different manufacturers do not always separate the subpopulations equally well.

Expected Results

Expected results for the B-cell subpopulations are described in the paragraph 'Reports' and shown in tables 7 and 9.

Required Time

Preparation of MNCs takes approximately 1.5 h. The same amount of time is required for surface staining, while 2 h should be scheduled if combined with intracellular staining. Data acquisition for 2,000–5,000 B-cells requires up to 5 min per staining.

Summary

The different B-cell subpopulations in peripheral blood and precursor cells in bone marrow can be analyzed by four-color flow cytometry and commercially available fluorescence-labeled antibodies against differentiation-dependent antigens. Mature naïve B-cells, IgM memory cells, class-switched memory cells, plasmablasts and other atypical or not conclusively defined subpopulations can be distinguished in peripheral blood. Flow-cytometry can assess the complete range of B-cell pre-stages in the bone marrow. This method has proven useful for the elucidation of scientific and diagnostic questions concerning primary and secondary immunodeficiencies, autoimmune diseases and hematological tumors.

References

1 Janeway CA, Jr, Travers P, Walport MJ: Immunologie, 5. Aufl. Heidelberg, Spektrum Akademischer Verlag, 2002.
2 Duchosal MA: B-cell development and differentiation. Semin Hematol 1997;34:2–12.
3 Burrows PD, Cooper MD: B cell development and differentiation. Curr Opin Immunol 1997;9:239–244.
4 Ghia P, ten Boekel E, Rolink AG, Melchers F: B-cell development: a comparison between mouse and man. Immunol Today 1998;19:480–485.
5 Carsetti R, Rosado MM, Wardmann H: Peripheral development of B cells in mouse and man. Immunol Rev 2004;197:179–191.
6 Nunez C, Nishimoto N, Gartland GL, Billips LG, Burrows PD, Kubagawa H, et al: B cells are generated throughout life in humans. J Immunol 1996;156:866–872.
7 Ghia P, ten Boekel E, Sanz E, de la Ha, Rolink A, Melchers F: Ordering of human bone marrow B lymphocyte

precursors by single-cell polymerase chain reaction analyses of the rearrangement status of the immunoglobulin H and L chain gene loci. J Exp Med 1996;184:2217–2229.

8 Nomura K, Kanegane H, Karasuyama H, Tsukada S, Agematsu K, Murakami G, et al: Genetic defect in human X-linked agammaglobulinemia impedes a maturational evolution of pro-B cells into a later stage of pre-B cells in the B-cell differentiation pathway. Blood 2000;96:610–617.

9 Noordzij JG, Bruin-Versteeg S, Comans-Bitter WM, Hartwig NG, Hendriks RW, de Groot R, et al: Composition of precursor B-cell compartment in bone marrow from patients with X-linked agammaglobulinemia compared with healthy children. Pediatr Res 2002;51:159–168.

10 Klein U, Rajewsky K, Kuppers R: Human immunoglobulin (Ig)M(+)IgD(+) peripheral blood B cells expressing the CD27 cell surface antigen carry somatically mutated variable region genes: CD27 as a general marker for somatically mutated (memory) B cells. J Exp Med 1998;188:1679–1689.

11 Klein U, Goossens T, Fischer M, Kanzler H, Braeuninger A, Rajewsky K, et al: Somatic hypermutation in normal and transformed human B cells. Immunol Rev 1998;162:261–280.

12 Odendahl M, Jacobi A, Hansen A, Feist E, Hiepe F, Burmester GR, et al. Disturbed peripheral B lymphocyte homeostasis in systemic lupus erythematosus. J Immunol 2000;165:5970–5979.

13 Deneys V, Mazzon AM, Marques JL, Benoit H, De Bruyere M: Reference values for peripheral blood B-lymphocyte subpopulations: a basis for multiparametric immunophenotyping of abnormal lymphocytes. J Immunol Methods 2001;253:23–36.

14 Agematsu K, Futatani T, Hokibara S, Kobayashi N, Takamoto M, Tsukada S, et al: Absence of memory B cells in patients with common variable immunodeficiency. Clin Immunol 2002;103:34–42.

15 Warnatz K, Denz A, Drager R, Braun M, Groth C, Wolff-Vorbeck G, et al: Severe deficiency of switched memory B cells ($CD27^{+}IgM^{-}IgD^{-}$) in subgroups of patients with common variable immunodeficiency: a new approach to classify a heterogeneous disease. Blood 2002;99:1544–1551.

16 Kruetzmann S, Rosado MM, Weber H, Germing U, Tournilhac O, Peter HH, et al: Human immunoglobulin M memory B cells controlling *Streptococcus pneumoniae* infections are generated in the spleen. J Exp Med 2003;197:939–945.

17 Gaspar HB, Conley ME: Early B cell defects. Clin Exp Immunol 2000;119:383–389.

18 Schiff C, Lemmers B, Deville A, Fougereau M, Meffre E: Autosomal primary immunodeficiencies affecting human bone marrow B-cell differentiation. Immunol Rev 2000;178:91–98.

19 de Vries E, Noordzij JG, Kuijpers TW, van Dongen JJ: Flow cytometric immunophenotyping in the diagnosis and follow-up of immunodeficient children. Eur J Pediatr 2001;160:583–591.

20 Noordzij JG, Bruin-Versteeg S, Verkaik NS, Vossen JM, de Groot R, Bernatowska E, et al.: The immunophenotypic and immunogenotypic B-cell differentiation arrest in bone marrow of RAG-deficient SCID patients corresponds to residual recombination activities of mutated RAG proteins. Blood 2002;100:2145–2152.

21 Noordzij JG, Bruin-Versteeg S, Hartwig NG, Weemaes CM, Gerritsen EJ, Bernatowska E, et al: XLA patients with BTK splice-site mutations produce low levels of wild-type BTK transcripts. J Clin Immunol 2002;22:306–318.

22 Noordzij JG, Verkaik NS, van der Burg M, van Veelen LR, Bruin-Versteeg S, Wiegant W, et al: Radiosensitive SCID patients with Artemis gene mutations show a complete B-cell differentiation arrest at the pre-B-cell receptor checkpoint in bone marrow. Blood 2003;101:1446–1452.

23 Liu YJ, Malisan F, De Bouteiller O, Guret C, Lebecque S, Banchereau J, et al: Within germinal centers, isotype switching of immunoglobulin genes occurs after the onset of somatic mutation. Immunity 1996;4:241–250.

24 Bohnhorst JO, Bjorgan MB, Thoen JE, Natvig JB, Thompson KM. Bm1-Bm5 classification of peripheral blood B cells reveals circulating germinal center founder cells in healthy individuals and disturbance in the B cell subpopulations in patients with primary Sjögren's syndrome. J Immunol 2001;167:3610–3618.

25 Ferry BL, Jones J, Bateman EA, Woodham N, Warnatz K, Schlesier M, et al: Measurement of peripheral B cell subpopulations in common variable immunodeficiency (CVID) using a whole blood method. Clin Exp Immunol 2005;140:532–539.

26 Paramithiotis E, Cooper MD: Memory B lymphocytes migrate to bone marrow in humans. Proc Natl Acad Sci U S A 1997;94:208–212.

27 Gratama JW, van der LR, van der HB, Bolhuis RL, van de Winkel JG: Analysis of factors contributing to the formation of mononuclear cell aggregates ('escapees') in flow cytometric immunophenotyping. Cytometry 1997;29:250–260.

Sack U, Tárnok A Rothe G (eds): Cellular Diagnostics. Basics Methods and Clinical Applications of Flow Cytometry. Basel Karger 2009 pp 230–260

Characterization of Natural Killer Cells

Roland Jacobs · Reinhold E. Schmidt

Clinic for Immunology and Rheumatology Hanover, Germany

Introduction

Characteristics of NK Cells

Natural killer cells (NK cells) bear this name because they are able to eliminate tumors and virus-infected cells without prior sensitization [1]. Thus, they are allocated to the innate immune system. They present large granular lymphocyte (LGL) morphology with a high cytoplasm-to-nucleus ratio and numerous granules in the cytoplasm (fig. 1) [2]. NK cells account for 10–15% of all lymphocytes in human peripheral blood. Phenotypically, NK cells are characterized by the expression of T-cell-associated markers (CD2 and in part CD8 at low density) and myeloid markers such as CD11b and CD16 in most NK cells. In general, all mature human NK cells express CD56, an isoform of the neural cell adhesion molecule (N-CAM) which mediates homophilic binding between neural and muscle cells. In lymphocytes, however, no specific function of CD56 has been determined so far. Furthermore, NK cells express opioid and adrenergic receptors. Although the functional impact of opioid receptors on NK cells is widely unknown, adrenergic receptors have been shown to be responsible for rapid recruitment of NK cells under acute stress situations [3–6].

Although NK cells also intracellularly express the CD3-associated ζ-chain, they do not express any molecules of the T-cell receptor (TCR) and the TCR-associated CD3 complex on the cell surface. Thus, human NK cells are defined as CD3– CD56+ lymphocytes [7, 8].

Activating and Inhibitory Receptors Expressed by NK Cells

In the past few years, several new NK cell receptors have been characterized. A family of molecules has been described and termed 'natural cytotoxicity receptors' (NCRs)

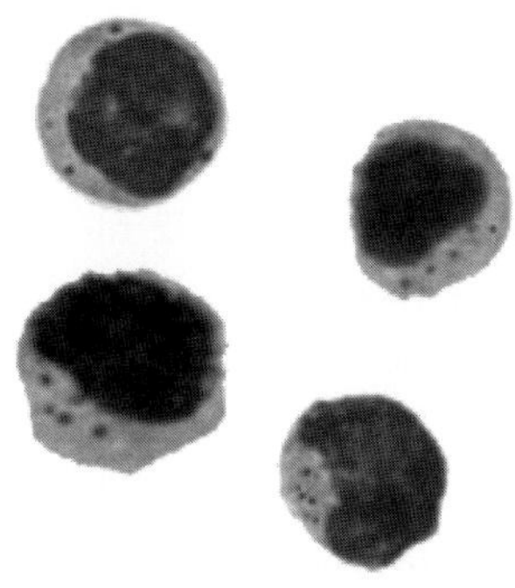

Fig. 1. NK cell morphology. Sorted NK cells from peripheral blood present as relatively large lymphocytes under the microscope. Typical granules in the cytoplasm become visible by Pappenheim staining. They are constitutively present in NK cells and contain cytolytic molecules such as perforin and granzymes.

[9]. This family comprises the molecules NKp30, NKp44 and NKp46 whose expression strongly correlates with the cytolytic activity of NK cells. Hemagglutinin expressed by virus-infected cells has been shown to serve as one ligand for NKp44 and NKp46. On tumor cells, however, corresponding ligands are still unknown. Although both NKp30 and NKp46 are constitutively expressed by all NK cells, NKp44 is only present on activated NK cells. NKp30 is associated with the intracellular ζ-chain and acts as signal modulator for NKp44 and NKp46 during cytolysis [10]. NKp80 is exclusive to NK cells in primates and acts as a stimulatory receptor. However, it lacks activation motifs in its cytoplasmic domain as well as charged amino acids in its transmembrane domain and is unlikely to be associated with activating adaptor proteins such as CD3ζ DAP12, DAP10 or FcRγ [11]. The ligand for NKp80 is AICL, which is a myeloid-specific receptor expressed by monocytes, macrophages and granulocytes [11]. CD244 (2B4) represents another activating receptor on NK cells whose ligand has been identified as CD48 [12, 13].

Several NK cell receptors have been to shown to exhibit MHC-I specificity [14]. In humans, they either belong to the immunoglobulin superfamily as killer cell immunoglobulin receptors (KIRs) or to the family of C-type lectin-like molecules as killer cell lectin-like receptors (KLRs). Each NK cell expresses one or more different MHC-I-specific receptors (on average 2–9 KLRs and/or KIRs) [15, 16]. The nomenclature of KIR molecules is based on their protein structure, particularly the number of immunoglobulin domains, and the length of the cytoplasmic tail. KIR molecules with two immunoglobulin domains (KIR2D) and those with three domains (KIR3D) contain either a long (L) inhibitory cytoplasmic tail with one or two immunoreceptor tyrosin-based motifs (ITIMs) or a short (S) cytoplasmic tail which associates with activating adaptor molecules containing immunoreceptor-tyrosin-based activating motifs (ITAMs). Hence, KIR2DL3 represents an inhibitory KIR with two domains and KIR3DS1 is an activating KIR with three extracellular domains. The KIR family is assigned to CD158 and the different members are named by alphabetical indices. This nomenclature, however, lacks information on structure and function of the respective receptors. Expression patterns of the receptors are genetically determined, but follow a stochastic scheme on the individual NK cells. The expression density of a given receptor in an individual NK cell can be either very high or low independently of the presence or absence of one or more ligands [15].

KLRs and KIRs differ considerably in their structure and specificity, though they are functionally homologous. In humans, heterodimers of CD94 and NKG2 molecules can form various KLRs. The functional features of these receptors depend on the associated NKG2 isoform. NKG2A and NKG2B bear intracellular ITIMs and hence they mediate inhibition of cellular functions.

NKG2C, NKG2E, and NKG2H isoforms associate with DAP-12 adaptor proteins like KIR molecules with short cytoplasmic tails and thus they induce stimulation of the cell. CD94:NKG2 heterodimers specifically bind HLA-E, which is a nonclassical HLA-I molecule representing a leader sequence of other HLA-I molecules and is as such coexpressed with different HLA proteins.

NKG2D *(KLRK1)* shows limited sequence identity to other NKG2 molecules and is expressed as a homodimer. For activation, NKG2D signals by associating with the adaptor protein DAP10. Human NKG2D binds UL-16-binding protein (ULBP) and MHC class-I-chain-related protein (MIC) A and B, which are upregulated in tumors and virus-infected cells [17].

Another member of the lectin-like receptor family is NKR-P1A (CD161) which is not well investigated so far. The molecule contains neither an ITIM nor a charged transmembrane amino acid. It is supposed to mediate negative signal via a currently unknown mechanism. A specific ligand for CD161 is also unknown so far.

Cytotoxic Function of NK Cells

The cytolytic activity is triggered by a balance of all inhibitory and activating signals received by the NK cell. In case of the predominance of negative signals, the NK cell remains inactive. However, if more activating signals are received, the cytolytic program of the NK cell will be started [18]. This cytolytic function towards the target can be mediated directly by cell-to-cell contact (natural cytotoxicity) or indirectly via antibodies coating the target cells (antibody-dependent cellular cytotoxicity; ADCC). NK cells can bind Fc portions via the FcγR (CD16) thereby becoming activated.

After successful binding to the target cell and a predominantly activating balance of received signals all the involved receptor-ligand pairs are physically concentrated by cytoskeleton structures. This establishes a narrow and limited contact zone between the effector and target cells which is called the 'immunological synapse'. Concomitantly, the intracellular granules of the effector cells are directed towards the synapse. Here they fuse with the cell membrane, thus releasing cytolytic substances such as granzymes and perforin. Formation of a synapse strengthens the complexes between effector and target cells, facilitates pairing of involved receptors and ligands, and ensures directed release of cytolytic molecules to the limited contact zone. Thus, only the touched target cell is killed and other cells in close proximity are spared from lysis [19].

In contrast to ADCC, the ζ-chain plays an essential role in directly mediated NK cytotoxicity [10, 20, 21]. Hence, the two mechanisms are initiated by different signals, but end in secreting the same cytolytic molecules from intracellular granules, thus killing the targets.

Another mechanism which enables NK cells to eliminate target cells is the induction of apoptosis. Activated NK cells express Fas ligand (CD95L) that binds to CD95 on target cells. Ligation of the molecules initializes a cascade of proteolytic enzymes in the target cell, leading to apoptotic death [22]. In addition to CD95L, NK cells also express TRAIL (TNF-related apoptosis-inducing ligand), which accordingly induces apoptosis via TRAIL receptors on the target cells [23, 24].

Production of Immune Mediators by NK Cells

In addition to their cytolytic function, NK cells are also capable of producing a considerable variety of chemokines and cytokines such as interferon-γ (IFN-γ) tumor necrosis factor (TNF), interleukin-1 (IL-1), IL-5, IL-8, IL-10, granulocyte-macrophage-colony-stimulating factor (GM-CSF), and macrophage inflammatory protein-1α (MIP-1α) and MIP-1β[1, 25–29]. By secreting these substances, NK cells are able to regulate immune responses and hematopoiesis.

NK cells, in turn, are sensitive to stimulation via IL-2, IL-12, IL-15, IL-21, and IFN-γ. The fact that NK cells are also able to produce the latter cytokines implies an autocrine activation pathway via IFN-γ [30–32]. In general, NK cells respond to the factors listed above with proliferation or enhanced cytotoxic capacity.

NK Cell Subpopulations

According to the surface density of CD56 expression, NK cells can be subdivided into $CD56^{dim}$ (weak expression) and $CD56^{bright}$ (5–10 times higher expression) NK cells, with the $CD56^{bright}$ population comprising about 1–10% of all NK cells in peripheral blood (fig. 2) [33–36]. The two populations differ considerably in terms of phenotype and function. In contrast to $CD56^{dim}$ NK cells, $CD56^{bright}$ NK cells do not express KIRs except for KIR2DL4 (fig. 3A). Both NK cell populations are positive for KLRs. Whilst the density of CD161 is similar on $CD56^{dim}$ and $CD56^{bright}$ NK cells, CD94 has a higher expression on the surface of $CD56^{bright}$ NK cells (fig. 3B) [37]. Functionally, $CD56^{dim}$ NK cells comply with the 'classical' type of NK cells with strong cytolytic capacity. $CD56^{bright}$ NK cells are clearly inferior in this regard.

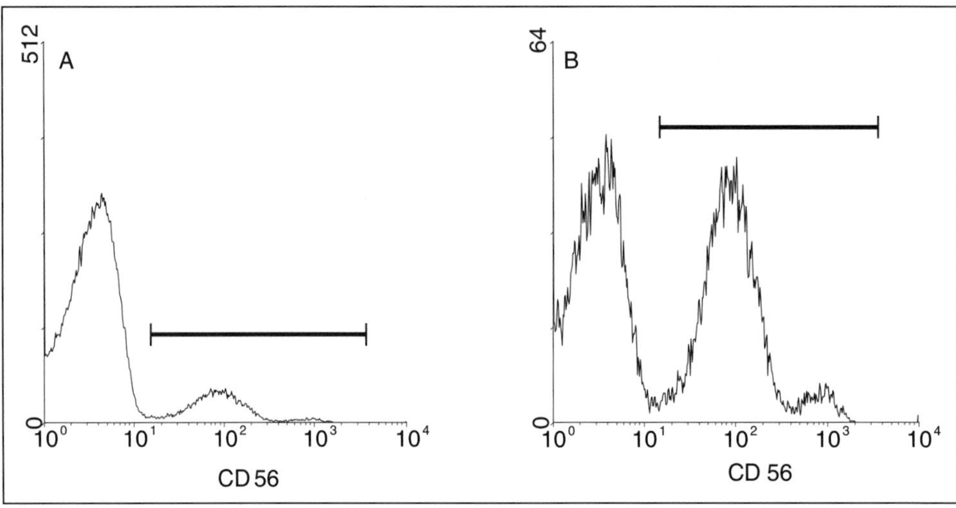

Fig. 2. CD56 expression by NK cells. **A** CD56^{bright} NK cells comprise only about 1–10% of all NK cells in human peripheral blood. The small proportion of T cells that also coexpresses CD56 has to be excluded from analysis by appropriate gating. The remainder CD56+ cells represent NK cells. **B** By exclusion of T cells, the CD56^{bright} NK cells become emphasized in the histogram plot.

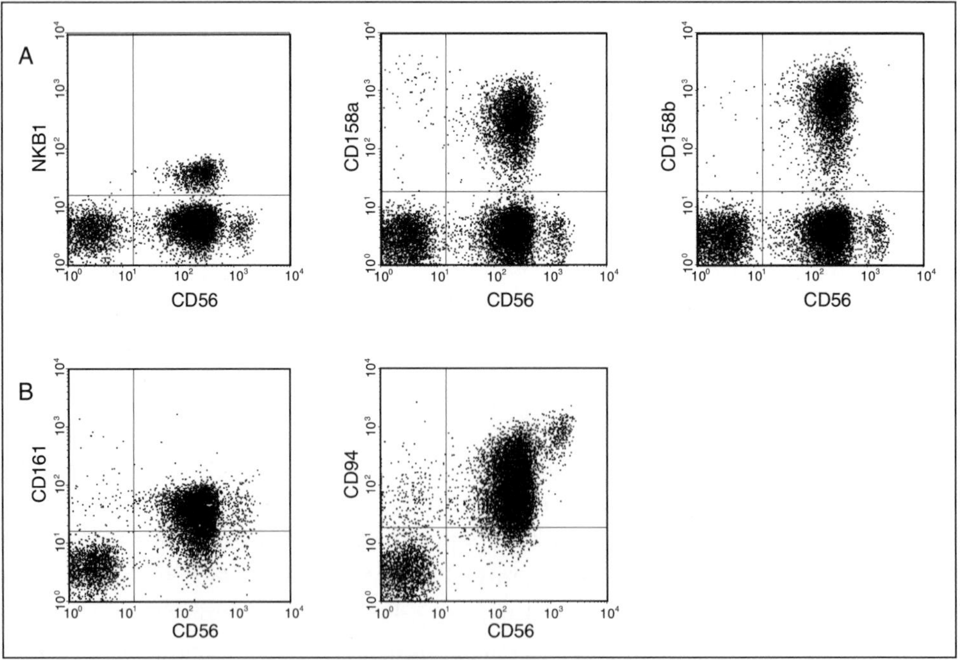

Fig. 3. KIR and KLR expression by NK cell subpopulations. **A** Almost all KIRs on NK cells are expressed by the CD56^{dim} population. **B** KLRs are expressed by both CD56^{dim} and CD56^{bright} NK cells. The cells of the analyses were gated on CD3– lymphocytes.

Thus, the two cell populations are differently armed for cytolysis. $CD56^{bright}$ NK cells are less active in binding tumor cells, which is also called 'conjugate formation' and is mandatory for performing effective cytolysis. Moreover, the granules in $CD56^{bright}$ NK cells contain lower concentrations of cytolytic substances such as perforin, serin esterases, granzymes A and B but, interestingly, higher amounts of granzyme K [38].

In contrast, $CD56^{bright}$ NK cells are superior in producing cytokines such as IFN-γ and TNF-α as compared to $CD56^{dim}$ NK cells. Thus $CD56^{bright}$ represent immunoregulatory rather than cytolytic lymphocytes. Hence, the NK cell compartment comprises a heterogeneous cell fraction, and specific functions can be attributed to distinct NK cell populations.

Protocols

Although most of the following methods can be performed using whole blood, the protocols are optimized for analyses of isolated peripheral blood mononuclear cells (PBMCs).

Preparation of Mononuclear Cells from Peripheral Blood

In order to functionally analyze lymphocytes (NK cells), it is necessary to separate granulocytes and erythrocytes from PBMCs. This is easily achieved by density gradient centrifugation.

Materials

- 50-ml tubes
- PBS = Phosphate-buffered saline (without Mg^{2+} and Ca^{2+}, as these cations support cell aggregation)
- Lymphocyte separation solution (density 1.077 g/ml).

Methods

- Dilute heparinized blood 1:2 with PBS in a 50-ml tube (e.g. 20 ml blood + 20 ml PBS).
- Carefully layer 10 ml of lymphocyte separation solution (density = 1.077 g/ml) underneath the blood dilution (using a thin pipette or preferably a syringe with an extra long injection needle). Alternatively, the blood suspension can be carefully layered on top of the separation solution. In both cases, mixing of the solutions must be prevented.
- Centrifuge the tube at 1,000 × *g* for 20 min.
- Remove the uppermost plasma-PBS mix.

- Carefully harvest the small opaque layer (PBMCs) with a pipette and transfer it into a fresh tube.
- Fill the tube containing the recovered PBMCs with PBS.
- Spin at 1,000 × *g* for 10 min.
- Remove the majority of the supernatant leaving approximately 100–200 μl.
- Resuspend the cell pellet in the remaining buffer by using a pipette with a (blue) tip.
- Add 10 ml PBS.
- Spin at 250 × *g* for 10 min.
- Remove the supernatant and resuspend the pellet in 2 ml PBS or medium of choice.

Comment

- In contrast to the instructions of several other protocols, we activate the brake function of the centrifuge during gradient centrifugation. This considerably shortens the procedure and has, at least in our experience, no negative influence on cell yield and viability.
- The last centrifuge step at 250 × *g* is important to remove the platelets, which are enriched thereby in the supernatant.
- Underlayering the separation solution beneath the blood is easier, particularly with high sample throughput than with the more common method of overlaying blood on the preloaded separation solution.
- Resuspension of the pellet in a small volume (100–200 μl) by using a 200-μl pipette with a blue tip is very effective and prevents possible clotting, as is frequently observed after vortexing.

Troubleshooting

In some cases, erythrocytes may remain in the interphase, possibly hampering subsequent analyses. In most cases, the erythrocytes can be eliminated by osmotic lysis. For this purpose, the pellet is first resuspended in 2 ml distilled water for 30 s and then well mixed with 2 ml of double-concentrated PBS. Alternatively, CD45 can be used in each sample allowing to gate on all leukocytes without erythrocyte contamination.

Expected Results

The cell yield should be around 1×10^6 PBMCs/ml of fresh blood and cell viability should be 100%.

Time Required

If the brake option of the centrifuge is used, the entire procedure will take about 1–1.5 h.

Phenotyping

Since NK cells are not characterized by a single surface molecule, multicolor analysis is essential. CD56 is a characteristic NK cell marker and is expressed by nearly all NK cells. However, in HIV patients, for example, there are quite frequently CD56– cells. These cells have to be ascribed to the NK cell compartment due to their remainder phenotype (CD3–, CD94+ and/or CD161+). A considerable fraction of T-cells also coexpresses CD56 and these T-cells must be excluded for proper NK cell analysis. This can be achieved by two-color staining using differently labeled CD3 and CD56 antibodies, depicting NK cells as CD3– CD56+ lymphocytes in a dot plot graph (fig. 4). In order to increase the quality of analysis, unspecific binding of antibodies must be avoided. All proteins including antibodies are sticky, meaning that antibodies can unspecifically bind to cells giving false-positive results. This unspecific binding can be minimized by adding proteins such as 0.1% BSA to all involved buffers. Furthermore, most NK cells express FcγRIII (CD16). As CD16 binds the Fc portion of antibodies, it can also capture any antibody used for phenotyping, independent of its specificity, also resulting in false-positive results. This can be avoided by adding either inactivated human serum or commercial IgG solution, which will saturate the binding capacity of the Fc receptors (fig. 5).

Phenotyping Cultured NK Cells

Human NK cells can quite easily be maintained in culture as clones or cell lines for several weeks. The cells need to be stimulated by cytokines such as IL-2 and irradiated feeder cells. During culture, they will increase in size and, depending on the longevity of the culture, a certain proportion of the cells will lose viability. Thus, cytometric analysis of these NK cells requires consideration of culture-related parameters. Changes in cell size and the presence of feeder cells can be dealt with by appropriate gating (fig. 6). Cell death can be visualized by adding dyes such as propidium iodide (PI) or 7AAD, which specifically stain dead cells and enable the gating on viable cells.

Materials

- 96-well round-bottom plate
- Sealing tape
- Multichannel pipette
- PBS/0.1% BSA
- Human immunoglobulin solution or human pool serum
- Antibodies of intended specificities and corresponding isotype control antibodies.

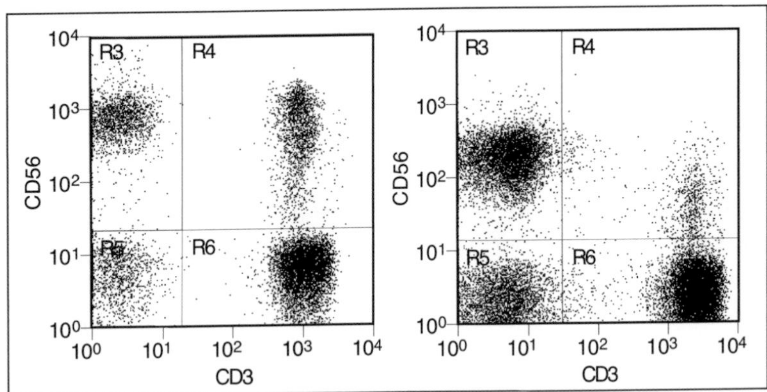

Fig. 4. Definition of NK cells by phenotype. NK cells are defined as CD3– CD56+ lymphocytes. Accordingly, they are depicted as dots in the upper left quadrant (R3). CD3+ CD56+ cells (R4) represent T cells including NKT cells. The proportions of the different CD56+ lymphocytes vary between individuals as illustrated in the two examples.

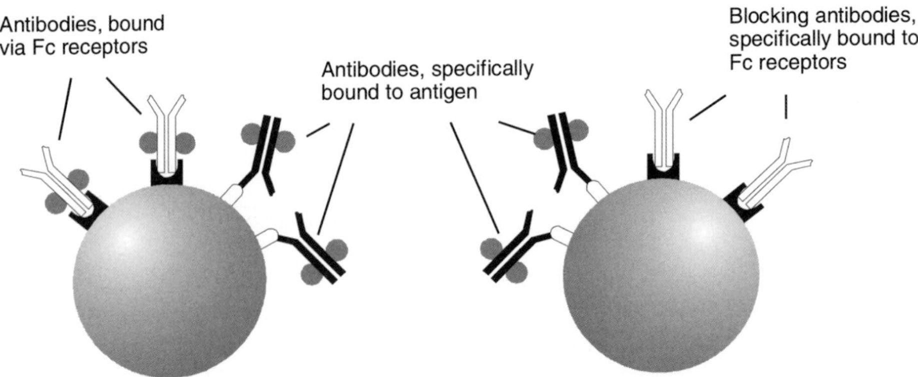

Fig. 5. Fc receptor blockade. Unspecific binding of test antibodies can be minimized by saturating Fc receptors with unlabeled antibodies or pool serum.

Working Steps

- Resuspend 3×10^5 cultured cells in 100 µl PBS/BSA into the planned number of wells of a 96-well round-bottom plate.
- Add 10 µl of human IgG solution (or pool serum).
- Add the differently labeled appropriate antibody combinations (e.g. CD3-FITC and CD56-PE) to the dedicated wells.
- Prepare additional wells with isotype control antibodies covering all applied isotypes labeled with the appropriate dye.
- Seal the plate with tape to prevent drying out.
- Incubate for 20 min at 4 °C in the dark.
- 3 times washing:
  - Centrifuge the plate at $300 \times g$ for 3 min.
  - Decant the supernatant vigorously with one movement.

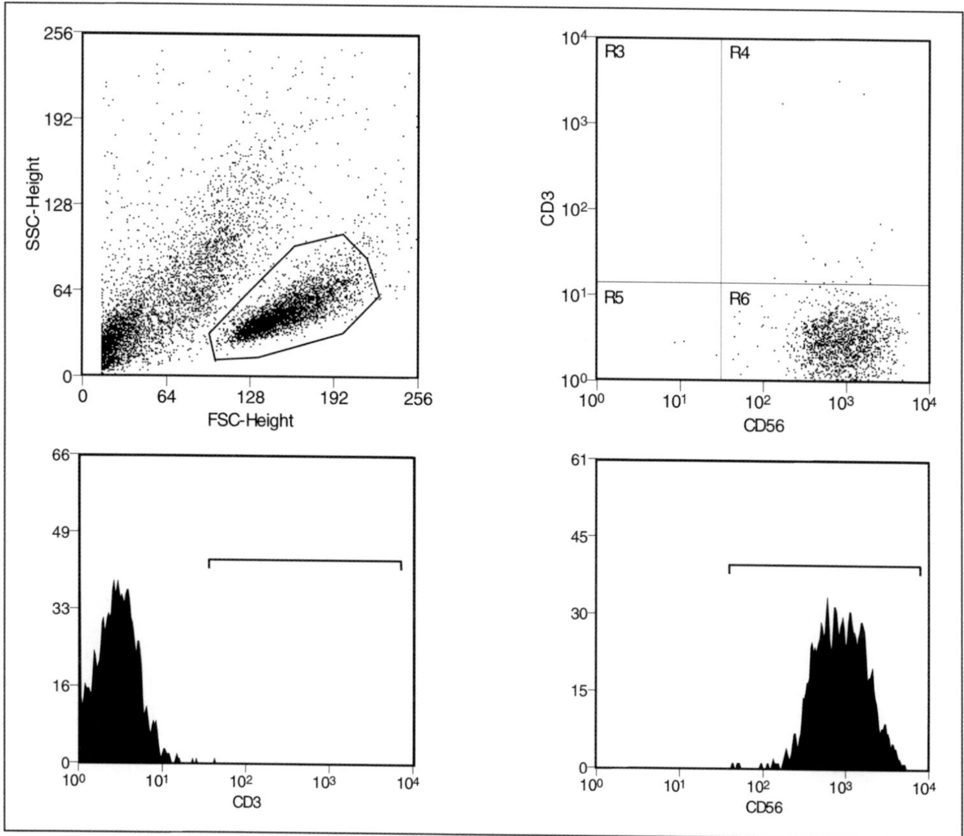

Fig. 6. Phenotyping of cultured NK cells. Due to the supplementation of culture media with growth factors such as IL-2, cells alter their properties in the FSC vs. SSC plot. Most of the cells increase in size, thus the gate has to be adjusted in order to include all cultured NK cells. The events outside the gate comprise debris and feeder cells which are essential for the culture of NK cells. The two-color dot plot and the two histograms confirm the NK phenotype of the cells.

- Dab the surface of the plate with a tissue.
- Resuspend the pellet on a shaker.
- Add 200 µl PBS/BSA to each well.

– Transfer the cells from the wells into 600-µl tubes that can be placed into empty wells of the plate.

– Top up the tube to 300 µl each with PBS/BSA.

Cytometric Analysis

The FACS machine must be set up for lymphocyte acquisition, and proper compensation must be performed. Gate on the population with a high forward scatter (FSC) and a low side scatter (SSC) (i.e. in the lower right position in an FSC vs. SSC dot plot). The activated cells are bigger than lymphocytes and so the FSC amplification might possibly need to be reduced. The events in the upper position are caused by dead feeder cells and must be excluded by appropriate gating. Acquire at least 10,000 events of the chosen gate. If differences in expression densities of some molecules are of interest, acquisition speed should be kept low.

Viability Controls

Commercial viability controls (e.g. ViaProbe) are available but PI can be used as well. Similar to trypan blue, which is a reliable indicator of dead cells in light microscopy, PI can enter only dead cells. There it binds to DNA, thus staining the cells red. PI signals equally in FL2 and FL3 and therefore positive (= dead) cells present as a narrow diagonal in the FL2 vs. FL3 plot. One of these channels (either FL2 or FL3) can be used additionally for antibody staining (fig. 7).

Working Steps

Add 20 μl PI (50 μg/ml PBS) to each sample at least 1 min before cell acquisition. The samples are instantly ready for analysis, no further washing is necessary.

Data Analysis

For data analysis, an appropriate gate is applied according to FSC and SSC properties. Analysis is then restricted to the cells in the gate and displayed, for example, in a CD3 vs. CD56 plot. The quadrants of the plot separate four cell populations:

- Lower left: CD3– CD56– cells (= all non-NK and all non-T-cells)
- Lower right: CD3+ T-cells
- Upper left: CD3– CD56+ NK cells
- Upper right: CD3+ CD56+ double-positive T-cells (= non-MHC-restricted T-cells including NKT cells, which are additionally characterized by a limited TCR repertoire (same VαJα chain and only a few different Vβ chains: VαQ preferentially paired with Vβ11). NKT cells recognize antigens (glycolipids) via CD1 and not MHC-I presentation.

If coexpression analysis of further molecules is wanted, the appropriate antibodies (other than CD3 and CD56) with different labels can be added according to the cytometer equipment. For analysis of complex multicolor phenotyping, additional gates must be introduced. Therefore, a region in a CD3 vs. CD56 plot is determined including only CD56+ and excluding all CD3+ cells. The combination of the lymphocyte and the CD3– CD56+ region in a new gate comprises only NK cells. Based on this NK cell gate, all further coexpression patterns can be clearly determined.

Comment

The reason for the acquisition speed limitation for distinction of expression densities of certain molecules is based on the technology of the cytometer. During acquisition, the cells are lined up in the center of the tube (= focus) by hydrodynamic forces. If acquisition speed exceeds a certain threshold, the sample stream becomes less focused, which results in blurring of distinct populations.

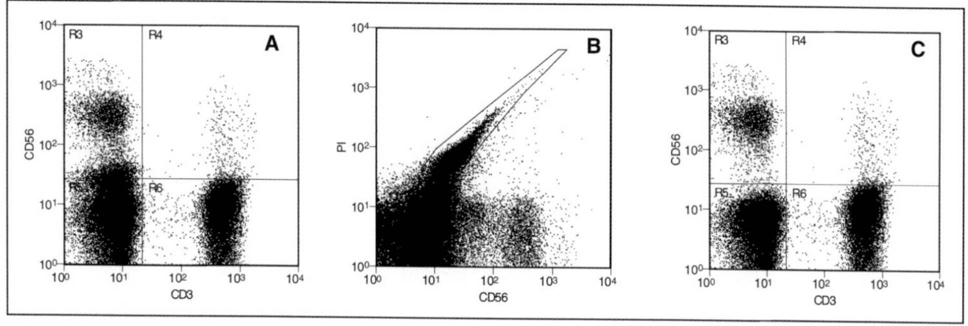

Fig. 7. Viability control. **A** Autofluorescence signals emitted by dead cells hamper proper analysis. This can be overcome by excluding dead cells. PI penetrates only into dead cells lacking membrane integrity and binds to DNA. As a result, the cells exhibit equal fluorescence in both FL2 and FL3. Thus dead cells are plotted as a narrow diagonal and can easily be excluded by gating as shown in **B**, allowing an exact analysis of living cells only as shown in **C**.

For analysis of very rare cells the number of events should be significantly increased in order to obtain reliable statistics and usable plots.

Measuring of the cells should be performed in a timely manner in respect of cell preparation. However, samples can be analyzed the next day without notable loss in quality if the cells were stored at 4 °C in the dark. If a longer lapse of time between preparation and measuring is inevitable (e.g. technical problems), cells can be acquired up to 1 week after staining if they have been fixed by resuspension in 0.1% paraformaldehyde-supplemented PBS/BSA. Note: the quality of the cell analysis might be reduced with time.

The use of 96-well plates, as opposed to several 5-ml tubes, has proved to be convenient in practice. The positions of the samples have simply to be logged on a protocol sheet and all washing steps can be performed using multichannel pipettes, thus shortening the time for preparing the staining considerably. Leaving blank at least one well between two samples significantly minimizes the risk of contaminating the samples with each other. Before measuring, the samples can finally be transferred into 600-μl tubes which precisely fit into the empty wells of the plate, thus also serving as a tube rack. If corresponding positions for the tubes are selected, the sequence of the samples will not change and the protocol sheet will still be valid and no tube labeling is necessary, again saving time (fig. 8). For acquisition, the 600-μl tubes are vortexed and placed into a standard 5-ml cytometer tube. Be sure that the acquisition capillary enters the small inner tube containing the sample.

Quality Controls

The cytometer has to be set up and compensated properly by using beads and cell samples.

For each antibody an identically labeled isotype control must be measured in parallel. The isotype control antibodies are of the same isotype (e.g. IgG1) and derive

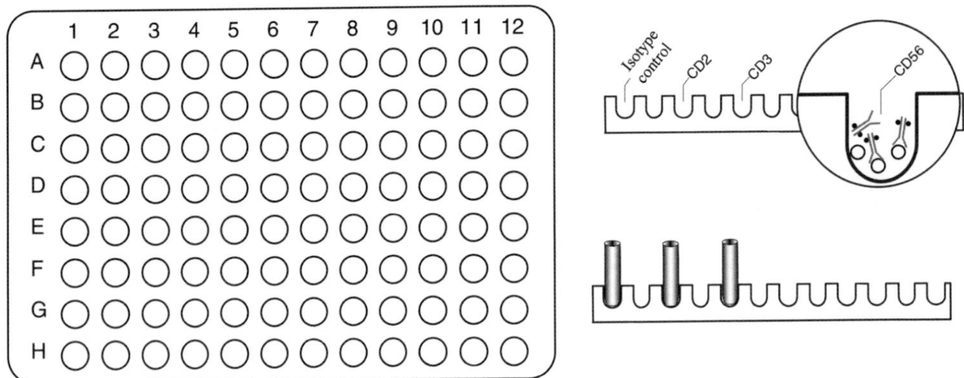

Fig. 8. Phenotyping in a microtiter plate. Phenotyping can be simplified and time can be saved by using a 96-well U-bottom plate and 600-μl tubes which perfectly fit into the wells.

from the same species (e.g. mouse) as the detection antibodies, but are specific for irrelevant antigens, and thus the samples should be completely negative.

Troubleshooting

Dead cells adulterate FACS analyses by their autofluorescent properties. Therefore, only living cells may be considered for analysis. This can be achieved by performing viability controls (see above) and a corresponding gating strategy.

Clotting of cells, which is particularly observed when measuring thawed cell samples, may block the orifices or tubings of the cytometer. To remove clumps, cells can be rinsed through fine-pored (e.g. 30 μm) meshes. DNAse may be added to prevent clotting, which is mainly caused by DNA-associated proteins sticking to the cells. Cutting the long DNA molecules does not affect the adherence of proteins, but does avoid the agglutination of cells.

Time Required

Preparation of PBMCs will take about 1–1.5 h.

Staining of the cells will also take about 1–1.5 h.

Time for measuring the samples will strongly depend on cell numbers and the frequency of the molecules of interest. Normally, each sample will be acquired within seconds to few minutes. Analysis of acquired data and especially of multicolor data will depend on the software and even more on the experience of the operator.

Determination of Intracellular Molecules

Constitutively expressed and inducible molecules can be analyzed on a single-cell level by intracellular staining. Relevant constitutively expressed intracellular proteins of NK cells include perforin, granzymes, and the ζ-chain (CD247), which is commonly

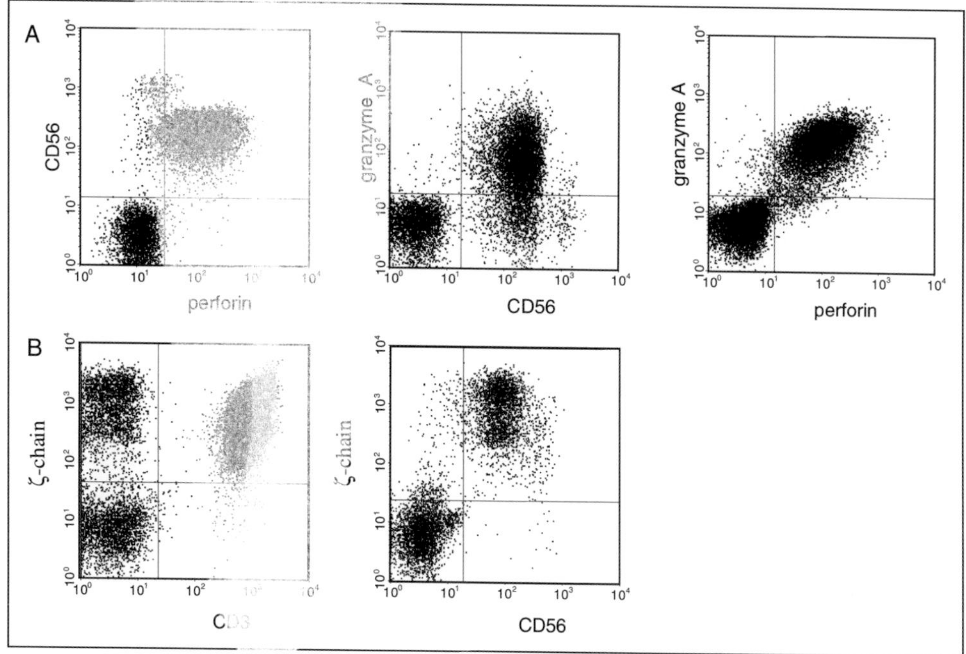

Fig. 9. Intracellular staining of constitutive molecules in NK cells. **A** Perforin and granzyme A are essential for lysing target cells and are constitutively expressed in almost all CD56^{dim} NK cells. The two molecules are basically stored in the same cells. **B** The CD3 ζ-chain is expressed in all T cells. However, the ζ-chain is also expressed in NK cells, as this signal-transducing molecule is also associated with CD2, CD16, and NCR (in this plot, CD3+ T-cells were excluded by gating).

known as a TCR-associated signal transducer (fig. 9). CD247 also functions as signal transducer in NK cells, but here the ζ-chain is associated with FcγRIII (CD16) and CD2, both of which can be used independently to stimulate NK cells. In addition, the ζ-chain mediates activating signals initiated via NKp30 and DAP12 (two molecules that function as coreceptors of activating NK cell receptors). The protocol for staining the intracellular ζ-chain of NK cells can be found below. Staining protocols for other intracellular molecules (e.g. perforin and granzymes) are identical, except that 4% paraformaldehyde is substituted for the 1% solution.

Materials

- PBS/0.1% BSA.
- R10 medium: RPMI-1640 medium supplemented with 10% FCS, glutamine (2 mmol/l) penicillin-streptomycin (100 U/ml) and sodium pyruvate (1 mmol/l).
- Permeabilizing reagent:
  - Saponin stock solution (100 mg/ml) in distilled water.
  - HEPES (238 mg/ml) in distilled water.
  - Working solution: mix 500 μl saponin solution with 500 μl HEPES solution and fill up with 49 ml distilled water.

- Paraformaldehyde solution (1 and 4%):
  - Stir 4% paraformaldehyde (weight percentage) in warm PBS (ca. 80 °C, do not boil!). Store the completely dissolved solution at 4 °C.
  - Prepare a 1% paraformaldehyde solution by diluting the above solution 1 : 4 with PBS.
- Human immunoglobulin solution or human inactivated pool serum.
- Antibodies of the intended specificity.

Method

- Adjust cells to 3×10^6 /ml in R10.
- Transfer 100 μl for each sample into a well of a round-bottom plate.
- Add 10 μl human IgG solution (or pool serum) to each sample.
- Supplement the appropriate number of wells with differently labeled CD3 (e.g. PerCP) and CD56 (e.g. PE) and the other well with corresponding isotype control antibodies.
- Incubate for 20 min at 4 °C.
- Wash 3 times:
  - Add 150 ml of PBS/BSA into each well.
  - Centrifuge at $300 \times g$ for 5 min.
  - Remove (or decant) the supernatant.
- Fixation:
  - Resuspend the cells in 50 μl 1% paraformaldehyde solution (for ζ-chain; most other molecules might tolerate 4% paraformaldehyde).
  - Incubate for 10 min in the dark at room temperature.
  - Wash once.
- Resuspend the cell pellets in 50 μl of saponin solution (permeabilization).
- Add the differently labeled anti-ζ-chain antibody (e.g. FITC) into one well and an appropriate isotype control antibody into the other well.
- Incubate for 20 min at 4 °C in the dark.
- Wash 3 times and resuspend the cell pellets each in 300 μl PBS/BSA.

Acquisition

Perform cytometer setup for lymphocytes and assure proper compensation for all utilized dyes. Set a gate on lymphocytes according to the FSC vs. SSC properties of the cells. The two samples can then be measured consecutively acquiring at least 10,000 events in the selected lymphocyte gate.

Data Analysis

As mentioned above, the ζ-chain is also expressed by T-cells. Thus, in order to analyze ζ-chain expression exclusively in NK cells, cells from the lymphocyte gate have to be limited to NK cells by further gating on CD56+ CD3– lymphocytes only. NK cells bearing the ζ-chain can then be depicted by plotting CD56 vs. ζ-chain in a dot plot graph.

Comment
For many surface molecules, paraformaldehyde treatment is harmless, which helps to save time since extra- and intracellular staining can be performed simultaneously after fixation and permeabilization.

Unfortunately, this is not the case for NK cells. The CD56 antigen is quite sensitive to paraformaldehyde treatment. Therefore, CD56 must be stained prior to fixation.

As the cells analyzed for intracellular cytokines must be fixed and permeabilized viability controls are not possible.

Most of the intracellular staining protocols use 4% paraformaldehyde. However, in case of the ζ-chain a lower concentration of paraformaldehyde (1%) is preferable. Therefore, optimal concentrations of paraformaldehyde should be titrated for every antigen when analyzed for the first time.

Required Time
Separation of PBMCs will take 1–1.5 h. Since the extra- and intracellular staining must be performed consecutively, due to the sensitivity of CD56 to paraformaldehyde treatment, the phenotyping procedure will take 1.5–2 h.

Intracellular Cytokine Staining

Similar to most effector cells, NK cells do not constitutively store mentionable amounts of cytokines. Thus, NK cells must be induced by stimulation in order to produce the different mediators (fig. 10). The protocol for staining intracellular proteins is explained below using IFN-γ as an example.

Materials
- PBS/0.1% BSA.
- R10 medium: RPMI-1640 medium supplemented with 10% FCS, glutamine (2 mmol/l), penicillin-streptomycin (100 U/ml), and sodium pyruvate (1 mmol/l)
- Phorbol myristate acetate (PMA)
- Ionomycin
- Brefeldin A: stock solution (5 mg/ml) in DMSO, working solution dilute stock solution with PBS (1 : 50)
- Permeabilizing reagent:
  - Saponin stock solution (100 mg/ml) in distilled water
  - HEPES (238 mg/ml) in distilled water
  - Working solution: mix 500 μl saponin solution with 500 μl HEPES solution and fill up with 49 ml distilled water

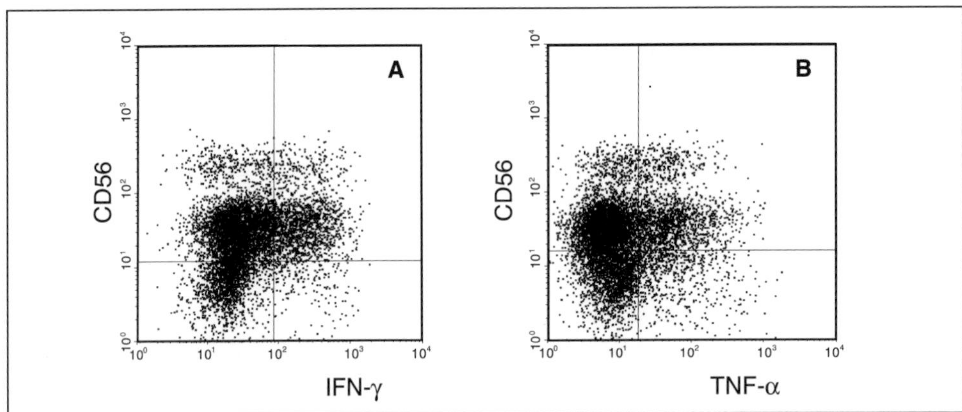

Fig. 10. Intracellular staining of cytokines in NK cells. Fixation and permeabilization enable the detection of cytokines such as IFN-γ and TNF-α in NK cells after prior stimulation. Expression of the cytokines by $CD56^{dim}$ and $CD56^{bright}$ NK cells is clearly visible. Cells were gated on CD3− lymphocytes.

- Paraformaldehyde solution (4%):
  - Stir 4% paraformaldehyde (weight percentage) in warm PBS (ca. 80 °C, do not boil!). Store the completely dissolved solution at 4 °C
- Human immunoglobulin solution or human inactivated pool serum
- Antibodies of the intended specificity.

Induction of Cytokine Production

- Adjust 1×10^6 PBMCs in 1 ml R10 in a 5-ml tube.
- Add 5 ng PMA and 500 ng ionomycin 1 min later.
- Incubate (37 °C, 5% CO_2) for 1 h.
- Add 20 μl brefeldin A working solution to prevent secretion and achieve intracellular accumulation of the produced cytokines.
- Incubate for an additional 3 h (37 °C, 5% CO_2).
- Centrifuge the cells at $300 \times g$ for 5 min.
- Remove the supernatant and resuspend the cells in 100 μl PBS/BSA.

Staining Procedure

- Transfer half (50 μl) of the cells into a different tube.
- Add 10 μl human IgG solution (or pool serum).
- Supplement one tube (A) with differently labeled CD3 (e.g. PerCP) and CD56 (e.g. PE) and the other tube (B) with appropriate isotype control antibodies.
- Incubate for 20 min at 4 °C.
- Washing:
  - Add 4 ml of PBS/BSA in each tube.
  - Centrifuge at $300 \times g$ for 5 min.
  - Remove (or decant) the supernatant.

- Fixation:
  - Resuspend the cell in 50 μl 4% paraformaldehyde solution.
  - Incubate for 10 min in the dark at room temperature.
  - Wash once.
- Resuspend the cell pellets in 50 μl of saponin solution (permeabilization).
- Add again differently labeled anti-IFN-γ antibody (e.g. FITC) into tube A and an appropriate isotype control antibody into tube B.
- Incubate for 20 min at 4 °C in the dark.
- Wash once and resuspend the cell pellets each in 300 μl PBS/BSA.

Analysis

Perform cytometer setup for lymphocytes and assure proper compensation for all utilized dyes. Set a gate on lymphocytes according to the FSC vs. SSC properties of the cells. The two tubes can then be measured consecutively acquiring at least 10,000 events in the selected lymphocyte gate.

Data Analysis

PMA is an unspecific stimulus and can also induce the production of cytokines such as IFN-γ in cells other than NK cells. Thus, in order to analyze IFN-γ-producing NK cells, cells from the lymphocyte gate have to be limited to NK cells by further gating on CD56+ CD3−lymphocytes. IFN-γ-producing NK cells can then be depicted by plotting CD56 vs. IFN-γ in a dot plot graph.

Comment

For many surface molecules, paraformaldehyde treatment is harmless, which helps to save time since extra- and intracellular staining can be performed simultaneously after fixation and permeabilization.

Unfortunately, this is not the case for NK cells. The CD56 antigen is sensitive to paraformaldehyde treatment. Therefore, CD56 must be stained prior to fixation.

Activation of the cells with PMA/ionomycin can also be performed overnight, making the work flow more flexible.

Since cells analyzed for intracellular cytokines have to be fixed and permeabilized, viability controls are not possible.

Certain molecules such as CD4 are strongly downregulated by PMA treatment. Although CD4 is irrelevant for NK cells since they normally do not express this antigen, it cannot be presumed that expression of other NK-cell-specific proteins would not be affected by PMA.

Required Time

Separation of PBMCs will take 1−1.5 h, stimulation of the cells will need at least 4 h.

Since the extra- and intracellular staining must be performed consecutively due to the sensitivity of CD56 to paraformaldehyde treatment, the phenotyping procedure will take 1.5−2 h.

Analysis of Signaling Molecules in NK Cells

Phosphorylation of intracellular molecules is a key event in many cell activation pathways. The standard method for investigating signal transduction is isolating cellular proteins by gel electrophoresis and detecting phosphorylated proteins on Western blots of the gels by using antibodies. Analyzing phosphorylated proteins by intracellular staining strongly simplifies the tracking of signaling pathways as compared to standard Western blot methods. The combination of extra- and intracellular staining allows direct flow-cytometric analysis of activation steps in single-cell populations (fig. 11). An example of how to measure phosphorylation of signal transducer and activator of transcription (STAT) 3 molecules in NK cells is given below.

Working Steps

- Resuspend PBMCs at 3×10^6/ml in R10 medium.
- Add 20 ng/ml IL-21.
- Incubate at 37 °C for 45 min.
- Add C56-PE antibodies.
- Incubate for 15 min.
- Add the same volume of 4% paraformaldehyde solution (final concentration 2%).
- Incubate for 10 min at 37 °C.
- Centrifuge (300 × *g*, 10 min).
- Resuspend the pellet in 1 ml of ice cold 90% methanol.
- Incubate the cells for 30 min on ice.
- Wash twice with PBS/1% FCS.
- Resuspend in 100 μl PBS/FCS.
- Add pSTAT3-ALEXAFluor 647 and CD3PerCP antibodies.
- Incubate for 1 h at room temperature.
- Wash once.

Data Acquisition

Load cytometer settings for lymphocytes and assure proper compensation for all utilized dyes. Set a gate on lymphocytes according to the FSC vs. SSC properties of the cells. The tubes can then be measured consecutively, acquiring at least 10,000 events in the selected lymphocyte gate.

Data Analysis

In order to restrict analysis to NK cells, cells from the lymphocyte gate have to be further gated on CD56+ CD3– lymphocytes. STAT3+ NK cells can then be depicted by plotting CD56 vs. STAT3 in a dot plot graph or as a STAT3 histogram.

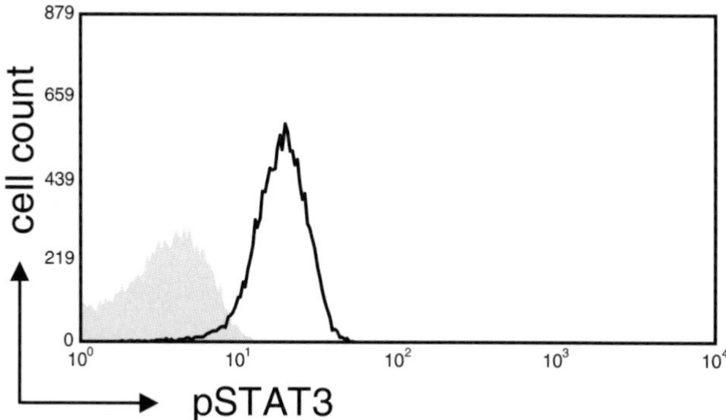

Fig. 11. Analysis of signal transduction by flow cytometry. Lymphocytes were activated with IL-21 (20 ng/ml) at 37 °C for 45 min. Phosphorylated STAT3 molecules were stained intracellularly and analyzed by gating on NK cells, isotype control is depicted in gray.

Comment

Standard protocols for intracellular staining using saponin will not work. The optimal concentration for methanol might need to be determined.

Analysis of Conjugate Formation

One basic function of NK cells is the ability to kill altered cells. A prerequisite for killing is the initial binding to the target cell, which is established via several adhesion molecules. The complexes formed can be identified by flow cytometry. Tumor cell lines, such as the standard target K562, exhibit a high degree of autofluorescence, so that further staining is unnecessary. The effector cells are stained as for normal phenotyping before being incubated with the targets. Careful selection of the surface markers is essential. For example MHC-I (all PBMCs) or a combination of CD3 and CD56 (NK cells) are feasible markers since these molecules are neither involved in cell binding nor does their cross-linking affect functional activity of the effectors. In cytometry, both effectors and the bigger targets can easily be distinguished according to their FSC vs. SSC properties. During analysis, proper gating must include both populations. Conjugates can be identified as double-positive events (fig. 12).

Materials

- PBS/0.1% BSA
- R10 medium: RPMI-1640 medium supplemented with 10% FCS, glutamine (2 mmol/l), penicillin-streptomycin (100 U/ml), and sodium pyruvate (1 mmol/l)
- Antibodies of appropriate specificities.

Working Steps

- Incubate 2×10^6 effectors with CD3-PercP and CD56-PE antibodies in 100 μl PBS/BSA for 20 min in a 5-ml tube at 4 °C in the dark.

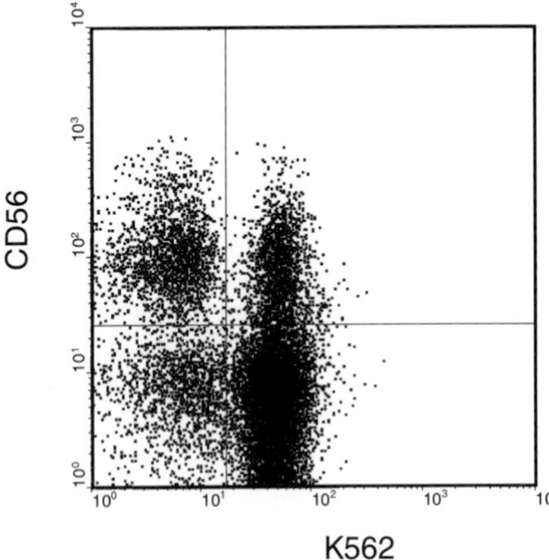

Fig. 12. Determination of conjugate formation between NK and target cells. Prior to cell lysis, effector cells must bind the targets, thereby forming conjugates. If the two cell types fluoresce differently, conjugates are detectable as double-positive events in the upper right quadrant. The dot plot depicts red fluorescence (y-axis) of effectors and green autofluorescence (x-axis) of K562 target cells.

- Washing:
  - Add 4 ml of PBS/BSA into the tube.
  - Centrifuge at 300 × *g* for 5 min.
  - Remove (or decant) the supernatant.
  - Resuspend the cells in 600 μl R10 medium and split the cells by transferring 300 μl to a different tube.
- Preparing K562 target cells:
  - Suspend 2×10^5 K562 cells in 4 ml R10 medium.
  - Centrifuge at 300 × *g* for 5 min.
  - Remove the supernatant.
  - Resuspend the cells in 600 μl R10 medium and split the cells by transferring 300 μl in a different tube.

Data Acquisition

After basic lymphocyte setup of the cytometer, target cells from one tube can be acquired. Normally, the FSC amplification will have to be decreased due to the large size of the targets. A gate can now be drawn around the cells and the fluorescent properties of the cells in FL1 (green) and FL2 (red) can be controlled in a second plot. In FL1 a clear green autofluorescence signal will be visible. A lower autofluorescence signal in FL2 needs to be corrected by reducing FL2 amplification.

The stained effector cells can be measured using the same instrument settings and applying a lymphocyte gate. Under these conditions the cells will signal in FL2 (red) but must be compensated properly in order to eliminate positive events in FL1.

After completing the instrument settings, the cells of both tubes used before can be mixed and immediately measured. In the FSC vs. SCC plot, a gate is drawn to include the effector and target cell populations. Based on these settings, the corresponding FL1 vs. FL2 plot should now depict FL1 (green) and FL2 (red) single-positive but no double-positive populations. Depending of the chosen antibodies, a double-negative effector cells may also be present.

The remaining samples are mixed together centrifuged (3 min, 50 × *g*) and incubated for at least 5 min at 37 °C. Carefully resuspend the cell sediment (rough handling might disrupt the conjugates) and measure on flow cytometer as described above. The dot plot of this sample should show a double-positive (FL1 and FL2) cell fraction. This population comprises conjugates exhibiting the green fluorescence properties of the targets and the red staining of the effectors. If kinetics of conjugate formation is of interest, any number of samples can be prepared and measured after different incubation times.

Data Analysis
The percentage of targets which have been bound by effector cells can easily be calculated by gating on all (green) targets and determining the percentage of red cells within this gate.

Comment
The numbers of effector cells should be 5–10 times higher than that of the target cells (E/T ratio: 5–10 : 1) to obtain reliable results. Since NK cells comprise about 10% of PBMCs, an E/T ratio of 50–100 : 1 has to be used when PBMCs are analyzed.

Troubleshooting
Resuspension can cause disruption of conjugates due to shearing forces. Thus, the cell suspension should be treated very carefully.

Required Time
PBMC preparation needs about 1–1.5 h. The conjugate assay including cell staining can be performed within 1 h.

Flow Cytometry-Based Cytotoxicity Assay

The golden standard for determining NK cell cytotoxicity is the chromium-51 release assay where target cells are radiolabeled. Different numbers of effector cells are added and after 4 h the release of radioactivity from lysed cells is measured in a γ-counter. Alternatively, cytotoxicity can be determined in a flow-cytometry-based assay, which is very similar to the conjugate analysis mentioned above (fig. 13). If K562 cells are used as targets, staining can be omitted due to the high autofluorescence signal. Other cells might need prior staining with labeled antibodies or dyes such as CFSE. The protocol given below is based on K562 cells used as targets.

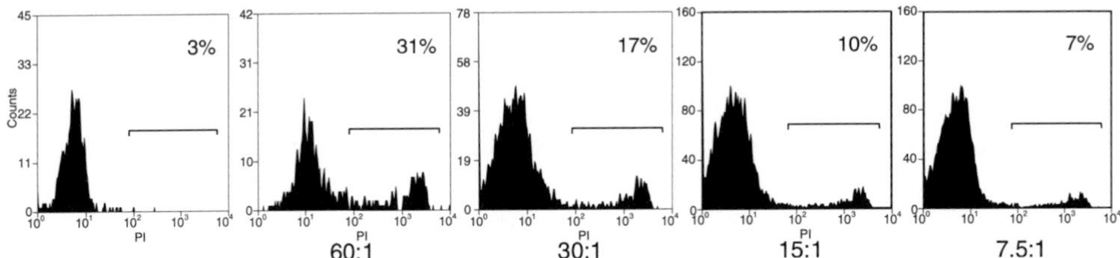

Fig. 13. Assessment of natural cytotoxicity by flow cytometry. Green fluorescent K562 target cells were incubated with effectors at the indicated E/T ratios for 4 h. The percentages of lysed target cells (inserts) were determined after adding PI. On the left hand side, spontaneous lysis is presented, which was obtained by incubating the target cells in medium alone. Specific lysis was calculated for each E/T ratio by subtracting the spontaneous from the measured values.

Materials

- PBS/0.1% BSA
- R10 medium: RPMI-1640 medium supplemented with 10% FCS glutamine (2 mmol/l) penicillin-streptomycin (100 U/ml) and sodium pyruvate (1 mmol/l)
- Trypan blue solution (0.5% trypan blue in saline)
- PI solution (50 µg/ml PI in PBS).

Working Steps

- Preparing targets:
  - Transfer targets from the culture flask into a 5-ml tube.
  - Centrifuge the cells (300 × *g*, 10 min).
  - Resuspend the pellet in 1 ml R10 medium.
  - Viability control by trypan blue exclusion staining: Mix 20 µl cell suspension and 20 µl trypan blue solution with a pipette.
  - Calculate the percentage of dead (blue) cells within all (clear and blue) cells, which should be lower than 3%.
  - Adjust the cells to 1 × 10^5/ml R10 medium.
  - Preparing effector cells.
  - Isolate PBMCs (see 'Preparation of Mononuclear Cells from Peripheral Blood', pp. 235).
  - Adjust the cells to 3 × 10^6/ml R10 medium.
- Cytotoxicity assay:
  - Effector and target cells are coincubated at four different E/T ratios (60 : 1, 30 : 1, 15 : 1 and 7.5 : 1) according to the pipetting protocol in table 1.
  - Centrifuge the tubes mildly (100 × *g*, 3 min).
  - Incubate the samples at 37 °C for 4 h.
  - Place the tubes on ice until analysis.
  - Add 20 µl PI as viability indicator.

Tab. 1. Pipetting scheme for effector and target cells in four different E/T ratios

E/T ratio	Effector cell suspension, µl	Target cell suspension, µl	Medium, µl
60 : 1	200	100	0
30 : 1	100	100	100
15 : 1	50	100	150
7.5 : 1	25	100	175
Minimal value	0	100	200

- Vortex briefly.
- Place the tubes again on ice for at least 5 min.
- Measure the samples within 30 min after adding PI.

Data Acquisition

Use the background sample in order to set a proper gate around all FL1+ (green) cells representing the target cells. All samples can consecutively be measured by acquiring 5,000–10,000 events per sample and analyzed using histogram plots of FL2 or FL3. PI+ (= dead) cells signal equally well in the two channels and thus lysed target cells appear positive. The fraction of PI+ cells in the background sample should be minimal. Percentages of PI+ target cells in the other samples are expected to increase with higher E/T ratios.

Data Analysis

Specific lysis can be calculated from histogram statistics by subtracting percentages of the background sample from the samples containing effector and target cells.

Troubleshooting

Due to lysis, cells might tend to increase in size and alter fluorescence properties. Thus, the initially defined target cell gate has to be controlled for each sample to ensure that all targets are included. Otherwise, the assessed values become unreliable.

An advantage of the flow-cytometric determination of cytotoxicity is the avoidance of radioactive material and hence connected risks and costs (equipment, storage, radioactive waste disposal). However, the disadvantage is the time-consuming analysis, lack of standardization, and a weak correlation compared with the chromium-51 release assay. For these reasons, the chromium release assay is still preferred.

Required Time

PBMC preparation needs about 1–1.5 h. Incubation time and cell preparation require an additional 4.5 h; acquisition and analysis take about 2–3 min per sample.

Identification of Cytotoxic NK Cells

The final step of cellular cytotoxicity is the release of cytolytic molecules from intracellular granules of the effector cells towards the targets. The membranes of the granules are lined by the molecule LAMP-1 (lysosomal associated protein, CD107a). Since the membranes are inside the cell, CD107a cannot be detected by antibodies until granule release is initiated, at which time CD107a becomes accessible at the surface of the effector cell. Thus CD107a can be used to identify NK cells which are killing the target cells (fig. 14).

Working Steps

- Resuspend PBMCs at 1×10^6/ml in R10 medium.
- Incubate the cells with 1×10^5/ml K562 target cells (E/T ratio = 10:1).
- Medium alone (negative control).
- PMA/ionomycin (positive control).
- Add 20 µl/ml CD107a-FITC antibody (1D4B; BD Pharmingen).
- Incubate samples for 1 h at 37 °C.
- Add monensin (6 µg/ml).
- Incubate for 3–5 h.
- Stain the samples with CD3-PerCP and CD56-PE for 20 min.
- Wash once.

Data Acquisition

Load cytometer settings for lymphocyte acquisition and ensure proper compensation. Acquire at least 50,000 events in the lymphocyte gate.

Data Analysis

Gate on NK cells by using the combination of region(R)1 (lymphocytes in an FSC vs. SSC plot) and R2 (CD3– CD56+ in FL2 vs. FL3). The two-color plot FL2 vs. FL1 displays the degranulated (CD107a+) NK cells.

Comment

Use of monensin is required as it prevents acidification of endocytic vesicles, thus inhibiting degradation of recycling CD107a molecules and enhancing its detectability.

Required Time

PBMC preparation needs about 1–1.5 h. Incubation time and cell preparation require an additional 4.5 h, acquisition and analysis take about 2–3 min per sample.

Direct Fluorochrome Labeling of NK Cells

Like other lymphocytes, NK cells can be labeled directly with various dyes. These fluorochromes should not interfere with cellular viability or function. There are sev-

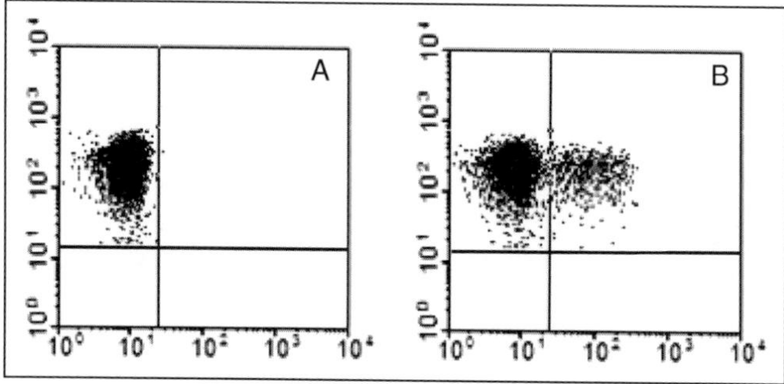

Fig. 14. Identification of target-cell-killing NK cells. Lymphocytes were incubated with NK-resistant L1210 in **A** and NK-sensitive K562 target cells in **B**. Cells were gated on NK cells and release of cytotoxic granules was analyzed by using CD107a antibodies.

eral useful substances currently available that fulfil these prerequisites. A commonly used molecule is CFSE, a green fluorescent dye. CFSE is taken up by cells into the cytoplasm where it binds to intracellular proteins. The dye is useful for staining target cells in cytometric cytotoxicity assays but also for analyzing the proliferative capacity of lymphocytes. Since the dye is also distributed to the daughter cells during mitosis, fluorescence intensity decreases with each cell division (fig. 15). By combining CFSE staining with appropriate surface markers, the proliferative responses of distinct populations within a cell suspension can be determined. CFSE signals very brightly and thus overstaining might occur, causing compensation problems in multicolor cytometric analyses. Therefore, optimal CFSE concentration and incubation time must be determined in earlier test stainings.

Materials

- PBS
- PBS/0.1% BSA
- R10 medium (RPMI-1640 supplemented with 10% FCS, 2 mmol/l glutamine, 100 U/ml penicillin/streptomycin, and 1 mmol/l pyruvate
- FCS
- CFSE
- Stock solution 25 mg CFSE in 10 ml DMSO (5 mmol/l; aliquots can be stored at −20 °C)
- Working solution: 1 : 100 dilution in PBS
- Example for a concentration: suspend 10 μl of the working solution into 1 ml cell suspension in protein free PBS. The optimal CFSE concentration is cell type dependent.

Working Steps

- Spin down the cells in a 5-ml tube at 250 × *g* for 5 min.
- Resuspend cells in PBS without any protein additives.

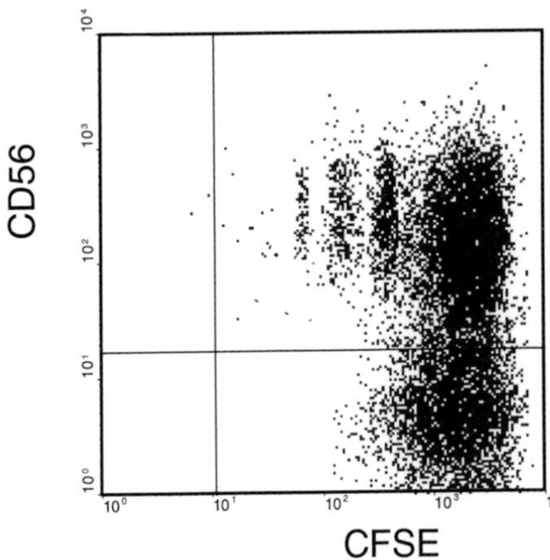

Fig. 15. Measuring proliferation of NK cells. Each cell leads to half the fluorochrome which was used to stain the initial cell sample. The plot presents the proliferative response of CFSE-labeled lymphocytes which were stimulated with IL-2. Analysis was restricted to CD3– cells by gating.

- Wash the cells (250 × *g*, 5 min).
- Adjust the cells to 3 × 10^6/ml in PBS without protein.
- Add CFSE (e.g. 10 µl) of the titrated concentration by placing the substance carefully on the tube wall above the cell suspension and vortex to achieve equal staining of all cells.
- Incubate for 10 min (optimal time for each cell type has to be determined) at 37 °C.
- Stop staining by adding 100 µl/ml FCS (or a different protein source).
- Fill the tube with PBS.
- Wash twice at 250 × *g* for 5 min with PBS.
- Suspend the cells in appropriate medium.
- Control fluorescence intensity of a cell aliquot on a cytometer.
- Stimulate cells at a concentration of 3 × 10^6/ml with IL-2 (500 U/ml) in R10 medium at 37 °C for several days.
- Harvest aliquots (100 µl/per sample) at different time points (days) into 5-ml tubes.
- Add 10 µl immunoglobulin (or human serum).
- Add differently labeled antibodies (e.g. CD3-APC and CD56-PE, FL1 channel is used by CFSE, FL3 is available for viability controls).
- Perform appropriate isotype controls in parallel.
- Incubate the samples for 20 min at 4 °C.
- Wash the cells once (4 ml, 250 ×*g*, 5 min).
- Resuspend the cells in 300 µl PBS/BSA.
- Cells are now ready for analysis.

Data Acquisition

Load instrument settings for lymphocytes and assure proper compensation. At least 10,000 cells of each sample in the chosen gate are then acquired. If the expression density of a marker is of specific interest (e.g. in order to distinguish $CD56^{dim}$ and $CD56^{bright}$ NK cells), the acquisition speed should be minimized in order to ensure optimal hydrodynamic focusing of the cells.

Viability Control

Since the cells are taken from culture, a viability control should be performed. There are several commercially available viability probes (e.g. ViaProbe or 7AAD), but PI can be used as well. Similar to trypan blue, which is commonly used in light microscopy, PI can only enter dead cells where it stains the DNA. The resulting red fluorescence signals equally strong in FL2 and FL3, thus the dots appear in a very narrow diagonal if plotted in FL2 vs. FL3 graph. This allows the exclusion of the double-positive events while still using fluorochromes which signal in either FL2 or FL3.

Working Steps

Add 20 µl PI (50 µg/ml PBS) at least 1 min before sample acquisition. No further washing is required.

Data Analysis

Set a region (R1) on lymphocytes according to the FSC vs. SSC properties of the cells. Define a second region (e.g. CD3 vs. CD56; R2) including only CD3– CD56+ cells. A third region (R3 FL2 vs. FL3) is useful for excluding all dead cells. By combining the regions to an NK cell gate (R1*R2 AND NOT R3) proliferating NK cells can be analyzed in a CFSE histogram plot (FL1 green fluorescence) where nonproliferating cells exhibit the highest fluorescence intensity since cell division leads to half the CFSE concentration.

Media and Buffers

- Wash buffer: PBS containing 0.1% BSA
- R10 medium: RPMI-1640 medium supplemented with 10% FCS, glutamine (2 mmol/l), penicillin streptomycin (100 U/ml), sodium pyruvate (1 mmol/l)
- CFSE: Stock solution 5 mmol/l (25 mg CFSE solved in 10 ml DMSO), store as aliquots at −20 °C
- Working solution: diluted stock solution in PBS (1 : 100)
- *Note:* add 10 µl to 1 ml cell suspension. Since staining efficacy is strongly cell type specific, the optimal concentration must be individually determined.
- Saponin stock solution: 100 mg/ml in distilled water.
- Working solution: mix 500 µl saponin solution with 500 µl HEPES solution (238 mg/ml in distilled water) and add up with 49 ml distilled water.
- Brefeldin A: stock solution (5 mg/ml) in DMSO, working solution dilute stock solution with PBS (1 : 50).
- Paraformaldehyde solution (1 or 4%)

- Stir 4% paraformaldehyde (weight percentage) in warm PBS (ca. 80° C, do not boil!). Store the completely dissolved solution at 4 °C
- Prepare a 1% paraformaldehyde solution by diluting the above solution 1 : 4 with PBS.

Summary

Flow cytometry is a very useful tool for analyzing NK cells. Multiparameter analyses allow the definition of NK cell subpopulations and determination of specific functional properties such as NK cell-mediated cytotoxicity, proliferative capacity, and cytokine production. Functional capacities can also be attributed to particular NK cell populations by appropriate gating. Without flow cytometry, comparable results would require complex cost- and time-consuming cell isolation and sorting steps. Thus, flow cytometry constitutes a precise, elegant, time-saving and relatively inexpensive method to analyze phenotype and function on a single (NK) cell level.

Acknowledgment

The authors would like to thank Rachel Thomas for carefully revising the manuscript.

References

1 Trinchieri G: Biology of natural killer cells. Adv Immunol 1989;47:187–376.

2 Timonen T, Ortaldo JR, Herberman RB: Characteristics of human large granular lymphocytes and relation to natural killer and K cells. Int J Cancer 1975;15: 596–605.

3 Benschop R J, Schedlowski M, Wienecke H, Jacobs R, Schmidt RE: Adrenergic control of natural killer cell circulation and adhesion. Brain Behav Immun 1997; 11: 321–332.

4 Schedlowski M, Falk A, Rohne A, Wagner TO F, Tewes U, Jacobs R, Schmidt R E: Catecholamine effects on NK cells and NK function in humans. 2nd Int Congr ISNIM Paestum 1993.

5 Schedlowski M, Jacobs R, Alker J, Prohl F, Stratmann G, Richter S, Hadicke A, Wagner TO, Schmidt RE, Tewes U: Psychophysiological neuroendocrine and cellular immune reactions under psychological stress. Neuropsychobiology 1993;28: 87–90.

6 Jetschmann, JU, Benschop RJ, Jacobs R, Kemper A, Oberbeck R, Schmidt RE, Schedlowski M: Expression and invivo modulation of alpha- and beta-adrenoceptors on human natural killer (CD16+) cells. J Neuroimmunol 1997;74:159–164.

7 Hercend T, Schmidt RE: Characteristics and uses of natural killer cells. Immunol Today 1988;9:291–293.

8 Ritz J, Schmidt RE, Michon J, Hercend T, Schlossman SF: Characterization of functional surface structures on human natural killer cells. Adv Immunol 1988;42:181–211.

9 Moretta A, Vitale M, Sivori S, Bottino C Morelli L, Augugliaro R, Barbaresi M, Pende D, Ciccone E, Lopez-Botet M: Human natural killer cell receptors for HLA-class I molecules. Evidence that the Kp43 (CD94) molecule functions as receptor for HLA-B alleles. J Exp Med 1994;180:545–555.

10 Pende D, Parolini S, Pessino A, Sivori S, Augugliaro R, Morelli L, Marcenaro E, Accame L Malaspina A, Biassoni R, et al: Identification and molecular characterization of NKp30, a novel triggering receptor involved in natural cytotoxicity mediated by human natural killer cells. J Exp Med 1999;190:1505–1516.

11 Welte S, Kuttruff S, Waldhauer I, Steinle A: Mutual activation of natural killer cells and monocytes mediated by NKp80-AICL interaction. Nat Immun 2006;7:1334–1342.

12 Mathew PA, Garni-Wagner BA, Land K, Takashima A, Stoneman E, Bennett M, Kumar V: Cloning and characterization of the 2B4 gene encoding a molecule associated with non-MHC-restricted killing mediated by activated natural killer cells and T cells. J Immunol 1993;151:5328–5337.

13 Nakajima H, Colonna M: 2B4: An NK cell activating receptor with unique specificity and signal transduction mechanism. Hum Immunol 2000;61:39–43.

14 Lanier LL, NK cell receptors. Annu Rev Immunol 1998;16:359–393.

15 Valiante NM, Uhrberg M, Shilling HG., Lienert-Weidenbach K, Arnett KL, D'Andrea A, Phillips JH, Lanier LL, Parham P: Functionally and structurally distinct NK cell receptor repertoires in the peripheral blood of two human donors. Immunity 1997;7:739–751.

16 Uhrberg M, Valiante NM, Young NT, Lanier LL, Phillips JH, Parham P: The repertoire of killer cell iG-like receptor and CD94:NKG2A receptors in T cells: Clones sharing identical alpha beta TCR rearrangement express highly diverse killer cell Ig-like receptor patterns. J Immunol 2001;166:3923–3932.

17 Bauer S, Groh V, Wu J, Steinle A, Phillips JH, Lanier LL, Spies T: Activation of NK cells and T cells by NKG2D, a receptor for stress-inducible MICA. Science 1999;285:727–729.

18 Long EO: Regulation of immune responses through inhibitory receptors. Annu Rev Immunol 1999;17:875–904.

19 Stinchcombe JC, Bossi G, Booth S, Griffiths GM: The immunological synapse of CTL contains a secretory domain and membrane bridges. Immunity 2001;15:751–761.

20 Wu J, Cherwinski H, Spies T, Phillips JH, Lanier LL: DAP10 and DAP12 form distinct, but functionally cooperative, receptor complexes in natural killer cells. J Exp Med 2000;192:1059–1068.

21 Vivier E, Morin P, O'Brien C, Druker B, Schlossman SF, Anderson P: Tyrosine phosphorylation of the Fc gamma RIII(CD16):zeta complex in human natural killer cells. Induction by antibody-dependent cytotoxicity but not by natural killing. J Immunol 1991;146:206–210.

22 Vujanovic NL, Nagashima S, Herberman RB, Whiteside TL: Nonsecretory apoptotic killing by human NK cells. J Immunol 1996;157:1117–1126.

23 Cretney E Takeda K, Yagita H, Glaccum M, Peschon JJ, Smyth MJ: Increased susceptibility to tumor initiation and metastasis in TNF-related apoptosis-inducing ligand-deficient mice. J Immunol 2002;168:1356–1361.

24 Takeda K, Smyth MJ, Cretney E, Hayakawa Y, Kayagaki N, Yagita H, Okumura K: Critical role for tumor necrosis factor-related apoptosis-inducing ligand in immune surveillance against tumor development. J Exp Med 2002; 195:161–169.

25 Levitt LJ, Nagler A, Lee F, Abrams J, Shatsky M, Thompson D: Production of granulocyte/macrophage-colony-stimulating factor by human natural killer cells. Modulation by the p75 subunit of the interleukin 2 receptor and by the CD2 receptor. J Clin Invest 1991;88:67–75.

26 Basu S, Sodhi A: Increased release of interleukin-1 and tumour necrosis factor by interleukin-2-induced lymphokine-activated killer cells in the presence of cisplatin and FK-565. Immunol Cell Biol 1992; 70(Pt 1):15–24.

27 Fehniger TA, Shah MH, Turner MJ, VanDeusen JB, Whitman SP, Cooper MA, Suzuki K, Wechser M, Goodsaid F, Caligiuri MA: Differential cytokine and chemokine gene expression by human NK cells following activation with IL-18 or IL-15 in combination with IL-12: implications for the innate immune response. J Immunol 1999;162:4511–4520.

28 Warren HS, Kinnear BF, Phillips JH, Lanier LL: Production of IL-5 by human NK cells and regulation of IL-5 secretion by IL-4, IL-10, and IL-12. J Immunol 1995;154:5144–5152.

29 Saito S, Kasahara T, Sakakura S, Enomoto M, Umekage H, Harada N, Morii T, Nishikawa K, Narita N, Ichijo M: Interleukin-8 production by CD16-CD56bright natural killer cells in the human early pregnancy decidua. Biochem Biophys Res. Commun 1994;200:378–383.

30 Chiorean EG, Miller JS: The biology of natural killer cells and implications for therapy of human disease. J Hematother Stem Cell Res 2001;10:451–463.

31 Matera L, Contarini M, Bellone G, Forno B, Biglino A: Up-modulation of interferon-gamma mediates the enhancement of spontanous cytotoxicity in prolactin-activated natural killer cells. Immunology 1999;98:386–392.

32 Wendt K, Wilk E, Buyny S, Schmidt RE, Jacobs R: Interleukin-21 differentially affects human natural killer cell subsets. Immunology 2007;122:486–495.

33 Jacobs R, Stoll M, Stratmann G, Leo R, Link H, Schmidt RE: CD16− CD56+ natural killer cells after bone marrow transplantation. Blood 1992;79:3239–3244.

34 Caligiuri MA, Zmuidzinas A, Manley TJ, Levine H, Smith KA, Ritz J: Functional consequences of interleukin 2 receptor expression on resting human lymphocytes. Identification of a novel natural killer cell subset with high affinity receptors. J Exp Med 1990;171:1509–1526.

35 Carson WE, Fehniger TA, Caligiuri MA: CD56bright natural killer cell subsets: characterization of distinct functional responses to interleukin-2 and the c-kit ligand. Eur J Immunol 1997;27:354–360.

36 Cooper MA, FehnigerTA, Caligiuri MA: The biology of human natural killer-cell subsets. Trends Immunol 2001;22:633–640.

37 Jacobs R, Hintzen G, Kemper A, Beul K, Behrens G, Kempf S, Sykora KW, Schmidt RE: $CD56^{bright}$ cells differ in their KIR repertoire and function from $CD56^{dim}$ NK cells. Eur J Immunol 2001;31:3121–3126.

38 Wendt K, Wilk E, Buyny S, Buer J, Schmidt RE, Jacobs R: Gene and protein characteristics reflect functional diversity of $CD56^{dim}$ and $CD56^{bright}$ NK cells. J Leukoc Biol 2006;80:1529–1541.

Sack U, Tárnok A, Rothe G (eds): Cellular Diagnostics. Basics, Methods and Clinical Applications of Flow Cytometry. Basel, Karger, 2009, pp 261–271

Lymphocyte Subsets in the Peripheral Blood of Healthy Children

Ulrich Sack[a] · Fee Gerling[a] · Attila Tárnok[b]

[a] Institute of Clinical Immunology and Transfusion Medicine, Medical Faculty,
[b] Department of Pediatric Cardiology, Cardiac Center Leipzig GmbH, University of Leipzig, Leipzig, Germany

Introduction

Developmental changes from neonatal to adult age massively influence the composition of peripheral blood leukocytes and lymphocytes. Immunophenotyping is an important diagnostic tool for decision-making in various diseases in children. Typical indications are:

- diagnosis of immunodeficiencies [1–3],
- monitoring of systemic and chronic diseases [4, 5],
- cellular diagnostics of allergies [6, 7],
- functional characterization of cells [8, 9],
- genetic investigation [10, 11],
- differential diagnosis of lymphocytosis [12],
- diagnosis of leukemia [13–18],
- investigation of T-cell receptor expression in Kawasaki syndrome [19, 20].

Furthermore, there are several indications that are not yet so important for routine diagnostics [21].

Clinical decision-making relies on measurable aberrations of a patient's values from 'normal.' In children correct decision-making heavily depends on the availability of such normal ranges for a given age group. It is of great concern that no reliable data exist to date that take into account age-dependent changes. The normal values used for diagnostics in children are often based on adult values. Only a handful of published studies reports on the composition of the lymphocyte subset in different age groups of children. As outlined below, these studies have drawbacks, making their application difficult in everyday clinical diagnostic settings:

- limited number of children enrolled,
- different age groups or a limited age range investigated,

- technical equipment, namely flow cytometers and software,
- reagents including lysis and antibodies,
- different parameters analyzed,
- one- or two-platform technologies.

Aim

The present study was aimed at combining existing studies (as far as they were combinable) in a meta-analysis describing age-dependent changes from neonatal to adult age. These values might be more useful than those that are available now. Furthermore, this investigation also aimed at defining valid data even for short periods in children's lives and studying the influence of technical approaches, sample preparation, antibody selection, instrumentation, and data analysis.

Material and Methods

The present work is a meta-analysis compiling earlier data published by others. For this study, we screened databases for all publications reporting normal values in children. As the primary raw data were mostly not at hand, all available data (individual and calculated values) were included. We tried to contact all authors of these publications in order to ask them for the primary data. Unfortunately, this was successful only in few cases. We exclusively included studies based on flow cytometry and staining with monoclonal antibodies. Technical options such as isolation of lymphocytes prior to staining, simultaneous staining with multiple fluorescent dyes or data analysis options were investigated for their influence on data, but data from these studies were not excluded.

Identification of References and Raw Data

We focused on the following lymphocyte populations:
- T-cells (CD3+),
- T-helper cells (CD3+ CD4+ CD8−),
- cytotoxic T-cells (CD3+ CD4− CD8+),
- B-cells (CD19+),
- NK cells (CD16+ CD56+ CD3−).

We could identify 17 original publications investigating lymphocyte subpopulations in children by flow cytometry. Most of these papers were found in reference databases; additional ones were identified in literature searches and textbooks. We added data from our own laboratory to improve the database by further normal values.

Although a number of other subpopulations were described in several of these publications, we decided not to consider rarely reported findings.

In 8 of the publications the data were reported as means and standard deviations. Such data were entered into the calculation of age-dependent normal values. This was not feasible for the other 9 papers a they published data as median and percentiles. Raw data could not be obtained for most studies.

Data Collection

Data were collected in tables and investigated by data analysis and comparison methods. In particular the following data were taken into consideration:

- age, gender,
- sample collection, anticoagulation, preanalytical procedures, time to workbench,
- definition of normal group, ethnicity,
- applied immunophenotyping method, direct or indirect staining, monoclonal antibodies used,
- cell separation or whole-blood protocols,
- instrumentation and software,
- definition of cell populations, data analysis.

Statistical Analysis and Visual Presentation

Data analysis, statistical examination and definition of age-related values were based on the calculation of patient groups across relevant age ranges. As a consequence of this approach the patient counts were initially small. However, this method opened the opportunity to merge the values from other studies that did not work with equal time spans. Thereby, mean values and standard deviations could be recalculated for these combined data. Nonparametric data were analyzed for fundamentally fitting values. It is noteworthy that there were no two studies that grouped the children into identical age groups.

Data handling, statistical analysis and graphic presentation were performed using SPSS V12 (SPSS Science, Erkrath, Germany) and SigmaPlot 10 (Systat, Erkrath, Germany).

Results

Based on the identified data we generated a data set describing normal values in short time periods during childhood. Nearly all available publications could be included, except selected ones. Excluded were data generated using

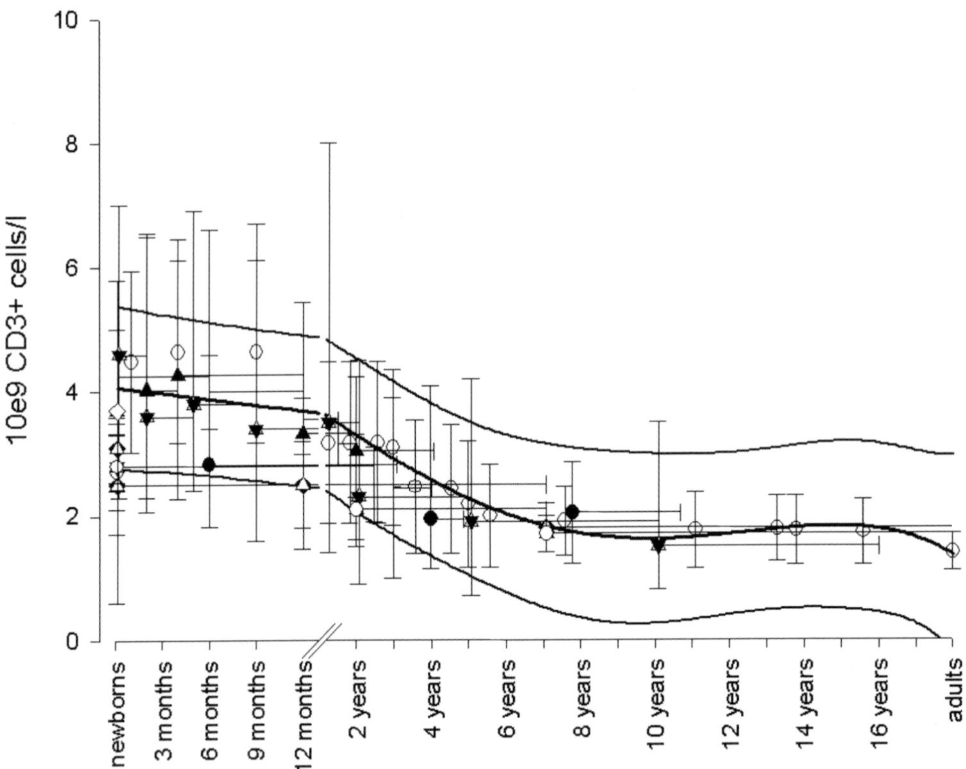

Fig. 1. T-cell count (CD3+) in peripheral blood from children from birth to adolescence. Lines represent mean ± 1 SD. Calculation was based on 6 publications [22–27] and is in accordance with 7 further available papers, added with median, percentiles, and time span as published [28–34].

- analysis by cell sorters: such data did not fit with other publications, done using flow cytometer analyzers,
- analysis by microscopic counting, and
- rosetting techniques instead of monoclonal antibodies.

All other data could be included into our study. We found no influence of the following variables on the results:

- gender,
- ethnicity,
- anticoagulant used,
- preparation of sample (Ficoll vs. whole blood technology),
- monoclonal antibodies used,
- selected staining protocols,
- technical platform.

The results of the 16 included papers were statistically homogeneous. Cell counts are presented in the following figures.

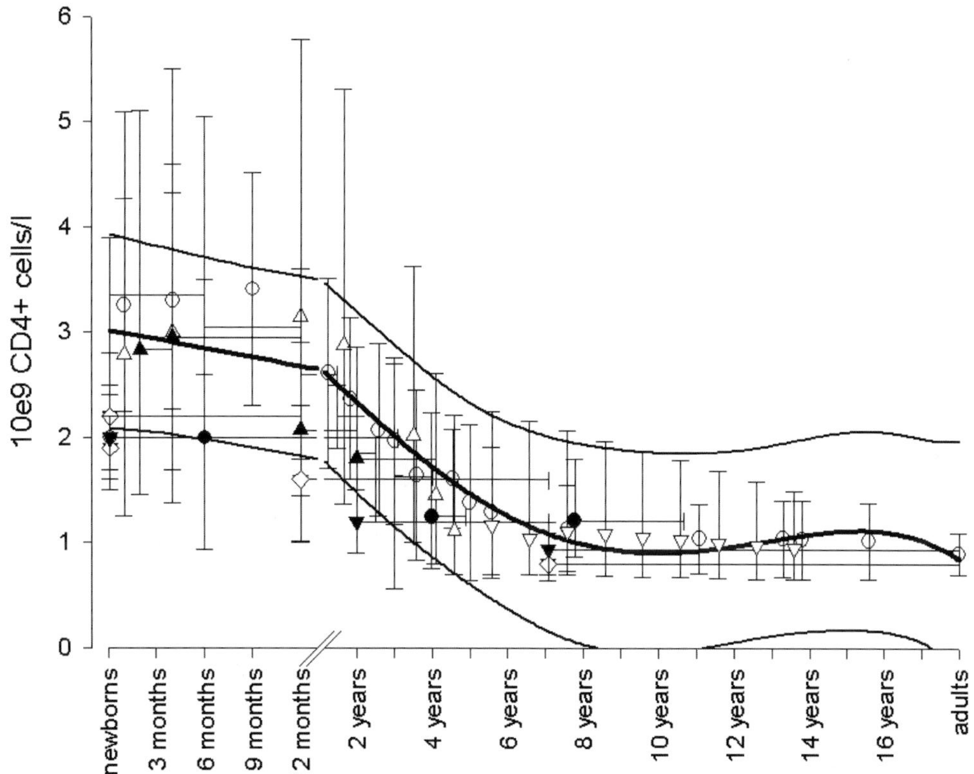

Fig. 2. Number of T-helper cells (CD3+ CD4+ CD8−) in peripheral blood from children from birth to adolescence. Data represent mean ± 1 SD. Calculation was based on 6 publications [23–27, 35, 36] and is in accordance with 7 further available papers, added with median, percentiles, and time span as published [28–32, 37, 38].

All lymphocyte subsets exhibited a common feature i.e. a substantial dynamic change during the first year of life. This finding is missing from previous publications because they were based on too long age ranges. Especially around the fourth to sixth month of life, artificial T lymphocytosis could cause misinterpretation.

During the first year of life T-cell count is clearly higher than at older ages. The number of T-cells was higher than expected especially during the second half of the first year of life (fig. 1). Based on most accessible papers, there is general consensus on this intermediate peak. All published data fitted well with our own findings, but the common practice of classifying children into one 'first-year group' sometimes obscured this fact and thereby caused underestimation of expected values.

T-helper cell counts exhibited an intermediate peak around the 6th month of life, similarly to T-cells counts (fig. 2). All publications report homogeneous findings, with a tendency to find too low T-helper cell counts around the 6th month due to the fact that the time periods selected were too long.

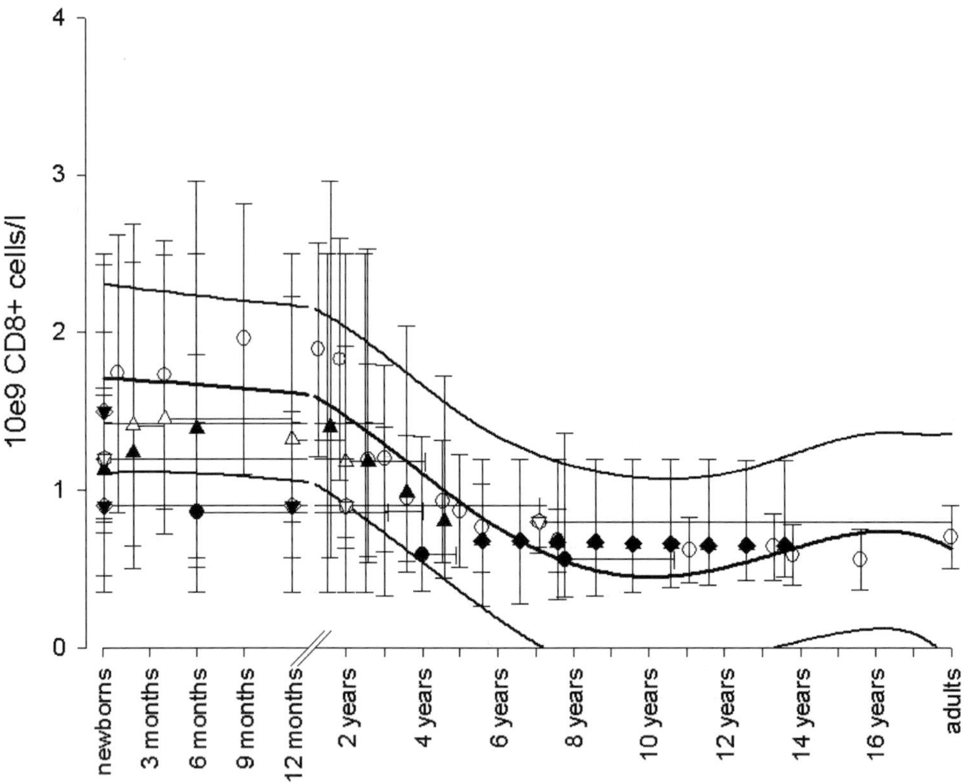

Fig. 3. Number of cytotoxic T-cells (CD3+ CD4− CD8+) in peripheral blood from children from birth to adolescence. Data represent mean ± 1 SD. Calculation was based on 7 publications [23–27, 35, 36] and is in accordance with 7 further available papers, added with median, percentiles, and time span as published [28–32, 37, 38].

Similarly to T-helper cell counts, cytotoxic T-cell counts showed a slightly delayed intermediate peak in the first year of life (fig. 3). This finding is also consistent with all included publications.

B-cell counts were lower than T-cell counts, but the kinetics was similar (fig. 4). This is consistent with nearly all included publications.

The time-course of NK cell counts obviously differs from that of other lymphocytes. From birth on, there were a decreasing number of NK cells in the peripheral blood (fig. 5). Not all publications recognised this tendency. Frequently, NK cells were reported with smaller counts.

Discussion

For a multitude of reasons, cytometric analysis is an ideal tool for immunophenotyping, especially in children. It is a high-throughput technology that allows rapid and accurate quantification of even minute cell subsets such as stem or progenitor

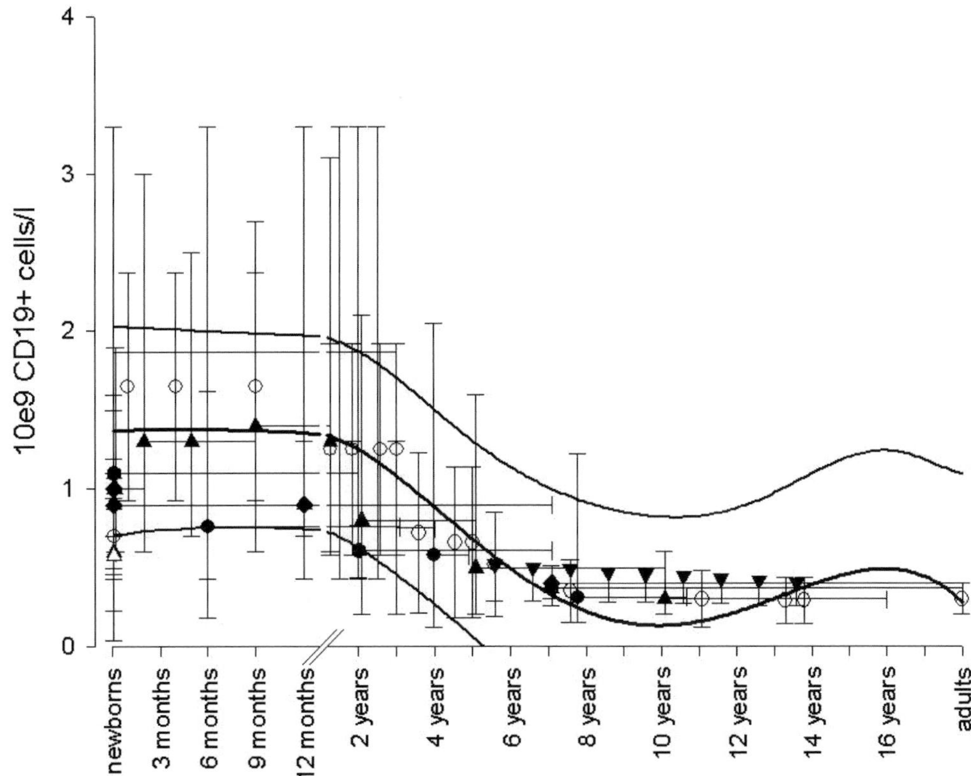

Fig. 4. Number of B-cells (CD19+) in peripheral blood from children from birth to adolescence. Data represent mean ± 1 SD. Calculation was based on 6 publications [22, 24–27] and is in accordance with 7 further available papers, added with median, percentiles, and time span as published [28, 29, 31–34, 37].

cells [39] in a few seconds to minutes by measuring tens of thousands to millions of cells. By measuring multiple colors different epitopes may easily be detected on the same cell, thus enabling precise phenotyping. Present technologies already allow the measurement of 6 colors for routine immunophenotyping and future developments may push this further to 17 or more colors [40, 41] by applying sophisticated technology and novel fluorescent colors derived from nanotechnology [4, 42]. There are two additional effects of multiplexed cell analysis:

i) The required blood volume for diagnosis is drastically reduced. This is important particularly in critically ill neonates. Future progress in instrument and computer development [43] will enable cytometric analysis on slides with even smaller samples in specially designed microscopes [44, 45], whichmay be substantially cheaper than the instruments we are using now [46].
ii) The complex data pattern that results from multiplexed measurements will allow computer-aided systemic analysis [47], which may lead to improved risk assessment and hopefully to an individualized medicine [48].

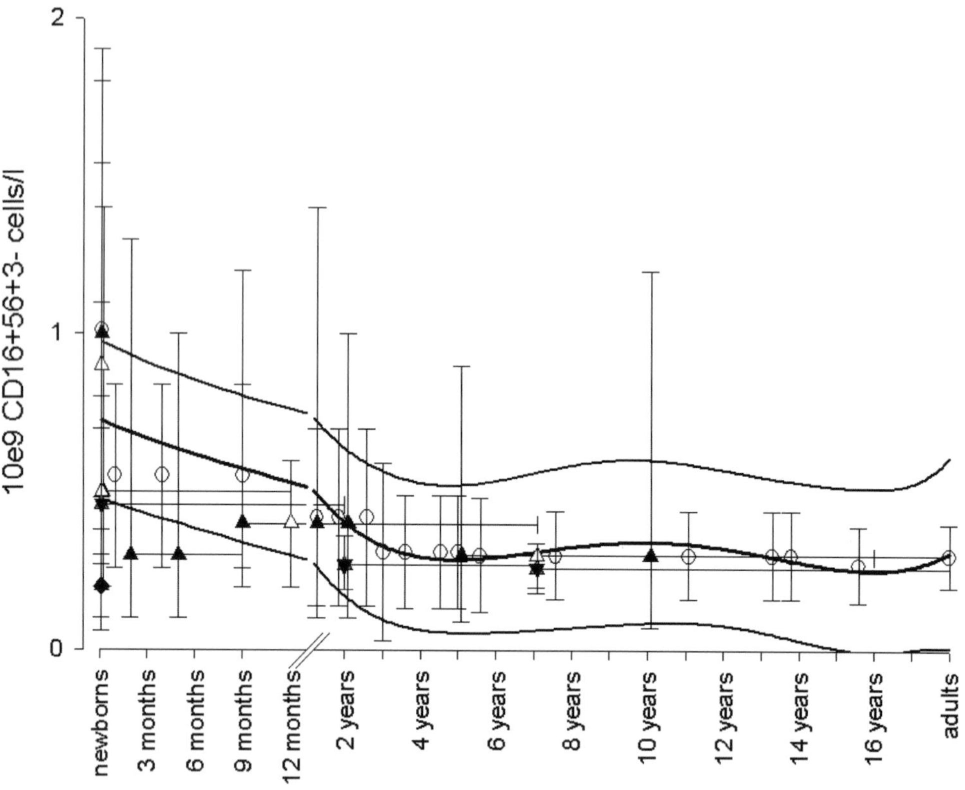

Fig. 5. NK cell count (CD3− CD16+ CD56+) in peripheral blood from children from birth to adolescence. Data represent mean ± 1 SD. Calculation was based on 2 publications [26, 27] and is in accordance with further available papers, added with median, percentiles, and time span as published [28, 32–34].

Nevertheless, all these multiparameteric investigations require reliable normal values obtained in healthy children. Our results show that when strictly adhering to validated protocols for clinical laboratory diagnostics, flow-cytometric data are largely independent of technological platforms, selected antibodies, or laboratory protocols. Furthermore this allows the merging of data and findings from several studies into one data set as a base for decision-making. High stability need not be true under all conditions. In fact, in immunodeficiencies most detection systems (labeling, instrument, and analysis) must be validated under conditions representing realistic symptoms of such diseases. In healthy persons this aspect is less critical. Time courses of investigated cell populations underline the necessity to compare flow-cytometric diagnostic findings with strictly age-matched normal values. Failure to take into account changes in early childhood may lead to misinterpretation and misleading diagnostic reports.

Conclusions

Development-related alterations of lymphocyte subset counts in children could be gathered from literature data for diagnostic use. The results were mostly independent of gender, ethnic factors, and sampling procedure, anticoagulation, preanalytical procedures, time to workbench, immunophenotyping method, staining procedure, selected monoclonal antibodies, technical devices and software.

Our data indicate that previous normal values are not sufficiently precise for the interpretation of lymphocyte subsets in children [49]. Mainly during the first year of life, count and subset distribution of lymphocytes are different from those of adults. Therefore, a close-meshed data set for normal values is required to guarantee adequate diagnostic interpretation.

References

1 Noroski LM, Shearer WT: Screening for primary immunodeficiencies in the clinical immunology laboratory. Clin Immunol Immunopathol 1998;86:237–245.

2 de Vries E, Noordzij JG, Kuijpers TW, Van Dongen JJ: Flow cytometric immunophenotyping in the diagnosis and follow-up of immunodeficient children. Eur J Pediatr 2001;160:583–591.

3 Siegrist CA: The child with recurrent infections: which screening for immunodeficiency?. Arch Pediatr 2001;8:205–210.

4 Bocsi J, Mittag A, Sack U, Gerstner AO, Barten MJ, Tarnok A: Novel aspects of systems biology and clinical cytomics. Cytometry A 2006;69:105–108.

5 Seipp MT, Erali M, Wies RL, Wittwer C: HLA-B27 typing: evaluation of an allele-specific PCR melting assay and two flow cytometric antigen assays. Cytometry B Clin Cytom 2005;63B: 10–15.

6 Ebo DG, Hagendorens MM, Bridts CH, Schuerweg AJ, De Clerck LS, Stevens WJ: Flow cytometric analysis of in vitro activated basophils, specific IgE and skin tests in the diagnosis of pollen-associated food allergy. Cytometry B Clin Cytom 2005;64:28–33.

7 Milovanova TN: Comparative analysis between CFSE flow cytometric and tritiated thymidine incorporation tests for beryllium sensitivity. Cytometry B Clin Cytom 2007;72:265–275.

8 Liu J, Roederer M: Differential susceptibility of leukocyte subsets to cytotoxic T cell killing: Implications for HIV immunopathogenesis. Cytometry A 2007;71:94–104.

9 Gille C, Orlikowsky TW: Flow cytometric methods in the detection of neonatal infection. Transf Med Hemother 2007;34:157–163.

10 Hamurcu Z, Demirtas H, Kumandas S: Flow cytometric comparison of RNA content in peripheral blood mononuclear cells of Down syndrome patients and control individuals. Cytometry B Clin Cytom 2006;70:24–28.

11 Rockenbauer E, Petersen K, Vogel U, Bolund L, Kolvraa S, Nielsen KV, Nexo BA: SNP genotyping using microsphere-linked PNA and flow cytometric detection. Cytometry A 2005;64:80–86.

12 Shim YK, Vogt RF, Middleton D, Abbasi F, Slade B, Lee KY, Marti GE: Prevalence and natural history of monoclonal and polyclonal B-cell lymphocytosis in a residential adult population. Cytometry B Clin Cytom 2007;72:344–353.

13 Qadir M, Barcos M, Stewart CC, Sait SN, Ford LA, Baer MR: Routine immunophenotyping in acute leukemia: role in lineage assignment and reassignment. Cytometry B Clin Cytom 2006;70:329–334.

14 Habib LK, Finn WG: Unsupervised immunophenotypic profiling of chronic lymphocytic leukemia. Cytometry B Clin Cytom 2006;70:124–135.

15 Lamb LS Jr, Neuberg R, Welsh J, Best R, Stetler-Stevenson M, Sorrell A: T-cell lymphoblastic leukemia/lymphoma syndrome with eosinophilia and acute myeloid leukemia. Cytometry B Clin Cytom 2005;65:37–41.

16 Langebrake C, Brinkmann I, Teigler-Schlegel A, Creutzig U, Griesinger F, Puhlmann U, Reinhardt D: Immunophenotypic differences between diagnosis and relapse in childhood AML: implications for MRD monitoring. Cytometry B Clin Cytom 2005;63:1–9.

17 Shankey TV, Forman M, Scibelli P, Cobb J, Smith CM, Mills R, Holdaway K, Bernal-Hoyos E, Van Der HM, Popma J, Keeney M: An optimized whole blood method for flow cytometric measurement of ZAP-70 protein expression in chronic lymphocytic leukemia. Cytometry B Clin Cytom 2006;70:259–269.

18 Sutherland DR, Kuek N, Davidson J, Barth D, Chang H, Yeo E, Bamford S, Chin-Yee I, Keeney M: Diagnosing PNH with FLAER and multiparameter flow cytometry. Cytometry B Clin Cytom 2007;72:167–177.

19 Lina G, Cozon G, Ferrandiz J, Greenland T, Vandenesch F, Etienne J: Detection of staphylococcal superantigenic toxins by a CD69-specific cytofluorimetric assay measuring T-cell activation. J Clin Microbiol 1998;36:1042–1045.

20 Reichardt P, Lehmann I, Sierig G, Borte M: Analysis of T-cell receptor V-beta 2 in peripheral blood lymphocytes as a diagnostic marker for Kawasaki disease. Infection 2002;30:360–364.

21 Sack U, Rothe G, Barlage S, Gruber R, Kabelitz D, Kleine TO, Lun A, Renz H, Ruf A, Schmitz G: Flow cytometry in clinical diagnostics. J Lab Med 2000;24:277–297.

22 Hicks MJ, Jones JF, Minnich LL, Weigle KA, Thies AC, Layton JM: Age-related changes in T- and B-lymphocyte subpopulations in the peripheral blood. Arch Pathol Lab Med 1983;107:518–523.

23 Yanase Y, Tango T, Okumura K, Tada T, Kawasaki T: Lymphocyte subsets identified by monoclonal antibodies in healthy children. Pediatr Res 1986;20:1147–1151.

24 Babcock GF, Taylor AF, Hynd BA, Sramkoski RM, Alexander JW: Flow cytometric analysis of lymphocyte subset phenotypes comparing normal children and adults. Diagn Clin Immunol 1987;5:175–179.

25 Wiener D, Shah S, Malone J, Lowell N, Lowitt S, Rowlands DT Jr: Multiparametric analysis of peripheral blood in the normal pediatric population by flow cytometry. J Clin Lab Anal 1990;4:175–179.

26 Beck R, Lam-Po-Tang PR: Comparison of cord blood and adult blood lymphocyte normal ranges: a possible explanation for decreased severity of graft versus host disease after cord blood transplantation. Immunol Cell Biol 1994;72:440–444.

27 Shahabuddin S, Al Ayed I, Gad El-Rab MO, Qureshi MI: Age-related changes in blood lymphocyte subsets of Saudi Arabian healthy children. Clin Diagn Lab Immunol 1998;5:632–635.

28 Erkeller-Yuksel FM, Deneys V, Yuksel B, Hannet I, Hulstaert F, Hamilton C, Mackinnon H, Stokes LT, Munhyeshuli V, Vanlangendonck F: Age-related changes in human blood lymphocyte subpopulations. J Pediatr 1992;120:216–222.

29 Heldrup J, Kalm O, Prellner K: Blood T and B lymphocyte subpopulations in healthy infants and children Acta Paediatr 1992;81:125–132.

30 Denny T, Yogev R, Gelman R, Skuza C, Oleske J, Chadwick E, Cheng SC, Connor E: Lymphocyte subsets in healthy children during the first 5 years of life. JAMA 1992;267:1484–1488.

31 Kotylo PK, Fineberg NS, Freeman KS, Redmond NL, Charland C: Reference ranges for lymphocyte subsets in pediatric patients. Am J Clin Pathol 1993;100:111–115.

32 Hulstaert F, Hannet I, Deneys V, Munhyeshuli V, Reichert T, De Bruyere M, Strauss K: Age-related changes in human blood lymphocyte subpopulations. II. Varying kinetics of percentage and absolute count measurements. Clin Immunol Immunopathol 1994;70:152–158.

33 Comans-Bitter WM, de Groot R, van den Beemd R, Neijens HJ, Hop WC, Groeneveld K, Hooijkaas H, Van Dongen JJ: Immunophenotyping of blood lymphocytes in childhood Reference values for lymphocyte subpopulations. J Pediatr 1997;130:388–393.

34 O'Gorman MR, Millard DD, Lowder JN, Yogev R: Lymphocyte subpopulations in healthy 1–3-day-old infants. Cytometry 1998;34:235–241.

35 Likanonsakul S, Wasi C, Thepthai C, Sutthent R, Louisirirotchanakul S, Chearskul S, Vanprapa N, Lebnark T: The reference range of CD4+ and CD8+ T-lymphocytes in healthy non-infected infants born to HIV-1 seropositive mothers: a preliminary study at Siriraj Hospital. Southeast Asian J Trop Med Public Health 1998;29:453–463.

36 Lisse IM, Aaby P, Whittle H, Jensen H, Engelmann M, Christensen LB: T-lymphocyte subsets in West African children: impact of age, sex, and season. J Pediatr 1997;130:77–85.

37 Robinson M, O'Donohoe J, Dadian G, Wankowicz A, Barltrop D, Hobbs JR: An analysis of the normal ranges of lymphocyte subpopulations in children aged 5–13 years. Eur J Pediatr 1996;155:535–539.

38 The European Collaborative Study: Age-related standards for T lymphocyte subsets based on uninfected children born to human immunodeficiency virus 1-infected women. Pediatr Infect Dis J 1992;11: 1018–1026.

39 Kamprad M, Kindler S, Schuetze N, Emmrich F: Flow cytometric immunophenotyping of umbilical cord and peripheral blood haematopoietic progenitor cells by different CD34 epitopes, CD133, P-glycoprotein expression and rhodamine-123 efflux. Transfus Med Hemother 2007;34:195–203.

40 Perfetto SP, Chattopadhyay PK, Roederer M: Seventeen-colour flow cytometry: unravelling the immune system. Nat Rev Immunol 2004;4: 648–655.

41 Mittag A, Lenz D, Gerstner AO, Tarnok A: Hyperchromatic cytometry principles for cytomics using slide based cytometry. Cytometry A 2006;69: 691–703.

42 Chattopadhyay PK, Price DA, Harper TF, Betts MR, Yu J, Gostick E, Perfetto SP, Goepfert P, Koup RA, De Rosa SC, Bruchez MP, Roederer M: Quantum dot semiconductor nanocrystals for immunophenotyping by polychromatic flow cytometry. Nat Med 2006;12:972–977.

43 Parks DR, Roederer M, Moore WA: A new 'Logicle' display method avoids deceptive effects of logarithmic scaling for low signals and compensated data. Cytometry A 2006;69:541–551.

44 Tarnok A: Slide-based cytometry for cytomics – a minireview. Cytometry A 2006;69:555–562.

45 Laffers W, Schlenkhoff C, Pieper K, Mittag A, Tárnok A, Gerstner AOH: Concepts for absolute immunophenosubtyping by slide based cytometry. Transfus Med Hemother 2007;34:188–194.

46 Shapiro HM, Perlmutter NG: Personal cytometers: slow flow or no flow? Cytometry A 2006;69:620–630.

47 Tarnok A, Bocsi J, Brockhoff G: Cytomics – importance of multimodal analysis of cell function and proliferation in oncology. Cell Prolif 2006;39:495–505.

48 Valet G: Human cytome project, cytomics, and systems biology: the incentive for new horizons in cytometry. Cytometry A 2005;64:1–2.

49 Sack U, Gerling F, Tarnok A: Age-related lymphocyte subset changes in the peripheral blood of healthy children – a meta-study. Transfus Med Hemother 2007;34:176–181.

Sack U, Tarnok A, Rothe G (eds): Cellular Diagnostics. Basics, Methods and Clinical Applications of Flow Cytometry. Basel, Karger, 2009, pp 272–285

Quantitative Assessment of Plasmacytoid and Myeloid Dendritic Cells

Rüdiger V. Sorg

Institute for Transplantation Diagnostics and Cell Therapeutics, Heinrich Heine University Hospital Düsseldorf, Germany

Introduction

Dendritic cells (DCs) are leukocytes which, as highly specialized antigen-presenting cells, play a pivotal role in the initiation and polarization of T-cell responses, thereby controlling most aspects of cellular and humoral adaptive immunity [1, 2]. As immature cells, they reside in most nonlymphoid tissues and express chemokine receptors like CCR6 (CD196) which guide them to sites of infection or inflammation where the corresponding chemokine, CCL20, is produced [3]. Antigen uptake activity of immature DCs using receptors such as the C-type lectins CD205 (DEC205) or CD209 (DC SIGN) is well developed [4]. This antigen sampling, however, only results in effective antigen presentation and induction of T-cell immunity when the immature DCs are triggered at the same time by danger signals like the presence of pathogens, signs of inflammation or tissue damage [5]. Danger-triggered DCs start to mature. They lose antigen uptake activity. CCR6 is downregulated and CCR7 (CD197) is upregulated, which guides them to the draining lymph nodes where the respective chemokine ligands, CCL19 and CCL21, are produced [3]. Furthermore, the maturation process results in induction of molecules relevant to T-cell stimulation, including HLA class II molecules, adhesion molecules such as CD54 (ICAM-1), costimulatory molecules such as CD40, CD80 (B7.1) and CD86 (B7.2) as well as additional accessory molecules such as interleukin-12 (IL-12) or CD252 (OX40L) [1, 6]. Mature DCs are also discriminated from immature DCs by expression of CD83 [7].

In the lymph node, mature DCs present peptides generated from the antigenic material taken up in the periphery on HLA class I and class II molecules in an immunostimulatory context of accessory molecules, allowing for specific recognition of those peptides by antigen-specific cytotoxic (CTL) and helper T-cells (Th), respectively. The accessory molecules on DCs mediate adhesive and costimulatory interactions which are essential for activation of the peptide-specific T-cells. Additional membrane-bound or secreted accessory molecules such as IL-12 or CD252 constitute polarizing signals which are not essential for activation per se, but determine the qualitative type of T-cell response (e.g. Th1 vs. Th2), thereby indirectly controlling the development of cellular or humoral immunological effector mechanisms [8, 9]. The polarizing signals are critical for the generation of an immune response appropriate for a given pathogen. Thus, DCs not only carry the antigen to the lymph nodes but also information on the pathogen itself and on the microenvironment it was derived from. Several pattern recognition factors which recognize conserved pathogen structures such as the toll-like receptors (TLRs) contribute to this function [4]. Furthermore, DCs are capable of reacting to a multitude of microenvironmental signals, including the presence of various proinflammatory cytokines [10]. The integration of all of these signals at the time of antigen uptake decides whether a DC will migrate and mature, thus, engage in T-cell activation, and also via a resulting distinct repertoire of polarizing signals, which type of response will develop (e.g. induction of Th1 response by IL-12-secreting DCs).

In the absence of danger signals or in the presence of anti-inflammatory signals like IL-10, immature DCs fail to mature and migration to lymph nodes is not efficient. T-cells recognizing their antigenic peptide on such an immature DC or on a so-called regulatory DC, generated in the presence of IL-10, either become apoptotic or differentiate to antigen-specific, immunosuppressive regulatory T-cells [11, 12]. Thus, DCs contribute to the maintenance of peripheral tolerance. Moreover, since they are also involved in negative selection of thymocytes during T-cell development [13], DCs appear to be an integral part of the mechanisms preventing autoimmunity.

Distinct subtypes of DC develop from lymphoid and myeloid cell lineages [14]. Myeloid DCs (MDCs) are closely related to monocytes. Indeed, monocytes can be regarded as pro-DCs because in the presence of granulocyte-macrophage-colony-stimulating factor (GM-CSF), IL-4 and CD154 (CD40L) they become mature, Th1 responses inducing DCs, which are also referred to as DC1 [15]. In peripheral blood (PB), HLA-DR+ and CD11c+ immature MDCs have been identified, which express the myeloid antigens CD13 and CD33, but no antigens typical for other cell lineages (lineage markers, lin) [16–20]. This population has also been described as ILT1+/ILT3+ (CD85h+/CD85k+) [21] or CD68^{dim} [22] subtype of DC and resembles DCs in the germinal centers of lymph nodes [23]. Following stimulation with tumor necrosis factor-α (TNF-α), they mature to Th1 responses inducing DCs [24], whereas remaining immature they stimulate the development of regulatory T-cells [25, 26].

At least one additional subtype has been identified in PB as lin−, HLA-DR+, CD11c− and CD123+ DCs [16–20, 27]. These cells lack myeloid markers such as CD13 but express preTα and show additional features typical for the lymphoid lineage, including partial recombination of the loci for the immunoglobulin heavy chain. Therefore, this type of DC is thought to be of lymphoid origin [28, 29]. They have also been described as an ILT1−/ILT3+ (CD85h−/CD85k+) [21], CD68^{bright}

[22] or BDCA-2+ (CD303+)/BDCA-4+ (CD304+) [30] subset. They resemble so-called plasmacytoid cells (pDCs – plasmacytoid DCs) in the T-cell zones of secondary lymphoid tissues [21, 31] as well as the major population of DCs in the thymus [32, 33]. This subtype of DC also represents the natural-interferon-producing cells (NIPCs) in PB which, following viral infection, produce large amounts of interferon-α and -β [21, 34], an activity which has meanwhile also been described for MDCs [35]. In contrast to MDCs, pDCs appear not to respond to inflammatory signals with migration via the afferent lymphatic vessels, but to enter lymph nodes in a CD62L-dependent process via the endothelia [21, 27]. In the presence of IL-3 and CD154, they mature to DCs which induce Th2 responses and are therefore referred to as DC2 [15, 24]. Similar to MDCs, pDCs have also been reported to induce regulatory T-cells depending on their state of maturation [36, 37].

The initial concept of myeloid DCs (or DC1) which induce Th1 responses and lymphoid plasmacytoid DCs (or DC2) which induce Th2 responses had to be modified. MDCs as well as pDCs can stimulate Th1, Th2 and regulatory T-cell responses, depending on the maturation stimuli [8, 9]. The two DC subtypes, however, are receptive to distinct activation stimuli and since their expression profile of pattern recognition receptors is different [38], they may be specialized for distinct pathogens. To complicate things even more, additional DC subsets appear to exist [20, 30, 39].

Although the precise function of MDC and pDC subsets is not well understood, there is evidence that frequencies, numbers and the proportional contribution of the two subsets to the DC compartment (pDC:MDC ratio) in PB may vary. Immediately after surgery, the total DC count in the blood increases [40]. Patients suffering from chronic graft-versus-host disease (GvHD) show an increased frequency of pDCs [41]. There appears to be an increased risk of relapse together with a reduction in chronic GvHD following bone marrow transplantation, depending on the amount of pDC in the graft [42]. There is a shift in the pDC:MDC ratio during pregnancy [43] and an age-dependent decline in pDC [44, 45]. Both DC subsets are reduced in patients suffering from liver disease [46], rheumatoid arthritis [47] and during active tuberculous infection [48], whereas in psoriatic arthritis [47], in asthma [49] as well as in prostate cancer [50] pDCs are reduced. Moreover, in HIV-infected patients, pDC counts may have predictive value [51]. Thus, it appears to be expedient to determine blood DC counts and the contribution of DC subsets to the DC compartment; a protocol is introduced describing two distinct methods each for identification and quantification of MDCs and pDCs in PB.

Protocol

Principle of Detection

Currently, no marker is available which by itself is sufficiently specific to identify MDC or pDC subsets. CD1a, CD83 or CMRF-44, all of which are used to identify

DCs, have limitations. Despite being detectable on myeloid DCs after in vitro generation (DCs differentiated from CD34+ stem cells or CD14+ monocytes), CD1a is expressed in vivo only on a specific DC subpopulation in the epidermis, the Langerhans cells. CD83 and CMRF-44 expression in vivo depends on the state of maturation; only mature DCs are positive for these markers. Moreover, their expression is not restricted to DCs. Only BDCA-2 (CD303) has been shown to allow specific identification and enumeration of pDCs in PB [30], but it is not clear yet whether expression of this marker alone is sufficient to identify pDCs in other tissue specimens as well. Therefore, combinations of monoclonal antibodies (mAbs), all of which are not specific by themselves, are used in a three-color protocol to detect MDCs as well as pDCs. DC subset specificity results from the simultaneous fulfillment of several criteria only. Furthermore, for each DC subset, two independent protocols are introduced. They may be used individually or in parallel. Particularly if tissue samples other than PB and cord blood are to be analyzed, both protocols should be used in parallel, until the specificity of the individual protocols is proven.

pDCs are identified as:
- CD303+ leukocytes with side scatter (SSC) characteristics lower than granulocytes, lacking CD14 and CD19 expression.
- CD303+ leukocytes with SSC characteristics lower than granulocytes, coexpressing CD123 and HLA-DR.

MDCs are identified as:
- CD1c+ leukocytes with forward scatter (FSC) characteristics of larger lymphocytes, lacking CD14 and CD19 expression.
- CD11c+ leukocytes, coexpressing HLA-DR and lacking or showing only weak expression of the lineage markers CD3, CD11b, CD14, CD16, CD19, CD34, CD56 und CD235a.

The respective target populations are successively narrowed down by consecutive gating (fig. 1). In the final dot plot, they should show as homogeneous populations with FSC and SSC characteristics of larger lymphocytes.

Materials

In table 1 possible sources for the mAbs conjugated with phycoerythrin (PE), fluorescein isothiocyanate (FITC) or PE-cyanin-5 (PC5) are listed. The Blood Dendritic Cell Enumeration kit (anti-CD303-FITC, anti-CD1c-PE, anti-CD14/CD19-PC5) can be purchased from Miltenyi Biotec. To avoid adhesive interactions, stainings are performed in 5-ml round-bottom polypropylene tubes. Lysis of erythrocytes as well as washing steps are performed in 15-ml conical polypropylene tubes which allow

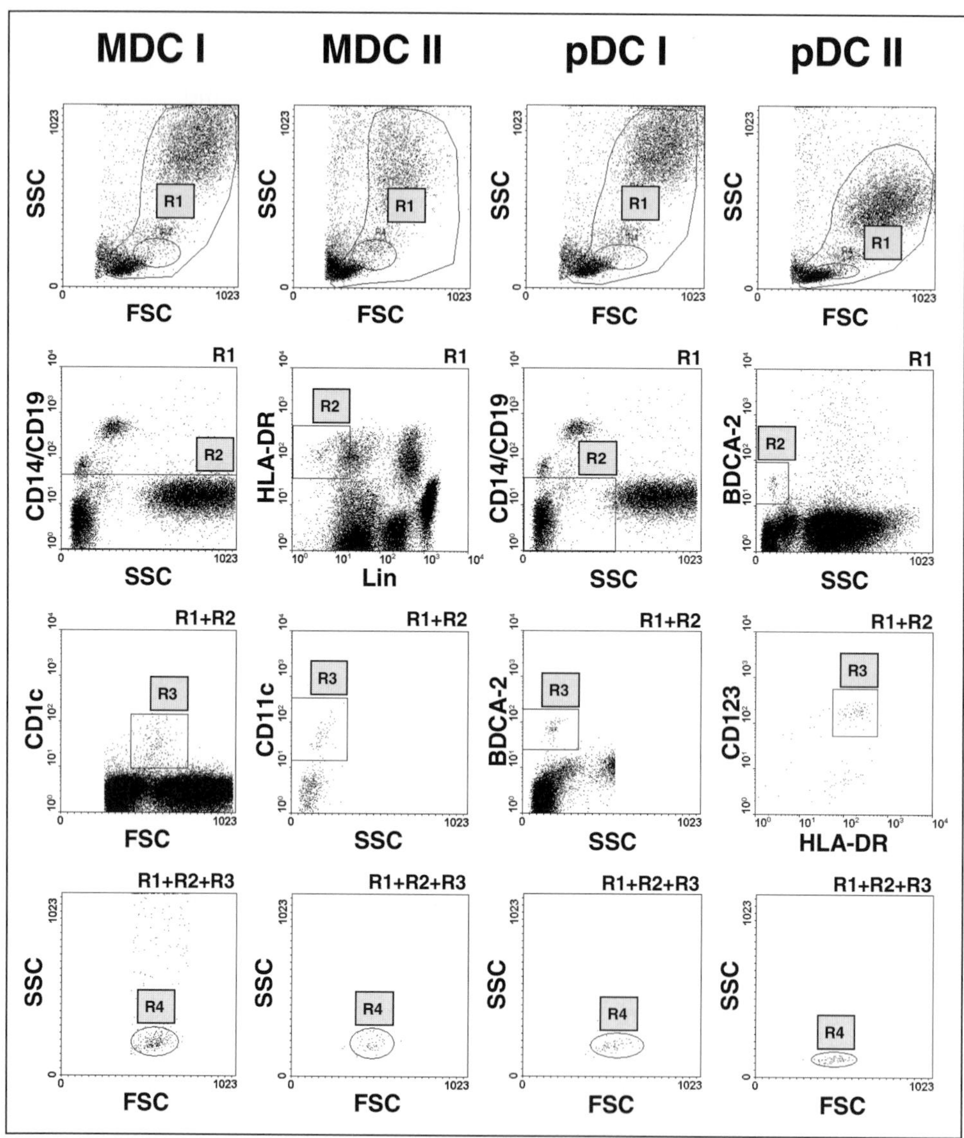

Fig. 1. Identification of myeloid (MDC) and plasmacytoid (pDC) subtypes of DCs using two different protocols each successively narrowing down (R1–R4) the target population.

more precise removal of supernatants following sedimentation of cells. Tubes can be purchased from BD Biosciences or Greiner Bio-One. Washing buffer is phosphate-buffered saline (PBS; pH 7.4) supplemented with human serum albumin (HSA) which is available as a 20 % stock solution from Octapharma. Alternatively, bovine serum albumin (BSA) may be used as a protein supplement of PBS. To lyse erythrocytes, ammonium chloride solution is used (8.3 g/l NH_4Cl, 1 g/l $KHCO_3$, 0.0375 g/l EDTA-Na_2).

Table 1. Monoclonal antibodies

Specificity/conjugate	Clone	Isotype	Source
CD1c-PE	AD5-8E7 (BDCA-1)	IgG2a	Miltenyi Biotec
CD3-PE	UCHT1	IgG1	Beckman-Coulter
CD11b-PE	Bear 1	IgG1	Beckman-Coulter
CD11c-FITC	3.9	IgG1	Serotec
CD14-PE/PC5	RMO52	IgG2a	Beckman-Coulter
CD16-PE	3G8	IgG1	Beckman-Coulter
CD19-PE/PC5	J4.119	IgG1	Beckman-Coulter
CD34-PE	8G12	IgG1	BD Biosciences
CD56-PE	NKH-1	IgG1	Beckman-Coulter
CD123-PE	9F5	IgG1	BD Biosciences
CD235a-PE	11E4B7.6	IgG1	Beckman-Coulter
CD303-FITC	AC144 (BDCA-2)	IgG1	Miltenyi Biotec
HLA-DR-PC5	Immu-357	IgG1	Beckman-Coulter

Sources
BD Biosciences – *www.bd.com*
Beckman Coulter – *www.beckman.com*
Miltenyi Biotec – *www.miltenyibiotec.com*
Serotec – *www.serotec.com*
Greiner Bio-One – *www.gbo.com*
Octapharma – *www.octapharma.de.*

Procedure: Identification and Quantification of Peripheral Blood Dendritic Cells

PB should be processed within 24 h. It can be stored at 20 °C until analysis [52]. For other types of specimens or longer storage, stability of samples should be determined first (time course analyses). Ethylene diamine tetraacetate (EDTA), acid-citrate-dextrose (ACD), citrate or citrate-phosphate-dextrose (CPD) can be used as anticoagulant [52]. Depending on the anticoagulant used, it may be necessary to use different dilution factors.

The following protocol describes two procedures each for analysis of pDCs and MDCs in PB. It can be applied directly for cord blood as well. For other tissue types, it has to be verified first that both procedures for pDCs and both procedures for MDCs produce comparable results, and that other characteristics of the identified DCs also resemble typical pDCs or MDCs. In doubt, cells should be characterized further by flow cytometry to confirm their MDC or pDC nature. A selection of markers for such analyses is given in table 2.

- Determine sample volume.
- Use about 300 µl of PB to determine the leukocyte number on a blood cell analyzer (e.g. Abbott CellDyn 3500). The volume required may vary with the analyzer used.

Table 2. Phenotype of DCs

Molecule	pDC	MDC	Molecule	pDC	MDC
CD1a	–	–	CD44	+	+
CD1c	–	+	CD45RA	+	–
CD2	+/–	+	CD45R0	–	+
CD3	–	–	CD54	+	+
CD4	++	+	CD56	–	–
CD5	–	+/–	CD58	+	++
CD8	–	–	CD61	–	–
CD10	+/–	–	CD62L	++	+
CD11a	+	+	CD64	–	+
CD11b	–	+/–	CD69	–	+
CD11c	–	+	CD71	–	–
CD13	–	+	CD80	–	–
CD14	–	+/–	CD83	–	–
CD15	–	–	CD85a	–	+
CD16	–	–	CD85d	–	+/–
CD18	+	++	CD85h	–	+
CD19	–	–	CD85j	+	+
CD20	–	–	CD85k	+	+
CD23	–	–	CD86	+/–	+
CD25	–	–	CD98	+	++
CD27	–	–	CD116	+	++
CD32	+/–	+	CD120a	+	+
CD33	+/–	+	CD120b	+	++
CD34	–	–	CD123	++	+
CD36	+	+	HLA-DR	+	++
CD38	+	++	HLA-ABC	+	+
CD40	+/–	+/–			

- Document cell count/μl.
- Transfer 1 ml PB (250 μl PB/labeling) into one conical polypropylene 15-ml tube.
- Add 10 ml of ice-cold ammonium chloride solution.
- Incubate for 10 min at 4 °C.
- Sediment cells for 5 min at 550 × *g* and 4 °C.
- Discard supernatant and resuspend cells in 10 ml 0.5 % HSA/PBS.
- Sediment cells for 5 min at 550 × *g* and 4 °C.
- Discard supernatant.
- Resuspend cells in 880 μl (210 μl/labeling) 0.5 % HSA/PBS.
- Transfer 200 μl/tube in 4 round-bottom polypropylene 5-ml tubes.
- Prepare the following labelings by adding the mAbs:
  - *MDC I:* 20 μl of Blood Dendritic Cell Enumeration kit mAb mix (relevant here: anti-BDCA-1-PE, anti-CD14/CD19-PC5) (control: IgG2a-PE, anti-CD14/CD19-PC5)*,

- *MDC II:* 10 µl each of anti-CD11c-FITC, anti-CD3-PE, anti-CD11b-PE, anti-CD14-PE, anti-CD16-PE, anti-CD19-PE, anti-CD34-PE, anti-CD56-PE, anti-CD235a-PE and anti-HLA-DR-PC5 mAbs (control: IgG1-FITC, anti-Lin-PE, anti-HLA-DR-PC5)*,
- *pDC I:* 20 µl of Blood Dendritic Cell Enumeration kit mAb mix (relevant here: anti-BDCA-2-FITC, anti-CD14/CD19-PC5) (control: IgG1-FITC, anti-CD14/CD19-PC5)*,
- *pDC II:* 10 µl each of anti-BDCA-2-FITC, anti-CD123-PE and anti-HLA-DR-PC5 mAbs (control: IgG1-FITC, anti-CD123-PE, anti-HLA-DR-PC5)*.

  *At least at the beginning, while establishing the procedures, use the controls shown in parentheses for each labeling.

- Incubate for 20 min at 4 °C in the dark.
- Fill tubes with 0.5 % HSA/PBS.
- Sediment cells for 5 min at 550 × *g* and 4 °C.
- Discard supernatant.
- Resuspend cells in 400 µl 0.5 % HSA/PBS and analyze.

Acquisition and Data Analysis

Due to the expected low frequency of pDCs and MDCs in PB (about 0.1 % of leukocytes), at least 100,000 leukocytes should be acquired (corresponding to about 100 target cells). If during data acquisition a region is used (*'live gate'*), care should be taken that neither leukocytes nor target cells are excluded.

For data analysis, 4 dot plots are required for each procedure (MDC I, MDC II, pDC I, pDC II; fig. 1).

MDC I

The 1st dot plot shows FSC vs. SSC. Region R1 encompasses all leukocytes representing 100 % of the cells. The 2nd dot plot shows SSC vs. CD14/CD19 expression for all events in R1. Region R2 encompasses all CD14−/CD19− events. To define the upper limit of R2, the corresponding fluorescence of granulocytes is used. The 3rd dot plot shows FSC vs. CD1c expression for all events in regions R1 + R2 (CD14−/CD19− leukocytes). Region R3 encompasses all CD1c+ events which display an FSC of large lymphocytes. The 4th dot plot shows FSC vs. SSC for all events in regions R1 + R2 + R3 (CD14−/CD19−/CD1c+ large lymphocytes). A homogeneous population should appear which is included in region R4. The number of events in R4 represents the results for the target population. Region R4 is also depicted in the 1st dot plot to control for the relative position (large lymphocytes).

MDC II

The 1st dot plot shows FSC vs. SSC. Region R1 encompasses all leukocytes representing 100 % of the cells. The 2nd dot plot shows expression of lineage antigens vs.

expression of HLA-DR for all events in R1. Region R2 encompasses all lin–/dim/HLA-DR+ events. To define the lower limit of HLA-DR, the corresponding fluorescence of monocytes is used, for the upper limit of lin–/dim, the mean fluorescence of B cells. The 3rd dot plot shows SSC vs. CD11c expression for all events in regions R1 + R2 (lin–/HLA-DR+ leukocytes). Region R3 encompasses all CD11c+ events. The 4th dot plot shows FSC vs. SSC for all events in regions R1 + R2 + R3 (lin–/HLA-DR+/CD11c+ leukocytes). A homogeneous population should appear which is included in region R4. The number of events in R4 represents the results for the target population. Region R4 is also depicted in the 1st dot plot to control for the relative position (large lymphocytes).

pDC I

The 1st dot plot shows FSC vs. SSC. Region R1 encompasses all leukocytes representing 100% of the cells. The 2nd dot plot shows SSC vs. expression of CD14/CD19 for all events in R1. Region R2 encompasses all CD14–/CD19– events which display an SSC lower than granulocytes. To define the upper limit of R2 for CD14–/CD19–, the corresponding fluorescence of granulocytes is used. The 3rd dot plot shows SSC vs. expression of BDCA-2 (CD303) for all events in regions R1 + R2 (CD14–/CD19–/$SSC^{<gran}$ leukocytes). Region R3 encompasses all BDCA-2+ events. The 4th dot plot shows FSC vs. SSC for all events in regions R1 + R2 + R3 (CD14–/CD19–/BDCA-2+/$SSC^{<gran}$ leukocytes). A homogeneous population should appear which is included in region R4. The number of events in R4 represents the results for the target population. Region R4 is also depicted in the 1st dot plot to control for the relative position (large lymphocytes).

pDC II

The 1st dot plot shows FSC vs. SSC. Region R1 encompasses all leukocytes representing 100% of the cells. The 2nd dot plot shows SSC vs. expression of BDCA-2 (CD303) for all events in R1. Region R2 encompasses all BDCA-2+ events which display an SSC lower than granulocytes. The 3rd dot plot shows expression of HLA-DR vs. expression of CD123 for all events in regions R1 + R2 (BDCA-2+/$SSC^{<gran}$ leukocytes). Region R3 encompasses all HLA-DR+/CD123+ events. The 4th dot plot shows FSC vs. SSC for all events in regions R1 + R2 + R3 (BDCA-2+/$SSC^{<gran}$/HLA-DR+/CD123+ leukocytes). A homogeneous population should appear which is included in region R4. The number of events in R4 represents the results for the target population. Region R4 is also depicted in the 1st dot plot to control for the relative position (large lymphocytes).

The frequency of pDC and MDC is calculated accordingly:

$$\text{Frequency, \%} = \frac{\text{Events in region R4} \times 100}{\text{Events in region R1}} \qquad (1)$$

The pDC and MDC cell count/μl is calculated from cell numbers/μl and frequencies accordingly:

$$\mathrm{DC}/\mu\mathrm{l} = \text{Cell number}/\mu\mathrm{l} \times \text{frequency} \times \text{dilution factor} \tag{2}$$

The dilution factor due to the anticoagulant is calculated accordingly:

$$\text{Dilution factor} = \frac{\text{Sample volume}}{\text{Sample volume} - \text{volume anticoagulant}} \tag{3}$$

The pDC:MDC ratio can be calculated from pDC and MDC cell numbers/μl or from the respective frequencies.

Report

In the report on the identification and quantification of pDCs and mDCs, the following parameters should be listed:

- Method used
- Frequency of pDCs and MDCs*
- Anticoagulant and dilution factor
- Cell counts/μl for pDCs and MDCs*
- pDCs:MDCs ratio*.

*When normal values become available, deviations should be documented.

Troubleshooting

With the exception of BDCA-2 (CD303), the currently used mAbs are not specific for the DC subsets. The protocol becomes specific for pDCs and MDCs only by the combination of several parameters. It has been validated for PB and cord blood of healthy individuals. For other tissue types or patient samples, specificity has to be confirmed before the protocol can be used. It has to be ascertained that the tissue specimen does not contain any non-DCs, which are absent in PB or cord blood, but share distinct immunophenotypic characteristics with DCs, which would cause a false-positive identification and overestimation of target cell numbers. If there is a specimen type specific difference in the immunophenotype of DCs, particularly downregulation of certain markers will cause an underestimation of cell numbers. Therefore, preferably both procedures for MDC and pDC enumeration should be used in parallel and when in doubt, additional parameters should be tested for the target cell population. Examples of mAbs which may be used to further characterize pDCs and MDCs are listed in table 2.

Expected Results

Normal ranges for pDCs and MDCs have not been established yet. Frequencies between 0.1 and 0.5% have been reported for PB. MDCs vary from 10 to 40 cells/μl, pDC from 5 to 20 cells/μl [24, 44, 52–55].

Time Required

From receiving the samples to the final report, the procedure will require approximately 90 min.

Summary

DCs play a central role in initiation and polarization of T-cell responses, thereby controlling most aspects of adaptive immunity. Distinct subtypes of DC – pDCs and MDCs – develop from lymphoid and myeloid progenitors, respectively. They show differential responsiveness to activating signals and appear to represent specialized antigen-presenting cells for distinct groups of pathogens. The frequencies, cell numbers and pDC:MDC ratio may vary depending on pathological conditions. Therefore, enumeration of DCs appears to be expedient, and a protocol for such analyses is presented.

Acknowledgment

I would like to thank my colleagues Zakir Özcan, Zhongbo Jin, Nina Blasberg, Gesine Kögler and Peter Wernet for their contribution. I am grateful for the financial support from the German José Carreras Leukemia Foundation and the EU EUROCORD III grant QLRT-2001-01918.

References

1 Banchereau J, Briere F, Caux C, Davoust J, Lebecque S, Liu YJ, Pulendran B, Palucka K: Immunobiology of dendritic cells. Annu Rev Immunol 2000;18:767–811.

2 Ueno H, Klechevsky E, Morita R, Aspord C Cao T, Matsui T, Di Pucchio T, Connolly J, Fay JW, Pascual V, Palucka AK, Banchereau J: Dendritic cell subsets in health and disease. Immunol Rev 2007;219:118–142.

3 Caux C, Ait-Yahia S, Chemin K, de Bouteiller O, Dieu-Nosjean MC, Homey B, Massacrier C, Vanbervliet B, Zlotnik A, Vicari A: Dendritic cell biology and regulation of dendritic cell trafficking by chemokines. Semin Immunopathol 2000;22:345–369.

4 Geijtenbeek TB, van Vliet SJ, Engering A, t Hart, BA, van Kooyk, Y: Self- and nonself-recognition by C-type lectins on dendritic cells. Annu Rev Immunol 2004;22:33–54.

5 Matzinger P: Tolerance, danger, and the extended family. Annu Rev Immunol 1994;12:991–1045.

6 Lipscomb MF, Masten BJ: Dendritic cells: immune regulators in health and disease. Physiol Rev 2002;82:97–130.
7 Zhou LJ, Tedder TF: Human blood dendritic cells selectively express CD83, a member of the immunoglobulin superfamily. J Immunol 1995;154:3821–3835.
8 Kalinski P, Hilkens CM, Wierenga EA, Kapsenberg ML: T-cell priming by type-1 and type-2 polarized dendritic cells: the concept of a third signal. Immunol Today 1999;20:561–567.
9 Kapsenberg ML: Dendritic-cell control of pathogen-driven T-cell polarization. Nature Rev Immunol 2003;3:984–993.
10 Jonuleit H, Kuhn U, Muller G, Steinbrink K, Paragnik L, Schmitt E, Knop J, Enk AH: Pro-inflammatory cytokines and prostaglandins induce maturation of potent immunostimulatory dendritic cells under fetal calf serum-free conditions. Eur J Immunol 1997;27:3135–3142.
11 Steinman RM, Hawiger D, Nussenzweig MC: Tolerogenic dendritic cells. Annu Rev Immunol 2003;21:685–711.
12 Steinman RM, Hawiger D, Liu K, Bonifaz L, Bonnyay D, Mahnke K, Iyoda T, Ravetch J, Dhodapkar M, Inaba K, Nussenzweig M: Dendritic cell function in vivo during the steady state: a role in peripheral tolerance. Ann NY Acad Sci 2003;987:15–25.
13 Brocker T, Riedinger M, Karjalainen K: Targeted expression of major histocompatibility complex (MHC) class II molecules demonstrates that dendritic cells can induce negative but not positive selection of thymocytes in vivo. J Exp Med 1997;185:541–550.
14 Shortman K, Liu YJ: Mouse and human dendritic cell subtypes. Nature Rev Immunol 2002;2:151–161.
15 Rissoan MC, Soumelis V, Kadowaki N, Grouard G, Briere F, de Waal Malefyt R, Liu YJ: Reciprocal control of T helper cell and dendritic cell differentiation. Science 1999;283:1183–1186.
16 O'Doherty U, Peng M, Gezelter S, Swiggard WJ, Betjes M, Bhardwaj N, Steinman RM: Human blood contains two subsets of dendritic cells, one immunologically mature and the other immature. Immunology 1994;82:487–493.
17 Kohrgruber N, Halanek N, Groger M, Winter D, Rappersberger K, Schmitt-Egenolf M, Stingl G, Maurer D: Survival, maturation, and function of CD11c− and CD11c+ peripheral blood dendritic cells are differentially regulated by cytokines. J Immunol 1999;163:3250–3259.
18 Robinson SP, Patterson S, English N, Davies D, Knight SC, Reid CD: Human peripheral blood contains two distinct lineages of dendritic cells. Eur J Immunol 1999;29:2769–2778.
19 Ito T, Inaba M, Inaba K, Toki J, Sogo S, Iguchi T, Adachi Y, Yamaguchi K, Amakawa R, Valladeau J, Saeland S, Fukuhara S, Ikehara S: A CD1a+/CD11c+ subset of human blood dendritic cells is a direct precursor of Langerhans cells. J Immunol 1999;163:1409–1419.
20 MacDonald KP, Munster DJ, Clark GJ, Dzionek A, Schmitz J, Hart DN: Characterization of human blood dendritic cell subsets. Blood 2002;100:4512–4520.
21 Cella M, Jarrossay D, Facchetti F, Alebardi O, Nakajima H, Lanzavecchia A, Colonna M: Plasmacytoid monocytes migrate to inflamed lymph nodes and produce large amounts of type I interferon. Nat Med 1999;5:919–923.
22 Strobl H, Scheinecker C, Riedl E, Csmarits B, Bello-Fernandez C, Pickl WF, Majdic O, Knapp W: Identification of CD68+lin− peripheral blood cells with dendritic precursor characteristics. J Immunol 1998;161:740–748.
23 Grouard G, Durand I, Filgueira L, Banchereau J, Liu YJ: Dendritic cells capable of stimulating T cells in germinal centres. Nature 1996;384:364–367.
24 Arpinati M, Green CL, Heimfeld S, Heuser JE, Anasetti C: Granulocyte-colony stimulating factor mobilizes T helper 2-inducing dendritic cells. Blood 2000;95:2484–2490.
25 Jonuleit H, Schmitt E, Schuler G, Knop J, Enk AH: Induction of interleukin 10-producing, nonproliferating CD4(+) T cells with regulatory properties by repetitive stimulation with allogeneic immature human dendritic cells. J Exp Med 2000;192:1213–1222.
26 Jonuleit H Adema G, Schmitt E: Immune regulation by regulatory T cells: implications for transplantation. Transpl Immunol 2003;11:267–276.
27 Olweus J, BitMansour A, Warnke R, Thompson PA, Carballido J, Picker LJ, Lund Johansen F: Dendritic cell ontogeny: a human dendritic cell lineage of myeloid origin. Proc Natl Acad Sci USA 1997;94:12551–12556.
28 Res PC, Couwenberg F, Vyth-Dreese FA, Spits H: Expression of pT alpha mRNA in a committed dendritic cell precursor in the human thymus. Blood 1999;94:2647–2657.
29 Corcoran, L Ferrero I, Vremec D, Lucas K, Waithman J, O'Keeffe M, Wu L, Wilson A, Shortman K: The lymphoid past of mouse plasmacytoid cells and thymic dendritic cells. J Immunol 2003;170:4926–4932.
30 Dzionek A, Fuchs A, Schmidt P, Cremer S, Zysk M, Miltenyi S, Buck DW, Schmitz J: BDCA-2, BDCA-3, and BDCA-4: three markers for distinct subsets of dendritic cells in human peripheral blood. J Immunol 2000;165:6037-6046.

31 Grouard G, Rissoan MC, Filgueira L, Durand I, Banchereau J, Liu YJ: The enigmatic plasmacytoid T cells develop into dendritic cells with interleukin (IL)-3 and CD40-ligand. J Exp Med 1997;185:1101–1111.

32 Vandenabeele S, Hochrein H, Mavaddat N, Winkel K, Shortman K: Human thymus contains 2 distinct dendritic cell populations. Blood 2001;97:1733–1741.

33 Bendriss-Vermare N, Barthelemy C, Durand I, Bruand C, Dezutter-Dambuyant C, Moulian N, Berrih-Aknin S, Caux C, Trinchieri G, Briere F: Human thymus contains IFN-alpha-producing CD11c(–), myeloid CD11c(+), and mature interdigitating dendritic cells. J Clin Invest 2001;107:835–844.

34 Siegal FP, Kadowaki N, Shodell M, Fitzgerald-Bocarsly PA, Shah K, Ho S, Antonenko S, Liu YJ: The nature of the principal type 1 interferon-producing cells in human blood. Science 1999;284:1835–1837.

35 Diebold SS, Montoya M, Unger H, Alexopoulou L, Roy P, Haswell LE, Al-Shamkhani A, Flavell R, Borrow P, Reis e Sousa C: Viral infection switches non-plasmacytoid dendritic cells into high interferon producers. Nature 2003;424:324–328.

36 Kuwana M, Kaburaki J, Wright TM, Kawakami Y, Ikeda Y: Induction of antigen-specific human CD4(+) T cell anergy by peripheral blood DC2 precursors. Eur J Immunol 2001;31:2547–2557.

37 Kuwana, M: Induction of anergic and regulatory T cells by plasmacytoid dendritic cells and other dendritic cell subsets. Hum Immunol 2002;63:1156–1163.

38 Kadowaki N, Ho S, Antonenko S, de Waal Malefyt R, Kastelein RA, Bazan F, Liu YJ: Subsets of human dendritic cell precursors express different toll-like receptors and respond to different microbial antigens. J Exp Med 2001;194:863–870.

39 Schakel K, Mayer E, Federle C, Schmitz M, Riethmuller G, Rieber EP: A novel dendritic cell population in human blood: one-step immunomagnetic isolation by a specific mAb (M-DC8) and in vitro priming of cytotoxic T lymphocytes. Eur J Immunol 1998;28:4084–4093.

40 Ho CS, Lopez JA, Vuckovic S, Pyke CM, Hockey RL, Hart DN: Surgical and physical stress increases circulating blood dendritic cell counts independently of monocyte counts. Blood 2001;98:140–145.

41 Clark FJ, Freeman L, Dzionek A, Schmitz J, McMullan D, Simpson P, Mason J, Mahendra P, Craddock C, Griffiths M, Moss PA, Chakraverty R: Origin and subset distribution of peripheral blood dendritic cells in patients with chronic graft-versus-host disease. Transplantation 2003;75:221–225.

42 Waller EK, Rosenthal H, Jones TW, Peel J, Lonial S, Langston A, Redei I, Jurickova I, Boyer MW: Larger numbers of CD4(bright) dendritic cells in donor bone marrow are associated with increased relapse after allogeneic bone marrow transplantation. Blood 2001;97:2948–2956.

43 Darmochwal-Kolarz D, Rolinski J, Tabarkiewicz J, Leszczynska-Gorzelak B, Buczkowski J, Wojas K, Oleszczuk J: Myeloid and lymphoid dendritic cells in normal pregnancy and pre-eclampsia. Clin Exp Immunol 2003;132:339–344.

44 Teig N, Moses D, Gieseler S, Schauer, U: Age-related changes in human blood dendritic cell subpopulations. Scand J Immunol 2002;55:453–457.

45 Perez-Cabezas B, Naranjo-Gomez M, Fernandez MA, Grifols JR, Pujol-Borrell R, Borras FE: Reduced numbers of plasmacytoid dendritic cells in aged blood donors. Exp Gerontol 2007;42:1033–1038.

46 Wertheimer AM, Bakke A, Rosen HR: Direct enumeration and functional assessment of circulating dendritic cells in patients with liver disease. Hepatology 2004;40:335–345.

47 Jongbloed SL, Lebre MC, Fraser AR, Gracie JA, Sturrock RD, Tak PP, McInnes IB: Enumeration and phenotypical analysis of distinct dendritic cell subsets in psoriatic arthritis and rheumatoid arthritis. Arthritis Res Ther 2006;8:R15.

48 Lichtner M, Rossi R, Mengoni F, Vignoli S, Colacchia B, Massetti AP, Kamga I, Hosmalin A, Vullo V, Mastroianni CM: Circulating dendritic cells and interferon-alpha production in patients with tuberculosis: correlation with clinical outcome and treatment response. Clin Exp Immunol 2006;143:329–337.

49 Matsuda H, Suda T, Hashizume H, Yokomura K, Asada K, Suzuki K, Chida K, Nakamura H: Alteration of balance between myeloid dendritic cells and plasmacytoid dendritic cells in peripheral blood of patients with asthma. Am J Respir Crit Care Med 2002;166:1050–1054.

50 Sciarra A, Lichtner M, Autran GA, Mastroianni C, Rossi, R, Mengoni F, Cristini C, Gentilucci A, Vullo V, Di Silverio F: Characterization of circulating blood dendritic cell subsets DC123+ (lymphoid) and DC11C+ (myeloid) in prostate adenocarcinoma patients. Prostate 2007;67:1–7.

51 Lichtner M, Rossi R, Rizza MC, Mengoni F, Sauzullo I, Massetti AP, Luzi G, Hosmalin A, Mastroianni CM, Vullo V: Plasmacytoid dendritic cells count in antiretroviral-treated patients is predictive of HIV load control independent of CD4+ T-cell count. Curr HIV Res 2008;6:19–27.

52 Vuckovic S, Gardiner D, Field K, Chapman GV, Khalil D, Gill D, Marlton P, Taylor K, Wright S, Pinzon-Charry A, Pyke, CM, Rodwell R, Hockey RL, Gleeson M, Tepes S, True D, Cotterill A, Hart DN: Monitoring dendritic cells

in clinical practice using a new whole blood single-platform TruCOUNT assay. J Immunol Methods 2004;284:73–87.

53 Szabolcs P, Park KD, Reese M, Marti L, Broadwater G, Kurtzberg J: Absolute values of dendritic cell subsets in bone marrow, cord blood, and peripheral blood enumerated by a novel method. Stem Cells 2003;21:296–303.

54 Ueda Y, Hagihara M, Okamoto A, Higuchi A, Tanabe A, Hirabayashi K, Izumi S, Makino T, Kato S, Hotta, T: Frequencies of dendritic cells (myeloid DC and plasmacytoid DC) and their ratio reduced in pregnant women: comparison with umbilical cord blood and normal healthy adults. Hum Immunol 2003;64:1144–1151.

55 Ma L, Scheers W, Vandenberghe P: A flow cytometric method for determination of absolute counts of myeloid precursor dendritic cells in peripheral blood. J Immunol Methods 2004;285:215–221.

Sack U, Tárnok A, Rothe G (eds): Cellular Diagnostics. Basics, Methods and Clinical Applications of Flow Cytometry. Basel, Karger, 2009, pp 286–304

Peripheral Blood Stem Cells

Stefan Fruehauf

Paracelsus-Klinik, Osnabrück, Germany

Introduction

Over the last years, the transplantation of circulating stem cells has become a routine procedure for the treatment of patients with hemato-lymphatic illnesses and in some cases for patients with solid tumors as well. It has largely replaced bone marrow transplantation and enables intensive chemo- and radiotherapies, for which bone marrow toxicity was previously dose-limiting.

The first part of this chapter gives an introduction to peripheral blood stem cells and their meaning for transplantation medicine. The second part is concerned with the challenge of mobilizing hematopoietic stem cells. It is followed by an overview of current standard protocols for the assessment of CD34+ cells. A current example then demonstrates how flow cytometry can aid in predicting the success of cell mobilization prior to transplantation. The chapter concludes with a standard protocol for the measurement of mobilized hematopoietic stem cells

What differentiates stem cells from hematopoietic stem cells and peripheral blood stem cells? In the current literature, differing definitions can frequently be found, which often leads to confusion. We use the definition of Weissman [1]: Stem cells are clonogenic cells which can renew themselves and which possess the ability to further differentiate into multiple cell lines. Progenitor cells, however, have lost this ability; they already belong to a specific cell line; their ability to differentiate is reduced and they cannot renew themselves. CD34 antigen expression as well as cell lineage affiliation is often used in clinical practice in order to characterize hematopoietic stem cells and progenitor cells.

Finally, those cells which are found in the marrow are called hematopoietic stem cells. Hematopoietic stem cells which are circulating in the body or in other organs are called peripheral blood stem cells.

The definition of a threshold value for peripheral blood stem cells has contributed to the clinical success of stem cell therapy. Implementation of the threshold value ensures a quick and lasting recovery after transplantation. A dose effect has already been described in 1987 for colony-forming progenitor cells in the transplant [2].

With the availability of a directly fluorescence-marked antibody against hematopoietic progenitor cells, the immunologic quantification of the hematopoietic reconstitution capacity in the mobilized peripheral blood was successfully achieved for the first time [3]. Quantification can be carried out within 1 h and can therefore support decision-making of when and how many leukaphereses are to be performed. Data from many transplantation studies reveal that a recovery of all three cell lines, i.e. granulopoiesis, megakaryopoiesis and erythropoiesis, occurs within 2 weeks following transplantation of peripheral blood stem cells if the amount of progenitor cells applied is greater than the threshold dose [4]. This advantage could be shown in randomized studies on autologous bone marrow transplantation, for which the reconstitution period amounted to approximately 3 weeks [5]. A shorter hematopoietic reconstitution period results in shorter treatment with intravenous antibiotics and thus in a shorter duration of dependency on transfusions and a transplantation-associated mortality of under 2%. The increased safety and the increased number of collectable peripheral blood stem cells offer new therapeutic options. Patients over 60 years and a spectrum of oncological illnesses, for example testicular tumors, are now treated with high-dose chemotherapy and subsequent transplantation of peripheral blood stem cells – this was not possible previously with bone marrow transplantations or possible only in a limited number of cases due to increased mortality risks [6]. A further important development is the use of multiple cycles of high-dose chemotherapy and transplantation of peripheral blood stem cells in order to be able to administer a higher total dose at a higher dose rate.

This ground-breaking development was made possible by the characterization and diagnostic use of the CD34 antigen. In 1984, Curt Civin at John Hopkins Hospital in Baltimore presented a new antigen called My-10 which was isolated from the Kg1a cell lines. His antibody HPCA-1 was used in subsequent years by many groups for leukemia classification and stem cell therapy. HPCA-1 bound with comparatively low affinity to the My-10 antigen, and despite many attempts there was no success in producing a reagent coupled directly with fluorescence staining. Towards the end of the 1980s, it became more difficult as one-dimensional FACS analyses were used as the standard and small cell populations are difficult to detect in histogram tests. In principle, the My-10 findings were not reproducible in different laboratories.

Stefan Serke, who sadly died much too young, was the first to multidimensionally depict CD34+ cells. He managed to confirm the intermediate side scatter (SSC) of CD34+ cells from bone marrow and peripheral blood. Functional investigations and correlation analyses of colony-forming cells as well as of HPCA-1+ cells validated the blood stem cell population he described [7].

In the early 1990s, when it became generally accepted that peripheral blood stem cells were superior to bone marrow with regard to their repopulation ability, a feverish search began, as reflected in the large number of publications, for the immunophenotypically characterizable peripheral blood stem cells in order to prove a correlation with hematopoietic reconstitution after transplantation. The HPCA-1 antibody separated out the CD34+ cell population; however it did not delineate clearly enough from the negative fraction.

The next milestone was the introduction of the directly fluorescence-coupled CD34+ antibody 8G12 by P. Lansdorp and his working group. They succeeded, for the first time, in characterizing CD34+ cell populations with very small volumes, e.g. in the peripheral blood of quiescent hematopoiesis [8].

Clinical research groups found a threshold value of $2.0–2.5 \times 10^6$ CD34+ cells in the transplant, which must be exceeded for recovery of blood reconstitution within 14 days after stem cell injection [4, 9]. This greatly favored the breakthrough in stem cell transplantation in subsequent years. Almost immediately, quality assurance initiatives were launched to draft methods for reproducible CD34+ cell assays available to interested working groups.

In the beginning, the Mulhouse meeting was organized by E. Wunder and P. Henon in 1992 and brought together many experts in this field. The *Stem Cell Manual* elaborated as a result of this meeting represented the reference work in many transplantation and flow cytometry labs for many years [10].

Quality assurance initiatives in the German-speaking countries were coordinated by Kai Gutensohn and on the European level by David Barnett from Sheffield. The International Society of Hematotherapy and Graft Engineering (ISHAGE) intervened on the international level for the standardization of CD34+ cell measurement. These initiatives were started by D.R. Sutherland from Toronto, who, under the auspices of the International Society for Cellular Therapy (ISCT – former ISHAGE), helped reproducible CD34+ cell measurement to become established [11].

The proportion of CD34+ cells in peripheral blood in homeostasis amounts to only approximately 0.06% and to approximately 1.1% in the marrow [12, 13]. There are indications that there are also CD34– hematopoietic cells in humans and in mice [14]. However, their clinical meaning remains unclear. In the 1990s, attempts were also made to further divide the hematopoietic cell population into different primitive or lineage-determined subtypes by combining the CD34 antigen with further markers.

Leon Terstappen has defined CD34+ CD38– cells; this definition enables distinction between stem cells with the ability to repopulate and those without this ability. The CD34+ CD38– cells were not adopted for clinical use as their identity could not be determined because continuous CD38 expression on CD34+ cells could not be reliably detected in various laboratories due to varying threshold values for the CD34+ CD38– population, which made data comparison difficult.

It was similarly hoped that CD34+ Thy-1^{bw}+ (CDw90) cells could be reliably defined as these cells exhibit a more primitive phenotype [15]. There were more CD34+ CDw90+ cells in peripheral blood than in the bone marrow transplant both proportionally and absolutely [16]. However, with 30% of the CD34+ cells, the CD34+ CDw90+ population is too large to exclusively contain repopulating cells. Further antigen combinations such as CD34+ HLA-DR have not become generally accepted clinically due to their low reproducibility or insufficient sensitivity and specificity. Interestingly enough, bone marrow CD34 cells differ from those of peripheral blood also in gene expression [17]. This is due to their different functions: the cell cycle in bone marrow progresses more rapidly, while peripheral stem cells remain in G_0/G_1 phase. Finally, in peripheral blood stem cells apoptosis-associated genes are highly regulated compared with bone marrow cells, maybe to prevent hematopoietic stem cells from multiplying in an uncontrolled fashion once they have exited the bone marrow, and thus the principal site of new blood formation in adults.

How Do CD34+ Cells Get in the Peripheral Blood?

We would first like to address the question of how hematopoietic stem cells can be mobilized. The classical mobilization strategies for peripheral blood stem cells include mobilization based primarily on the use of granulocyte colony-stimulating factor (G-CSF) alone or in combination with other cytokines or chemotherapy. G-CSF has become generally accepted in routine clinical use because it produces a 20- to 100-fold rise in stem cells in peripheral blood and because there are no unwanted side effects in the sense of the WHO classification [18, 19]. The very different responses of patients administered G-CSF may possibly be attributed to interindividual genetic factors and have not been clarified yet.

Thus yields of precursor cells differ between individual patients by a factor of more than 100 following G-CSF-supported chemotherapy [4]. Factors which decrease the yield of precursor cells include prior cytotoxic chemotherapy and irradiation [4, 20, 21]. A statistic analysis of these clinical data does not, however, allow prediction of individual patient outcome.

As alternatives to G-CSF there are other cytokines which induce the mobilization of stem cells, so-called 'stem cell factors' [22]; these include interleukin 3 (IL-3) [23] and granulocyte-macrophage colony-stimulating factor (GM-CSF). However, in comparison with G-CSF, these agents have the disadvantage that their action onset is slow, that they are less effective and in part exhibit strong side effects.

The exact biological mechanism of mobilization has not yet been satisfactorily explained. Thus it was already mentioned that peripheral stem cells appear to be ready at any time to enter apoptosis. Figure 1 summarizes current knowledge on mobilization [24]. As shown in the figure, it appears that the G1a protein is important in telling the CXCR4 receptor whether to live, to die or to exit the bone marrow. The 'pathways' of the different agents coalesce on this protein.

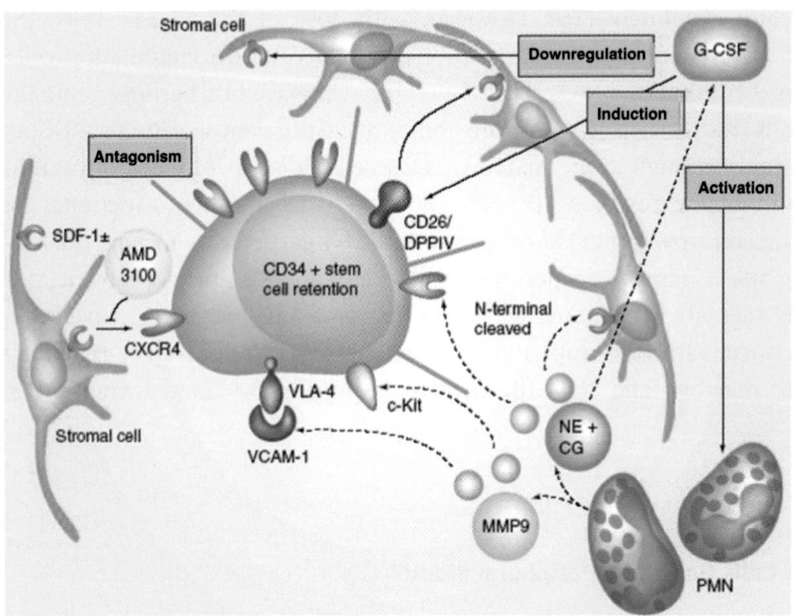

Fig. 1. Role of the SDF-1/CXCR4 interaction. Hematopoietic stem cells are mobilized through: (1) downregulation of SDF-1 or reduction of RNA stability; (2) induction of CD26 and the following N-terminal cleavage of SDF-1; (3) neutrophil elastase (NE)-, cathepsin G (CG)- and matrix metalloproteinase (MMP) 9-induced cleavage of adhesion molecules. A new substance directly interrupts stem cell retention in the bone marrow: AMD3100, a competitive CXCR4 antagonist which interferes with the SDF-1α/CXCR4 interaction. CXCR4 = CXC motif receptor 4; DPPIV = dipeptidyl peptidase IV; G-CSF = granulocyte colony stimulating factor; PBPCs = peripheral blood progenitor cells; PMN = polymorphonuclear leukocyte; SDF = stromal-cell-derived factor; VCAM = vascular cell adhesion molecule; VLA = very late antigen. According to Fruehauf and Seeger [24].

Hematopoietic stem cells are bound in the bone marrow stroma via the SDF-1/CXCR4 receptor complex. G-CSF ensures that expression is downregulated by SDF-1 in the bone marrow stromal cells. At the same time, metalloproteinases which bind the CXCR4 receptor agonistically and cause separation are stimulated. This in turn stimulates G1a protein expression and drives the cell to exit the bone marrow.

AMD-3100 is a very promising molecule which can be used in patients categorized as 'poor mobilizers', that is patients who respond poorly to G-CSF and the conventional cytokines. AMD-3100 is a bicyclam, a small, synthetic molecule which functions as a CXCR4 receptor agonist. It induces detachment of the cell from the bone marrow and at the same time highly upregulates G1a expression so that the cell is stimulated to start 'the chemotaxis pathway' and enter the circulating blood [25].

Methods of CD34+ Analysis

In the past few years, considerable efforts have been undertaken to standardize, CD34+ analyses between different laboratories in order to obtain reproducible and comparable results across centers. Three important protocols were developed for this purpose.

They can be divided first into 'dual-platform' and 'single-platform' assays. 'Dual-platform' assays require a hematology analyzer and a flow cytometer. In these protocols, the proportion of CD34+ cells in the blood is always determined. In order to be able to draw conclusions regarding the absolute number of CD34+ cells, however, the total number of leukocytes is also required. This must be determined separately. The advantage of a 'single-platform' system is that it is possible to work with only a single flow cytometer and thus analysis with the hematology analyzer becomes superfluous.

Three 'dual-platform' protocols for the analysis of CD34+ regulation have been developed. The first protocol, ('Mulhouse protocol') was presented in 1989 by Siena et al. [3]. With this procedure, simple forward scatter (FSC) – as opposed to SSC, analyzes all selected cells, thus providing the denominator for the calculation of percentages. A gate is then created for the CD34+ cell fraction. In addition, CD34-PE-stained samples are used. All events that are measured above these isotype controls, and at the same time exhibit a low to intermediate SSC, become the denominator in the computation of the proportional fraction of CD34+ cells.

'The Nordic protocol' is a modification of this procedure. In this protocol, a gate is determined by selecting the events that are visible as individual, strongly stained accumulations but display low to intermediate SSC. These gating regions enable isotype control measurements. This protocol promises a somewhat more accurate numerator than the original Mulhouse procedure.

The Sihon protocol developed by a Dutch working group goes one step further and pursues another strategy to select all nucleus-containing cells (denominators): the laser dye LDS-751 specifically stains DNA (and to a certain extent RNA). To obtain the numerator, the monocytes and granulocytes are first filtered through an FITC-conjugated antibody against CD14 and CD66e. The CD34+ cells are detected with the assistance of a class III CD34 antibody which is coupled to PE.

The ISHAGE protocol is the most widely used protocol. It was introduced to reduce interlaboratory variability in the determination of CD34+ levels.

The method is based on a multiparameter procedure: an FITC-coupled CD45 antibody reliably identifies the nucleus-containing cell population of leukocytes (fig. 2). A very reliable and stable denominator can be obtained this way. Thereupon, a Boolean gating strategy is used in order to separate PE-stained cells from nonspecific staining events (fig. 2). Now the cells of this gate are compared with the SSC (fig. 2). The CD34-expressing cells (peripheral stem cells), which are also CD45+

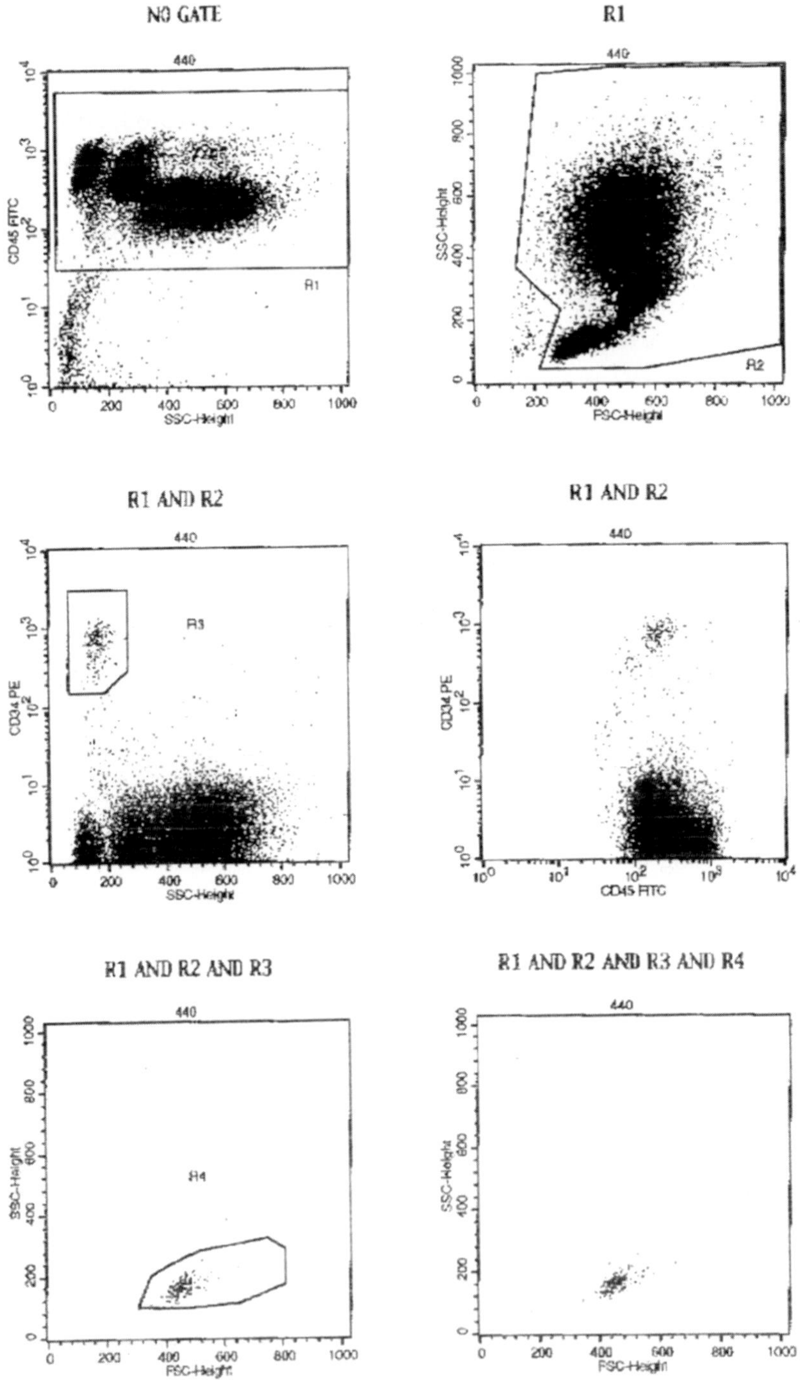

Fig. 2. Example a CD34+ measurement: at first, the core containing leukocytes is determined in the CD45+ vs. SSC image. This is used as gate for other analyses (SSC vs. FSC, CD34-PE vs. SSC, CD34-PE vs. CD45-RTC). The cells from these gates are carried over into the SSC vs. FSC representation. This representation contains the CD34+ cells.

(white blood cells) are captured in this way, and as a 'security' exhibit an appropriate intermediate SSC value.

In all the above-mentioned protocols, the percentage of CD34+ cells is computed first by calculating the quotient of the numerator and denominator, and multiplying this by the total number of leukocytes counted with the hemocytometer. The limitations of this method are thus evident as well. Dual protocols are prone to inaccuracies, above all if the sample examined is not fresh. In addition, the presence of nucleus-containing erythrocytes can cause the measured CD34+ cell levels to appear to be too low when using DNA dyes. These restrictions and problems led to the development of a single-platform protocol.

According to this protocol, the sample is incubated with magnetic CD34+ beads. The absolute number of CD34+ cells can now be determined solely with a flow cytometer, which is an advantage.

The absolute CD34+ cell number can be calculated directly by comparing the counted beads and the measured CD34+ cells with a washing buffer sample.

The first commercial kit for single-platform measurement of CD34+ cells was the ProCount® reagent from BD Biosciences. It contains a three-color assay: a commercial DNA dye is measured in the cytometer against FL1, CD34-PE and CD45-PerCP. The addition of Trucount® fluorescence then permits direct CD34+ cell counts on the cytometer.

The addition of fluorescent beads transforms the original dual-platform protocol of the ISHAGE protocol into a single-platform protocol. Addition of the 7-AAD dye serves to assess cell vitality. This single-platform protocol elaborated by ISHAGE was validated by an international group of 36 participating centers [26]. The observed deviations have a narrow distribution so that the ISHAGE protocol meets its original objective of inter-institutional reproducibility.

An example of the ISHAGE protocol on which standard operation procedures (SOPs) are based is given in 'Standard Operating Procedure: Flow-Cytometric Assessment of CD34+ Cells and CD3+ Cells' (pp. 295).

Applications: The Relationship between CD34+ Levels in Steady State and Subsequent Successful Mobilization

The levels of CD34+ cells in quiescent hematopoiesis display marked differences among patients and may greatly differ in bone marrow aspirates (up to 168-fold) [27].

From flow-cytometric measurements, the calculated proportional frequency of peripheral CD34+ cells is found to be reversed as compared with the counts in the preceding chemotherapy cycles. These data are valid for patients with Hodgkin's disease, multiple myeloma, and non-Hodgkin's lymphomas, but not for patients with

Table 1. Gates statistics from the FACS plots from figure 2[a]

'Gate'	Events	'Gated' %	Total%
G1	29,234	100,00	97,45
G2	29,234	100,00	97,45
G3	196	0,67	0,65
G4	6896	23,59	22,99
G5	39,234	100,00	97,45
G6	196	0,67	0,65
G7	196	0,67	0,65

[a] The results: 97.45% of all cells of the sample were analyzed for CD34. Of those, 0.67% are CD34+. This equals 0.65% of all cells of the blood sample (including the cells that, because of the gate analysis, were not carefully examined anymore).

solid tumors. In each chemotherapy cycle, the pool of circulating CD34+ cells decreases further by a defined amount.

No statistical relationship could be found between previous irradiation and chemotherapy. Thus it may be assumed that both therapies influence the number of circulating peripheral stem cells – however through independent mechanisms.

No connection was found between CD34+ cells in the marrow and the preceding cytotoxic chemotherapy.

A relationship was established between the number of mobilizable CD34+ cells and the CD34+ cells on 'day 0' of treatment [27]; this observation applies to patients with hemato-oncologic illnesses (multiple myeloma, non-Hodgkin's lymphoma).

A very large number of events is required for the highly sensitive measurement of CD34+ cells (in our case over 184,000 events and 100 patients).

On average, there were 152 $\pm$ 7 CD34+ cells (mean $\pm$ 1 SEM; corrected values for nonspecific events) in the CD34+ measuring window. The measured coefficient of variation was 13 $\pm$ 6% and was thus quite close to the expected coefficient of variation of 12% $\pm$ 3% [28].

At the same time, the minimum forerunner cell yield with a leukapheresis can be predicted with 95% probability. During leukapheresis, patients are connected to a cell separator which extracts the mononuclear cells and the subset of hematopoietic stem cells and returns the remaining blood to the patients. Table 1 shows the connection between CD34+ levels in quiescent hematopoiesis ('steady state') and the later yield for the example in non-Hodgkin's lymphomas. Based on the number of CD34+ cells in quiescent hematopoiesis, it is possible to measure the number of circulating peripheral stem cells which can be expected before a mobilization chemotherapy.

Standard Operating Procedure: Flow-Cytometric Assessment of CD34+ and CD3+ Cells

This SOP is from our daily laboratory practice.

Range of Application

These working instructions apply to the flow-cytometric assessment of CD34+ cells in peripheral blood (EDTA blood), leukapheresis products, selection products (and their intermediate products) and bone marrow.

A 'dual-platform' method will be used, in which the proportion of CD34+ cells in a sample (flow-cytometrically determined) is combined with the counts of the same sample on the hematology analyzer in order to determine the number of CD34+ cells per unit volume sample.

Safety

Biologic danger: All samples should be considered as potentially infectious. It is recommended that laboratory coats and protective gloves be worn all the time.

Material and Equipment

- Flow cytometer (e.g. FACSCalibur, BD Biosciences); software (e.g. CellQuest pro) and printer
- Adequate laboratory centrifuge for Eppendorf tubes
- Refrigerator for storage of reagents at 2–8 °C
- 1.5-ml reaction tubes (e.g. Eppendorf)
- Round-ended tubes (FACS tubes, e.g. BD Biosciences)
- Adequate MultiPipetter with Combi-tips (e.g. Eppendorf)
- Adequate pipettes with tips (i.e. Eppendorf Research)
- Vortex equipment
- Single-use gloves (e.g. VWR, Bruchsal Saveskin)
- Distilled water for dilution from 7.12 (e.g. Braun Melsungen)
- Phosphate-buffered saline (PBS; 1× concentration for dilution of samples and to wash cells (e.g. Gibco)
- Ammonium chloride lysis reagent for erythrocyte lysis, 10× concentrations, 100 ml (e.g. BD Pharmingen)
- Monoclonal antibodies:

- CD45-fluorescein isothiocyanate (FITC; e.g. clone 2D1, mouse IgG1; BD Biosciences)
- CD34-phycoerythrin (PE; e.g., clone 8G12, mouse IgG1; BD Biosciences)
- CD3-PE (e.g., clone SK7, mouse IgG1; BD Biosciences)
- mouse IgG1-PE (e.g. clone X40; BD Biosciences)

– FACS Flow (sheet fluid, e.g. BD Biosciences)
– FACS Rinse (e.g. BD Biosciences)
– FACS Clean (e.g. BD Biosciences).

Execution

Double staining consists of double assays in each case (also see the pipetting scheme below).

Preparation of Custom Solutions

Ammonium chloride lysis solution: ammonium chloride lysis reagent (10× concentration) with distilled water. Dilute 1:10 (e.g. 10 ml ammonium chloride lysis reagent + 90 ml distilled water). The date of the preparation and the expiration date should be noted on the container (durability 1 month, storage at 2–8 °C). Warm up to room temperature (RT: 15–25 °C) before use.

Leukocyte Counting in a Sample

White blood cells (WBCs) are enumerated, e.g. with the Coulter® $A^C \cdot T$ diff™ Analyzer.

For allogeneic leukaphereses, allogeneic selection products (intermediate products) and the negative fraction after enrichment, the following applies: the reaction containers are marked from 1 to 5. If several samples from several patients are analyzed at the same time, reaction containers of different colors are used (one color per patient). If this is not enough, the reaction containers can additionally be marked with symbols. The pipetting scheme is depicted in table 2.

For autologous leukaphereses, autologous bone marrow, autologous selection products (intermediate products) as well as the negative fraction following enrichment, the following applies: the reaction containers are marked from 1 to 3. If several samples from several patients are analyzed at the same time, reaction containers of different colors are used (one color per patient). If this is not enough, the reaction containers can additionally be marked with symbols. The pipetting scheme is depicted in table 3.

Cell Staining and Erythrocyte Lysis

– Administration and calculation of the sample (cells).
  Important: Pipette cells always directly to the bottom of the tube in order to ensure even mixing with the antibody.

Table 2. Allogeneic leukaphereses, allogeneic selection products (intermediate products) and negative fraction after enrichment[a]

Tube No.	CD45-FITC, μl	CD34-PE, μl	IgG1-PE, μl	CD3-PE; μl
1	10	10	–	–
2	10	10	–	–
3	10	–	10	–
4	10	–	–	10
5	10	–	–	10

[a] *Important:* Check after pipetting and mixing whether liquid is at the top margin of the cover. With such small volumes, it is important that the entire liquid is located at the bottom of the reaction container.

Table 3. For autologous leukaphereses, autologous bone marrow, autologous selection products (intermediate products) as well as the negative fraction following enrichment

Tube No.	CD45-FITC, μl	CD34-PE, μl	IgG1-PE, μl
1	10	10	–
2	10	10	–
3	10	–	10

- EDTA blood is added undiluted to the sample (20 μl per reactions tube/antibody).
- For leukapheresis products, selection products and their intermediate products as well as bone marrow, 50 μl of cell suspension is pipetted per reaction container/antibody. This corresponds to a target leukocyte count of 1×10^6 cells. This must be calculated beforehand with the hematology analyzer.
  The necessary cell volume for dilution is always computed for 10 assays otherwise the volume which can be inferred is too small.
  Factor 10 which is used as a multiplier corresponds to the number of assays. Calculation of the necessary cell volume: i.e. 300×10^3 WBCs:

$$\frac{\text{Estimated leukocyte number} \times \text{number of assays}}{\text{Current leukocyte numbers}} = \frac{1 \times 10^6 \times 10}{300 \times 10^3/\mu\text{l}} = 33.3\ \mu\text{l sample} \quad (1)$$

  Calculation of the required amount of PBS (target volume = 50 μl /assay ⇒ 500 μl for 10 assays)

$$500\ \mu\text{l} - \mu\text{l sample} = \text{PBS}\ (\mu\text{l})\ \text{to be pipetted} = 500\ \mu\text{l} - 33.3\ \mu\text{l} = 466.7\ \mu\text{l PBS} \quad (2)$$

  If the volume after the decimal is ≥ 5, the value is rounded up; if it is <5 it rounded down! The amount of PBS to be pipetted will be compared.
- From the calculated and prepared sample dilutions, 50 μl is now added to each antibody.
- After addition of the cells to the antibody, tubes are vortexed and incubated for 15 min at RT in the dark (stained).

- 1,000 µl ammonium chloride lysis solution, warmed to RT, is added to every sample tube, vortexed and incubated for 15 min at RT in the dark.
- Centrifuge for 3 min at 500 × *g* at RT and decant (i.e. pour off the top of liquid so that there is almost no liquid left with the pellet in the tube) the rest of the liquid.
- The cell sediment is resuspended in 1 ml PBS, vortexed and centrifuged again (3 min at 500 × *g*, RT).
- The liquid is decanted, and the cells are resuspended in 500 µl PBS. If the analysis cannot be carried out immediately, the cells can be stored at 2–8 °C in the dark. The maximal storage time is 4 h.

Instrument Settings

- First turn on the FACSCalibur, then the computer. Then calibrate the machine (see manufacturer's instructions).
- Select 'APPLE' ? 'CD34 Measurement/Analysis' and confirm with 'OK'.
- A window with 6 dot plots will appear:

  (1) SSC-Height vs. CD45-FITC NO GATE
  (2) FSC-Height vs. SSC-Height R1
  (3) SSC-Height vs. CD34-PE* R1 + R2
  (4) CD45-FITC vs. CD34-PE* R1 + R2
  (5) FSC-Height vs. SSC-Height R1 + R2 + R3
  (6) FSC-Height vs. SSC-Height R1 + R2 + R3 + R4

  *Instead of CD34-PE, mouse IgG1-PE or CD3-PE can be displayed.
- The computer is connected to the FACSCalibur through 'Acquire' 'Connect to Cytometer'. A window opens 'Browser: CD34 Measurement/Analysis'. Open the window to its maximum size to see all of the information.
- In order to start measuring, activate the tab 'ACQUISITION':

  ☑ R Setup: with checkmark means that the result of the measurement will not be saved; the words adjacent to Directory and File are in italics.

  ☐ Setup: without checkmark means that the result of the measurement will be saved; the words adjacent to 'Directory' and 'File' are in italics.

  There must be no checkmark in the box when you start a measurement!

Marking the Sample

- The sample designation is entered under 'File', which can be a letter and a number (maximum 999!). The letter/number combination (see below) is sequential. Click in addition 'Change', under 'Custom Prefix' the letter and register the number under 'file count'; confirm with 'OK'.
- Assignment of letter/number combinations: after the 999th measurement, the equipment starts to count from 1 again, therefore letters must be assigned. The letters start from A and end at Z and then go over into double letters AA-ZZ. Example: 1–999, A1-Z999, AA1-ZZ999.
- A window displays the registered letter number combination.

Entering the Samples to Be Measured into the Computer

- The material is entered under 'SAMPLE ID', e.g. EDTA for peripheral blood from EDTA tubes, L5 for the 5th leukapheresis of a patient, L5 ref. tube for the thawed safety sample of the leukapheresis patient, BM for bone marrow.
- For allogeneic samples, enter the donor data under 'SAMPLE ID': Name, first name and ID.
- Under 'PATIENT ID' enter surname and first name of the patient.
- The current date is indicated automatically.
- Under 'Experiment Components' in the register map select 'Acquisition' panel.
- On the right, besides the current panel names, click and hold the symbol (triangle in a rectangle); select 'Load Tubes from Panel'; a further window will appear; select the appropriate new panel. The name is adopted and appears in the window 'Browser CD34 measurement/analysis'.
- The panel name appears with the tubes (antibody designation). An arrow on the left of the tube label indicates the current measurement; after each measurement, this arrow jumps to the next tube; in front of tubes that have already been measured, a green checkmark appears.
- To save all the information on the screen, other windows must be opened:
  (1) Open the window 'Counters' via 'Acquire' on the menu;
  (2) In order to download the saved instrument settings, proceed as follows (if this step is not carried out, erroneous settings from the previous measurement may be loaded or even the previous calibration values!):
  (3) Click on 'Cytometer' on the menu bar and select 'Instrument settings'. The settings of the equipment are stored there. A window will open; click on 'OPEN'; a new window will appear. In this window, select 'Desktop' → 'Facsstation G4' → 'Open'; then 'BD files' → 'Open' → 'Instrument Setting Files' → 'Open'; select 'CD34 settings' ('open'). In order to save the selected setting, click on 'Set' and then on 'Done', so the window will close again.
  (4) Next 'Cytometer' is selected from the menu and the appropriate window from the following is chosen: 'Detectors/Amps' and 'Compensation'.
- Examine the settings:
  (1) Select the field 'ACQUIRE' from the menu list then choose 'Acquisition' and 'Storage'.
  (2) At 'ACQUISITION GATE': 'ACCEPT, ALL events'.
  (3) At 'COLLECTION CRITERIA': acquisition will stop when 30,000 of all events are counted.
  (4) At 'STORAGE GATE': data file will contain all events.
  (5) At 'RESOLUTION': 1,024.

Data Acquisition/Measurement

Note: Holding arm must be positioned forward during the measurement.

- After being turned on, the machine is on stand by; switch to 'run' for measuring and switch the flow rate to 'high'.

- Check the settings using the control:
  - First measure the control sample (tube No. 3) with activated setup mode (checkmark is set). Vortex the sample and place it in the tube rack of the cytometer. The equipment settings as well as the results of the plots will be double-checked with this control sample. If the cells lie outside of the plots, the complete settings are verified and corrected. Regular training course are organized to teach the staff how to optimally correct the photomultiplier tension and compensation.
  - Once the verifications are completed, the samples are successively measured with the setup mode switched off (no checkmarks in the box).
- Measuring the samples:
  - Tube 1 is vortexed and placed on the tube rack of the cytometer. The measurement is started by clicking 'ACQUIRE' once. After the measurement of 30,000 events, the data are stored automatically into the previously selected file. If necessary, correct all gates, and afterwards print out the measuring protocol; for this, click 'File' from the menu and select 'Print'; a window appears and ' Print' must be clicked again.
  - After each printed file, the window 'Browser CD34 measurement/analysis' closes. For each new measurement, the window must again be opened through Windows (Menu) → 'Browser CD34 Measurement/Analysis'.
  - In the browser window, a green checkmark indicating that the measurement is finished appears below the selected panel.
  - Carry out the same procedure with tubes 2 and 3.
    Important: from tube 2, 'Print' must be clicked each time to obtain a printout of the measuring protocol.
  - For allogeneic products, additional measurements of tubes 4 and 5 are required.

    In principle, in case of double analyses, the gates set after the first measurement are not modified anymore. For this reason, the gates for CD34+ cells should be set rather 'generously'.

Evaluation of CD34+ Cells

After the acquisition the cells are evaluated. All gates must be adapted to each individual sample.

- The flow-cytometric data are represented as dot plots. For the evaluation of the data, four regions are set as shown in table 4.
- Description of the statistics table (table 5): following each measurement, the document is printed out and the relevant values (bold in the document) are manually circled and the document is signed by the operator with date and initials.

Criteria for a Valid Analysis

- The flow cytometer FACSCalibur needs to be successfully calibrated. If this is not the case, no further measurements must be taken. If the cytometer cannot be properly calibrated after repeated attempts, the person in charge is notified.

Table 4. List of regions

Dot plot no.	Number of region	Dot plot axis label	Enclosed cell population	Region linkage
1	R1	SSC-height vs. CD45-FITC	leukocytes (CD45+)	no gate
2	R2	FSC-height vs. SSC-height	cells without debris, without pairs	R1
3	R3	SSC-height vs. CD34-PE	CD34+ cells, with low SSC	R1 and R2
4		CD45-FITC vs. CD34-PE	control plot for setting R1	R1 and R2
5	R4	FSC-height vs. SSC-height	cells with low to middle FSC and low SSC	R1, R2, and R3
6		FSC-height vs. SSC-height	cells with middle FSC and low SSC ⇒ true CD34+ cells	R1, R2, R3, and R4

Table 5. Description of statistics table[a]

Gate	Events	% Gated	% Total	Linked region
G1	not applicable	not applicable	not applicable	R1
G2	not applicable	not applicable	not applicable	R2
G3	not applicable	not applicable	not applicable	R3
G4	not applicable	not applicable	not applicable	R4
G5	not applicable	not applicable	leukocytes	R1 and R2
G6	not applicable	not applicable	not applicable	R1, R2, and R3
G7	not applicable	CD34+ cells and control	not applicable	R1, R2, R3, and R4

[a] The relevant values are in italics. Values calculated from dot plot No. 3.

- The deviation when carrying out double assay of CD45 and CD34 (and CD3 in the case of allogenic products) is must not be greater than 20%.

  Calculation of the deviation according to the formula

$$\text{Deviation} = \frac{[(\text{1st value}) - (\text{2nd value})] \times 200}{[(\text{1st value}) - (\text{2nd value})]} \tag{3}$$

  If in a double assay both values lie under 0.5%, the deviation is allowed to be 20%. If the allowed tolerance (>20%) is exceeded, the staining must be repeated.

Documentation

Documentation on the Flow Cytometry Protocol (Print from the FACSCalibur)

- The mean values of both CD34 measurements are calculated and registered in the section 'Remarks'. In the case of leukapheresis products and bone marrow, the mean value of the CD45 measurements is calculated and registered as well. The deviation of both values and the control value should be registered.

- For further documentation, the control value is subtracted from the CD34 mean value and/or the CD3 mean value.
- Sign the printouts of EDTA blood with date and signature.
- Sign and date the printouts pertaining to leukapheresis products and bone marrow and file along with the production and test protocols and the patients' files after obtaining the signatures of the individuals in charge.

Documentation of the Data from the Calculation Protocol FACSCalibur (Valid only for Stem Cell Preparations, Not for EDTA Blood and the Corresponding Production and Test Protocols)

- Enter the following values (take the values from the printout of the hematology analyzer. Use WBCs for the calculation):
  - WBC $\times$ 10^8 absolute (calculated entire cell number/result from the hematology analyzer)
  - CD45% (mean of the double assay)
  - CD34% (CD34 mean – control).
- Values that are to be calculated:
  - WBC $\times$ 10^8 absolute

$$\text{WBC} \times 10^8 \text{ absolute} = \frac{\text{CD45 (mean)} \times \text{WBC} \times 10^8}{100} \tag{4}$$

  - The WBC value for bone marrow is *not* calculated anew for CD45
  - CD34 $\times$ 10^6 absolute

$$\text{CD34} \times 10^6 \text{ absolute} = \frac{\text{CD34\%} \times \text{WBC} \times 10^8}{100} \tag{5}$$

  - CD34 $\times$ 10^6/kg body weight = CD34 $\times$ 10^6 abs./kg KG.
- After the calculation, the values are recorded on the production and test protocols.

Documentation on the 'Reported Results of the CD34+ Monitoring from EDTA Blood' form

- Enter the following values:
  - CD34% (CD34 mean – control)
  - WBC and thrombocytes (unit: $\times$ 10^3/µl; the value is taken directly from the hematology analyzer printout).
- Values that are to be calculated:
  - CD34/µl

$$\text{CD34}/\mu\text{l} = \frac{\text{CD34\%} \times \text{leukocytes (WBC)} \times 10^3}{100} \tag{6}$$

Summary

Due to the ability to calculate CD34+ levels during quiescent hematopoiesis by flow cytometry, it is possible, in many hemato-oncological illnesses, to determine the latest possible mobilization success and the blood stem cell yield during leukapheresis. Flow cytometry assists in the reliable determination of cell levels with use of immunological markers. Due to standardized and validated protocols, the results of different hospitals and institutions can be compared. Flow cytometry has thus found acceptance in routine clinical methods for peripheral stem cell transplantation.

Acknowledgements

We would like to thank Bernhard Berkus for setting up the FACS plots from figure 2.

References

1 Weissman IL: Stem cells: units of development, units of regeneration, and units in evolution. Cell 2000;100:157–168.

2 To LB, Russell J, Moore S, Juttner CA: Residual leukaemia cannot be detected in very early remission peripheral blood stem cell collections in acute non-lymphoblastic leukaemia. Leuk Res 1987;11:327–329.

3 Siena S, Bregni M, Brando B, Belli N, Ravagnani F, Gandola L, Stern AC, Lansdorp PM, Bonadonna G, Gianni AM: Flow cytometry for clinical estimation of circulating hematopoetic progenitors for autologous transplantation in cancer patients. Blood 1991;77:400–409.

4 Haas R, Mohle R, Fruhauf S, Goldschmidt H, Witt B, Flentje M, Wannenmacher M, Hunstein W: Patient characteristics associated with successful mobilizing and autografting of peripheral blood progenitor cells in malignant lymphoma. Blood 1994;83:3787–3794.

5 Beyer J, Schwella N, Zingsem J, Strohscheer I, Schwaner I, Oettle H, Serke S, Huhn D, Stieger W: Hematopoietic rescue after high-dose chemotherapy using autologous peripheral-blood progenitor cells or bone marrow: a randomized comparison. J Clin Oncol 1995;13:1328–1335.

6 Peters WP, Ross M, Vredenburgh JJ, Meisenberg B, Marks LB, Winer E, Kurtzberg J, Bast RC Jr, Jones R, Shpall E, Wu K, Rosner G, Gilbert C, Mathias B, Coniglio D, Petros W, Henderson IC, Norton L, Weiss RB, Budman D, Hurd D: High-dose chemotherapy and autologous bone marrow support as consolidation after standard-dose adjuvant therapy for high-risk primary breast cancer. J Clin Oncol 1993;11:1132–1143.

7 Serke S, Sauberlich S, Abe Y, Huhn D: Analysis of CD34-positive hemopoietic progenitor cells from normal human adult peripheral blood: flow-cytometrical studies and in-vitro colony (CFU-GM, BFU-E) assays. Ann Hematol 1991;62:45–53.

8 Fruehauf S, Haas R, Zeller WJ, Hunstein W: CD34 selection for purging in multiple myeloma and analysis of CD34+ B cell precursors. Stem Cells 1994;12:95–102.

9 Ho AD, Young D, Maruyama M, Corringham RE, Mason JR, Thompson P, Grenier K, Law P, Terstappen LW, Lane T: Pluripotent and lineage-committed CD34+ subsets in leukapheresis products mobilized by G-CSF, GM-CSF vs. a combination of both. Exp Hematol 1996;24:1460–1468.

10 Wunder E, Sovalat H, Fritsch G, Silvestri F, Henon P, Serke S: Report on the European Workshop on Peripheral Blood Stem Cell Determination and Standardization – Mulhouse, France, February 6–8 and 14–15, 1992. J Hematother 1992;1:131–142.

11 Keeney M, Sutherland DR: Stem cell enumeration by flow cytometry: current concepts and recent developments in CD34+ cell enumeration. Cytotherapy 2000;2:395–402.

12 Fruehauf S, Haas R, Conradt C, Murea S, Witt B, Mohle R, Hunstein W: Peripheral blood progenitor cell (PBPC) counts during steady-state hematopoiesis allow to estimate the yield of mobilized PBPC after filgrastim (R-metHuG-CSF)-supported cytotoxic chemotherapy. Blood 1995;85:2619–2626.
13 Koerbling M, Anderlini P: Peripheral blood stem cell versus bone marrow allotransplantation: does the source of hematopoietic stem cells matter? Blood 2001;98:2900–2908.
14 Goodell MA, Rosenzweig M, Kim H, Marks DF, DeMaria M, Paradis G, Grupp SA, Sieff CA, Mulligan RC, Johnson RP: Dye efflux studies suggest that hematopoietic stem cells expressing low or undetectable levels of CD34 antigen exist in multiple species. Nat Med 1997;3:1337–1345.
15 Craig W, Kay R, Cutler RL, Lansdorp PM: Expression of Thy-1 on human hematopoietic progenitor cells. J Exp Med 1993;177:1331–1342.
16 Haas R, Mohle R, Pforsich M, Fruehauf S, Witt B, Goldschmidt H, Hunstein W: Blood-derived autografts collected during granulocyte colony-stimulating factor-enhanced recovery are enriched with early Thy-1+ hematopoietic progenitor cells. Blood 1995;85:1936–1943.
17 Steidl U, Kronenwett R, Martin S, Haas R: Molecular biology of hematopoietic stem cells. Vitam Horm 2003;66:1–28.
18 Duhrsen U, Villeval JL, Boyd J, Kannourakis G, Morstyn G, Metcalf D: Effects of recombinant human granulocyte colony-stimulating factor on hematopoietic progenitor cells in cancer patients. Blood 1988;72:2074–2081.
19 Seggewiss R, Buss EC, Herrmann D, Goldschmidt H, Ho AD, Fruehauf S: Kinetics of peripheral blood stem cell mobilization following G-CSF-supported chemotherapy. Stem Cells 2003;21:568–574.
20 Prince HM, Toner GC, Seymour JF, Blakey D, Gates P, Eerhard S, Chapple P, Wall D, Quinn M, Juneja S, Wolf M, Januszewicz EH, Richardson G, Scarlett J, Briggs P, Brettell M, Rischin D: Docetaxel effectively mobilizes peripheral blood CD34+ cells. Bone Marrow Transplant 2000;26:483–487.
21 Fruehauf S, Klaus J, Huesing J, Veldwijk MR, Buss EC, Topaly J, Seeger T, Zeller LW, Moehler T, Ho AD, Goldschmidt H: Efficient mobilization of peripheral blood stem cells following CAD chemotherapy and a single dose of pegylated G-CSF in patients with multiple myeloma. Bone Marrow Transplant 2007;39:743–50.
22 Stiff P, Gingrich R, Luger S, Wyres MR, Brown RA, LeMaistre CF, Perry J, Schenkein DP, List A, Mason JR, Bensinger W, Wheeler C, Freter C, Parker WRL, Emmanouilides C: A randomized phase 2 study of PBPC mobilization by stem cell factor and filgrastim in heavily pretreated patients with Hodgkin's disease or non-Hodgkin's lymphoma, Bone Marrow Transplant 2000;26:471–481.
23 Orazi A, Cattoretti G, Schiro R, Siena S, Bregni M, Di Nicola M, Gianni AM: Recombinant human interleukin-3 and recombinant human granulocyte-macrophage colony-stimulating factor administered in vivo after high-dose cyclophosphamide cancer chemotherapy: effect on hematopoiesis and microenvironment in human bone marrow. Blood 1992;79:2610–2619.
24 Fruehauf S, Seeger T: New strategies for mobilization of hematopoietic stem cells. Future Oncol 2005;1:375–383.
25 Fruehauf S, Seeger T, Maier P, Li L, Weinhardt S, Laufs S, Wagner W, Eckstein V, Bridger G, Calandra G, Wenz F, Zeller WJ, Goldschmidt H, Ho AD: The CXCR4 antagonist AMD3100 releases a subset of G-CSF-primed peripheral blood progenitor cells with specific gene expression characteristics. Exp Hematol 2006;34:1052–1059.
26 Gratama JW, Kraan J, Keeney M, Sutherland DR, Granger V, Barnett D: Validation of the single-platform ISHAGE method for $CD34^+$ hematopoietic stem and progenitor cell enumeration in an international multicenter study. Cytotherapy 2003;5:55–65.
27 Fruehauf S, Schmitt K, Veldwijk MR, Topaly J, Benner A, Zeller WJ, Ho AD, Haas R: Peripheral blood progenitor cell (PBPC) counts during steady-state haemopoiesis enable the estimation of the yield of mobilized PBPC after granulocyte colony-stimulating factor supported cytotoxic chemotherapy: an update on 100 patients. Br J Haematol 1999;105:786–794.
28 Rümke CL: The statistically expected variability in differential leukocyte counting; in Koepke J (ed): Differential Leukocyte Counting. Skokie, CAP Today 1978, p 39.

Sack U, Tárnok A, Rothe G (eds): Cellular Diagnostics. Basics, Methods and Clinical Applications of Flow Cytometry.
Basel, Karger, 2009, pp 305–316

Endothelial Progenitor Cells

Johannes C. Fischer

Institute for Transplantation Diagnostics and Cellular Therapeutics, Heinrich Heine University Düsseldorf, Germany

Introduction

Cardiovascular disease remains a primary cause of morbidity and mortality throughout industrialized countries [1, 2]. Recent evidence suggests that circulating endothelial progenitor cells (EPCs) are mobilized from bone marrow and are crucial for maintaining vascular integrity and normal endothelial function throughout adult life [3, 4].

Therefore EPCs have been extensively studied as biomarkers to assess the severity of cardiovascular disease and as a source for cell-based therapy in several human cardiovascular disorders.

The concept of bone-marrow-derived EPCs is based on the work of Ashara et al. [5] in 1997. They were the first to describe putative EPCs in peripheral blood. Their work was based on the evidence that peripheral blood can endothelialize artificial graft material with blood-borne [6–8] or bone-marrow-borne cells [8]. Prior to these findings, neovascular formation in adults was considered to result exclusively from proliferation, migration, and remodeling of preexisting endothelial cells, a process referred to as angiogenesis. Angiogenesis was believed to be responsible for reducing regional hypoxia [9]. In contrast, vasculogenesis, a process defined as the formation of new blood vessels from EPCs, was thought to be restricted to embryogenesis [10].

Based on the concept that stem cells serve as an important defense against aging as they replace lost organ cells [11], the role of bone-marrow-derived stem cells in endothelial repair has been the focus of research over the past decade [12–14]. In most studies to date, mainly two methods were used for the detection of EPCs. EPCs were obtained and quantified either by in vitro cell culture or they were identified and enumerated by flow cytometry.

Flow-Cytometric Definition of Endothelial Progenitor Cells

While no single unique cell surface molecule indicates circulating EPCs, CD34 expression serves as a fundamental marker in humans [5, 15, 16]. In addition, several other signs of endothelial progeny have been proposed and among them the expression of certain surface molecules that allow characterization of the cells by flow cytometry.

A widely used definition characterized CD34+ AC133+ vascular endothelial growth factor receptor (VEGFR)-2+ cells as potential precursors for endothelial cell colony-forming units (CFU-EC) [4, 17, 18].

Besides this definition, other marker combinations have been used to identify those putative EPCs in blood such as total CD34+, CD34+ CD133+, CD133+, CD133+ VEGFR-2+, CD133+ VE-cadherin+, CD133+ CXCR4+; others included CD45 and designated 'putative precursors' as CD45+ CD133+, CD45+ CD13+ VEGFR-2+, CD45+ CD34+ CD146+, or CD45+ CD34+ CD133+ VEGFR-2+ [14, 19–23].

As a result of these heterogeneous phenotypes, flow-cytometric data may be difficult to interpret; they may also not be comparable between laboratories, which may result in discrepancies regarding their significance for cardiovascular disease [24].

Most of the suggested precursor phenotypes cited above have not been directly tested in vivo or in vitro for true endothelial capacity, thus leading to measurement artifacts [25].

Definition of Endothelial Progenitor Cells in Cultures

Mainly three different methods for the isolation of EPCs using in vitro cultures have been described.

The original method used by Asahara et al. [5] can now be performed with a commercially available kit (Endocult, StemCell Technologies, Vancouver, BC, Canada).

In brief, peripheral blood mononuclear cells (MNCs) are plated on fibronectin-coated dishes. After 48 h, the nonadherent cells are removed and replated on fibronectin-coated dishes. From these cells, colonies emerge within 5–9 days. Typically, they show round cells centrally, with spindle-shaped cells sprouting at the periphery. Colonies of this type are widely referred to as CFU-ECs) [4].

In another widely employed and methodologically similar approach, whole unfractionated MNCs are cultured in supplemented endothelial growth medium for 4 days, whereupon the nonadherent fraction of cells is removed, resulting in a target adherent-cell population referred to as circulating angiogenic cells (CACs) [26–28].

In both methods, the resulting cultured cells display features of an endothelial phenotype. They bind the endothelial-specific lectin Ulex Europeus Agglutinin-1 (UEA-1), show uptake of acetylated low-density liproprotein (acLDL), express von

Willebrand factor (vWF), platelet-endothelial cell adhesion molecule (PECAM-1 or CD31), VEGFR-2, vascular endothelial cadherin (VE-cadherin or CD144), Tie-2/TEK (angiopoietin-1 receptor precursor or tunica intima endothelial cell kinase [29–31]. These adherent cultured cells promote neovascularization in animal models of critical limb ischemia or myocardial infarction [26, 32]. CACs do not display the colony morphology of CFU-ECs though the sprouting spindle-shaped cells of CFU-ECs are similar in morphology to CACs in culture [26, 33]. Furthermore, CACs can be collected from culture in far greater numbers than CFU-ECs, comprising approximately 2% of the whole MNC population [28].

The third method produces 'endothelial colony-forming cells' (ECFCs). ECFCs are obtained using methods similar to those originally reported for endothelial outgrowth from peripheral blood [27]. MNCs are collected and plated onto collagen-I-coated plates in endothelial-specific growth medium [34, 35]. Nonadherent cells are discarded after 24–48 h. ECFC colonies emerge 10–21 days after plating (5–7 days for cord blood MNCs). These cells display a cobblestone appearance typical of endothelial cells. Thus far, ECFCs have proven to be phenotypically indistinguishable from cultured ECs, and have de novo vessel-forming ability [34–37].

Because ECFCs emerge much later in culture as compared to both CFU-ECs and CACs, ECFCs have been called 'late outgrowth' EPCs, while CFU-ECs and CACs have been called 'early outgrowth' EPCs [38, 39]. Furthermore, ECFCs are the only type of EPCs that display a clear hierarchy of proliferative potentials and show proliferation both on a clonal level. Thus ECFCs contain EPCs whereas CAC and CFU-EC do not.

Dissecting Flow-Cytometrically Defined Populations Containing Cultured EPCs

Upon writing this chapter only two studies analyzed the progeny of CD34+ CD133+ VEGFR-2+ cells as well as CD34+ VEGFR-2+ cells in hematopoietic and ECFC assays. These studies revealed that after flow-cytometric sorting these cells yielded hematopoietic cell colonies at nearly the same frequency as primitive hematopoietic stem cells, but did not form endothelial cell colonies [37, 40]. These results are underscored by the strong CD45 expression of those cells (fig. 1, 2). Thus, there are no convincing data demonstrating that CD34+ CD133+ VEGFR-2+ or CD34+ VEGFR-2+ cells differentiate into endothelial cells in vitro [40, 41] or form new blood vessels [42]. Most likely, these cells are hematopoietic cells which promote new blood vessel growth by paracrine effects [33, 43, 44]. In contrast, ECFCs could be grown at high frequencies from the sorted CD34+ CD45– population [40, 41].

Accordingly, embryonic CD34 + CD45– hemato-endothelial precursors differentiating along the hematopoietic lineage acquire the hematopoietic marker CD45 or CD43 and lose their endothelial capacity [45, 46].

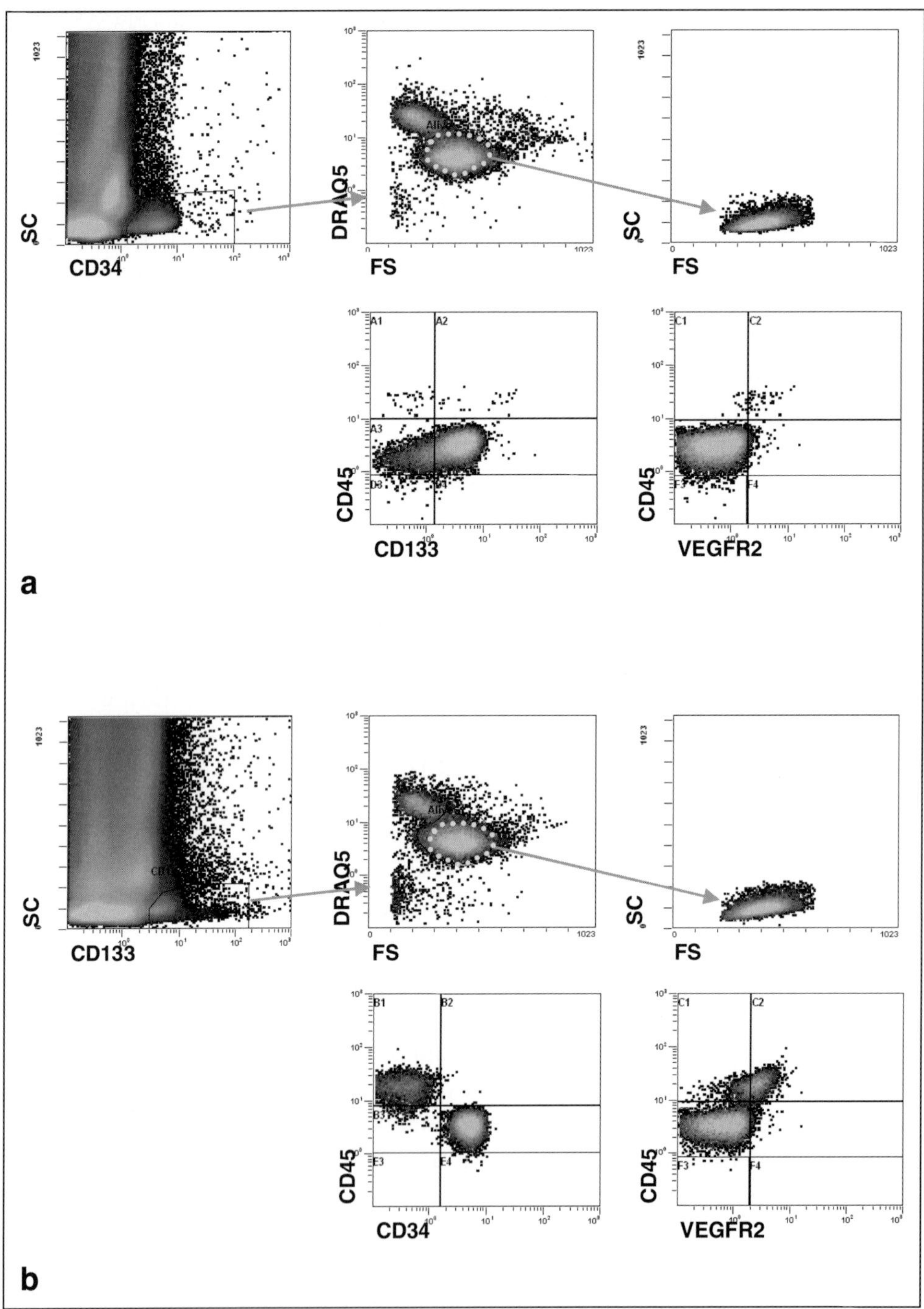

SC
CD34
DRAQ5
FS
SC
FS
A1
A2
A3
CD45
CD133
C1
C2
CD45
VEGFR2
a
SC
CD133
DRAQ5
FS
SC
FS
B1
B2
B3
E3
E4
CD45
CD34
C1
C2
F3
F4
CD45
VEGFR2
b

Nevertheless, additional studies are needed to address the in vivo endothelial potential of both CD34+ CD45− and CD34+ CD45+ cell subsets in relevant vasculogenesis models and to dissect the ontogeny of the CD34+ CD45− cell(s) that gives rise to ECFCs in vitro, i.e., hemangioblast, endothelial committed precursor and/or a high proliferative vessel wall endothelial cells.

In summary, recent experiments have ascertained that these CD133+ CD34+ VEGFR-2+ cells do not represent endothelial progenitors, but are hematopoietic or monocyte/macrophage progenitors showing cell properties in culture that are shared with endothelial cells [37, 41]. The fraction of CD34+ cells most likely containing true EPCs are CD45− [40, 47] and seems not to reside within the CD133+ fraction [40, 47].

Clinical Relevance

Despite these uncertainties [48] or controversies [41, 49] concerning the definition of EPCs, research suggests a role for these bone-marrow-derived CD34+ cells in atherosclerotic disease. Peripheral blood cell counts of progenitor cells were inversely associated with scores in the Framingham risk algorithm and positively correlated with endothelial function [4]. Moreover, regular physical exercise increased the number of circulating progenitor cells [50]. Progenitor cell counts also increased after medication with statins [51], and after brief ischemic events [52]. Finally, higher progenitor cell counts predicted better survival in 519 patients with coronary artery disease as confirmed by angiography [18] although, using a different progenitor cell definition, an increase in concentration of angiogenic cells in relation to the severity of coronary artery disease could be seen [53].

◀

Fig. 1. Measurement of endothelial progenitors from peripheral G-CSF mobilized blood (apheresis product). Acquisition and analysis was performed on a dual-laser five-color FC 500 flow cytometer (Beckman Coulter, Krefeld, Germany) with an adapted optical filter setting. Filter setting was optimized for simultaneous measurement of the following flourochromes or conjugates: CD45 FITC, CD34 ECD (Beckman Coulter), CD133 PE (Miltenyi Biotec, Bergisch Gladbach, Germany) VEGFR-2 APC (R&D Systems, Wiesbaden, Germany) and DRAQ5 (Alexis/Axxora, Lörrach, Germany). In Brief 5,000,000 nucleated cells were stained and analysed after lysing erythrocytes (Versa Lyse, Beckman Coulter). CD34+ cells (**a**) or CD133+ cells (**b**) are displayed according to their DRAQ5 staining. DRAQ5 binds covalently to DNA. When used in concentrations of 1 mmol/l it stains all nucleated cells. Cells with membrane defects (dead cells stain brighter (see region, alive). Alive CD34+ (**a**) or CD133+ (**b**) cells are displayed for CD45, VEGFR-2 and CD133/ CD34 expression, respectively. According to Timmermans et al. [40] and Case et al. [41], cells containing clonable CFU-ECs could be found in the lower CD45− CD34+ quadrant (here 12 events out of 5,000,000 recorded). Of note is that all CD133+ VEGFR-2+ cells express CD45.

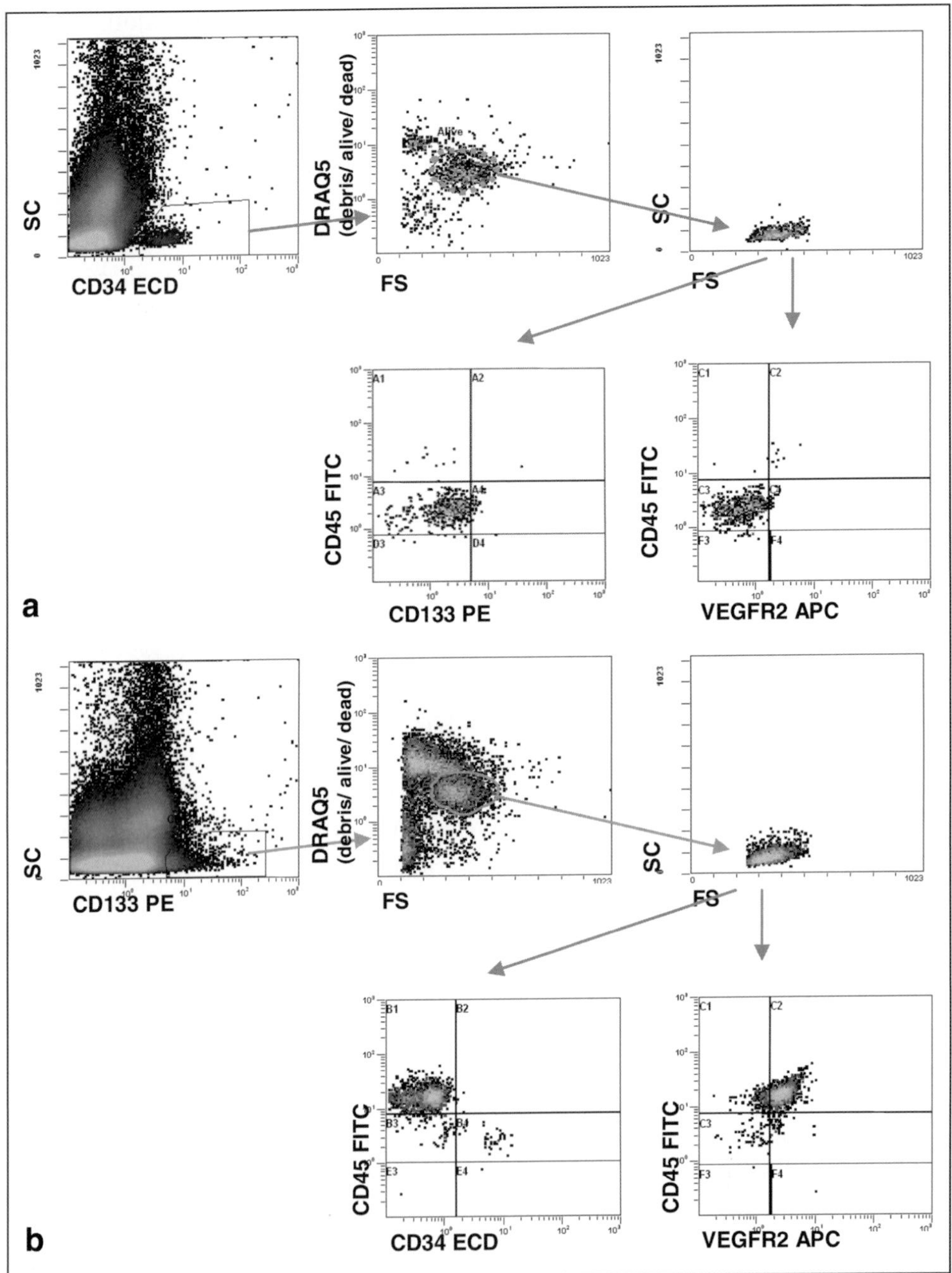

Fig. 2. Measurement of EPCs from peripheral nonmobilized blood (apheresis product). Staining and measurement procedures were done as described for figure 1. Alive CD34+ (**a**) or CD133+ cells (**b**) are displayed for CD45, VEGFR-2 and CD133/ CD34 expression, respectively. Only a few CD45− CD34+ cells could be found (here 6 events out of 2,500,000 MNCs recorded). Of note again is that all CD133+ VEGFR-2+ cells express CD45, but do not express CD34. In addition there is only a small amount of cells double positive for CD34 and CD133 (negative for VEGFR-2).

Practical Considerations

There are several different definitions for EPCs by flow cytometry, but one is always dealing with rare-event analysis!

Rare-event flow cytometry is not regular flow cytometry. It is impossible to perform on the same machine in between regular flow-cytometric acquisitions. Rare-event analysis requires dedicated machines and procedures [54, 55], even in EPC diagnostics [48]. Most recent data suggest that the population containing 'true' EPCs consists of CD45− nucleated CD34+ cells [40, 47, 56] resulting in concentrations of 0.01% of all MNCs in peripheral blood.

Thus, the flow-cytometric measurement of those cells is rare-event analysis.

As a practical definition, rare events are all events of interest that are as frequent or less frequent as spurious events. Spurious events are defined as all infrequent artificial flow events that have nothing to do with unlabeled or specifically (fluorescent-) labeled single viable cells. Expected measurement range should enable detection of frequencies from <0.1% to as low as practically feasible.

But what is feasible? Assuming a normal WBC count of 5,000 μl and 100% recovery

- 1%: 5,000 found within 100 μl of blood
- 0.1%: 500 found within 100 μl of blood
- 0.01%: 50 found within 100 μl of blood
- 0,001%: 5 found within 100 μl of blood.

Thus a data file of 5 million events is required; acquisition will take 34 min for 2,500 events/s.

For rare-event analysis as described [57] elsewhere, accuracy and precision of the flow cytometer should be maximized by:

- signal to noise improvement,
- increasing the frequency of the rare event of interest (signal),
- pre-enrichment techniques,
- decreasing the frequencies of spurious events (noise)
  - by preventing their occurrence, and
  - by excluding them during analysis.

Rare-event analysis by flow cytometry is followed by two main problems:

- A rare-event problem requires exclusion of the spurious event.
- Spurious-event exclusion requires multiparametric polychromatic flow cytometry: multicolor assays are needed.

After all spurious events have been minimized or excluded (table 1), an appropriate negative control should be used with an

- isotype-matched irrelevant antibody with the same concentration and same fluorochrome-protein ratio,
- acquisition of the same amount of total events,
- addition of a fluorochrome minus one (FMO) control.

Table 1. Causes of spurious events and strategies to avoid them

Causes of spurious events	Solutions to spurious events
Debris	clean the cytometer, use filtered solutions
Nonspecific labeled cell with nonspecific binding of fluorochromes or antibody	use blocking serum, exclude dead cells
Highly autofluorescent cells	include at least one negative marking or an unstained fluorescence parameter) for the rare event, label all potentially interfering cells using a fluorescence 'dump' channel
Be aware of nonnucleated cells	exclude by a nuclear stain
Aggregates	exclude by pulse shape analysis (area versus peak or width)
Dead cells	exclude by viability stain (membrane integrity stain) when possible
Random (uncorrelated) noise and spikes from PMTs and electronics	exclude by increasing the number of parameters that positively define the rare event define rare event with the brightest and tightest positive marking
Transient flow disturbances (partial clogging; clumps ...)	excluded by time gating

After checking what was found within the rare event gate, the frequency of spurious events is the actual detection limit of rare-event analysis, e.g. with a frequency of spurious events of 0.02% (1 in 5,000), one cannot try to detect the real rare event of interest at a frequency of 0.01% (1 in 10,000).

The resulting theoretical precision can be calculated by the Poisson approximation, which can be considered as a Bernoulli process, mathematically characterized by a binomial distribution [58]. An approximation for the mathematical precision of a measured event count (95% confidence interval for the 'true' event count) is given below:

- $5{,}000 \pm \text{sqrt}(5{,}000) = 5{,}000 \pm 70.7$ (1%)
- $500 \pm \text{sqrt}(500) = 500 \pm 22.4$ (4%)
- $50 \pm \text{sqrt}(50) = 50 \pm 7.1$ (14%)
- $5 \pm \text{sqrt}(5) = 5 \pm 2.2$ (45%).

One should thus realize that results based on 4 counted events could either be 2 or 6 true events.

Summary

Nowadays, the model of adult circulating remodeling appears more complex: besides angiogenesis and arteriogenesis, the recruitment not only of endothelial progenitor cells (EPCs) but also of a multiplicity of other cells is involved [44, 59–61]. Proliferative endothelium may line the vessel lumen while other cells support vessel formation through secretion of cytokines and modification of the extracellular space [36, 62, 63]. Despite their fundamentally different origin, all the cells involved in postnatal vasculogenesis have been lumped into the single term 'EPC'. Therefore, most relationships found between EPCs and cardiovascular disease may have been focused on cells that do not give rise to the vessel by themselves, but which recruit or facilitate angiogenic cells [33].

In summary, there is broad evidence that progenitor counts obtained by flow cytometry are inversely correlated with various cardiovascular risk factors although the biologic mechanism for this observation remains unclear [56], especially considering the possible limitations as concerns the precise functional phenotype of ECPs [40, 44, 48]. Thus flow-cytometric enumeration of EPCs is far from being a standardized procedure and many efforts are still needed to develop easy-to-use appropriate assays.

References

1 Anderson RN, Smith BL: Deaths: leading causes for 2002. Natl Vital Stat Rep 2005;53:1–89.

2 Hoyert DL, Heron MP, Murphy SL, Kung HC: Deaths: final data for 2003. Natl Vital Stat Rep 2006;54:1–120.

3 Vasa M, Fichtlscherer S, Adler K, Aicher A, Martin H, Zeiher AM, Dimmeler S: Increase in circulating endothelial progenitor cells by statin therapy in patients with stable coronary artery disease. Circulation 2001;103:2885–2890.

4 Hill JM, Zalos G, Halcox JP, Schenke WH, Waclawiw MA, Quyyumi AA, Finkel T: Circulating endothelial progenitor cells, vascular function, and cardiovascular risk. N Engl J Med 2003;348:593–600.

5 Asahara T, Murohara T, Sullivan A, Silver M, van der Zee R, Li T, Witzenbichler B, Schatteman G, Isner JM: Isolation of putative progenitor endothelial cells for angiogenesis. Science 1997;275:964–967.

6 Shi Q, Wu MH, Hayashida N, Wechezak AR, Clowes AW, Sauvage LR: Proof of fallout endothelialization of impervious Dacron grafts in the aorta and inferior vena cava of the dog. J Vasc Surg 1994;20:546–556; discussion 556–547.

7 Wu MH, Shi Q, Wechezak AR, Clowes AW, Gordon IL, Sauvage LR: Definitive proof of endothelialization of a Dacron arterial prosthesis in a human being. J Vasc Surg 1995;21:862–867.

8 Noishiki Y, Tomizawa Y, Yamane Y, Matsumoto A: Autocrine angiogenic vascular prosthesis with bone marrow transplantation. Nat Med 1996;2:90–93.

9 Heil M, Eitenmuller I, Schmitz-Rixen T, Schaper W: Arteriogenesis versus angiogenesis: similarities and differences. J Cell Mol Med 2006;10:45–55.

10 Risau W: Mechanisms of angiogenesis. Nature 1997;386:671–674.

11 Beausejour C: Bone marrow-derived cells: the influence of aging and cellular senescence. Handb Exp Pharmacol 2007:67–88.

12 Urbich C, Dimmeler S: Endothelial progenitor cells: characterization and role in vascular biology. Circ Res 2004;95:343–353.

13 Dignat-George F, Sampol J, Lip G, Blann AD: Circulating endothelial cells: realities and promises in vascular disorders. Pathophysiol Haemost Thromb 2003;33:495–499.

14 Blann AD, Woywodt A, Bertolini F, Bull TM, Buyon JP, Clancy RM, Haubitz M, Hebbel RP, Lip GY, Mancuso P, Sampol J, Solovey A, Dignat-George F: Circulating endothelial cells. Biomarkers of vascular disease. Thromb Haemost 2005;93:228–235.

15 Peichev M, Naiyer AJ, Pereira D, Zhu Z, Lane WJ, Williams M, Oz MC, Hicklin DJ, Witte L, Moore MA, Rafii S: Expression of VEGFR-2 and AC133 by circulating human CD34(+) cells identifies a population of functional endothelial precursors. Blood 2000;95:952–958.

16 Reyes M, Dudek A, Jahagirdar B, Koodie L, Marker PH, Verfaillie CM: Origin of endothelial progenitors in human postnatal bone marrow. J Clin Invest 2002;109:337–346.

17 Urbich C, Dimmeler S: Endothelial progenitor cells functional characterization. Trends Cardiovasc Med 2004;14:318–322.

18 Werner N, Kosiol S, Schiegl T, Ahlers P, Walenta K, Link A, Bohm M, Nickenig G: Circulating endothelial progenitor cells and cardiovascular outcomes. N Engl J Med 2005;353:999–1007.

19 Murga M, Yao L, Tosato G: Derivation of endothelial cells from CD34– umbilical cord blood. Stem Cells 2004;22:385–395.

20 Shaw JP, Basch R, Shamamian P: Hematopoietic stem cells and endothelial cell precursors express Tie-2, CD31 and CD45. Blood Cells Mol Dis 2004;32:168–175.

21 Romagnani P, Annunziato F, Liotta F, Lazzeri E, Mazzinghi B, Frosali F, Cosmi L, Maggi L, Lasagni L, Scheffold A, Kruger M, Dimmeler S, Marra F, Gensini G, Maggi E, Romagnani S: CD14+ CD34low cells with stem cell phenotypic and functional features are the major source of circulating endothelial progenitors. Circ Res 2005;97:314–322.

22 Fadini GP, Coracina A, Baesso I, Agostini C, Tiengo A, Avogaro A, de Kreutzenberg SV: Peripheral blood CD34+KDR+ endothelial progenitor cells are determinants of subclinical atherosclerosis in a middle-aged general population. Stroke 2006;37:2277–2282.

23 Shaffer RG, Greene S, Arshi A, Supple G, Bantly A, Moore JS, Mohler ER, 3rd: Flow cytometric measurement of circulating endothelial cells: the effect of age and peripheral arterial disease on baseline levels of mature and progenitor populations. Cytometry B Clin Cytom 2006;70:56–62.

24 Leor J, Marber M: Endothelial progenitors: a new Tower of Babel? J Am Coll Cardiol 2006;48:1588–1590.

25 Strijbos MH, Kraan J, den Bakker MA, Lambrecht BN, Sleijfer S, Gratama JW: Cells meeting our immunophenotypic criteria of endothelial cells are large platelets. Cytometry B Clin Cytom 2007;72:86–93.

26 Kalka C, Masuda H, Takahashi T, Kalka-Moll WM, Silver M, Kearney M, Li T, Isner JM, Asahara T: Transplantation of ex vivo expanded endothelial progenitor cells for therapeutic neovascularization. Proc Natl Acad Sci U S A 2000;97:3422–3427.

27 Lin Y, Weisdorf DJ, Solovey A, Hebbel RP: Origins of circulating endothelial cells and endothelial outgrowth from blood. J Clin Invest 2000;105:71–77.

28 Dimmeler S, Aicher A, Vasa M, Mildner-Rihm C, Adler K, Tiemann M, Rutten H, Fichtlscherer S, Martin H, Zeiher AM: HMG-CoA reductase inhibitors (statins) increase endothelial progenitor cells via the PI 3-kinase/Akt pathway. J Clin Invest 2001;108:391–397.

29 Asahara T, Masuda H, Takahashi T, Kalka C, Pastore C, Silver M, Kearne M, Magner M, Isner JM: Bone marrow origin of endothelial progenitor cells responsible for postnatal vasculogenesis in physiological and pathological neovascularization. Circ Res 1999;85:221–228.

30 Dimmeler S, Zeiher AM: Endothelial cell apoptosis in angiogenesis and vessel regression. Circ Res 2000;87:434–439.

31 Gehling UM, Ergun S, Schumacher U, Wagener C, Pantel K, Otte M, Schuch G, Schafhausen P, Mende T, Kilic N, Kluge K, Schafer B, Hossfeld DK, Fiedler W: In vitro differentiation of endothelial cells from AC133-positive progenitor cells. Blood 2000;95:3106–3112.

32 Cho HJ, Kim HS, Lee MM, Kim DH, Yang HJ, Hur J, Hwang KK, Oh S, Choi YJ, Chae IH, Oh BH, Choi YS, Walsh K, Park YB: Mobilized endothelial progenitor cells by granulocyte-macrophage colony-stimulating factor accelerate reendothelialization and reduce vascular inflammation after intravascular radiation. Circulation 2003;108:2918–2925.

33 Rehman J, Li J, Orschell CM, March KL: Peripheral blood 'endothelial progenitor cells' are derived from monocyte/macrophages and secrete angiogenic growth factors. Circulation 2003;107:1164–1169.

34 Ingram DA, Mead LE, Tanaka H, Meade V, Fenoglio A, Mortell K, Pollok K, Ferkowicz MJ, Gilley D, Yoder MC: Identification of a novel hierarchy of endothelial progenitor cells using human peripheral and umbilical cord blood. Blood 2004;104:2752–2760.

35 Yoder MC, Ingram DA: Isolation, expansion and use of clonogenic endothelial progenitor cells. Indiana University Research and Technology Corporation, Indianapolis, 2006.

36 Ingram DA, Mead LE, Moore DB, Woodard W, Fenoglio A, Yoder MC: Vessel wall-derived endothelial cells rapidly proliferate because they contain a complete hierarchy of endothelial progenitor cells. Blood 2005;105:2783–2786.

37 Yoder MC, Mead LE, Prater D, Krier TR, Mroueh KN, Li F, Krasich R, Temm CJ, Prchal JT, Ingram DA: Redefining endothelial progenitor cells via clonal analysis and hematopoietic stem/progenitor cell principals. Blood 2007;109:1801–1809.

38 Gulati R, Jevremovic D, Peterson TE, Chatterjee S, Shah V, Vile RG, Simari RD: Diverse origin and function of cells with endothelial phenotype obtained from adult human blood. Circ Res 2003;93:1023–1025.

39 Hur J, Yoon CH, Kim HS, Choi JH, Kang HJ, Hwang KK, Oh BH, Lee MM, Park YB: Characterization of two types of endothelial progenitor cells and their different contributions to neovasculogenesis. Arterioscler Thromb Vasc Biol 2004;24:288–293.

40 Timmermans F, Van Hauwermeiren F, De Smedt M, Raedt R, Plasschaert F, De Buyzere ML, Gillebert TC, Plum J, Vandekerckhove B: Endothelial outgrowth cells are not derived from CD133+ cells or CD45+ hematopoietic precursors. Arterioscler Thromb Vasc Biol 2007;27:1572–1579.

41 Case J, Haneline LS, Yoder MC, Ingram DA: Reply to Fadini et al: Critical assessment of putative endothelial progenitor phenotypes. Exp Hematol 2007;35:1481–1482.

42 Rohde E, Malischnik C, Thaler D, Maierhofer T, Linkesch W, Lanzer G, Guelly C, Strunk D: Blood monocytes mimic endothelial progenitor cells. Stem Cells 2006;24:357–367.

43 Massberg S, Schaerli P, Knezevic-Maramica I, Kollnberger M, Tubo N, Moseman EA, Huff IV, Junt T, Wagers AJ, Mazo IB, von Andrian UH: Immunosurveillance by hematopoietic progenitor cells trafficking through blood, lymph, and peripheral tissues. Cell 2007;131:994–1008.

44 Rohde E, Bartmann C, Schallmoser K, Reinisch A, Lanzer G, Linkesch W, Guelly C, Strunk D: Immune cells mimic the morphology of endothelial progenitor colonies in vitro. Stem Cells 2007;25:1746–1752.

45 Vodyanik MA, Thomson JA, Slukvin II: Leukosialin (CD43) defines hematopoietic progenitors in human embryonic stem cell differentiation cultures. Blood 2006;108:2095–2105.

46 Vodyanik MA, Slukvin II: Hematoendothelial differentiation of human embryonic stem cells. Curr Protoc Cell Biol 2007;chapter 23:Unit 23 26.

47 Case J, Mead LE, Bessler WK, Prater D, White HA, Saadatzadeh MR, Bhavsar JR, Yoder MC, Haneline LS, Ingram DA: Human CD34+ AC133+ VEGFR-2+ cells are not endothelial progenitor cells but distinct, primitive hematopoietic progenitors. Exp Hematol 2007;35:1109–1118.

48 Khan SS, Solomon MA, McCoy JP Jr: Detection of circulating endothelial cells and endothelial progenitor cells by flow cytometry. Cytometry B Clin Cytom 2005;64:1–8.

49 Fadini GP, Avogaro A, Agostini C: Critical assessment of putative endothelial progenitor phenotypes. Exp Hematol 2007;35:1479–1480; author reply 1481–1472.

50 Laufs U, Werner N, Link A, Endres M, Wassmann S, Jurgens K, Miche E, Bohm M, Nickenig G: Physical training increases endothelial progenitor cells, inhibits neointima formation, and enhances angiogenesis. Circulation 2004;109:220–226.

51 Walter DH, Rittig K, Bahlmann FH, Kirchmair R, Silver M, Murayama T, Nishimura H, Losordo DW, Asahara T, Isner JM: Statin therapy accelerates reendothelialization: a novel effect involving mobilization and incorporation of bone marrow-derived endothelial progenitor cells. Circulation 2002;105:3017–3024.

52 Adams V, Lenk K, Linke A, Lenz D, Erbs S, Sandri M, Tarnok A, Gielen S, Emmrich F, Schuler G, Hambrecht R: Increase of circulating endothelial progenitor cells in patients with coronary artery disease after exercise-induced ischemia. Arterioscler Thromb Vasc Biol 2004;24:684–690.

53 Guven H, Shepherd RM, Bach RG, Capoccia BJ, Link DC: The number of endothelial progenitor cell colonies in the blood is increased in patients with angiographically significant coronary artery disease. J Am Coll Cardiol 2006;48:1579–1587.

54 Jones EA, English A, Kinsey SE, Straszynski L, Emery P, Ponchel F, McGonagle D: Optimization of a flow cytometry-based protocol for detection and phenotypic characterization of multipotent mesenchymal stromal cells from human bone marrow. Cytometry B Clin Cytom 2006;70:391–399.

55 Donnenberg AD, Donnenberg VS: Rare-event analysis in flow cytometry. Clin Lab Med 2007;27:627–652, viii.

56 Prater DN, Case J, Ingram DA, Yoder MC: Working hypothesis to redefine endothelial progenitor cells. Leukemia 2007;21:1141–1149.

57 Donnenberg VS, O'Connell PJ, Logar AJ, Zeevi A, Thomson AW, Donnenberg AD: Rare-event analysis of circulating human dendritic cell subsets and their presumptive mouse counterparts. Transplantation 2001;72:1946–1951.

58 Rosenblatt JI, Hokanson JA, McLaughlin SR, Leary JF: Theoretical basis for sampling statistics useful for detecting and isolating rare cells using flow cytometry and cell sorting. Cytometry 1997;27:233–238.

59 Fujiyama S, Amano K, Uehira K, Yoshida M, Nishiwaki Y, Nozawa Y, Jin D, Takai S, Miyazaki M, Egashira K, Imada T, Iwasaka T, Matsubara H: Bone marrow monocyte lineage cells adhere on injured endothelium in a monocyte chemoattractant protein-1-dependent manner and accelerate reendothelialization as endothelial progenitor cells. Circ Res 2003;93:980–989.

60 Ingram DA, Caplice NM, Yoder MC: Unresolved questions, changing definitions, and novel paradigms for defining endothelial progenitor cells. Blood 2005;106:1525–1531.

61 Hur J, Yang HM, Yoon CH, Lee CS, Park KW, Kim JH, Kim TY, Kim JY, Kang HJ, Chae IH, Oh BH, Park YB, Kim HS: Identification of a novel role of T cells in postnatal vasculogenesis: characterization of endothelial progenitor cell colonies. Circulation 2007;116:1671–1682.

62 Bompais H, Chagraoui J, Canron X, Crisan M, Liu XH, Anjo A, Tolla-Le Port C, Leboeuf M, Charbord P, Bikfalvi A, Uzan G: Human endothelial cells derived from circulating progenitors display specific functional properties compared with mature vessel wall endothelial cells. Blood 2004;103:2577–2584.

63 Urbich C, Aicher A, Heeschen C, Dernbach E, Hofmann WK, Zeiher AM, Dimmeler S: Soluble factors released by endothelial progenitor cells promote migration of endothelial cells and cardiac resident progenitor cells. J Mol Cell Cardiol 2005;39:733–742.

Sack U, Tárnok A, Rothe G (eds): Cellular Diagnostics. Basics, Methods and Clinical Applications of Flow Cytometry. Basel, Karger, 2009, pp 317–324

Mesenchymal Progenitor Cells

Johannes C. Fischer

Institute for Transplantation Diagnostics and Cellular Therapeutics, Heinrich Heine University Düsseldorf, Germany

Introduction

The primary role of adult stem cells is to maintain and repair tissues. To conserve this capacity over a long period of time, the stem cell pool must be kept relatively constant. Thus, apart from their ability to differentiate, adult stem cells possess self-renewing capabilities. Therefore during recent years stem cells have emerged as a promising source for the development of new cellular therapies. Since multipotent mesenchymal stem cells (MSCs) can easily be obtained from different donor tissues and expanded ex vivo, they appear to be a suitable cell source in regenerative medicine.

Friedenstein and Kuralesova [1] and Friedenstein et al. [2] were the first to characterize MSCs. They identified a plastic-adherent, fibroblast-like population in adult bone marrow that could regenerate rudiments of bone in vivo (colony-forming units-fibroblasts; CFU-F). Although there is no consistent definition for MSCs, MSC-like cells have recently been isolated from different organs and tissues, including adipose, cord blood, fetal liver, blood, bone marrow and lung [3–9].

After expansion, they can be used for in vitro research, or clinical application. In general, MSCs have the ability to differentiate into adipocytes, osteoblasts, and chondrocytes [10]. However, according to some reports, MSC-like cells, e.g. multipotent adult progenitor cells (MAPCs) [11, 12] or unrestricted somatic stem cells (USSC) derived from cord blood [13] can also give rise to skeletal and cardiac muscle cells and to nonmesodermic derivatives such as hepatocytes (endoderm) and neurons or astrocytes (neuroectoderm). These findings suggest that at least some of the MSC-like cells can differentiate across the germ layer borders, while others are more restricted in their differentiation potential. Additionally, clonal analyses revealed that only a percentage of MSCs of a given population retained multipotent capabilities and were able to regenerate long-term cell cultures [4, 9, 14]. This, together with the fact that antigens have been identified that appear to be present on MSC subsets only [15], suggests that, in analogy to the hematopoietic system, MSCs

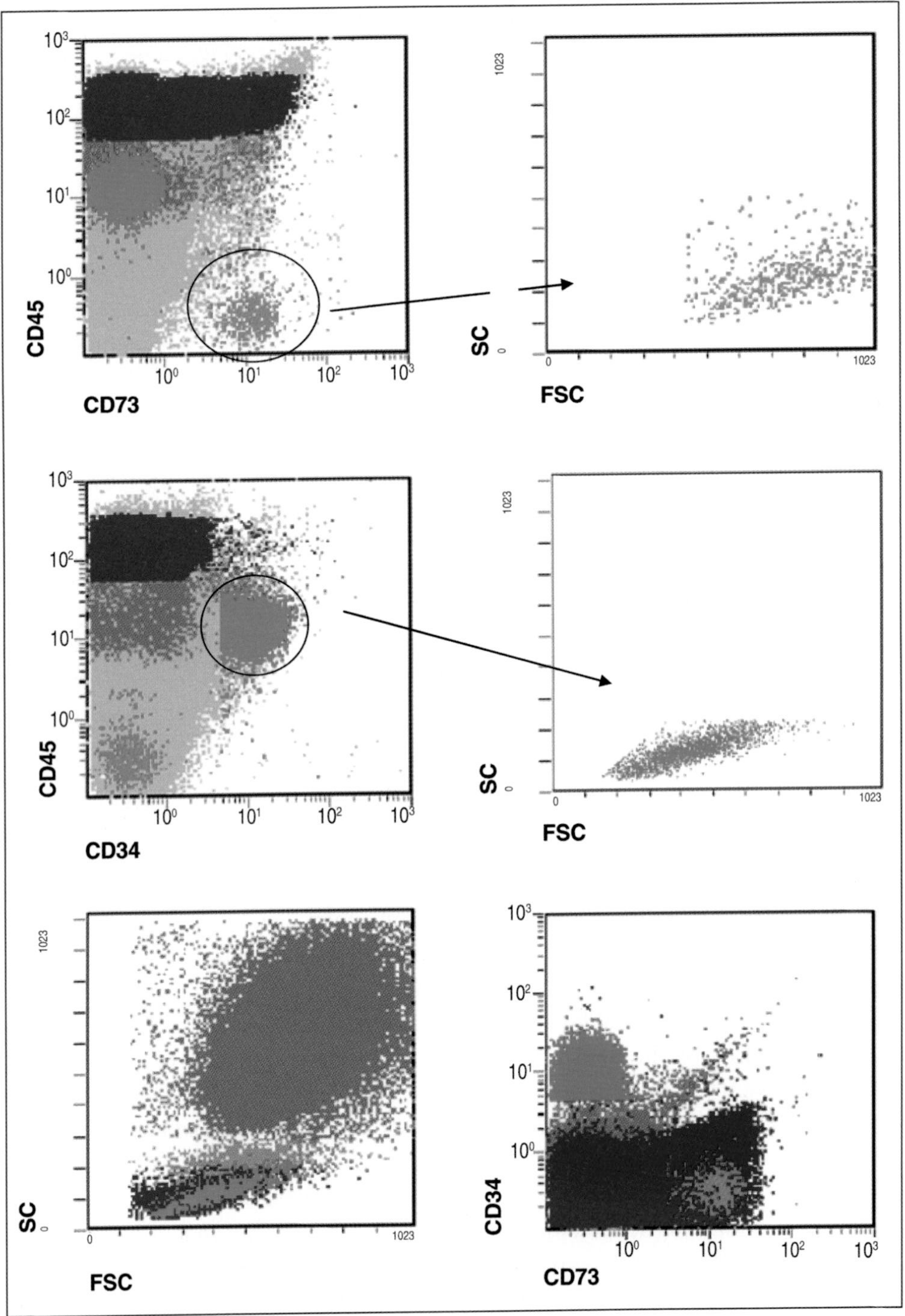

CD45
CD73
SC
FSC
CD45
CD34
SC
FSC
SC
FSC
CD34
CD73

might be organized in a hierarchical mode. However, due to the lack of qualified markers like CD34 and CD133, which discriminate more primitive and more mature hematopoietic cells in cord blood [16], there is so far no efficient phenotypic way to discriminate MSCs of different hierarchical levels.

Characterization and Isolation

Due to the rarity of the cells in bone marrow and the lack of a marker that specifically isolates MSCs, cells are commonly isolated by adherence to plastic and consecutive passage in tissue culture. MSCs may be expanded several fold in vitro, for instance after aspiration of bone marrow. But their proliferation rate as well as their multipotency is limited to several cell culture passages [17].

So far, cultured MSCs are well characterized as plastic-adherent, spindle-shaped cells that express a panel of key markers, including CD105 (endoglin, SH2), CD73 (ecto-5 nucleotidase, SH3, SH4), CD166 (ALCAM), CD29 (β1-integrin), CD44 (H-CAM), CD90 (Thy-1), and STRO-1, and CD166 [4, 18–20]. Off note, MSCs are also positive for CD56 [21].

Heterogeneous expression of other molecules such as the low-affinity nerve growth receptor (CD271) [22] or CD109 [15] has been reported. So far, CD271 is one of the most specific markers for bone marrow-derived MSCs [23]. But only $CD271^{bright}$ populations, positive for CD10, CD13, CD73, and CD105 but not $CD271^{dim}$ populations contain CFU-F [21].

The isolation of MSCs from primary tissue is therefore hampered by the limited selectivity of available markers. Markers that meet established criteria for their positive selection include the markers cited above. CD45 and glycophorin A (CD235) are used for the negative selection of MSCs [11].

Whether culture-expanded MSCs differ from their in vivo progeny remains uncertain as proliferation on plastic surfaces could induce both phenotypic and functional changes [24, 25]. Certain subtypes of MSCs may also have a survival advantage in culture [26]. Even MSC clones derived from single cells vary in their gene

---

◄

Fig. 1. Simultaneous measurement of hematopoietic progenitors and mesenchymal progenitors from human bone marrow. Acquisition and analysis were performed on a dual-laser five-color FC 500 flow cytometer (Beckman Coulter, Krefeld, Germany) The following fluorochromes or conjugates were used: CD45 FITC, CD34 ECD (Beckman Coulter), CD73 PE (BD Pharmingen, Heidelberg, Germany) In brief, 2,000,000 nucleated cells were stained and analyzed after lysing the erythrocytes (VersaLyse, Beckman Coulter). CD34 hematopoietic cells are shown in green ($CD34+$, $CD45^{dim}$), whereas mesenchymal progenitor cells are shown in red. Lymphocytes are displayed in blue. Note the different scatter characteristics of the mesenchymal and hematopoietic progenitor cells. This staining could be expanded for more markers (e.g. live/dead staining).

expression, ability to differentiate, expansion potential and phenotype [4, 14, 25]. Although MSCs from all sources are negative for the hematopoietic markers CD14, CD45 as well as CD34, the definition of MSCs generated ex vivo is a composite of morphological, phenotypic and functional characteristics.

In addition to bone marrow, MSCs or MSC-like cells have also been isolated from diverse adult and fetal tissues [3–9, 13]. Therefore, it appears that MSCs reside within the connective tissue of most organs as predicted by early studies with chick embryos [27]. These populations are not functionally equivalent with respect to their differentiation potential [19], particularly when assayed using stringent in vivo assays [28]. In addition, depending on the age of the donor [30] or the site of bone marrow collection [29], differences in the starting material could be observed, especially in the frequencies of the obtained MSC progenitors.

Thus, characterizing populations as MSC or MSC-like also depends, in part, on the methods used to evaluate their differentiation potential as proliferation on plastic surfaces could induce both phenotypic and functional changes [24, 25, 31]. As said before, some of those MSC-like cells could be further differentiated into cells expressing ectodermal, mesodermal, and endodermal markers as the USSC derived from cord blood [13, 32]. However, it is important to realize that no single isolation method is regarded as a standard in the field.

The varied approaches used to culture, expand and select for MSCs make it difficult to compare the experimental results directly. Moreover, some isolation schemes introduce epigenetic and genetic changes in cells that may dramatically affect their therapeutic utility [31]. Adult human MSCs can undergo spontaneous immortalization [33]. Independent laboratories have confirmed these data for MSCs derived from human or murine bone marrow [34–38]: following approximately 20 population doublings in vitro, MSC cultures may enter a senescence phase. Cells that are able to continue to divide until they reach a crisis phase can undergo tumorigenic transformation.

Animal and Clinical Models for Therapeutic Use of Mesenchymal Stem Cells

MSCs show a strong propensity to attenuate tissue damage in response to injury and disease. In concordance with the first findings, MSCs were originally evaluated for their ability to repair skeletal defects first in experimental animal models and subsequently in human patients suffering from osteogenesis imperfecta, a genetic defect in bone and other tissues caused by mutations in the genes for type 1 collagen [39]. Furthermore, MSCs were effectively used as therapeutic vectors in animal models of lung injury [40, 41], kidney disease [42], myocardial infarction [43], and various neurological disorders [44–47]. However, in a number of such reports, MSCs affected tissue repair although only low and/or transient levels of engraftment could be seen in vivo. In trials on children with osteogenesis imperfecta receiving cellular therapy measurable improvements in growth velocity, bone mineral density, and ambulation were observed although the levels of engrafted donor MSCs in bone, skin, and other tissues reached less than 1%. In addition, improvement in cardiac function after infusion of human MSCs into immunodeficient mice with acute myocardial infarction could be observed although no engrafted donor cells could be detected 3 weeks following injection [48]. These and other studies suggest that the ability of MSCs to secrete soluble factors that alter the tissue microenvironment may play a more prominent role than their transdifferentiation in effecting tissue repair [49]. This is also true with respect to an immunosuppressive effect of MSCs, e.g. modulated by indoleamine 2,3-dioxygenase [50]. MSCs induce little, if any, proliferation of allogeneic lymphocytes [50]. MSCs appear to be immunosuppressive in vitro. They inhibit T-cell proliferation to alloantigens and mitogens and prevent the development of cytotoxic T-cells [51]. In vivo, MSCs prolong skin allograft survival [51] and can obliterate graft-versus-host disease after allogeneic transplantation [52], MSCs have several immunomodulatory effects: they stimulate antibody secretion by B-cells [53], but suppress allospecific antibody production [54].

Summary

Due to the lack of a single definitive marker and precise knowledge regarding the anatomical location and distribution of mesenchymal stem cells (MSCs) in vivo, the demonstration of their existence has relied primarily on retrospective assays. The gold standard assay utilized to identify MSCs is the CFU-F assay which, at minimum, identifies adherent, spindle-shaped cells that proliferate to form colonies [2]. The general approach to the culture of MSCs involves the isolation of MSC-containing mononuclear cells from bone marrow aspirates and seeding these cells on tissue culture plates at a standard plating density in a minimum essential medium base containing fetal bovine serum. Nonadherent hematopoietic cells are removed within 24–48 h, and the adherent cells are cultured and passaged to expand the MSC population [28, 55]. Under this condition, cells can typically be expanded to 40 population doublings. These cells show characteristics of adherent-derived MSCs in that they share a similar phenotypic profile (CD45− CD34− CD14− CD42+ CD105+ CD73+) [15, 19–21, 56] and have the ability to differentiate to a variety of mesenchymal tissues (i.e. bone, cartilage and adipose tissue) both in vitro and in vivo [4, 6, 13, 25].

References

1 Friedenstein A, Kuralesova AI: Osteogenic precursor cells of bone marrow in radiation chimeras. Transplantation 1971;12:99–108.

2 Friedenstein AJ, Chailakhjan RK, Lalykina KS: The development of fibroblast colonies in monolayer cultures of guinea-pig bone marrow and spleen cells. Cell Tissue Kinet 1970;3:393–403.

3 Kopen GC, Prockop DJ, Phinney DG: Marrow stromal cells migrate throughout forebrain and cerebellum, and they differentiate into astrocytes after injection into neonatal mouse brains. Proc Natl Acad Sci U S A 1999;96:10711–10716.

4 Pittenger MF, Mackay AM, Beck SC, Jaiswal RK, Douglas R, Mosca JD, Moorman MA, Simonetti DW, Craig S, Marshak DR: Multilineage potential of adult human mesenchymal stem cells. Science 1999;284:143–147.

5 De Bari C, Dell'Accio F, Tylzanowski P, Luyten FP: Multipotent mesenchymal stem cells from adult human synovial membrane. Arthritis Rheum 2001;44:1928–1942.

6 Zuk PA, Zhu M, Mizuno H, Huang J, Futrell JW, Katz AJ, Benhaim P, Lorenz HP, Hedrick MH: Multilineage cells from human adipose tissue: implications for cell-based therapies. Tissue Eng 2001;7:211–228.

7 Seo BM, Miura M, Gronthos S, Bartold PM, Batouli S, Brahim J, Young M, Robey PG, Wang CY, Shi S: Investigation of multipotent postnatal stem cells from human periodontal ligament. Lancet 2004;364:149–155.

8 Sabatini F, Petecchia L, Tavian M, Jodon de Villeroche V, Rossi GA, Brouty-Boye D: Human bronchial fibroblasts exhibit a mesenchymal stem cell phenotype and multilineage differentiating potentialities. Lab Invest 2005;85:962–971.

9 da Silva Meirelles L, Chagastelles PC, Nardi NB: Mesenchymal stem cells reside in virtually all post-natal organs and tissues. J Cell Sci 2006;119:2204–2213.

10 Delorme B, Chateauvieux S, Charbord P: The concept of mesenchymal stem cells. Regen Med 2006;1:497–509.

11 Jiang Y, Vaessen B, Lenvik T, Blackstad M, Reyes M, Verfaillie CM: Multipotent progenitor cells can be isolated from postnatal murine bone marrow, muscle, and brain. Exp Hematol 2002;30:896–904.

12 Reyes M, Dudek A, Jahagirdar B, Koodie L, Marker PH, Verfaillie CM: Origin of endothelial progenitors in human postnatal bone marrow. J Clin Invest 2002;109:337–346.

13 Kogler G, Sensken S, Airey JA, Trapp T, Muschen M, Feldhahn N, Liedtke S, Sorg RV, Fischer J, Rosenbaum C, Greschat S, Knipper A, Bender J, Degistirici O, Gao J, Caplan AI, Colletti EJ, Almeida-Porada G, Muller HW, Zanjani E, Wernet P: A new human somatic stem cell from placental cord blood with intrinsic pluripotent differentiation potential. J Exp Med 2004;200:123–135.

14 Muraglia A, Cancedda R, Quarto R: Clonal mesenchymal progenitors from human bone marrow differentiate in vitro according to a hierarchical model. J Cell Sci 2000;113(Pt 7):1161–1166.

15 Vogel W, Grunebach F, Messam CA, Kanz L, Brugger W, Buhring HJ: Heterogeneity among human bone marrow-derived mesenchymal stem cells and neural progenitor cells. Haematologica 2003;88:126–133.

16 Giebel B: Cell polarity and asymmetric cell division within human hematopoietic stem and progenitor cells. Cells Tissues Organs 2008;188:116–126.

17 Siddappa R, Licht R, van Blitterswijk C, de Boer J: Donor variation and loss of multipotency during in vitro expansion of human mesenchymal stem cells for bone tissue engineering. J Orthop Res 2007;25:1029–1041.

18 Caplan AI: Mesenchymal stem cells. J Orthop Res 1991;9:641–650.

19 Jones EA, Kinsey SE, English A, Jones RA, Straszynski L, Meredith DM, Markham AF, Jack A, Emery P, McGonagle D: Isolation and characterization of bone marrow multipotential mesenchymal progenitor cells. Arthritis Rheum 2002;46:3349–3360.

20 Jones EA, English A, Kinsey SE, Straszynski L, Emery P, Ponchel F, McGonagle D: Optimization of a flow cytometry-based protocol for detection and phenotypic characterization of multipotent mesenchymal stromal cells from human bone marrow. Cytometry B Clin Cytom 2006;70:391–399.

21 Buhring HJ, Battula VL, Treml S, Schewe B, Kanz L, Vogel W: Novel markers for the prospective isolation of human MSC. Ann N Y Acad Sci 2007;1106:262–271.

22 Campioni D, Lanza F, Moretti S, Ferrari L, Cuneo A: Loss of Thy-1 (CD90) antigen expression on mesenchymal stromal cells from hematologic malignancies is induced by in vitro angiogenic stimuli and is associated with peculiar functional and phenotypic characteristics. Cytotherapy 2008;10:69–82.

23 Quirici N, Soligo D, Bossolasco P, Servida F, Lumini C, Deliliers GL: Isolation of bone marrow mesenchymal stem cells by anti-nerve growth factor receptor antibodies. Exp Hematol 2002;30:783–791.

24 Bianco P: Life in plastic is fantastic. Blood 2007;110:3090.

25 Wagner W, Feldmann RE, Jr., Seckinger A, Maurer MH, Wein F, Blake J, Krause U, Kalenka A, Burgers HF, Saffrich R, Wuchter P, Kuschinsky W, Ho AD: The heterogeneity of human mesenchymal stem cell preparations – evidence from simultaneous analysis of proteomes and transcriptomes. Exp Hematol 2006;34:536–548.

26 Javazon EH, Beggs KJ, Flake AW: Mesenchymal stem cells: paradoxes of passaging. Exp Hematol 2004;32:414–425.

27 Young HE, Mancini ML, Wright RP, Smith JC, Black AC Jr, Reagan CR, Lucas PA: Mesenchymal stem cells reside within the connective tissues of many organs. Dev Dyn 1995;202:137–144.

28 Kuznetsov SA, Krebsbach PH, Satomura K, Kerr J, Riminucci M, Benayahu D, Robey PG: Single-colony derived strains of human marrow stromal fibroblasts form bone after transplantation in vivo. J Bone Miner Res 1997;12:1335–1347.

29 Veyrat-Masson R, Boiret-Dupre N, Rapatel C, Descamps S, Guillouard L, Guerin JJ, Pigeon P, Boisgard S, Chassagne J, Berger MG: Mesenchymal content of fresh bone marrow: a proposed quality control method for cell therapy. Br J Haematol 2007;139:312–320.

30 Tokalov SV, Gruner S, Schindler S, Wolf G, Baumann M, Abolmaali N: Age-related changes in the frequency of mesenchymal stem cells in the bone marrow of rats. Stem Cells Dev 2007;16:439–446.

31 Larson BL, Ylostalo J, Prockop DJ: Human multipotent stromal cells undergo sharp transition from division to development in culture. Stem Cells 2008;26:193–201.

32 Sensken S, Waclawczyk S, Knaupp AS, Trapp T, Enczmann J, Wernet P, Kogler G: In vitro differentiation of human cord blood-derived unrestricted somatic stem cells towards an endodermal pathway. Cytotherapy 2007;9:362–378.

33 Rubio D, Garcia-Castro J, Martin MC, de la Fuente R, Cigudosa JC, Lloyd AC, Bernad A: Spontaneous human adult stem cell transformation. Cancer Res 2005;65:3035–3039.

34 Li H, Fan X, Kovi RC, Jo Y, Moquin B, Konz R, Stoicov C, Kurt-Jones E, Grossman SR, Lyle S, Rogers AB, Montrose M, Houghton J: Spontaneous expression of embryonic factors and p53 point mutations in aged mesenchymal stem cells: a model of age-related tumorigenesis in mice. Cancer Res 2007;67:10889–10898.

35 Miura M, Miura Y, Padilla-Nash HM, Molinolo AA, Fu B, Patel V, Seo BM, Sonoyama W, Zheng JJ, Baker CC, Chen W, Ried T, Shi S: Accumulated chromosomal instability in murine bone marrow mesenchymal stem cells leads to malignant transformation. Stem Cells 2006;24:1095–1103.

36 Tolar J, Nauta AJ, Osborn MJ, Panoskaltsis Mortari A, McElmurry RT, Bell S, Xia L, Zhou N, Riddle M, Schroeder TM, Westendorf JJ, McIvor RS, Hogendoorn PC, Szuhai K, Oseth L, Hirsch B, Yant SR, Kay MA, Peister A, Prockop DJ, Fibbe WE, Blazar BR: Sarcoma derived from cultured mesenchymal stem cells. Stem Cells 2007;25:371–379.

37 Wang Y, Huso DL, Harrington J, Kellner J, Jeong DK, Turney J, McNiece IK: Outgrowth of a transformed cell population derived from normal human BM mesenchymal stem cell culture. Cytotherapy 2005;7:509–519.

38 Zhou YF, Bosch-Marce M, Okuyama H, Krishnamachary B, Kimura H, Zhang L, Huso DL, Semenza GL: Spontaneous transformation of cultured mouse bone marrow-derived stromal cells. Cancer Res 2006;66:10849–10854.

39 Horwitz EM, Gordon PL, Koo WK, Marx JC, Neel MD, McNall RY, Muul L, Hofmann T: Isolated allogeneic bone marrow-derived mesenchymal cells engraft and stimulate growth in children with osteogenesis imperfecta: implications for cell therapy of bone. Proc Natl Acad Sci U S A 2002;99:8932–8937.

40 Ortiz LA, Gambelli F, McBride C, Gaupp D, Baddoo M, Kaminski N, Phinney DG: Mesenchymal stem cell engraftment in lung is enhanced in response to bleomycin exposure and ameliorates its fibrotic effects. Proc Natl Acad Sci U S A 2003;100:8407–8411.

41 Ortiz LA, Dutreil M, Fattman C, Pandey AC, Torres G, Go K, Phinney DG: Interleukin 1 receptor antagonist mediates the antiinflammatory and antifibrotic effect of mesenchymal stem cells during lung injury. Proc Natl Acad Sci U S A 2007;104:11002–11007.

42 Kunter U, Rong S, Boor P, Eitner F, Muller-Newen G, Djuric Z, van Roeyen CR, Konieczny A, Ostendorf T, Villa L, Milovanceva-Popovska M, Kerjaschki D, Floege J: Mesenchymal stem cells prevent progressive experimental renal failure but maldifferentiate into glomerular adipocytes. J Am Soc Nephrol 2007;18:1754–1764.

43 Minguell JJ, Erices A: Mesenchymal stem cells and the treatment of cardiac disease. Exp Biol Med (Maywood) 2006;231:39–49.

44 Jin HK, Carter JE, Huntley GW, Schuchman EH: Intracerebral transplantation of mesenchymal stem cells into acid sphingomyelinase-deficient mice delays the onset of neurological abnormalities and extends their life span. J Clin Invest 2002;109:1183–1191.

45 Acosta FL Jr, Lotz J, Ames CP: The potential role of mesenchymal stem cell therapy for intervertebral disc degeneration: a critical overview. Neurosurg Focus 2005;19:E4.

46 Goya RL, Kuan WL, Barker RA: The future of cell therapies in the treatment of Parkinson's disease. Expert Opin Biol Ther 2007;7:1487–1498.

47 Kim M, Lee ST, Chu K, Kim SU: Stem cell-based cell therapy for Huntington disease: A review. Neuropathology 2008;28:1–9.

48 Iso Y, Spees JL, Serrano C, Bakondi B, Pochampally R, Song YH, Sobel BE, Delafontaine P, Prockop DJ: Multipotent human stromal cells improve cardiac function after myocardial infarction in mice without long-term engraftment. Biochem Biophys Res Commun 2007;354:700–706.

49 Prockop DJ: Stemness does not explain the repair of many tissues by mesenchymal stem/multipotent stromal cells (MSCs). Clin Pharmacol Ther 2007;82:241–243.

50 Meisel R, Zibert A, Laryea M, Gobel U, Daubener W, Dilloo D: Human bone marrow stromal cells inhibit allogeneic T-cell responses by indoleamine 2,3-dioxygenase-mediated tryptophan degradation. Blood 2004;103:4619–4621.

51 Bartholomew A, Sturgeon C, Siatskas M, Ferrer K, McIntosh K, Patil S, Hardy W, Devine S, Ucker D, Deans R, Moseley A, Hoffman R: Mesenchymal stem cells suppress lymphocyte proliferation in vitro and prolong skin graft survival in vivo. Exp Hematol 2002;30:42–48.

52 Ringden O, Uzunel M, Rasmusson I, Remberger M, Sundberg B, Lonnies H, Marschall HU, Dlugosz A, Szakos A, Hassan Z, Omazic B, Aschan J, Barkholt L, Le Blanc K: Mesenchymal stem cells for treatment of therapy-resistant graft-versus-host disease. Transplantation 2006;81:1390–1397.

53 Rasmusson I, Le Blanc K, Sundberg B, Ringden O: Mesenchymal stem cells stimulate antibody secretion in human B cells. Scand J Immunol 2007;65:336–343.

54 Comoli P, Ginevri F, Maccario R, Avanzini MA, Marconi M, Groff A, Cometa A, Cioni M, Porretti L, Barberi W, Frassoni F, Locatelli F: Human mesenchymal stem cells inhibit antibody production induced in vitro by allostimulation. Nephrol Dial Transplant 2008;23:1196–1202.

55 Ohgushi H, Caplan AI: Stem cell technology and bioceramics: from cell to gene engineering. J Biomed Mater Res 1999;48:913–927.

56 Bruder SP, Horowitz MC, Mosca JD, Haynesworth SE: Monoclonal antibodies reactive with human osteogenic cell surface antigens. Bone 1997;21:225–235.

Sack U, Tárnok A, Rothe G (eds): Cellular Diagnostics. Basics, Methods and Clinical Applications of Flow Cytometry. Basel, Karger, 2009, pp 325–342

Determination of Cell Physiological Parameters: pH, Ca^{2+}, Glutathione, Transmembrane Potential

Attila Tárnok[a] · Gregor Rothe[b]

[a] Pediatric Cardiology, Cardiac Center Leipzig GmbH, University of Leipzig,
[b] Laborzentrum Bremen, LADR Group, Bremen, Germany

Introduction/Background

The determination of physiological properties of cells will be exemplified using phagocytic cells. Professional phagocytes such as neutrophils, eosinophils, basophils and monocytes/macrophages play a pivotal role in the nonspecific or innate immune response. These cells remove cell and tissue debris, apoptotic cells and foreign particles. Furthermore, phagocytes are involved in inflammatory reactions and inhibit tumor growth. Monocytes exert their influence via the expression of tissue factors and their interactions with thrombocytes, homeostasis and wound healing [1]. Phagocytes can engulf microorganisms and thereby induce their destruction. In addition, they can present bacterial antigens via the major histocompatibility (MHC) system and induce a specific adaptive immune response. Via this mechanism, phagocytes link the innate with the adaptive immune response. The cascade of phagocytic responses to microorganisms consists of a complex sequence of graded processes: chemotaxis, actin polymerization, migration, aggregation, phagocytosis, degranulation, release of reactive oxygen species (O_2^-, H_2O_2) and changes in the intracellular pH (pH_i), cytosolic free calcium concentration ($[Ca_i^{2+}]$), and transmembrane potential (MP).

Chemotaxis and migration are the initial responses of phagocytes to activation. They are characterized by initial adhesion processes, reorganization of the cytoskeleton and involvement of chemokine receptors. The binding of uncoated or opsonized particles to specific surface receptors initiates phagocytosis. Equipped with pattern recognition receptors, these cells can bind opsonized cells as well as specific components of the cell membrane. The receptors characterized best are opsonin-dependent receptors such as immunoglobulin receptors FcγRI, FcγRII, FcγRIII, complement receptor-3 and integrin receptors $\alpha_5\beta_1$ (VLA-5) and $\alpha_v\beta_3$ and opsonin-independent receptors such as the N-formyl-Met-Leu-Phe (fMLP) receptor, mannose receptor, β-glucane receptor and CD14 (lipopolysaccharide receptor) [2].

The initial event induced by the binding of ligands is the accumulation and association of receptors on the cell surface. This enhanced presentation of receptors is a major property of phagocyte activation and includes signal transduction with, e.g., tyrosin kinases, GTPases and GTPkinases. Phosphatidylinositol 3-kinase is particularly important for the response of phagocytes to chemical signals. These stimulus-dependent activation steps can be significantly enhanced by priming (prestimulation). Priming is induced by treatment with low concentrations of agonists such as interleukin-8 (IL-8), lipopolysaccharide, leukotriene B4 and platelet-activating factor. The priming of neutrophils accelerates responses such as the production of reactive oxygen species. The last step of the activation cascade of phagocytes is the intracellular destruction of microorganisms and ingested foreign materials.

Dysfunction of phagocytes and the innate defense system can be congenital or acquired. Congenital defects can affect phagocyte functions such as chemotaxis, motility or adhesion. Examples are Chediak-Higashi syndrome, Schwachman's syndrome and chronic granulomatous disease. Cellular defects include gp110 defect, deficient endocytosis and cytotoxicity, and defects in myeloperoxidase, gp150 or G6-PD. The more frequent acquired immune defects include chronic diseases such as diabetes, autoimmune diseases, kidney or liver dysfunctions, alcoholism, immunosuppression during surgery, burns and viral infection. In these syndromes a reduced immune response is associated with increased susceptibility to infections and recurrent infections. By contrast, exaggerated responses of neutrophils, e.g. during systemic inflammatory response syndrome, may also lead to tissue destruction even in the absence of infection.

Abnormalities of the ligand-induced oxidative burst or phagocytosis reaction (see 'Oxidative Burst, Phagocytosis and Expression of Adhesion Molecules', pp. 343) as well as of changes in pH_i, $[Ca_i^{2+}]$ or MP can be indicators of decreased or abnormally increased responses of phagocytes [3–5].

Basic Principles of the Cytometric Analysis of Cell Physiological Responses

Cytometric methods (flow- or slide-based cytometry) are standard assays for the analysis of $[Ca^{2+}]_i$, pH and MP of cells. Other assays enable the estimation of additional intracellular ions such as K^+, Na^+, Cl^-. By cytometry, many individual cells can be measured with high throughput. The results are obtained in a standardized manner by establishing a calibration curve and analyzing the responses to defined stimuli. Fluorescent dyes that are sensitive to changes in cellular ion concentration, pH or MP enable cost effective measurements. All reagents mentioned hereafter are available from Invitrogen Inc./Molecular Probes (Eugene, OR, USA) and partially from other providers.

In general, cell physiological analyses are performed on living cells as measurements of time-dependent responses. The analyzed responses such as stimulation-induced changes in [Ca_i^{2+}] or MP are often rapid, and time may be recorded as an additional parameter during flow-cytometric data acquisition. Following a prerun with unstimulated cells, the sample is taken from the cytometer, the stimulus is added, mixed, and measurements are continued. Alternatively, the stimulus can be directly injected during the measurement [3, 4, 6] or can be placed in a drop of liquid above the cell suspension and mixed with the cells after the prerun by vortexing, a procedure which requires some dexterity. In case of sustained responses, the analysis may also be performed at a defined time after stimulation or as an end-point determination. Negative controls using the solvent in the absence of a stimulus should always be included. Such negative controls are necessary because mechanical stimulation by addition of a stimulus and mixing may by themselves lead to cell activation in some cell types [6]. For data analysis, the time course of the cell physiological changes is determined and the data are calibrated using a calibration curve. Calibration, e.g. of fluorescence ratios, can be performed directly using specific software packages available as commercial products or for free.

Cell physiological analysis requires gentle standardized cell treatment. Staining, storage, and measurement should always be done under identical conditions. Changes in temperature should be avoided as these can stimulate cells [7]. Cells should not be stored on ice. Ideally, they should be measured at 37 °C if enabled by the instrument. Damaged and 'dead' cells can impair the measurements as such cells may suggest unresponsive cell populations. Dead cells can be discriminated from live cells using propidium iodide (PI) or 7-amino-actinomycin D (7AAD) (both at 5 µg/ml final concentration). Using a blue light source for excitation, these dyes enable clear distinction of dead cells as extremely bright cells with orange or red emission. Other dead-cell dyes are available if dead cells need to be discriminated using a different excitation source. The use of calcium chelators such as EDTA or citrate should be avoided. Keeping cells in calcium-free medium is not physiological. Intracellular calcium pools are depleted, and cells may respond less well to the same stimulus. In general, cells can be simultaneously stained for a cell physiological parameter and for cell surface antigens using antibodies. This multiparametric staining can yield useful additional information on the response of cell subsets. However, it should be remembered that the binding of antibodies to surface antigens may by itself lead to cell activation or modify cellular responses. For example, anti-CD3 will stimulate a calcium flux in T cells. If this type of analysis has to be performed, appropriate control experiments need to be done.

There are two principles for dyes to indicate changes in cell physiology. Indicator dyes can either change their fluorescence intensity (example: following an increase in intracellular free calcium, the dye Fluo3 increases in fluorescence whereas Fura Red fluorescence decreases) or their spectrum of emission (example: Indo-1). In general, ratiometric dyes that change their spectrum of emission (i.e. color) are best

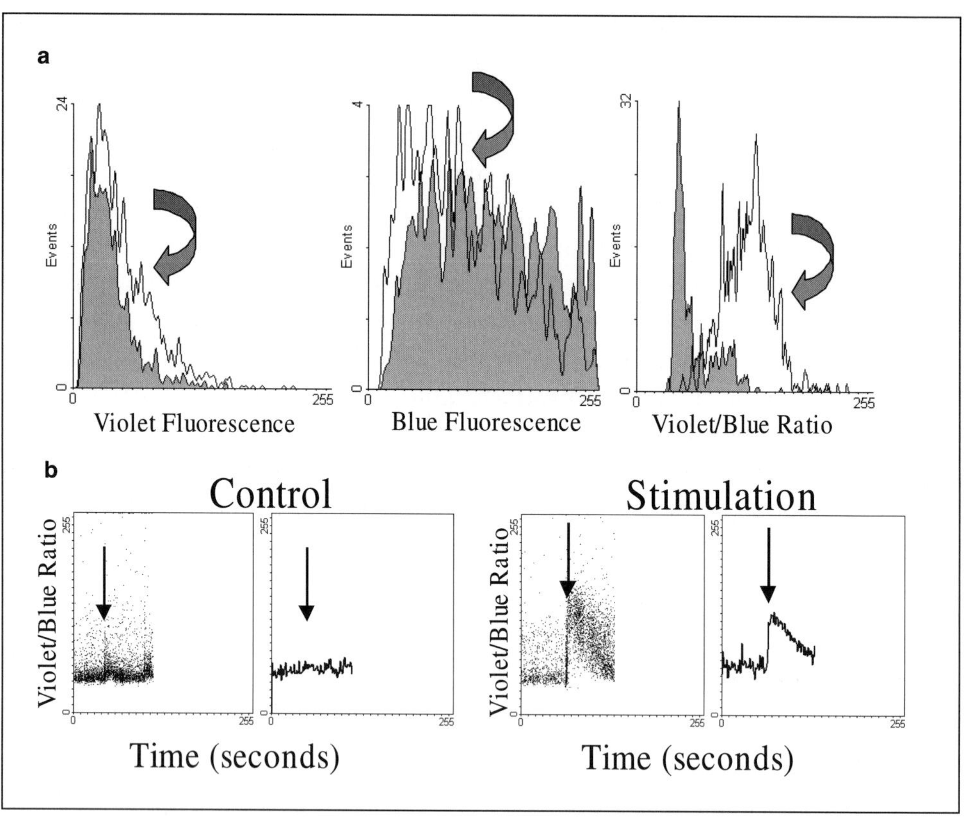

Fig. 1. Measurement of $[Ca^{2+}]_i$ with Indo-1.
a Unstimulated cells (gray) and stimulated cells (white) stained with Indo-1 are shown in their respective violet and blue fluorescence. Upon calcium release (straight lines) stimulated cells (gray shaded) show a slight shift towards violet and a reduction of their blue fluorescence. An unequivocal discrimination between responding and nonresponding cells is only possible using the violet to blue ratio (top right).
b Time course of a measurement following addition of a buffer solution as a control (left) or a stimulus (right). Arrows indicate the time point of the injection of the stimulus into the cell suspension during the measurement [22]. Following stimulation, the violet to blue ratio is dramatically increased. This leads to the clear discrimination of responding and nonresponding cells. Data are shown pairwise as dot plots or as mean calcium concentrations (lines).

suited for the measurement of cell physiological parameters. By the simultaneous measurement of two emission wavelengths (colors) the intensity of both colors is determined ratiometrically. This approach enables a calibrated analysis of different cells also in case of differences in dye loading (fig. 1). To this end, the cytometer should be able to record on line or analyze fluorescence ratios off line as additional parameters.

For the measurement, cells have to be loaded with the appropriate indicator. Labeling for MP is passive, i.e. after addition, the dye associates to cells shortly before the measurement. In contrast, staining for other dyes depends on active intracellular enrichment through metabolic activity. These dyes include, among others, indicators for Ca^{2+} and pH. As they are mostly hydrophilic in their active state, i.e. when they can bind ions, they cannot penetrate the hydrophobic (lipophilic) cell membrane and accumulate in the cell. In order to label cells, these dyes are usually added as acetoxymethyl ester (AM). Due to ester groups, the dyes are lipophilic and can pass the cell membrane. Inside the cells these ester groups are removed hydrolytically by intracellular esterases. Thereby, the dyes are converted into their active forms and become hydrophilic. As hydrophilic dyes cannot escape from the cells, they accumulate there. Dye accumulation can be increased by applying a nonionic detergent, such as Pluronic F127. If the cells lack intracellular esterases, the dye does not accumulate due to the absence of hydrolysis. In this case, low cellular fluorescence does not change following stimulation. If this is the case, the active dye (without AM) can be used. Pluronic F127 or hypotonic shock may enable loading of the cells with the dye. Effective dye loading should be verified using a potent or maximal stimulus.

Principles of Measurement

Intracellular Free Calcium Concentration

Ca^{2+} is a universal second messenger. It appears in the cell at increasing concentrations only fractions of seconds to seconds after cell activation. In quiescent cells, calcium concentration is actively maintained at a level of 100–150 nmol/l by intracellular Ca^{2+}-dependent ATPase. By contrast, the concentration of Ca^{2+} in the physiologic extracellular space is as high as 1.3 mmol/l. Stimulation of surface receptors typically leads to a three-step increase in the calcium concentration of nonmuscular cells. In the first step, the sources of calcium release are calciosomes (intracellular calcium pools). In the second phase, the calcium increase depends on an influx from the extracellular space. This second phase is followed by an increase in the calcium level in the calciosomes and the export of membrane bound Ca^{2+} through the cell membrane with membrane-bound ATPase. This export leads to a reduction in $[Ca_i^{2+}]$.

Ca^{2+}-mobilized signals are regulated by the following mechanisms:

i) the intracellular second messengers inositol-1,4,5-triphosphate, cyclic ADP-ribose and nicotinic acid dinucleotide phosphate;
ii) Ca^{2+} channels in the membrane of the endoplasmatic reticulum or in the plasma membrane, and
iii) calcium sensors, including the ubiquitous calcium-binding protein calmodulin.

$[Ca_i^{2+}]$ can be intracellularly determined by a shift of Indo-1 fluorescence from blue (absence of calcium) to violet (presence of calcium). This method enables precise quantification independent of cell size and total fluorescence intensity as

Table 1. Calcium indicators in cytometry

Indicator	Emission response upon Ca^{2+} increase	Excitation wavelength nm	Emission wavelength nm	Calcium affinity K_d, nmol/l	
				22 °C	37 °C
Indo-1	ratio increase	325–360	390/520	~230	~250
Fluo-3	ilncrease	488	526	~330	~860
Fluo-4	increase	488	516	~350	n.i.
Calcium Green-1[a]	ncrease	488	530	~250	n.i.
Calcium Green-2	n.i.	488	536	~550	n.i.
Calcium Green-5N	n.i.	488	532	~14,000	n.i.
Calcium Orange	increase	550	575	~330	n.i.
Calcium Crimson	increase	550	610	~200	n.i.
Oregon Green 488 BAPTA-1[b]	n.i.	488	523	~170	n.i.
Oregon Green 488 BAPTA-2 [c]	n.i.	488	523	~580	n.i.
Oregon Green 488 BAPTA-5N[b]	n.i.	488	521	~20,000	n.i.
Fura Red[c]	ratio increase	405/488	597	~400	n.i.
Fluo-3/Fura Red[d]	ratio increase	488	530/660[e]	~400	n.i.

n.i. = No information.

[a] Calcium Green-1 fluoresces stronger than Fluo-3; both contain Ca^{2+} in a bound and unbound form. The effect of unbound Ca^{2+} for the increase in fluorescence is greater for Fluo-3

[b] Molar absorption of Oregon Green BAPTA-indicators at 488 nm is approximately twice as high as for the respective Calcium Green-indicator.

[c] Fura Red can be used ratiometrically regarding excitation, showing an increase in emission at 597 nm with calcium at 405 nm excitation and a decrease at 488 nm excitation.

dSimultaneous staining with Fluo-3 and FuraRed can be used for ratiometric measurement with blue excitation.

[e] Emission maximum of Fura Red is at 597 nm. In order to minimize spectral overlap between Fluo-3 and Fura Red, it is recommended to detect Fura Red fluorescence in the deep red channel.

the ratio of violet to blue fluorescence in the cells represents the calcium concentration. A calibration curve can be constructed using cells in a defined extracellular calcium concentration and the presence of calcium ionophores such as ionomycin or the brom/calcium ionophore A23187. Intracellular buffering of calcium can impair this calibration.

The kinetics of the $[Ca_i^{2+}]$ response to a stimulus is determined in three steps:

i) preparation of cells with Indo-1 and measurement at 37 °C without stimulation;
ii) addition of a stimulus and measurement of the changes in the calcium concentration until an equilibrium is reached, and

Table 2. Indicators for ratiometric measurement of the pH in cytomery

Dye	pK_a	Fluorescence, nm	
		excitation	emission
Fluorescein diacetate (FDA)	6.3	ratio 436/495	525
Carboxyfluorescein diacetate (CFDA)	6.4	ratio 441/488	535
Bis-carboxyethylcarboxyfluorescein-acetox methyl ester (BCECF-AM)	7.0	ratio 439/490	535
		488	ratio 520/620
Hydroxycoumarin (4-methylumbelliferone; 4-MU)	7.8	~350	ratio 430/470
		~350	ratio 450/560
Diacetoxydicyanobenzene (ADB) (yields dicyanohydrochinone (DCH) after hydrolysis)	8.0	~350	ratio 425/540
Carboxy SNARF-1 acetoxymethyl ester	7.5	488, 514 or 530	ratio 575/670

iii) addition of ionomycin or brom/calcium ionophore A23187 in order to determine the maximal Indo-1 ratio (positive control).

If ultraviolet excitation (365 nm) is not available, the indicators Fluo-3 or Fluo-4 can be used with blue excitation. This method also enables the sensitive detection of a calcium increase, and can be used in a ratiometric manner when combined with Fura Red. Additional fluorescent dyes to determine the calcium concentration are shown in table 1.

Alkalization of the Cytosol

The pH of eukaryotic cells varies in the range of 7.0–7.4. Regulation of the pH strongly correlates with leukocyte function. In phagocytes, intracellular pH also differs in various cell organelles. Furthermore, pH regulation is strongly linked to cell viability and apoptosis. The best-known mechanism of pH regulation is via amiloride-sensitive Na^+/H^+ transport. Alkalization of the intracellular pH is obtained by stimulating this antiport, e.g. with cytokines. Regulation of the intracellular pH is of particular importance for the activation of cells, as an increase in the intracellular pH indicates elevated metabolic activity.

The ratiometric indicator SNARF-1 is the dye of choice to measure cytoplasmic pH. The dye is excited by a blue laser and shifts in emission from green to red with increasing pH. Therefore, fluorescence is simultaneously measured in the green and the red fluorescence channels. pH changes are determined based on the red/green fluorescence intensity ratio. More pH indicators are shown in table 2. Ratiometric measurements with other indicators, e.g. FDA, CFDA or BCECF, are possible by determining the green fluorescence intensity following excitation at two different wavelengths. These are frequently used in microscopy but are difficult to use in flow or slide-based cytometry.

Lysosomal Proteinases

Cellular endopeptidases are classified into four groups based on their inhibitors [8]:

i) serine proteinases, inhibited by diisopropyl fluorophosphate (DFP);

ii) cysteine proteinases, inhibited by E-64;

iii) asparagine proteinases, inhibited by pepstatin;

iv) metalloproteinases, inhibited by phenanthroline.

Due to their broad substrate specificity, these enzymes are involved in intracellular protein turn-over as well as extracellular tissue destruction during inflammation. Lysosomal proteases are differentially expressed in various cell lines and cell types. Therefore, they are relevant for cell typing. One specific function of phagocytes is the activity of elastase linked to the lysosomal destruction of bacteria. Extracellular release of elastase by phagocytes is associated with tissue destruction. Cysteine proteases expressed by monocytes and macrophages play an important role in antigen presentation.

The activity of lysosomal proteases can be detected intracellularly using R110 peptide derivatives. To this end, cells are incubated with the appropriate substrates, e.g. $(Z\text{-}Arg\text{-}Arg)_2$-R110. The specificity of these methods is verified by preincubation with specific inhibitors, e.g. Z-Phe-Ala-CHN_2 for cathepsin B and L [9, 10].

Membrane Potential

Quiescent cells maintain large ion gradients between the intra- and extracellular space. As an example, potassium ions are accumulated in the cell via the activity of Na^+K^+ATPase. The efflux of K^+ ions establishes an electron counter-gradient and the cytoplasm becomes electron negative as compared to the extracellular medium. This electrochemical K^+ gradient is the major contributor to the MP of mammalian cells. The maintenance of a large negative MP was found to be a control mechanism that keeps cells in a resting state. In addition, metabolically active cells maintain a strong MP across the mitochondrial membrane system. The binding of ligands to transmembrane receptors in various cell types leads to rapid changes in the MP and subsequently to a physiological cell response.

The detailed investigation of MP in small cells became possible after the development of MP-sensitive indicator dyes (table 3). These probes are charged lipophilic molecules. They migrate between cytosol, cell membrane, and the extracellular medium according to the Nernst equation. For cationic (positively loaded)

Table 3. Indicator dyes for MP

Indicator	Excitation wavelength[a], nm	Emission wavelength[a], nm
Carbocyanines (mitochondrial membrane selective)		
$DiOC_2(3)$	488	500–510
$DiOC_5(3)$	488	520–530
$DiOC_6(3)$	488	520–530
$DiIC_1(3)$	488	575–585
$DiIC_1(5)$	633–647[b]	660–680
$DiIC_5(3)$	488, 514	540–580
$DiSC_3(5)$	568, 633	$>590 > 680$
JC-1[c]	488	527/590
Oxonols (plasma membrane selective)		
$DiBAC_4(3)$	488	520–530
$DiBAC_4(5)$	568–595	610–640
$DiSBAC_2(3)$	568	590–630
$DiTBAC_4(3)$	488	575–585
$DiTBAC_4(5)$	633–647	670–680
Rhodamines (mitochondrial membrane selective)		
Rhodamine 123[c]	488	530
CMTMRos[c, d]	488	600
CMXRos[c, d]	488	600

[a] Common laser and emission regions.
[b] Depending on laser.
[c] Mitochondrial MP.
[d] Potential measurement also possible after fixation of cells.

indicators such as cyanine dyes, cellular concentration decreases when the cell depolarizes (i.e. MP drops to zero) and increases when the cell hyperpolarizes (becomes negative). Using the negatively charged oxonol dye, the response is in the opposite direction.

MP indicators for flow-cytometric analysis can be divided into three major types: carbocyanine, oxonol and rhodamine/rosamine dyes (table 3). All of them respond to MP changes by alterations in their transmembrane distribution. They can be used to measure changes in the MP of nonexcitatory and stimulated cells induced, e.g. by respiratory activity, ion channel permeability, receptor activation, or binding of substances with pharmacological activity. With some limitations, within cells oxonols generally respond to the plasma MP while carbocyanine and rhodamine dyes respond to mitochondrial MP differences in a concentration-dependent manner. [11].

Carbocyanines

Membrane hyperpolarization leads to increased intracellular uptake of moderately lipophilic cationic carbocyanine dyes such as $DiOC_6(3)$ and $DiOC_5(3)$. Protocols for flow cytometry apply low extracellular dye concentrations (<0.1 μmol/l) in order to minimize the toxic effect of carbocyanine dyes [11, 12]. Staining at low extracellular dye concentrations mainly reflects the mitochondrial MP, and to a lesser extent the plasma MP [12, 13]. Potential-dependent enrichment of the carbocyanine dye JC-1 in mitochondria leads to a pronounced shift in its emission wavelength (from ~527 to 590 nm) due to formation of J-aggregates [14]. The ratio of green to red JC-1 fluorescence at 488 nm excitation can be used to determine the mitochondrial MP [15]. Alternative dyes include $DiOC_2(3)$ (blue excitation) and $DiIC_1(5)$ (red excitation). Loss of mitochondrial MP gradients is a very early step in the cascade of programmed cell death (apoptosis). Thus, JC-1 is a good indicator of apoptosis.

Oxonols

Oxonols like $DiBAC_4(3)$, the most popular of these dyes in flow cytometry, are lipophilic anions. Hyperpolarization of the membrane leads to reduced cell staining by oxonols. This is in contrast to the increase observed using cationic carbocyanines. Oxonols are more specific indicators of the plasma MP than carbocyanines. Because of their anionic character, no labeling of mitochondria occurs. For the same reason oxonols are substantially less toxic than carbocyanines. In addition, oxonols are better suited to detecting the viability of bacteria than carbocyanines and rhodamines [16].

Rhodamine and Rosamine

Rhodamine 123 is a lipophilic cation and is particularly well suited for the labeling of mitochondria in living cells [17]. Its spectral properties resemble those of fluorescein. As labeling is dependent on mitochondrial function, staining intensity directly reflects cell viability [18]. Furthermore, it enables discrimination of cell populations based on their level of mitochondrial respiratory activity [19, 20]. The rosamine derivatives CMTMRos and CMXRos are also suitable for measuring mitochondrial MP. Both dyes bind to intracellular proteins. Fixation after stimulation therefore enables one to retain the staining of cells.

Protocols

Isolation of Leukocytes

For the determination of cell physiological parameters, gentle primary isolation of leukocytes from erythrocytes is preferred over secondary erythrocyte lysis.

Work Steps

- Pour whole blood into heparin tubes (10 U/ml heparin; do not use EDTA or citrate tubes as cells may not respond physiologically following calcium depletion).
- Overlay 5 ml Ficoll (density: 1.077 g/ml) carefully with 3 ml blood. Avoid mixing of blood with Ficoll (Ficoll activates leukocytes).
- Keep samples for approximately 40 min at room temperature; erythrocytes aggregate and sediment into Ficoll; leukocytes remain above the layer and can be seen as an opaque zone.
- Carefully withdraw approximately 800 μl of the supernatant with leukocytes. Avoid contact with Ficoll.

The cell suspension thus obtained will contain $\sim 2 \times 10^7$ unseparated leukocytes/ml in autologous plasma.

If necessary, Ficoll separation can be avoided. In this case, leukocytes are specifically labeled and thus discriminated from abundant erythrocytes during analysis. Antibodies (CD45, pan-leukocyte antigen), a viable dye for DNA (Hoechst 33258 (excitation: UV, emission: 490 nm); SYTO 16 dye (excitation: 488 nm, emission: 518 nm) or DRAQ5 (excitation: 488 nm or 633 nm, emission: 670 nm) can be used for this purpose. One has to make sure that these additional dyes do not lead to artifactual stimulation or modify the response to a stimulus. Thus DRAQ5 is at present the only cell permeant DNA dye that is excitable by a red light source. Unfortunately, it has some toxicity and may thus alter physiological responses.

Measurement of Intracellular Free Calcium Concentration

Staining

Indo-1

- 5×10^6 cells/ml are incubated in HEPES-buffered medium (containing a physiological concentration of Ca^{2+}) with 0.5–5 μmol/l Indo-1/AM at 37 °C (water bath). Staining is completed after 20–30 min. Depending on cell type, the lowest possible concentration that results in a homogeneous fluorescence ratio should be selected.
- Add PI at a final concentration of 5 μg/ml. Assess the fluorescence of Indo-1/calcium complexes by using band-pass filters in the 390- and 440-nm range. For the PI fluorescence of dead cells, a 620-nm band-pass filter is required. Both dyes are excited by a UV light source ($\sim$360 nm).

Fluo-3/AM

- 5×10^6 cells/ml are incubated in HEPES-buffered medium (containing a physiological concentration of Ca^{2+}) with 0.5–2 μmol/l Fluo-3/AM at 37 °C (water bath). Staining is completed after 20–30 min. Depending on cell type, the low-

est possible concentration that yields homogeneous fluorescence should be selected.
- Add PI (final concentration 5 μg/ml). Assess Fluo-3 fluorescence using a 535-nm band-pass filter (excitation: 488 nm); PI fluorescence: 620-nm band-pass filter.

Measurement
- Perform an initial measurement of the unstimulated cells for 30 s in order to determine the basal value.
- Add a stimulus, e.g. fMLP (10^{-8} mol/l).
- Negative control: addition of the solvent of your stimulus.
- Positive control: addition of ionomycin or Br/Ca ionophore A23187 (2 μmol/l) for a maximal calcium response.
- Continue the measurement for a few minutes after stimulation depending on the kinetics of the response; acquire time as an additional parameter.

The absolute [Ca_i^{2+}]can be determined using a calibration curve generated with known extracellular calcium concentrations [21, 22]. This is substantial extra work, however.

Data Analysis
- Data (ratio for Indo-1 or intensity for Fluo-3) are plotted as dot plots versus side scatter. The leukocyte subsets of neutrophils, monocytes and lymphocytes can be discriminated based on scatter.
- Gate on the cells of interest (e.g. neutrophils) and show their PI fluorescence as a histogram. Gate on the PI-negative (viable) cells.
- Show the Indo-1 ratio (violet/blue) or the Fluo-3 fluorescence intensity vs. time. Set a cut-off for a positive reaction based on the negative control. Cells with values above this limit are regarded as responding.
- Determine the percentage of responding cells after stimulation. Positive control should be 100%.
- The mean calcium concentration (absolute values if a calibration curve has been established, otherwise relative values) can now be reported as a function of time. This can be performed with commercial and shareware software.

Intracellular pH

Staining

SNARF-1
- 5×10^6 cells/ml are incubated in HEPES-buffered medium with 0.2–1 μmol/l SNARF-1/AM at 37 °C (water bath). Staining is completed after 20–30 min. Depending on cell type, the lowest possible concentration that yields a homogeneous fluorescence ratio should be selected.
- Add PI (final concentration 5 μg/ml).

Measurement

- For the orange SNARF-1 fluorescence select a band-pass filter of 575 nm; for the red SNARF-1 and PI fluorescence, a 620-nm long-pass filter. (Although both PI and SNARF-1 red are measured in the same fluorescence channel, PI fluorescence intensity is orders of magnitude higher and easily distinguishable. Dead cells will show low or no SNARF-1 orange fluorescence.)
- If feasible on your instrument, also directly measure the SNARF-1 red/orange ratio. Turn on time as a parameter.
- Initial baseline measurement for 30 s without stimulation.
- Add a stimulus, e.g. fMLP (10^{-8} mol/l), or TNF-α (1 ng/ml) for neutrophils.
- Negative control: addition of the solvent of the stimulus.

Calibration Curve

- After SNARF-1/AM incubation, split sample into six aliquots (see staining 1).
- Centrifuge (60 × *g*, 4 °C, 5 min).
- Resuspend and incubate for 5 min in 140 mmol/l KCl buffer (10 mmol/l Mes, 10 mmol/l HEPES, pH 6.4; 6.8; 7.2; 7.6; 8.0; 8.4; each with 10 µmol/l nigericin).
- Measure.

Data Analysis

- See protocol for calcium.
- When a calibration curve is measured, the SNARF-1 ratio values can be transformed into absolute pH values.

Proteinases

Staining

- Incubate 5×10^6 cells/ml in HEPES-buffered medium with or without specific proteinase inhibitors (e.g. 100 µmol/l Z-Phe-Ala-CHN_2 or 1 mmol/l DFP) at 37 °C for 10 min.
- 20 min incubation with appropriate proteinase substrate, e.g. 4 µmol/l (Z-Arg-Arg)$_2$-R110.
- Add PI (final concentration 5 µg/ml).
- Assess R110 fluorescence with a 515- to 535-nm band-pass filter; PI fluorescence, e.g. with a 620-nm long-pass filter (excitation: 488 nm).

Measurement and Analysis

The intracellular turnover is determined by the amount of fluorescence acquired during the incubation.

Staining

JC-1

- Transfer 5×10^6 cells/ml into cell culture medium with 10% serum and mix with JC-1 (final concentration: 20 μmol/l; note: titration is recommend as ratiometric response very much depends on JC-1 loading into cells.). Mix thoroughly during addition and the following 20 s.
- Wash twice with PBS; centrifuge for 5 min at $200 \times g$.
- Incubate cells for 15 min at room temperature.
- Add PI (final concentration 5 μg/ml).

DiBAC$_4$(3)

- Transfer 5×10^6 cells/ml into HBSS. Incubate with DiBAC$_4$(3) for 10–20 min at 37 °C (final concentration: 100 nmol/l). Staining is completed after 10 min. Depending on cell type, select the lowest possible concentration yielding homogeneous fluorescence.
- Add PI (final concentration 5 μg/ml).

Measurement

- For JC-1, use 530-nm (green) and 570-nm (orange) band-pass filters; for DiBAC$_4$(3), a 525-nm band-pass filter, for PI fluorescence a 620-nm long-pass filter (excitation: 488 nm).
- Initial baseline measurement for 30 s without stimulation.
- Add a stimulus, e.g. fMLP (10^{-8} mol/l), or TNF-α (1 ng/ml) for neutrophils.
- Negative control: addition of the solvent of your stimulus.
- Positive control: addition of valinomycin (10 μmol/l in ethanol); valinomycin is an ionophore and leads to cell depolarization.

Calibration Curve

- Stain 250 μl cell suspension for 10 min at 37 °C in HBSS with 250 μl DiBAC$_4$(3) solution (400 nmol/l) and 500 μl HBSS or K-HBSS (K^+ concentration in 5 mmol/l steps from 5 to 37.5 mmol/l).
- Measure fluorescence.

Data Analysis

- See protocol for calcium.
- By means of a calibration curve, the measured DiBAC$_4$(3) values can be converted to absolute MP values according to a modified Nernst equation:

$$E_K + (\text{mV}) = -59 \log [K^+]_i/[K^+]_o \,. \tag{1}$$

$E_K + (\text{mV})$ is the K^+ MP, the intracellular K^+ concentration $[K^+]_i$ is estimated at 100 mmol/l [3], $[K^+]_o$ is the concentration of the extracellular potassium of the cali-

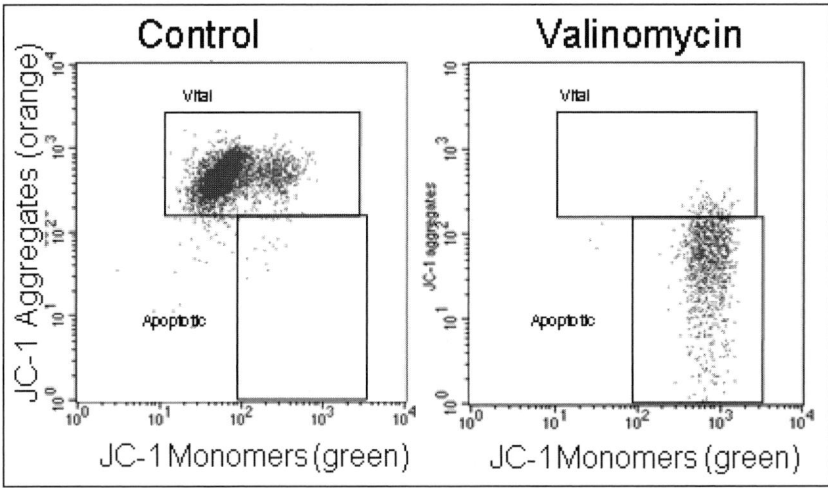

Fig. 2. Measurement of the MP with JC-1. Left figure shows a control measurement without stimulation. Right figure depicts stimulation with valinomycin (2 μmol/l). Valinomycin leads to cell depolarization and apoptosis.

bration solution. In the range of 5–40 mmol/l extracellular K^+ the DiBAC4(3) fluorescence intensity and the E_K should have a linear relationship.

Expected Results

A transient increase in $[Ca_i^{2+}]$ is typically achieved with low concentrations of fMLP (see typical examples for maximum response in figures 1 and 2, and in Tárnok et al. [3, 5, 6]. However, low concentrations are insufficient to evoke an oxidative burst. These experiments are of some use for determining the reaction of blood cells to a specific ligand, but are not well suited for dose-response determinations.

Determination of intracellular elastase activity in neutrophils or of cathepsin B in monocytes yields information about altered cell maturation and differentiation cells. This type of information can be important for inflammatory processes or neoplastic diseases of the hematopoietic system. Increased cathepsin B activity correlates with activation of monocytes in vivo. Degranulation can be observed in vivo during inflammatory processes, but is hardly detectable in peripheral blood cells. One can assume that this process is associated with cellular adhesion.

Troubleshooting

Artificial activation of cells, such as spontaneous calcium burst or nonspecific labeling, are crucial problems in the analysis of phagocytic cells. Both problems can be minimized if samples are quickly and carefully obtained and rapidly analyzed.

Transient Calcium

- Insufficient staining with Indo-1 or Fluo-3 can be improved by the addition of Pluronic F-127.
- Minor changes in the Indo-1 fluorescence ratio (even after stimulation with calcium ionophore) can be due to incomplete hydrolysis of Indo-1/AM. Complete hydrolysis can be obtained by extended incubation in a substrate-free environment. Inactivation of the calcium ionophore can be an additional reason. In solution, the ionophore is only stable for a limited period of time (even at −20 °C). It should be stored in lyophilized aliquots at −80 °C until use.
- The lack of a cellular response can also be due to the absence of extracellular calcium. Check that the cells are not kept in a medium containing a calcium chelator. The lowest possible concentration of fluorochrome that yields homogeneous signals should be applied in order to avoid intracellular buffering of calcium by the probe.

Alkalinization of the Cytosol

The absence of a cellular response in the presence of pH-sensitive cellular fluorescence may indicate the simultaneous activation of hydrogen-exporting antiport and the metabolic generation of H^+ in the cells. Both processes can be separately addressed by the addition of the antiport inhibitor amiloride, which enables analysis of the metabolic generation of H^+ without the compensatory effect of the antiport.

Lysosomal Proteinases

Selective cell death may occur in samples incubated with proteinase substrates. At the same time addition of an inhibitor preserves cellular viability. A reason for this phenomenon can be a high intracellular metabolic accumulation of R110. This effect can be avoided by reducing the incubation time or the substrate concentration.

The sensitivity of fluorescent peptide substrates to enzymes depends on factors such as pH, ion strength or oxidative potential. Therefore, it is advisable to test the specificity of such experiments under unknown intracellular conditions with specific inhibitors.

Membrane Potential

Failure of cells to respond to stimulation may be due to staining toxicity. This may happen especially when using cyanine dyes. One should avoid prolonged cell incubation and reduce the dye concentration in a titration experiment. Absence of mitochondrial staining as checked by microscopy confirms a correct concentration range. In general, oxonol dyes are preferable for the measurement of cellular MP.

Summary

The analysis of cell physiological parameters such as pH, Ca^{2+} or membrane potential enables sensitive detection of functional responses to stimulation. The functional characterization of phagocytic blood cells is an interesting application. Neutrophils and monocytes are involved, among others, in acute and chronic inflammatory processes. Dysfunctions of these cells can be accompanied by increased susceptibility to infections. The functional repertoire of phagocytes correlates with their state of differentiation. Ligands, e.g. bacterial products, cytokines, complement, can activate phagocytes and induce specific cellular reactions like chemotaxis. Phagocytosis comprises a complex cascade of stimulatory actions and functional responses. Cellular activation also involves processes on the cell surface such as expression of adhesion antigens which determine an interaction with other blood cells or the endothelium. Artifactual activation can already occur during the preparation of neutrophils, and this can be demonstrated using sensitive cytometric assays.

References

1 Osterud B: Tissue factor expression by monocytes: regulation and pathophysiological roles. Blood Coagul Fibrinolysis 1998; 9(suppl 1):9–14
2 Kindt TJ, Goldsby RA, Osborne BA, Kuby J: Immunology, ed 6. New York, Palgrave Macmillan, 2006.
3 Tárnok A, Schlüter T, Berg I, Gercken G: Silica induces changes in cytosolic free calcium, cytosolic pH, and plasma membrane potential in bovine alveolar macrophages. Anal Cell Pathol 1997;15:61–72.
4 Tárnok A, Dörger M, Berg I, Gercken G, Schlüter T: Rapid screening of possible cytotoxic effects of particulate air pollutants by measurement of changes in cytoplasmic free calcium, cytosolic pH, and plasma membrane potential in alveolar macrophages by flow cytometry. Cytometry 2001;43:204–210.
5 Tárnok A, Bocsi J, Rössler H, Schlykow V, Schneider P, Hambsch J: Low degree of activation of circulating neutrophils. Determination by flow cytometry during cardiac surgery with cardiopulmonary bypass. Cytometry 2001;46:41–49.
6 Tárnok A, Ulrich H: Detection and purification of rare Ca^{2+} responders by fixed-time flow cytometry; in Radbruch A (ed): Flow Cytometry and Sorting. Springer Lab Manual. Berlin, Springer, 2000, pp 140–158.
7 Forsyth KD, Levinsky RJ. Preparative procedures of cooling and re-warming increase leukocyte integrin expression and function on neutrophils. J Immunol Methods 1990;128:159–163.
8 Hooper NM: Proteases: a primer. Essays Biochem 2002;38:1–8.
9 Rothe G, Klouche M: Phagocyte function; in Darzynkiewicz Z, Roederer M, Tanke HJ (eds): Methods in Cell Biology. San Diego, Elsevier, 2004, vol 75: Cytometry, ed 4, pp 679–798.
10 Rothe G, Assfalg-Machleidt I, Machleidt W, Klingel S, Zirkelbach C, Banati R, Mangel WF, Valet G: Flow cytometric analysis of protease activities in vital cells. Biol Chem Hoppe-Seyler 1992;373:547–554.
11 Shapiro HM: Cell membrane potential analysis. Methods Cell Biol 1994;41:121–133.
12 Petit PX, Glab N, Marie D, Kieffer H, Métézeau P: Discrimination of respiratory dysfunction in yeast mutants by confocal microscopy, image, and flow cytometry. Cytometry 1996;23:28–38.

13 Wilson HA, Seligmann BE, Chused TM: Voltage-sensitive cyanine dye fluorescence signals in lymphocytes: plasma membrane and mitochondrial components. J Cell Physiol 1985;125:61–71.

14 Chen LB, Smiley ST: Probing mitochondrial membrane potential in living cells by a J-aggregate forming dye; in Mason WT (ed): Fluorescent and Luminescent Probes for Biological Activity. San Diego, Academic Press, 1993, pp 124–132.

15 Cossarizza A, Baccarani-Contri B, Kalashnikova G, Franceschi C: A new method for the cytofluorimetric analysis of mitochondrial membrane potential using the J-aggregate forming lipophilic cation 5,5′,6,6′-tetrachloro-1,1′,3,3′-tetraethylbenzimidazolcarbocyanine iodide (JC-1). Biochem Biophys Res Commun 1993;197:40–45.

16 Mason DJ, López-Amorós, Allman R, Stark JM, Lloyd D: The ability of membrane potential dyes and calcafluor white to distinguish between viable and non-viable bacteria. J Appl Bacteriol 1995;78:309–315.

17 Chen LB, Fluorescent labeling of mitochondria. Methods Cell Biol 1989;29:103–123.

18 Kaprelyants AS, Kell DB: Rapid assessment of bacterial viability and vitability by rhodamine 123 and flow cytometry. J Appl Bacteriol 1992;72:410–422.

19 Skowronek P, Krummeck G, Haferkamp O, Rödel G: Flow cytometry as a tool to discriminate respiratory-competent and respiratory-deficient yeast cells. Curr Genet 1990;18:265–267.

20 Spangrude GJ, Johnson GR: Resting and activated subsets of mouse multipotent hematopoietic stem cells. Proc Natl. Acad Sci U S A 1990; 87:7433–7437.

21 Robinson JP, Darzynkiewicz Z, Hyun W, Orfao A, Rabinovitch P (eds): Current Protocols in Cytometry. New York, Wiley, 2003.

22 Tárnok A: Improved kinetic analysis of cytosolic free calcium in pressure sensitive cells by fixed time flow-cytometry. Cytometry 1996;23:82–89.

Sack U, Tárnok A, Rothe G (eds): Cellular Diagnostics. Basics, Methods and Clinical Applications of Flow Cytometry. Basel, Karger, 2009, pp 343–376

Oxidative Burst, Phagocytosis and Expression of Adhesion Molecules

Andreas Lun[a] · Ilka Schulze[b] · Joachim Roesler[c] · Volker Wahn[d]

[a] Central Institute for Laboratory Medicine and Pathobiochemistry, Charité, Berlin,
[b] University Hospital of Freiburg, Center for Pediatric and Adolescent Medicine, Freiburg i.Br.,
[c] Clinic and Polyclinic for Pediatric Medicine, Carl Gustav Carus University Hospital, Dresden,
[d] Charité Center for Immunodeficiency, Berlin, Germany

Introduction

Innate Immune Defense – Mechanisms of Granulocytes and Macrophages

Innate defense mechanisms of neutrophils comprise the following functional steps: adhesion to the endothelium following expression of adhesion molecules on the cell surface and on the endothelium, migration guided by chemoattractants, phagocytosis of microbial particles, activation of enzymes needed for degradation, production of reactive oxygen species (ROS), and other mechanisms to neutralize intruding pathogens [1]. Some of these defense mechanisms can be monitored in vitro by flow cytometry. For further short descriptions of important defense mechanisms, see 'Defects in the Phagocytic System' (pp. 345).

Diagnostic Approach

Neutrophils provide important components of the immune defense system. Therefore, innate or acquired defects in their functions lead to increased susceptibility to special opportunistic infectious agents such as *Staphylococcus aureus* and *Aspergillus* spp. Thus, neutrophil function should be scanned if a patient presents with the following symptoms:

– recurrent infections with abscess formation in lymph nodes (especially fistulating) or abscesses in internal organs, particularly in the liver, but also in the spleen, the kidneys, the lungs, and all otherwise unexplained invasive *Aspergillus* infections including hypersensitivity pneumonitis due to *Aspergillus*;

- granuloma formation even if no infection can be detected;
- pneumonia, gingivitis, stomatitis, lymphadenitis, osteomyelitis, delayed wound healing, especially if combined with fistula formation;
- disease manifestations which include the following differential diagnoses: tuberculosis, exogenous allergic alveolitis, allergic bronchopulmonary aspergillosis, idiopathic pulmonary fibrosis, Crohn's disease, colitis, obstructive uropathy or autoimmune diseases;
- particular coincidences that do not fit with more common illnesses considered first (e.g. atypical young age in the presence of manifestation suggestive of Crohn's disease), and
- delayed separation of the umbilical cord (more than 21 days postnatally).

As an example, patients with chronic granulomatous disease (CGD) suffer from recurrent severe life-threatening infections, especially with *S. aureus* and *Aspergillus* spp.

Besides the clinical symptoms and the isolated pathogens, the gender and the age of the patient at the first manifestation of disease, and the incidence of the illness in the population have to be taken into account when trying to determine through differential diagnosis whether a phagocyte defect is implicated:

- CGD, an innate deficiency of one of the subunits of NADPH oxidase, is the most frequent defect in granulocyte function with an incidence of 1 in 200,000–250,000 live births [2];
- $gp91^{phox}$ deficiency, one of the common causes of CGD, and the G6PDH defect are inherited in an X-linked mode;
- most of the time, CGD first manifests at an age of up to 2 years. In rare cases, the disease can manifest for the first time at any age, however [3].

Diagnostic Procedure When Suspecting a Granulocyte Deficiency

The urgency of applying special phagocyte diagnostics may vary according to the following situations:

- if pathogens, symptoms and/or case history point to a granulocyte disorder, a specialized diagnostic procedure should be performed immediately as described below;
- as an initial screening, if the patient is suspected of suffering from an ambiguous immunodeficiency, we recommend, a full blood count with leukocyte differentiation and microscopic examination of leukocytes and an assay of the IgG, IgA and IgM concentrations. These examinations rule out other reasons for the patient's symptoms such as neutropenia (<1,000 neutrophils/µl), hemato-oncological disease or lowered immunoglobulin concentrations.

Due to a lower prevalence and incidence of complement deficiencies in comparison to granulocyte deficiencies, a test of the hemolytic complement activity CH50 and APH50 (CH100 and AH100) is not recommended initially if the symptoms are not indicative of a complement deficiency. If a granulocyte function deficiency is not ruled out after these analyses, the following tests should all be performed in parallel: examination of the expression of adhesion molecules (CD18, CD15s), and a burst test, or – if fresh blood is available – a nitroblue tetrazolium (NBT) test. Simultaneous performance of these tests allows for a more confident diagnosis.

A primary granulocyte function defect (e.g. CGD) is a severe condition. Therefore, a verification of the results by examining a second blood sample of the patient is indicated to confirm the diagnosis prior to informing the patient.

Defects in the Phagocytic System

Innate immunity fends off infectious agents in the first minutes, hours and days following exposure, during the lag period of 24–36 h required to generate a specific adaptive immune response. Innate immune responses to septic injury comprise three basic mechanisms: the induction of proteolytic cascades that are aimed at walling off the infection, the recognition of microbial patterns and engulfment of the infectious agent by phagocytic cells, and the induction of a humoral response in the guise of antimicrobial peptide release (complement, kallikrein, clotting system) [4].

The cellular defense against an infectious agent includes the following steps: extravasation of cells from the blood stream to tissue using adhesion molecules [1], recognition of infectious agents by pattern recognition molecules such as mannan-binding lectin (MBL) [5] and engulfment and killing of the infectious agent by phagocytic cells [6, 7]. The main phagocyte defects or disorders which can be detected by flow cytometry are listed in table 1.

Recognition of Microbial Patterns by Macrophages

Polymorphonuclear neutrophils (PMNs) recognize microbial products via pattern recognition receptors such as MBL or the Toll-like receptors (TLR) [8, 9]. Human PMNs express all TLRs except TLR-3 [8]. Polymorphisms in the genes coding for TLR-2 (Arg753Gln) and TLR-4 (Asp299Gly) are associated with increased susceptibility to bacterial infections, especially sepsis, meningitis, osteomyelitis and respiratory infections [10, 11].

MBL gene polymorphisms are significantly associated with susceptibility to infections [5, 12–18].

Leukocyte Adhesion Deficiency

The trafficking of PMNs from the bloodstream to tissues is important for the defense against bacterial infections, rapid accumulation of leukocytes at the sites of

Table 1. Defects of phagocyte function

Disease	Defect / inheritance / affected function	Diagnostics or additional diagnostics
Congenital defects of phagocyte function		
Leukocyte adhesion deficiency type I	integrin β_2 adhesion protein / AR / adherence, chemotaxis, endocytosis	flow cytometry, CD11a, b, c, CD18 (migration and phagocytosis)
Leukocyte adhesion deficiency type II LAD II	fucosidase regulator I, GDP fucose transporter / AR / rolling, chemotaxis	flow cytometry CD15s, hh-blood group
Specific granule deficiency in neutrophils 5 cases worldwide	enhancer-binding protein epsilon is missing (myeloid-specific transcription factor, which is expressed in granulocyte differentiation CEBPE gene in the region 14q11.2) / AR /	microscopy and special staining: absence of the neutrophil-specific granula, neutrophils with bilobar nuclei (disturbed migration, chemotaxis test, and receptor-expression)
Chronic granulomatous disease X-chromosomal autosomal recessive	membrane component gp91^{phox} (X-linked) / membrane component p22^{phox} cytoplasmic components p67^{phox}, p47^{phox} / AR / disturbed killing because of a decreased formation of ROS by the NADPH oxidase	DHR 123 test with PMA for maximum stimulation (detection of cytochrome $b_{558'}$ confirmation of CGD diagnosis through a second independent test, for example chemoluminescence, NBT test)
Glucose 6-phosphate dehydrogenase deficiency	neutrophil G6PDH deficiency (G6PDH below 1% of normal) (X-linked) insufficient supply of NADPH and decreased formation of ROS in the NADPH oxidase reaction insufficient supply of NADPH in erythrocytes resulting in hemolytic anemia	determination of G6PDH in erythrocytes, Hint: hematological crisis after oxidative strain, search for signs for hemolysis, if G6PDH is under 1% of the normal values → dihydro-rhodamine 123 test as with CGD with rest activity
Immune dysregulation Chediak-Higashi syndrome	CHS1/Lysosomal trafficking regulator (LYST1) gene1q42.1-q42.2; protein is involved in either vesicle fusion or fission / AR / defect results in the formation of phago-lysosomes, decreased pigmentation of hairs and eyes; neutrophils are deficient in chemotactic and bactericidal activities	microscopy, Pappenheim staining → large lysosomal granules, deficiency of cathepsin G and elastase, also in bone marrow, flow-cytometric phagocytosis test, migration

Table 1 continued on next page

Table 1. Continued

Disease	Defect / inheritance / affected function	Diagnostics or additional diagnostics
Enzyme defect with unknown clinical relevance		
Myeloperoxidase deficiency (unknown)	myeloperoxidase gene (MPO) / AR / no detectable activity of the lysosomal enzyme myeloperoxidase in neutrophils and monocytes	microscopy, peroxidase staining or hematology analyzer with peroxidase staining (Bayer, Technicon H3, Advia) burst test like CGD with a residual activity
Acquired		
Burn trauma, diabetes mellitus, sepsis	chemotaxis	Flow-cytometric migration test phagocytosis test
Hemodialysis, diabetes mellitus	leukocyte adhesion	Flow-cytometric phagocytosis test expression of CD11a,b,c, CD18
Leukemia, sepsis, diabetes mellitus, malnutrition	phagocytosis, microbicide properties	phagocytosis test

AR = Autosomal recessive.

inflammatory response and tissue injury [19]. Leukocyte interaction with vascular endothelial cells is mediated by adhesion molecules. The crucial role of the β_2-integrins in leukocyte emigration is obvious in leukocyte adhesion deficiency (LAD) I. LAD I patients suffer from life-threatening bacterial infections (pneumonia, omphalitis) without the formation of pus and markedly delayed wound healing. The severe form usually causes death in early childhood unless bone marrow transplantation is performed. LAD I results from a congenital deficiency of the leukocyte β_2-integrin receptor complex CD11/CD18 on the cell surface [20] (table 1). In patients who suffer from LAD I, separation of the umbilical cord is delayed. Furthermore, they present with persistent leukocytosis.

LAD II shows the role of selectin receptors and their fucosylated ligands. Patients with LAD II suffer from a less severe form. This form resembles mild LAD I, but comprises early progressive neurodegeneration. Neutrophils from LAD II patients do not express sialyl Lewis X (CD15s), a fucose-containing glycoconjugate ligand for E-, P-, and L-selectins [21] and fail to bind to recombinant E-selectin, to E-selectin on endothelial cells and to platelet P-selectin [22]. Flow-cytometric examination shows significantly reduced expression of CD15 on the surface of neutrophils.

Leukocyte attachment on vascular endothelium needs rapid integrin activation on the surface of leukocytes by chemoattractants. LAD III leukocytes express intact integrins but with an impaired ability to generate a high avidity for endothelial ligands [23]. LAD III neutrophils are unable to mediate leukocyte attachment on TNF-α-stimulated endothelium despite normal selectin-mediated rolling. The activation of the integrins on the surface of leukocytes and platelets is disturbed. The integrins cannot rapidly alter their affinity and adhesiveness in response to chemokine signals. Chemokine activation of leukocyte-expressed chemokine receptors and their associated G protein machinery upregulates the avidity of integrin for endothelial ligands. The LAD III defect is in the CalDAG-GEFI gene, which is a key Rap-1/2 guanine exchange factor (GEF) [24]. Diagnosis is not possible by flow cytometry.

Phagocytosis

Binding of the opsonized particles to the receptors induces cell activation and triggers phagocytosis through a signal cascade. Pseudopodia are formed that engulf the particles and form a phagocytic vacuole. The membrane of the vacuole fuses with the membrane of lysosomal granules; as a consequence, their content spreads into the evolving phagolysosome. Cellular activation is triggered through the inositol-3-phosphate (IP_3) and the diacylglycerol (DAG) pathway. To put it simple, the cascade is characterized by the following steps [6, 7, 25–27]: the receptor-coupled G protein activates phospholipase C. This enzyme cleaves phosphatidylinositol-bisphosphate to IP_3 and DAG. Both compounds route the activation cascade to a separate pathway. IP_3 triggers the release of Ca^{2+} from the endoplasmic reticulum and the extracellular space into the cytosol. Ca^{2+} mediates activation through actin and myosin motility and chemotaxis. DAG activates protein kinase C. This enzyme phosphorylates and activates NADPH oxidase and triggers other processes like the degranulation of the secretory granules and the synthesis of arachidonic acid metabolites. Further, protein kinase C modulates the expression of adhesion molecules and chemotaxis.

Measuring phagocytosis is of little clinical relevance and can be omitted in most cases.

Oxidative Burst, Chronic Granulomatous Disease

CGD is a rare genetic disease characterized by the inability of phagocytes to generate ROS via NADPH oxidase activation (table 1). A deficiency of NADPH activation results in proliferation of intracellular microorganisms, life-threatening bacterial or fungal infections (e.g., with *S. aureus*, *Burkholderia cepacia*, mycobacteria, *Aspergillus* spp.), and inflammatory granulomas [2, 6, 28–31]. PMNs directly recognize microbial products via pattern recognition receptors such as TLRs [32]. TLR engagement activates MAP kinases, JNK and PI3K in PMNs [33]. PMN stimulation through TLRs causes an immediate defensive response, including the production of antimicrobial molecules like ROS by NADPH oxidase and cytokines [8, 34]. NADPH oxidase is a multicomponent enzyme system that catalyzes the NADPH-dependent reduction of oxygen to superoxide anions. It is the precursor of the other ROS and consists of the flavochrome b558 component, which is composed of the two subunits: $gp91^{phox}$ and $p22^{phox}$, and is located in the plasma membrane and in specific granules. The other components of the NADPH complex ($p47^{phox}$, $p67^{phox}$, $p40^{phox}$, and the G protein Rac2) are cytosolic proteins [28]. The activation of NADPH oxidase is tightly regulated at infectious and inflammatory sites by cytokines such as IL-8, IFN-γ, IL-1α, and TNF-α to avoid tissue and vascular lesions [35]. In most CGD cases (70%), the defect is located in the X-linked CYBB gene that encodes the $gp91^{phox}$ subunit. The production of reactive oxygen metabolites can be measured by flow cytometry using the dihydrorhodamine 123 test (DHR test). Carriers of the X-chromosomal form will display two granulocyte populations in this test: a normal population and a functionally defective subset.

The α- and β-subunits of NADPH oxidase produce cytochrome b558, which can be measured by flow cytometry [36, 37]. In about 25% of CGD patients, the disease is caused by a mutation in the neutrophil cytosolic factor 1 gene on chromosome 7q11.23, which encodes $p47^{phox}$ [38].

The exact knowledge of the subunit of the NADPH oxidase involved does not play an important role for therapy. It is however necessary for genetic counseling. It is also the prerequisite for prenatal diagnostics (index patient) and maybe for gene therapy in the future. Furthermore molecular diagnosis is of benefit prior to bone marrow transplantation as only genomic DNA is available after the transplantation. This complicates the localization of the involved gene.

Examination of Granulocyte Function

Adhesion Molecules

Principle

A blood sample is incubated at room temperature with FITC- and phycoerythrin (PE)-tagged antibodies, which are directed against CD11c, CD18 and CD15 surface antigens. The cells are repeatedly washed with phosphate buffer, the erythrocytes are disintegrated and antigen expression on the surface of granulocytes is measured by flow cytometry.

Materials

- Flow cytometer with an excitation of 488 nm (argon laser), forward (FSC) and sideward (SSC) scatter; fluorescence 1, wavelength for FITC: 515–545 nm; fluorescence 2, wavelength for PE: 546–602 nm;

- software for measuring, data accumulation and analysis (Cell Quest or equivalent);
- centrifuge equipped for FACS tubes with lids; roller mixer for the gentle resuspension of the sedimented blood cells, and vortex mixer to disperse the cells;
- storage space at 4 °C for the test kits;
- transfer pipettes (2–20, 20–200 and 100–1,000 μl), dispenser and dispenser tips and twelve 75-mm tubes (Falcon or equivalent) with an appropriate rack.

Reagents

- Calibrite^{TM} beads should be performed immediately (Becton Dickinson; catalogue No. 340486; polymethylmethacrylate microparticles of approximately 6 μm in diameter labeled with FITC, PE or peridinin-chlorophyll-protein complex (PerCP) or unlabeled) for the calibration of the flow cytometer and verification of the settings prior to the measurement;
- BD LeucoGate CD45/CD14 (FITC/PE) (Becton Dickinson; catalogue No. 342408; applied 1 : 10 diluted with acidic PBS);
- BD Simulset^{TM} control$_{\gamma 1/\gamma 2}$ (FITC/PE) as negative control (Becton Dickinson; catalogue No. 342409; applied 1 : 10 diluted with acidic PBS);
- CD11a FITC, e.g. clone G-25.2 BD (applied undiluted);
- CD11b PE, e.g. clone D12 BD (applied 1 : 10 diluted with acidic PBS);
- CD11c PE, e.g. clone S-HCL-3 BD (applied 1 : 10 diluted with acidic PBS);
- CD18 FITC, e.g. clone L-130 BD (applied 1 : 10 diluted with acidic PBS);
- CD15s, e.g. clone CSLEX1 BD (applied undiluted);
- anti-mouse IgM FITC e.g. clone DS-1 BD (applied undiluted).

The antibody dilution remains stable for 3 weeks at 4 °C.

- Dulbecco's PBS (1×) without calcium or magnesium, tested for endotoxin, filtered under sterile conditions (PAA Laboratories GmbH, Pasching, Austria; catalogue No. H15–002, 500 ml catalogue No. H22–002, 2,000 ml);
- acidic PBS: 0.2 g NaN_3 and 0.037 g EDTA dissolved in 1 liter PBS;
- lysing stock solution: (82.9 g NH_4Cl, 10.2 g $KHCO_3$ and 292 mg EDTA disolved in 1 liter distilled water; for use, the stock solution is diluted 1 : 10 in distilled water. The dilution is prepared fresh every day.

Procedure

An EDTA blood sample is drawn from the patient and a healthy individual.

A full blood count and 2 blood smears are performed. One smear is stained with the Pappenheim method, the other is kept in store for peroxidase staining. Table 2 summarizes the schedule of setup and incubation.

Measurement

The flow cytometer has to be calibrated before measuring. As the system settings generally remain stable over time, a control with beads of every color is sufficient.

Table 2. Schedule for setup and incubation – adhesion molecules

Tube 1	Tube 2	Tube 3	Tube 4	Tube 5	Tube 6	Tube 7	Tube 8
–	–	–	–	–	–	–	1 µl CD15s
–	–	–	–	–	–	50 µl blood	50 µl blood
Vortex and incubate at room temperature in the dark for 10 min; afterwards, add 3 ml PBS azide and vortex, centrifuge and discard supernatant.							
25 µl Leucogate FITC/PE 50 µl blood	25 µl negative control FITC/PE 50 µl blood	25 µl CD11a FITC 50 µl blood	25 µl CD11b PE 50 µl blood	25 µl CD11c PE 50 µl blood	25 µl CD18 FITC 50 µl blood	25 µl anti-mouse IgM FITC	25 µl anti-mouse IgM FITC
Vortex and incubate at room temperature in the dark for 10 min. Add 3 ml lysing solution, vortex and incubate at room temperature in the dark for 10 min. Centrifuge and discard supernatant. Afterwards, add 3 ml PBS azide and vortex, centrifuge and discard supernatant. Measure with the flow cytometer.							

Compensation within the panel is consistent for well-titrated antibodies. The granulocytes in the FSC lin scattergram are gated versus SSC lin; 10,000–15,000 cells from each sample are measured and the results are saved in a file.

Data Analysis

Granulocytes (as other leukocytes) are gated in an FSC-SSC diagram and checked with the surface markers CD45+/CD14$-^{\mathrm{dim}}$. The negative control sets the limit for the positive expression of surface markers; 95% of the cells are negative (fig. 1). Tube No. 7 serves as negative control for CD15 detection in tube No. 8; it is filled with FITC-labeled anti-IgM. The results are given in percentage of cells with a higher surface fluorescence than the negative control.

Quality Control

In order to verify the settings of the flow cytometer, Calibrite Beads are measured with adjustments for surface molecule detection. The results are compared with representative data from previous measurements. If the results do not match with previous data, it is recommended to rinse the flow cytometer before modifying the system settings. To monitor the whole process, which includes drawing the blood sample, storage of the specimen and transport, the sample from a healthy individual is treated likewise.

Using a lin FSC versus lin SSC scattergram allows for a survey of the size and correct classification of the cell population and comparison with the results of the leukocyte differentiation.

If a LAD I or II is detected on granulocytes, it has to be detected as well on lymphocytes and monocytes in order to verify the diagnosis. A LAD II is always related

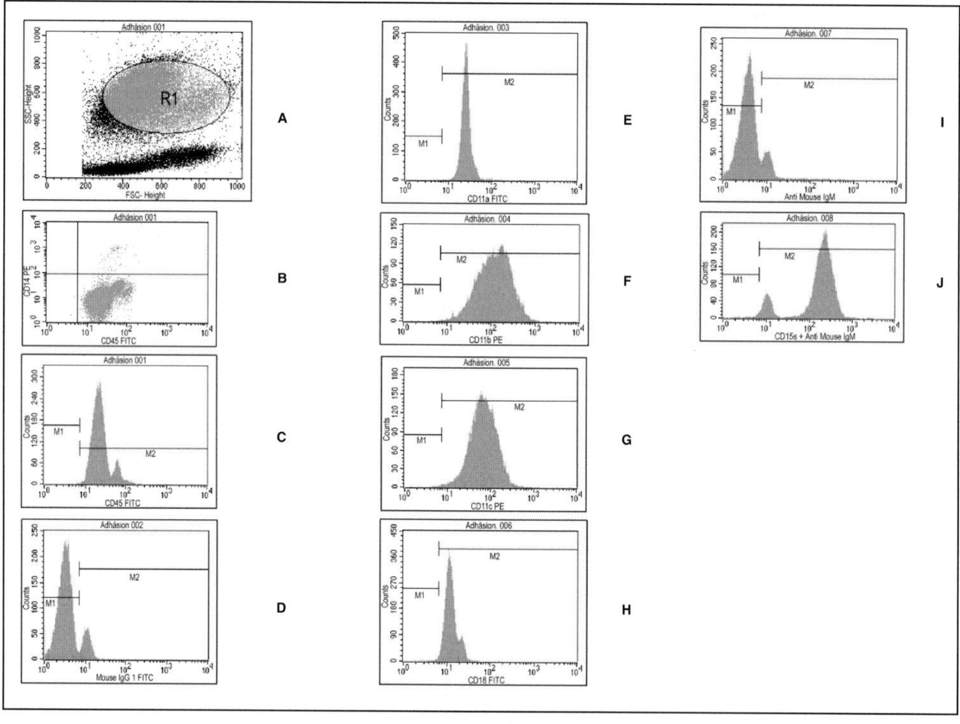

Fig. 1. Expression of adhesion molecules, analysis document. **A** Granulocyte gate FSC. **B** CD14PE CD45 FITC control of granulocyte gate. **C** Histogram, positive control by CD45. **D** Histogram, negative control by anti mouse-IgG1. **E–H** Histogram, expression of adhesion molecules CD11a, CD11b, CD11c und CD18. **I, J** Histogram, expression of adhesion molecule CD15s.

to the H-deficient Bombay red blood cell blood group, and this has to be checked (table 3).

Expected Results, Interpretation and Pitfalls

Deficiencies of adhesion are known as LAD I and LAD II. The three integrin molecules associated with LAD I are coded on gene locus 21q22.3. These are CD18/CD11a, also known as LFA-1; CD18/CD11b also known as complement receptor 3 or Mac-1, and CD18/CD11c, also known as Leu-CAMc.

LAD II (chromosome 11) derives from a defect of the GDP-fucose transporter (FUCT1) which leads to absence of sialyl Lewis X, an oligosaccharide epitope on the surface of leukocytes.

This epitope is responsible for the binding of the leukocytes to activated endothelial cells and permits leukocyte rolling (a loose binding of leukocytes to the endothelium surface which is required for transmigration).

In inconspicuous samples, 95–99% of the granulocytes express adhesion molecules. The cells of patients with LAD, however, do not express the corresponding adhesion molecule or a combination of adhesion molecules on the surface. A defi-

Table 3. Troubleshooting

Problem	Cause	Solution
High fluorescence of all cells (even lymphocytes, without stimulation)	measuring chamber contaminated with fluorescent dye unstained cell fluorescence	rinse measurement chamber
Monocytes not distinguishable	old blood sample, monocyte-poor sample	mark the monocytes with CD14
Too few granulocytes acquired	Neutropenia	new sample with more whole blood according to the differential blood count
Monocytes and granulocytes are hard to separate	sample was stored too long precursor of granulopoiesis in peripheral blood	examine fresh sample examine in an interval without acute infection, prior test for immature precursors, e.g. meta-myelocytes
Two-peak population of granulocytes in the histogram of an adhesion molecule	blood pipetted to the tube wall which therefore was not incubated with antibodies	repeat and as a guideline remove blood from the tube wall with a cotton wool wad so all leukocytes are incubated with antibodies

ciency of CD18 also leads to the absence of CD11a, b, c as CD18 serves as an anchor for the other integrins and embeds them in the cell membrane. The consequences of this deficiency are the disruption of the adsorption of the leukocytes to the endothelium, the migration of the granulocytes, phagocytosis and oxidative burst when a bacterial stimulation takes place [39, 40].

LAD II is characterized by the absence of CD15s, a molecule on the surface of granulocytes.

Patients suffering from LAD II lack substance H on their erythrocytes. This is a fucosylated glycoprotein which is a precursor of blood group proteins A, B and O. Therefore these patients have the blood group 'Bombay' and cannot secrete the blood antigens (A, B, H) in their saliva. When confronted with infections, they react with a steep increase in their leukocyte numbers (30,000–150,000/µl) [41].

Further LADs which clearly do not belong to LAD I or LAD II have been studied recently.

Several antibiotics exert an influence on granulocyte adhesion: imipenem and rifampicin increase adhesion whereas colistin, polymyxin B, chinin and chloroquin inhibit adhesion [42]. Penicillin G, cephaloxin, bacitracin, minociclin, doxycyclin, chloramphenicol, erythromycin, licomycin, neomycin, streptomycin, acetylspiramycin, fluctosin, primaquin, tinidazol and sulfhisoxazol do not influence adhesion [42].

Pharmaceuticals which inhibit the expression of the adhesion molecules and are applied in high doses can decrease the expression of such molecules. This may lead to a false diagnosis of LAD.

The expression of adhesion molecules 2 h after a surgery is also significantly decreased, probably due to inflammatory reactions [43].

Caution: Partial deficiencies or deficiencies with expression of nonfunctional proteins or decreased function can be missed in the process of analysis. Therefore functional tests (phagocytosis, motility) can be important!

The only patient [44] thus far having a Rac2 deficiency displayed a clinical picture similar to LAD in combination with CGD. Her case history showed severe bacterial infections and poor wound healing.

Rac2 is a regulator of granulocyte activation. The expression of the adhesion molecules cannot be upregulated when stimulated (see 'Burst' for the modifications of stimulated ROS formation).

An abnormal positioning in the FSC or SSC gives reason to examine the granulocytes microscopically (e.g. Chediak-Higashi syndrome).

A mutation in the enhancer binding protein epsilon (CEBPE gene 14q11.2) leads to the very rare (5 cases worldwide) neutrophil-specific granule deficiency (SGD) which causes defective granulocyte differentiation and maturation [30]. Phenotypically, the disease is characterized by bilobular nuclei, reduced granularity of the cytoplasm and significantly reduced superoxide production in phorbolmyristate acetate (PMA) medium, and malfunctioning chemotaxis and bactericidal activity. The expression of the surface molecules CD45, CD11b, CD14, CD15 and CD16 is peculiar. Even if suffering from sepsis, SGD patients produce no or very little IL-6 whereas IL-8 is produced in an adequate amount.

Inaccurate pipetting: if part of the cells was pipetted to the tube wall and therefore had no or insufficient contact with the FITC-labeled anti-CD18 antibodies, the histogram displays two peaks.

Required Time

The test, including the measurements, has to be performed in one session. An interruption for hours and a finalization afterwards is not acceptable. Only data analysis may be postponed.

Measuring the patient specimen in parallel with a control sample and acquiring the data takes approximately 1.5–2 h. Processing of up to 10 patient samples takes only a little longer (30–60 min).

Data analysis requires about 10 min for the first sample; any subsequent specimen takes approximately 5 min. Usually, phagocytosis, expression of adhesion molecules and formation of ROS are determined from one patient sample and a control sample. The time required for the analysis of a patient sample with its necessary control is then approximately 3–4 h.

Phagocytosis Test

Principle

Heparinized full blood is incubated with FITC-labeled, opsonized inactivated *Escherichia coli* bacteria at 37 °C for 10 min; a control sample is treated likewise and incubated at 0 °C for 10 min. Phagocytosis is stopped by the addition of iced quenching solution. The sample is washed twice; the erythrocytes are then lysed and removed. In this step, the leukocytes are simultaneously fixed. This operation is followed by staining with DNA staining solution to exclude artifacts of bacterial aggregates and thrombocytes. In the following measurement, the granulocytes are selected according to the FSC and SSC light signal (FSC-SSC diagram). This cell fraction is then checked for contamination with bacterial aggregates by DNA fluorescence and for fluorescence of the absorbed FITC-labeled *E. coli*.

When incubated at 37 °C, granulocytes usually show distinct internalization of the fluorescence-marked *E. coli*, which is not the case at 0 °C.

Materials

- Flow cytometer with an excitation of 488 nm (argon laser), FSC and SSC light; fluorescence 1, wavelength for FITC: 515–545 nm; fluorescence 2, wavelength for propidium iodide (PI; DNA staining solution): 546–600 nm
- Software for measuring, data accumulation and analysis (Cell Quest or equivalent)
- Centrifuge equipped for FACS tubes with lids, roller mixer for the gentle resuspension of the sedimented blood cells, vortex mixer to disperse the cells
- Water bath (37 °C chemotaxis), ice bath (0 °C)
- Storage space at 4 °C for the test kits
- Variable transfer pipettes (2–20, 20–200 and 100–1,000 µl), dispenser and dispenser tips and twelve 75-mm tubes (Falcon or equivalent) with an appropriate rack.

Reagents

- Test kit Phagotest® from Orpegen Pharma (Heidelberg, Germany), consisting of:
  - FITC-marked opsonized and stabilized *E. coli* suspension, approximately 1×10^9 bacteria/ml;
  - Quenching solution (similar to trypan blue), ready to use;
  - DNA staining solution (PI), ready to use;
  - lysing solution (diethylene glycol, formaldehyde);
  - washing solution, (PBS with EDTA and preservative), diluted preportioned salts in 1,000 ml distilled water;
  - Calibrite™ beads (Becton Dickinson; catalogue No. 340486; polymethylmethacrylate microparticles of approximately 6 µm in diameter labeled with FITC, PE or PerCP or unlabeled) for the primary calibration of the flow cytometer and the verification of the settings prior to the measurement. A sufficient amount of solution prior to use is important.

Procedure

- Prearrangements:
  - set up ice bath (0 °C), and water bath (37 °C);
  - prepare washing solution (dilute preportioned salts in 1,000 ml distilled water), and lysing solution, (it is diluted 1 : 10 from the stock solution with distilled water; the lysis solution has to be prepared fresh);
  - precool quenching solution and opsonized FITC-marked *E. coli* suspension (approximately 1×10^9 bacteria/ml) in an ice bath;
- blood count with differentiation of the leukocytes in order to be able to modify the sample (400,000–950,000 neutrophil granulocytes/sample) (table 4).

Measurement

The cells are measured with the flow cytometer at a wave length of 488 nm and the Cell-Quest software. During data accumulation, a live-gate is placed on the cells in the red fluorescence histogram (DNA content of a diploid cell) to exclude bacterial aggregates which display the same scatter light characteristics as leukocytes. 10,000–15,000 cells from each sample are measured and the results are saved in a file.

During data acquisition, the green fluorescence of the granulocytes in the control sample which was incubated at 37 °C should be higher than that of granulocytes incubated on ice; this is due to phagocytosis of the FITC-marked bacteria.

Data Analysis

The percent fraction of cells (monocytes and granulocytes) which have phagocytosed bacteria is analyzed as well as their mean green fluorescence intensity, which is proportional to the number of internalized bacteria per cell. For this purpose, the cells are selected in the SSC versus fluorescence 2 (DNA fluorescence) (gate 1) dot plot. Afterwards, the limits for the granulocytes/monocytes are set in FSC versus lin SSC (gate 2).

The green fluorescence of the cells being gated through gate 1 and gate 2 is observed in the SSC versus fluorescence 1 dot plot and then analyzed in the FL1 histogram.

For phagocytosis and burst test analysis, similar data analysis is utilized (fig. 2).

In the green fluorescence 1 histogram of the control sample (0 °C incubation) a marker is set in such a way that less than 1% of all cells are positive. The marker position in the test sample determines the percent fraction of phagocytosing cells. The mean fluorescence activity equals the quantity of bacteria per single leukocyte. In addition to measured values, a comment summarizes all results including the burst test. To evaluate the opsonization ability of the plasma, the use of nonopsonized FITC-marked *E. coli* is possible.

Quality Control

In order to verify the settings of the flow cytometer, Calibrite Beads are measured with adjustments for surface molecule detection. The results are compared with

Table 4. Scheme for setup and incubation – phagocytosis

Patient sample		Control sample (healthy individual)	
Test sample	control sample	test sample	control sample
100 μl blood	100 μl blood	100 μl blood	100 μl blood
Incubate for 10 min on ice			
20 μl precooled *E. coli* bacteria suspension	20 μl precooled *E. coli* bacteria suspension	20 μl precooled *E. coli* bacteria suspension	20 μl precooled *E. coli* bacteria suspension
Vortex			
Incubate for 10 min 37 °C	incubate for 10 min in an ice bath	incubate for 10 min 37 °C	incubate for 10 min in an ice bath
Stop the phagocytosis by placing the test sample in an ice bath			
100 μl ice cold quenching solution	100 μl ice cold quenching solution	100 μl ice cold quenching solution	100 μl ice cold quenching solution
Vortex Wash twice with 3 ml washing solution, centrifuge for 5 min at 250 × *g* and 4 °C; decant supernatant. 2 ml lysing solution (preheated to room temperature); vortex. Incubate for 20 min at room temperature in the dark. Centrifuge for 5 min at 250 × *g* and 4 °C; decant supernatant. Add 3 ml washing solution, mix and centrifuge (5 min 250 × *g*, 4 °C).			
200 μl DNA staining solution	200 μl DNA staining solution	200 μl DNA staining solution	200 μl DNA staining solution
Store sample till measured in the dark and on ice. Perform measurement within 1 h after addition of the DNA staining			

representative data from previous measurements. If the results do not match with previous data, it is recommended to rinse the flow cytometer before modifying the system settings. To monitor the whole process, which includes drawing the blood sample, storage of the specimen and transport, the sample of a healthy individual is treated likewise.

Using a lin FSC versus lin SSC scattergram allows for a survey of the size and the correct classification of the cell population and comparison with the results of the leukocyte differentiation.

Ideally, the frequency peak of the control granulocyte fluorescence is positioned at 0–20 arbitrary fluorescence units for the sample incubated on ice and at 37 °C, at 1,000 fluorescence units for the sample incubated. The curves of the frequency

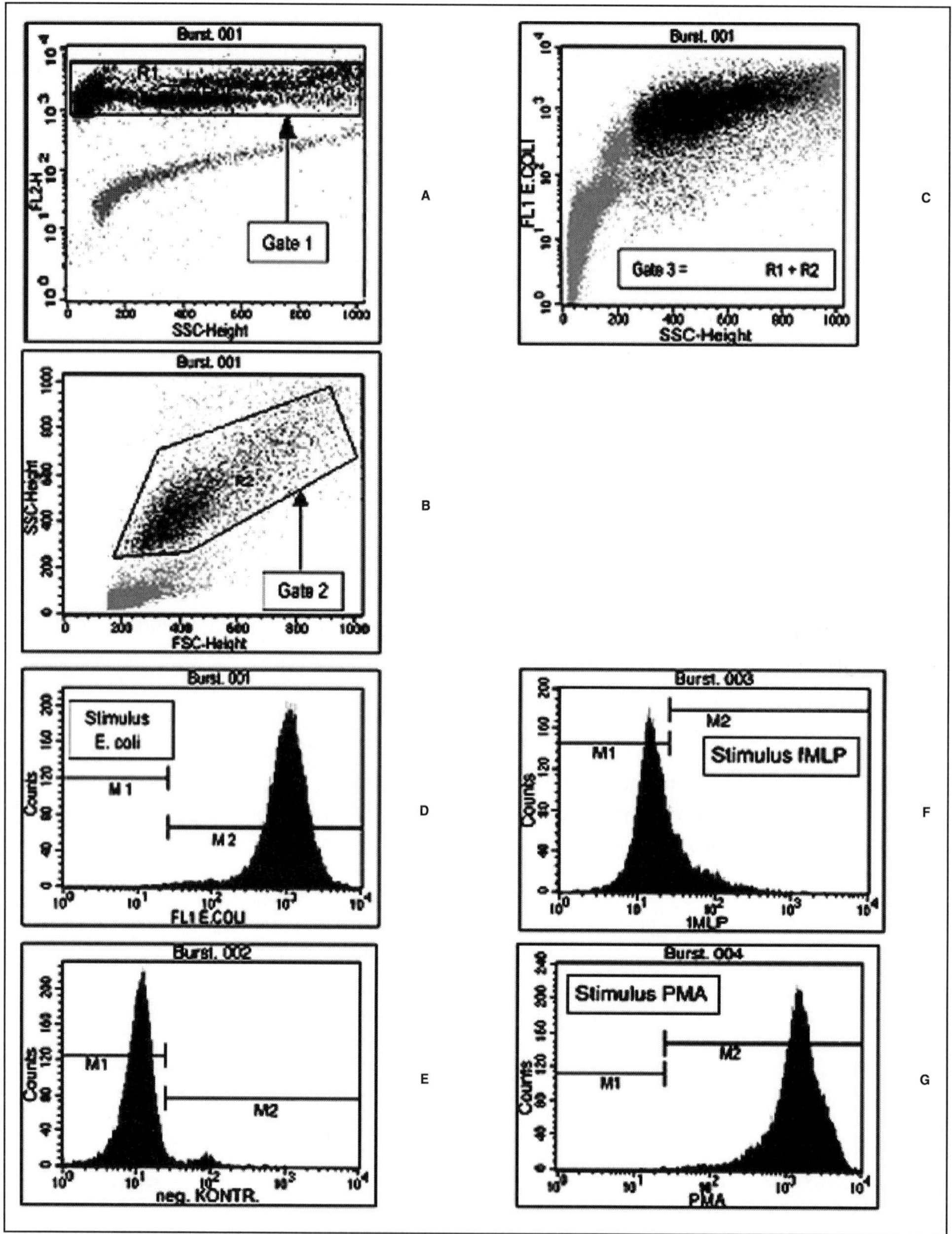

distribution do not overlap. Since a phagocytosis dysfunction makes a stimulus from *E. coli* in the burst test ineffective, combined determination of burst and phagocytosis is recommended.

A complex diagnostic finding which summarizes and interprets all test results is more meaningful than separate single findings (table 5).

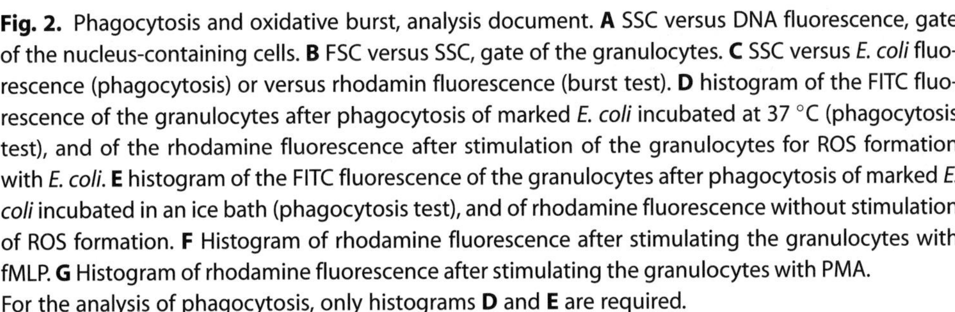

◄

Fig. 2. Phagocytosis and oxidative burst, analysis document. **A** SSC versus DNA fluorescence, gate of the nucleus-containing cells. **B** FSC versus SSC, gate of the granulocytes. **C** SSC versus *E. coli* fluorescence (phagocytosis) or versus rhodamin fluorescence (burst test). **D** histogram of the FITC fluorescence of the granulocytes after phagocytosis of marked *E. coli* incubated at 37 °C (phagocytosis test), and of the rhodamine fluorescence after stimulation of the granulocytes for ROS formation with *E. coli*. **E** histogram of the FITC fluorescence of the granulocytes after phagocytosis of marked *E. coli* incubated in an ice bath (phagocytosis test), and of rhodamine fluorescence without stimulation of ROS formation. **F** Histogram of rhodamine fluorescence after stimulating the granulocytes with fMLP. **G** Histogram of rhodamine fluorescence after stimulating the granulocytes with PMA.
For the analysis of phagocytosis, only histograms **D** and **E** are required.

Expected Results, Interpretation and Pitfalls

Normally, more than 90% of the granulocytes and 65–95% of the monocytes fulfill phagocytosis. The mean fluorescence intensity of granulocytes ranges around 800–1,800 fluorescence units, the fluorescence of the monocytes ranges between 300 and 900 fluorescence units. The nonphagocytosing cells in the ice bath reach 10 fluorescence units.

Inherited defects comprise actin dysfunction; tuftin deficiency, complement fragment C3bi receptor deficiency and LAD (see 'Defects in the Phagocytic System', pp. 345) which can all result in impaired phagocytosis.

Acquired defects of phagocytosis are observed in patients suffering from sepsis, diabetes mellitus, renal failure and infections. Reduced phagocytosing capacity is observed in patients suffering from AIDS and burns, and in newborns and older people [45, 46]. Some cytokines (IL-2, IFN-γ) and lactobacilli increase phagocytosis [46]. Phagocytosis of granulocytes is reduced after surgery [43]. Hypocalcemia leads to reduced phagocytosis [47]. The quinoline derivates pefloxacin, ofloxacin and fleroxacin and the antibiotics nitrofurantoin, rifampicin, chloramphenicol, amikacin, gentamycin, tetracycline and chloroquine inhibit phagocytosis if administered in high doses [42, 48]. Such inhibitory concentrations were not seen in blood but in urine.

There is no disease entity with an isolated dysfunction of phagocytosis. Diagnoses which include a permanent phagocytosis dysfunction may have severe implications. Therefore analysis of a second blood sample from the patient for verification is in queue before the diagnosis is definite and the patient informed.

Required Time

The test, including the measurement, has to be performed in one session. An interruption for hours and a finalization afterwards is not acceptable. Only the analysis can be postponed. Up to 4 patient samples can be processed at once. Measuring the patient sample together with a control sample and acquiring the data takes approximately 2.5–3 h. Processing of 4 patient samples takes only a bit longer (30 min). Data analysis requires about 10 min for the first sample; a subsequent specimen takes approximately 5 min.

Table 5. Troubleshooting phagocytosis measurement

Problem	Cause	Solution
High fluorescence of all cells (nonstimulated lymphocytes)	measuring chamber contaminated with fluorescent dye unstained cell fluorescence	rinse measuring chamber
Wrong test tubes (citrate, EDTA, serum)	Ca^{2+} bound in complexes, not available for the activation	new sample and briefing on the right sample drawing conditions
Too few granulocytes acquired	neutropenia	new sample with more whole blood according to the differential blood count
Monocytes and granulocytes are hard to separate	sample was stored too long precursor of the granulopoiesis in peripheral blood	examine fresh sample examine in an interval without acute infection, prior test for immature precursors, e.g. metamyelocytes
Sample contains too few vital cells	wrong sample, fluoride was added	new sample, use heparin test tube
High fluorescence in the control sample at 0 °C	eosinophils, increased auto-fluorescence. Incompletely anticoagulated samples can simulate phagocytosis-positive cells in the 0 °C setup. This is caused by thrombocyte aggregates and dead cells with released DNA, which unspecifically bind and lock up bacteria, so that their fluorescence cannot be suppressed.	optimize gating draw a new sample paying attention to the proper mixing of blood and heparin in the tube, invert several times
Second population of non-phagocytosing granulocytes	blood pipetted to the tube wall which then was not incubated with antibodies	repeat, remove blood from the tube wall so that all granulocytes have contact to *E. coli*

Usually, phagocytosis, expression of adhesion molecules and formation of ROSare determined from one patient sample and a control sample. The time needed for a patient sample including the necessary controls is then approximately 3–4 h.

Dihydrorhodamine Test for the Determination of Reactive Oxygen Species – Formation following a Stimulus (Burst Test)

Principle

The formation of ROS in granulocytes can be triggered in three different ways: i) by opsonized *E. coli* as particle stimulus, ii) through the protein kinase C activator PMA, which serves as maximum stimulus, or iii) through the chemotactic peptide N-formyl-MetLeuPhe (fMLP), which serves as a weak stimulus. Added DHR 123 is oxidized by H_2O_2 to rhodamine, which is used as indicator for the emerging ROS. The increase in fluorescence is approximately proportional to NADPH oxidase activity. The reaction is stopped through the addition of FACS 'lysing solution'; the erythrocytes are lysed and in this step the leukocytes are simultaneously fixed and permeabilized. The addition of DNA staining solution permits the discrimination of leukocyte and bacterial aggregates. Rhodamine fluorescence as a criterion for ROS formation is measured by flow cytometry. The rhodamine fluorescence histograms of the stimulated and nonstimulated samples are compared. With an intact NADPH oxidase the mean fluorescence without stimulation is clearly lower than after moderate stimulation and drastically lower than after maximal stimulation.

The first description [49–51] and the numerous following descriptions [41, 52] of the diagnosis of CGD using DHR 123 and flow cytometry show a clear advantage of this method over older testing methods such as the NBT slide test.

Materials

- Flow cytometer with an excitation of 488 nm (argon laser), FSC and SSC light and fluorescent light; fluorescence 1, wavelength for FITC: 515–545 nm; fluorescence 2, wavelength for DNA measurement: 546–600 nm
- centrifuge equipped for FACS tubes with lids, roller mixer for the careful resuspension of the sedimented blood cells and vortex mixer to disperse the cells;
- ice bath (0 °C), water bath (37 °C) and storage space at 4 °C for the test kits;
- transfer pipettes (2–20, 20–200 and 100–1,000 µl) and manual dispenser;
- burst test (Phagoburst®) test kit (Orpegen Pharma, Heidelberg, Germany) and burst test according to Rothe et al. [53].

In this section, the burst test from Orpegen Pharma and a test according to Rothe et al. [53] are presented; they do not differ in test principle, apparatus, measurement, analysis or influence of preanalytical factors. Reagents and incubation schedule of the two tests will be labeled with A or B. The test according to Rothe et al. [53] makes it possible to prepare the reagents oneself.

Reagents

- Burst test (Phagoburst®) test kit or else purchase from Becton Dickinson;
- heparin blood vacutainer or tubes from Becton Dickinson or Sarstedt;

- kit components: all 7 components are stored at 4 °C (watch the storage time);
  - 1 vial (2 ml) opsonized not marked *E. coli* suspension, 1 × working solution;
  - chemotactic peptide fMLP (200 × stock solution, 1 mmol/l – by dilution with washing solution, a 1 × working solution is set up, e.g. 5 μl for 1,000 μl);
  - PMA (200 × stock solution, 1 mmol/l – by dilution with washing solution, a 1 × working solution is set up, e.g. 5 μl for 1,000 μl);
  - vial with substrate disk contains the fluorescence dye DHR123; it has to be reconstituted 30 min prior to use by adding 1 ml of washing solution, discard after use!
  - DNA staining solution (PI) 1 × working solution;
  - washing solution (Instamed-Salts PBS + EDTA + preservative) dissolved in 1,000 ml distilled water;
  - FACS 10 × lysing solution (diethylene glycol + formaldehyde) to produce a 1 × solution, dilute 1 : 10 with bidistilled water or BD lysing solution;
- distilled water to reconstitute the washing solution and to dilute the lysing solution;
- Calibrite™ beads (polymethylmethacrylate microparticles of approximately 6 μm in diameter labeled with FITC, PE or PerCP or unlabeled) for the calibration of the flow cytometer and verification of the settings prior to the measurement. A sufficient amount of solution prior to use is important.

Drawing the Sample, Transport and Storage

The blood sample should not contain any clots, should not be overly hemolytical and reach the laboratory without cooling within 4 h drawing. These requirements are vital for the detection of a carrier status.

For the analysis or the exclusion of CGD, longer storage or transport times (postage) is acceptable. The temperature range should not exceed 15–22 °C.

Procedure

The burst test from Orpegen Pharma is performed according to the manufacturer's manual. The manufacturer has performed the tests to verify the accuracy, specificity and sensitivity and thus the test kit is a certified product. Therefore these tests do not have to be performed anymore when holding to the manufacturer's manual.

- Preliminary steps:
  - washing solution, lysing solution, fMLP work solution, PMA work solution and substrate solution are produced by diluting the preportioned salts or diluting the stock solutions;
  - set up the ice water bath (0–4 °C) and switch on the water bath (37 °C);
  - check the temperature;

- precool the bacteria suspension in the ice water bath;
- singularize the cells by vigorous mixing with a vortex mixer;
- turn on and check the flow cytometer.

Test Setup

The heparin blood samples have to be mixed for at least 5 min with the Coulter mixer. The samples are pipetted to the bottom of 4 Falcon tubes (5 ml). Blood traces on the tube walls should be avoided; any traces must be removed at once (table 6).

The flow-cytometric measurement should be performed within 30 min after this preparation. Longer storage leads to a poorer cut-off of the granulocyte gates and a smaller difference between the histogram peaks of the rhodamine fluorescence with and without stimulation.

Burst Test According to Rothe et al. [53]

Reagents

- DHR 123, MW 346 (Molecular Probes, Eugene, OR, USA)
  Use dimethylformamide (DMF) to rinse the glass ampoule and obtain the DHR 123 (10 mg content), dissolve DHR 123 in 26.2 ml DMF. Discard the broken-off ampoule head as it contains oxidized DHR at the neck. The resulting pale-pink stock solution of 1.1 mmol/l DHR (381 μg/ml) in DMF is portioned into 100-μl aliquots. Aliquotation is done in a cool environment using Eppendorf tubes (volume of 1.5 ml, with Safe-Lock and brown light protection, catalogue No. 0030 120.191). The solution can be stored at −20 °C with a stability >12 months, decay is noticeable by an orange coloring.
- DMF (Aldrich-Chemie, Sternheim, Germany; MW 73.1);
- fMLP (Sigma, Deisenhofen, Germany; MW 437.6);
  A 1 mmol/l stock solution of fMLP (436.7 μg/ml) is prepared in DMF and stored in portions of 10 μl in Eppendorf tubes (volume of 1.5 ml, with Safe-Lock) at −20 °C with a stability of >60 months.
- Hanks balanced salt solution (HBSS) is produced using 10 mmol/l Hepes solution (pH 7.40). Dry powdered HBSS medium is dissolved in 1 liter bidistilled water while adding 5 ml 2 mol/l Hepes solution and adjusted to pH 7.40 with approximately 2 ml 2 N sodium hydroxide solution. The pH at 37 °C is controlled in an aliquot. At 4 °C the shelf life is <1 week.
- Hepes (Serva, Heidelberg, Germany; MW 238.3):
  A 2 mol/l stock solution (476.6 g/l) is prepared in bidistilled water and adjusted to pH 7.00 by the addition of 14 g/l NaOH. Storage at 4 °C results in a shelf life of approximately 5 months.
- Perhydrol 30% (H_2O_2) (Merck, Darmstadt, Germany; catalogue No. 7210):

Table 6. Incubation scheme for the measurement of stimulated formation of ROS in granulocytes

Without stimulation negative control	fMLP stimulation *low stimulus,* chemotactical peptide	*E. coli* stimulation particular stimulus	PMA stimulation *high stimulus,* protein kinase C ligand
100 μl blood	100 μl blood	100 μl blood	100 μl blood
Incubate 10 min on ice			
20 μl washing solution	20 μl fMLP	20 μl *E.coli*	20 μl PMA
Close tubes with lid while in ice bath			
Vortex, incubate at 37 °C for 10 min			
20 μl dihydrorhodamine	20 μl dihydrorhodamine	20 μl dihydrorhodamine	20 μl dihydrorhodamine
Vortex, incubate at 37 °C for 10 min			
Stop the reaction by placing the test sample in an ice bath			
3 ml lysing solution	3 ml lysing solution	3 ml lysing solution	3 ml lysing solution
Incubate for 20 min at room temperature in the dark			
Vortex; samples are centrifuged at 300 × *g* for 5 min at 4 °C; decant supernatant			
Stir up sedimented cells by vortexing			
3 ml washing solution	3 ml washing solution	3 ml washing solution	3 ml washing solution
Centrifuge at 300 × *g* for 5 min at 4 °C; decant supernatant			
3 ml washing solution	3 ml washing solution	3 ml washing solution	3 ml washing solution
Centrifuge at 300 × g for 5 min at 4 °C; decant supernatant			
200 μl DNA staining solution	200 μl DNA staining solution	200 μl DNA staining solution	200 μl DNA staining solution
Vortex and incubate samples in the dark for 10 min in an ice bath			
Measure within 30 min			

An approximately 100 mmol/l fresh working solution of H_2O_2 is prepared daily by diluting hydrogen peroxide 1:100 (v/v) with 5 mmol/l Hepes-buffered saline solution (0.15 mol/l NaCl, pH 7.35).

- LDS-741 (Exciton, Dayton, OH, USA; MW 445.5):
  A 0.4 mmol/l stock solution (178 μg/ml) is produced using DMF and stored at –20 °C; stable for ≫12 months.
- *E. coli* suspension:

A suspension of *E. coli* K-12 (Sigma) is cultured in RMPI-1640 (Sertomed, Biochrom, Berlin) for 48 h at 37 °C and 5% CO_2; 5 ml of the culture suspension are centrifuged at 5,000 × *g*, washed twice with 10 ml 5 mmol/l Hepes-buffered saline solution (HBSS, 0.15 mol/l NaCl, pH 7.35) and resuspended in HBSS to a concentration of 3×10^9 bacteria/ml HBSS. The washed bacteria are portioned into Eppendorf tubes and stored at −20 °C; shelf life >60 months.
– Histopaque 1,077 (Sigma; catalogue No. 1977–1) or equivalent polysaccharide-containing separation media.

Test Setup for Leukocyte-Rich Plasma
3 ml Histopaque 1077 is overlayed with 3 ml of heparin blood. The erythrocytes sediment by aggregation without centrifugation. After 40 min, a clear plasma supernatant has formed which contains approximately 2 ml spontaneously separated leukocytes and platelets. 0.8 ml of the top of the plasma supernatant are pipetted and stored on ice. Table 7 displays the incubation schedule for the measurement of ROS formed in granulocytes after stimulation according to Rothe et al. [53].

The burst test with isolated granulocytes showed a slightly better discriminatory power than the conventional whole blood method. Granulocytes from a CGD patient with substantial residual activity were more reliably discerned as well using the granulocyte preparation instead of the whole blood test [Roesler, personal observations].

The vitality of isolated unfixed cells can be judged after activation by PMA according to shifts in the FSC-SSC diagram because of the Na^+ and H_2O influx in viable cells. The population of dead or extensively damaged cells does not shift towards larger cells (FSC).

Multichannel pipettes are used to transfer 1.0 ml HBSS into each of 12 polypropylene tubes (75 mm). After heating of the HBSS in a water bath at 37 °C for 10 min, 20 μl leukocyte suspension in autologous plasma is added, as well as 10 μl DHR working solution with or without the appropriate stimulating substances. After each step, all tubes are briefly shaken. After incubating the leukocytes for 25 min, they are incubated for another 5 min at 37 °C with 1 μl 0.4 mmol/l LDS-751 to stain the DNA. After incubation, the samples are placed on ice. The stained samples are stable for approximately 90 min when kept on ice. If these 90 min need to be exceeded, the samples can be treated with 1% paraformaldehyde in PBS to fix them. For the reagent controls, inhibition of the PMA-induced increase in fluorescence can be checked by 20 min preincubation on ice. For the negative control, 4 μl NADPH oxidase inhibitor DPL with a final concentration of 400 μmol/l is used. For the positive control, 4 μl exogenous 100 mmol/l H_2O_2 with a final concentration of 400 μmol/l is used.

Measurement
The parameters for the measurements with the FACScan for the Orpegen Test are: FSC, cell form and size; SSC, granularity; fluorescence 1, rhodamine: 515–545 nm

Table 7. Incubation schedule for the measurement of stimulated formation of ROS in granulocytes according to Rothe et al. [53]

Time, min	Negative control	fMLP	E. coli	PMA
−10 min, 37 °C	1 ml HBSS	1 ml HBSS	1 ml HBSS	1 ml HBSS
0 min	+20 μl leukocyte-rich plasma +10 μl DHR	+20 μl leukocyte-rich plasma +10 μl DHR +10 μl FMLP	+20 μl leukocyte-rich plasma +10 μl DHR +10 μl *E. coli*	+20 μl leukocyte-rich plasma +10 μl DHR +10 μl PMA
25 min	+1 μl LDS-751	+1 μl LDS-751	+1 μl LDS-751	+1 μl LDS-751
30 min	stop reaction by placing the samples in an ice bath			

and fluorescence 2, DNA dye: 550–600 nm. The parameters for the measurements with the FACScan for the test modification according to Rothe are: FSC, cell form and size; SSC, granularity; fluorescence 1, rhodamine: 515–545 nm and fluorescence 3, DNA dye: >650 nm.

For a preliminary setup, the Calibrite^{TM} beads (Becton Dickinson) are measured. Measurement of the cells is performed in the flow cytometer (FACScan, CELLQuest software). The results contain the histogram display of fluorescence 2 and the dot plot depiction: lin FSC versus lin SSC, SSC-H versus FL1-H. In histogram FL2, a gate has to be set on the nucleus-containing cells; in the dot plot lin FSC versus lin SSC, a gate has to be set on the granulocytes.

With the first sample, the scatter diagram (FSC versus SSC) is checked; if differences with the normal appearance are detected, the detectors for FSC and SSC might have to be adjusted. Then the gates are set for the nucleus-containing cells and the granulocytes.

Thereafter, the four samples per patient are measured in consecutive order. Thereafter, the next patient samples can be measured likewise.

Data Analysis

In the data analysis document (fig. 2), the nucleus-containing cells are isolated (gate 1) in the dot plot lin SSC versus fluorescence 2. Following that the granulocytes are isolated (gate 2) in the dot plot lin FSC versus lin SSC of gate 1. The resulting histogram of fluorescence 1 of gate 3 (gate 1 and gate 2) displays the rhodamine fluorescence, that is ROS formation of the single granulocytes. The additional statistics shows the mean fluorescence intensity (data not shown). The 95th percentile is defined in the F1 histogram of the unstimulated sample.

A double peak in the histogram or a tail in the regions with high fluorescence intensity points to a primary stimulated granulocyte population. In this case, the marks at the borders of both populations have to be repositioned. The mark of the 95th percentile and that of the unstimulated population are added to the histograms of stimulated cells. The positioning of the granulocyte gate in the FSC-SSC diagram has to be adjusted for stimulated granulocytes as PMA stimulation leads to displacement towards the stronger FSC signal as the granulocytes swell [51].

The percentage of granulocytes in the PMA-, *E. coli*- or fMLP-stimulated samples showing an oxidative burst activity above the mentioned mark (see above) and the geometrical mean value of the fluorescence activity of all cells are recorded.

These parameters correspond to cell activity. If damaged cells are present in the sample, a 'tail' can be observed in the low-fluorescence region. For the evaluation of the mean fluorescence of these samples, the limits for positive and negative ROS formation have to be placed at the inflection points of the curve. Next to that, one has to note whether the peak fluorescence following PMA stimulation has shifted towards high fluorescence.

Criteria to Generate Diagnostic Findings

In the DHR fluorescence histogram, a normal burst displays a clear shift of the cells towards high fluorescence (*E. coli* 1.5–2 log steps, PMA 2–3 log steps, fMLP minor shift) in the stimulated cells, which is not the case in the nonstimulated control cells. If distinct fluorescence or a population with distinct rhodamine fluorescence is found in an unstimulated sample, the increased fluorescence after stimulation is not so clear. If, after a long storage period or any other in vitro influences, a sample contains many damaged cells, a functionally normal cell subset can be observed. This usually excludes CGD except for some very rare special forms. After stimulation with *E. coli* or PMA, granulocytes from CGD patients do not show any shift or only a very low shift of the fluorescence peak to high fluorescence. After PMA stimulation, carriers of an X-chromosomal CGD display a two-peaked distribution curve: a functionally normal granulocyte population with a peak of high fluorescence intensity and a pathological granulocyte population with a peak comparable to that of an unstimulated control sample with no increase or a very low increase in fluorescence after stimulation. (It has to be taken into account that H_2O_2 can diffuse from intact cells to damaged cells and thus mimic a function. This effect can be avoided by diluting the cell suspension by mixing it with neutral cells and/or by the addition of catalase.) An in vitro impairment of cells and two peaks in the histogram can also mimic a carrier status. With some experience, however, the in vitro impairment can be distinguished ('smeared' cell population, missing shift in the FSC-SSC diagram, uniform fluorescence of activated eosinophil granulocytes). The two peaks can be checked by an antibody staining of cytochrome b558. The relation of the two populations differs between carriers ('Lyon pattern', X-inactivation pattern). Figure 3 displays the fluorescence histograms of a family whose members suffer from CGD. A carrier status is usually not associated with CGD. In rare cases, however, the ratio of damaged granulocytes to granulocytes with normal ROS formation is shifted, resulting in CGD [3].

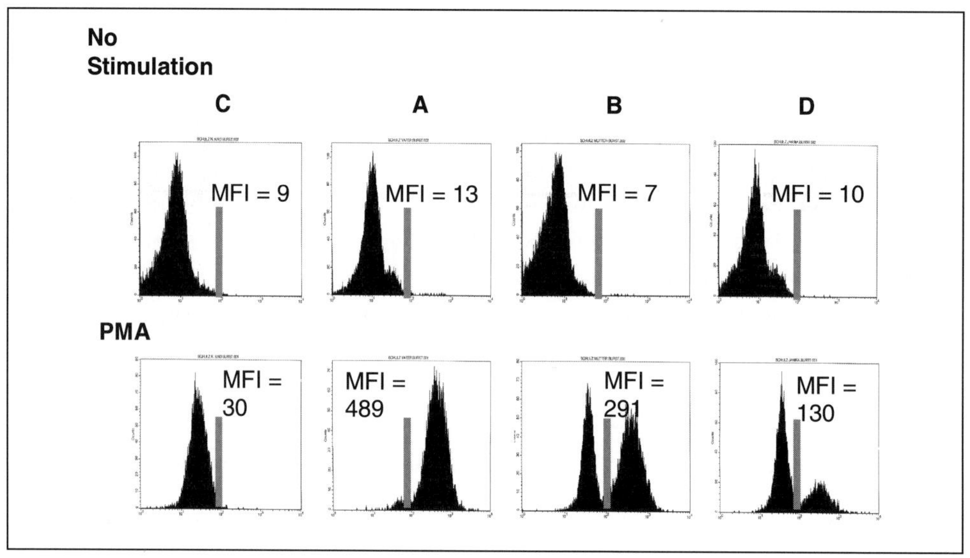

Fig. 3. Histograms of unstimulated and PMA-stimulated ROS formation in granulocytes of a family suffering from CGD. **A** Healthy father. **B, D** Heterozygous carriers (mother and sister of the index patient). **C** Index patient.

Patients with a dysfunction of phagocytosis are characterized by a lacking reaction to *E. coli* stimulation, but display a clear reaction following PMA stimulation (fig. 4). For the diagnosis of CGD, the result of PMA stimulation (the maximum stimulation) is crucial. fMLP and *E. coli* stimulation support the diagnostic finding.

A reduced oxidative burst can be observed in the presence of a glucose-6-phosphate dehydrogenase (G6DPH) dysfunction (G6DPH in granulocytes <1% of normal) (gene locus Xq28) [6, 28]. The neutrophils of these patients are not able to provide NADPH through the pentose phosphate pathway. In addition, these patients suffer from hemolytic nonspherocytic anemia [7] caused by the deficiency in G6PDH in the erythrocytes. The family doctor has to be notified of the first diagnostic suspicion of CGD by phone to make sure that this diagnostic finding is not lost in the abundance of other diagnostic findings.

Unfortunately, one case ended tragically because the suspected diagnosis and the prescribed control examination were not taken seriously. The patient died of the next manifestation of the disease due to the unknown diagnosis.

Genetic Analysis

Informed consent from the patient and/or the parents is always necessary for genetic evaluation and the results must always be made available to them. After the diagnosis of CGD, a molecular examination for clarification, especially as a prerequisite for prenatal diagnostics, should be recommended. Furthermore, the molecular diagnosis is of benefit prior to bone marrow transplantation as after the transplantation only genomic DNA is available. This complicates the determination of the affected gene.

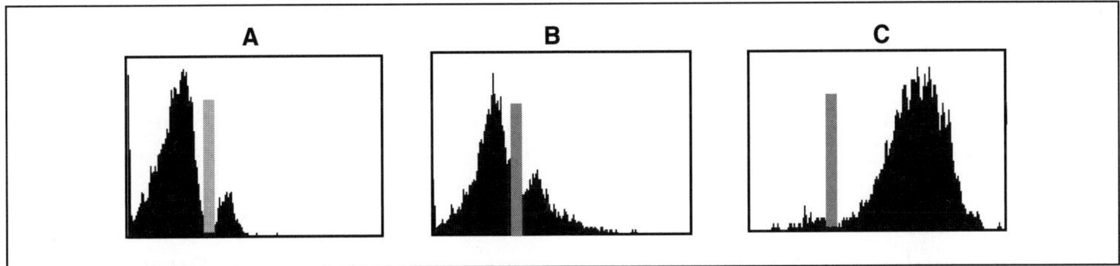

Fig. 4. Histograms of ROS formation in granulocytes of a patient suffering from a temporary drug-induced disturbance of phagocytosis. **A** Unstimulated. **B** *E. coli*-stimulated (center). **C** PMA-stimulated.

The first steps to identify the gene in question are a thorough family history and a flow-cytometric examination of the parents, grandparents and siblings including determination of cytochrome b558 expression. It is not permitted to evaluate underage girls for the sole purpose of determining their carrier status. However, it is important to know whether their X-inactivation pattern puts them at risk. This can justify the DHR assay in underage girls and these considerations should be discussed with the parents. (We encountered several X-linked CGD carriers with severe symptoms who were misdiagnosed and received wrong and even detrimental treatments because their carrier status was unknown.)

The goal of flow cytometry is to narrow down the molecular search to one specific gene. Postponing these examinations until the time of projected parenthood may result in the loss of important information and thus in missing the opportunity of early prenatal diagnosis. Table 8 summarizes the most important indicators.

All patients that suffer from $p47^{phox}$ deficiency and some patients suffering from certain missense mutations exhibit residual activity of the NADPH oxidase and therefore, to some extent, have a better prognosis [54].

Due to the existing residual activity, infections are less frequent (but can still be life-threatening); as a consequence, the deficiency is mostly diagnosed at a later date than in patients without residual activity [2, 29].

Up to date, no defect of $p40^{phox}$ has been reported and only one patient was identified in whom the the *Rac2* gene mutation has triggered CGD and phagocytosis dysfunction.

Quality Control

Before starting the measurement, the performance data of the laser have to be checked.

Device stability is easily verified by measuring a sample of the previous day.

A control with Calibrite beads has to be performed at least once a week; if the sample of the previous day displays awkward results, an additional control with Calibrite beads is necessary (table 9).

Table 8. Classification of chronic granulomatous disease via NADPH oxidase activity and immunologic test for subunits[a]

Defect chromosomal localization	Frequency (100% are all CGD patients)	Immunological test for the subunits				Flow-cytometric evidence pointing to the affected gene
		Gp91^{phox}	P22^{phox}	P47^{phox}	P67^{phox}	
Gp91^{phox} CYBB	approximately 60%	missing	trace	normal	normal	in approximately 80% cell population with two peaks in DHR test of the mother [24] cytochrome b_{558} reduced proportionally to gp91^{phox}
	<10%	reduced	reduced	normal	normal	
	<2%	normal	normal	normal	normal	
Xp21.1 P22^{phox} CYBA 16q24	<5%	missing or trace	missing	normal	normal	cytochrome b_{558} reduced proportionally to p22^{phox} never cell population with two peaks in DHR test of the mother[b]
	<1%	normal	normal	normal	normal	
P47^{phox} NCF-1 7q11.23	approximately 25%	normal	normal	missing	normal	characteristic residual activity in DHR-test cytochrome b_{558} positive [38, 51, 58]
P67^{phox} NCF-2 1q25	<5%	normal	normal	normal	missing	cytochrome b_{558} positive
Rac2 = cytosolic GTPase 22q12.3-q13.2	only case	normal	normal	normal	normal	ROS production reduced after Zymosan particle stimulus, negative with fMLP stimulus, in addition disturbed chemotaxis [60]

Table 8 continued on next page

Table 8. Continued

Defect chromosomal localization	Frequency (100% are all CGD patients)	Immunological test for the subunits				Flow-cytometric evidence pointing to the affected gene
		$Gp91^{phox}$	$P22^{phox}$	$P47^{phox}$	$P67^{phox}$	
Glucose-6-phosphate-dehydrogenase defect chromosome?		normal	normal	normal	normal	residual activity variable, mostly normal; if enzyme in granulocytes below 1% of standard level, typical CGD symptoms occur [61] in addition, hemolytic nonspherocytic anemia [6, 7, 28]

NCF = Neutrophil cytosolic factor.

[a] More constellations, unobserved until now, are possible.

[b] Most of the time, cases of CGD with '$gp91^{phox}$ missing' and '$P22^{phox}$ missing' can only be safely distinguished through mutational analysis. With females, a $p22^{phox}$ deficiency is most probable.

Control of the current measurement: Is the scattergram of the cell population in the lin FSC to lin SSC as usual or is it shifted? Is the 'live gate' set on the granulocytes? Does the sample of the healthy control show increased rhodamine fluorescence after stimulation? Are the histograms of the nuclear fluorescence and the rhodamine fluorescence within the measurement range?

Influences of Drugs on Reactive Oxygen Species Formation

Nitric oxide (NO) interferes with the formation of ROS; following fMLP stimulation, ROS formation is reduced, PMA stimation does not display any effect [55]. NO inhibits the signal cascade, but not NADPH oxidase.

In therapeutic dosage, thiopentane and propofol inhibit ROS formation [56].

GM-CSF highly stimulates the formation of ROS, whereas G-CSF, IL-6 and IL-2 exhibit only moderate stimulatory effects [57]. This is accompanied by an increase in the size of the granulocytes. Cytokines which do not stimulate the formation of ROS cause no increase in size.

Local anesthetics (propofol, bupivacain), barbiturates (thiopental) and beta-blockers (propranolol, metoprolol, carvedilol) inhibit the formation of ROS depending on concentration [42]. Imipenem, quinolone, cephalexin, penicillin, chlor-

Table 9. Troubleshooting – determination of oxidative burst

Problem	Cause	Solution
Signal too weak	measuring chamber contaminated with fluorescent dye or flow cytometer main settings wrong laser, not adjusted	rinse measuring chamber → OK, if not OK, new calibration with beads, if not OK, call service
No stimulation in healthy control sample and patient sample	Usage of EDTA- or citrate blood	order new sample in correct tube [62]
Blood sample coagulated	serum tube	order new sample in correct tube [62]
Discrepancy between results of different stimuli	stimulus forgotten	repeat test
Patient sample and healthy control are conspicuous in FSC-SSC dot plot; both display only minimal or no ROS formation after stimulation	samples too old, major part of cells unable to function	order new sample
Patient sample already displays histogram of normal to reduced ROS formation after stimulation	high amount of immature granulocyte precursors	examine blood smear, draw new sample in an interval without stimulated granulopoiesis
Cell population with insufficient ROS formation	high amount of eosinophil granulocytes	consider in calculations, exclude eosinophils
Complete myeloperoxidase deficiency can lead to a false positive result	MPO deficiency can lead to a strongly decreased DHR signal	the addition of recombinant human MPO enhances the DHR signal of MPO deficient cells but not of NADPH oxidase-deficient cells [63]

amphenicol, streptomycin, cefamanodol and metronidazol in therapeutic dosage do not influence ROS formation.

Cresol in concentrations as they are observed in patients with renal insufficiency inhibits the production of ROS by inhibiton of protein kinase C and NADPH oxidase [58].

Expected Results, Interpretation and Pitfalls

The evaluation of the burst test is based on the comparison of a stimulated and an unstimulated sample. Reference values of the fluorescence intensity after stimulation are helpful for quick decision-making in uncomplicated cases. Samples with high background fluorescence (e.g. high content of eosinophils or application of fluorescent drugs) lead to high unspecific fluorescence intensity. This can be misinterpreted if only holding to reference values. Measured values of the unstimulated sample enable the physician to recognize and consider such problems. In uncomplicated samples, unstimulated granulocytes display a mean fluorescence intensity ranging between 8 and 40 arbitrary fluorescence units. Granulocytes prestimulated in vivo may exhibit higher values. Following PMA stimulation, a distinct shift in the histograms towards high rhodamine fluorescence intensity (2 steps on a logarithmic scale) is considered to be normal.

Severe sepsis can cause reduced ROS production (metabolic depletion of the cells). After laparoscopic surgery, patients may display a significantly diminished production of ROS as an effect of the inflammatory reaction [43].

Hypocalcemia leads to reduced ROS production [47].

Patients suffering from Wegener's granulomatosis show reduced production of ROS after PMA stimulation [42, 59].

Required Time

The test has to be performed according to the instructions up to the measurement. An interruption for hours and delayed resumption of the test is only acceptable after fixation of the cells with 1% paraformaldehyde in PBS. The analysis of the measured data can be postponed if needed.

Up to 4 patient samples can be processed at once. Measuring the patient sample together with the control sample and acquiring the data takes approximately 2.5–3 h. Processing of 4 patient samples takes insignificantly longer (30 min). The data analysis of a patient sample or control sample takes about 10 min for the first sample; any subsequent sample takes approximately 5 min. Usually, phagocytosis, expression of adhesion molecules and formation of ROS are determined from one patient sample and the control sample. The time required for a patient sample with its necessary control sample is then approximately 3–4 h.

Summary

The Indication and diagnostic procedure in case of a suspected defective granulocyte function are described. Furthermore, the methods for the determination of adhesion molecule expression, the phagocytosis of opsonized fluorescent-labeled *E. coli* and the stimulated formation of ROS are detailed for reproduction. The instructions describe possible pitfalls and their elimination and provide an estimation of the required working time and assistance with the interpretation of results.

References

1 Albelda SM, Smith CW, Ward PA: Adhesion molecules and inflammatory injury. FASEB J 1994;8:504–512.

2 Winkelstein JA, Marino MC, Johnston RB, Jr., Boyle J, Curnutte J, Gallin JI, Malech HL, Holland SM, Ochs H, Quie P, Buckley RH, Foster CB, Chanock SJ, Dickler H: Chronic granulomatous disease. Report on a national registry of 368 patients. Medicine (Baltimore) 2000;79:155–169.

3 Lun A, Roesler J, Renz H: Unusual late onset of X-linked chronic granulomatous disease in an adult woman after unsuspicious childhood. Clin Chem 2002;48:780–781.

4 Hoffmann JA, Kafatos FC, Janeway CA, Ezekowitz RA: Phylogenetic perspectives in innate immunity. Science 1999;284:1313–1318.

5 Ezekowitz RA: Role of the mannose-binding lectin in innate immunity. J Infect Dis 2003;187(suppl 2):S335–S339.

6 Domachowske JB, Malech HL: Phagocytes; in Rich RR, Fleisher TA, Schwartz BD, Shearer WT, Strober W (eds): Clinical Immunology – Principles and Practice. St Louis, Mosby, 1996, vol 1, pp 392–407.

7 Mitchel NR, Cotran RS: Acute and chronic inflammation; in Kumar V, Cotran RS, Robbins SL (eds): Basic Pathology. Philadelphia, Saunders, 1996, pp 33–59.

8 Hayashi F, Means TK, Luster AD: Toll-like receptors stimulate human neutrophil function. Blood 2003;102:2660–2669.

9 Medzhitov R, Janeway CA, Jr.: Innate immune induction of the adaptive immune response. Cold Spring Harb Symp Quant Biol 1999;64:429–435.

10 Akashi-Takamura S, Miyake K: Toll-like receptors (TLRs) and immune disorders. J Infect Chemother 2006;12:233–240.

11 Schroder NW, Schumann RR: Single nucleotide polymorphisms of Toll-like receptors and susceptibility to infectious disease. Lancet Infect Dis 2005;5:156–164.

12 Zhang H, Zhou G, Zhi L, Yang H, Zhai Y, Dong X, Zhang X, Gao X, Zhu Y, He F: Association between mannose-binding lectin gene polymorphisms and susceptibility to severe acute respiratory syndrome coronavirus infection. J Infect Dis 2005;192:1355–1361.

13 Terai I, Kobayashi K, Matsushita M, Miyakawa H, Mafune N, Kikuta H: Relationship between gene polymorphisms of mannose-binding lectin (MBL) and two molecular forms of MBL. Eur J Immunol 2003;33:2755–2763.

14 Cedzynski M, Szemraj J, Swierzko AS, Bak-Romaniszyn L, Banasik M, Zeman K, Kilpatrick DC: Mannan-binding lectin insufficiency in children with recurrent infections of the respiratory system. Clin Exp Immunol 2004;136:304–311.

15 Dean MM, Minchinton RM, Heatley S, Eisen DP: Mannose binding lectin acute phase activity in patients with severe infection. J Clin Immunol 2005;25:346–352.

16 Takahashi K, Shi L, Gowda LD, Ezekowitz RA: Relative roles of complement factor 3 and mannose-binding lectin in host defense against infection. Infect Immun 2005;73:8188–8193.

17 Neth O, Hann I, Turner MW, Klein NJ: Deficiency of mannose-binding lectin and burden of infection in children with malignancy: a prospective study. Lancet 2001;358:614–618.

18 Ezekowitz RA: Mannose-binding lectin in prediction of susceptibility to infection. Lancet 2001;358:598–599.

19 Campbell JJ, Butcher EC: Chemokines in tissue-specific and microenvironment-specific lymphocyte homing. Curr Opin Immunol 2000;12:336–341.

20 Anderson DC, Springer TA: Leukocyte adhesion deficiency: an inherited defect in the Mac-1, LFA-1, and p150.95 glycoproteins. Annu Rev Med 1987;38:175–194.

21 Polley MJ, Phillips ML, Wayner E, Nudelman E, Singhal AK, Hakomori S, Paulson JC: CD62 and endothelial cell-leukocyte adhesion molecule 1 (ELAM-1) recognize the same carbohydrate ligand, sialyl-Lewis x. Proc Natl Acad Sci USA 1991;88:6224–6228.

22 Phillips ML, Schwartz BR, Etzioni A, Bayer R, Ochs HD, Paulson JC, Harlan JM: Neutrophil adhesion in leukocyte adhesion deficiency syndrome type 2. J Clin Invest 1995;96:2898–2906.

23 Alon R, Aker M, Feigelson S, Sokolovsky-Eisenberg M, Staunton DE, Cinamon G, Grabovsky V, Shamri R, Etzioni A: A novel genetic leukocyte adhesion deficiency in subsecond triggering of integrin avidity by endothelial chemokines results in impaired leukocyte arrest on vascular endothelium under shear flow. Blood 2003;101:4437–4445.

24 Pasvolsky R, Feigelson SW, Kilic SS, Simon AJ, Tal-Lapidot G, Grabovsky V, Crittenden JR, Amariglio N, Safran M, Graybiel AM, Rechavi G, Ben Dor S, Etzioni A, Alon R: A LAD-III syndrome is associated with defective expression of the Rap-1 activator CalDAG-GEFI in lymphocytes, neutrophils, and platelets. J Exp Med 2007;204:1571–1582.

25 Kuijpers TW, van Lier RA, Hamann D, de Boer M, Thung LY, Weening RS, Verhoeven AJ, Roos D: Leukocyte adhesion deficiency type 1 (LAD-1)/variant. A novel immunodeficiency syndrome characterized by dysfunctional beta-2-integrins. J Clin Invest 1997;100:1725–1733.

26 Quie P, Mills EL, Roberts RL, Noya FJD: Disorders of the polymorphonuclear phagocytic system; in Stiehm ER (ed): Immunologic Disorders in infants and children. Philadephia, Saunders, 1996, pp 446–447.

27 Wild MK, Luhn K, Marquardt T, Vestweber D: Leukocyte adhesion deficiency II: therapy and genetic defect. Cells Tissues Organs 2002;172:161–173.

28 Heyworth PG, Cross AR, Curnutte JT: Chronic granulomatous disease. Curr Opin Immunol 2003;15:578–584.

29 Segal BH, Leto TL, Gallin JI, Malech HL, Holland SM: Genetic, biochemical, and clinical features of chronic granulomatous disease. Medicine (Baltimore) 2000;79:170–200.

30 Gombart AF, Shiohara M, Kwok SH, Agematsu K, Komiyama A, Koeffler HP: Neutrophil-specific granule deficiency: homozygous recessive inheritance of a frameshift mutation in the gene encoding transcription factor CCAAT/enhancer binding protein-epsilon. Blood 2001;97:2561–2567.

31 Shiflett SL, Kaplan J, Ward DM: Chediak-Higashi syndrome: a rare disorder of lysosomes and lysosome related organelles. Pigment Cell Res 2002;15:251–257.

32 Akira S, Sato S: Toll-like receptors and their signaling mechanisms. Scand J Infect Dis 2003;35:555–562.

33 Yum HK, Arcaroli J, Kupfner J, Shenkar R, Penninger JM, Sasaki T, Yang KY, Park JS, Abraham E: Involvement of phosphoinositide 3-kinases in neutrophil activation and the development of acute lung injury. J Immunol 2001;167:6601–6608.

34 Strassheim D, Asehnoune K, Park JS, Kim JY, He Q, Richter D, Kuhn K, Mitra S, Abraham E: Phosphoinositide 3-kinase and Akt occupy central roles in inflammatory responses of Toll-like receptor 2-stimulated neutrophils. J Immunol 2004;172:5727–5733.

35 Yuo A, Kitagawa S, Suzuki I, Urabe A, Okabe T, Saito M, Takaku F: Tumor necrosis factor as an activator of human granulocytes. Potentiation of the metabolisms triggered by the Ca^{2+}-mobilizing agonists. J Immunol 1989;142:1678–1684.

36 Becker R, Pfluger KH: Myeloperoxidase deficiency: an epidemiological study and flow-cytometric detection of other granular enzymes in myeloperoxidase-deficient subjects. Ann Hematol 1994;69:199–203.

37 Ward DM, Shiflett SL, Kaplan J: Chediak-Higashi syndrome: a clinical and molecular view of a rare lysosomal storage disorder. Curr Mol Med 2002;2:469–477.

38 Roos D, de Boer M, Kuribayashi F, Meischl C, Weening RS, Segal AW, Ahlin A, Nemet K, Hossle JP, Bernatowska-Matuszkiewicz E, Middleton-Price H: Mutations in the X-linked and autosomal recessive forms of chronic granulomatous disease. Blood 1996;87:1663–1681.

39 Allende LM, Hernandez M, Corell A, Garcia-Perez MA, Varela P, Moreno A, Caragol I, Garcia-Martin F, Guillen-Perales J, Olive T, Espanol T, Arnaiz-Villena A: A novel CD18 genomic deletion in a patient with severe leucocyte adhesion deficiency: a possible CD2/lymphocyte function-associated antigen-1 functional association in humans. Immunology 2000;99:440–450.

40 Abramson JS, Mills EL, Sawyer MK, Regelmann WR, Nelson JD, Quie PG: Recurrent infections and delayed separation of the umbilical cord in an infant with abnormal phagocytic cell locomotion and oxidative response during particle phagocytosis. J Pediatr 1981;99:887–894.

41 Etzioni A, Sturla L, Antonellis A, Green ED, Gershoni-Baruch R, Berninsone PM, Hirschberg CB, Tonetti M: Leukocyte adhesion deficiency (LAD) type II/carbohydrate deficient glycoprotein (CDG) IIc founder effect and genotype/phenotype correlation. Am J Med Genet 2002;110:131–135.

42 Stahl D, Hänsch GM: Granulozyten-Funktionsprüfung; in Thomas L (Hrsg): Labor und Diagnose. Frankfurt, TH-Books Verlagsgesellschaft, 1998, pp 804–811.

43 Sietses C, Wiezer MJ, Eijsbouts QA, van Leeuwen PA, Beelen RH, Meijer S, Cuesta MA: The influence of laparoscopic surgery on postoperative polymorphonuclear leukocyte function. Surg Endosc 2000;14:812–816.

44 Gu Y, Williams DA: RAC2 GTPase deficiency and myeloid cell dysfunction in human and mouse. J Pediatr Hematol Oncol 2002;24:791–794.

45 Dobmeyer TS, Raffel B, Dobmeyer JM, Findhammer S, Klein SA, Kabelitz D, Hoelzer D, Helm EB, Rossol R: Decreased function of monocytes and granulocytes during HIV-1 infection correlates with CD4 cell counts. Eur J Med Res 1995;1:9–15.

46 Schiffrin EJ, Rochat F, Link-Amster H, Aeschlimann JM, Donnet-Hughes A: Immunomodulation of human blood cells following the ingestion of lactic acid bacteria. J Dairy Sci 1995;78:491–497.

47 Ducusin RJ, Uzuka Y, Satoh E, Otani M, Nishimura M, Tanabe S, Sarashina T: Effects of extracellular Ca2+ on phagocytosis and intracellular Ca^{2+} concentrations in polymorphonuclear leukocytes of postpartum dairy cows. Res Vet Sci 2003;75:27–32.

48 Wenisch C, Parschalk P, Graninger W: Effect of ciprofloxacin and other quinolones on granulocyte function assessed by flow cytometry. Drugs 1995;49(suppl 2):301–303.
49 Rothe G, Emmendorffer A, Oser A, Roesler J, Valet G: Flow cytometric measurement of the respiratory burst activity of phagocytes using dihydrorhodamine 123. J Immunol Methods 1991;138:133–135.
50 Roesler J, Hecht M, Freihorst J, Lohmann-Matthes ML, Emmendorffer A: Diagnosis of chronic granulomatous disease and of its mode of inheritance by dihydrorhodamine 123 and flow microcytofluorometry. Eur J Pediatr 1991;150:161–165.
51 Emmendorffer A, Hecht M, Lohmann-Matthes ML, Roesler J: A fast and easy method to determine the production of reactive oxygen intermediates by human and murine phagocytes using dihydrorhodamine 123. J Immunol Methods 1990;131:269–275.
52 Emmendorffer A, Nakamura M, Rothe G, Spiekermann K, Lohmann-Matthes ML, Roesler J: Evaluation of flow cytometric methods for diagnosis of chronic granulomatous disease variants under routine laboratory conditions. Cytometry 1994;18:147–155.
53 Rothe G, Kellermann W, Schaerer B, Valet G: Messung der phagosomalen Wasserstoffperoxidproduktion von Granulozyten und Monozyten mit Dihydrorhodamin 123; in Schmitz G, Rothe G (Hrsg): Durchflußzytometrie in der klinischen Zelldiagnostik. Stuttgart, Schattauer, 1994, pp 331–350.
54 Vowells SJ, Fleisher TA, Sekhsaria S, Alling DW, Maguire TE, Malech HL: Genotype-dependent variability in flow cytometric evaluation of reduced nicotinamide adenine dinucleotide phosphate oxidase function in patients with chronic granulomatous disease. J Pediatr 1996;128:104–107.
55 Frohlich D, Rothe G, Wittmann S, Schmitz G, Schmid P, Taeger K, Hobbhahn J: Nitrous oxide impairs the neutrophil oxidative response. Anesthesiology 1998;88:1281–1290.
56 Frohlich D, Rothe G, Schwall B, Schmitz G, Hobbhahn J, Taeger K: Thiopentone and propofol, but not methohexitone nor midazolam, inhibit neutrophil oxidative responses to the bacterial peptide FMLP. Eur J Anaesthesiol 1996;13:582–588.
57 Yuan L, Inoue S, Saito Y, Nakajima O: An evaluation of the effects of cytokines on intracellular oxidative production in normal neutrophils by flow cytometry. Exp Cell Res 1993;209:375–381.
58 Vanholder R, De Smet R, Waterloos MA, Van Landschoot N, Vogeleere P, Hoste E, Ringoir S: Mechanisms of uremic inhibition of phagocyte reactive species production: characterization of the role of p-cresol. Kidney Int 1995;47:510–517.
59 Riecken B, Gutfleisch J, Schlesier M, Peter HH: Impaired granulocyte oxidative burst and decreased expression of leucocyte adhesion molecule-1 (LAM-1) in patients with Wegener's granulomatosis. Clin Exp Immunol 1994;96:43–47.
60 Williams DA, Tao W, Yang F, Kim C, Gu Y, Mansfield P, Levine JE, Petryniak B, Derrow CW, Harris C, Jia B, Zheng Y, Ambruso DR, Lowe JB, Atkinson SJ, Dinauer MC, Boxer L: Dominant negative mutation of the hematopoietic-specific Rho GTPase, Rac2, is associated with a human phagocyte immunodeficiency. Blood 2000;96:1646–1654.
61 Gray GR, Stamatoyannopoulos G, Naiman SC, Kliman MR, Klebanoff SJ, Austin T, Yoshida A, Robinson GC: Neutrophil dysfunction, chronic granulomatous disease, and non-spherocytic haemolytic anaemia caused by complete deficiency of glucose-6-phosphate dehydrogenase. Lancet 1973;ii:530–534.
62 Prince HE, Lape-Nixon M: Influence of specimen age and anticoagulant on flow cytometric evaluation of granulocyte oxidative burst generation. J Immunol Methods 1995;188:129–138.
63 Mauch L, Lun A, O'gorman MR, Harris JS, Schulze I, Zychlinsky A, Fuchs T, Oelschlagel U, Brenner S, Kutter D, Rosen-Wolff A, Roesler J: Chronic granulomatous disease (CGD) and complete myeloperoxidase deficiency both yield strongly reduced dihydrorhodamine 123 test signals but can be easily discerned in routine testing for CGD. Clin Chem 2007;53:890–896.

Sack U, Tárnok A, Rothe G (eds): Cellular Diagnostics. Basics, Methods and Clinical Applications of Flow Cytometry. Basel, Karger, 2009, pp 377–389

Platelet Function and Thrombopoietic Activity

Andreas Ruf[a] · Stefan Barlage[b]

[a] Center for Laboratory Medicine, Microbiology and Transfusion Medicine, Städtisches Klinikum Karlsruhe gGmbH, Karlsruhe,
[b] Laboratory Medicine, MVZ Leverkusen, Leverkusen, Germany

Introduction

Platelets are key players in primary hemostasis. Furthermore, platelets play an essential role in the pathogenesis of atherothrombotic diseases such as myocardial or cerebral ischemia. A reliable determination of platelet function and thrombopoietic activity is therefore of major clinical interest. Flow-cytometric methods have significantly broadened the range of methods for the determination of platelet function and thrombopoietic activity [1–6]. Today, they rank among the standard methods for the laboratory diagnosis of hereditary disorders of platelet function [1, 2, 5]. The measurement of reticulated platelets by flow cytometry is an established screening tool for the differentiation of thrombocytopenia [7–9]. Determination of platelet function is an established approach to control the quality of platelet concentrates [10]. Furthermore, flow cytometry can reliably assess platelet activation in vivo or suppression of platelet function ex vivo upon antiplatelet therapy. However, the diagnostic significance of flow-cytometric methods for monitoring platelet activation or antiplatelet therapy in various diseases must be clarified in prospective clinical studies [6, 11–14].

A large number of different protocols for the flow-cytometric analysis of platelets have been described. These differ in substance in the type of anticoagulation, sample preparation and sample stabilization [2, 4]. In the following, the authors report protocols for the analysis of platelet function and the measurement of reticulated platelets which are based on a consensus of the European Working Group on to Clinical Cell Analysis and the German Society for Thrombosis and Haemostasis [15]. They do not claim general validity; however, they enable users to establish reliable methods to address their respective tasks.

Protocols

Platelet Function

Principle
Immune flow-cytometric analysis of thrombocyte function is based on the immune staining of adhesion receptors or activation-dependent expressed antigens with fluorochrome-coupled antibodies. Subsequently, the specific fluorescence of the thrombocyte is measured in a flow cytometer and antigen expression, i.e. the antigen-positive thrombocytes, is quantified. The most important target antigens are shown in tables 1 and 2.

Materials
- Antibodies
  - fluorochrome-coupled antibodies directed against adhesion receptors (table 1), i.e. activation-dependent antigens (table 2)
  - unconjugated specific antibodies or control antibodies (see 'Negative Control', p. 382)
- Phosphate buffer: $Na_2HPO_4 \times 2\ H_2O$ 0.15 mol/l; $NaH_2PO_4 \times H_2O$ 0.15 mol/l dissolved in distilled water; the Na_2HPO_4 solution is mixed and titrated with the NaH_2PO_4 solution to a pH of 7.4; the osmolarity is adjusted to 300 mosmol/l with distilled water
- Formaldehyde solution: 10% paraformaldehyde in distilled water
- Glyoxal: 40% glyoxal in distilled water
- Stabilization solution: 1 part glyoxal solution and 8 parts formaldehyde solution are added to phosphate buffer to a total of 200 parts
- Dilution solution: glycine (0.2% w/v) dissolved in phosphate buffer.

Methods
Stabilization (Optional) and Dilution
- The platelet preparation is fixed with stabilization solution: the proportion of the mix is: 2 parts platelet preparation and 1 part stabilization solution; entire volume i.e. 1.5 ml;
- Incubation for 10 min at room temperature;
- Dilution of the sample in 1:10 relation with the dilution solution.

Immunostaining
- FITC-coupled antibodies against the target antigens are added to 50 μl of the stabilized, diluted and/or simply diluted blood samples in saturated concentration. If a negative control is needed (see 'Negative Control', p. 382), the unconjugated antibody in approximately twenty times molecular excess is added; the conjugated antibody should only be added afterwards. Alternatively, a negative antibody can be used as a negative control.

Table 1. Platelet adhesion receptors

Receptor	CD nomenclature	Function
Complex of		
GpIIb	CD41 (α subunit of $\alpha_{IIb}/\beta 3$ integrin)	aggregation/adhesion
GpIIIa	CD 61 (β subunit of $\alpha_{IIb}/\beta 3$ integrin)	
Complex of		
GpIX	CD42a	adhesion
GpIbα	CD42b	
GpIbβ	CD42c	
GpV	CD42d	

Gp = Glycoprotein.

Table 2. Activation-dependent antigens [modified from ref. 2]

Mechanism of expression	Antigen/epitope
Conformational change induced by	
Activation	ligand binding sites of GpIIb/IIIa
Ligand binding	LIBS on GpIIb/IIIa
Receptor binding	RIBS on fibrinogen
Secretion	
α-Granules	CD62P (P-selectin), GMP 33
Lysosomes	CD63, CD107a (LAMP-1), CD107b (LAMP-2)
Binding to the platelet surface	CD40 ligand
	thrombospondin
	multimerin
	fibrinogen
	factor Va
	factor VIII
	factor Xa
Redistribution	
From SCS to surface	GpIIb/IIIa
From surface to SCS	Gp Ib

Gp = Glycoprotein; LIBS = ligand-induced binding sites; RIBS = receptor-induced binding sites; LAMP-1 = lysosomal associated membrane protein-1; SCS = surface connected canalicular system.

- When double staining: addition of PE-conjugated antibody for platelet identification (see 'Comments', pp. 381).
- Incubation for 15 min at room temperature in the dark.
- Addition of 450–1,000 μl phosphate buffer (see 'Troubleshooting', pp. 384).

Instrument Settings

- The light scatter and fluorescence detectors should be set to logarithmic amplification.
- The amplifier for the light scatter detectors is adjusted in such a way that the median light scatter intensity of the platelets lies approximately at the end of the 2nd decade of the measuring range.
- The amplifier for the fluorescence detectors is adjusted in such a way that the median autofluorescence of the platelets lies approximately in the center of the 1st decade of the measuring range.

Measurements

After double immune staining, PE antibody fluorescence is used as the trigger for platelet identification. After single staining, the platelets can be identified on the basis of their light scatter characteristics. At least 5,000 platelets per sample are measured.

Analysis

The type of data analysis depends on the antigen density on the individual thrombocytes. For antigens with a high density such as glycoprotein Ib or the glycoprotein IIb/IIIa complex, the number of antibody binding sites can be quantitatively determined. Calibration is carried out with latex particles which carry a defined number of specific antimouse immunoglobulin binding sites [16]. If the goal is to determine an adhesion receptor defect rather than activation-dependent differences in expression density, comparison of the specific fluorescence intensities of the patient's thrombocytes with a healthy control is sufficient.

The differences even in heterozygous carriers of the phenotype are so large that quantification with calibrated beads is not necessary (fig. 1). The results can be represented as a relationship (%) related to the fluorescence intensity in normal controls. If, however, activation-dependent antigens expressed in low densities (e.g. CD62P) and only on subpopulations of thrombocytes (e.g. circulating activated thrombocytes in vivo) need to be quantified, the following data analysis should be used.

First the data to be measured are represented in a light scatter cytogram, then the platelet cluster is encircled (fig. 2). Afterwards, the selected platelets are displayed in an FITC fluorescence histogram. The proportion of antigen-positive platelets is determined by setting a discriminator. The discriminator is set in such a way that 99–99.5% of the measured values of the negative control are situated on the left of the discriminator (fig. 3a). Subsequently, using specific antibodies, the percentage of activated (positive) platelets whose fluorescence intensity lies above the discriminator can be determined (fig. 3 below).

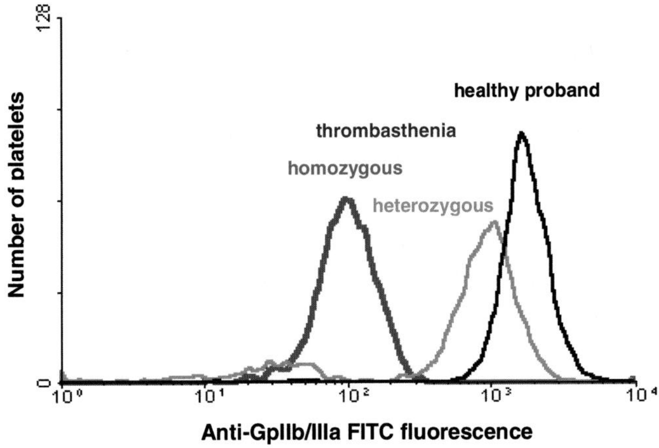

Fig. 1. Fluorescence histogram of thrombocytes from a healthy control and patients with thrombasthenia following immune staining with an anti-GpIIb/IIIa-antibody.

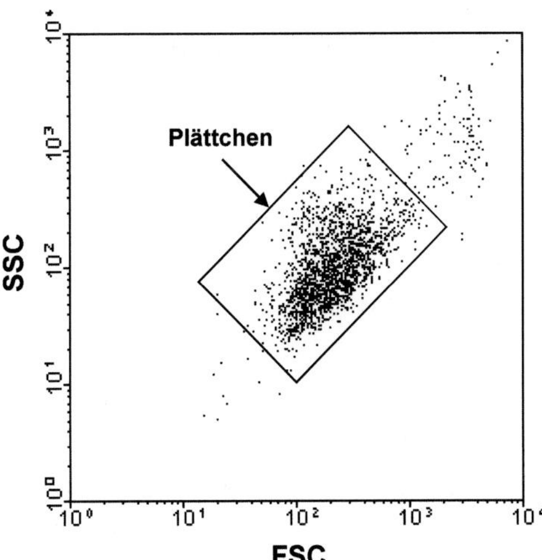

Fig. 2. Light scatter cytogram of thrombocytes.

Storage Stability

Samples fixed in accordance with section 'Stabilization (Optional) and Dilution' (p. 378) are stable for storage for up to 24 h at 4 °C [17].

Comments

Double Immune Staining

If cell types other than platelets are present in the preparation as well, a double immune staining is necessary since in these cases discrimination of the cells is not possible with light scatter parameters. The platelets should be identified with antibodies against strongly expressed antigens such as CD41a. Furthermore, the antibodies used for platelet identification should be applied as PE conjugates and the antibodies against a rarely expressed activation-dependent epitope should be applied as an

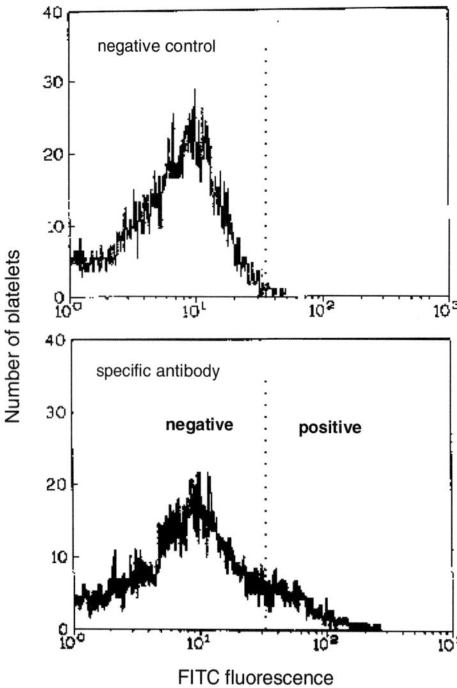

Fig. 3. Fluorescence histograms of thrombocytes after a negative-control analysis (**a**) and after immune staining with a specific antibody (**b**).

FITC conjugate. If this rule is not followed, compensation will not be successful with many flow cytometers.

Negative Control

If antigen-positive thrombocytes are to be analyzed, a proper negative control is crucial to ensure the quality of the results. The best way to obtain a negative control is to quantitate nonspecific binding through displacement with unconjugated specific antibody. Alternatively, a nonspecific control antibody can be used as a negative control. This must be used at the same concentration and with the same F/P ratio as the specific antibody. The fluorochrome conjugation of the specific antibody and the control antibody should have been produced with the same procedure.

Choice of a Target Antigen for Detection of Platelet Activation

Secretion-dependent antigens such as CD62P, for example, are actually less sensitive in indicating activation of thrombocytes than fibrinogen-binding or other activation parameters [18]. The expression of secretion-dependent antigens, however, is irreversible in vitro [19, 20]. CD62P and other secretion-dependent antigens can therefore be used as memory markers for the determination of platelet activation during the production and storage of thrombocyte concentrate. Experimental data from animal models confirm that CD62P is very rapidly separated from the platelet surface in vivo [21]. CD62P is therefore less well suited to confirm the presence of circulating activated platelets in vivo than other secretion-dependent antigens such as

CD63 [22]. The duration of secretion-dependent antigen expression in vivo is not yet clarified in humans.

Quality Control
Control materials are currently not available. Samples from healthy adults should therefore be used as controls.

Supporting Protocols

Determination of Thrombocyte Reactivity in vitro

The protocols listed above can be modified in order to determine thrombocyte reactivity. It is only necessary to stimulate the thrombocytes before sample stabilization and immune staining. Applicable platelet agonists include ADP, stable thromboxane A_2 mimetics, thrombin-receptor-activating peptide or thrombin. If thrombin is used as agonist, fibrin accumulation in the sample must be prevented. This can be done using the synthetic tetrapeptide glycyl-prolyl-arginyl proline [23].

To select the proper agonist concentration, the following is to be considered: the respective agonists should be used at least at *two different* concentrations:

- at a concentration that causes a minimal effect (change in antigen expression) in the thrombocytes of healthy persons (threshold concentration);
- at the lowest concentration that causes a maximum effect in the thrombocytes of healthy persons.

When using threshold concentrations, thrombocyte hyperreactivity can be detected. With the lowest concentration that causes the maximum effect, hyporeactivity can be detected.

The relationship between concentration and the effect of the agonists depends to a considerable degree on the matrix of the thrombocyte preparation. For example, the threshold concentration for ADP is a great deal higher in whole blood than in suspensions of washed thrombocytes [24]. In order to prevent aggregation of the thrombocytes after stimulation, the platelet preparation must be spiked with coagulation inhibitors (GpIIb/IIIa inhibitors) or strongly diluted before the stimulation. Dilution of the sample lowers the collision rate of the thrombocytes, which prevents their aggregation. However, this presupposes that the diluted samples are not agitated. The dilution factor depends on the presence of other cells in the platelet preparation. With a whole-blood sample, a dilution of 1:10 is usually sufficient since the erythrocytes likewise lower the collision rate of the thrombocytes. As dilution medium, an isotonic pH 7.4 HEPES buffer is suitable [25].

Expected Results

Clarification of Adhesion Receptor Defects

Both thrombasthenia (defect in the GpIIb/IIIa complex) and Bernard-Soulier syndrome (defect in the GpIb/IX/V complex) are extremely rare diseases. In typical heterozygous carriers, following immunostaining of the thrombocytes, the specific fluorescence is <60% compared with healthy test persons; and <10% compared with typical homozygous carriers. Both with thrombasthenia and Bernard Soulier syndrome there are variants with normal density adhesion receptors, which, however, exhibit functional defects. These variants are detectable with flow-cytometric methodology only if antibodies are used which recognize the appropriate adhesion receptor that causes the functional defect [1, 5].

Quality Control of Thrombocyte Concentrates

At present, there are no obligatory upper limits for activated thrombocytes in thrombocyte concentrates, not the least because it has not yet been ascertained whether thrombocyte activation under conditions of storage unfavorably affects thrombocyte function after transfusion. Normally, the portion of activated thrombocytes (CD62P+) after 5-day storage is below 30%.

Thrombocyte Activation in vivo and Thrombocyte Reactivity

Through the analysis of collected patient data, thrombocyte activation and changed reactivity have been proven in vivo. The diagnostic meaning of these findings is still unclear, however [2, 11–14].

Troubleshooting

Artificial Activation

Artificial thrombocyte activation in vitro is the most frequent preanalytic interference. This can be prevented by proper sample stabilization. A sample stabilizer should meet the following criteria:

- immediate blocking of cell activity;
- no antigen destruction;
- morphologic stabilization, and
- no increase in cellular autofluorescence.

Sample Dilution

If the sample is insufficiently diluted, inaccuracies in the measurements may arise. In order to avoid this, the final concentration of all cells in the sample and the flow rate should be coordinated in such a way that not more than 1,000–1,500 cells/s are measured.

Reticulated Platelets

Principle

The method described here essentially corresponds to that of Matic et al. [26] and permits the quantification of reticulated thrombocytes from unseparated whole blood. The fact that no platelet-rich plasma needs to be prepared shortens the time required for the test and prevents i) the potentially selective loss of individual thrombocyte populations through centrifugation or ii) inadvertent thrombocyte activation, which could contribute to faster RNA degradation. The protocol achieves i) a selective thrombocyte analysis from whole blood by staining the thrombocytes with an antibody against GpIb; ii) a short incubation period with high RNA staining specificity by a high thiazole orange concentration; iii) stable staining by subsequent fixation, and iv) minimization of the influence of aggregates and/or different thrombocyte sizes by selection of the analysis region in a two-dimensional representation of thiazole orange fluorescence versus forward light scatter.

Materials

EDTA anticoagulated whole blood – sample processing should take place within 2–4 h after probe collection.

Thiazole orange (e.g. Sigma Aldrich) is dissolved at a concentration of 1 mg/ml in methanol. This stock solution can be stored at −20 °C for several months. Further dilution to prepare the working solution is carried out with PBS.

For the specific labeling of the thrombocytes, an antibody against glycoprotein Ib (CD42b) conjugated with PE or another fluorochrome spectrally separated from thiazole orange is used.

Methods

Staining and Fixation

- 5 µl undiluted EDTA-anticoagulated whole blood is incubated with 5 µl of a stock solution of the CD42b antibody and 50 µl thiazole orange (1 µg/ml final concentration in PBS) solution for 15 min at room temperature in polypropylene tubes.
- Subsequently, the sample is fixed by addition of 1 ml formaldehyde (1% w/v in PBS). This sample is used directly for the flow-cytometric measurement, which should take place within 45 min.
- Parallel to testing the patient's sample, a sample from a healthy control with normal thrombocyte counts and without any signs of a quantitative or qualitative thrombocyte disturbance should be analyzed.

Measurements
Data acquisition is performed with logarithmic amplification for light scatter and fluorescence signals. Thiazole orange fluorescence is excited at 488 nm, with an emission maximum around 533 nm, and can thus be analyzed similarly to fluorescein. For the specific collection of thrombocytes, the threshold value is defined based on the fluorescence signal of the GpIb antibody (e.g. PE fluorescence). The exclusion of thiazole-orange-positive leukocytes and erythrocytes, and in particular reticulocytes, can be achieved by this method. A further gating, e.g. on the typical scatter profile of thrombocytes, is not done to prevent large thrombocytes from being excluded from the measurement. At least 5,000 thrombocytes are analyzed.

Analysis
For analyzing the fraction of reticulated thrombocytes, a region is defined within the FSC/FL1 dot plot on the basis of samples from healthy controls; this region runs at the upper border of the thrombocyte population and is defined in such a way that it includes 1% of the thrombocytes on average (fig. 4). Following validation of this region for normal probands, the daily control samples as well as the patient samples are analyzed. Additionally, the absolute counts of reticulated thrombocytes can be determined, e.g. by extrapolation using thrombocyte counts determined on a hematology analyzer.

Expected Results
The results of the daily normal controls should lie within the range determined for the group of normal probands (<5% reticulated thrombocytes). If values exceed the normal range, the measurement should be repeated in order to exclude technical problems (i.e. extreme incubation periods, poor fixation).

With repeated measurements of healthy control probands, the inter-assay variation coefficient should be <10% and the intra-assay variation coefficient <5%.

Patients with increased elimination of thrombocytes as well as those in the regeneration phase, e.g. following chemotherapy or hematopoietic stem cell transplantation, may show values above the range of normal controls [27–30]; however, with the analysis described here, the percentages of reticulated thrombocytes generally vary between 5 and 20%. In patients with substantial or massive thrombocytopenia, however, substantially higher or extremely high values may occasionally be found.

Troubleshooting
In vitro activation of thrombocytes, e.g. due to inappropriate sample storage, may lead to degranulation. The resulting reduced signal intensity may in turn lead to a falsely low proportion of reticulated thrombocytes in the measurement. Parallel measurements of the expression of activation antigens can reveal the activation in these cases.

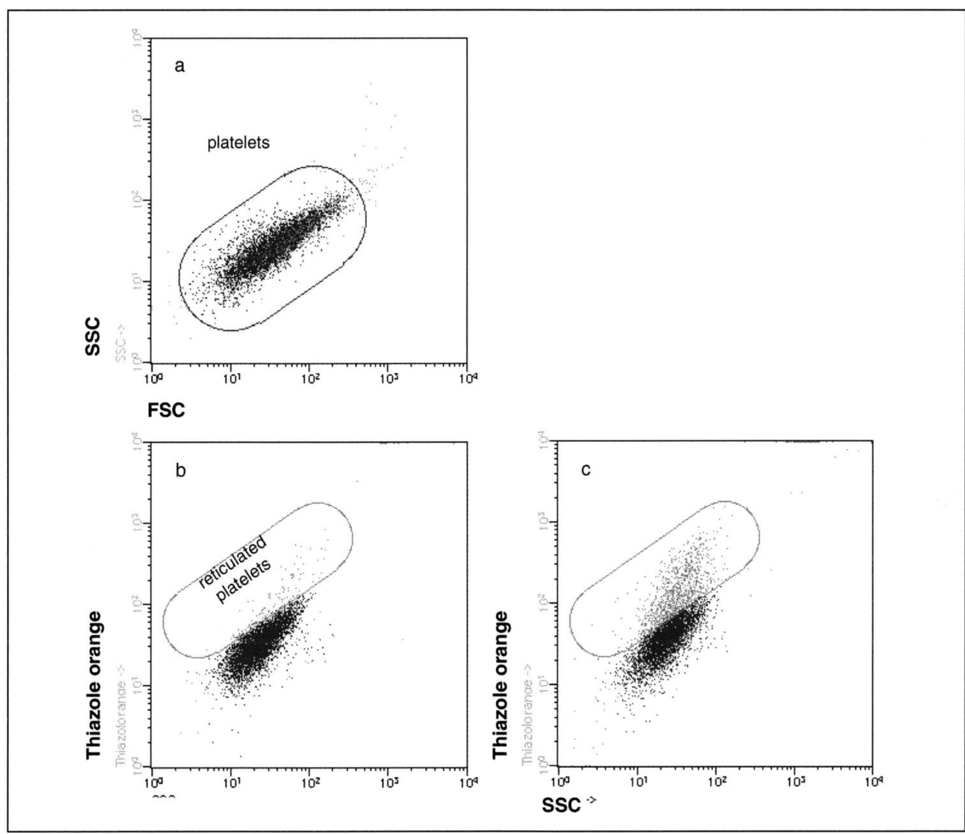

Fig. 4. a Representation of the thrombocyte on the basis of light scatter signals in the FSC/SSC dot plot. After setting the threshold value on the basis of the fluorescence of the bound PE-conjugated GpIb antibody, erythrocytes and leukocytes in the measurement are excluded from the analysis. **b** Representation of thiazole orange-positive reticulated thrombocytes from a control patient and **c** from a patient after chemotherapy who has an increased portion of reticulated thrombocytes. Reticulated thrombocytes are quantitated using a two-dimensional region in order to minimize the influence from thrombocyte aggregates.

Summary

Flow-cytometric platelet assays comprise immunocytometric methods for the determination of platelet function and methods based on RNA dyes for the determination of reticulated platelets. Some of these assays such as the analysis of platelet adhesion receptors, determination of platelet activation and measurement of reticulated platelets have become clinically validated. Analysis of platelet adhesion receptors is used for the clarification of rare inherited disorders of platelet function. Determination of platelet activation is also routinely used to characterize the quality of stored platelet concentrates. Measurement of reticulated platelets is very helpful in the diagnostic process of the differentiation of thrombocytopenia. Determination of platelet activation in vivo or reactivity ex vivo to plan therapeutic strategies or to monitor antiplatelet therapy should be used in the scope of clinical studies only until their diagnostic value is established.

References

1 Michelson AD: Flow cytometry: A clinical test of platelet function. Blood 1996;87:4925–4936.

2 Michelson AD, Furman MI: Laboratory markers of platelet activation and their clinical significance. Curr Opin Hematol 1999;6:342–348.

3 Ault KA: The clinical utility of flow cytometry in the study of platelets. Semin Hematol 2001;38:160–168.

4 Hickerson DH, Bode AP: Flow cytometry of platelets for clinical analysis. Hematol Oncol Clin North Am 2002;16:421–454.

5 Peerschke EI: The laboratory evaluation of platelet dysfunction. Clin Lab Med 2002;22:405–420.

6 Michelson AD. Evaluation of platelet function by flow cytometry. Pathophysiol Haemost Thromb 2006;35:67–82.

7 Kienast J, Schmitz G: Flow cytometric analysis of thiazole orange uptake by platelets: a diagnostic aid in the evaluation of thrombocytopenic disorders. Blood 1990;75:116–121.

8 Rinder HM, Munz UJ, Ault KA, Bonan JL, Smith BR: Reticulated platelets in the evaluation of thrombopoietic disorders. Arch Pathol Lab Med 1993;117:606–610.

9 Salvagno GL, Montagnana M, Degan M, Marradi PL, Ricetti MM, Riolfi P, Poli G, Minuz P, Santonastaso CL, Guidi GC: Evaluation of platelet turnover by flow cytometry. Platelets 2006;17:170–177.

10 Rand ML, Leung R, Packham MA: Platelet function assays. Transfus Apheresis Sci 2003;28:307–317.

11 Tsiara S, Elisaf M, Jagroop IA, Mikhailidis DP: Platelets as predictors of vascular risk: is there a practical index of platelet activity? Clin Appl Thromb Hemost 2003;9:177–190.

12 Michelson AD, Frelinger AL 3rd, Furman MI: Current options in platelet function testing.Am J Cardiol 2006;98:4N–10N.

13 Gurbel PA, Becker RC, Mann KG, Steinhubl SR, Michelson AD: Platelet function monitoring in patients with coronary artery disease. J Am Coll Cardiol 2007;50:1822–1834.

14 Cattaneo M: Resistance to antiplatelet drugs: molecular mechanisms and laboratory detection. J Thromb Haemost 2007;5(suppl 1):230–237.

15 Schmitz G, Rothe G, Ruf A, Barlage S, Tschope D, Clemetson KJ, Goodall AH, Michelson AD, Nurden AT, Shankey TV, European Working Group on Clinical Cell Analysis: Consensus protocol for the flow cytometric characterisation of platelet function. Thromb Haemost 1998;79:885–896.

16 Schwarzt A, Repollet EF, Vogt R, Gratama JW: Standardizing flow cytometry: construction of a standardized fluorescence calibration plot using matching spectral calibrators. Cytometry 1996;26:22–31.

17 Grau AJ, Ruf A, Vogt A, Lichy C, Buggle F, Patscheke H, Hacke W: Increased fraction of circulating activated platelets in acute and previous cerebrovascular ischemia. Thromb Haemost 1998;80:298–301.

18 Ruf A, Patscheke H: Flow cytometric detection of activated platelets: comparison of determining shape change, fibrinogen binding, and P-selectin expression. Semin Thromb Hemost 1995;21:146–151.

19 McEver RP: Properties of GMP-140, an inducible granule membrane protein of platelets and endothelium. Blood Cells 1990;16:73–80.

20 Michelson AD, Benoit SE, Kroll MH, Li JM, Rohrer MJ, Kestin AS, Barnard MR: The activation-induced decrease in the platelet surface expression of the glycoprotein Ib-IX complex is reversible. Blood 1994;83:3562–3573.

21 Michelson AD, Barnard MR, Hechtman HB, MacGregor H, Connolly RJ, Loscalzo J, Valeri CR: In vivo tracking of platelets: circulating degranulated platelets rapidly lose surface P-selectin but continue to circulate and function. Proc Natl Acad Sci U S A 1996;93:11877–11882.

22 Marquardt L, Ruf A, Mansmann U, Winter R, Schuler M, Buggle F, Mayer H, Grau AJ: Course of platelet activation markers after ischemic stroke. Stroke 2002;33:2570–2574.

23 Michelson AD, Ellis PA, Barnard MR, Matic GB, Viles AF, Kestin AS: Downregulation of the platelet surface glycoprotein Ib-IX complex in whole blood stimulated by thrombin, adenosine diphosphate, or an in vivo wound. Blood 1991;77:770–779.

24 Ruf A, Frojmovic M.M, Patscheke H: Platelet aggregation; in von Bruckhausen F, Walter U (eds): Handbook of Experimental Pharmacology. Platelet and Their Factors. Berlin, Springer, 1997, pp 84–97.

25 Shattil SJ, Hoxie JA, Cunningham M, Brass LF: Changes in the platelet membrane glycoprotein IIb/IIIa complex during platelet activation. J Biol Chem 1985;260:11107–11114.
26 Matic GB, Chapman ES, Zaiss M, Rothe G, Schmitz G: Whole blood analysis of reticulated platelets: improvements of detection and assay stability. Cytometry 1998;34:229–234.
27 Consolini R, Calleri A, Bengala C, Legitimo A, Conte PF: Evaluation of thrombopoiesis kinetics by measurement of reticulated platelets and CD34(+) cell subsets in patients with solid tumors following high dose chemotherapy and autologous peripheral blood progenitor cell support. Haematologica 2001;86:959–964.
28 Richards EM, Jestice HK, Mahendra P, Scott MA, Marcus RE, Baglin TP: Measurement of reticulated platelets following peripheral blood progenitor cell and bone marrow transplantation: implications for marrow reconstitution and the use of thrombopoietin. Bone Marrow Transplant 1996;17:1029–1033.
29 Jiménez MM, Guedán MJ, Martín LM, Campos JA, Martínez IR, Vilella CT: Measurement of reticulated platelets by simple flow cytometry: An indirect thrombocytopoietic marker. Eur J Intern Med 2006;17:541–544.
30 Takami A, Shibayama M, Orito M, Omote M, Okumura H, Yamashita T, Shimadoi S, Yoshida T, Nakao S, Asakura H: Immature platelet fraction for prediction of platelet engraftment after allogeneic stem cell tansplantation. Bone Marrow Transplant 2007;39:501–507.

Sack U, Tárnok A, Rothe G (eds): Cellular Diagnostics. Basics, Methods and Clinical Applications of Flow Cytometry. Basel, Karger, 2009, pp 390–425

DNA and Proliferation Analysis by Flow Cytometry

Gero Brockhoff

Clinic for Gynecology and Obstetrics, Caritas Hospital St. Josef, University of Regensburg, Germany

'Discussion about CV have been known to bring out rather primitive competitive instincts within groups of flow cytometrists, who are usually friendly, well-adjusted people.'
Longobardi Givan A: Flow Cytometry – First Principles.New York, Wiley-Liss, 1992

Introduction

DNA analysis belongs to the earliest applications in flow cytometry. The use of fluorescence dyes that interact in a stoichiometrically linear relationship with DNA allow determination of the absolute DNA content, i.e. the ploidy of individual cells as well as the cell cycle distribution of a cell population.

The main applications of clinical DNA cytometry are currently found in tumor biology diagnostics and oncology. Degenerate cells can exhibit malignant chromosomal abnormalities that lead to DNA material which deviates from the normal diploid state. This genetic instability is considered to be the central pathological molecular characteristic of cancer cells that is responsible not only for the emergence but also for the progression of neoplastic growth. Manifestations of the genetic instability of tumor cells can be demonstrated both on the chromosomal and DNA levels. On the chromosomal level, a disturbance in chromosome distribution leads to aneuploidy in the context of cell division with the daughter cells, which is also responsible for increased mutability of the aneuploid cells. This chromosomal instability represents the main form of genetic instability in human tumors and has been demonstrated in a multiplicity of malignant illnesses. The underlying mechanisms are only partially understood, but phenomena such as deletion, unequal translocation, errors in replication and subsequent repair, endomitotic processes may cause chromosomal aberrations. Mitosis spindle checkpoint proteins play a central role in these processes: during cell division, they regulate both the accumulation of chromosomes at the mitotic spindle and their correct distribution in daughter cells. In the presence of dysfunctions in mitosis-regulating molecules, cells can develop whose abnormal DNA content can be detected by flow cytometry.

Quantification of the absolute DNA content is particularly important in pathology diagnostics as aneuploidy frequently accompanies malignancy. However, the assessment of ploidy is valuable not only for diagnosis but also for therapy monitoring. The diagnosis of both abnormal DNA content as a marker of chromosome instability and a high proliferation fraction can contribute to the prognostic evaluation and to individual optimization of therapy and monitoring of the disease process. DNA quantification can be carried out in multiparametric assays with simultaneous cell selection and/or identification and phenotyping by immunological staining.

Cell cycle is essentially investigated in basic research, but the results shed light on our understanding of cell proliferation. The proportional distribution of a population of cells in the G_1, S and G_2/M phases of the cell cycle is determined based on fluorescence intensity. In clinical applications, the S phase fraction (SPF) expresses the proliferation activity of a (tumor) cell population. With the help of dynamic proliferation assays, additional information can be obtained regarding the regulation of cell proliferation under defined conditions.

Static DNA Analysis

Static DNA analysis is widely used in research as well as for diagnostic purposes. DNA dyes, which interact quantitatively with DNA, enable the determination of the proportional distribution of a cell population or subpopulation that is in the G_1, S and G_2/M phases. If a fluorescence dye is stoichiometrically incorporated into double-stranded DNA, the measured fluorescence intensity will be directly proportional to the DNA content (fig. 1). This is represented in the DNA histogram of a flow cytometry analysis. G_1 and/or interphase cells usually have a simple, i.e. diploid DNA content, which is doubled in the G_2 phase by DNA replication in the S phase. In mitosis, the chromosomes are divided into the two daughter cells, so that following cytokinesis the cells are once again diploid for DNA content. The duration of this cyclic process is cell type specific; differences are due in particular to an unequally long G_1 phase. In contrast, the duration of the S phase is usually constant and relatively short: if the DNA replication process has already begun, complete doubling of the genome occurs without interruption. The G_2/M phase is usually very short (a few hours). A typical DNA histogram, shown in figure 1, reflects the unequal duration of the cell cycle phases, but does not quantitatively reflect the duration of the individual cell cycle phases. However, the fact that most cells are in G_1 phase shows that this phase must be the longest.

There is a vast spectrum of dyes which are suitable for static DNA analysis; there is a difference between base-pair-specific and nonspecific dyes. Propidium iodide

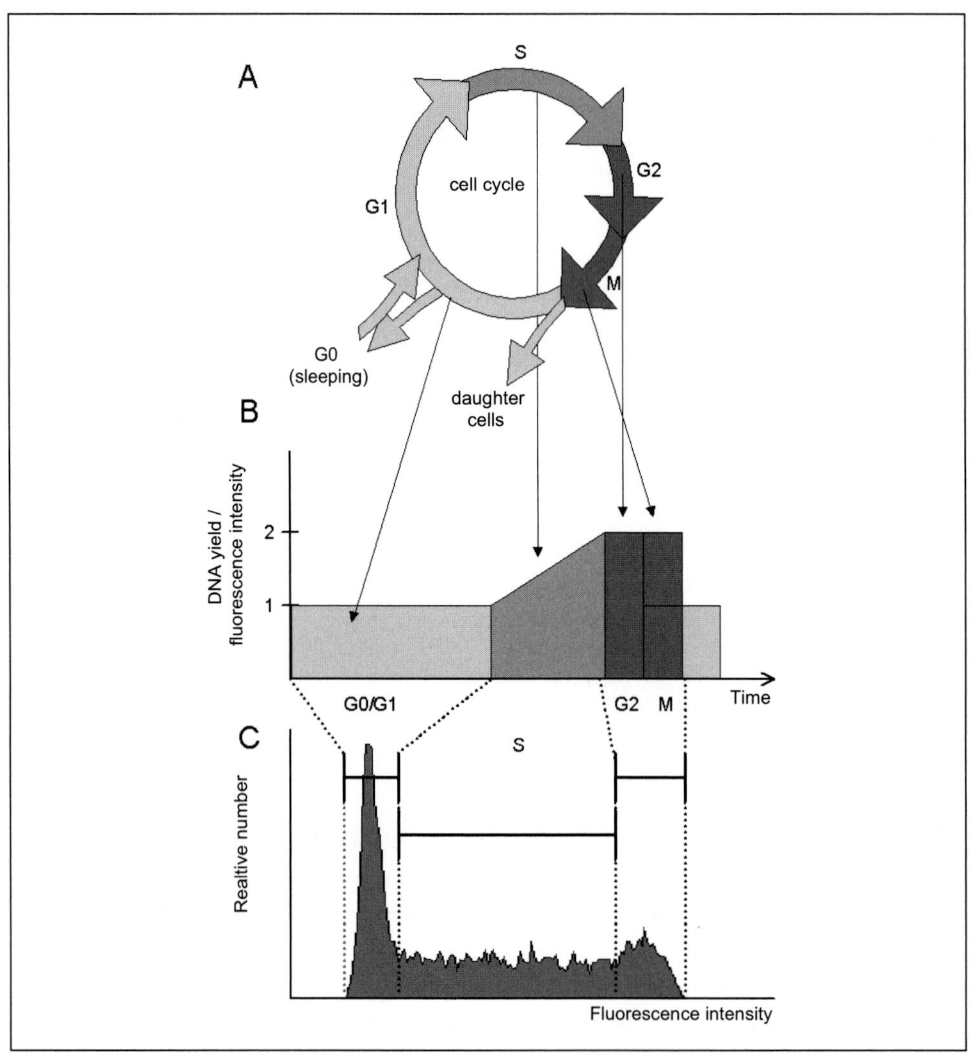

Fig. 1. The DNA content in the cell cycle process. **A** Cell multiplication is typically based on cells passing more or less continuously through the G_1, S and G_2/M cell cycle phases. Mammalian cells require permanent growth stimulus by growth factors to increase their numbers. Without external stimulus, the cells reduce their metabolism and enter a resting phase (G_0 phase). Growth factors, which function as competence factors, recruit cells from the G_0 phase to re-enter the G_1 phase, whereby they participate actively in cell multiplication. **B** In interphase and/or G_1 phase (gap-1) somatic, nontransformed cells display a diploid (2n) DNA content. In the DNA synthesis phase (S phase) the DNA content is increased by DNA replication, until the genome is doubled in the G_2 phase (gap-2) (4n). In the mitosis phase, the duplicated chromosomes are distributed evenly in two daughter cells, which again exhibit a DNA content of 2n. The fluorescence intensity of fluorochromes, which interact in a stoichiometrically linear relationship with the DNA, is proportional to the DNA content in the cell cycle process and reflects the cyclical processes of cell multiplication. **C** In the DNA histogram of a flow-cytometric measurement, fluorescence intensity can be assigned to individual cell cycle phases; however, as shown here, G_0 and G_1 phase cells with identical DNA content cannot be differentiated with a simple static DNA analysis. The number of cells in the respective cell cycle phases is shown on the y-axis of the histogram.

Table 1. Some examples of fluorochromes which can be used in DNA analysis[a]

	Vital dye	Non-vital dye
Base pair-nonspecific binding	Vybrant DyeCycle (diverse)	ethidium bromide* PI* LDS-751 acridine orange (metachromatic: DNA binding green emission; RNA binding red emission) TO-PRO-3*
Base pair-specific binding (preferential binding to AT and GC base pairs)	Hoechst 33258* (AT) Hoechst 33342* (AT) DAPI* (AT) DRAQ5 (AT)	7-AAD* (GC) chromomycin* (GC) mithramycin* (GC)

[a] Only a subset of the many frequently used dyes with different biophysical characteristics are listed here. Those fluorochromes characterized with '*' demonstrate multiple amplification of emission after binding to nucleic acid. 'AT' and/or 'GC' indicates preferred base pair binding. This is only a very small and arbitrary selection of frequently used nucleic acid dyes. A large selection of dyes with detailed biophysical characterization may be found e.g. on the web pages of Invitrogen (*http://probes.invitrogen.com*).

(PI) or ethidium bromide, for example, do not discriminate in binding to double-stranded nucleic acid while others such as 7-aminoactinomycin-D (7-AAD), Hoechst dye or 4′,6-diamidino-2-phenylindol (DAPI) preferentially bind to AT or GC base pairs. There is also a difference between vital (live-cell) and nonvital (fixed-cell) dyes, which reach the target DNA either without permeabilization or only after permeabilization of the cell and nuclear membrane. Cells can be permeabilized with short-chain alcohols such as methanol or ethanol, or with detergents. The choice of the fluorochrome primarily depends on the available excitation source (e.g. UV wavelengths, 488 or 635 nm laser) and the detection possibilities (filters and fluorescence channels). A small selection of widely used DNA dyes is found in table 1. There are actually dozens of dyes characterized by different biophysical properties which bind nucleic acid:

- preferential DNA or RNA binding;
- base pair-nonspecific or preferential binding (AT or GC preference);
- membrane permeable or impermeable (called vital or non-vital dyes);
- emission intensification through nucleic acid binding;
- different spectral characteristics depending upon DNA or RNA binding (example acridine orange);
- specific excitation maxima (from ultraviolet to infrared), and
- accessibility of the chromatin (euchromatin, heterochromatin).

DNA and proliferation analyses can include one parameter (only DNA staining) or two or more parameters. The advantage of one-parameter analyses is that a nuclear preparation can be carried out, which usually ensures a high-quality DNA measurement (fig. 5) [1–4] (see also 'Internal Quality Control of DNA Analysis: The Coefficient of Variation', pp. 397). Multiparametric analyses, however, fundamentally require (if not just nuclear antigens need to be detected) intact cells. In the case of heterogeneous cell populations, this enables the identification of subpopulations or an additional phenotype through the detection of antigen expression.

Static DNA Quantification/Analysis of Ploidy

Fluorochromes, which quantitatively bind in a linear relationship to DNA, emit fluorescence by an interaction whose intensity is proportional to the DNA content (fig. 1). Thus the DNA content can be quantified in relation to a standard. The DNA content of a cell is species specific and varies periodically in the course of the cell cycle. Somatic, untransformed cells have usually a diploid (2n) DNA content, which increases through DNA replication in the S phase to twice the amount (4n) in the G_2 phase. The doubled genome is distributed by cell division into two daughter cells which are again diploid. Chromosomal aberrations lead to a total deviation from diploid DNA content in at least 4% of the entire genome to be analyzed by flow-cytometric DNA measurement. This approximately corresponds to the quantity of DNA in a large chromosome. That is, those genetic and/or chromosomal changes that do not affect the entire DNA content (e.g. mutations, balanced translocations), or losses and/or increases in DNA quantity <4%, cannot be measured by a simple flow-cytometric ploidy analysis (figs. 2, 3).

The term 'DNA index' has become generally accepted in flow-cytometric DNA quantification; diploid cells have a DNA index of 1, tetraploid cells have a DNA index of 2 and triploid cells, an index of 1.5. An overview of current terminology is shown in table 2.

Protocol for a Single-Parametric DNA Measurement with Propidium Iodide

- Adherent cultured cells are harvested as usual (e.g. trypsinized, scraped off, incubated with EDTA) and washed twice with PBS. (Cells growing in suspension are washed without a separate harvesting step.)
- After washing, the cells are fixed and the cell membrane is permeabilized; thereafter, the cells are suspended in 70% methanol or ethanol. (To avoid clumping, cells can be suspended in PBS, and 100% alcohol can be slowly added under cooled conditions to make up a final concentration of 70%.) 1 million cells/ml 70% methanol represents the optimal cell density.

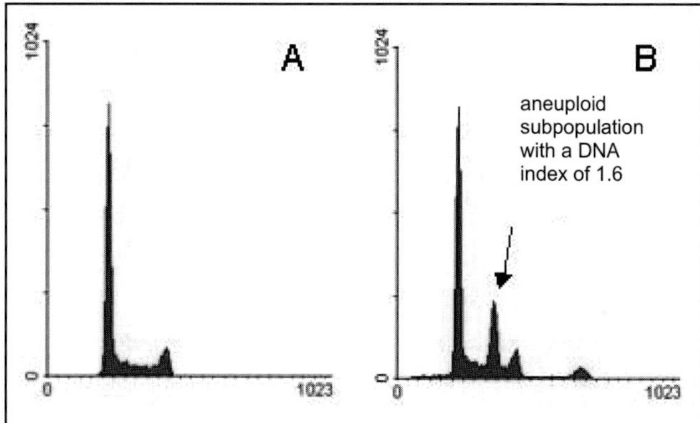

Fig. 2. Diploidy and aneuploidy. **A** DNA analysis with an exclusively diploid population. **B** DNA analysis of a mainly diploid and a smaller aneuploid subpopulation with a DNA index of 1.6.

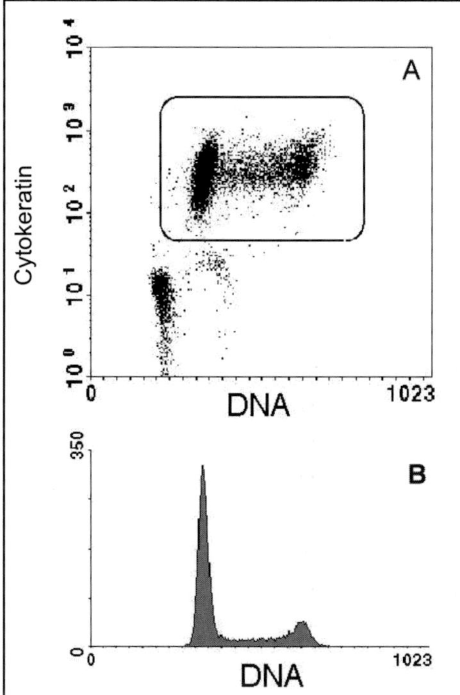

Fig. 3. DNA cytokeratin analysis of a heterogenic cell population consisting of cytokeratin-18-positive, aneuploid and cytokeratin-18-negative diploid cells. **A** After gating, the purely aneuploid population can be displayed in a DNA histogram (**B**).

Table 2. Terminology of DNA-quantification in flow cytometry

Ploidy	DNA content	DNA index
Diploid	simple (normal)	1
Hypodiploid	DNA loss	<1
Hyperdiploid	DNA increase	>1
Euploid	there is a simple chromosome set or a whole number multiple of it	i.e. 1, 2, 3, or 4
Aneuploid	deviation from a simple chromosome set, but not a multiple of a whole number	$\neq 1$, $\neq 2$, $\neq 3$, $\neq 4$
Triploid	quantitatively, half of the genome is doubled	1.5
Tetraploid	there is a doubling of the genome	2

- The cells should be fixed for at least 1 h; they can be kept for several days (weeks) in 70% alcohol.
- For the following DNA staining, the cells are again pelleted, washed twice with PBS and resuspended in 425 μl PBS (1 million cells per aliquot). Finally 50 μl of RNAse (from a concentrated stock solution kept at 1 mg/ml) is added, and the cells are incubated for about 20 min at 37 °C. (PI is integrated into double-stranded nucleic acid, so the RNA should be digested.)
- Subsequently, 25 μl PI is added from a stock solution (PI concentration 1 mg/ml; the final concentration amounts to 50 μg/ml, and the final volume of the sample is 500 μl).
- The cells need to be incubated for only a few minutes and can then be analyzed by flow cytometry.
- Instrument settings: the fluorescence of the DNA stain is measured with linear amplification! The pulse-processed signals are likewise applied linearly (see 'Pulse Processors and Doublet Discrimination', p. 397). The voltage of the amplifier should be adjusted in such a way that both diploid and aneuploid cells are on the linear x-axis of a histogram, without the left-sided diploid G_1 cells or right-sided aneuploid G_2 phase cells falling outside of the scale. The choice of the detection channel depends on the staining material.

Comment – Error Check – Troubleshooting

Interacting Nucleic Acid Dyes

PI and many other DNA dyes bind double-stranded nucleic acid, – not only DNA, but also as far as present in the cell, double-stranded RNA. This additional take-up of dye by the cell, which is of course independent of DNA content and cell cycle, decreases the quality of a DNA histogram: The G_0/G_1 peak becomes broader, which results in an increased coefficient of variation (CV; see 'Internal Quality Control of DNA Analysis: The Coefficient of Variation', pp. 397), and the computation of the distribution of cell cycle phases increases the inaccuracy. Digestion of the RNA before DNA staining is therefore recommended.

The Stoichiometry of Dye Binding to DNA

Since PI interacts with DNA in a stoichiometric relationship, the cell amounts/concentrations of DNA and of dye are mutually dependent on each other. The goal is to attain a saturated DNA staining with PI because if the cell number is too high in relation to the dye quantity, there is a risk of understaining. On the other hand, it is also possible to 'overstain' the cells with a too high concentration of dye, resulting in a nonspecific staining. In particular if the proportional distribution of the cell cycle phases as well as calculation of the entire DNA content of the cells (ploidy estimation) is important, a constant cell number and dye concentration for each attempt is very important. Therefore one should be sure to maintain the approximate proportion of 0.5 million cells/ml PBS to 50 μg/ml PI! With excessive amounts of dye and low a DNA quantity, a DNA histogram is obtained that is shifted to the right on the intensity axis, leading to overestimation of DNA content whereas insufficient amounts of dye lead to intensity shifts to the left.

Pulse Processors and Doublet Discrimination

With DNA analyses, it is particularly important to identify single cells and doublets and to discriminate larger aggregates. This may be achieved by using pulse processor technology and laser excitation, which enable time-dependent fluorescence signals to be recorded. With the commercially available state-of-the art devices, additional calculation of intensity signal (height), one (area signal) or two time-dependent pulse-processed signals (surface and width) is possible.

The signal intensity (height) therefore supplies information on fluorescence density and thus the maximally emitted brightness under fully excited conditions. From the signal integral and thus the surface (area), the entire surface and cell-bound 'fluorescence content' can be derived while the pulse width (width) allows conclusions to be drawn regarding particle size, and therefore single cells and cell pairs. With respect to the combined representation, two signals clearly allow differentiation between single cells and cell aggregates. In figure 4, the fluorescence signals of an individual cell in G_1 and G_2/M phase as well as a G_1 cell pair, which are registered as a singular event, are directly compared. With the flow of a fluorescent particle through the illumination point, if the cell does not exceed the dimension of the laser beam, maximum excitation and emission can be observed. However, a cell pair which e.g. consists of two G_1 phase cells is too large to be encompassed by the laser. Thus, a cell pair may be distinguished from an individual cell in G_2 phase which does not exceed the width of the laser beam. The technical aspects and consequences of cell pair discrimination have been discussed by Wersto et al. [5]. With the use of mercury vapor lamps, a comprehensive cell and particle illumination is possible in contrast to the one-sided excitation with coherent laser light. This all-round irradiation enables the discrimination of large and small particles and aggregates without pulse processors. In addition, high-resolution DNA histograms may be obtained with such illumination [1, 2, 6].

Internal Quality Control of DNA Analysis: The Coefficient of Variation

In particular the ploidy assessment and the identification of even a few subpopulations with a DNA index that deviates only slightly from diploidy (i.e. near-diploid cells) make flow cytometry-based quality control of DNA analysis necessary. The CV of the G_0/G_1 peaks of a DNA histogram represents a substantial quality criterion of static DNA measurements. The CV is defined as the quotient derived from the standard deviation of this peak divided by the mean fluorescence intensity of this

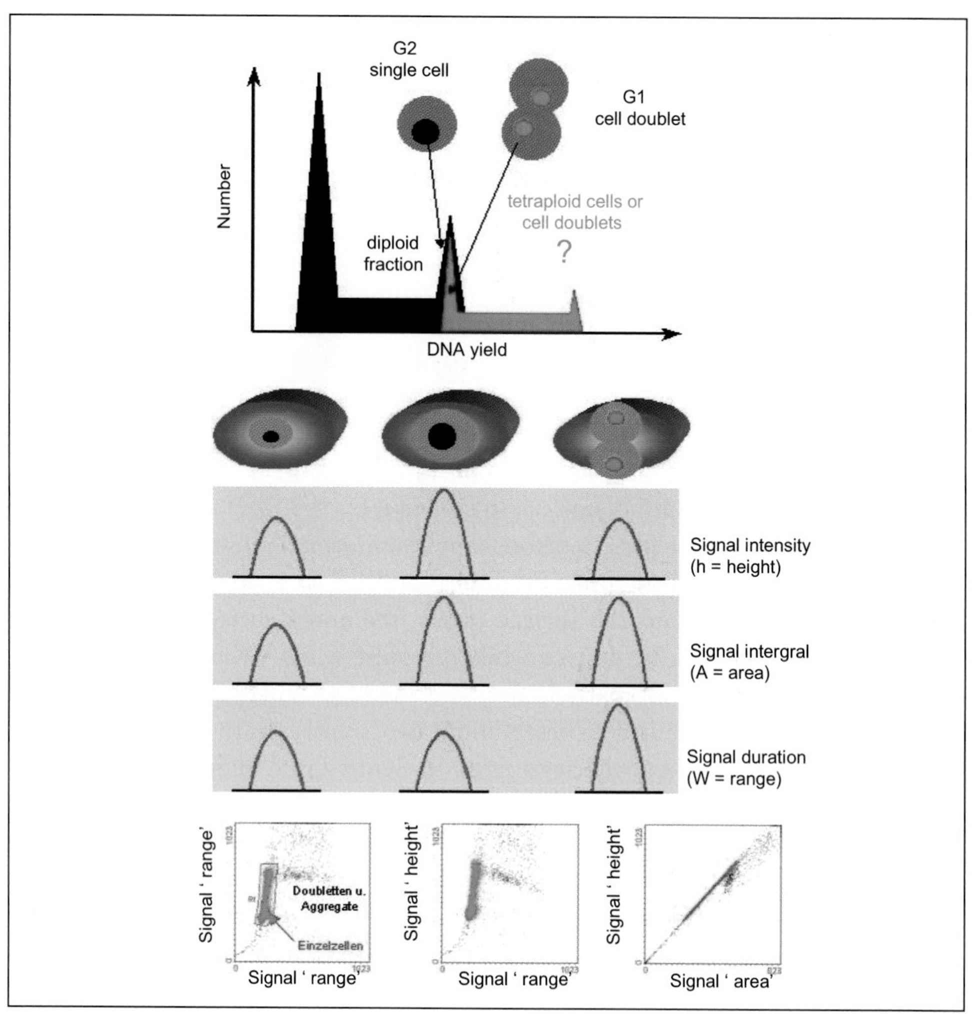

Fig. 4. Discrimination of cell doublets on the basis of pulse processor technology.
Cell pairs which could be falsely read as G_2 phase cells or tetraploid cells due to double fluorescence intensity are identified by time-dependent signals (area, width). Cell aggregates whose diameter is larger than the cross-section of the laser beam do not become maximally activated. Moreover, it takes longer for large, aggregated particles, to cross the illumination point compared with single cells, which is the reason why the width signal of a cell pair is larger than that of a single cell. The applications of pulse width versus surface area, width versus intensity (height) or surface area versus intensity are equivalent and, with adequate gating, allow the discrimination of debris, cell pairs and larger aggregates. See Wersto et al. [5].

peak and describes its width. Multiplication of the CV by 100 yields a value which can be expressed in percent.

Subpopulations which differ only slightly in DNA content (i.e. DNA index of 1.0 und 1.2) are obviously easier to detect and identify the smaller the G_0/G_1 peak. Therefore, DNA cytometry experts always aim at performing measurement that gen-

Table 3. Parameters that influence the quality of a DNA analysis (CV)

Instrumentation technology	Preparation	Biological material
– Instrument comparison, adjustment of flow system and optical system (Illumination point) – Flow rate	– exploration of whole cells with intact cytoplasm or analysis of a nuclei preparation – single (only DNA) or multi-parametric measurement (DNA- and i.e. antibody staining) – DNA dye – RNA digestion	– homogeneous cell lines or heterogenic material from tissue disaggregation – cell and/or tissue type – Primary material: fresh, frozen or isolated from paraffin – diploid or aneuploid population

erate an as small as possible CV of the G_0/G_1 peaks. This tendency is so strong that it may even cause 'competitive' tensions between cytometry experts within or between laboratories, as was well described by A. Longobardi Givan (see motto at the beginning of this chapter). However, one frequently forgets that the CV depends on many different parameters based both on instrumentation and preparation, each of which exerts differential influences based on the users' needs. The quality of a DNA measurement is affected by the parameters listed in table 3.

Thus the CV depends on the particular flow cytometer, the adjustment of the illumination points and the adjusted flow rate. In general, smaller CVs are obtainable with private instruments equipped with a flow cuvette, than with public systems, which must be optimally adjusted. Furthermore, aneuploid tumor cells usually generate a larger CV than diploid investigation material or homogeneous cell lines due to their greater heterogeneity. Of course, the dye used plays a crucial role too. Thus a dye such as 7-AAD is incorporated less efficiently into double-stranded DNA from cells in G_0 phase than from cells in G_1 phase and thus provides slightly different fluorescence intensities for resting, metabolically inactive G_0 cells with more strongly condensed chromatin than for metabolic more active G_1 cells with less condensed chromatin [7]. To obtain high-resolution DNA histograms it is best to use a dye excited by UV light such as DAPI with cell or complete nuclear excitation (in contrast to lateral laser activation) with a mercury vapor lamp (fig. 5).

It is advisable to verify the optimal calibration of flow cytometers in regular and short time intervals (e.g. weekly) with reference particles (e.g. beads or chicken erythrocytes) in order to ensure the optimum quality of the DNA analysis from the technical side. For most instruments, a DNA measurement should not exceed a flow rate of 300 events/s. Beyond that, reproducible data can be achieved only at constant temperatures (laser actual working time, room temperature). The following can serve as reference values: a CV of <6% is a good value if heterogeneous primary tumor material is examined and if, following tissue digestion, whole cells are present and PI is used. With human epithelial tumor cell cultures a CV of <5% can be

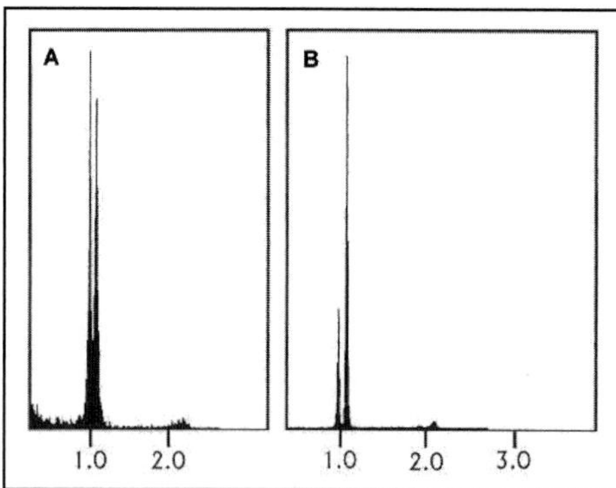

Fig. 5. High-resolution DNA cytometry applied to a disk epithelium carcinoma [8]. The DAPI DNA measurements of the tumor cells were generated with a PAS II flow cytometer from Partec (Münster, Germany). **A** Histogram of diploid tumor cells with an aneuploid subpopulation with a DNA index of 1.08. **B** Re-analysis of a locally relapsed cell line that mainly consists of aneuploid cells. This example clarifies that the aneuploid fraction was identifiable only with the high-resolution DNA analysis of a nuclear preparation. On the x-axis, the relative DNA content and the fluorescence intensity is indicated in each case by the author.

reached with whole intact cells – also when carrying out multiparametric approaches (double or multiple staining). When isolating nuclei or when analyzing lymphocytes, CV values between 1 and 2% can be obtained using DAPI and excitation with a mercury vapor lamp.

Clinical Uses of Static DNA Analysis

Flow-cytometric DNA analyses are as old as flow cytometry [9–13], and it has taken many years to be able to employ them in a clinically useful fashion. Today, the central clinical interest in DNA analytics lies in oncology, particularly because cancer can be regarded as an illness of uncontrolled cell proliferation, in which the degenerate cells frequently, although not always, have an abnormal DNA content (aneuploidy). With the help of flow-cytometric DNA analysis both the proliferation fraction and the DNA content (i.e. ploidy) can be quantified and, dependent on tumorigenicity, relevant additional prognostic and diagnostic information can be obtained. The knowledge that a sample has a high proliferation fraction and an abnormal DNA content, which serve as rough markers of chromosomal instability, can contribute to specification of the prognosis, individual therapy optimization as well as the monitoring of the disease process.

DNA analysis can be applied in both hematologic diseases [14–18] and in the analysis of solid tumors [19–35]. In contrast to leukemic illnesses, solid tumor analytics require either cell nuclear isolation or a cell type with optimum conditions of cell integrity [31, 36]. A single-parameter investigation of cell nuclei in suspension permits the generation of high-resolution DNA histograms for the detection of malignant subpopulations with an only slightly abnormal DNA content [1–4, 6]. Nuclear isolation was originally developed by Vindelov [37] and later optimized by Hedley et al. [38] and Vindelov et al. [39]. Similar associated protocols can be found in these publications. Corver et al. [95], however, optimized protocols which describe the intact preservation of cells and their deparaffinization. In multiparametric analysis, however, tumor heterogeneity is taken into account by cell selection and discrimination. Additionally, DNA analysis following the confirmation of tumor-biologically relevant antigens can be combined with further cell phenotyping [20, 32–34, 40–43]. That is, in multi-parametric assessments, e.g. epithelial cells can be identified through antibody staining against cytokeratine; however, these same cells can be identified as carcinoma cells by macroscopic and histological testing by a pathologist. Accordingly, other cell types (e.g. stromal cells, inflammatory cells, endothelial cells) which may impair proliferation and ploidy calculation, or even falsify them, can also be differentiated [31, 32, 35, 41, 44, 45].

Thus, for example nondifferentiated, barely proliferating stromal cells do not lead to a falsely low SPF with diploid tumors. Other multiparametric applications aim at characterizing the (tumor) cells through the confirmation of specific antigens (oncogenes) [20, 34, 40, 41, 43] or to examine in more detail the regulation of proliferation through the confirmation of the expression and function of cell cycle regulating molecules (cyclin, cyclin-dependent kinases, PCNA, Ki67, p120 and many more [6, 9, 11, 46, 47]). In comparison to single modal analysis, multiparametric approaches allow more precise measurement of the cell cycle fraction of different cell populations. Each of these approaches requires an adequate standardization concerning material preparation as well as data acquisition and handling, in order to ensure that external laboratories can be assured of an independent interpretation and utilization of clinical DNA data.

It would exceed the framework of this paper to comprehensively discuss the value and use of flow cytometric DNA analysis in tumor diagnostics and oncology. The reader is directed to the literature [48] and to numerous scientific essays which deal in detail with the possibilities and limits of clinical DNA cytometry. It should be mentioned, however, that the parameters which require standardization in order to produce externally reproducible DNA analyses have already been defined very clearly and in detail at a consensus meeting of scientists and clinicians in Toronto in 1992 [38, 49, 50]. Since then, numerous works have been published which have made a contribution to this essential standardization, regarding the following:

- use of fresh, frozen and paraffin-embedded tissues [35, 45, 51, 52];
- possibilities, pro and cons of purely mechanical, fully automatic and enzymatic tissue digestion and cell disaggregation [31, 36, 53–55];
- the fixation and permeabilization of single-cell suspensions [36];
- use of different DNA dyes with regard to the excitation, emission, binding specificity and stoichiometry [42, 43];
- data acquisition, processing and evaluation [21–23, 25, 26, 28–30, 56, 57] as well as

- clinical data evaluation from single and multiparametric analyses [14–18, 32, 33.40–43, 57, 58].

The work specified above forms the foundation of the clinical application of flow-cytometric DNA analysis. In previous years it could be demonstrated that multiparametric approaches in which cells are identified and phenotyped on the basis of antibody staining can be very valuable as supplemental clinical diagnostics. This approach takes into account conclusions reached already in 1992 [42]. Data were interpreted taking into account the numerous types of neoplasias and were newly defined at the European level in 1998 [59]. With strict consideration of these comprehensively compiled consensus guidelines, flow-cytometric DNA analysis – along with ploidy and proliferation assessment – can be an informative component in complementary oncology diagnostics and therefore can make a substantial contribution to an individualized tumor therapy [60]. DNA cytometry is also regarded as a valuable component [61, 62] in predictive medicine.

Proliferation Analysis by Flow Cytometry

The analysis of proliferation can be carried out using a static or dynamic method. With 'static analysis', the cell cycle fraction of a given cell population can be quantified while dynamic techniques used in applied research provide additional information about the absolute duration of the individual cell cycle phases and the regulation of cell proliferation under defined cell culture conditions. Moreover, the fraction of resting cells, i.e. cells in G_0 phase can be quantified with dynamic measuring procedures. Table 4 gives a comparative overview of the possibilities, including the pros and cons, of static and dynamic proliferation analyses using flow cytometry.

Software for DNA Data Evaluation

Computer-aided DNA data evaluation has become quite complex in the course of the past years; however, it is based on fundamental theoretical papers on the computation of cell cycle fractions, which had already been published in the 1970s and early 1980s [63–65].

If the variables T_{G_1}, T_S and T_{G_2M} are defined as the duration of the individual cell cycle phases then, the phase fractions $[G_1]$, $[S]$ and $[G_2M]$ can be calculated based on the equations 1–3 by integration over the appropriate time periods.

$$[G_1] = 2\left(1 - \exp\left(-\frac{T_{G_1} \ln 2}{T_c}\right)\right) \tag{1}$$

Table 4. Comparison of static and dynamic proliferation analysis in flow cytometry

Static DNA analysis	Anti-BrdU/EdU technique	BrdU/Hoechst quenching technique
– Ploidy assessment – Identification of lines with different ploidy – Quantification of the cell cycle fraction of active proliferating cells (G_1, S, G_2/M phase – Fresh-frozen and paraffin-embedded material – Relatively simple cell or nuclei preparation – Analysis in routine-flow cytometer – Use in research and in supplementary diagnostic	– calculation of the length of the cell cycle phases G_1, S and G_2/M – analysis on routine flow cytometer possible (single-laser excitation)	– length of the cell cycle phase G_1, S und G_2/M can be estimated – identification of resting G_0 phase cells – high resolution assessment of the regulation of cell proliferation (including the crossing of restriction points) – description of replication history on a single cell basis (compare to fig. 15)
No kinetic information; only rough indications regarding the relative length of the individual cell cycle phases <f	– relatively difficult and sensitive preparation (DNA-denaturing/intact double strands) – detection of resting cells only possible with additional analysis following continuous BrdU application	– UV activation (is required, not every routine machine is equipped with this) – careful optimization of preparation and staining possible (dye relation)

$$[S] = 2\left(1 - \exp\left(-\frac{T_{G_1}\ln 2}{T_c}\right) - \exp\left(-\frac{(T_{G_1} + T_s)\ln 2}{T_c}\right)\right) \tag{2}$$

$$[G_2M] = 2\left(\exp\left(\frac{-(T_{G_1} + T_s)\ln 2}{T_c}\right) - \frac{1}{2}\right) \tag{3}$$

In the past few years, computer programs whose mathematical algorithms contain e.g. aggregate and debris correctives and also consider phenomena such as 'cellular aging' and exponentially proliferating cultures, have become generally accepted [48]. The term 'cellular aging' is somewhat misleading, because it does not refer to biological cell aging, but rather to the fact that the cells are not evenly distributed over the whole cell cycle, but, due to repetitive cell divisions, there is an increased occurrence of 'younger' cells (cells in the early G_1 phase) as compared with 'older' cells (cells in G_2 and/or M phases). A given cell population is reduced through 'cellular aging' according to the following equation, based on the assumption that all cells are found in the active growth fraction (G_1, S, G_2/M).

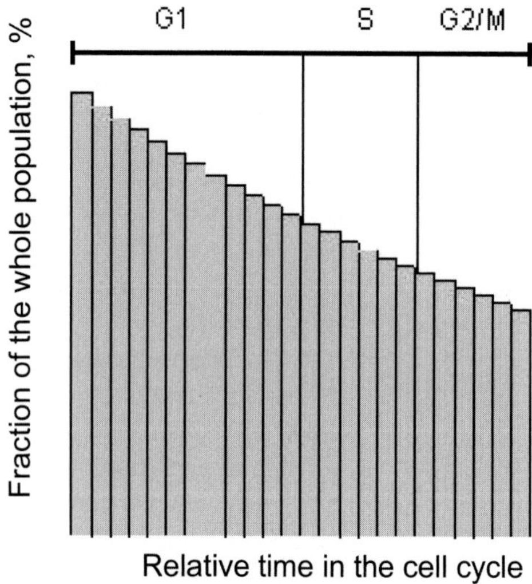

Fig. 6. The term 'cellular aging' describes a statistical phenomenon in which, due to cell division and the subsequent increase in cells, the number of young cells entering the cell cycle is larger than that of cells leaving it.

$$\left[\frac{dN}{N_{ges}(t)}\right] = \ln 2 \frac{dt_c}{T_c} \exp\left(\left(1 - \frac{t_c}{T_c}\right) \ln 2\right) \tag{4}$$

dN designates the number of cells in the time interval dt_c within the cell cycle, t_c is the time in the cell cycle after the preceding mitosis, N_{ges}, the total number of the cells at time t and T_c the length of the cell cycle.

The calculation thus deals with a statistical phenomenon (fig. 6) in which a proportion of a given cell population is reduced relative to the total population in accordance with equation 4. Software options are available which consider the dynamics of cell increase through cell division, and allow the definition of diploid and apoptotic subpopulations as well as standards with known DNA content. Through the ability of the software options to consider occurrences of nuclear fragments with DNA loss ('sliced nuclei'), both clinical and basic research applications are included. This is so because this ability concerns the investigation of primary tissue material which has been isolated mechanically and/or automatically. Only a fraction of the various available programming functions which are currently required for accurate DNA proliferation and evaluation are mentioned here. The most popular and most frequently used programs are likely ModFit and WinCycle for Windows (fig. 7), which have been developed by Bagwell et al. [56] and Rabinovitch [66] and are currently offered by Verity software House (Topsham, ME, USA; *www.vsh.com*) and Phoenix Flow Systems (San Diego, CA, USA; *www.phnxflow.com*).

Static Proliferation Analysis

Static flow-cytometric DNA analysis is a snapshot of the distribution of a cell population within a cell cycle phase. Since the fluorescence intensity of cells with

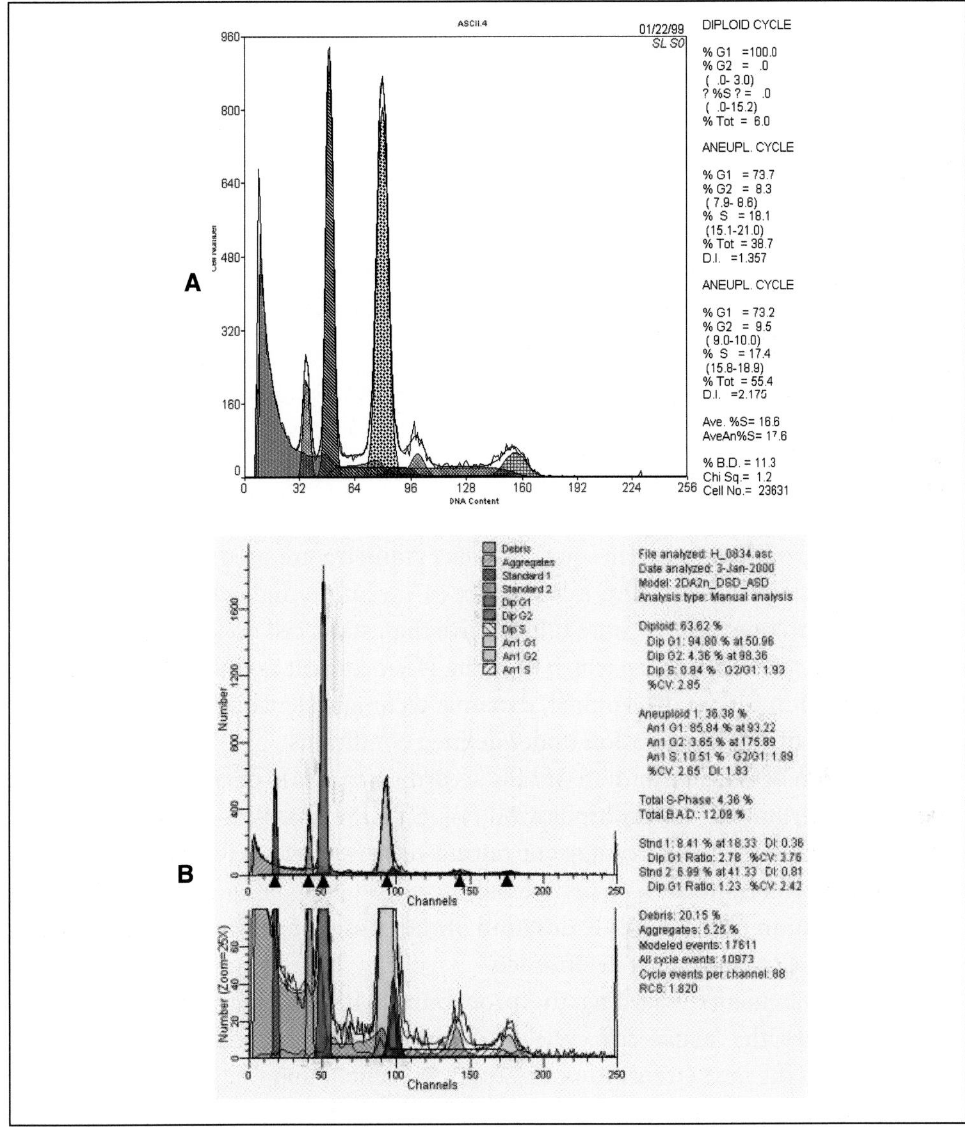

Fig. 7. Example of the calculation of cell division including DNA quantification with WinCycle for Windows from Phoenix Flow Systems (left) and ModFit LT from Verity Software House (right). Both programs allow the detection of up to three subpopulations with different DNA amounts and calculate the cell cycle of each subpopulation. Additionally, debris and cell aggregates are taken into account as so-called 'BADs' (background, aggregate, debris).

DNA stains is detectable proportionally to the absolute DNA content, the ploidy of a cell population as well as the distribution of the cell population in specific cell cycle phases can be calculated on the basis of the staining (fig. 1). In many applications, the size of the SPF can thus be used as a proxy for the proliferation activity of the population of interest. A high SPF reflects a strong proliferation activity. The SPF is assumed to have a prognostic (rather than a diagnostic) meaning in various malignant illnesses [49]. In the framework of applied research, a simple DNA measurement can be used to observe to what extent the cell cycle distribution of different cell populations changes compared with control conditions. Thus questions can be explored such as to what extent the SPF is increased by the influence of growth factors or which tumor treatments potentially reduce SFP. Furthermore, static analytics allow the identification of cell cycle arrest (e.g. in G_1 or G_2 phase). However, a detailed analysis of proliferation is possible only with dynamic techniques.

Dynamic Proliferation Analyses

The dynamic proliferation techniques in flow cytometry are used primarily in basic research. They are somewhat more complex in execution and interpretation; however, these techniques are far more informative than static cell cycle analyses.

In contrast to a snapshot in which only the DNA content and the momentary cell cycle distribution can be determined, dynamic techniques enable investigations into the regulation of cell proliferation under defined conditions:

- calculation of the distribution of cells according to phase of cell cycle, which actively contribute to cell multiplication (G_1, S, G_2/M phase);
- quantification of the percentage of resting cells, i.e. cells not actively contributing to cell multiplication (G_0 phase cells);
- determination of the absolute duration of the phases of cell cycle; up to three sequential cell cycles can be determined.
- making statements regarding the probability with which resting cells are recruited into the 'active cell cycle' and with which other cells cross from one phase into the next (transition of restriction points), and
- retracing of the replication history of individual cells.

Dynamic proliferation assays are based on the incorporation of the synthetic thymidine analogue bromodeoxyuridine (BrdU) into the DNA during its replication stage, by which cells are exposed to the base analogue either continuously or during a brief pulse of about 30 min. The proof of this substitution is made either by means of immunological antibody staining (anti-BrdU technology) and/or 'click chemistry', or by using a fluorochrome whose fluorescence characteristics are dependent on BrdU incorporation (BrdU/Hoechst quenching technology).

Dynamic Proliferation Analysis by a Short BrdU Pulse: The Anti-BrdU Technique

A substantial advantage of anti-BrdU technology compared with BrdU/Hoechst quenching technology is the fact that the widespread 488 nm excitation source can

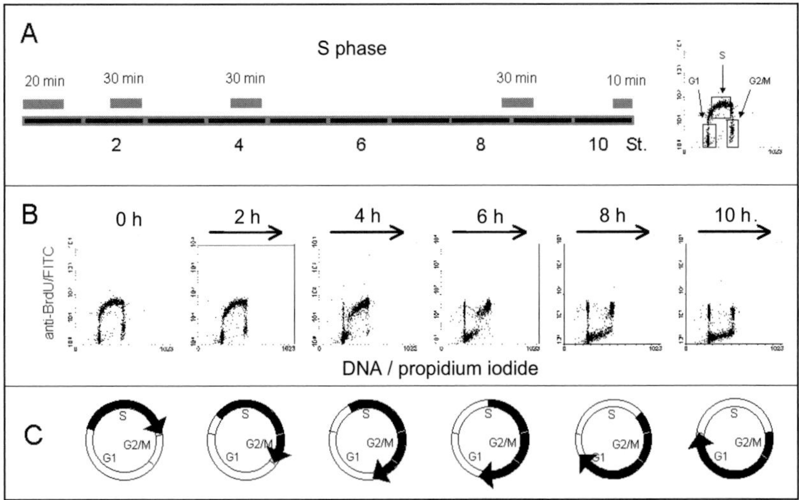

Fig. 8. A Schematic representation of BrdU incorporation during S phase over 10 h. The cells are placed in BrdU for 30 min. All cells in S phase at this time will incorporate BrdU into their DNA. Cells entering S phase only 10 min after BrdU addition will incorporate BrdU for only 20 min. Cells which have already nearly completely DNA replication at the time of BrdU addition cross from S phase into G_2 phase during the 30-min BrdU incubation and will replace thymidine with BrdU for only about 10 min. **B** Snapshots of assays which have been incubated in parallel and which in each case have been harvested after 0, 2, 4, 6, 8 and 10 h. The cells are stained with an anti-BrdU/FITC antibody (y-axis) and PI (x-axis). Immediately after the BrdU pulse (0 h). all S phase cells are BrdU-positive, the cells of other cell cycle phases are negative. Cells which were at the beginning or end of the S phase during BrdU incubation are more weakly BrdU-positive than cells which have incorporated BrdU over the whole 30 min. This information results in the typical two-parametric, horseshoe-shaped representation in the dot plot. At later times, after 2, 4, 6 h, the entry of BrdU-positive cells into G_2/M phase, and the exit of BrdU-negative cells from the G_1 phase into S phase can be observed. The S phase is completed after 10 h, so that all BrdU-positive cells have reached the G_2/M phase and can be found in the G_1-phase of the second cell cycle following cell division. BrdU-negative cells are distributed in all cell cycle phases of the first cell cycle after 10 h. **C** BrdU-stained cells (in black) in the first 10 h of the cell cycle.

be used because BrdU incorporation into the DNA is detected with fluorochrome-conjugated antibodies. Anti-BrdU antibodies that are conjugated with a dye and excitable at 488 nm, e.g., fluorescein isothiocyanate (FITC), Alexa Fluor-488, R-PE) are commercially available and well known. With anti-BrdU technology, the cells are exposed to a 30-min BrdU pulse. All cells in S phase incorporate BrdU instead of thymidine into their DNA and are thus marked (fig. 8A). After 30 min, the BrdU is removed from the culture and the cells are incubated under usual conditions in the absence of BrdU. In principle, many parallel cultures are incubated; they are harvested and analyzed at different time intervals or at 2-hour intervals after BrdU withdrawal. The detection of BrdU-positive cells following antibody staining (or 'click chemistry') and simultaneous representation of the cell cycle phases with an-

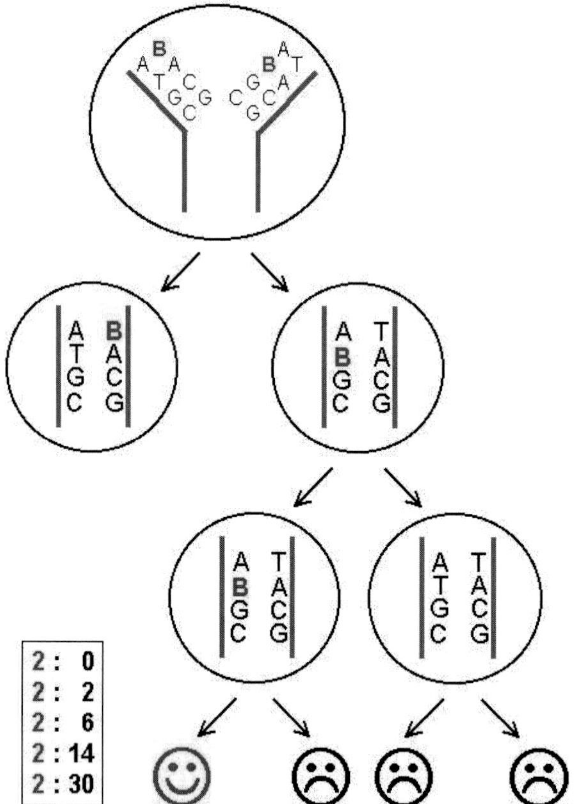

Fig. 9. Replacement of thymidine by BrdU with brief BrdU exposure (BrdU pulse): All cells which are in the S phase during a brief BrdU incubation incorporate BrdU into one of their two DNA strands instead of thymidine. Thus the daughter cell generation is marked with BrdU. After the next division of each BrdU-positive cell, a BrdU-positive and a BrdU-negative cell are born. Thereafter no more BrdU can be incorporated because it has been removed from the medium. The BrdU-positive strand, however, has become transferred from one generation to the next. The number of BrdU-positive cells remains constant in the culture while the number of BrdU-negative cells increases. After three generations the relationship is 2:14. The BrdU-positive cells can be identified with antibody staining.

other dye which penetrates into the DNA (e.g. PI) enables observation of the progress of BrdU-marked cells and detection of BrdU-negative cells in the cell cycle [67–74] (fig. 8).

Figure 9 shows schematically how BrdU is incorporated into cells if it is applied with a brief pulse. Only one DNA strand is incompletely marked by the 30-min exposure to BrdU. This incomplete marking is sufficient to detect it with antibody staining. Because only one of the two DNA strands is marked with BrdU, in the next generation 50% of the daughter cells are positive for BrdU and to the 50% are BrdU negative. This relationship persists in the following cell generations, so that the rela-

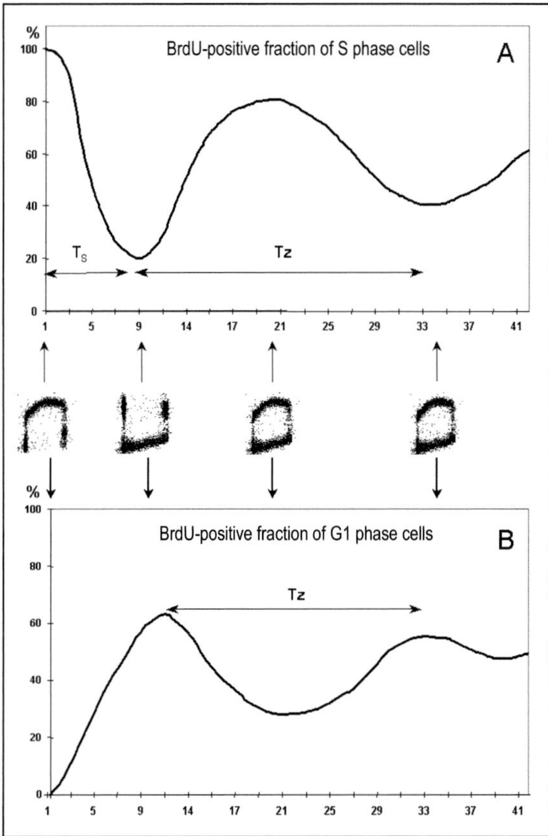

Fig. 10. Proportional distribution of BrdU-positive and BrdU-negative cells over time. **A** Immediately after the 30-min BrdU pulse, S-phase cells are 100% BrdU positive. In the course of the first 9 h, these cells reach the G_2/M phase of the first cell cycle and after cell division they reach the G_1 phase of the second cell cycle. At this time, almost no BrdU-positive cells are in S phase, in other words, the proportion of BrdU-positive cells has decreased to a minimum. Following the G_1 phase, the cells enter into the S phase of the second cell cycle; thus the fraction of BrdU-positive cells achieves a relative maximum after approximately 21 h. After about 33 h, the next relative minimum can be recognized, and so on. **B** After BrdU application, all G_1 phase cells are BrdU-negative. Only after conclusion of the S and G_2/M phases of the first cell cycle do the first BrdU-positive cells emerge in the G_1 phase of the second cell cycle. Subsequently, these cells enter the S phase of the second cell cycle, and after 21 h a relative minimum of BrdU-positive cells has reached the G_1 phase. Before the last cells leave the G_1 phase of the second cell cycle, other cells have already finalized the S and G_2/M-phase of this cell cycle and after further cell division, fall back into the G_1 phase of the third cell cycle: accordingly, a second relative maximum of cells in this cell cycle phase is reached after approximately 33 h. Two additional observations are to be noted: (a) The proportional distribution of the respective cell cycle phases over time follows a sine-similar process. The amplitude of this curve flattens over time. The reason for this is that with an increasing number of cell divisions the proportion of BrdU-positive cells in the total population decreases (compare to fig. 9). b) The cells do not pass the cell cycle phases in constant synchronous succession: they proliferate asynchronously over time. As a consequence, the relationship between BrdU-positive and BrdU-negative cell cycle fractions does not remain constant over time, so that the relative maxima and minima for the respective cell cycle phases become smaller from generation to generation.

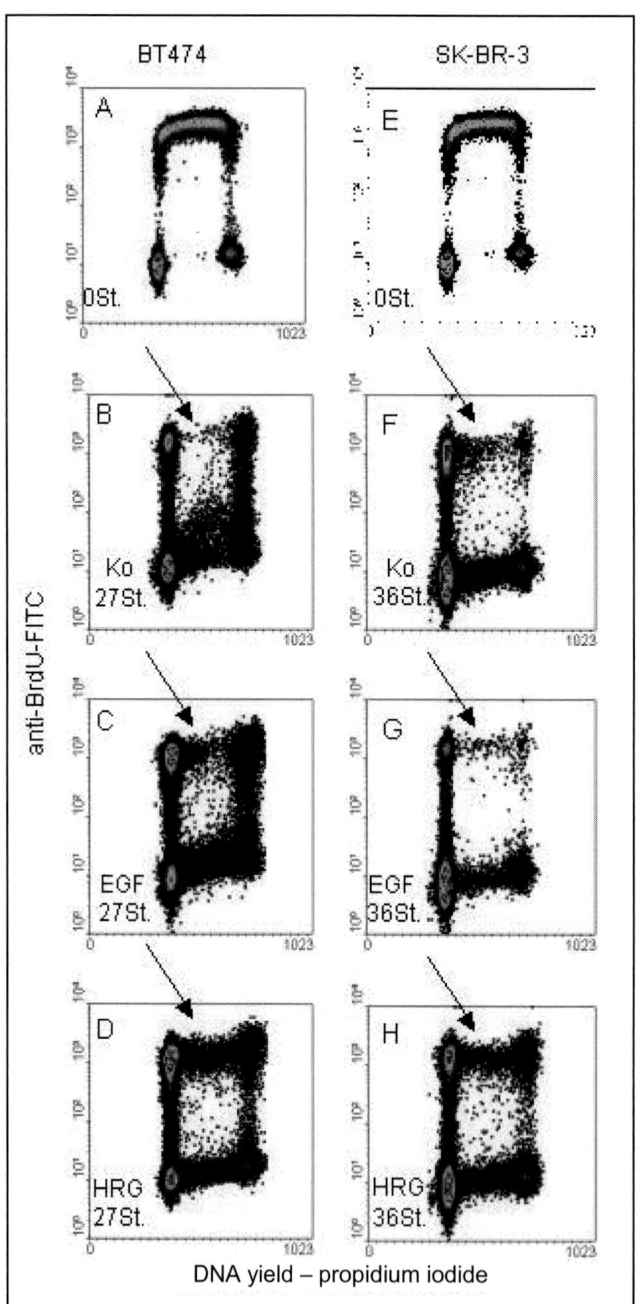

Fig. 11. Example of anti-BrdU technology. The proliferation of the two breast carcinoma cell lines BT474 and SK-BR-3 is compared with controls (**B, F**) in the presence of the growth factors EGF (**C, G**) and HRG (**D, H**) [75]. The proliferation of BT474 cells is stimulated both in the presence of EGF (**C**) and HRG (**D**): After 27 h in the presence of EGF, a cohort of cells enters the S phase of the second cell cycle, which would not be observed without addition of growth factors (**B**). Stimulated by HRG, the first cells have already completely the S phase of the second cell cycle (**D**). Compared with controls (**F**), the proliferation of SK-BR-3 cells is inhibited in the presence of EGF

tive proportion of BrdU-positive cells in the total culture is reduced as the total cell number increases; however, the absolute number of BrdU-positive cells (apart from dying cells) remains constant. In figure 10, the cyclic distribution of BrdU-positive and BrdU-negative cells is shown in the cell cycle phases of following cell cycles. Thus, as BrdU-positive cells gradually go through the individual cell cycle phases, maximum and minimum BrdU-positive fractions in the S-phase emerge cyclically (fig. 10A). The same principle is valid in the anti-cyclical sense for the G_1 phase (fig. 10B). With an increasing number of cell divisions an increasing asynchronous mixing of BrdU-positive and BrdU-negative cells becomes evident in the cell cycle phases, thus, after 2 or 3 cell divisions, it is no longer possible to distinguish between cell cycle phases.

As an example of the use of anti-BrdU technology, figure 11 shows the influence of epidermal growth factor (EGF) and heregulin (HRG) on the proliferation of SK-BR-3 and BT474-breast carcinoma cells. Passage of a cohort of cells from the G_1 phase of the second cell cycle into S phase after 27 and 36 h is shown. While proliferation of BT474 cells is stimulated by EGF and HRG in comparison to control (without addition of growth factors), only HRG has an influence on SK-BR-3 cells while EGF inhibits the proliferation of this cell line [75]. From figure 11 one can judge in each case a growth-factor-induced extension and/or shortening of the G_1 phase.

Protocol for Dynamic Proliferation Analysis following a Brief BrdU Pulse

- Numerous parallel cultures are prepared. The number of plated culture flasks depends on the total period of time during which the cell kinetic analysis is intended to be run. A goal should be to observe initially positive BrdU S phase cells until they enter the S phase of the second cell cycle which follows the G_1 phase. Thus the duration of all cell cycle phases can be determined in the order G_2/M phase (1st cycle) to S phase (1st cycle), G_1 phase (2nd cycle) and thus the duration of the entire cell cycle (fig. 11) can be determined. The total duration of the cell cycle varies depending on the specific cell. For many cell types it is meaningful and sufficient to assess one period of 40 h. It is important to make sure that cell-plating density is such that the culture is still in a subconfluent state at the time of harvest and that cell-cell contact-induced inhibition of proliferation is prevented.
- BrdU pulse: simultaneous addition of 20 μmol/l BrdU and 10 μmol/l desoxycytidin (DC); incubation time: 30 min.

◀

(**G**). Only a few isolated cells are found in the S phase of the second cell cycle. HRG, however, stimulates the proliferation/multiplication of SK-BR-3 cells (**H**). **A** and **E** are the control measurements at time t = 0 h after a BrdU pulse. The measurements of the two cell lines took place at different times (after 27 and 36 h) because the duration of the entire cell cycle is longer for SK-BR-3 cells than BT474 cells.

- The cells are washed carefully three times with PBS, in order to completely remove BrdU. – The cells are harvested as usual at 2-hour intervals after the BrdU pulse (e.g. after 0, 2, 4, 6, 8 h, depending upon cell type; trypsinize; scrape off; incubate with EDTA).
- Wash cells with PBS (containing 2% BSA), fix each sample in 1 ml 65% methanol and store at 4 °C. (The samples can be collected in this way and later further processed together.)
- Staining: remove methanol by washing twice in PBS.
- RNA digest: add 1 mg/ml RNAse and incubate for 20 min at 37 °C.
- Pellet cells and resuspend in PBS, 0.01 N HCl, 5 mg/ml pepsin. Incubate for 5 min at 37 °C. (Make up pepsin solution only just before use because of autolytic digestion!)
- Stop reaction by addition of ice-cold PBS; wash cells.
- Denature DNA with 2 N HCl/PBS and incubate for 10 min at room temperature.
- Wash cells thoroughly with PBS, which neutralizes pH.
- Stain with anti-BrdU antibody directly or indirectly (e.g. anti-BrdU IgG; DAKO Cytomation, Glostrup, Denmark) 10 μg/ml, 30 min.
- Wash sample and incubate with the secondary reagent (e.g. 'rabbit anti-mouse'-FITC IgG; DAKO Cytomation) 10 μg/ml, 30 min.
- Wash sample again with PBS.
- Add PI, working concentration 2.5 μg/ml.
- The preparation is finished and ready for analysis with the flow cytometer.
- Instrument settings: PI staining is measured again linearly; anti-BrdU staining intensity is logarithmically strengthened. The choice of the detection channels depends on the flow cytometer and on the dye which is conjugated to the antibody.

Comment – Error Check – Troubleshooting

- Anti-BrdU technology is based on a critical double staining: the first stain is BrdU which is detected with antibodies; with the second stain, PI is stoichiometrically and covalently incorporated into the entire DNA. The critical aspect of this double staining lies in denaturing the DNA for the antibody staining while making sure the DNA remains double stranded for the PI staining (denaturation makes the BrdU accessible to the antibody; PI binds only to double-stranded nucleic acid.) The preparation of the DNA must be optimized for this compromise and possibly be slightly varied depending on cell type. Of course, the application of other protocols is possible. It is useful at this point to refer to Hammers et al. [76, 77], who devised a successful preparation, which represents an alternative to enzymatic digestion when it is based on the use of UV radiation to denature the DNA. Before starting detailed proliferation trials, it is advisable to test the preparation and make slight modifications to optimize each trial for the specific cells to be examined. Such modifications can involve incubation periods, enzyme concentrations and incubation periods.

- After a 30-min pulse, the BrdU must be completely removed. It is recommended to wash the cultures at least three to four times! BrdU remnants in the medium become visible in cultures harvested later because cells can insert small quantities of BrdU in the DNA and accordingly during the antibody staining there is a weakly positive reaction. These cells distort the image in the two-dimensional dot plot representation and the accurate temporal resolution of the cell cycle phases.
- Instead of BrdU, 5-ethynyl-2-deoxyuridine (EdU) may be used as an alternative thymidine analogue. Unlike BrdU assays which require harsh treatments to denature the DNA, with the Click-iT EdU assay (Invitrogen), standard aldehyde-based fixation and detergent permeabilization are sufficient for the Click-iT detection reagent to gain access to the DNA. As a consequence, the Click-iT EdU cell proliferation kit is not only easy to use, but is also compatible with cell cycle dyes and multiplexing for the detection of antibody-based surface and intracellular markers. We have compared in detail the results of EdU with BrdU and obtained the best results on the basis of a permeabilization with saponin [98]. Detailed suggestions regarding preparation and staining can be found under *http://probes.invitrogen.com.*

Dynamic Proliferation Analysis by Continuous BrdU-Cell Staining: The BrdU/Hoechst Quenching-Technique

The inclusion of cell cycle kinetics with BrdU/Hoechst quenching technology is somewhat more fastidious and more complex, but unique as concerns the information that can be gained on the regulation of cell proliferation. The basis of this proliferation assay had already been described in the late 1970s and early 1980s [13, 78–82]. Later the use of this technology was published only in a single article [47, 83–85, 96, 97]. The strength of the BrdU/Hoechst quenching method lies in the fact that the regulation of cell proliferation and cell cycle distribution can be observed over three sequential cell cycles (fig. 13). Thus not only can the duration or a change in duration of a cell cycle phase be determined (e.g. an agent-induced extension or a growth-factor-induced shortening of the G_1 phase), but, additionally, the cell cycle in which the change arose may also be identified from the beginning of the observation [86, 87].

In contrast to the anti-BrdU technology, with the BrdU/Hoechst quenching technology the cells are exposed continuously to the nucleotide analogue BrdU, which is incorporated into the DNA instead of thymidine in several sequential S phases. Figure 12 makes clear that the more Hoechst dye molecules are displaced from the DNA, the more BrdU is incorporated into the DNA. The dynamic proliferation assay is based on the simultaneous use of DNA staining materials such as PI and Hoechst33258, whereby Hoechst33258 binds to AT base pairs and, by BrdU inser-

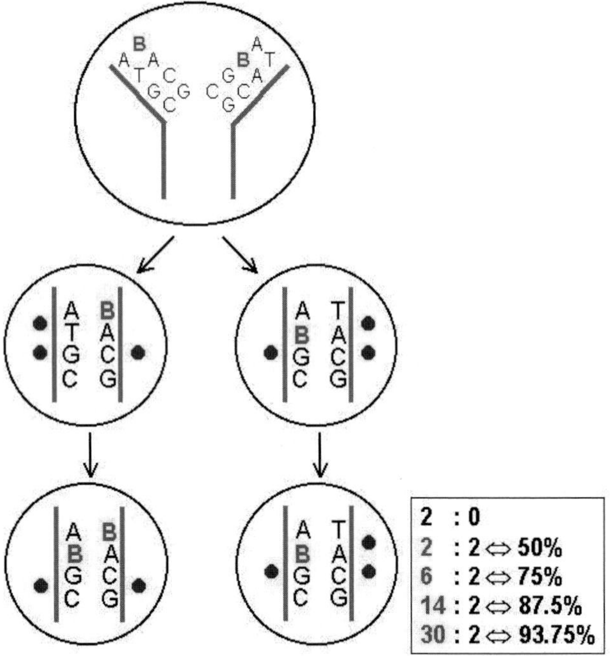

Fig. 12. Substitution of BrdU for thymidine. With continuous BrdU incubation, BrdU is incorporated into one strand in the daughter generation after the first cell division. The cells of this generation will substitute BrdU for thymidine in the next S-phase so that in the third generation there will be one cell with two BrdU-positive strands and one with one positive strand. With continuous BrdU incubation, the number of BrdU-positive cells rises from generation to generation. With each cell division, half of the BrdU-negative cells will likewise incorporate BrdU, so that after the second cell division 75% of the cells will be positive for BrdU, and after the third division, 87.5% positive and so on. The Hoechst dye (points) which is bound to the DNA strands is quenched by incorporated BrdU.

tion into the DNA (quenching), results in a reduction of the fluorescence intensity while the fluorescence intensity of PI, a dye that is not incorporated into the DNA, is not affected by the presence of BrdU. A two-dimensional application of the fluorescence intensity of Hoechst33258 against PI enables the identification of all cell cycle phases of up to three sequential cell cycles.

Figure 13 schematically shows the fluorescence intensities of Hoechst- and PI-stained, asynchronously proliferating cells in three cell cycles. The fluorescence intensity of PI during the first cell cycle increases linearly according to the doubling of the DNA content from the G_1 phase to the G_2/M phase. The Hoechst fluorescence, however, decreases due to the BrdU incorporated in the S phase: it is 'quenched'. The resulting shift to the left of the 'trace' of cells in the first cell cycle depends on the relationship between the concentrations of the two staining materials Hoechst and PI. During cell division, DNA is evenly distributed to the daughter cells, which halves the fluorescence intensity of both staining materials. Since the increase of

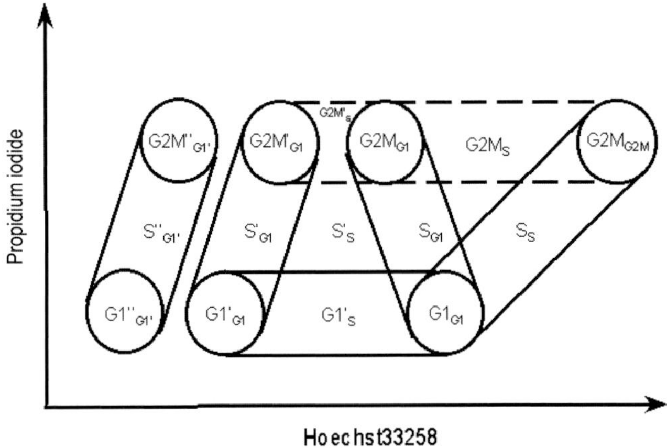

Fig. 13. Hoechst33258 versus PI fluorescence in the process of three sequential cell cycles with continuous BrdU exposure. The relative fluorescence intensities of the G_1, S and G_2/M phases are shown on an asynchronously proliferating population. The cell cycle phases of the second and third cell cycle are marked with "'" and "''". The original cell cycle phases are indicated as subscripts.

BrdU substitution in the second cell cycle no longer reaches the extent of the first cell cycle, but with a rise of 50–75% now reaches 25%, the quenching effect in the second cell cycle turns out to be smaller. From the second cell cycle, the increased Hoechst dye binding outweighs the BrdU-induced quenching effect due to the increased DNA content. The fluorescence intensity of the Hoeschst dye even increases in both the second and the following cell cycles with an increase in the DNA content; thus, the traces of the cell cycle phases in the second cell cycle run parallel to each other. The Hoechst/PI fluorescence intensities of cells in the first and second cell cycle form a mirror image in the two-dimensional dot plot. The trace of the third cell cycle runs parallel to the second.

Figure 14 shows the Hoechst and ethidium bromide fluorescence of phytohemagglutinin-stimulated, peripheral blood lymphocytes in three sequential cell cycles. The cells were synchronized before so that the observation starting point is found in the G_0/G_1 phase of the first cell cycle. The measurements are performed over a period of 35–96 h starting from continuous BrdU addition. The lymphocytes shown in figure 13 gradually start proliferating from the G_0/G_1 phase of the first cell cycle and gradually trace along the track of the Hoechst, PI and ethidium bromide fluorescence as represented in figure 14. But even after 96 h a few cells can be found that have not left the G_0/G_1 phase of the first cell cycle. At 65 h the first apoptotic cells emerge whose origin can be clearly assigned to the G_1 phase of the third cell cycle. The proportion of these cells increases up to the image taken after 96 h.

In addition to detailed cell cycle analysis, BrdU/Hoechst quenching technology also allows the quantification of the fraction of resting G_0 cells. Advantage is taken of the fact that in contrast to the G_0 fraction, all proliferating cells follow the course of the scheme shown in figure 13. G_0 cells are at the starting point of the first cell

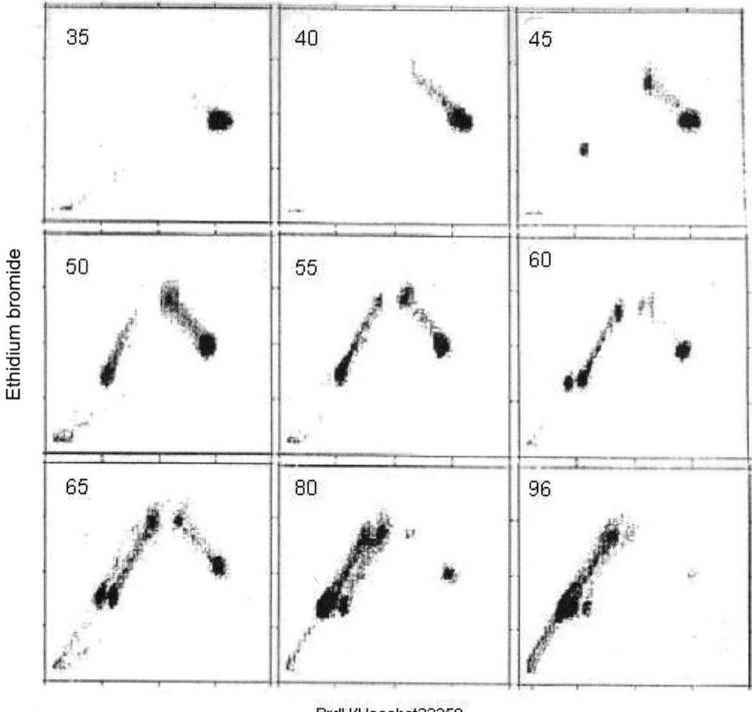

Fig. 14. Hoechst/ethidium bromide fluorescence of synchronized (proliferation starts in G_1 of the first cell cycle) and then proliferating peripheral blood lymphocytes, imaged after BrdU incubation for 35, 40, 45, 55, 55, 60, 65, 80 and 96 h. The phytohemagglutinin-activated lymphocytes can be observed as they advance through three sequential cell cycles. After 45 h, the first cells have reached the G_1 phase of the second cycle, after 60 h the same cells have reached the G_1 phase of the third cell cycle. After 65, 80 and 96 h, it can be clearly recognized that the first cells are dying. The origin of the apoptotic cells is the G_1 phase of the third cell cycle. According to Poot et al. [84].

cycle after an extended incubation with BrdU – e.g. after 96 h of BrdU exposure. By appropriate 'gating' it is possible to exactly calculate at every time point the duration of all cell cycle phases, the proportional distribution of the cells in each phase, and the proportion of resting cells of the total population. The kinetics of the cell proliferation of three sequential cell cycles can be represented in cell cycle phases exit kinetics, in which it is considered that the cells of the second cell cycle have divided once and those of the third cell cycle, twice [88]. From this, the proportional distribution of the cell cycle fractions of the three cell cycles can be read over time, and an overview of the duration of the phases and the associated phase transitions can be gained. Figure 15 gives an example of a semi-logarithmic application of such exit kinetics. For further details, the original literature can be consulted [78, 79, 84, 85].

In contrast to the example given in figure 14, figure 16 represents asynchronously proliferating cells; following BrdU application they have their starting point in all

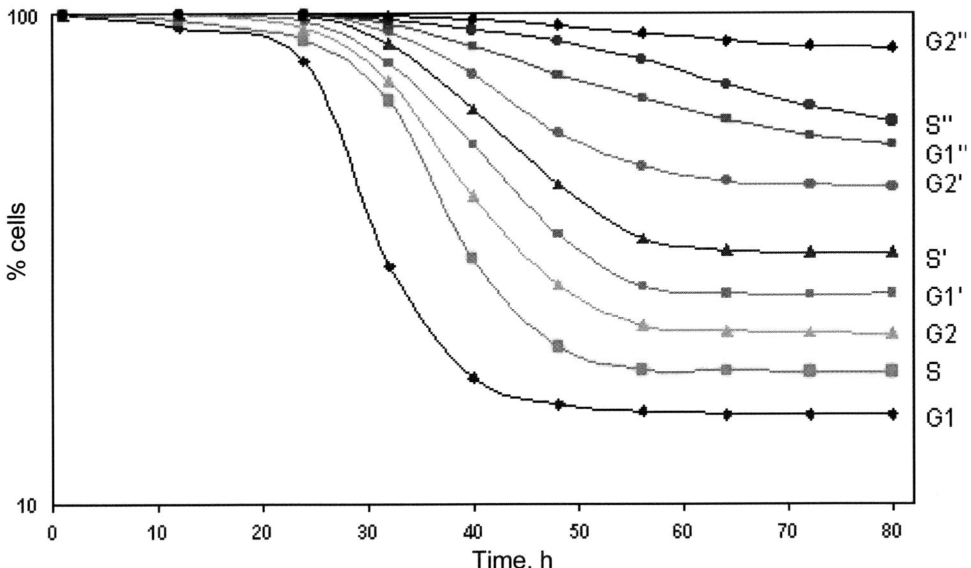

Fig. 15. Example of cell cycle phase exit kinetics on the basis of the computations of Smith and Martin [88]. The time (x-axis) is plotted against the percentage of cells, which did not yet withdraw from the respective cell cycle compartment (y-axis). The cell cycle phases are indicated on the right side.

cell cycle phases (not only in the G_0/G_1 phase). Figure 16 shows the cell cycle distribution of J82 carcinoma cells in each case after 0, 6, 12 and 42 h. After 12 h, the first cells are found in the G_1 phase of the second cell cycle; after 42 h, most of the cells already have passed the second cell cycle and are distributed over the individual cell cycle phases of the third cycle. After 42 h, a very small population is still found in the G_1 phase of the first cell cycle.

Based on the proliferation analysis using the BrdU/Hoechst quenching method (in contrast to anti-BrdU technology), it seems legitimate to affirm that continuous cell exposure to BrdU is necessary. The substitution of this halogenated base analogue for thymidine during replication may influence cell metabolism and physiology and exert an unwanted inhibition of proliferation. It is described in detail in the literature that BrdU addition to the medium or BrdU incorporation into the DNA has effects on nucleotide metabolism and 3D DNA structure; as a consequence, it may also trigger reciprocal DNA-protein interactions, which in turn may influence DNA replication and transcription, leading to cytotoxic effects and disturbance of 'natural' proliferation [13, 89–91]. Inhibition effects can be attributed to a single and/or bifiliary thymidine substitution; they have been demonstrated e.g. for NIH-3T3 cells and peripheral blood lymphocytes [78, 79, 84, 85, 92]. These effects could be reproduced in vitro in human stomach carcinoma cells [93] as well as breast and bladder carcinoma cell lines in vitro. Different p53-dependent and p53-independent proliferation-inhibiting mechanisms have also been described [93]. This makes clear

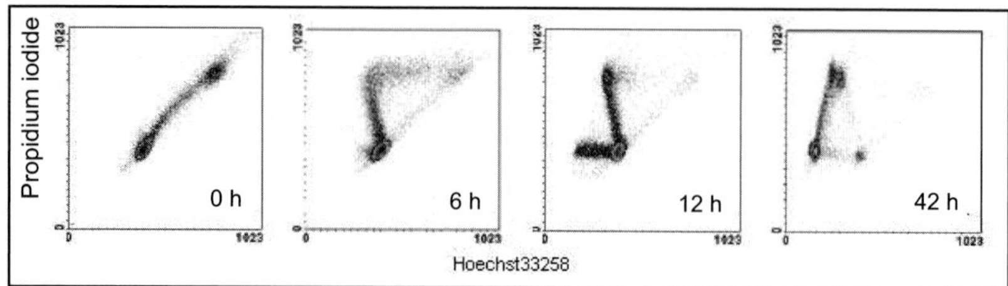

Fig. 16. Example of BrdU/Hoechst quenching measurements of J82 bladder carcinoma cells. Asynchronously proliferating cell cultures progress through three cell cycles: At the beginning (0 h) PI and Hoechst fluorescence are linear proportionally to each other. After 6 h, the first cells are found in G_2 phase. Their Hoechst fluorescence is 'quenched'. After 12 h, the first cells after cell division can be observed in G_1 phase, and after 42 h only few cells are in the G_2 phase of the first cell cycle. Most cells are already distributed over the third cell cycle at this time. By courtesy of Elmar Endl, Institute for Molecular Medicine and Experimental Immunology, University of Bonn).

that not only the proliferation of 'normal' cells, but also the proliferation of human tumor cell lines, which are normally considered to proliferate in an uncontrolled fashion, can be inhibited or at least slowed down by the presence of BrdU [83]. Even the addition of DC – which should contribute to a balanced nucleotide metabolism [13, 89, 90] – does not inevitably attenuate the inhibitory effect of BrdU. In any case, BrdU-induced potentially inhibitory effects are cell-type dependent and specific and must be clarified and excluded on a case-by-case basis before each experimental proliferation assay.

Figure 17 shows how the BrdU/Hoechst quenching technology was used in order to examine the interference of EGF with the cell cycle of J82 bladder carcinoma cells. Growth factors can in principle influence the cell cycle as competence or progression factors. Command factors typically recruit resting cells from the G_0 phase into the cell cycle (transition of G_0 to G_1 phase), whereas progression factors stimulate the passage of G_1 to S phase. Thus these two classes of growth factors act at different points of the cell cycle on a molecular level. The fraction of J82 cells entering S phase in the presence of EGF is larger than without addition of this growth factor, but the cells progress into S phase at the same pace S phase, one can conclude that EGF influences J82 cells as a command factor rather than a progression factor.

Protocol for the Dynamic Proliferation Analysis on Breast Carcinoma Cells through Continuous BrdU Incubation (BrdU/Hoechst Quenching Technology)

- Seed many cell culture samples in T75 culture flasks (e.g. Greiner, Frickenhausen, Germany); the density of the seeded cells is selected in such a way that the culture is in a subconfluent state after 7 days (depending upon specific cell type doubling time, e.g. 2×10^5 cells per bottle).
- The cells should be harvested all on the same day, but should be incubated with BrdU for different amounts of time, this means that BrdU addition must take place at different times to make sure that the cells have been in culture for equal

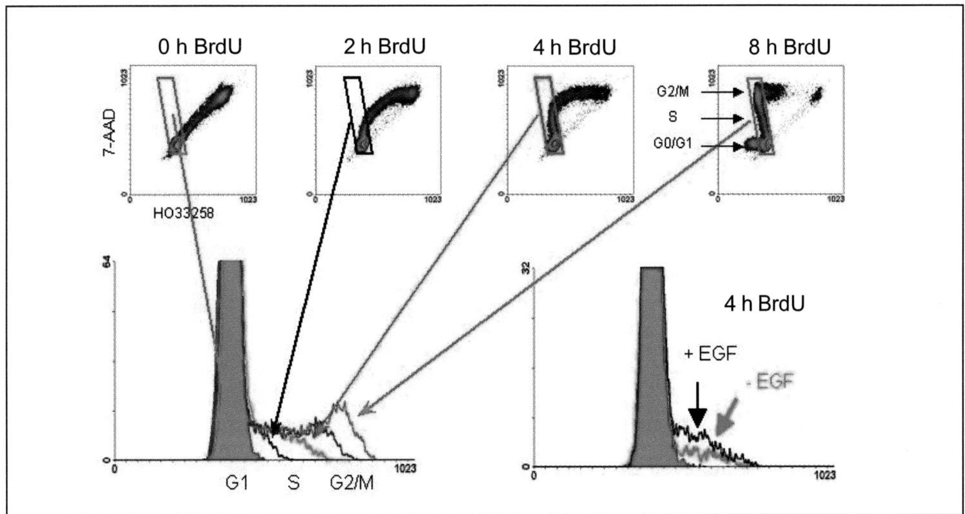

Fig. 17. Influence of EGF on the proliferation of J82 bladder carcinoma cells. Through suitable gating, the fraction of cells which enter S phase from G_1 phase following 2, 4, 6 and 8 h of continuous exposure to BrdU can be proportionally computed. Moreover, progression of the cells in the S phase of the cell cycle at any given time can also be determined. Comparison of the cultural conditions of J82 cells with and without the EGF shows that the cells progress at the same pace in the S phase of the first cell cycle both with and without growth factor. In the presence of EGF, however, the cell fraction entering S phase is greater than without addition of the growth factor. By courtesy of E. Endl, Institute for Molecular Medicine and Experimental Immunology, University of Bonn).

amounts of time at the time of harvesting (apart from specific treatments, e.g. with growth factors, without additive of inhibitors) and have the same growth density.

- Cell kinetics are done at 2-hour intervals, so that BrdU (e.g. Sigma, Deisenhofen, Germany) is accepted in each case 2, 4, 6, 8, 10 h at a concentration of 60 µmol/l before cell harvesting. DC is supplemented in half equimolar concentration.
- Following BrdU addition, the cells are incubated in the dark. Cultures incubated with and without BrdU and DC additive serve as controls to establish whether continuous BrdU incubation has an inhibitory effect on cell proliferation. These controls can be examined as a single-parameter; the distribution of the cell cycle phases is not changed by the BrdU incorporation into the DNA.
- At the times indicated in each case, the cells are trypsinized as usual, washed once with PBS and placed in 2 ml medium with the following additives: DMEM (without phenol red, 10% DMSO, 10% FCS).
- The cells are frozen in suitable vials at −20 °C, and can be stained at a later time with DNA dyes and measured on the flow cytometer.
- Thaw cells and wash once with the following DNA buffer: 100 mmol/l Tris pH 7.4, 154 mmol/l NaCl, 1 mmol/l $CaCl_2$, 0.5 mmol/l $MgCl_2$, 0.2% BSA.

- Permeabilize: place cells in 1 ml DNA buffer containing 0.1% NP40 (= Igepal, e.g. Sigma), and incubate on ice for 30 min.
- Addition of the fluorochrome (Hoechst33258, e.g. from Sigma): final concentration: 1.5 μg/ml (use 7.5 μl from a 200 μg/ml aqueous stock solution).
- PI: 10 μg/ml (10 μl from a 1 mg/ml stock solution; e.g. Sigma). Cells (nuclei) incubate for a further 15 min and then measure with the FACS. (The cells are not to be washed again.)

Comment – Error Check – Troubleshooting

- If measuring artifacts develop during signal detection, reduce the flow velocity through the FACS because detergent is present in the suspension and can disturb the current streaming through the FACS.
- BrdU is a photosensitizer. In the presence of BrdU, light radiation can photochemically cause DNA damage, release repair mechanisms and as a consequence impede proliferation. Therefore all cultures should be incubated protected from direct exposure to light. If (due to longer duration experiments) a change of the culture medium is necessary, this should take place in a dark environment without allowing the culture bottles to cool down.

Alternative Techniques for Dynamic Proliferation Analysis

In addition to the anti-BrdU and the BrdU/Hoechst quenching technology which were established many years ago, an alternative method of analysis of cell proliferation has been published in 1999 by Beisker et al. [94]. The two DNA dyes PI and TO-PRO-3 are used simultaneously; they interact with one another through fluorescence resonance energy transfer (FRET). TO-PRO-3-fluorescence is enhanced by BrdU incorporation into the DNA. This method requires excitation by a double laser; instead of the 488-nm excitation of a UV laser, which is required with BrdU/Hoechst quenching technology, a 635-nm excitation of a diode laser is employed. Like the anti-BrdU technology, this method is applicable with a routine bench-top flow cytometer. Furthermore, preparation of cells and/or nuclei requires very little effort and similar to the quenching technique, three cell cycles can be observed. For more details, the reader is referred to the original work [94].

Conclusion

The potential of flow cytometry for the analysis of DNA and cell proliferation is extremely varied. No other technology (MTT, ^{3}H-thymidine and others) provides comparable information content in and of itself. The great strengths of proliferation analysis by flow cytometry lie in its ability to carry out multiparametric assays (in combination with selection and phenotyping markers) and the possibility to observe the regulation of cell proliferation, i.e. the regulation of the cell cycle (dynamic analyses, kinetics measurements), in detail under defined conditions. Since a substantial characteristic of malignant tumors is uncontrolled cell proliferation, which is frequently found in combination with chromosomal instability (aneuploidy), DNA and proliferation analyses represent a very valuable contribution in supplemental diagnostics for oncology. In addition, flow-cytometric methods provide us with incomparable information content, thus contributing uniquely to the advancement of our understanding of cell cycle regulation and dysregulated cell proliferation. This creates not only an essential basis for the specification of diagnosis and prognosis, but lays the foundation for therapies which purposefully intervene in cell proliferation.

References

1 Otto FJ: High-resolution analysis of nuclear DNA employing the fluorochrome DAPI. Methods Cell Biol 1994;41:211–217.

2 Otto F: DAPI staining of fixed cells for high-resolution flow cytometry of nuclear DNA. Methods Cell Biol 1990;33:105–110.

3 Hemmer J, Kraft K: High-resolution DNA flow cytometry in oral verrucous carcinoma. Oncol Rep 2000;7:433–435.

4 Hemmer J, van Heerden WF, Polackova J, Kraft K: High-resolution DNA flow cytometry in papillary cystadenoma lymphomatosum (Warthin's tumour). J Oral Pathol Med 1998;27:405–406.

5 Wersto RP, Chrest FJ, Leary JF, Morris C, Stetler-Stevenson MA, Gabrielson E: Doublet discrimination in DNA cell-cycle analysis. Cytometry 2001;46:296–306.

6 Zellner A, Meixensberger J, Roggendorf W, Janka M, Hoehn H, Roosen K: DNA ploidy and cell-cycle analysis in intracranial meningiomas and hemangiopericytomas: a study with high-resolution DNA flow cytometry. Int J Cancer 1998;79:116–120.

7 Ferlini C, Biselli R, Scambia G, Fattorossi A: Probing chromatin structure in the early phases of apoptosis. Cell Prolif 1996;29:427–436.

8 Hemmer J, Thein T, Van Heerden WF: The value of DNA flow cytometry in predicting the development of lymph node metastasis and survival in patients with locally recurrent oral squamous cell carcinoma. Cancer 1997;79:2309–2313.

9 Bruno S, Crissman HA, Bauer KD, Darzynkiewicz Z: Changes in cell nuclei during S phase: progressive chromatin condensation and altered expression of the proliferation-associated nuclear proteins Ki-67, cyclin (PCNA), p 105, and p 34. Exp Cell Res 1991;196:99–106.

10 Crissman HA, Steinkamp JA: Rapid, simultaneous measurement of DNA, protein, and cell volume in single cells from large mammalian cell populations. J Cell Biol 1973;59:766–771.

11 Danova M, Riccardi A, Giordano M, Girino M, Mazzini G, Dezza L, Ascari E: Cell cycle-related proteins: a flow cytofluorometric study in human tumors. Biol Cell 1988 64:23–28.

12 Dittrich W, Goehde W: Impulse fluorometry of single cells in suspension. Z Naturforsch B 1969;24:360–361.

13 Latt SA, George YS, Gray JW: Flow cytometric analysis of bromodeoxyuridine-substituted cells stained with 33258 Hoechst. J Histochem Cytochem 1977;25:927–934.

14 Nowak R, Oelschlagel U, Range U, Bergmann S, Bornhauser M, Holig C, Schuler U, Krebs U, Gunther H, Kroschinsky F, Ehninger G: Flow cytometric DNA quantification in immunophenotyped cells as a sensitive method for determination of aneuploid multiple myeloma cells in peripheral blood stem cell harvests and bone marrow after therapy. Bone Marrow Transplant 1999;23:895–900.

15 Nowak R, Oelschlagel U, Range U, Molle M, Ehninger G: The incidence of DNA aneuploidy in multiple myeloma does not correlate with stage of disease. Am J Clin Pathol 1998;109:226–232.

16 Nowak R, Oelschlaegel U, Schuler U, Zengler H, Hofmann R, Ehninger G, Andreeff M: Sensitivity of combined DNA/immunophenotype flow cytometry for the detection of low levels of aneuploid lymphoblastic leukemia cells in bone marrow. Cytometry 1997;30:47–53.

17 Nowak R, Oelschlagel U, Hofmann R, Zengler H, Huhn R: Detection of aneuploid cells in acute lymphoblastic leukemia with flow cytometry before and after therapy. Leuk Res 1994;18:897–901.

18 Oelschlaegel U, Freund D, Range U, Ehninger G, Nowak R: Flow cytometric DNA-quantification of three-color immunophenotyped cells for subpopulation specific determination of aneuploidy and proliferation. J Immunol Methods 2001;253:145–152.

19 Brockhoff G, Endl E, Minuth W, Hofstädter F, Knuechel R: Options of flow cytometric three-colour DNA measurements to quantitate EGFR in subpopulations of human bladder cancer. Anal Cell Pathol 1996;11:55–70.

20 Brockhoff G, Wieland W, Woelfl G, Hofstädter F, Knuechel R: Evaluation of flow-cytometric three-parameter analysis for EGFR quantification and DNA assessment in human bladder carcinomas. Virchows Arch 1998;432:77–84.

21 Bagwell CB, Clark GM, Spyratos F, Chassevent A, Bendahl PO, Stal O, Killander D, Jourdan ML, Romain S, Hunsberger B, Wright S, Baldetorp B: DNA and cell cycle analysis as prognostic indicators in breast tumors revisited. Clin Lab Med 2001;21:875–895.

22 Bagwell CB, Clark GM, Spyratos F, Chassevent A, Bendahl PO, Stal O, Killander D, Jourdan ML, Romain S, Hunsberger B, Baldetorp B: Optimizing flow cytometric DNA ploidy and S-phase fraction as independent prognostic markers for node-negative breast cancer specimens. Cytometry 2001;46:121–135.

23 Bergers E, Baak JP, van Diest PJ, Willig AJ, Los J, Peterse JL, Ruitenberg HM, Schapers RF, Somsen JG, van Beek MW, Bellot SM, Fijnheer J, van Gorp LH: Prognostic value of DNA ploidy using flow cytometry in 1301 breast cancer patients: results of the prospective Multicenter Morphometric Mammary Carcinoma Project. Mod Pathol 1997;10:762–768.

24 Bergers E, van Diest PJ, Baak JP: Reliable DNA histogram interpretation. Number of nuclei requiring measurement with flow cytometry. Anal Quant Cytol Histol 1997;19:277–284.

25 Bergers E, Baak JP, van Diest PJ, van Gorp LH, Kwee WS, Los J, Peterse HL, Ruitenberg HM, Schapers RF, Somsen JG, van Beek MW, Bellot SM, Fijnheer J: Prognostic implications of different cell cycle analysis models of flow cytometric DNA histograms of 1,301 breast cancer patients: results from the Multicenter Morphometric Mammary Carcinoma Project (MMMCP). Int J Cancer 1997;74:260–269.

26 Bergers E, van Diest PJ, Baak JP: Comparison of five cell cycle analysis models applied to 1,414 flow cytometric DNA histograms of fresh frozen breast cancer. Cytometry 1997;30:54–60.

27 Bergers E, van Diest PJ, Baak JP: Tumour heterogeneity of DNA cell cycle variables in breast cancer measured by flow cytometry. J Clin Pathol 1996;49:931–937.

28 Bergers E, Montironi R, van Diest PJ, Prete E, Baak JP: Interlaboratory reproducibility of semiautomated cell cycle analysis of flow cytometry DNA-histograms obtained from fresh material of 1,295 breast cancer cases. Hum Pathol 1996;27:553–560.

29 Bergers E, van Diest PJ, Baak JP: Cell cycle analysis of 932 flow cytometric DNA histograms of fresh frozen breast carcinoma material. Correlations between flow cytometric, clinical, and pathologic variables. MMMCP Collaborative Group. Multicenter Morphometric Mammary Carcinoma Project Collaborative Group. Cancer 1996;77:2258–2266.

30 Bergers E, van Diest PJ, Baak JP: Reproducibility of semi-automated cell cycle analysis of flow cytometric DNA-histograms of fresh breast cancer material. Anal Cell Pathol 1995;8:1–13.

31 Brockhoff G, Fleischmann S, Meier A, Wachs FP, Hofstädter F, Knüchel R: Use of a mechanical dissociation device to improve standardization of flow cytometric cytokeratin DNA measurements of colon carcinomas. Cytometry 1999;38:184–191.

32 Corver WE, Koopman LA, van der Aa J, Regensburg M, Fleuren GJ, Cornelisse CJ: Four-color multiparameter DNA flow cytometric method to study phenotypic intratumor heterogeneity in cervical cancer. Cytometry 2000;39:96–107.

33 Corver WE, Bonsing BA, Abeln EC, Vlak-Theil PM, Cornelisse CJ, Fleuren GJ: One-tube triple staining method for flow cytometric analysis of DNA ploidy and phenotypic heterogeneity of human solid tumors using single laser excitation. Cytometry 1996;25:358–366.

34 Leers MP, Hoop JG, Nap M: Her2/neu analysis in formalin-fixed, paraffin-embedded breast carcinomas: comparison of immunohistochemistry and multiparameter DNA flow cytometry. Anticancer Res 2003;23:999–1006.

35 Leers MP, Theunissen PH, Schutte TB, Ramaekers FC: Bivariate cytokeratin/DNA flow cytometric analysis of paraffin-embedded samples from colorectal carcinomas. Cytometry 1995;21:101–107.

36 Corver WE, Cornelisse CJ, Hermans J, Fleuren GJ: Limited loss of nine tumor-associated surface antigenic determinants after tryptic cell dissociation. Cytometry 1995;19:267–272.

37 Vindelov LL: Flow microfluorometric analysis of nuclear DNA in cells from solid tumors and cell suspensions. A new method for rapid isolation and straining of nuclei. Virchows Arch B Cell Pathol 1977;24:227–242.

38 Hedley DW, Friedlander ML, Taylor IW, Rugg CA, Musgrove EA: Method for analysis of cellular DNA content of paraffin-embedded pathological material using flow cytometry. J Histochem Cytochem 1983;31:1333–1335.

39 Vindelov LL, Christensen IJ, Nissen NI: Standardization of high-resolution flow cytometric DNA analysis by the simultaneous use of chicken and trout red blood cells as internal reference standards. Cytometry 1983;3:328–331.

40 Corver WE, Koopman LA, Mulder A, Cornelisse CJ, Fleuren GJ: Distinction between HLA class I-positive and -negative cervical tumor subpopulations by multiparameter DNA flow cytometry. Cytometry 2000;41:73–80.

41 Corver WE, Fleuren GJ, Cornelisse CJ: Improved single laser measurement of two cellular antigens and DNA-ploidy by the combined use of propidium iodide and TO-PRO-3 iodide. Cytometry 1997;28:329–336.

42 Corver WE, Cornelisse CJ, Fleuren GJ: Simultaneous measurement of two cellular antigens and DNA using fluorescein-isothiocyanate, R-phycoerythrin, and propidium iodide on a standard FACScan. Cytometry 1994;15:117–128.

43 Plander M, Brockhoff G, Barlage S, Schwarz S, Rothe G, Knuechel R: Optimization of three- and four-color multiparameter DNA analysis in lymphoma specimens. Cytometry 2003;54A:66–74.

44 Brockhoff G, Knüchel R: Flow cytometric DNA-analysis in oncology: from single to multiparametric measurements. J Lab Med 2003;5–6:167–174.

45 Leers MP, Theunissen PH, Koudstaal J, Schutte B, Ramaekers FC: Trivariate flow cytometric analysis of paraffin-embedded lung cancer specimens: application of cytokeratin subtype specific antibodies to distinguish between differentiation pathways. Cytometry 1997;27:179–188.

46 Sramkoski RM, Wormsley SW, Bolton WE, Crumpler DC, Jacobberger JW: Simultaneous detection of cyclin B1, p105, and DNA content provides complete cell cycle phase fraction analysis of cells that endoreduplicate. Cytometry 1999;35:274–283.

47 Endl E, Steinbach P, Knuechel R, Hofstaedter F: Analysis of cell cycle-related Ki-67 and p120 expression by flow cytometric BrdUrd-Hoechst/7AAD and immunolabeling technique. Cytometry 1997;29:233–241.

48 Bauer KD, Duque RE, Shankey TV: Clinical Flow Cytometry: Principles and Application. Baltimore, Williams & Wilkins, 1993.

49 Shankey TV, Rabinovitch PS, Bagwell B, Bauer KD, Duque RE, Hedley DW, Mayall BH, Wheeless L, Cox C: Guidelines for implementation of clinical DNA cytometry. International Society for Analytical Cytology. Cytometry 1993;14:472–477.

50 Hedley DW, Shankey TV, Wheeless LL: DNA cytometry consensus conference. Cytometry 1993;14:471.

51 Leers MP, Schutte B, Theunissen PH, Ramaekers FC, Nap M: Heat pretreatment increases resolution in DNA flow cytometry of paraffin-embedded tumor tissue. Cytometry 1999;35:260–266.

52 Leers MP, Theunissen PH, Ramaekers FC, Schutte B: Multi-parameter flow cytometric analysis with detection of the Ki67-Ag in paraffin embedded mammary carcinomas. Cytometry 1997;27:283–289.

53 Molnar B, Bocsi J, Karman J, Nemeth A, Pronai L, Zagoni T, Tulassay Z: Immediate DNA ploidy analysis of gastrointestinal biopsies taken by endoscopy using a mechanical dissociation device. Anticancer Res 2003;23:655–660.

54 Nap M, Brockhoff G, Brandt B, Knüchel R, Leers MP, Schmidt H, De Angelis G, Eltze E, Semjonow A: Flow cytometric DNA and phenotype analysis in pathology. A meeting report of a symposium at the annual conference of the German Society of Pathology, Kiel, Germany, 6–9 June 2000. Virchows Arch 2001;438:425–432.

55 Novelli M, Savoia P, Cambieri I, Ponti R, Comessatti A, Lisa F, Bernengo MG: Collagenase digestion and mechanical disaggregation as a method to extract and immunophenotype tumour lymphocytes in cutaneous T-cell lymphomas. Clin Exp Dermatol 2000;25:423–431.

56 Bagwell CB, Mayo SW, Whetstone SD, Hitchcox SA, Baker DR, Herbert DJ, Weaver DL, Jones MA, Lovett EJ 3rd: DNA histogram debris theory and compensation. Cytometry 1991;12:107–118.

57 Corver WE, Fleuren GJ, Cornelisse CJ: Software compensation improves the analysis of heterogeneous tumor samples stained for multiparameter DNA flow cytometry. J Immunol Methods 2002;260:97–107.

58 Nowak R, Oelschlagel U, Heider T, Naumann R, Ehninger G: Some limitations in the detection of residual aneuploid cells with DNA quantification on immunophenotyped cells by flow cytometry. Br J Haematol 2000;110:751–753.

59 Ormerod MG, Tribukait B, Giaretti W: Consensus report of the task force on standardisation of DNA flow cytometry in clinical pathology. DNA Flow Cytometry Task Force of the European Society for Analytical Cellular Pathology. Anal Cell Pathol 1998;17:103–110.

60 Valet G: Past and present concepts in flow cytometry: a European perspective. J Biol Regul Homeost Agents 2003;17:213–222.

61 Valet GK, Tarnok A: Cytomics in predictive medicine. Cytometry 2003;53B:1–3.

62 Valet G: Predictive medicine by cytomics: potential and challenges. J Biol Regul Homeost Agents 2002;16:164–167.

63 Baisch H, Beck HP, Christensen IJ, Hartmann NR, Fried J, Dean PN, Gray JW, Jett JH, Johnston DA, White RA, Nicolini C, Zeitz S, Watson JV: A comparison of mathematical methods for the analysis of DNA histograms obtained by flow cytometry. Cell Tissue Kinet 1982;15:235–249.

64 Dean PN, Jett JH: Mathematical analysis of DNA distributions derived from flow microfluorometry. J Cell Biol 1974;60:523–527.

65 Brunsting A, Collins JM, Kane FR, Bagwell CB: An examination of some basic assumptions of DNA distribution analysis using biological data. Cell Tissue Kinet 1979;12:123–134.

66 Rabinovitch PS: DNA content histogram and cell-cycle analysis. Methods Cell Biol 1994;41:263–296.

67 Lacombe F, Belloc F, Bernard P, Boisseau MR: Evaluation of four methods of DNA distribution data analysis based on bromodeoxyuridine/DNA bivariate data. Cytometry 1988;9:245–253.

68 Beisker W, Dolbeare F, Gray JW: An improved immunocytochemical procedure for high-sensitivity detection of incorporated bromodeoxyuridine. Cytometry 1987;8:235–239.

69 Dean PN, Dolbeare F, Gratzner H, Rice GC, Gray JW: Cell-cycle analysis using a monoclonal antibody to BrdUrd. Cell Tissue Kinet 1984;17:427–436.

70 Dolbeare F, Gratzner H, Pallavicini MG, Gray JW: Flow cytometric measurement of total DNA content and incorporated bromodeoxyuridine. Proc Natl Acad Sci USA 1983;80:5573–5577.

71 Dolbeare F, Beisker W, Pallavicini MG, Vanderlaan M, Gray JW: Cytochemistry for bromodeoxyuridine/DNA analysis: stoichiometry and sensitivity. Cytometry 1985;6:521–530.

72 Gray JW, Dolbeare F, Pallavicini MG, Beisker W, Waldman F: Cell cycle analysis using flow cytometry. Int J Radiat Biol Relat Stud Phys Chem Med 1986;49:237–255.

73 Langer EM, Hemmer J, Kleinhans G, Goehde W: The applications of the BrdUrd-technique for the estimation of cycling S-phase cells in human renal cell carcinoma. Urol Res 1988;16:303–307.

74 Yanagisawa M, Dolbeare F, Todoroki T, Gray JW: Cell cycle analysis using numerical simulation of bivariate DNA/bromodeoxyuridine distributions. Cytometry 1985;6:550–562.

75 Brockhoff G, Heiss P, Schlegel J, Hofstädter F, Knüchel R: Epidermal growth factor receptor, c-erbB2 and c-erbB3 receptor interaction, and related cell cycle kinetics of SK-BR-3 and BT474 breast carcinoma cells. Cytometry 2001;44:338–348.

76 Hammers HJ, Saballus M, Sheikzadeh S, Schlenke P: Introduction of a novel proliferation assay for pharmacological studies allowing the combination of BrdU detection and phenotyping. J Immunol Methods 2002;264:89–93.

77 Hammers HJ, Kirchner H, Schlenke P: Ultraviolet-induced detection of halogenated pyrimidines: simultaneous analysis of DNA replication and cellular markers. Cytometry 2000;40:327–335.

78 Rabinovitch PS, Kubbies M, Chen YC, Schindler D, Hoehn H: BrdU-Hoechst flow cytometry: a unique tool for quantitative cell cycle analysis. Exp Cell Res 1988;174:309–318.

79 Kubbies M, Schindler D, Hoehn J, Rabinovitch PS: BrdU-Hoechst flow cytometry reveals regulation of human lymphocyte growth by donor age-related growth fraction and transition rate. J Cell Physiol 1985;125:229–234.

80 Latt SA, Sahar E, Eisenhard ME: Pairs of fluorescent dyes as probes of DNA and chromosomes. J Histochem Cytochem 1979;27:65–71.

81 Latt SA, Stetten G: Spectral studies on 33258 Hoechst and related bisbenzimidazole dyes useful for fluorescent detection of deoxyribonucleic acid synthesis. J Histochem Cytochem 1976;24:24–33.

82 Latt SA, Wohlleb JC: Optical studies of the interaction of 33258 Hoechst with DNA, chromatin, and metaphase chromosomes. Chromosoma 1975;52:297–316.

83 Diermeier S, Schmidt-Brücken E, Kubbies M, Kunz-Schughart LA, Brockhoff G: Continuous bromodeoxyuridine (BrdU) exposition differentially affects cell cycle progression of human breast and bladder cancer cell lines. Cell Prolif 2004;37:195–206.

84 Poot M, Hoehn H, Kubbies M, Grossmann A, Chen Y, Rabinovitch PS: Cell-cycle analysis using continuous bromodeoxyuridine labeling and Hoechst 33358-ethidium bromide bivariate flow cytometry. Methods Cell Biol 1994;41:327–340.

85 Poot M, Schindler D, Kubbies M, Hoehn H, Rabinovitch PS: Bromodeoxyuridine amplifies the inhibitory effect of oxygen on cell Proliferation. Cytometry 1988;9:332–338.

86 Kubbies M: High-resolution cell cycle analysis: The flow cytometric bromodeoxyuridine-Hoechst quenching technique; in Radbruch A (ed): Flow Cytometry and Cell Sorting, 2nd ed. Heidelberg, Springer, 1999, pp 112–124.

87 Ormerod MG, Kubbies M: Cell cycle analysis of asynchronous cell populations by flow cytometry using bromodeoxyuridine label and Hoechst-propidium iodide stain. Cytometry 1992;13:678–685.
88 Smith JA, Martin L: Do cells cycle? Proc Natl Acad Sci U S A 1973;70:1263–1267.
89 Cleaver JE: Thymidine metabolism and cell kinetics; in Neuberger A, Tatum EL (eds): Frontiers of Biology. Amsterdam, North-Holland Publishing Company, 1967, vol 6, p 92.
90 Goz B: The effects of incorporation of 5-halogenated deoxyuridines into the DNA of eukaryotic cells. Pharmacol Rev 1977;29:249–272.
91 Kaufmann ER, Davidson RL: Biological and biochemical effects of bromodeoxyuridine and deoxycytidine on Syrian hamster melanoma cells. Somat Cell Genet 1978;4:587–601.
92 Kubbies M, Hoehn H, Schindler D Chen YC, Rabinovich PS: Cell cycle analysis via BrdUrd/Hoechst flow cytometry: principles and applications; in Yen A (ed): Flow Cytometry: Advanced Research and Clinical Application. Boca Raton, CRS-Press, 1989, vol 2, pp 5.
93 Peng DF, Sugihara H, Hattori T: Bromodeoxyuridine induces p53-dependent and -independent cell cycle arrests in human gastric carcinoma cell lines. Pathobiology 2001;69:77–85.
94 Beisker W, Weller-Mewe EM, Nuesse M: Fluorescence enhancement of DNA-bound TO-PRO-3 by incorporation of bromodeoxyuridine to monitor cell cycle kinetics. Cytometry 1999;37:221–229.
95 Corver WE, Ter Haar NT, Dreef EJ, Miranda NF, Prins FA, Jordanova ES, Cornelisse CJ, Fleuren GJ: High-resolution multi-parameter DNA flow cytometry enables detection of tumour and stromal cell subpopulations in paraffin-embedded tissues. J Pathol 2005;206:233–241.
96 Diermeier S, Horváth G, Knuechel-Clarke R, Hofstaedter F, Szöllosi J, Brockhoff G. Epidermal growth factor receptor coexpression modulates susceptibility to herceptin in HER2/neu overexpressing breast cancer cells via specific erbB-receptor interaction and activation. Exp Cell Res 2005;304(2):604–19.
97 Brockhoff G, Heckel B, Schmidt-Bruecken E, Plander M, Hofstaedter F, Vollmann A, Diermeier S. Differential impact of cetuximab, pertuzumab and trastuzumab on BT474 and SK-BR-3breast cancer cell proliferation. Cell Prolif 2007;40:488–507.
98 Diermeier-Daucher S, Clarke ST, Bradford JA, Vollmann-Zwerenz A, Hill D, Brockhoff G: Applicability of EdU for dynamic proliferation assessment in Flow Cytometry, Submitted, 2008.

Sack U, Tárnok A, Rothe G (eds): Cellular Diagnostics. Basics, Methods and Clinical Applications of Flow Cytometry. Basel, Karger, 2009, pp 426–458

Multidrug Resistance

Alexandra Dorn-Beineke · Cornelia Keup

Institute of Clinical Chemistry, Faculty for Clinical Medicine, University of Heidelberg, Germany

Introduction

Historical Background

Multidrug resistance (MDR) has been a subject of experimental and clinical research for over three decades now. Early reports in the field of MDR appeared in the 1970s [1, 2]. Juliano and Ling [2] identified a protein of 170 kDa (P-glycoprotein; Pgp) which works as an efflux pump for several cytotoxic substances in resistant cell lines. Laboratories in North America and the Netherlands described the amino acid sequence and structure of Pgp, which exhibit homologies to bacterial transport proteins [3]. An important evolutionary antitoxic principle was found. The other proteins discussed in this chapter were described in the 1990s.

The original definition of MDR is purely descriptive: 'MDR describes the cross-resistance in cells or microorganisms against diverse, structurally or chemically unrelated cytotoxic substances' [4]. The affected substances have in common that they are natural products, e.g. alkaloids or antibiotics of herbal or fungal origin ('MDR-related drugs'). Nevertheless, after more than three decades of research, it is clear that the mechanisms of MDR are more complex than expected. Among other pharmacological and cellular MDR mechanisms, changes in drug targets, increased DNA repair mechanisms and inhibition of apoptosis are well known.

Indications

In the context of MDR, mainly indications with an oncological or pharmacological background such as those listed below are of interest:

– exploration of processes at physiological barriers, e.g. the blood-brain barrier [5];

- phenotype-genotype correlations to evaluate functionally active mutations/polymorphisms of MDR transport proteins [6];
- investigation of altered substrate-binding properties of MDR transport proteins caused by mutations/polymorphisms [7];
- correlation of expression and function of MDR transport proteins with the plasma levels of their MDR-related drugs [8];
- investigation of resistance and cross-resistance of cytotoxic drugs in cell lines with well-known MDR expression patterns or directly in patient material (e.g. 'in vitro cytostasis') [9];
- evaluation of primary and secondary therapy failures, e.g. in antiviral therapies in HIV [10] or imatinib resistance in antineoplastic therapies [11, 12];
- implementation of clinical studies, including the administration of modulators of MDR transport proteins in new therapeutic schedules and observation of the clinical outcome [13];
- determination of prognostic parameters in diseases of interest [14, 15], and
- development of converting substances for MDR transport proteins [16].

MDR transport proteins became well known to the general public with the discovery of Hoffmeyer et al. [6], who were the first to describe the association of MDR1 polymorphism with Pgp expression in humans. They showed that the silent C3435T polymorphism in exon 26 of the MDR1 gene influences the level of intestinal Pgp expression in Caucasians. Therefore, individuals with the CC genotype (wild type) exhibit an approximately twofold higher Pgp expression in the intestine compared to individuals with the TT genotype. Heterozygous individuals with the CT genotype show intermediate intestinal Pgp expression. Correlating with the lower intestinal Pgp expression, higher peak steady-state plasma concentrations of dioxin could be measured in TT individuals than in the CC group. Hoffmeyer et al. thus described a functionally active MDR1 polymorphism. A discussion about the implementation of a 'genetic passport' arose. In addition, Wang et al. [17] reported that the 3435C → T substitution affects ABCB1 mRNA stability and appears to be a main factor in allelic variation of ABCB1 mRNA expression in the liver. An overview of sequence variations within Pgp affecting ATP binding and hydrolysis has recently been published by Lawson et al. [18].

Studies by different working groups who performed flow-cytometric expression analysis and functional tests in patients suffering from leukemia show that patients with acute myelogenous leukemia (AML) with Pgp+ blasts have a worse course of disease than patients with Pgp− blasts [19–21]. Patients with Pgp+ blasts showed a more progressive increase in Pgp expression with advanced age and a significant correlation of Pgp expression with an impaired rate of complete remissions as well as an elevated incidence of resistant disease [21]. Another important clinical impact of Pgp in patients with acute leukemia was observed by Wuchter et al. [15], who showed that the function of Pgp measured in 106 adults with AML using the rhodamine efflux assay correlated with response to induction chemotherapy, relapse rate, overall survival and cytogenetic risk groups. In 92 children with acute lymphoblastic leukemia (ALL), the functional Pgp activity was lower than in AML patients and was not correlated with immunological subgroups, responses to induction therapy, relapse or overall survival.

In a cell model, Honjo et al. [7] observed that an amino acid exchange in amino acid position 482 in the MDR transporter breast cancer resistance protein (BCRP) causes differences in substrate specificity. Substitution of glycine or threonine for arginine (wild-type) by single nucleotide polymorphism (SNP) in exon 12 causes enhanced anthracycline resistance, and increased rhodamine 123 transport and ATP hydrolytic activity. Nevertheless, in contrast to BCRP wild-type cells, the mutants were not able to transport methotrexate. Honjo et al. [7] called them 'gain-of-function mutants'. Gain-of-function mutants may compromise the success of treatments with such substances.

A summary of ABCB1, ABCC and ABCG2 genetic variants involved in cancer therapy is given in a paper by Huang [22]. A comprehensive list of genetic polymorphisms in ABC transporter genes is available in several databases (Pharmacogenetics Research network website: *www.pharmGKB.org*; National Center for Biotechnology Information (NCBI): *www.ncbi.nlmnih.gov/SNP*; JSNP: *http://snp.ims.u-tokyo.ac.jp/*).

MDR analysis is performed with different methods. Besides flow cytometry and immunological methods, molecular biological methods including microarray-based methods [23] as well as MALDI-TOF-MS-based genotyping [24] are used. As may be gathered from table 1, MDR diagnosis became important not only in vitro, but also in vivo. ^{99m}Tc-disofenin and N-[^{11}C]-acetyl-LTE4 are substrates of MRP2 and are used for the diagnosis of the Dubin-Johnson syndrome [25].

ABC Transporters and Their Function

Pgp, its homologous proteins MDR3, MRP1, cMOAT (MRP2) and the other known MRP homologous proteins MRP3–9, BCRP as well as TAP1/2 belong to the supergene family of ABC transporters (ATP-binding cassette) [26]. ABC transporters described in prokaryotes and eukaryotes are involved in energy-consumptive transport of diverse substrates across biological membranes. These substrates include ions, heavy metals, amino acids, peptides, sugars, steroids and toxic compounds. ABC transporters are evolutionary preserved membrane-anchored polypeptides. Up to now, 49 ABC transporters have been identified in the human genome; they are divided into seven subclasses (A–G), based on their genomic organization, the arrangement of their domains and their sequence. For members of four subclasses (A, B, C and G), it could be shown that they induce drug resistance in cell cultures. The minimum structure required for an active ABC transporter is 2 transmembrane (TMD) and 2 ABC domains. The 'half-transporters' dimerize to a functionally

Table 1. In vitro und in vivo detection methods for MDR transporters

In vitro detection methods	In vivo detection methods
RNA detection	*single photon emission tomography*
RT-PCR	^{99m}Tc-Sestamibi
Northern blot	^{99m}Tc-Tetrofosmin
Dot blot	^{99m}Tc-Furifosmin
RNAse protection assay	^{99m}Tc-Disofenin
In situ hybridization	other ^{99m}Tc(III) complexes
Protein detection	*positron emission tomography*
Western blot	radioactive-labeled MDR substrates
Immunohistochemistry	N-[^{11}C]-acetyl-LTE4
Flow cytometry	
Functional assays using fluorescent or radioactive-labeled substrates	
Influx assays	
Efflux assays	
Microarray-based detection methods	
MALDI-TOF-MS-based genotyping	

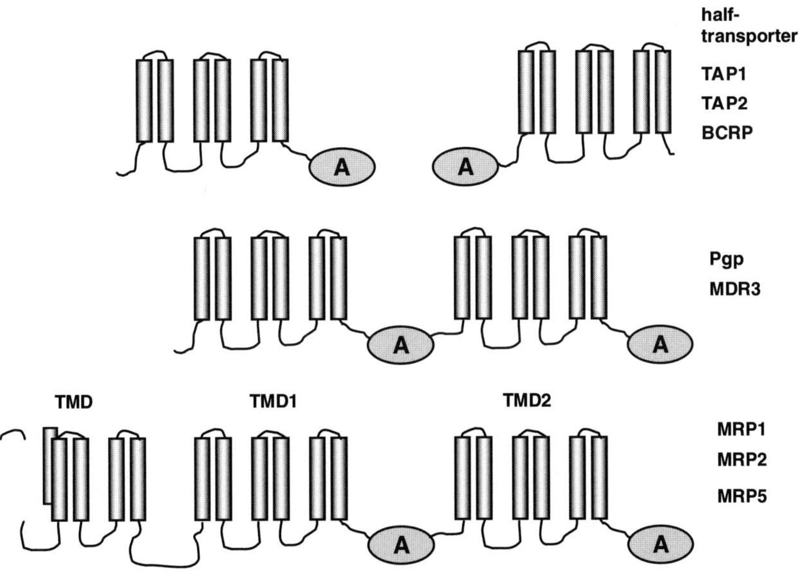

Fig. 1. The figure gives an overview of the domain arrangements of ABC transporter proteins according to Klein et al. [26]. A = ABC domain; TMD = transmembrane domain. The ABC domain consists of a mini protein of 200–250 amino acids, which contains two short conserved peptide motifs, Walker A and Walker B. Both motifs are involved in the ATP-binding process. The third conserved region, situated between Walker A and Walker B motif, is the ABC signature. TMD1 and TMD2 consist of six, TMD0 of five transmembrane α-helical segments.

Table 2. Overview of ABC transporters modified according to Litman et al. [43] and Scheffer et al. [83]

Protein	Symbol	Amino acids	Chromosome	Domain	Phenotype
MDR1	ABCB1	1280	7q21	(TMD-ABC)2	MDR
TAP1	ABCB2	686	6p21.3	TMD-ABC	MDR
TAP2	ABCB3	748	6p21.3	TMD-ABC	BLS
MDR3	ABCB4	1279	7q21	(TMD-ABC)2	PFIC3
MRP1	ABCC1	1531	16p13.1	TMD0(TMD-ABC)2	MDR
MRP2	ABCC2	1545	10q24	TMD0(TMD-ABC)2	DJS
MRP3	ABCC3	1527	17q21.3	TMD0(TMD-ABC)2	MDR
MRP4	ABCC4	1325	13q32	(TMD-ABC)2	?
MRP5	ABCC5	1437	3q27	(TMD-ABC)2	?
MRP6	ABCC6	1503	16p13.1	TMD0(TMD-ABC)2	PXE
MRP7	ABCC10	1513	6p21	TMD0(TMD-ABC)2	?
MRP8	ABCC11	1382	16q12.1	(TMD-ABC)2	PKC?
MRP9	ABCC12	1359	16q12	(TMD-ABC)2	PKC?
BCRP	ABCG2	655	4q22	ABC-TMD	MDR

BLS = Bare lymphocyte syndrome; DJS = Dubin-Johnson syndrome; PFIC = progressive familial intrahepatic cholestasis (Byler disease); PKC = paroxysmal kinesigenic choreoathenosis; PXE = pseudoxanthoma elasticum.

active protein [27]. Furthermore, ABC transporters play important roles in physiological processes. Mutations of different ABC transporters could be associated with certain diseases (fig. 1, table 2).

Current databases for ABC transporters are as follows:

- *http://nutrigene.4t.com/humanabc.htm* and
- *www.gene.ucl.ac.uk/nomenclature/genefamily/abc.html.*

Characterization and Resistance Profiles of Multidrug Resistance Transporters

In this chapter, we limit the description to a few MDR transport proteins and flow cytometric applications. The protocols were selected according to methodological aspects. One critical issue is the selection of antibodies for the immunological detection of Pgp. Work on vital (not fixed) and nonvital (fixed) cells represents a further issue. Protocols for direct labeling and indirect labeling of cells are described with specific modifications for work on peripheral blood as well as cell lines. Finally, gating strategies for the analysis of immunological and functional tests are discussed in detail considering the requirements of the different fluorochromes.

P-Glycoprotein (ABCB1)

Localization and Function of P-Glycoprotein

Pgp is an ABC transporter located in the cell membrane transporting hydrophobic substrates out of the cell, thus causing subcompartmentalization of the substrates within the cell [28]. Human Pgp is encoded by the MDR1 gene, which is located on chromosome 7 (7q21.1). Pgp is a transmembrane glycoprotein of 170 kDa, consisting of 1,280 amino acids. The protein has two hydrophobic domains each containing six transmembrane α-helical segments. Two ATP-binding sites with a (TMD ABC)2 structure have a cytoplasmic localization [29].

Pgp is not only expressed in resistant cells but also in normal cells [30]. Physiologically, Pgp is localized at the apical side of the cell membrane in polarized epithelium cells. These include the brush border of intestinal epithelium cells, the canalicular membrane of hepatocytes and the luminal membrane of proximal renal tubular cells [31]. Furthermore, Pgp is expressed on epithelial cells of the blood-brain barrier as well as on choroid plexus cells [32, 33]. Finally, in particular CD34+ precursor cells in the bone marrow exhibit a functionally active Pgp [34].

One physiological function of Pgp is the detoxification of the organism [35]. Pgp also plays an important role in the regulation of cell death and cell differentiation, immunoregulation, the regulation of chloride channels and cholesterol metabolism [36–39].

To date, the functional mechanism of the transmembrane protein is not completely clarified. Several hypotheses exist, e.g. the 'flippase model' [40], Pgp as a 'hydrophobic vacuum cleaner' [41], or the 'catalytic cycle model', which is favored nowadays [42]. Localization and function of Pgp are summarized in figure 2.

Resistance Profile of P-Glycoprotein

In the literature, MDR is grouped into Pgp-mediated 'classical MDR' and 'non-Pgp MDR'. The following compilation of the resistance profile of Pgp was modified according to Litman et al. [43].

Pgp-mediated MDR-related drugs:

- anthracyclines (daunorubicin, doxorubicin, epirubicin);
- vinca alkaloids (vinblastin, vincristin, vinorelbin, vindesin);
- epipodophyllotoxins (etoposide, teniposide);
- camptothecin derivatives (CPT-11, topotecan);
- microtubule-interacting substances (colchicine, paclitaxel, docetaxel);
- actinomycin D;
- cyclic or linear peptides (valinomycin, gramicidin D);
- anthracene (bisantrene, mitoxantrone);
- HIV-1 protease inhibitors (ritonavir, saquinavir, indinavir);
- imatinib (Gleevec®), and
- homoharringtonine.

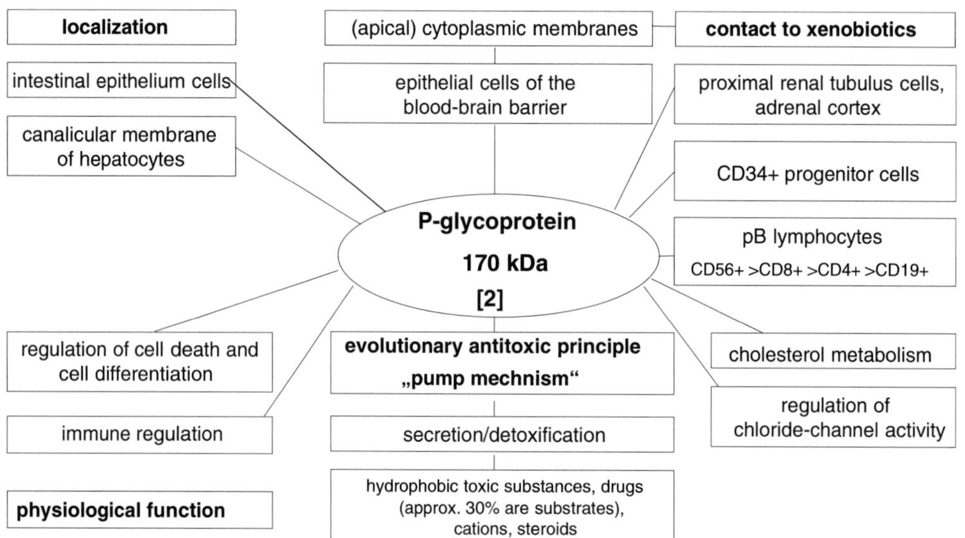

Fig. 2. Localization and physiological function of Pgp. PB = Peripheral blood; CD = cluster of differentiation.

Non-Pgp-MDR-related substances:

- platinum derivatives;
- antimetabolites, and
- alkylating agents.

Multidrug-Resistance-Associated Protein (MRP1, ABCC1)

Localization und Function of MRP1

MRP1 belongs to the C branch of the ABC transporter superfamily. This protein was first described by Cole et al. [44] in a resistant lung carcinoma cell line. MRP1 is a 190-kDa protein consisting of a sequence of 1,531 amino acids; it is located in region 16p13,112–13 of chromosome 16. Its structural formula is TMD0(TMD-ABC)2. In contrast to Pgp, it holds an additional TMD (TMD0) consisting of 5 α-helical segments. MRP1 causes a resistance phenotype almost identical to that caused by Pgp although both proteins only exhibit 14% homology in their amino acid sequence [45].

MRP1 is located in polarized epithelial cells at the basolateral membrane, the endoplasmic reticulum and cytoplasmic vesicles [46]. MRP1 expression was detected ubiquitously with different methods: in the lung, skeletal muscles, testis, colon, adrenal gland and kidney. In addition, MRP1 expression was described in many human tumors [47]. MRP1 was identified as a transporter of lipophilic anions ('GSX pump') which transports GSH conjugates, glucuronates, sulfates or other organic anions. Physiological substrates are metallic anions, conjugated steroid hormones, leukotriene C4 (LTC4) and sulfated bile salts [48]. The close relationship between MRP1 transport and intracellular glutathione (GSH) levels was described by Lorico et al. [49].

MRP1, like Pgp, causes the sequestration of substrates away from the cellular target, as well as the removal of substrates from the cell. The key function of MRP1 is the mediation of LTC-dependent inflammation processes. It is the main transporter of endogenous LTC4, a mediator of hypersensitivity reactions, shock and inflammation. In addition, MRP1 plays a physiological role in oxidative stress [50].

Resistance Profile of MRP1

MRP1 is overexpressed in most Pgp-expressing cell lines [51]. Transfection of MRP1-DNA in sensitive cells causes an MDR phenotype [46]. MRP1 overexpression was described to be associated with resistance against anthracyclines, epipodophyllotoxins and vinca alkaloids. Resistance against taxol was rather weak [52]. No resistances were shown against antimetabolites or platinum derivatives [53]. In short-term exposure assays in MRP1-expressing cells, resistance against methotrexate was found [54]. Additionally, the antiviral drugs saquinavir and ritonavir are known to be substances transported by MRP1 [55].

Breast Cancer Resistance Protein (BCRP, ABCG2)

Localization und Function of BCRP

Doyle et al. [56] were the first to describe BCRP in 1998. These scientists identified overexpression of a 2.4-kB mRNA, which codes for a 655-amino-acid transmembrane protein in the human MDR breast carcinoma cell line MCF-7/AdrVp. The coding gene was mapped on chromosome 4q22. In human tissues, BCRP is highly expressed in the placenta [56]. Its physiological function is closely related to the maintenance of the maternal-fetal placental barrier. Jonker et al. [57] stated that BCRP is strongly induced during lactation, thus causing accumulation of clinically and toxicologically important substrates by active secretion into the milk. Other human tissues with lower mRNA expression of ABCG2 are the brain, the prostate, the small intestine, the testes and the ovaries, the colon and the liver [56]. Comparison of mRNA levels of MDR efflux pumps in peripheral blood cells reveals that BCRP was the predominant form in monocytes whereas MDR1 was mainly found in lymphocytes [58]. Observations of ABCG2 overexpression in mitoxantrone-selected human cell lines of different origin followed [59]. There is no consensus to date as to the expression of BCRP on blood cell membranes. Maliepaard et al. [60] found no significant protein expression on peripheral blood cells whereas Scharenberg et al. [61] observed relatively high ABCG2 transcript levels in hematopoietic progenitor cells, decreasing with maturation and lineage commitment. In contrast, they found higher ABCG2 expression in CD56+ natural killer (NK) cells and glycophorin A+ erythroblasts. The subcellular

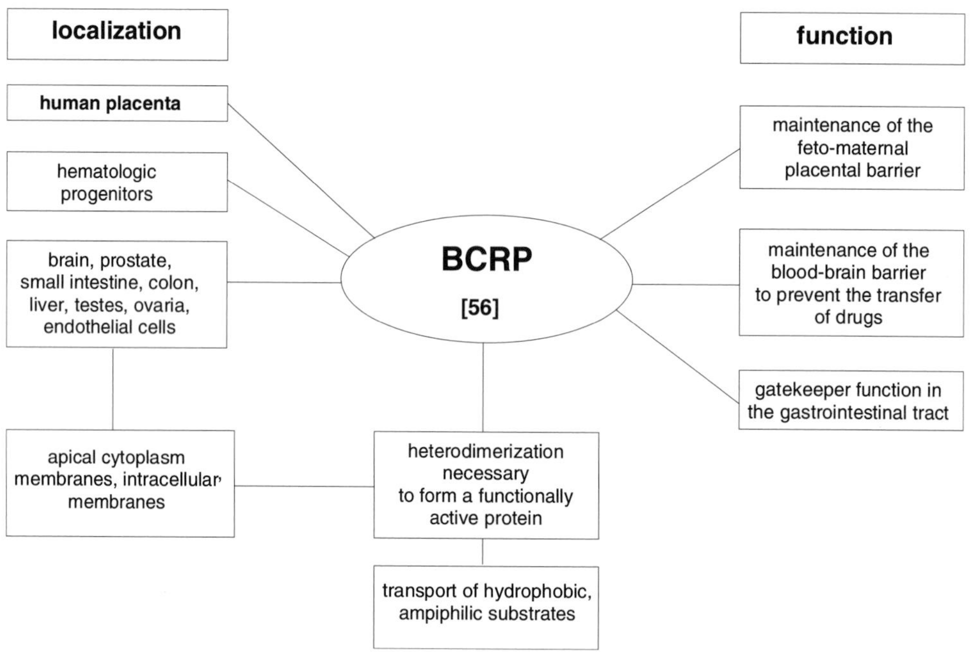

Fig. 3. Localization and physiological function of BCRP.

localization of BCRP was described on apical cytoplasm membranes of the intestine, the colon and syncytiotrophoblasts [60]. Other authors found BCRP expression also on intracellular membranes [62]. In contrast to the proteins of the MRP family mentioned above, BRCP is a half-transporter of the ABCG/white subfamily. It was shown that heterodimerization is necessary to form a functionally active protein [27]. Localization and function of BCRP are summarized in figure 3.

Resistance Profile of BCRP

The resistance profile of BCRP includes the anthraquinone mitoxantrone, the anthracycline derivatives doxorubicin and daunorubicin, etoposide, flavopiridol as well as the topoisomerase-I inhibitors topotecan and SN-38 [9, 63–66]. In a CD4+ T-cell line highly expressing BCRP wild-type, it could be demonstrated that high-level BCRP expression in CD4+ T-cells leads to reduced anti-HIV-1 activity of nonnucleoside reverse transcriptase inhibitors such as zidovudine and lamivudine [67].

Lung-Resistance-Related Protein (LRP/MVP)

Localization and Function of LRP/MVP

Besides the ABC transporter proteins, the 110-kDa LRP/MVP was described as a protein involved in non-Pgp MDR [68]. The LRP/MVP gene is located in the 16p13.1–16p11.2 region of chromosome 16. LRP/MVP was identified as the human

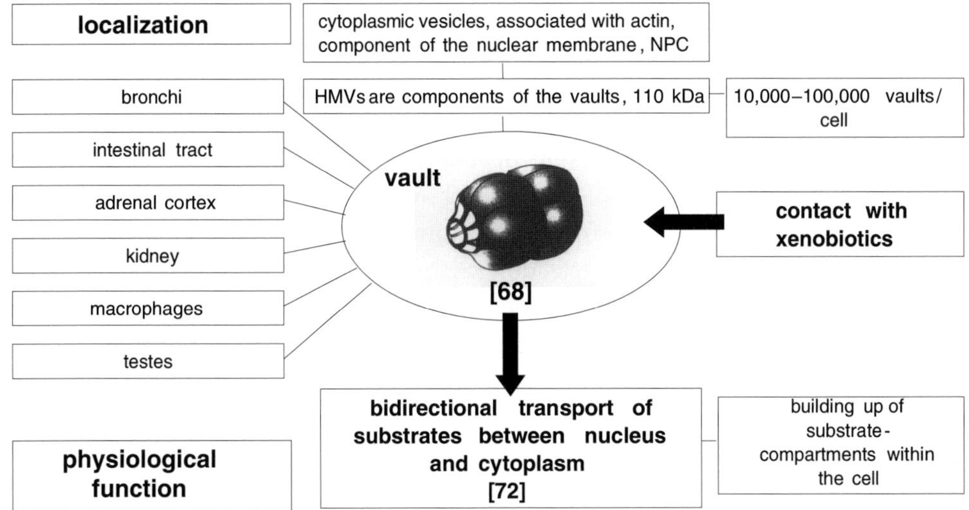

Fig. 4. Localization and physiological function of LRP/MVP.

major vault protein (MVP) [69]. Vaults are ribonucleoprotein particles, consisting of MVP, three smaller proteins, and an RNA molecule. MVP forms more than 70% of the vault ribonucleoprotein particles (multi-subunit particles). These particles are present in all eukaryotes [70].

Vaults are mainly localized in the cytoplasm, only 5% were found in the nuclear pore complex of human cells [69]. Additionally, an association with actin was shown [68]. LRP/MVP was found ubiquitously in normal human tissues. High LRP/MVP expression was described in the bronchi, the digestive tract, the adrenal glands and the macrophages [71]. The mechanism of action of LRP/MVP is the bidirectional transport of substrates between the nucleus and the cytoplasm. This leads to compartments of substrates within the cell [72] (fig. 4).

Resistance Profile of LRP/MVP

LRP/MVP was found in several Pgp-MDR cell lines of different tissues. LRP/MVP could be induced by sodium butyrate in SW-620 human rectal cancer cells, and it could be shown using LRP/MVP-specific ribozymes that resistances arose against doxorubicin, vincristin, VP-16, taxol and gramicidine D. Furthermore, LRP/MVP plays an important role for the doxorubicin transport between nucleus and cytoplasm [73]. In contrast to the classical MDR phenotype induced by Pgp, the total cytoplasmic substrate concentration did not change. It was shown that LRP/MVP decreases the nucleus/cytoplasm ratio of doxorubicin and leads to an increased redistribution in vesicles [74]. The statements about the role of LRP/MVP concerning induction of resistance against platinum derivatives are contradictory [75–77]. A recent study revealed that DNA damage enhances MVP promoter activity [77].

Protocols

The current in vivo and in vitro detection methods of MDR transport proteins are listed in table 1. To improve the standardization of the detection methods, consensus recommendations based on the results of multicenter studies were elaborated especially for the detection of Pgp ('Memphis workshop', 'French multi-center trial', Amsterdam-Rotterdam trial') [78–80]. It is recommended to use a combination of immunological and functional Pgp detection methods. It is also recommended to work with well-characterized cell lines in order to ensure a sufficient quality control. To the best of our knowledge, no similar recommendations exist for the remaining transporter proteins regarding the selection of methodology and technical aspects of the measurements, and/or for the interpretation of the data.

Substrates and Modulators of Multidrug Resistance Transporter Proteins

The resistance profiles of Pgp, MRP1, BCRP and LRP/MVP are described above. These resistance profiles partially overlap. Functional in vitro investigations are based among other things on the use of radioactively or fluorescently labeled substances and are performed either as influx or efflux measurements. Modulators (converters) of MDR transporters more or less specifically inhibit the various proteins (table 3). The classification of MDR substrates and MDR modulators into different groups of substances is shown in table 4.

The sensitivity and specificity of functional tests can be influenced by the combination of substrates and converters which competitively or allosterically inhibit the pumping of a protein. The exclusion of nonviable cells is obligatory in functional tests [81]. Suitable cell lines for use as positive and negative controls are specified below.

Immunochemical Methods

P-Glycoprotein

Materials

Primary antibodies, isotype controls and conjugates

Antibodies, isotype controls and conjugates are listed in table 5.

Reagents

- Ficoll density gradient: density = 1.007 g/ml;
- immunoglobulin G solution: 1 ml contains 50 mg protein with a purity of at least 95% human immunoglobulin G;
- mouse serum: Sigma, order No. M-5905;
- donkey serum: Sigma, order No. D-9663;

Table 3. MDR substrates and modulators of Pgp, MRP1 and BCRP used in flow cytometry*

Pgp	MRP1	BCRP
Substrates		
Anthracyclines	anthracyclines	*mitoxantrone*
Rhodamine 123	rhodamine 123	BODIPY FL prazosin
Calcein-AM	*calcein-AM*	Hoechst 33342
Hoechst 33342		
$DiOC_2(3)$		
Modulators		
Vp (Isoptine®)	*Vp (Isoptine®)*	*fumitremorgin C*
CsA (Sandimmune®)	*CsA (Sandimmune®)*	tryprostatin A
PSC 833 (Valspodar®)	*PSC 833 (Valspodar®)*	GF120918
UIC-2	*PROB*	CI1033
GEN	Ko134	
NEM		
BSO		

* To the best of our knowledge, a functional test to measure LRP/MVP activity has not been described in the literature. Italics show the mainly used substances.
BODIPY = Dipyrromethene boron difluoride; BSO = DL-buthionine-(S,R)-sulfoximine; calcein-AM = calcein-acetoxymethylester; $DiOC_2(3)$ = 3,3′-diethyloxacarbocyanin; NEM = N-ethylmaleinimide.

- 30% bovine serum albumin (BSA): Serva, order No.: 1193702;
- sodium azide;
- 7-actinomycin D (7-AAD): Molecular Probes, order No. A 1310;
- Cellwash: BD Biosciences, order No. 349524.

Solutions and buffers

- Dilution of antibodies: PBS + 1% BSA + 0.1% sodium azide;
- NH_4Cl lysis 10×: 41.5 g NH_4Cl, 5 g $KHCO_3$, 0.19 g EDTA to 500 ml distilled water; dilute for the working solution to 1× with distilled water;
- PBS (Dulbecco) (w/o Ca^{2+}, Mg^{2+}): Biochrom KG, order No.: L 1825;
- Washing solution: Cellwash + 1% BSA;
- Dilution of 7-AAD: 2 µl 7-AAD ad 1,000 µl PBS.

Consumables

- Polypropylene round-bottom tubes (5 ml) '*FACS tubes*': BD Biosciences, Falcon, order No. 2053;
- polypropylene round-bottom tubes (14 ml): BD Biosciences, Falcon, order No. 2059;
- Blue Max™ Jr. polypropylene conical tube (15 ml): BD Biosciences, order No. 2096;
- Blue Max™ polypropylene conical tube (50 ml): BD Biosciences, order No. 2070;
- reaction tube, Eppendorf AG, order No. 0030120.086;
- 10-ml heparin Monovettes, Sarstedt Nürmbrecht, order No. 02.1064.

Table 4. Classification of MDR substrates and modulators

Classification	Substrates und modulators
Pgp antibodies (human)	UIC-2
Calcein ester	Calcein-AM
Carbocyanine	$DiOC_2(3)$
Diketopiperazine	tryprostatin A
DNA dyes	Hoechst 33342
Fluorescein derivatives	FDA, CMFDA, BCECF-AM
Inhibitors of GSH biosynthesis	BSO
Immunosuppressive agents	CsA (Sandimmune), PSC 833 (Valspodar)
Inhibitors of organic anion transporter	PROB
Mitochondrial dyes	rhodamine 123
Inhibitors of phosphodiesterases	sildenafil (Viagra®), zaprinast, trequinsin
Inhibitors of proteinkinases	GEN
Tyrosine kinase inhibitors	CI1033
Cytostatic drugs	anthracyclines, mitoxantrone

Table 5. Primary antibodies, isotype controls and conjugates for Pgp measurement

Primary antibody	Isotype control	Manufacturer/supply	Order No.
Unconjugated primary antibodies			
Monoclonal anti-human Pgp			
Clone 4E3	mouse IgG2a	Dako Cytomation	M 3523
Clone MRK16	mouse IgG2a	Alexis Biochemicals	801-008-C150
Clone MM4.17	mouse IgG2a	M. Cianfriglia, Rome	–
Conjugated primary antibodies			
Monoclonal anti-human Pgp			
Clone UIC2-PE	mouse IgG2a	Beckman Coulter	IM 2370
Clone 15D3-PE	mouse IgG1	BD Biosciences	340555
Unconjugated isotype controls			
Anti-human			
Mouse IgG2a		Caltag Laboratories	MG2a00
Conjugated isotype controls			
Anti-human			
Mouse IgG2a-PE		Caltag Laboratories	MG2a04
Mouse IgG1-PE		Caltag Laboratories	MG104
Secondary antibodies/conjugates			
Donkey anti-mouse PE conjugated $F(ab')_2$ (DAM-PE)		Jackson ImmunoResearch	715-116-151

Technical equipment

– Flow cytometer FACSCalibur, BD Biosciences.

Immunochemical Methods for Peripheral Blood Cells

Cell separation by sedimentation:

– for each patient, prepare 5 ml of Ficoll in a 15-ml Falcon tube and cover carefully with a layer of 4 ml heparinized blood;
– allow the erythrocytes to sediment to the bottom (fast sedimentation after approximately 20–40 min): leukocytes and thrombocytes remain in the supernatant on the Ficoll phase;
– discard approximately 2–3 ml of the leukocyte-enriched plasma and transfer to a new 15-ml Falcon tube;
– refill the cell suspension with washing solution to approximately 10 ml, wash ($275 \times g$, 5 min, 15 °C) and discard the supernatant;
– lysis of erythrocytes: add 4 ml of NH_4Cl lysis working solution (1×), incubate for 10 min at room temperature and gently vortex now and then;
– refill the cell suspension with Cellwash + 1% BSA and wash once again;
– discard the supernatant and adjust the cells to 1×10^6/ml with washing solution;

(1) Indirect cell staining:

– prepare in FACS tubes 1–6 100 µl cell suspension each (contains 1×10^5 cells);
– background reduction: by using IgG2a antibodies and the respective isotype controls and conjugates reduce unspecific binding with the immunoglobulin G solution → add 100 µl solution (50 mg IgG/ml) to each tube and incubate for 30 min on ice in the dark;
– wash with Cellwash + 1% BSA ($275 \times g$, 5 min, 4 °C);
– pipette the appropriate amount of diluted antibody into each tube according to the scheme given below and vortex;
– the antibody concentrations given are the final antibody concentrations:
  - tube 1 – conjugate control;
  - tube 2 10 µg/ml sotype control IgG2a;
  - tube 3 10 µg/ml 4E3 (isotype IgG2a);
  - tube 4 10 µg/ml MRK16 (isotype IgG2a);
  - tube 5 50 µg/ml IgG2a (isotype IgG2a);
  - tube 6 50 µg/ml MM 4.17 (isotype IgG2a);
– incubate on ice for 30 min in the dark;
– wash 2 × with Cellwash + 1% BSA ($275 \times g$, 5 min, 4 °C);
– saturate with 10 µl of donkey serum in all tubes and vortex;
– give 20 µl of DAM-PE (1 : 10 dilution) in each tube, vortex;
– incubate on ice for 30 min in the dark;
– wash 2 × with Cellwash + 1% BSA ($275 \times g$, 5 min, 4 °C);
– saturate with 10 µl of mouse serum in all tubes and vortex;
– counterstaining to identify the different cell populations, e.g.;

- CD14-FITC (Caltag Laboratories, order No. MHCD1401), 20 μl, 1 : 16;
- CD34-FITC (Caltag Laboratories, order No. CD34-581-01-4), 10 μl, 1 : 4;
- place on ice for 20 min in the dark;
- wash 1 × with Cellwash + 1% BSA (275 × g, 5 min, 4 °C);
- pipette 50 μl of 7-AAD dilution to each tube and vortex;
- incubate on ice for 10 min in the dark and measure at once.

(2) Direct cell staining:
- prepare in FACS tubes 7–10 100 μl cell suspension each (contains 1×10^5 cells);
- reduce background in tubes 7 and 9 with Immunoglobulin G solution (50 mg/ml; see indirect cell staining);
- pipette the appropriate amount of diluted antibody in each tube according to the scheme given below and vortex;
- the antibody concentrations given are the final antibody concentrations:
  - tube 7 10 μg/ml isotype control IgG2a-PE;
  - tube 8 10 μg/ml isotype control IgG1-PE;
  - tube 9 10 μg/ml UIC2-PE (isotype IgG2a);
  - tube 10 10 μg/ml 15D3-PE (isotype IgG1);
- incubate on ice for 30 min in the dark:
- wash 2 × with Cellwash + 1% BSA (275 × g, 5 min, 4 °C);
- saturate all tubes 5 min with 10 μl mouse serum and vortex;
- counterstain (see indirect cell staining);
- incubate all tubes on ice for 20 min in the dark;
- wash 1 × with Cellwash + 1% BSA (275 × g, 5 min, 4 °C);
- pipette 50 μl of 7-AAD dilution into each tube and vortex;
- incubate on ice for 10 min in the dark and measure at once.

Measurements by Flow Cytometry
- FSC, lin; SSC, lin; Fl1 (counterstaining), Fl2 (anti-Pgp) und Fl3 (7-AAD); log over 4 decades;
- calibration und compensation: Calibrite Beads; see under 'Quality Control' (p. 452);
- trigger on FSC to exclude cell debris;
- collect 20,000 events in a gate including all 7-AAD– (vital) cells;
- flow rate: low to medium.

MRP1 and LRP/MVP

Materials

This section shows the protocol for the immunochemical detection of MRP1 and LRP/MVP in cell lines. Additional material required for analysis of patient peripheral blood cells is identified with '(→ pB)'.

Table 6. Primary antibodies, isotype controls and conjugates for MRP1 and LRP/MVP measurement

Antibodies	Isotype controls	Manufacturer/supply	Order No.
Unconjugated primary antibodies			
Monoclonal anti-human MRP1			
Clone MRPm6	mouse IgG1	Alexis Biochemicals	801-013-C500
Clone MRPr1	rat IgG2a	Alexis Biochemicals	801-007-C125
Monoclonal anti-human LRP/MVP			
Clone LRP56	mouse IgG2b	Alexis Biochemicals	801-007-C100
Antibodies against ds-DNA			
Clone DNA42	mouse IgG2a	Dr. R. Smeenk, CLB	–
Unconjugated isotype controls			
Anti-human			
Mouse IgG1		Caltag Laboratories	MG100
Mouse IgG2b		Caltag Laboratories	MG2b00
Rat IgG2a		Southern Biotechnology Ass., Inc.	0117-01
Secondary antibodies/conjugates			
Immunoglobulins			
Goat anti-rat FITC-conjugated, $F(ab')_2$ (GAR-FITC)		Southern Biotechnology Ass., Inc.	3052-02
Rabbit anti-mouse FITC-conjugated, $F(ab')_2$ (RAM-FITC)		Dako Cytomation	F0313

Primary antibodies, isotype controls and conjugates

Antibodies, isotype controls and conjugates are listed in table 6.

Reagents

- Ficoll density gradient, immunoglobulin G solution (see 'Materials' Pgp, pp. 436) (→ pB);
- BSA, sodium azide, 7-AAD, Cellwash (see 'Materials' Pgp, pp. 436);
- trypsin EDTA: Gibco, order No. 25300-054;
- trypan blue solution 0.4%;
- formaldehyde solution min. 37% analytical grade;
- acetone;
- FBS;

Solutions and buffers

- NH_4Cl lysis 1 × (see 'Materials' Pgp, pp. 436) (→ pB);
- dilution for antibodies, PBS (Dulbecco) (see 'Materials' Pgp, pp. 436);
- washing solution: PBS + 0.1% BSA + 0.1% sodium azide, stored at 4 °C;
- fixation: 2% (v/v) formaldehyde in 100% acetone, stored at 4 °C;
- dilution for conjugates: PBS + 1% BSA + 0.1% sodium azide + 2% pooled human plasma;

Consumables and equipment

- see 'Materials' Pgp (pp. 436);
- cell culture flasks, BD Biosciences, order No. 353136.

Methods

Immunochemical methods for adherent cell lines

Preparation of cells:

- culture cell lines in flasks (incubator, 37 °C, 5% CO_2) with the appropriate culture medium;
- rinse each flask 2× with 15 ml PBS and trypsinize with 3 ml trypsin EDTA per flask;
- incubate for a few minutes in a water bath or incubator at 37 °C, then transfer the cells to a 50-ml Falcon tube and refill with PBS/5% FCS;
- wash cells (175 × *g*, 10 min, 15 °C), discard the supernatant and wash once again with PBS;
- discard the supernatant, vortex the cell pellet, refill with approximately with 5–10 ml PBS and count the cells in a Neubauer chamber;
- determine the vitality of the cells using trypan blue exclusion (>90% for experiments).

Fixation:

- prepare a 14-ml round-bottom tube with the volume of cell suspension required for all tubes (1×10^5 cells/tube);
- wash cells with 1 ml cold washing solution (460 × *g*, 5 min, 4 °C);
- discard the supernatant with a water jet pump and vortex the pellet;
- pipette 100 µl of fixation solution on the cell pellet, vortex for several seconds and refill with 2 ml cold washing solution;
- centrifuge all tubes (460 × *g*, 5 min, 4 °C);
- discard the supernatant with a water jet pump and repeat the washing step with 2 ml cold washing solution (460 × *g*, 5 min, 4 °C);
- discard the supernatant with a water jet pump and resuspend the cells with the respective volume of washing solution (e.g. for 9 tubes, 90 µl cell suspension are required);
- background reduction: (→ pB, (see 'Immunochemical Methods for Peripheral Blood Cells', pp. 439).

Staining with primary antibodies:

- prepare 10 µl cell suspension/FACS tube;
- add 10 µl diluted antibody according to the scheme given below (step 1) and vortex;
- the antibody concentrations given are the final antibody concentrations:

		1st step	*2nd step*
• tube 1	–	conjugate control RAM-FITC	RAM-FITC
• tube 2	2.5 µg/ml	DNA 42	RAM-FITC

- tube 3 – conjugate control GAR-FITC GAR-FITC
- tube 4 10 μg/ml isotype control IgG1 RAM-FITC
- tube 5 10 μg/ml MRPm6 (isotype IgG1) RAM-FITC
- tube 6 1.7 μg/ml isotype control IgG2a GAR-FITC
- tube 7 1.7 μg/ml MRPr1 (isotype IgG2a) GAR-FITC
- tube 8 0.6 μg/ml isotype control IgG2b RAM-FITC
- tube 9 0.6 μg/ml LRP56 (isotype IgG2b) RAM-FITC

– incubate for 45 min at room temperature;
– wash all tubes with 1 ml of washing solution (475 × g, 5 min, 4 °C), discard the supernatant and vortex;
– dilute goat anti-rat FITC (GAR-FITC) 1 : 500 and rabbit anti-mouse FITC (RAM-FITC) 1 : 50 with dilution for conjugates;
– add 40 μl diluted conjugate in step 2 according the scheme given above and vortex;
– incubate for 30 min at room temperature in the dark;
– wash with 1 ml washing solution (475 × g, 5 min, 4 °C), decant and vortex;
– counterstaining with monoclonal antibodies: (→ pB) (see 'Immunochemical Methods for Peripheral Blood Cells', pp. 439) (e.g. CD14-PE, BD Biosciences, order No. 345785, 20 μl, 1 : 8 or CD34-PE, BD Biosciences, order No. 345802, 20 μl, 1 : 2);
– vortex, place for 20 min on ice in the dark and wash once again;
– pipette 50 μl of 7-AAD dilution into each tube and vortex;
– incubate on ice for 10 min in the dark and measure at once.

Measurements by Flow Cytometry

– FSC, lin; SSC, lin; Fl1 (anti-MRP1, anti-LRP/MVP), (Fl2 pB, counter-staining) and Fl3 (7-AAD), log over 4 decades;
– calibration and compensation: Calibrite Beads, see under 'Quality Control' (p. 452);
– trigger on FSC to exclude cell debris;
– collect 20,000 7-AAD+, permeabilized cells in the appropriate gate;
– flow rate: low to medium.

Functional Multidrug Resistance Tests

Rhodamine 123 Efflux Assay

The protocol was modified according to Marie et al. [81].

Materials

Reagents

– Ficoll density gradient: Ficoll separating solution, density = 1.007 g/ml;
– sodium azide.

Solutions and buffers

- DMSO;
- RPMI-1640 (w 2.0 g/l $NaHCO_3$, w/o phenol red, w/o L-glutamine): Biochrom KG, order No. F 1275;
- dilution for antibodies: PBS + 1% BSA + 0.1% sodium azide;
- PBS (Dulbecco, w/o Ca^{2+}, Mg^{2+}): Biochrom KG, order No. L 1825;

Consumables and equipment

- See 'Materials' Pgp, (pp. 436).

Substrate

Rhodamine 123 (R123): Molecular Probes, order No. F1275;

stock solution: dissolve 25 mg rhodamine 123 in 4 ml of DMSO (= 16.4 mmol/l); stir at 2–8 °C, stability: several months;

solution A: 50 µl RPMI-1640 + 4 µl rhodamine 123 stock solution (= 1.31 µmol/l), always prepare just before use.

Converters

Isoptine®: Knoll;

active component: verapamil hydrochloride (Vp);

5.1 mmol/l (2.5 mg/ml) working solution, store at room temperature;

Sandimmune®: Novartis ;

active component: cyclosporin A (CsA);

41.6 mmol/l (50 mg/ml) stock solution; store at room temperature;

4.16 mmol/l working solution: dilute the 41.6 mmol/l stock solution 1 : 10 with RPMI-1640, store at 2–8 °C; stability: 2 weeks;

Valspodar® (PSC 833): Novartis;

Prepare a 8.23 mmol/l stock solution by dissolving 10 mg PSC 833 in 0.5 ml ethanol/tween 80 (9 : 1), filter under sterile conditions, dilute with 0.5 ml bidistilled water and make aliquots of 20 µl; store at –20 °C; stability: several months;

Prepare a 206 µmol/l working solution: dilute the 8.23 mmol/l stock solution 1 : 40 with RPMI-1640, store at 4 °C, stability: 2 weeks.

Methods

Functional MDR tests with cells of peripheral blood

Preparation of peripheral blood cells:

- See 'Immunochemical Methods for Peripheral Blood Cells' (pp. 439); do not use NH_4Cl lysis!
- refill the cell suspension after Ficoll separation with RPMI-1640 to approximately 10 ml, wash (275 × *g*, 5 min, 15 °C) and discard the supernatant;
- adjust cells with RPMI-1640 to 1×10^7/ml;

Preparation of reagents:

- dilute rhodamine 123 and converters; prepare the stock and working solutions described above;

- the following solutions were prepared according to the pipetting schemes given in tables 7 and 8;
- the final concentration of rhodamine 123 is approximately 1.2 µmol/l.

Procedure:

- pipette 100 µl cell suspension in labeled Falcon tubes (1×10^6 cells/tube);
- add 1 ml of the various solutions according to the first pipetting step (uptake phase) and vortex;
- incubate for 60 min at 37 °C in a water bath while shaking gently;
- centrifuge all tubes at $275 \times g$, 15 °C, 5 min;
- discard the supernatant and wash 1 × with RPMI-1640 ($275 \times g$, 15 °C, 5 min);
- add the solutions according to the second pipetting step (efflux phase) and vortex;
- tube 2: after addition of 100 µl RPMI-1640, place on ice at once, keep in the dark;
- incubate the remaining water bath for 60 min at 37 °C under gentle agitation;
- centrifuge tubes ($275 \times g$, 5 min, 15 °C) ;
- add 100 µl RPMI-1640/tube;
- counterstaining with the respective antibodies (e.g. CD14-PerCP, BD Biosciences, order No. 345786, 10 µl, 1 : 2 or CD 34-PerCP, BD Biosciences, order No. 340430, 20 µl, 1 : 4);
- vortex and place tubes on ice for 20 min in the dark; do not wash!
- cell lines: pipette 50 µl of 7-AAD dilution to each tube and vortex;
- incubate on ice for 10 min in the dark and measure at once.

Measurements by Flow Cytometry

- FSC, lin; SSC, lin; Fl1 (R123) und Fl3 (counterstaining); log over 4 decades;
- calibration and compensation: Calibrite™ Beads, see under 'Quality control' (p. 452):
- trigger on FSC to exclude cell debris (do not cut the lymphocytes!);
- collect 20,000 events in gate 1 (R1); when measuring cell lines without counterstaining: collect 20,000 events in a gate on 7-AAD cells;
- flow rate: low to medium.

Calcein Influx Assay

Calcein-acetoxymethylester (AM) is a nonfluorescent, hydrophobic compound that easily permeates intact, living cells by diffusion. The hydrolysis of calcein-AM by intracellular esterases produces calcein, a hydrophilic, strongly fluorescent compound that is well retained in the cell cytoplasm. Pgp exclusively transports the nonfluorescent calcein-AM out of the cell, thus preventing cleavage into the fluorescent, free calcein. Both calcein-AM and free calcein are substrates for MRP1. By selecting suitable MRP1 converters (e.g. probenecid; PROB), Pgp function can be differentiated from MRP1 function. The calcein influx assay was modified according to Feller et al. [82].

Table 7. Pipetting scheme 1 for the rhodamine 123 efflux assay

Solution	Pipetting step
Vp B	10 ml solution A + 20 μl 5.1 mmol/l Vp (= 10.2 μmol/l)
Vp C	10 ml RPMI-1640 + 20 μl 5.1 mmol/l Vp (= 10.2 μmol/l)
CsA B	10 ml solution A + 8 μl 4.16 mmol/l CsA (3.3 μmol/l)
CsA C	10 ml RPMI-1640 + 8 μl 4.16 mmol/l CsA (3.3 μmol/l)
PSC 833 B	1 ml solution A + 10 μl 206 μmol/l PSC 833 (= 2.03 μmol/l)
PSC 833 C	1 ml RPMI-1640 + 10 μl 206 μmol/l PSC 833 (= 2.03 μmol/l)

Table 8. Pipetting scheme 1 for the rhodamine 123 efflux assay

Tube (description)	Cell suspension	1st step (uptake phase)	2nd step (efflux phase)
Cells + RPMI-1640	100 μl	1 ml RPMI-1640	
Maximum uptake	100 μl	1 ml solution A	100 μl RPMI-1640 on ice
Efflux without converter	100 μl	1 ml solution A	1 ml RPMI-1640
Efflux with converter (final concentration)			
9.3 μmol/l Vp	100 μl	1 ml Vp B	1 ml Vp C
3.02 μmol/l CsA	100 μl	1 ml CsA B	1 ml CsA C
1.85 μmol/l PSC 833	100 μl	1 ml PSC 833 B	1 ml PSC 833 C

Materials

Reagents

- Ficoll density gradient, BSA, Cellwash (see 'Materials' Pgp, pp. 436) (→ pB);
- PBS, sodium azide (see 'Materials' Pgp, pp. 436);
- trypsin EDTA, trypan blue, FBS (see 'Materials' MRP1 and LRP/MVP, pp. 440).

Solutions and buffers

- Dilution for antibodies (see 'Materials' Pgp, pp. 436);
- DMSO (see 'Materials' Rhodamine 123 Efflux Assay, pp. 443);
- RPMI-1640 (w 2.0 g/l $NaHCO_3$, w/o phenol red, w/o L-glutamine): Biochrom KG, order No. F 1275;
- Hanks balanced salt solution (HBSS) w/o phenol red: Biochrom KG, order No. L2035;

Consumables

See 'Materials' MRP1 and LRP/MVP, (pp. 440).

Equipment

See 'Materials' MRP1 and LRP/MVP, (pp. 440).

Substrates

Calcein-AM): Molecular Probes, order No. C-3100:
prepare a 100 μmol/l stock solution: dissolve 50 μl calcein-AM in 500 μl DMSO and aliquot 20-μl portions; store at −20 °C; storage: several months;

prepare a 1.25 µmol/l working solution: dilute the 100 µmol/l stock solution 1 : 80 with HBSS, prepare just before use;
prepare a 0.0625 µmol/l working solution: dilute the 1.25 µmol/l working solution 1 : 20 with HBSS, prepare just before use.
Converters
PROB : Sigma, order No. P-8761:
prepare a 100 mmol/l stock solution by dissolving of 28.45 mg PROB in 1 ml DMSO, store at 2–8 °C, stability: 4 weeks;
N-Ethylmaleinimide (NEM): Sigma, order No. 12,828-7:
prepare a 1 mmol/l stock solution by dissolving of 1.25 mg NEM in 10 ml RPMI-1640, store at 2–8 °C; stability: 2 weeks;
Genisteine (GEN): Sigma, order No. G 6649:
prepare a 20 mmol/l stock solution by dissolving of 5.4 mg GEN in 1 ml DMSO; store at 2–8 °C; stability: 4 weeks:
DL-buthionine-(S,R)-sulfoximine (BSO): Sigma, order No. B 2640:
prepare a 2.5 mmol/l stock solution by dissolving of 5.56 mg BSO in 10 ml RPMI-1640; store at 2–8 °C; stability: 2 weeks;
Isoptine®: Knoll :
active component: Vp;
5.1 mmol/l (2.5 mg/ml) stock solution; store at room temperature;
1 mmol/l working solution: dilute the 5.1 mmol/l stock solution 1 : 5 with HBSS, store at 2–8 °C; stability 2 weeks;
Sandimmune®: Novartis:
active component: CsA ;
41.6 mmol/l (50 mg/ml) stock solution; store at room temperature;
prepare a 416 µmol/l working solution; dilute the 41.6 mmol/l stock solution 1 : 100 with HBSS; store at +2 bis +8 °C; stability: 2 weeks,
Valspodar® (PSC 833): Novartis:
prepare a 8.23 mmol/l stock solution by dissolving 10 mg PSC 833 in 0.5 ml ethanol/tween 80 (9 : 1); filter under sterile conditions, dilute with 0.5 ml bidistilled water and make aliquots of 20 µl; store at –20 °C; stability: several months;
prepare a 206 µmol/l working solution; dilute the 8.23 mmol/l stock solution 1 : 40 with RPMI-1640; store at 4 °C; stability: 2 weeks.

Functional Method for Adherent Cell Lines
Preparation of cells:
- See 'Methods' MRP1 and LRP/MVP (pp. 440).
- after each washing step with PBS, refill the cells with approximately 5–10 ml HBSS; determine cell count and viability using a Neubauer chamber (see 'Materials' Calcein Influx Assay, p. 446);
- adjust cells with RPMI-1640 to 1×10^7/ml.

Table 9. Pipetting scheme for the calcein influx assay

Tube	Cell suspension	1st step	2nd step
Cells + HBSS	100 µl cells + 700 µl HBSS	–	–
Influx without converter	100 µl cells + 700 µl HBSS	–	200 µl calcein
Influx with converter and control tube (final concentrations)			
1 mmol/l PROB	100 µl cells + 700 µl HBSS	10 µl 100 mmol/l PROB	200 µl calcein
9.9 µmol/l NEM	100 µl cells + 700 µl HBSS	10 µl 1 mmol/l NEM	200 µl calcein
9.9 µmol/l Vp	100 µl cells + 700 µl HBSS	10 µl 1 mmol/l Vp	200 µl calcein
198 µmol/l GEN	100 µl cells + 700 µl HBSS	10 µl 20 mmol/l GEN	200 µl calcein
24.8 µmol/l BSO	100 µl cells + 700 µl HBSS	10 µl 2.5 mmol/l BSO	200 µl calcein
3.3 µmol/l CsA	100 µl cells + 700 µl HBSS	8 µl 416 µmol/l CsA	200 µl calcein
2.04 µmol/l PSC 833	100 µl cells + 700 µl HBSS	10 µl 206 µmol/l PSC833	200 µl calcein
DMSO	100 µl cells + 700 µl HBSS	10 µl DMSO	200 µl calcein

Preparation of reagents:

– prepare the stock solutions and working solutions by diluting substrates and converters as described under 'Immunochemical methods for adherent cell lines' (pp. 442).

Pipetting scheme:

– the following solutions were prepared according to the pipetting schemes given in table 9;
– the final concentration of calcein-AM in the tubes is 12.4 nmol/l.

Procedure:

– pipette of 100 µl cell suspension in each of the labeled Falcon tubes;
– refill with HBSS to 800 µl and vortex;
– add 10 µl (or 8 µl) of the solutions mentioned above (see pipetting scheme: step 1);
– incubate for 5 min at 37 °C in a water bath under gentle agitation;
– add 200 µl calcein-AM to all tubes (except tube 1) and vortex;
– incubate for 2 h in a water bath under gentle agitation;
– centrifuge all tubes (275 × g, 5 min, 15 °C);
– add 100 µl HBSS;
– counterstaining with the appropriate antibodies (→ pB) (e.g. CD14-PE, BD Biosciences, order No. 345785) 20 µl, 1 : 8 or CD 34-PE, BD Biosciences, order No. 345802, 20 µl, 1 : 2);
– vortex and incubate for 20 min on ice in the dark; do not wash!
– cell lines: pipette 50 µl of 7-AAD dilution to each tube and vortex;
– incubate on ice for 10 min in the dark and measure at once.

Measurements by Flow Cytometry

– FSC, lin; SSC, lin; Fl1 (calcein), (Fl2 → pB: counterstaining) und Fl3 (7-AAD); log over 4 decades;

- calibration and compensation: Calibrite Beads, see under 'Quality control';
- trigger on FSC or SSC to exclude cell debris;
- collect 20,000 events in a gate including all 7-AAD– cells (vital cells);
- flow rate: low to medium.

Data Analysis

To analyze the flow cytometric data, the following approaches are suitable:
- analysis of percentage of positive cells;
- analysis of the arithmetic mean values of fluorescence intensities (MFI).

The latter shows higher sensitivity if antigens are weakly expressed.

Immunological methods:
- create a histogram of the fluorescence of the target antigen;
- place suitable marker regions;
- analysis of the MFI of isotype control and specific antibody of the same concentration;
- calculate the ratio:
  $MFI_{specific\ antibody}/MFI_{isotype\ control}$ (= ratio of MFI);
- define a cut-off value:
  - if ratio ≥ 1.5: positive for specific antibody expression;
  - if ratio < 1.5: negative for specific antibody expression [81].

Functional assays:
- create a histogram of the fluorescence of the substrate;
- place suitable marker regions;
- analysis of the MFI of substrate with converter and substrate of the same concentration without converter;
- calculate the ratio:
  $MFI_{substrate\ with\ converter}/MFI_{substrate\ without\ converter}$
- define a cut-off value:
  - if ratio > 1.0: positive (= functionally active protein and/or conversion is detectable);
  - if ratio $= 1.0$: negative (= no functionally active protein and/or conversion is detectable) [81].

Within each test series, quality controls (blank values, conjugate controls, isotype controls) of the same concentration compared to the specific antibody have to be prepared and evaluated. In functional assays, it has to be considered that some converters are dissolved in DMSO or distilled water; thus, additional control tubes must be included (see pipetting schemes). In case of a significant difference compared to the influx value or efflux value without converter, a higher dilution of the solvent must be selected. The arithmetic mean values of antibody fluorescence can be imported in Excel sheets with appropriate data base fields, e.g. via direct transmission

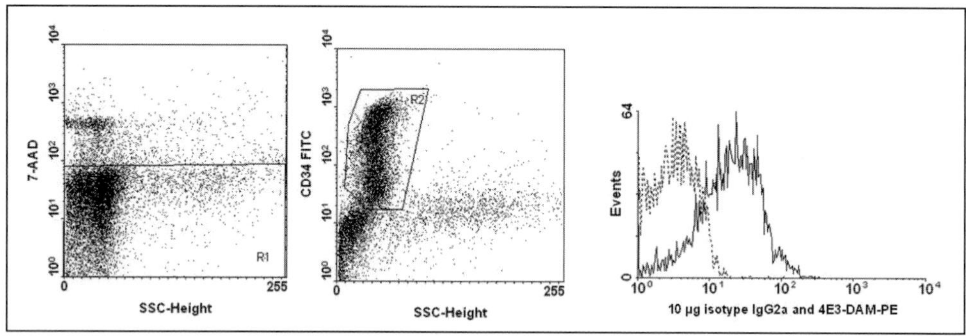

Fig. 5. Immunological Pgp detection on CD34+ leukemic blasts by flow cytometry using anti-(human) Pgp antibody 4E3.

from the flow cytometric software. In principle, a correlation of protein expression and function should be considered. In the following, the evaluation of different assays is demonstrated as an example.

Analysis of Staining Procedures with Anti-Pgp (Human) Antibodies

Figure 5 shows the staining procedure for peripheral blood cells with anti-Pgp-antibody 4E3. The first gating step is done in an SSC/FL3 plot. A gate (R1) including all 7-AAD-negative (vital) cells is created. In the SSC/FL1-plot, an additional gate (R2) is created including the target cell population (CD34+ blasts). The calculation of MFI occurs in a histogram over FL2. The gates R1 und R2 are connected. Thus, in the statistical evaluation only vital CD34+ cells are considered. Dotted lines: IgG2a isotype control; black line: anti-Pgp antibody 4E3. A ratio of MFI of 7.91 is calculated.

Analysis of Staining Procedures with Anti-MRP1 (Human) and Anti-LRP/MVP (Human) Antibodies

Figure 6 depicts the immunochemical staining of SK-Mel-23 cells derived from an intrinsically resistant melanoma cell line, with anti-MRP1 (human) antibody MRPm6. SK-Mel-23 cells are identified in an FSC/SSC plot (gate R1). A gate on all well-fixed and permeabilized 7-AAD+ cells is drawn in an SSC/FL3 plot (gate R2). The evaluation is done using a histogram displaying FL1. The gates R1 and R2 are connected. Only properly permeabilized SK-Mel-23 cells are included in the statistical evaluation. A ratio of MFI of 3.11 is calculated. Dotted line: IgG1 isotype control, black line: anti-MRP1-antibody MRPm6. The analysis of anti-MRP1- and anti-LRP/MVP-stained cells is identical.

Analysis of the Rhodamine Efflux Assay

Figure 7 shows the analysis of the rhodamine efflux assay performed with peripheral blood cells. Vital leukocytes are gated in an SSC/FL1 plot (gate R1) and separated from the debris and erythrocytes. CD34+ blasts are identified in an SSC/FL3 plot by

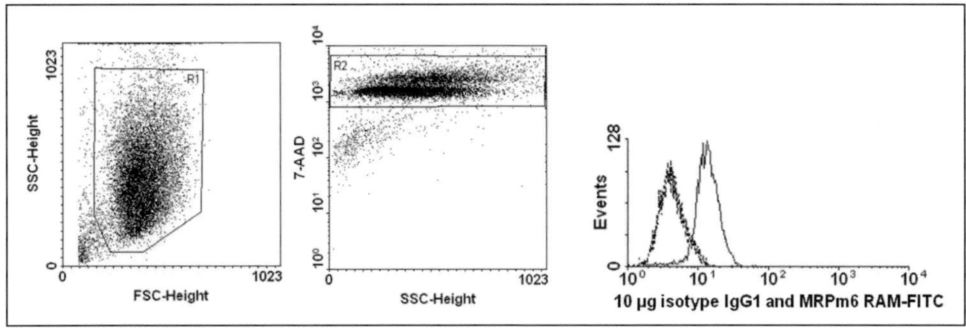

Fig. 6. Immunological MRP1 detection on SK-Mel-23 melanoma cells by flow cytometry using anti-(human) MRP1 antibody MRPm6.

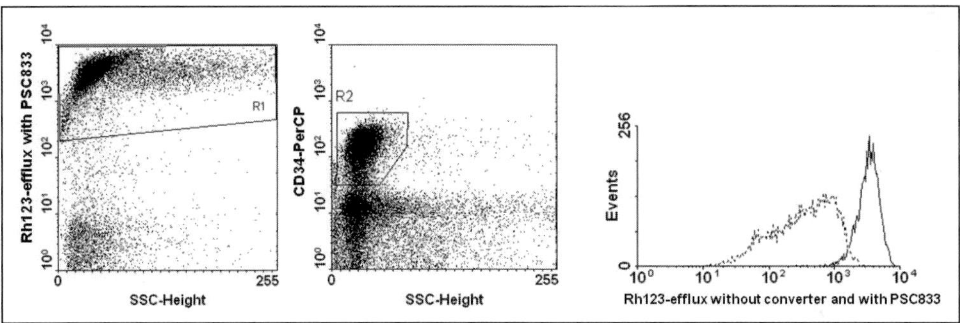

Fig. 7. Rhodamine 123 efflux monitored on CD34+ leukemic blasts.

counterstaining with CD34-peridinin-chlorophyll-protein (PerCP) (gate R2). The statistical analysis is done using a histogram depicting FL1. The efflux phase of CD34+ blasts without converter (dotted line) and with PSC 833 (black line) is shown. The results correspond to a functionally active protein with clear-cut conversion by PSC 833 (MFI = 5.3).

When analyzing cell lines, staining with 7-AAD has to be performed (see 'Analysis of Staining Procedures with Anti-MRP1 (Human) and Anti-LRP/MVP (Human) Antibodies', p. 450). In this case, Fl2 is not used for immunological labeling because of a strong spectral overlap with rhodamine. Counterstaining with an allophycocyanin (APC)-labeled antibody and viability staining with 7-AAD in peripheral blood are possible, but in our experience they are not necessary because of the high viability of these cells before measurements (see SSC/FL1 plot).

Analysis of the Calcein Influx Assay

The calcein influx assay in the melanoma cell line SK-Mel-23 is shown in figure 8. Cells are identified in an FSC/SSC plot (gate R1). The exclusion of nonviable, 7-AAD+ cells occurs in an SSC/FL3 plot (gate R2). The statistical analysis is done using a histogram depicting the FL1 whereas gate 1 and 2 are connected. Only vital

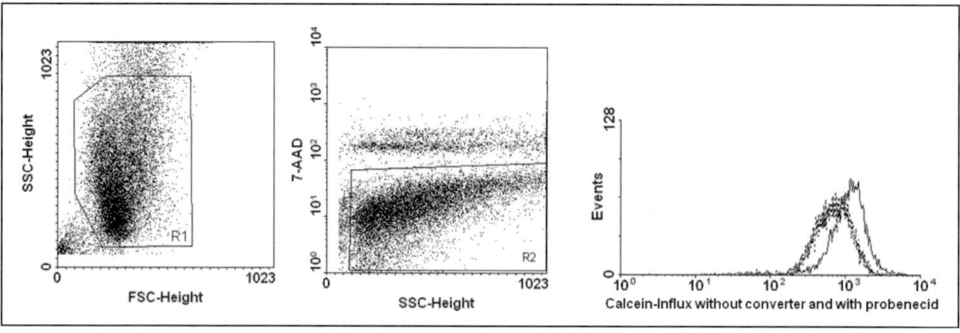

Fig. 8. Calcein influx monitored on SK-Mel-23 melanoma cells.

SK-Mel-23 cells are included. The influx phase without converter (dotted line) and with PROB (black line) is shown. The results correspond to a functionally active protein and a weak conversion (MFI = 1.88).

Quality Control

- Daily check of flow cytometric settings (compensation) and control of precision by measurement and calculation of the variation coefficient: Calibrite™ 3 Beads (Fa. BD Biosciences, order No. 340486) and Calibrite™ APC Beads (BD Biosciences, order No.: 340487);
- check antibodies by antibody titration and control cell lines;
- check unspecific binding of antibodies by use of isotype controls and conjugate controls;
- check cell permeabilization by staining of intracellular epitopes by ds-DNA antibodies, 7-AAD or propidium iodide (PI);
- check cell viability by vital dye (7-AAD, PI, trypan blue exclusion).

Troubleshooting

- Viability detection of cells should be >90% to avoid false-positive (immunochemical methods) or false-negative (functional assays) results;
- do not use of cell lysis in functional assays to avoid passive permeation through the cell membrane;
- perform background reducing in protocols with antibodies of IgG2a isotype in myeloid cells to avoid unspecific antibody binding to Fc receptors;
- titration of antibodies used in suitable cell lines to assure antibody excess;
- exclusion of nonviable cells with 7-AAD or PI to avoid false-positive results (immunological methods) or false-negative results (functional assays);

- selection of isotype controls according subclass and concentration of the specific primary antibody thus avoiding false-positive results by unspecific binding and calculation of MFI;
- exact identification of cell populations by counterstaining to augment the significance of results;
- work with red-labeled fluorochromes and indirect immunochemical methods in cases of weakly expressed antigens to augment sensitivity by enhanced quantum yield (amplification);
- use of anti-ds-DNA antibodies in staining protocols of intracellular antigens to check cell permeation by fixation;
- performance of flow-cytometric measurements with standardized settings to compare results from different experiments;
- select suitable positive and negative controls to continuously check the single components of the assays (e.g. antibodies, substrates, converters);
- calculate MFI ratios instead of absolute values in order to achieve higher sensitivity and reduce variations of substrate fluorescences influenced, e.g. by storage.

Time Required

The following time is needed from sample preparation to flow-cytometric measurements:

Immunochemical methods:

Detection of Pgp:

- direct staining: cell lines 165 min, pB 195 min;
- indirect staining: cell lines 205 min, pB 235 min;
- detection of MRP1 and LRP/MVP: cell lines 180 min, pB 210 min;

Functional MDR tests:

- rhodamine efflux assay: cell lines 215 min, pB 270 min;
- calcein influx assay: cell lines 215 min, pB 270 min.

Summary

In this chapter, we described special immunochemical and functional methods to detect proteins involved in multidrug resistance (MDR):

- P-glycoprotein (Pgp, ABCB1);
- multidrug resistance-associated protein 1 (MRP1, ABCC1);
- breast cancer resistance protein (BCRP/MXR/ABCP1/ABCG2) and
- lung resistance-related protein (LRP/MVP).

We also reviewed clinical aspects and basic knowledge on the physiological expression patterns and functions of the above-mentioned proteins. Their resistance profiles were described as well. Protocols show techniques using fixed and unfixed cells, direct and indirect staining methods, functional influx and efflux assays and give a theoretical and methodical survey of antibodies, substrates and MDR converters. Additionally, we summarized the main recommendations regarding the standardization of MDR-related methods.

References

1 Biedler JL, Riehm H: Cellular resistance to actinomycin D in Chinese hamster cells in vitro: cross-resistance, radioautographic, and cytogenetic studies. Cancer Res 1970;30:1174–1184.

2 Juliano RL, Ling V: A surface glycoprotein modulating drug permeability in Chinese hamster ovary cell mutants. Biochim Biophys Acta 1976;455:152–162.

3 Gerlach JH, Endicott JA, Juranka PF, Henderson G, Sarangi F, Deuchars KL, Ling V: Homology between P-glycoprotein and a bacterial haemolysin transport protein suggests a model for multidrug resistance. Nature 1986;324:485–489.

4 Watkins PB: The barrier function of CYP3A4 and P-glycoprotein in the small bowel. Adv Drug Deliv Rev 1997;27:161–170.

5 Cox DS, Scott KR, Gao H, Raje S, Eddington ND: Influence of multidrug resistance (MDR) proteins at the blood-brain barrier on the transport and brain distribution of enaminone anticonvulsants. J Pharm Sci 2001;90:1540–1552.

6 Hoffmeyer S, Burk O, von Richter O, Arnold HP, Brockmoller J, Johne A, Cascorbi I, Gerloff T, Roots I, Eichelbaum M, Brinkmann U: Functional polymorphisms of the human multidrug-resistance gene: multiple sequence variations and correlation of one allele with P-glycoprotein expression and activity in vivo. Proc Natl Acad Sci U S A 2000;97:3473–3478.

7 Honjo Y, Hrycyna CA, Yan QW, Medina-Perez WY, Robey RW, van de Laar A, Litman T, Dean M, Bates SE: Acquired mutations in the MXR/BCRP/ABCP gene alter substrate specificity in MXR/BCRP/ABCP-overexpressing cells. Cancer Res 2001;61:6635–6639.

8 Kurata Y, Ieiri I, Kimura M, Morita T, Irie S, Urae A, Ohdo S, Ohtani H, Sawada Y, Higuchi S, Otsubo K: Role of human MDR1 gene polymorphism in bioavailability and interaction of digoxin, a substrate of P-glycoprotein. Clin Pharmacol Ther 2002;72:209–219.

9 Allen JD, Brinkhuis RF, Wijnholds J, Schinkel AH: The mouse Bcrp1/Mxr/Abcp gene: amplification and overexpression in cell lines selected for resistance to topotecan, mitoxantrone, or doxorubicin. Cancer Res 1999;59:4237–4241.

10 Chaillou S, Durant J, Garraffo R, Georgenthum E, Roptin C, Dunais B, Mondain V, Roger PM, Dellamonica P: Intracellular concentration of protease inhibitors in HIV-1-infected patients: correlation with MDR-1 gene expression and low dose of ritonavir. HIV Clin Trials 2002;3:493–501.

11 Hamada A, Miyano H, Watanabe H, Saito H: Interaction of imatinib mesilate with human P-glycoprotein. J Pharmacol Exp Ther 2003;307:824–828.

12 Scheuring UJ, Pfeifer H, Wassmann B, Bruck P, Gehrke B, Petershofen EK, Gschaidmeier H, Hoelzer D, Ottmann OG: Serial minimal residual disease (MRD) analysis as a predictor of response duration in Philadelphia-positive acute lymphoblastic leukemia (Ph + ALL) during imatinib treatment. Leukemia 2003;17:1700–1706.

13 Dorr R, Karanes C, Spier C, Grogan T, Greer J, Moore J, Weinberger B, Schiller G, Pearce T, Litchman M, Dalton W, Roe D, List AF: Phase I/II study of the P-glycoprotein modulator PSC 833 in patients with acute myeloid leukemia. J Clin Oncol 2001;19:1589–1599.

14 Filipits M, Pohl G, Stranzl T, Suchomel RW, Scheper RJ, Jager U, Geissler K, Lechner K, Pirker R: Expression of the lung resistance protein predicts poor outcome in de novo acute myeloid leukemia. Blood 1998;91:1508–1513.

15 Wuchter C, Leonid K, Ruppert V, Schrappe M, Buchner T, Schoch C, Haferlach T, Harbott J, Ratei R, Dorken B, Ludwig WD: Clinical significance ofP-glycoprotein expression and function for response to induction chemotherapy, relapse rate and overall survival in acute leukemia. Haematologica 2000;85:711–721.

16 Lee BD, French KJ, Zhuang Y, Smith CD: Development of a syngeneic in vivo tumor model and its use in evaluating a novel P-glycoprotein modulator, PGP-4008. Oncol Res 2003;14:49–60.
17 Wang D, Johnson AD, Papp AC, Kroetz DL, Sadee W: Multidrug resistance polypeptide 1 (MDR1, ABCB1) variant 3435C → T affects mRNA stability. Pharmacogenet Genomics 2005;15:693–704.
18 Lawson J, O'Mara ML, Kerr ID: Structure-based interpretation of the mutagenesis database for the nucleotide binding domains of P-glycoprotein. Biochim Biophys Acta 2008;1778:376–391.
19 Legrand O, Simonin G, Perrot JY, Zittoun R, Marie JP: Both PGP and MRP1 activities using calcein-AM contribute to drug resistance in AML. Adv Exp Med Biol 1999;457:161–175.
20 Leith CP, Kopecky KJ, Chen IM, Eijdems L, Slovak ML, McConnell TS, Head DR, Weick J, Grever MR, Appelbaum FR, Willman CL: Frequency and clinical significance of the expression of the multidrug resistance proteins MDR1/P-glycoprotein, MRP1, and LRP in acute myeloid leukemia: a Southwest Oncology Group Study. Blood 1999;94:1086–1099.
21 Leith CP, Kopecky KJ, Godwin J, McConnell T, Slovak ML, Chen IM, Head DR, Appelbaum FR, Willman CL: Acute myeloid leukemia in the elderly: assessment of multidrug resistance (MDR1) and cytogenetics distinguishes biologic subgroups with remarkably distinct responses to standard chemotherapy. a Southwest Oncology Group Study. Blood 1997;89:3323–3329.
22 Huang Y: Pharmacogenetics/genomics of membrane transporters in cancer chemotherapy. Cancer Metastasis Rev 2007;26:183–201.
23 Gillet JP, Efferth T, Steinbach D, Hamels J, de Longueville F, Bertholet V, Remacle J: Microarray-based detection of multidrug resistance in human tumor cells by expression profiling of ATP-binding cassette transporter genes. Cancer Res 2004;64:8987–8993.
24 Humeny A, Rodel F, Rodel C, Sauer R, Fuzesi L, Becker C, Efferth T: MDR1 single nucleotide polymorphism C3435T in normal colorectal tissue and colorectal carcinomas detected by MALDDI-TOF mass spectrometry. Anticancer Res 2003;23:2735–2740.
25 Pinos T, Constansa JM, Palacin A, Figueras C: A new diagnostic approach to the Dubin-Johnson syndrome. Am J Gastroenterol 1990;85:91–93.
26 Klein I, Sarkadi B, Varadi A: An inventory of the human ABC proteins. Biochim Biophys Acta 1999;1461:237–262.
27 Kage K, Tsukahara S, Sugiyama T, Asada S, Ishikawa E, Tsuruo T, Sugimoto Y: Dominant-negative inhibition of breast cancer resistance protein as drug efflux pump through the inhibition of S-S dependent homodimerization. Int J Cancer 2002;97:626–630.
28 Bradley G, Juranka PF, Ling V: Mechanism of multidrug resistance. Biochim Biophys Acta 1988;948:87–128.
29 Pastan I, Gottesman MM: Multidrug resistance. Annu Rev Med 1991;42:277–286.
30 Beck WT, Danks MK: Mechanisms of resistance to drugs that inhibit DNA topoisomerases. Semin Cancer Biol 1991;2:235–244.
31 Thiebaut F, Tsuruo T, Hamada H, Gottesman MM, Pastan I, Willingham MC: Cellular localization of the multidrug-resistance gene product P-glycoprotein in normal human tissues. Proc Natl Acad Sci U S A 1987;84:7735–7738.
32 Cordon-Cardo C, O'Brien JP, Casals D, Rittman-Grauer L, Biedler JL, Melamed MR, Bertino JR: Multidrug-resistance gene (P-glycoprotein) is expressed by endothelial cells at blood-brain barrier sites. Proc Natl Acad Sci U S A 1989;86:695–698.
33 Rao VV, Dahlheimer JL, Bardgett ME, Snyder AZ, Finch RA, Sartorelli AC, Piwnica-Worms D: Choroid plexus epithelial expression of MDR1 P glycoprotein and multidrug resistance-associated protein contribute to the blood-cerebrospinal-fluid drug-permeability barrier. Proc Natl Acad Sci U S A 1999;96:3900–3905.
34 Chaudhary PM, Roninson IB: Expression and activity of P-glycoprotein, a multidrug efflux pump, in human hematopoietic stem cells. Cell 1991;66:85–94.
35 Schinkel AH, Mayer U, Wagenaar E, Mol CA, van Deemter L, Smit JJ, van der Valk MA, Voordouw AC, Spits H, van Tellingen O, Zijlmans JM, Fibbe WE, Borst P: Normal viability and altered pharmacokinetics in mice lacking MDR1-type (drug-transporting) P-glycoproteins. Proc Natl Acad Sci U S A 1997;94:4028–4033.
36 Johnstone RW, Ruefli AA, Tainton KM, Smyth MJ: A role for P-glycoprotein in regulating cell death. Leuk Lymphoma 2000;38:1–11.
37 Raghu G, Park SW, Roninson IB, Mechetner EB: Monoclonal antibodies against P- glycoprotein, an MDR1 gene product, inhibit interleukin-2 release from PHA-activated lymphocytes. Exp Hematol 1996;24:1258–1264.
38 Idriss HT, Hannun YA, Boulpaep E, Basavappa S: Regulation of volume-activated chloride channels by P-glycoprotein: phosphorylation has the final say! J Physiol 2000;524:629–636.

39 Luker GD, Nilsson KR, Covey DF, Piwnica-Worms D: Multidrug resistance (MDR1) P-glycoprotein enhances esterification of plasma membrane cholesterol. J Biol Chem 1999;274:6979–6991.
40 Higgins CF, Gottesman MM: Is the multidrug transporter a flippase? Trends Biochem Sci 1992;17:18–21.
41 Raviv Y, Pollard HB, Bruggemann EP, Pastan I, Gottesman MM: Photosensitized labeling of a functional multidrug transporter in living drug-resistant tumor cells. J Biol Chem 1990;265:3975–3980.
42 Dey S, Ramachandra M, Pastan I, Gottesman MM, Ambudkar SV: Evidence for two nonidentical drug-interaction sites in the human P-glycoprotein. Proc Natl Acad Sci U S A 1997;94:10594–10599.
43 Litman T, Druley TE, Stein WD, Bates SE: From MDR to MXR: new understanding of multidrug resistance systems, their properties and clinical significance. Cell Mol Life Sci 2001;58:931–959.
44 Cole SP, Bhardwaj G, Gerlach JH, Mackie JE, Grant CE, Almquist KC, Stewart AJ, Kurz EU, Duncan AM, Deeley RG: Overexpression of a transporter gene in a multidrug-resistant human lung cancer cell line. Science 1992;258:1650–1654.
45 Grant CE, Bhardwaj G, Cole SP, Deeley RG: Cloning, transfer, and characterization of multidrug resistance protein. Methods Enzymol 1998;292:594–607.
46 Breuninger LM, Paul S, Gaughan K, Miki T, Chan A, Aaronson SA, Kruh GD: Expression of multidrug resistance-associated protein in NIH/3T3 cells confers multidrug resistance associated with increased drug efflux and altered intracellular drug distribution. Cancer Res 1995;55:5342–5347.
47 Barrand MA, Bagrij T, Neo SY: Multidrug resistance-associated protein: A protein distinct from P-glycoprotein involved in cytotoxic drug expulsion. Gen Pharmacol 1997;28:639–645.
48 Jedlitschky G, Leier I, Buchholz U, Barnouin K, Kurz G, Keppler D: Transport of glutathione, glucuronate, and sulfate conjugates by the MRP gene-encoded conjugate export pump. Cancer Res 1996;56:988–994.
49 Lorico A, Rappa G, Finch RA, Yang D, Flavell RA, Sartorelli AC: Disruption of the murine *MRP* (multidrug resistance protein) gene leads to increased sensitivity to etoposide (VP-16) and increased levels of glutathione. Cancer Res 1997;57:5238–5242.
50 Wijnholds J, Evers R, van Leusden MR, Mol CA, Zaman GJ, Mayer U, Beijnen JH, van der Valk M, Krimpenfort P, Borst P: Increased sensitivity to anticancer drugs and decreased inflammatory response in mice lacking the multidrug resistance-associated protein. Nat Med 1997;3:1275–1279.
51 Loe DW, Deeley RG, Cole SP: Biology of the multidrug resistance-associated protein, MRP. Eur J Cancer 1996;32A:945–957.
52 Lautier D, Canitrot Y, Deeley RG, Cole SP: Multidrug resistance mediated by the multidrug resistance protein (MRP) gene. Biochem Pharmacol 1996;52:967–977.
53 Sharp SY, Smith V, Hobbs S, Kelland LR: Lack of a role for MRP1 in platinum drug resistance in human ovarian cancer cell lines. Br J Cancer 1998;78:175–180.
54 Hooijberg JH, Broxterman HJ, Kool M, Assaraf YG, Peters GJ, Noordhuis P, Scheper RJ, Borst P, Pinedo HM, Jansen G: Antifolate resistance mediated by the multidrug resistance proteins MRP1 and MRP2. Cancer Res 1999;59:2532–2535.
55 Deeley RG, Cole SP: Substrate recognition and transport by multidrug resistance protein 1 (ABCC1). FEBS Letters 2006;580:1103–1111.
56 Doyle LA, Yang W, Abruzzo LV, Krogmann T, Gao Y, Rishi AK, Ross DD: A multidrug resistance transporter from human MCF-7 breast cancer cells. Proc Natl Acad Sci U S A 1998;95:15665–15670.
57 Jonker JW, Merino G, Musters S, van Herwaarden AE, Bolscher E, Wagenaar E, Mesman E, Dale TC, Schinkel AH: The breast cancer resistance protein BCRP (ABCG2) concentrates drugs and carcinogenic xenotoxins into milk. Nat Med 2005;11:127–129.
58 Moon YJ, Zhang S, Morris ME: Real-time quantitative polymerase chain reaction for bcrp, MDR1, and MRP1 mRNA levels in lymphocytes and monocytes. Acta Haematol 2007;118:169–175.
59 Roller A, Bahr OR, Streffer J, Winter S, Heneka M, Deininger M, Meyermann R, Naumann U, Gulbins E, Weller M: Selective potentiation of drug cytotoxicity by nsaid in human glioma cells: The role of COX-1 and MRP. Biochem Biophys Res Commun 1999;259:600–605.
60 Maliepaard M, Scheffer GL, Faneyte IF, van Gastelen MA, Pijnenborg AC, Schinkel AH, van De Vijver MJ, Scheper RJ, Schellens JH: Subcellular localization and distribution of the breast cancer resistance protein transporter in normal human tissues. Cancer Res 2001;61:3458–3464.
61 Scharenberg CW, Harkey MA, Torok-Storb B: The abcg2 transporter is an efficient Hoechst 33342 efflux pump and is preferentially expressed by immature human hematopoietic progenitors. Blood 2002;99:507–512.
62 Scheffer GL, Maliepaard M, Pijnenborg AC, van Gastelen MA, de Jong MC, Schroeijers AB, van der Kolk DM, Allen JD, Ross DD, van der Valk P, Dalton WS, Schellens JH, Scheper RJ: Breast cancer resistance protein is loca-

lized at the plasma membrane in mitoxantrone- and topotecan-resistant cell lines. Cancer Res 2000;60:2589–2593.

63 Robey RW, Medina-Perez WY, Nishiyama K, Lahusen T, Miyake K, Litman T, Senderowicz AM, Ross DD, Bates SE: Overexpression of the ATP-binding cassette half-transporter, ABCG2 (MXR/BCRP/ABCP1), in flavopiridol-resistant human breast cancer cells. Clin Cancer Res 2001;7:145–152.

64 Sargent JM, Williamson CJ, Maliepaard M, Elgie AW, Scheper RJ, Taylor CG: Breast cancer resistance protein expression and resistance to daunorubicin in blast cells from patients with acute myeloid leukaemia. Br J Haematol 2001;115:257–262.

65 Maliepaard M, van Gastelen MA, de Jong LA, Pluim D, van Waardenburg RC, Ruevekamp-Helmers MC, Floot BG, Schellens JH: Overexpression of the BCRP/MXR/ABCP gene in a topotecan-selected ovarian tumor cell line. Cancer Res 1999;59:4559–4563.

66 Kawabata S, Oka M, Shiozawa K, Tsukamoto K, Nakatomi K, Soda H, Fukuda M, Ikegami Y, Sugahara K, Yamada Y, Kamihira S, Doyle LA, Ross DD, Kohno S: Breast cancer resistance protein directly confers SN-38 resistance of lung cancer cells. Biochem Biophys Res Commun 2001;280:1216–1223.

67 Wang X, Nitanda T, Shi M, Okamoto M, Furukawa T, Sugimoto Y, Akiyama S, Baba M: Induction of cellular resistance to nucleoside reverse transcriptase inhibitors by the wild-type breast cancer resistance protein. Biochem Pharmacol 2004;68:1363–1370.

68 Scheper RJ, Broxterman HJ, Scheffer GL, Kaaijk P, Dalton WS, van Heijningen TH, van Kalken CK, Slovak ML, de Vries EG, van der Valk P, et al.: Overexpression of a M_r 110,000 vesicular protein in non-P-glycoprotein-mediated multidrug resistance. Cancer Res 1993;53:1475–1479.

69 Scheffer GL, Wijngaard PL, Flens MJ, Izquierdo MA, Slovak ML, Pinedo HM, Meijer CJ, Clevers HC, Scheper RJ: The drug resistance-related protein LRP is the human major vault protein. Nat Med 1995;1:578–582.

70 Kedersha NL, Miquel MC, Bittner D, Rome LH: Vaults. II. Ribonucleoprotein structures are highly conserved among higher and lower eukaryotes. J Cell Biol 1990;110:895–901.

71 Izquierdo MA, Scheffer GL, Flens MJ, Giaccone G, Broxterman HJ, Meijer CJ, van der Valk P, Scheper RJ: Broad distribution of the multidrug resistance-related vault lung resistance protein in normal human tissues and tumors. Am J Pathol 1996;148:877–887.

72 Chugani DC, Rome LH, Kedersha NL: Evidence that vault ribonucleoprotein particles localize to the nuclear pore complex. J Cell Sci 1993;106:23–29.

73 Kitazono M, Sumizawa T, Takebayashi Y, Chen ZS, Furukawa T, Nagayama S, Tani A, Takao S, Aikou T, Akiyama S: Multidrug resistance and the lung resistance-related protein in human colon carcinoma SW-620 cells. J Natl Cancer Inst 1999;91:1647–1653.

74 Schuurhuis GJ, Broxterman HJ, de Lange JH, Pinedo HM, van Heijningen TH, Kuiper CM, Scheffer GL, Scheper RJ, van Kalken CK, Baak JP, et al: Early multidrug resistance, defined by changes in intracellular doxorubicin distribution, independent of P-glycoprotein. Br J Cancer 1991;64:857–861.

75 Zurita AJ, Diestra JE, Condom E, Garcia Del Muro X, Scheffer GL, Scheper RJ, Perez J, Germa-Lluch JR, Izquierdo MA: Lung resistance-related protein as a predictor of clinical outcome in advanced testicular germ-cell tumours. Br J Cancer 2003;88:879–886.

76 Oguri T, Fujiwara Y, Ochiai M, Fujitaka K, Miyazaki M, Takahashi T, Yokozaki M, Isobe T, Ohune T, Tsuya T, Katoh O, Yamakido M: Expression of lung-resistance protein gene is not associated with platinum drug exposure in lung cancer. Anticancer Res 1998;18:4159–4162.

77 Shimamoto Y, Sumizawa T, Haraguchi M, Gotanda T, Jueng HC, Furukawa T, Sakata R, Akiyama S: Direct activation of the human major vault protein gene by DNA-damaging agents. Oncol Rep 2006;15:645–652.

78 Beck WT, Grogan TM, Willman CL, Cordon-Cardo C, Parham DM, Kuttesch JF, Andreeff M, Bates SE, Berard CW, Boyett JM, Brophy NA, Broxterman HJ, Chan HS, Dalton WS, Dietel M, Fojo AT, Gascoyne RD, Head D, Houghton PJ, Srivastava DK, Lehnert M, Leith CP, Paietta E, Pavelic ZP, Weinstein R: Methods to detect P-glycoprotein-associated multidrug resistance in patients' tumors: Consensus recommendations. Cancer Res 1996;56:3010–3020.

79 Marie JP, Huet S, Faussat AM, Perrot JY, Chevillard S, Barbu V, Bayle C, Boutonnat J, Calvo F, Campos-Guyotat L, Colosetti P, Cazin JL, de Cremoux P, Delvincourt C, Demur C, Drenou B, Fenneteau O, Feuillard J, Garnier-Suillerot A, Genne P, Gorisse MC, Gosselin P, Jouault H, Lacave R, Robert J, et al: Multicentric evaluation of the MDR phenotype in leukemia. French Network of the Drug Resistance Intergroup, and Drug Resistance Network of Assistance Publique-Hôpitaux de Paris. Leukemia 1997;11:1086–1094.

80 Broxterman HJ, Sonneveld P, Feller N, Ossenkoppele GJ, Wahrer DC, Eekman CA, Schoester M, Lankelma J, Pinedo HM, Lowenberg B, Schuurhuis GJ: Quality control of multidrug resistance assays in adult acute leukemia: Correlation between assays for P-glycoprotein expression and activity. Blood 1996;87:4809–4816.

81 Marie JP, Legrand O, Perrot JY, Chevillard S, Huet S, Robert J: Measuring multidrug resistance expression in human malignancies: Elaboration of consensus recommendations. Semin Hematol 1997;34:63–71.

82 Feller N, Kuiper CM, Lankelma J, Ruhdal JK, Scheper RJ, Pinedo HM, Broxterman HJ: Functional detection of MDR1/p170 and MRP/P190-mediated multidrug resistance in tumour cells by flow cytometry [published erratum appears in Br J Cancer 1996 dec;74(12):2042]. Br J Cancer 1995;72:543–549.

83 Scheffer GL, Scheper RJ: Drug resistance molecules: Lessons from oncology. Novartis Found Symp 2002;243:19–31; discussion 31–17, 180–185.

Sack U, Tárnok A, Rothe G (eds): Cellular Diagnostics. Basics, Methods and Clinical Applications of Flow Cytometry. Basel, Karger, 2009, pp 459–475

Flow-Cytometric Analysis of Intracellularly Stained Cytokines, Phosphoproteins, and Microbial Antigens

Irina Lehmann[a] · Jörg Lehmann[b]

[a] Helmholtz Center for Environmental Research – UFZ,
[b] Fraunhofer Institute for Cell Therapy and Immunology, Leipzig, Germany

Introduction

The Identification of functionally active cells or infected cells ex vivo and in vitro using monoclonal or polyclonal antibodies recognizing specifically expressed proteins (i.e., cytokines, chemokines, or enzymes), posttranslational protein modifications or specific epitopes of intracellular pathogens is commonly applied in research and diagnostics. Most often, the protein of interest or the pathogen is detected using immunohistochemical or immunocytochemical techniques on glass or plastic slides (i.e. tissue sections or cytospins), which also enable the investigator to recognize the topographical localization of the detected cells in situ. On the other hand, these techniques are not really useful to analyze larger cell numbers, which would require the preparation and staining of a lot of sections or cytospins. Moreover, in situ techniques are less suitable for the simultaneous staining of an intracellular pathogen and/or protein and additionally one or more specific surface markers to identify the infected and/or protein-expressing cell type. In contrast, flow cytometry enables the simultaneous analysis of several parameters, including size (forward scatter; FSC) and granularity (side scatter; SSC), as well as the expression of surface and intracellular molecules (via fluorochrome-labeled antibodies) on the single-cell level. Therefore, in many tasks flow-cytometric analysis is to be preferred over histological staining techniques if the cells to be studied are available as single-cell suspensions. Since this basic requirement is often considered to represent a big disadvantage of flow cytometry, the last remark should be discussed in more detail. Several sample materials are inherently single-cell suspensions and can therefore be stained without time-consuming preparative procedures (i.e. blood, lymphoid fluid, bone marrow, cerebrospinal fluid, or bronchoalveolar lavage fluid). However, cells from a range of

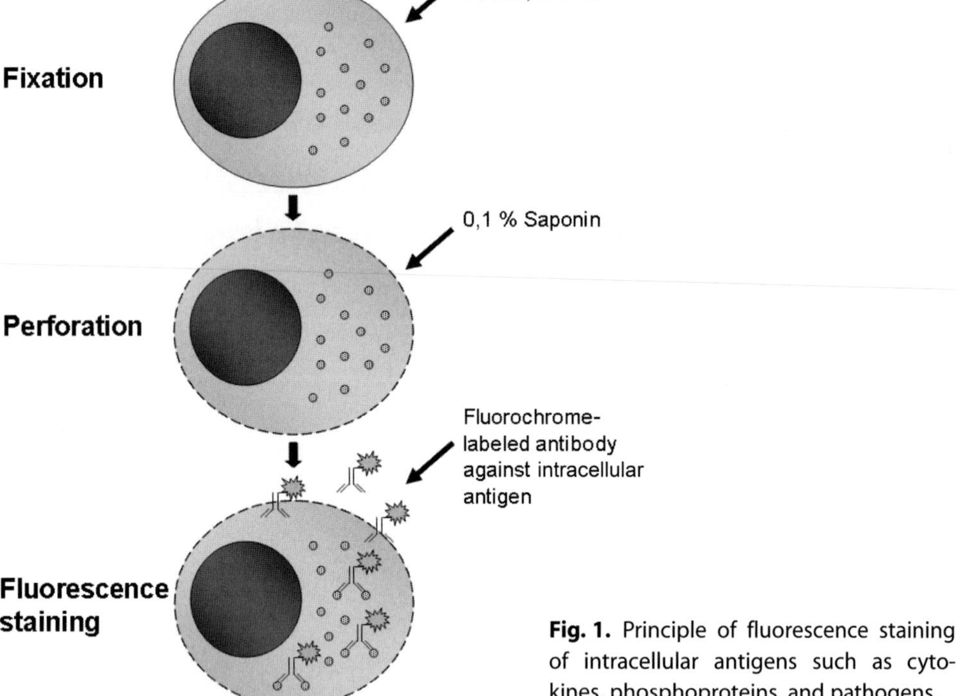

Fig. 1. Principle of fluorescence staining of intracellular antigens such as cytokines, phosphoproteins, and pathogens.

other tissues or organs can also be easily prepared for intracellular flow-cytometric analysis. For example, spleen or lymph node cells can be isolated only by mechanical disintegration (passage through a sieve of definite mesh size) without enzymatic dissociation [1]. However, epithelial cells, endothelial cells, connective tissue cells and tissue-infiltrating immune cells can also be isolated, but only by enzymatic dissociation (Lehmann J., unpublished) [2] or in certain cases also by growing-out methods [3]. Most of the isolation procedures for obtaining single-cell suspensions from tissues or organs described in the literature so far are appropriate for flow-cytometric analysis. Interestingly, a broad range of surface marker antigens has been shown to be unresponsive to treatment with trypsin or collagenase, both of which are often used for tissue disintegration or harvesting adherent cells in vitro [2, 3].

We describe an easy-to-handle and reproducible method for intracellular staining of cytokines, phosphoepitopes and pathogens, which was previously introduced in principle for cytokines by Sander et al. [4, 5] and improved by Jung et al. [6] and Picker et al. [7], who combined cytokine staining with the interruption of intracellular transport processes with monensin or brefeldin A, respectively; this results in accumulation of the cytokine in the Golgi complex and causes significant signal amplification (fig. 1).

Principle

The cells to be studied are fixed in ice-cold phosphate-buffered 4% paraformaldehyde (PFA) for exactly 10 min. This conserves the relevant epitopes of cell surface antigens and thus enables the use of most conventional cell surface marker antibodies for cell phenotyping. In contrast, fixation with formaldehyde is not recommended. Subsequently, the cells are permeabilized with saponin, which is a gentle detergent derived from the quillaja tree. Permeabilization of the cell membrane by saponin is reversible, thus, in the absence of saponin, the micropores in the cell membrane close again. For this reason, it is necessary to remove unbound antibodies from the inner cell compartment in the presence of saponin in the washing buffer. Otherwise, a high unspecific background will make the method unusable. The saponin-permeabilized cells can be incubated with appropriate antibodies (unlabeled or fluorochrome-labeled) directed against the cytokine, phosphoepitopes, or pathogen epitopes. The intracellularly stained target cells can be phenotyped simultaneously or successively by staining with antibodies against relevant cell surface molecules (e.g. certain CD antigens). This method may also be applied with heparinized whole blood as well as other heparinized body fluids or with cells derived from in vitro cell cultures.

In vitro stimulation of the target cells with the PKC activator phorbol-12-myristate-13-acetate (PMA) in combination with the Ca^{2+} ionophore ionomycin for 5 h in the presence of monensin or brefeldin A in order to inhibit the secretion of the produced cytokines is an important prerequisite for the intracellular staining of cytokines. Other stimulation protocols (e.g., anti-CD3/anti-CD28, concanavalin A, or antigen stimulation of T-cells for 24 h) yield suboptimal results.

For the intracellular staining of phosphoepitopes, special induction protocols, appropriate for the individual functional question, are needed.

The staining of intracellular pathogen epitopes does not require stimulation of the target cells.

In the following sections three typical protocols for the intracellular staining of (1) cytokines, (2) phosphoepitopes, and (3) intracellular bacterial or viral antigens are introduced as examples.

Intracellular Immunofluorescent Staining of Cytokines in Human Peripheral Blood T-Lymphocytes or Murine Splenic T-Lymphocytes ex vivo

To stain T-cell-derived cytokines (e.g. IFN-γ, TNF-α, IL-2, IL-4) human peripheral blood lymphocytes or murine splenic T-lymphocytes separated by Ficoll-Paque (GE Healthcare, Freiburg i. Br., Germany) or human whole blood (diluted 1:10 in RPMI-1640 cell culture medium) or erythrocyte-depleted murine splenocytes are

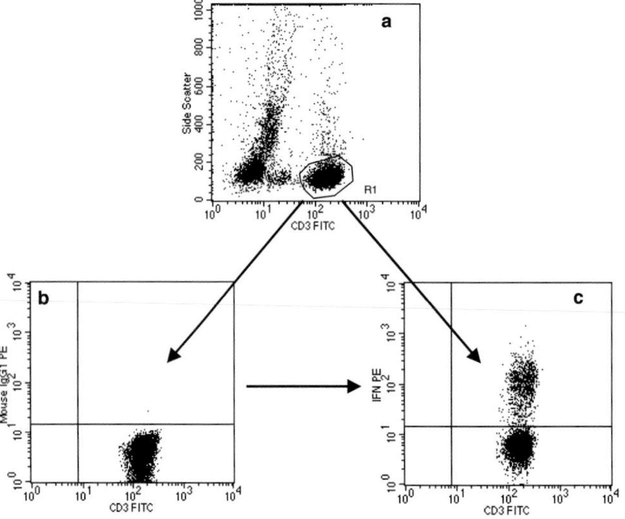

Fig. 2. Gating strategy for analysis of intracellular cytokine staining in T-lymphocytes. **a** First of all, CD3+ T-cells are discriminated in the SSC/FL1 plot. **b** To discriminate cytokine-positive T-cells, a marker set in the isotype control is then copied into the respective plot of the cytokine-stained samples (**c**).

stimulated with PMA (10 ng/ml) + ionomycin (1 μmol/l) (both reagents from Sigma, Taufkirchen, Germany) at 37 °C and 5% CO_2 for 5 h. Separated cells should have a density of (5×10^6/ml). During stimulation, cytokine secretion is inhibited by monensin (2.5 μmol/l) (e.g. GolgiStop^{TM}, BD Biosciences, Heidelberg, Germany). Afterwards, the cells are harvested, washed twice in PBS (1–5% FCS) and fixed for exactly 10 min in ice-cold phosphate-buffered 4% PFA (Serva, Heidelberg, Germany). In contrast to the majority of microbial or viral antigens, cytokines and phosphoepitopes are much more sensitive to protein degradation and should therefore be stored no onger than 24 h at 4 °C at this stage! In our experience, only IFN-γ remained stable for up to 48 h. To avoid loss of signal, it is strongly recommended to stain the cells within 24 h after fixation. Permeabilization has to be carried out as described for bacterial and viral antigens in this chapter. For the staining of T-cell cytokines, 5×10^5 permeabilized target cells are simultaneously stained with either anti-cytokine antibody plus anti-CD3 antibody plus anti-CD4 antibody or anti-cytokine antibody plus anti-CD3 antibody plus anti-CD8 antibody. To discriminate between Th1/Tc1 and Th2/Tc2 cells, for example, antibodies against IFN-γ or IL-4 should be used, respectively. Then, IFN-γ+ CD4+ and IFN-γ+ CD8+ or IL-4+ CD4+ and IL-4+ CD8+ cells are analyzed by flow cytometry gating onto CD3+ cells using the FSC or SSC mode against CD3+ PE (FL2) (fig. 2).

For the measurement of intracellular cytokines in monocytes/macrophages (e.g., IL-1β, IL-6, TNF-α, IL-10, IL-12), the cells are stimulated alternatively with LPS (1 μg/ml) or LPS (10 ng/ml) plus IFN-γ (100 U/ml) or heat-killed salmonellae (10^8–10^9 CFU/ml) for 5 h in the presence of monensin (2.5 μmol/l).

Typical staining patterns of cytokines in human peripheral T-cells and the corresponding normal values are shown in figure 3.

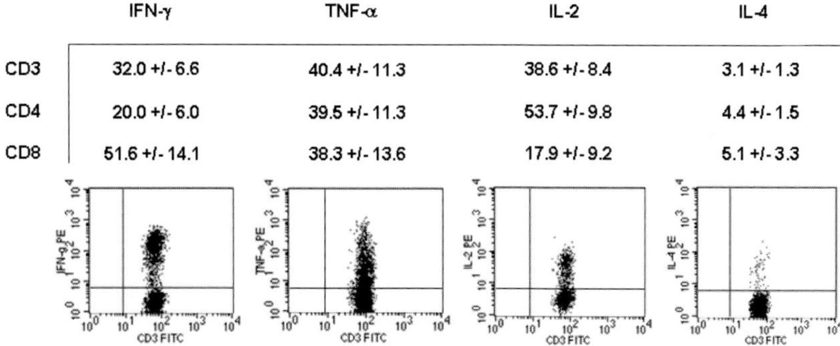

	IFN-γ	TNF-α	IL-2	IL-4
CD3	32.0 +/- 6.6	40.4 +/- 11.3	38.6 +/- 8.4	3.1 +/- 1.3
CD4	20.0 +/- 6.0	39.5 +/- 11.3	53.7 +/- 9.8	4.4 +/- 1.5
CD8	51.6 +/- 14.1	38.3 +/- 13.6	17.9 +/- 9.2	5.1 +/- 3.3

Fig. 3. Normal values ±SD (%) of IFN-γ-, TNF-α-, IL-2-, and IL-4-secreting peripheral human T-cells from healthy adults. The lower panel shows the typical staining patterns of these cytokines.

Intracellular Immunofluorescent Staining of Phosphoproteins in Human Peripheral Blood Mononuclear Cells or Murine Peritoneal Macrophages

Intracellular proteins can be phosphorylated on tyrosine, serine, or threonine residues by a battery of cellular kinases. On the other hand, cellular phosphatases can be activated to mediate dephosphorylation of phosphorylated proteins. In concert, both processes regulate many different cell-signaling pathways. For example phosphorylation of the p91 protein of signal transducer and activator of transcription 1 (Stat-1) in macrophages in response to receptor binding of IFN-γ results in dimerization of this protein and translocation into the nucleus which can be analyzed using p91-specific phosphotyrosine antibodies. Conventionally, this process can be measured by the electrophoretic mobility shift assay (EMSA) and/or Western blotting (WB). The EMSA uses a radioisotope-labeled (^{32}P) or chemiluminescence-labeled oligonucleotide representing the DNA response element in the promoter of corresponding genes (e.g. IFN-γ-inducible genes such as FcγRI) [8]. This oligonucleotide has to be incubated with the nuclear extract of the cells to be analyzed and is separated by polyacrylamide gel electrophoresis. For the identification of the phosphorylated transcription factor, the gel is blotted onto a nitrocellulose or a polyvinylidene fluoride membrane and then incubated with a phosphotyrosine-specific antibody. Alternatively, the phosphorylated transcription factor can directly be detected by a phosphotyrosine-specific antibody by WB if the whole lysate of the cells of interest is separated by SDS-PAGE and than blotted to a membrane. However, both techniques offer some disadvantages: they need at least 10^6 cells, are time-consuming, do not produce quantitative results, and are not conducive to multiparameter analysis. If the ^{32}P technique is used, the handling of radioactive isotopes is an additional disadvantage.

Several years ago, Perez and Nolan [9] introduced a novel approach for the analysis of phosphoproteins using flow cytometry. Flow cytometry requires fewer cells than conventional technology, can be better standardized and is suitable to perform quantitative multiparameter analyses of single cells and distinct cell populations, saving time-consuming cell separation processes.

Here a short protocol for the staining of phosphorylated intracellular Stat-1 in murine peritoneal macrophages which has been adapted from BD Biosciences [10] is presented as an illustration. Mouse peritoneal macrophages were harvested on day 4 after intraperitoneal injection of 1×10^6 attenuated salmonellae. As a positive control, the mice can be injected with recombinant IFN-γ 1 h before harvesting the peritoneal cells (resident or thioglycolate-elicited macrophages). The cells were washed twice with PBS, counted and adjusted to 5×10^6/ml; thereafter, the supernatant was completely removed and the pellet was fixed for exactly 10 min in prewarmed Lyse/Fix Buffer (BD™ Phosflow, Heidelberg, Germany). Other protocols recommend not to spin down the cells, but to add 20 vol of prewarmed 1 × Lyse/Fix buffer. The cells are then centrifuged ($300 \times g$) for 8 min and the supernatant is discarded. The cell pellet is resuspended by vortexing and the cells are permeabilized by adding 1 ml permeabilization buffer (BD™ Phosflow Perm/Wash Buffer I) to the cell pellet for 10 min at room temperature. Subsequently, the cells are spun down and washed once with Perm/Wash Buffer I (BD™ Phosflow; $300 \times g$, 8 min). The supernantant is removed again. The cell pellet is resuspended in 1 ml Stain Buffer (BD Pharmingen™) and 0.3 μg FcBlock (BD Pharmingen™) is added. After 15 min of incubation at room temperature, 100 μl of the cell suspension is transferred into 4-ml polystyrene tubes (BD Falcon™, Heidelberg, Germany) containing the antibody against phosphorylated Stat-1 (pY701-PE; BD™ Phosflow) and an antibody for the phenotyping of macrophages (CD11b-FITC). Alternatively, intracellular salmonellae can be co-stained (see next paragraph). The cells are stained for 30 min at room temperature in the dark followed by a final washing step with 2 ml/tube using Stain Buffer (BD Pharmingen™). After centrifugation, the cells are resuspended in 250 μl of the same buffer and can then be analyzed by flow cytometry. In order to analyze only the macrophage population, the cells are gated on CD11b+ cells.

Intracellular Immunofluorescent Staining of Bacterial or Viral Antigen in Murine Phagocytes and Dendritic Cells ex vivo or in vitro

Intracellular pathogens can be directly stained by specific antibodies following brief and gentle fixation of the target cells in ice-cold phosphate-buffered 4% PFA for exactly 10 min and subsequent permeabilization with saponin (0.1%; Sigma). 10^7 cells should be resuspended in 500 μl PFA. Following fixation, the cells are washed twice in 10 ml cold phosphate-buffered saline (PBS) containing 1–5% fetal calf serum (FCS; Biochrom, Berlin, Germany) and pelleted by centrifugation for 10 min at 1,200 rpm, 4 °C. Depending on the antigen, the cells can be stored as a pellet with

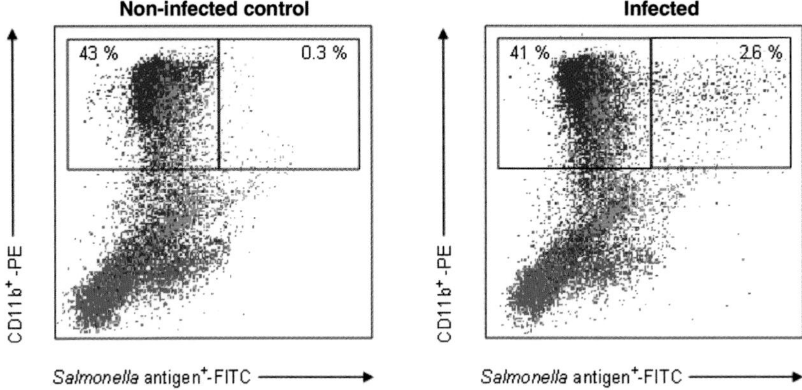

Fig. 4. Intracellular fluorescence staining of salmonella antigen in splenic phagocytes derived from a salmonella-infected BALB/c mouse using a primary polyclonal rabbit antibody directed against *Salmonella enterica* serovar Enteritidis and polyclonal FITC-labeled goat anti-rabbit IgG $F(ab')_2$ as secondary antibody. Simultaneously with the secondary staining, phagocytes were identified with a PE-labeled mouse anti-CD11b mAb.

about 100 μl supernatant under sterile conditions for up to 1 month at this state. Sometimes, this may be of interest for the storage of ready-prepared and fixed standard cells (e.g. a permanently infected cell line) as a positive control. However, the stability of an individual antigen has to be confirmed for any special application. Immediately before staining, the cells should be washed once again. After removal of the supernatant, the pellet is resuspended in 10 ml cold permeabilization buffer (PBS containing 0.1% saponin, 1% FCS, and 0.01 mol/l HEPES from Sigma) and centrifuged for 10 min at 1,200 rpm, 4 °C. This step is repeated twice. Finally, the cells are then adjusted to 5×10^6/ml in the required volume. Then, they are centrifuged again. In order to block nonspecific binding of fluorochrome-conjugated antibodies to FcγRII/III, the cell pellet is resuspended in 100 μl permeabilization buffer supplemented with purified 2.4G2 antibody (anti-CD16/CD32; 1 μg/10^6 cells; Fc Block, BD Pharmingen™) and preincubated for 15 min at room temperature. Then, the cell suspension is again adjusted to the required volume. The adequate antibody combination (e.g. FITC-conjugated anti-salmonella + PE-conjugated anti-CD11b) is given into 4-ml polystyrene tubes (BD Falcon™) following addition of 100 μl cell suspension. The individual antibody concentration should be tested for any application. In many cases, the optimal antibody concentration is 1–5 μg/ml. The cells are stained in permeabilization buffer for 30 min at 4 °C and should then be washed three times in permeabilization buffer to remove unbound antibody. Finally, the stained cells are resuspended in 250 μl fixation buffer containing 1% formaldehyde (CellFix™, BD™ Biosciences) for storage at 4 °C until flow-cytometric analysis.

The result is representatively shown for the staining of *Salmonella* Enteritidis in murine splenic phagocytes (fig. 4). Intracellular salmonella antigen (comprising vi-

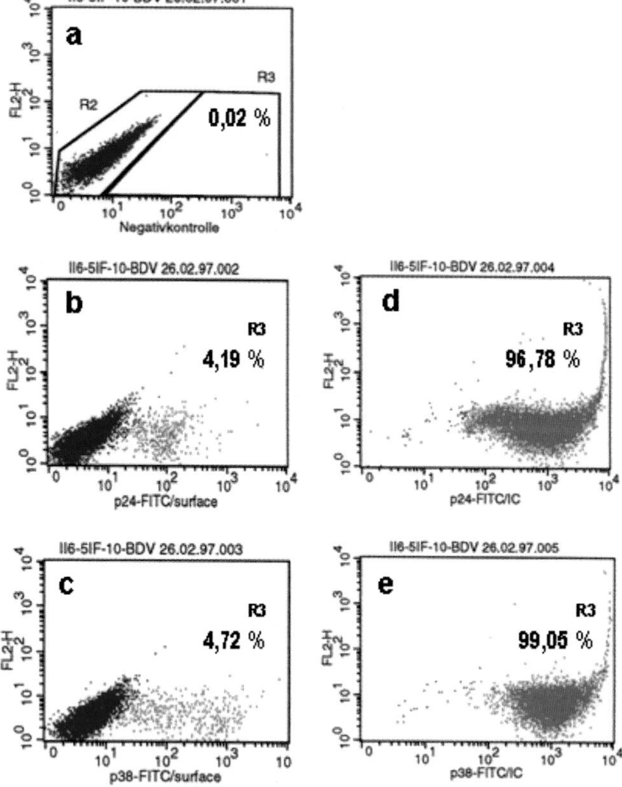

Fig. 5. Alternative surface (**a**–**c**) and intracellular staining (**d, e**) of viral proteins BDV-p24 (**b, d**) and BDV-p38 (**c, e**) in a murine dendritic cell line using primary mouse mAbs against BDV-p24 or BDV-p38 and polyclonal goat anti-mouse FITC as secondary antibody. **a** Negative control was stained with the secondary antibody only.

able and killed/processed intracellular salmonellae) was indirectly fluorescently stained using a polyclonal rabbit anti-*Salmonella* Enteritidis antiserum as primary antibody and an FITC-labeled goat anti-rabbit immunoglobulin $F(ab)_2$ fragment (Jackson Immuno Research, West Grove, PA, USA) as secondary antibody [1]. Phagocytes, including macrophages and granulocytes, were identified by simultaneous staining with the PE-labeled anti-CD11b mAb M1/70.15 (Caltag) and salmonella antigen as the secondary staining. The secondary staining step is carried out as described for the primary staining.

Intracellular viral antigens can be stained according to the same procedure as that described for the intracellular staining of bacterial antigen. In our hands, this protocol was shown to be valid for the staining of borna disease virus (BDV) proteins BDV-p24 or BDV-p38 in immortalized murine dendritic cells infected in vitro (fig. 5) as well as in mononuclear cells derived ex vivo from CSF from horses with borna disease (not shown). Moreover, reproducible results were also obtained using this protocol or a slightly modified protocol for the intracellular staining of pestivirus antigen in peripheral mononuclear cells from cattle suffering from bovine virus diarrhea (Lehmann, J., unpublished) [11] or of HIV-p24 antigen in AIDS patients [12]. The latter report demonstrates that HIV-p24+ CD4+ T-lymphocyte counts are a reliable parameter for monitoring HIV infection.

Acquisition and Analysis

Per sample, 10,000–20,000 cells are measured in the flow cytometer. Signal acquisition should take place within 24 h after fluorescence staining. The measured signals are saved as listmode files. The optimum instrument settings can be obtained by individual staining of the target cells with different labeling variants of a relevant surface marker. For example, in case of threefold staining with FITC/PE/PE-Cy5, the respective surface marker (e.g. anti-CD3) should be available as conjugate with each of the three different fluorochromes (e.g. anti-CD3-FITC, anti-CD3-PE, anti-CD3-PE-Cy5). On the basis of the peripheral blood mononuclear cell (PBMC)setting or another predefined setting, the settings for FSC, SSC, FL1, FL2 and FL3 are adapted for the currently measured sample by modifying voltage, gain and compensation.

The sufficient discrimination of the cells of interest is an essential prerequisite for correct data analysis. For example, to analyze T-cells for intracellular cytokines, the SSC^{low}/CD3+ population has to be gated in the SSC/CD3 plot (fig. 2a). Subsequently, in the FL1/FL2 plot gated on CD3+ T lymphocytes, the cytokine-producing cells are discriminated from the cytokine-negative cells by strict comparison with the isotype control (fig. 2b/c). If further subpopulations have to be analyzed (e.g. CD4+ CD8+), these subpopulations have to be discriminated by appropriate gating (e.g. CD3+ CD4+ or CD3+ CD8+) again followed by comparison with the isotype control while gated on CD4+ or CD8+ T-lymphocytes.

If intracellular antigens are indirectly fluorescence-stained, positive and negative signals should be discriminated using unstained isotype control plus fluorescence-labeled secondary antibody or if an unstained isotype control is not available, at least with the fluorescence-labeled secondary antibody alone. Any positive background signal occurring in the isotype or secondary antibody control has to be subtracted from the positive value detected in the test sample.

Three appropriate quality controls for intracellular protein staining can be suggested: i) staining of the nonpermeabilized cells; ii) competition with an excess of soluble protein of interest simultaneously with intracellular staining, or iii) competition with an excess of unlabeled antibody of the same specificity simultaneously with intracellular staining.

Results

If cell stimulation, fixation, and staining have been carried out properly and the instrument settings during acquisition are optimal, the described method yields very reliable results. As for any other diagnostic method based on antigen-antibody binding, the specificity and the affinity of the antibodies used are crucial also for intracellular antigen staining. The antibodies to be used, in particular noncommercial antibodies, should be very carefully tested for cross-reactivities. The described method is favored if large numbers of cells have to be analyzed and low frequencies of positive cells are expected.

When interpreting the results, it has to be remembered that intracellular cytokine production varies widely in healthy donors. Therefore, normal values are needed from a large number of healthy donors. Figure 3 indicates normal values for some important human cytokines. However, aberrances from normal values should always be evaluated in the context of other diagnostic parameters. A pathological decrease of cytokine-producing T-cells may occur as a result of medication (e.g. analgesic or anti-inflammatory drugs). Alternatively, a significantly decreased cytokine production by peripheral T-cells may also be an indication for innate or acquired immune defects. This may affect all cytokines, but also individual cytokines [13].

When interpreting the results of intracellular pathogen staining, it has to be remembered that not only living intracellular pathogens but also already processed pathogen antigens or, in the special case of a viral infection, single virus proteins produced by the host cell can be detected. Thus, this method may yield a positive result even in the absence of a living pathogen.

Method Protocol

Antibodies, Reagents, and Equipment for Intracellular Cytokine Staining in Human Peripheral T-Lymphocytes or Mouse Leukocytes ex vivo or in vitro

Cells

Heparinized whole blood, PBMCs separated by Ficoll-Paque, erythrocyte-depleted splenocytes or lymph node cells, primary cultured T-lymphocytes or T-cell clones

Cytokine Antibodies

Human

IFN-γ-FITC (Mouse IgG1, Beckman Coulter, Krefeld, Germany)
IL-4-PE (Beckman Coulter)

Mouse

IFN-γ-FITC (Clone XMG1.2, Rat IgG1, BD Pharmingen™, Heidelberg, Germany)
IL-4-PE (Clone 11B11, Rat IgG1, BD Pharmingen™)

Surface Marker Antibodies

Human
CD3-APC (BD Pharmingen ™)
CD4-PE-Cy5 (Mouse IgG1, Beckman Coulter)
CD8-PE-Cy5 (Mouse IgG1, Beckman Coulter)

Mouse
CD3ε-APC (Rat IgG2a, clone KT3, AbD Serotec, Düsseldorf, Germany)
CD4-PE-Cy5 (clone GK1.5, Rat IgG2b, Beckman Coulter)
CD8a-PE-Cy5 (clone 53–6.7, Rat IgG2a, Beckman Coulter)

Isotype Controls
Mouse IgG1-FITC/Mouse IgG1-PE (control for mouse mAb; Beckman Coulter)
Rat IgG1-FITC/-PE (control for rat mAb; clone R3–34, BD Pharmingen™)

Culture Media
RPMI-1640 (10% FCS) (both Biochrom)

Washing Buffer (pH 7.2)
PBS (1% FCS, w/o Ca^{2+}/Mg^{2+}, pH 7.2)

Fixation Buffer
4% phosphate-buffered PFA in PBS (1% FCS, w/o Ca^{2+}/Mg^{2+}), pH 7.4 (stable for 4 weeks) ⇒ prepared from a stock solution of 40% (w/v) PFA in double-distilled water by 1 : 10 dilution in phosphate buffer.

Preparation of the Fixation Buffer
- Prepare phosphate buffer by mixing 83% (v/v) solution A (22.6 g $NaH_2PO_4 \cdot H_2O$ in 1 liter H_2O bidest.) + 17% (v/v) solution B (25.2 g NaOH in 1 liter double distilled water).
- Prepare 40% PFA (solution C) by dissolving 4 g PFA in 10 ml double-distilled water through heating for several hours at 70 °C in an Erlenmeyer flask covered with an aluminum foil within a fume hood (PFA is toxic and of low solubility!).
- Add a few drops of 2 mol/l NaOH until the solution becomes clear and subsequently add 0.54 g glucose.
- Finally, add 90 ml of the mixture (A + B) to 10 ml of solution C, supplement 1% (v/v) fetal calf serum and adjust pH to 7.4–7.6 with 1 mol/l HCl if necessary [4].

Permeabilization Buffer
0.1% saponin in PBS (1% FCS, 10 mmol/l HEPES, pH 7.2) ⇒ freshly prepared from a sterile stock solution of 10% (w/v) saponin (Serva or Sigma) in 1 × PBS (1 mol/l HEPES) by 1 : 100 dilution in 1 × PBS (1% FCS), keep sterile at 4 °C!

Preparation and Storage of the Permeabilization Buffer Stock Solution
- Prepare a 10% (w/v) saponin stock solution by disolving 50 g saponin (Serva or Sigma) in 500 ml PBS (w/o Ca^{2+}/Mg^{2+}, pH 7.2, 1 mol/l HEPES) with using a magnetic stirrer until the saponin powder is completely dissolved.
- Sterilize the saponin stock solution by filtrating through a 0.2 μm filter.
- Sterile saponin stock solution should be stored at −20 °C.

Stimulators
- PMA (1 μg/ml) in RPMI-1640 (10% FCS) ⇒ freshly prepared from a stock solution of (100 μg/ml) PMA (Sigma) in absolute ethanol (−20 °C).
- Ionomycin (100 μmol/l) in RPMI-1640 (10% FCS) ⇒ freshly prepared from a stock soution of (10 mmol/l) ionomycin (Sigma) in absolute ethanol (−20 °C).

Golgi Transport Inhibitor
Monensin (250 μmol/l) ⇒ freshly prepared from a stock solution of (10 mmol/l) monensin (Sigma or BD) in absolute ethanol (−20°C).

Carrier Fluid for the Flow Cytometer
FACS flow™ (BD™ Biosciences).

Fixation and Storage Buffer for Ready Prepared Samples
CellFix™ (BD™ Biosciences)

Disposable Material
- 24-well tissue culture plate (BD Falcon™)
- 1.5-ml reaction tube (Eppendorf)
- 4 ml tubes 12 × 75 mm (polystyrene, BD Falcon™)
- Combitips + adapter for reservoir pipette (Multipette™, Eppendorf)
- Pipette tips for multichannel pipette (Varipette™, Eppendorf)

Instruments
- Flow cytometer (FACSCalibur™, BD™ Biosciences or FACS Canto™, BD™ Biosciences or Cytomics FC500, Beckman Coulter)
- CO_2 Incubator (37 °C, 5% CO_2, Kendro)
- Centrifuge (4 °C, swing-out rotor, used RCF 350 *g*, Megafuge™, Kendro/Thermo)
- Vortex mixer.

Work Steps for the Discrimination of Human or Murine Th0/Tc0, Th1/Tc2, and Th2/Tc2 Cells Using Intracellular Immunofluorescent Staining of IFN-γ and IL-4

Cell Stimulation

Unseparated Human Peripheral Blood T-Cells

- Collect venous blood using a heparinized vacutainer (Monovette^{TM}, Sarstedt, Nümbrecht, Germany).
- Dilute 100 μl heparinized blood in 900 μl cell culture medium RPMI-1640 in each of 2 wells of a 24-well tissue culture plate (Cellstar^{TM}, Greiner Bio-one, Frickenhausen, Germany).

Separated Human Peripheral Blood T-Cells

- PBMCs are separated with Ficoll-Paque^{TM} (GE Healthcare) according to the manufacturer's instructions.
- PBMC fraction is harvested from the interphase and washed 3 times in RPMI-1640 cell culture medium.
- Residual red blood cells can be depleted through hypotonic lysis using ammonium chloride buffer (0.15 mol/l NH_4Cl; 0.1 mmol/l EDTA disodium salt dihydrate; 10 mmol/l $NaHCO_3$ or a commercial product such as Lysing Reagent Orthomune®, Ortho-Clinical Diagnostics, Neckargemünd, Germany).
- Pipette 1 ml cell suspension in culture medium (5×10^6/ml) in each of 2 wells of a 24-well tissue culture plate (Greiner).

Mouse Splenic T-Cells

- Sacrifice the mouse and isolate the whole spleen aseptically.
- Cut the spleen tissue into small pieces and isolate splenocytes by mincing and passing the tissue through a sieve of 100 μm mesh size (Cellstrainer^{TM}, BD Falcon^{TM}).
- Red blood cells can be depleted by density gradient (Ficoll-Paque^{TM} or Percoll^{TM}, both GE Healthcare) or simply by hypotonic lysis using ammonium chloride buffer (0.15 mol/l NH_4Cl; 0.1 mmol/l EDTA disodium salt dihydrate; 10 mmol/l $NaHCO_3$ or a commercial product such as Lysing Reagent Orthomune®, Ortho-Clinical Diagnostics).
- Pipette 1 ml cell suspension in culture medium (5×10^6/ml) in each of 2 wells of a 24-well tissue culture plate (Greiner).
- To 1 of 2 wells (stimulated cells) add:
  - 10 μl PMA (1 μg/ml) ⇒ final concentration: (10 ng/ml)
  - 10 μl ionomycin (100 μmol/l) ⇒ final concentration: (1 μmol/l)
  - 10 μl monensin (250 μmol/l) ⇒ final concentration: (2.5 μmol/l).
- Vortex briefly to mix and incubate the cells for 5 h in 5% CO_2 at 37 °C.

Fixation

- Resuspend and transfer cells into Falcon tubes.
- Wash cells 3 times in washing buffer (3 ml/wash).
- Pellet by centrifugation at $350 \times g$ (= 1,200 rpm in a Megafuge™, Kendro/Thermo) at 4 °C for 5 min.
- Resuspend the sedimented cells in 500 µl cold fixation buffer.
- Incubate for 10 min at 4 °C.
- Add 4 ml cold washing buffer immediately after incubation. Spin for 5 min at $350 \times g$.
- Wash cells twice (3 ml/wash, $350 \times g$, 4 °C, 5 min). Aspirate supernatant.
- At this stage, cells can be stored up to 24 h at 4 °C.

Multicolor Staining for Intracellular Cytokines and Cell Surface Markers

- Resuspend cells and wash twice in permeabilization buffer (at room temperature! 2 ml/wash). Incubate the cells with the permeabilization buffer for 5 min before spinning down (4 min at $350 \times g$). Aspirate supernatant.
- Resuspend the cell pellet in 200 µl permeabilization buffer.
- Add 10 µl anti-mouse CD16/CD32 (Fc Block™, BD Pharmingen™) to the cell suspension. Mix well (only necessary for mouse splenocytes).
- Incubate at room temperature for 15 min.
- During the incubation period, pipette the following antibody combinations into fresh 4-ml Falcon tubes:

Sample No.	Staining (cytokine mAb/surface marker mAb)
Unstimulated cells	
1	IgG1-FITC/-IgG1-PE/anti-CD4-PE-Cy5/anti-CD3-APC
2	IgG1-FITC/IgG1-PE/anti-CD8-Cy5/anti-CD3-APC
3	anti-IFN-γ-FITC/anti-IL-4-PE/anti-CD4-PE-Cy5/anti-CD3-APC
4	anti-IFN-γ-FITC/anti-IL4-PE/anti-CD8-PE-Cy5/anti-CD3-APC
Stimulated cells	
5–8	as for 1–4

- If there is more than one cell preparation (e.g. cells from different animals or donors), samples No. 1–8 have to be prepared separately for each cell preparation; the number of samples is thus multiplied by the number of cell preparations.
- Pipette 2.5–5 µl of the individual antibody per tube (the antibody optimum has to be determined for special applications and individual antibodies!).
- Fill the cell suspension up to 400 µl with permeabilization buffer, resuspend. Add 100 µl of cell suspension to every tube with antibodies ⇒ 1 : 40–1 : 20 dilu-

tion of antibody in permeabilization buffer (if more than one well of a 24-well plate has been stimulated for cytokine production per patient or animal, the cells from different wells should be pooled before addition to the antibodies).

- Vortex well.
- Incubate at 4 °C for 30 min in the dark.
- Wash 3 times with cold permeabilization buffer (3 ml/wash). Aspirate supernatant.
- Resuspend cells in 250 μl 1% formaldehyde in PBS (1% FCS) *or* CellFix™ (BD). The fixed cells should be stored at 4 °C in the dark until analysis.

Flow-Cytometric Analysis

- Acquisition and analysis on a flow cytometer (e.g. FACSCalibur™ or FACS Canto™, BD™ Biosciences) using the corresponding software.
- Analysis of samples 3/4 and 7/8 enables discrimination between Th1/Tc1 (IFN-γ+ IL-4−), Th2/Tc2 (IFN-γ^{-}+ IL-4+), and Th0/Tc0 cells (IFN-γ+ IL-4+), while gating onto CD3+/CD4+ or CD3+/CD8+ cells.

Troubleshooting

Several subjective failures can significantly influence the staining result. A negative result may have several causes. First of all, reagents and protocol for cell stimulation should be checked. Correct concentrations of the stimulators and protein transport inhibitor are essential for an optimal result. In some cases, it may be of advantage to modify the stimulation time. For example, T-cells should be stimulated only for 4 h with PMA/ionomycin. If antigen stimulation is required, the optimal stimulation time should be determined for the individual protocol. A negative result may also occur if EDTA or citrate blood was used instead of heparinized blood since Ca^{2+} ions are bound by EDTA and citrate, thus activation via the Ca^{2+} ionophore ionomycin is inhibited. Negative results may also be due to insufficient permeabilization. In our hands, saponins from different sources differ significantly in terms of cell permeabilization. It may be recommendable to test different saponins and/or different concentrations for a certain application. Last but not least, the fixation step is critical. It is recommended to strictly follow the conditions (PFA concentration, temperature, and time) listed in the above protocol.

False-positive results are often caused by too high concentrations of antibodies or biotin-fluorochrome conjugates. The optimal concentration of each reagent should be titrated for each application. Another reason may also be that the washing-out steps after antibody staining were done with normal washing buffer instead of saponin buffer as required.

In order to check all conditions for intracellular protein staining before starting the experiment, it is recommended to use appropriate control systems to verify the stimulation protocol (e.g. by RT-PCR) or to verify the staining procedure (e.g. using a cytokine-transfected cell line).

The stimulation of CD4+ T-cells with PMA leads to distinct downregulation of the CD4 molecule. Therefore, the relatively weak CD4 fluorescence signal on T-lymphocytes after PMA stimulation is not caused by a subjective failure.

Lead Time

The time needed significantly depends on the number of samples and on the staining method – direct versus indirect. Stimulation and fixation take about 6.5 h. The simple direct intracellular staining protocol takes about 1.5 h without cell preparation. The simple indirect intracellular staining protocol (e.g. intracellular salmonella staining) takes about 2.5 h without cell preparation. These time estimations apply to up to 10 samples. If more than 10 samples have to be stained, the time requirements increase proportionally to the number of samples.

The lead time for the acquisition and analysis also depends on the number of samples. Per sample, 10–15 min should be allowed.

Summary

During the past decade, flow cytometry has entered almost all fields of biomedical research and diagnostics. Whereas flow cytometry was originally strongly focused on the immunophenotyping of cells from diverse sources, more recent applications demonstrate that flow cytometry is also a useful tool to address a broad range of functional questions. Among these are the measurement of the rate of phagocytosis, the mobilization of intracellular Ca^{2+} or the induction of the respiratory burst. Moreover, the flow-cytometric identification of a certain cytokine profile on the single-cell level has significantly contributed to the better understanding of T-cell regulation processes and has promoted the Th1/Th2 concept.

More recently, the identification of posttranslational protein modifications such as phosphorylation has given flow cytometry access to the field of proteomics on the single-cell level. Last but not least, the detection of intracellular pathogens by flow cytometry belongs to the palette of procedures that characterize the functional state of a certain cell type.

In this chapter, we introduce the reader to the latter three procedures and recommend representative protocols for their practical application.

References

1 Lehmann J, Bellmann S, Werner, C, Schröder, R, Schütze N, Alber G: IL-12p40-dependent agonistic effects on the development of protective innate and adaptive immunity against *Salmonella enteritidis.* J Immunol 2001;167:5304–5315.

2 Zimmermann T, Kunisch E, Pfeiffer R, Hirth A, Stahl HD, Sack U, Laube A, Liesaus E, Roth A, Palombo-Kinne E, Emmrich F, Kinne RW: Isolation and characterization of rheumatoid arthritis synovial fibroblasts from primary culture – primary culture cells markedly differ from fourth-passage cells. Arthritis Res 2001;3:72–76.

3 Lehmann J, Jüngel A, Lehmann I, Busse F, Biskop M, Saalbach A, Emmrich F, Sack U: Grafting of fibroblasts isolated from the synovial membrane of rheumatoid arthritis (RA) patients induces chronic arthritis in SCID mice – a novel model for studying the arthritogenic role of RA fibroblasts in vivo. J Autoimmun 2000;15:301–313.

4 Sander B, Andersson J, Andersson U: Assessment of cytokines by immunofluorescence and the paraformaldehyde-saponin procedure. Immunol Rev 1991;119:65–93.

5 Sander B, Hoiden I, Andersson U, Möller E, Abrams JS: Similar frequencies and kinetics of cytokine producing cells in murine peripheral blood and spleen. Cytokine detection by immunoassay and intracellular immunostaining. J Immunol Methods 1993;166:201–214.

6 Jung T, Schauer U, Heusser C, Neumann C, Rieger C: Detection of intracellular cytokines by flow cytometry. J Immunol Methods 1993;159:197–207.

7 Picker LJ, Singh MK, Zdraveski Z, Treer JR, Waldrop SL, Bergstresser PR, Maino VC: Direct demonstration of cytokine synthesis heterogeneity among human memory/effector T cells by flow cytometry. Blood 1995;86:1408–1419.

8 Lehmann J, Seegert D, Strehlow I, Schindler C, Lohmann-Matthes M-L, Decker T: IL-10-induced factors belonging to the p91 family of proteins bind to interferon-γ-responsive promoter elements. J Immunol 1994;153:165–172.

9 Perez OD, Nolan GP: Simultaneous measurement of multiple active kinase states using polychromatic flow cytometry. Nat Biotechnol 2002;20:155–162.

10 BD Biosciences: Phosflow protocols for mouse cells. Protocol I (Detergent Method); in: Techniques for Phospho Protein Analysis. Franklin Lakes, BD Biosciences, 2005, p 31. *www.bdbiosciences.com/pdfs/manuals/05-790030-3A1.pdf.*

11 Qvist P, Aasted B, Bloch B, Meyling A, Rønsholt L, Houe H: Flow cytometric detection of bovine viral diarrhea virus in peripheral blood leukocytes of persistently infected cattle. Can J Vet Res 1990;54:469–472.

12 Holzer TJ, Heynen CA, Novak RM, Pitrak DL, Dawson GJ: Frequency of cells positive for HIV-1 p24 antigen assessed by flow cytometry. AIDS 1993;7(suppl 2):3–5.

13 Lehmann I, Borte M, Sack U: Diagnosis of immunodeficiencies. J Lab Med 2001;25:495–511.

Sack U, Tárnok A, Rothe G (eds): Cellular Diagnostics. Basics, Methods and Clinical Applications of Flow Cytometry. Basel, Karger, 2009, pp 476–502

Detection of Antigen-Specific T-Cells using Major Histocompatibility Complex Multimers or Functional Parameters

Alexander Scheffold[a] · Dirk H. Busch[b] · Florian Kern[c]

[a] Miltonyi Biotex GmbH, Bergisch Gladbach,
[b] Institute for Medical Microbiology, Immunology and Hygiene, Technical University Munich, Munich, Germany
[c] Brighton and Sussex Medical School, University of Sussex, Brighton, UK

Introduction

In order to be able to respond to different antigens in a specific way, T- and B-lymphocytes of the immune system carry specific receptors on their surface. Each lymphocyte has receptors for only one single antigenic determinant (epitope). Only cells that are specific for an antigen eventually determine the response against it as, for example, after infection or vaccination. Most antigens will have several or many determinants that may be recognized by antigen receptors.

In order to be able to assess the adaptive immune response of an individual towards an antigen of interest, it is important to selectively determine the characteristics of T-cells and B-cells responding to that particular antigen. This chapter only covers T-cells.

Until a few years ago, we only disposed of indirect methods to measure antigen-specific T-cells. These measurements were relatively imprecise, for example they determined the presence of antigen-specific T-cells by measuring their effector functions such as cytokines secreted into the supernatant (ELISA) or proliferation (^{3}H-thymidine incorporation). All of these measurements were thus accomplished at a cell population level. Over the past 10 years or so, several new methods were developed that allow the detection of antigen-specific T-cells using flow cytometry and, therefore, permit a very detailed description of each single detected cell. These methods are either based on direct binding of a labeled antigen by the cells of interest, for example by using major histocompatibility complex (MHC) multimers, or by their function, e.g., cytokine production, CD40 ligand (CD40L) upregulation, or degranulation [1, 2] following antigen-specific activation. Using these methods, the frequencies as well as surface or functional markers of antigen-specific T-cells can be assessed.

In the present chapter we will discuss the use of flow-cytometric methods for the detection of intracellular cytokines (see 'Flow-Cytometric Analysis of Intracellularly Stained Cytokines, Phosphoproteins, and Microbial Antigens', pp. 459), the analysis of proliferation, e.g. by bromodeoxyuridine (BrdU) uptake or carboxyfluorescein succinimidyl ester (CFSE) dilution, and the use of MHC multimers in order to identify antigen-specific T-cells.

Background

Significance of Antigen-Specific T-Cells

T-cells are the central regulators and effectors of the adaptive immune response. The response against a specific antigen is usually determined by a small number of antigen-specific T-cells. Their frequency and their functional capacity, therefore, are very important parameters for determining the specific immune status of an individual, for example after vaccination against a specific pathogen, but also in the event of undesired immune responses such as allergies or autoimmunity [3]. One of the main problems in detecting and characterizing antigen-specific T-cells is their low frequency. In the absence of acute infection, the frequencies of these T-cells are normally in the range of 0.01–1%. If frequencies are even lower than that, antigen-specific T-cells can still be detected in some cases by combination with enrichment technologies [4–7].

Methods of Detection

Major Histocompatibility Complex/Human Leukocyte Antigen Multimers

The most direct and most rapid method to detect antigen-specific lymphocytes is labeling them with fluorescent antigen. This approach has been known for a long time for antigen-specific B-cells, whose antigen receptors have high affinity and therefore allow for stable binding of the antigen [8]. Because of the low binding affinity of the T-cell antigen receptor for its ligand, a complex of MHC/human leukocyte antigen (MHC/HLA) binding the antigenic peptide, a parallel approach for T-cells had been hampered for a long time. This problem was not solved until 1996 by the introduction of so-called MHC multimers [9]. The basic principle behind this method lies in the multimerization of MHC/peptide complexes to increase the relative binding avidity of the reagent to surface T-cell receptors (TCRs) to a degree that allows epitope-specific binding to T-cells. MHC multimer reagents conjugated with fluorochromes can be used for identification of T-cells by flow cytometry, with staining characterized by high specificity and sensitivity [10]. The most successful system for ex vivo T-cell staining has been the multimerization of MHC/peptide complexes to tetrameric molecules by specifically biotinylating a sequence tag fused to the C-terminus of the MHC heavy chain, followed by oligomerization of the MHCs with streptavidin, which has four biotin binding sites [9, 11]. The increase in relative binding avidity to specific TCRs achieved by MHC multimerization is sufficient to enable detection of antigen-specific T-cells within a wide range of physiological binding strengths for their ligand. Even 'low-affinity' T-cells can be stained in many cases with MHC tetramers, and several studies have demonstrated that MHC tetramer staining allows identification of clearly more than 95% of a given epitope-specific cell population [11]. Dimerization of MHC molecules, usually generated as immunoglobulin fusion proteins, also allows epitope-specific T-cell staining, but 'low-affinity' T-cells might not be detected as well as with MHC tetramers [12]. The MHC multimer has revolutionized T-cell research over the last years. A large repertoire of MHC/HLA class I multimer reagents is currently available, and most of the reagents can be purchased from commercial providers. MHC/HLA multimers have been used for extensive phenotypic characterizations of antigen-specific T-cell populations in both animal models and humans. In order to further simplify the generation of a broad spectrum of multimer reagents, a peptide exchange technology has recently been developed [13]. These multimers can be charged with any MHC/HLA binding peptide in small-scale applications. This procedure not only guarantees easier accessibility of MHC/HLA multimers for a given antigen, the exchange technology has also proven to be useful for very rapid epitope mapping approaches [14]. In addition to flow cytometry applications, some first reports indicate that MHC/HLA class I multimers might also be suitable for in situ detection of antigen-reactive T-cells [15]. MHC/HLA class II multimer reagents are more difficult to generate, and in many cases the epitope sequences have not been precisely determined. However, several recent studies have demonstrated that the general approach of MHC multimer staining can be used effectively for the detection of T-helper cells [16, 17].

As long as MHC multimer staining is performed at temperatures below approximately 10 °C, T-cells can be identified and purified without altering their original phenotype. However, since MHC multimer reagents represent the natural ligand bound to the TCR, placement of purified MHC multimer-stained T-cell populations into an in vitro cell culture results in functional alterations such as TCR internalization, activation, overstimulation, and cell death [18–21]. This intrinsic shortcoming of conventional MHC multimer staining substantially limits the applications of the technology for further analysis of ex-vivo-purified T-cells as well as for clinical medicine. To address the problem, a modified MHC multimer technology has been developed [21]; so-called 'MHC streptamers' allow removal of surface-bound MHC multimer reagents after cell staining and purification, conserving the phenotypical and functional status of isolated cell populations. This approach might further broaden the applications of MHC multimer technologies for ex vivo T-cell analysis and clinical applications.

MHC/HLA multimer staining is currently the only available method to detect epitope-specific T-cell populations by function-independent surface staining (when performed at low temperatures), and the combination of multimer staining with polychromatic flow cytometry allows to determine the true in/ex vivo phenotype of antigen-specific T-cells. A major limitation of this approach is that only T-cells with selected epitope specificities can be studied and that precise knowledge of an individual's MHC/HLA haplotype is required to select applicable reagents.

Function-Dependent Detection Methods

Since T-cells cover a wide range of different (effector) functions, information on their frequencies and phenotypes alone is probably of limited significance [3]. Additional parameters are needed that describe the functional profile of individual T-cells and the overall 'flavor' of the response [22]. The most important T-cell (effector) functions include proliferation (proliferation leads to the increase of the number of specific cells, e.g. during infection), production of cytokines and costimulatory molecules like CD154 (these mediators regulate the function and recruitment of other cells of the immune system to the site of the immune response), and cytotoxicity (this may for example eliminate infected cells in order to prevent further multiplication of the infectious agent). One or the other function may be more important depending on the situation.

Generally, these functions are only transiently switched on following contact with the specific antigen in order to limit the response in time and space (for example to the site of infection). Therefore, the detection of these functions depends on prior stimulation with specific antigen. The flow-cytometric detection of cytotoxic activity by staining for cytotoxic effector molecules (for example granzyme or perforin) is also possible. In contrast to most other mediators, these molecules are found preformed in the cells and following antigen stimulation can be released immediately. The assessment of antigen-specific cytotoxic activity, as a result, is only possible by determining the number of granzyme-secreting cells, e.g. using the secretion assay (Miltenyi-Biotec) [23]. An alternative approach to measuring cytotoxicity is the so-called CD107 mobilization assay, which measures degranulation. CD107a and b, or Lamp1 and Lamp2 [2] form part of the membranes of cytotoxic granules and can be found transiently on the cell surface during granule exocytosis (which is why the staining antibody must be present during T-cell stimulation).

Antigen-Specific T-Cell Stimulation

There are differences between the methods used for stimulation of CD4 and CD8 T-cells (table 1). CD4 T-cells can be stimulated by antigens that are presented via the exogenous pathway of antigen presentation on class II MHC molecules [24]. As a result, CD4 T-cells can be stimulated by the simple addition of protein antigens to a

Table 1. The principal possibilities of stimulating CD4 and CD8 T-cells

	CD4 T-cells	CD8 T-cells
Cells infected or transfected with an infectious agent/proteins of interest (e.g. CMV-infected fibroblasts)	not suitable because of predominant class-I MHC presentation of the relevant antigens	possible, depends on protein expression and MHC expression
Complete pathogens (bacteria, viruses)	suitable[a]	low efficiency
Complex protein mixes (e.g. pathogen lysates)	suitable[a]	low efficiency
Recombinant proteins	suitable[a]	low efficiency
Long polypeptides (>30 amino acids)	moderate to high efficiency, APC may be required	low efficiency
Peptide of 15–30 amino acids	usually very efficient	mostly low efficiency
Peptides of 12–15 amino acids	usually very efficient (unless epitopes are longer)	reasonably efficient, depends on peptide
Peptides < 12 amino acids	reasonable or high efficiency	usually good
Peptides of 9–10 amino acids	potentially low efficiency (depends on epitope length)	generally optimum peptide length

[a] Given that a sufficient number of antigen-presenting cells is present in the cell suspension (e.g. PBMCs).

cell suspension containing these T-cells (whole blood, peripheral blood mononuclear cells (PBMCs)). These protein antigens may include single (also recombinant) proteins, pathogen lysates or complete viruses [25]. Stimulation of CD8 T-cells with complete protein antigens by contrast is not reliable because the effective presentation of exogenous antigens on class I MHC depends on whether the antigen-presenting cells in the assay can efficiently 'cross present' (i.e. channel exogenous antigens into the endogenous pathway of antigen presentation). This is the case in only a minority of individuals (approximately 20%), so that no response or only a much reduced frequency of antigen-specific CD8 T-cells is found [26]. The stimulation of CD8 T-cells using infected fibroblasts (or other suitable cells) is complicated and time-consuming (HLA match) but equally unreliable. In contrast to this, externally added antigenic peptides generally induce stimulation very effectively [27]. Some authors additionally use costimulatory antibodies binding CD28 or CD49d in order to maximize stimulation [28]. Experiments using stimulation with peptides in combination with MHC multimer staining indicate that only MHC multimer-positive cells can be stimulated to produce cytokines using the cognate peptide [29–31].

The use of peptides is highly flexible since they can be used individually or in pools, such pools being able to cover complete protein amino acid sequences (protein spanning peptide pools) [32, 33]. If the overlap between neighboring peptides is chosen to be sufficiently large so that all epitopes that can possibly exist in a given protein sequence will be considered. Since the length of the peptide binding groove is generally 9 amino acids for both class I MHC and class II MHC molecules a stretch of amino acids binding to a particular MHC binding groove cannot be missed if the overlap between neighboring peptides is chosen to be 8 amino acids [34]. However, since the amino acids flanking the area of the peptide binding to the MHC seem to play an important role for the binding of the peptides to the TCR (i.e. recognition by T-cells), in particular for CD4 T-cells, it is important to make the overlap slightly bigger [35, 36]. It is impossible to accommodate all variations of peptide length in such a peptide scan so that a compromise has to be made. Of interest, some peptide manufacturers offer scans containing peptides of 9, 10, and 11 amino acids in length. In our hands the use of peptides of 15 amino acids length and 11 overlaps has proven very successful. Peptide pools composed of such peptides stimulate CD4 T-cells at least as effectively as recombinant proteins and moreover have the advantage of stimulating CD8 T-cells at the same time [33, 37]. At present, such peptide scans seem to be the best method to measure T-cell responses against antigens at the whole-protein level [30].

It may appear that the use of peptides of 15 amino acids length for stimulating CD8 T-cells is in conflict with the concept that the binding groove of class I MHC molecules can only accommodate a peptide of 9 amino acids in length. Given some variability of how the peptide can bind in the binding groves, class I MHC binding peptides are typically 9–11 amino acids long. However, since approaches using longer peptides are indeed very successful, it must be assumed that mechanisms exist that shorten these peptides in the extracellular space (clipping or trimming) [38–40]. It is not known for sure if such peptides can be taken up into the cells so that intracellular peptidases may contribute to shortening them, but this possibility has certainly not been ruled out. On the other hand, it has been shown that extracellular clipping occurs.

Types of Cell Suspensions

Cell suspensions that are suitable for staining with MHC multimers or for intracellular cytokine staining can be natural cell suspensions (for example anticoagulated blood, cerebrospinal fluid, ascites or joint aspirates) or artificial ones (cell preparations after gradient centrifugation, manipulated by culture or other measures). It is

Table 2. Comparison of MHC multimer technology and intracellular cytokine staining (ICS)

	ICS	MHC multimers
Identification of antigen-specific cells	depends on functionality	independent of functionality
CD8 T-cells	yes	many reagents available
CD4 T-cells	yes	few reagents available
Precise knowledge of recognized peptide and presenting MHC	not required	mandatory
Parallel phenotyping	yes	yes
Sorting	yes, live sorting possible[a]	yes, live sorting possible

[a] Using e.g. MHC streptamers [24].

not always easy to decide whether whole blood or PBMCs should be used for such experiments. Because of the simplicity of the approach, the use of whole blood appears to be the most straightforward solution [41]. However, many components of whole blood are likely to influence the uptake of antigen and their presentation by antigen-presenting cells. Proteins may adsorb free antigen before it can be taken up by antigen-presenting cells or loaded on to MHC molecules. There may be sufficient proteolytic activity to destroy complex antigens or peptides. As a result, higher concentrations of antigen are usually necessary for stimulating T-cells in whole blood, and likewise, higher concentrations of antibody are needed for staining. A direct comparison between dose-response curves has shown that even when antigen concentrations are chosen to be much higher in whole blood, the same degree of T-cell activation as in PBMCs (in percent of the reference population) is frequently not achieved [42].

The characteristics of MHC multimer technology and intracellular cytokine staining are summarized in table 2.

Protocol for Human Leukocyte Antigen Multimer Staining Combined with Multiparameter Phenotyping of Effector and Memory T-Cell Subsets

Patient/Donor Samples

PMBCs are isolated by conventional density gradient centrifugation (e.g. Ficoll) using heparinized whole blood (lithium heparinate or sodium heparinate) or citrated blood; $1-2 \times 10^6$ cells per staining are commonly needed. Also PBMCs derived from frozen samples are commonly used for MHC multimer stainings.

Reagents

This particular staining panel was designed for analysis on a Cyan-XDP flow cytometer.

- Phycoerythrin (PE)-conjugated MHC multimers (self-made or purchased from commercial providers; appropriate reagents have to be selected according to the antigen/epitope target and HLA haplotype of analyzed individuals)
- FACS staining buffer (phosphate-buffered saline, PBS, 0.5% bovine serum albumin, BSA)
- Propidium iodide (PI; stock solution 2 μg/ml in PBS)
- Monoclonal antibody-fluorochrome conjugates:
  - anti-CCR7 FITC
  - anti-CD19-PE-A610
  - anti-CD3-Pacific Blue
  - anti-CD45RA-PE-Cy7
  - anti-CD38-APC
  - anti-CD8-AmCyan.

Staining Procedure

Keep cells permanently cooled on ice to prevent T-cell activation.

- Transfer 2×10^6 PBMCs per well of a U-bottom 96-well microtiter plate; in order to allow correct compensation of the multicolor sample, single-color controls for each fluorochrome should be included; for the initial setup of the staining panel, fluorescence-minus-one (FMO) controls are required.
- Pellet cells by centrifugation ($490 \times g$, 2 min, 4 °C).
- Resuspend cells by carefully pipetting up and down in 25 μl MHC multimer PE (reagents have to be titrated like all other antibodies to determine optimal staining concentrations); incubate on ice in the dark.
- 20 min later, add antibody mix to the MHC multimer-containing samples and single-color controls.
- Incubate for an additional 20 min on ice in the dark.
- Add 150 μl FACS buffer and spin down cells ($490 \times g$, 2 min, 4 °C).
- Wash 2× in 200 μl FACS buffer.
- Finally, resuspend cells in 200 μl FACS buffer and transfer in FACS tube.
- Add 200 μl PI solution (1:100 dilution from stock) immediately before flow cytometry acquisition.

A typical result of HLA multimer staining is shown in figure 1.

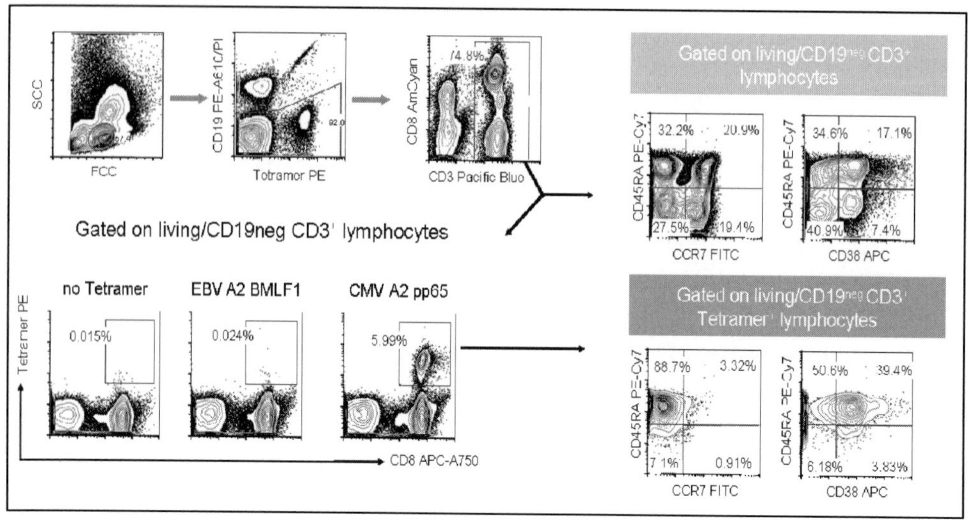

Fig. 1. Example for a phenotypical analysis of A2/CMV-pp65-specific CD8+ T-cells.

Aspects of Specific Relevance for Stainings with MHC Multimers

Controls

There is currently no consensus in the field what kind of staining controls should be added to each MHC multimer analysis to ensure correct interpretation of the staining results. Since some MHC/HLA multimers can be unstable and lose activity over time, a positive control should be included to document optimal performance of the reagents. For this purpose, T-cell lines or PBMC samples from known positive donors can be used. Especially when the frequencies of detected MHC multimer-positive T-cells are very low or when the staining intensities are dim, negative controls become crucial. HLA mismatch MHC multimers are sometimes used since the frequency of alloreactive T-cells stably binding to allo-MHC molecules is usually below the detection limit. Another procedure is to combine MHC multimer staining with markers for naïve and memory T-cells. Unspecific 'background' staining usually stains a population which contains large amounts of naïve T-cells whereas truly antigen-experienced T-cells usually exclude this subset.

Live/Dead Discrimination

MHC multimers – like most antibodies – tend to bind nonspecifically to dead cells. Especially when the frequencies of antigen-specific T-cells are very low, reliable staining results require the exclusion of dead cells. This can, for example, be achieved by PI staining (as it was used in the staining example) or by dyes that can even be combined with cell fixation (e.g. ethidium monoazide).

Fixation
If the target population consists of low-affinity T-cells, MHC multimers can dissociate upon staining. Especially when there is a longer time gap between staining and data acquisition, this phenomenon can result in false-negative results. In such cases, sample fixation (e.g. by paraformaldehyde; PFA) shortly after cell staining can prevent this problem.

Combination with Functional Assays
In vitro antigen stimulation, which is usually required to perform intracellular cytokine staining or CD107 expression analysis, can lead to strong downregulation of surface TCR, which might negatively interfere with MHC multimer staining. There are some reports showing that prestaining with MHC multimer prior to addition of antigen to cell cultures can partially circumvent this problem. However, it needs to be emphasized that extensive controls are needed for the correct interpretation of results derived from such a combination of interfering methods.

Protocol for Antigen-Specific Cytokine Production

Patient/Donor Samples

Whole-blood method: heparinized whole blood, 0.5–1 ml per staining panel (note: does not work with EDTA or citrated blood).

PMBC method: heparinized whole blood (lithium heparinate or sodium heparinate) or citrated blood, $1-2 \times 10^6$ cells per staining panel.

Reagents

Reagents mentioned here are those used in our laboratory. This is not an advertisement for any manufacturer.

Media/Buffers and Additives

- Sterile PBS (e.g. Gibco BRL, Berlin, Germany)
- VLE RPMI-1640 (Biochrom, Berlin, Germany)
- fetal calf serum (FCS; Biochrom)
- glutamine (L-glutamine; Biochrom)
- BSA (Serva, Heidelberg, Germany).

Chemicals

- Brefeldin A (BFA; Sigma, Munich, Germany)
- EDTA (Na_2EDTA; Sigma)

- sodium azide (NaN_3; Serva,)
  This substance is extremely toxic! Pertinent guidelines for handling and storage must be observed.
- PFA (Merck, Darmstadt, Germany)
- BD lysing solution (BD Biosciences, Heidelberg, Germany)
- permeabilizing solution (BD Biosciences)
- *Staphylococcus* enterotoxin B (SEB; Sigma)
  This substance is extremely toxic! Pertinent guidelines for handling and storage must be observed.

Monoclonal Antibodies
- Anti-IFN-γ-FITC
- anti-CD69-PE
- anti-CD8-PerCP
- anti-CD4-PerCP
- anti-CD3-APC.

Calibration

Traditionally, the calibration of flow cytometers has been achieved using an unstained sample (base fluorescence) and one single-stained sample for each fluro-chrome-labeled antibody used (i.e. FITC, PE, ECD/PerCP and Cy5/APC on standard 4-color machines). Photomultiplier (PMT) voltages are set using the unstained sample (compensations at this time are all 0). An analysis region for lymphocytes is set in a scatter light plot (SSC/FSC). PMTs are adjusted in such a way that the lymphocyte population does not extend over any of the axes for any of the fluorescence parameters and sits approximately in the center of the first logarithmic decade. Compensation is then set using a single-stained sample for each color/dye. It is important to observe that staining of the individual samples should be performed under the same conditions as the staining for the actual fully stained samples to be measured later. In many situations, beads coated with antimouse antibodies can be used to replace cells for the process of compensation, one of the advantages being that these beads tend to be very bright as they can be saturated with the antibodies to be used for the actual patient or donor sample. This is particularly useful if the antigen to be stained on the patient or donor cell is expressed at a very low level but may be expressed at a higher level following stimulation. When setting compensation manually, it should be observed that the mean fluorescent intensity (MFI) of the population measured as being positive in one fluorescent channel should have the same MFI in other fluorescent channels as the respective unstained samples. Depending on the software used, the MFI in the other fluorescence channels can be shown simultaneously. This method is in fact now quite old-fashioned. Modern flow cytometers featuring fully digitalized signal processing offer the advantage of fully automated digital compensation of each fluorescence channel against each other one (compensation matrix). This still requires the use of single-stained samples, but obviates the need for setting PMTs manually. Detailed descriptions are available in the documentation accompanying the reagents/machines in question.

Assay Quality Control

The most important control in the assay is the unstimulated sample. It is generally not necessary to use antibody isotype controls. Their purpose is the assessment of the unspecific binding of each of the used staining antibodies. In our experience this unspecific binding is negligible when staining standard surface and intracellular markers on lymphocytes. Regions for positive or negative events are best set using the unstimulated sample and so-called FMO controls. FMO controls are particularly useful when many fluorochromes are used at the same time. They can then be very helpful to delineate positive and negative events with regard to staining with the antibody in question. Many a time, the delineation between a positive and a negative population will not be a straight line when using multiple fluorochromes. More often than not FMO controls are mandatory rather than simply useful.

If isotype controls are used, it is important to ensure that they belong to the same immunoglobulin class (and indeed subclass) as the respective staining antibodies and that they have the same fluorescence/protein ratio (F/P ratio). Furthermore, it is mandatory to use exactly the same concentration of the isotype antibody as for the staining antibody (if this concentration is known) so that they can be used under exactly the same conditions. If an isotype control is necessary or not should be decided on a case-to-case basis. Not using isotype controls will usually lead to considerable savings both monetary and with regard to patients/donor material.

Technical Approach

Preparation of Buffers and Working Solutions

- Complete culture media: RMPI-1640 + FCS (10% final concentration, v/v) + glutamine (2 mmol/l final concentration) + penicillin/streptomycin or other applicable antibiotic;
- Wash buffer: PBS + NaN_3 (0.1% final concentration, w/v) + BSA (0.5% final concentration, w/v);
- EDTA solutions: PBS + EDTA (20 mmol/l final concentration for whole-blood method or 2 mmol/l final concentration for PBMC method);
- BFA stock solution: 5 mg BFA + 1 ml DMSO pure (5 mg/ml final concentration in stock solution);

- virus lysate: adjust to 0.5 mg/ml protein in distilled water, store in aliquots at −80 °C (for example human cytomegalovirus (HCMV) AD169 purified viral lysate; Advanced Biotechnology Incorporated, TEBU, Frankfurt/M., Germany);
- SEB is dissolved in PBS at 0.2 mg/ml to make the stock solution. Stock solution is stored in aliquots at −20 °C.

Incubation/Short-Time Culture

Whole-blood method: 1 ml of whole blood (undiluted) is pipetted into a sterile 4.5 ml Falcon tube (No. 2052). Large tubes or dishes can also be used (for example 50-ml Blue-Cap, conical). 10 μl of virus lysate stock solution is added per milliliter of whole blood (corresponding to 5 μg/ml of protein) or the required amount of peptide solution so that the end concentration of each peptide is 5 μg/ml. If peptides are dissolved in DMSO, it is important to make sure that the final concentration of DMSO does not exceed 1% (v/v). When dissolving peptides in DMSO it advisable to do this under nitrogen atmosphere since DMSO can rapidly oxidize cysteine-containing peptides. Tubes are incubated upright in a standard incubator (37 °C, 5% CO_2 and saturated H_2O atmosphere). Tubes are not to be closed tightly with a lid so that sufficient gas exchange can take place.

PBMC method: the cell concentration is adjusted to 1×10^6/ml in RPMI/FCS medium. 2 ml of cell suspension is transferred into a sterile tube (for example Falcon Tubes, No. 2052). 4 μl of virus lysate stock solution (corresponding to 2 μg of protein) is added or the amount of peptide solution necessary to achieve a peptide end concentration of 1 μg/ml/peptide. Tubes may be incubated upright or at an angle in a standard incubator (37 °C, 5% CO_2 and saturated H_2O atmosphere).

Whole-blood method and PBMC method: addition of 2 μl BFA stock solution per milliliter of sample (corresponding to 10 μg/l end concentration) at 2 h.

Notes:

- An SEB-stimulated ('positive') control can be run. For this purpose, 0.5 μl of stock solution is added to the positive control tube instead of the other stimulants.
- When working with PBMC, cells may be adjusted to 5×10^6/ml and 400 μl transferred into the reaction tubes (corresponding to 2 million cells.). Stimulants are added in volumes of 100 μl (dissolved in RPMI/FSC medium). After 2 h of incubation, BFA is added in 500 μl of RPMI/FSC media. The total reaction volume is now 1 ml. The addition of reagents is easy to handle this way and peptide concentrations are slightly higher during the first 2 h of incubation. It is also possible to start in a volume of 1 ml. Every investigator should adjust her/his own experimental protocol according to her/his particular needs. The above recommendations are purely for orientation.

Terminating the Short-Term Culture, Fixation, Permeabilization and Intracellular Staining

Note: When using the PBMC methods, surface staining can be performed prior to permeabilization; this might be useful if the antigens to be stained are fixation sensitive. When using the whole-blood method, staining of surface antigens has to be performed prior to the addition of lysis reagent if this contains a fixative like PFA. Proceed as with intracellular staining, followed by a wash step.

Whole-Blood Method

- Following an incubation time of 6 h total, ice-cold EDTA solution (EDTA in PBS, 20 mmol/l), corresponding to 11% of the total culture volume is added to each sample (final EDTA concentration will thus be 2 mmol/l) and tubes are incubated for 10 min at room temperature (if polypropylene tubes are used, additional EDTA is not necessary).
- Subsequently, BD lysing solution corresponding to at least 9 times the culture volume is added. It may be necessary for this purpose to divide the samples into several tubes. Incubation for 10 min at room temperature. If solutions from other manufacturers are used, the relevant instructions should be followed.
- Centrifugation (340 × *g*, 8 min, 4 °C); decant carefully or aspirate supernatant.
- Addition of 5 ml wash buffer.
- Centrifugation (340 × *g*, 8 min, 4 °C); decant carefully or aspirate supernatant.

PMBC Method

- After a total incubation time of 6 h, 2 ml of ice-cold PBS is added;
- centrifugation (340 × *g*, 8 min, 4 °C); decant carefully or aspirate supernatant;
- addition of 3 ml of cold EDTA solution (EDTA in PBS, 2 mmol/l) and incubation for 10 min in a water bath at 37 °C (if polypropylene tubes are used, addition of EDTA is not necessary);
- centrifugation (340 × *g*, 8 min, 4 °C); decant carefully or aspirate supernatant;
- addition of 0.7 ml of cold PBS, flush the inside wall of the tube repeatedly where cells are likely to adhere if tubes were incubated at an angle. Alternatively, vortex for approximately 30 s;
- pool samples if desired (for example, if a greater volume of cells with identical stimulation is required);
- addition of 3 ml of washed buffer;
- centrifugation (340 × *g*, 8 min, 4 °C); decant carefully, blot tubes dry.

Note: EDTA can be added directly after the incubation (instead of ice-cold PBS). In this case a volume of 20 mmol/l EDTA/PBS solution is added corresponding to 11% of the total sample volume so that the final EDTA concentration is 2 mmol/l (see whole-blood method). Continue as above.

Continued for Whole-Blood Method and PBMC Method

- Addition of permeabilizing solution, 0.5–1 ml/10^6 cells, incubation for 10 min at room temperature (an estimate of the cell number must be made when using whole blood or cells can be counted);
- addition of 3–4 ml of wash buffer;
- centrifugation (340 × *g*, 8 min, 4 °C);
- decant carefully, blot tubes dry or alternatively aspirate supernatant carefully;
- addition of monoclonal antibodies; the total staining volume is 100 µl; addition of antibodies as follows: anti-IFN-γ-FITC, anti-CD69-PE (or other antibody as desired), anti-CD8-PerCP, anti-CD3-APC; staining on melting ice for 30 min;
- addition of 3–4 ml of wash buffer;
- centrifugation (340 × *g*, 8 min, 4 °C);
- decant carefully, blot tubes dry or alternatively aspirate supernatant carefully;
- if desired, re-fix with 1% PFA/PBS for 5 min at room temperature (for sample acquisition at a later time);
- Add 3–4 ml of wash buffer.
- centrifugation (340 × *g*, 8 min, 4 °C).
- decant carefully, blot tubes dry or alternatively aspirate supernatant carefully;
- sample acquisition with flow cytometer (sample volume should not be less than 200 µl to avoid sample loss).

The principles of data analysis are shown in figure 2.

Remarks

Using Whole Blood versus PBMCs

It appears to us that using the whole-blood method is not equivalent to using the PBMC method [42]. It is striking in particular that when using whole blood higher concentrations of stimulating antigens have to be used. Sometimes, even multiples of the concentration of antigen used for the stimulation of PBMCs will not achieve the same degree of stimulation when using whole blood (in percent of responding T-cells). The method using whole blood should therefore be reserved for situations that can only be addressed using whole blood. The inference that the use of whole blood provides a more physiological environment than the use of PBMCs does not seem very rational. Whole blood contains a large number of factors that are not present at the physiological site of T-cell/antigen-presenting cell interaction (lymph nodes or tissue but not peripheral blood!). These factors may have a negative influence on the response. For example, stimulating antigens could be bound by plasma proteins.

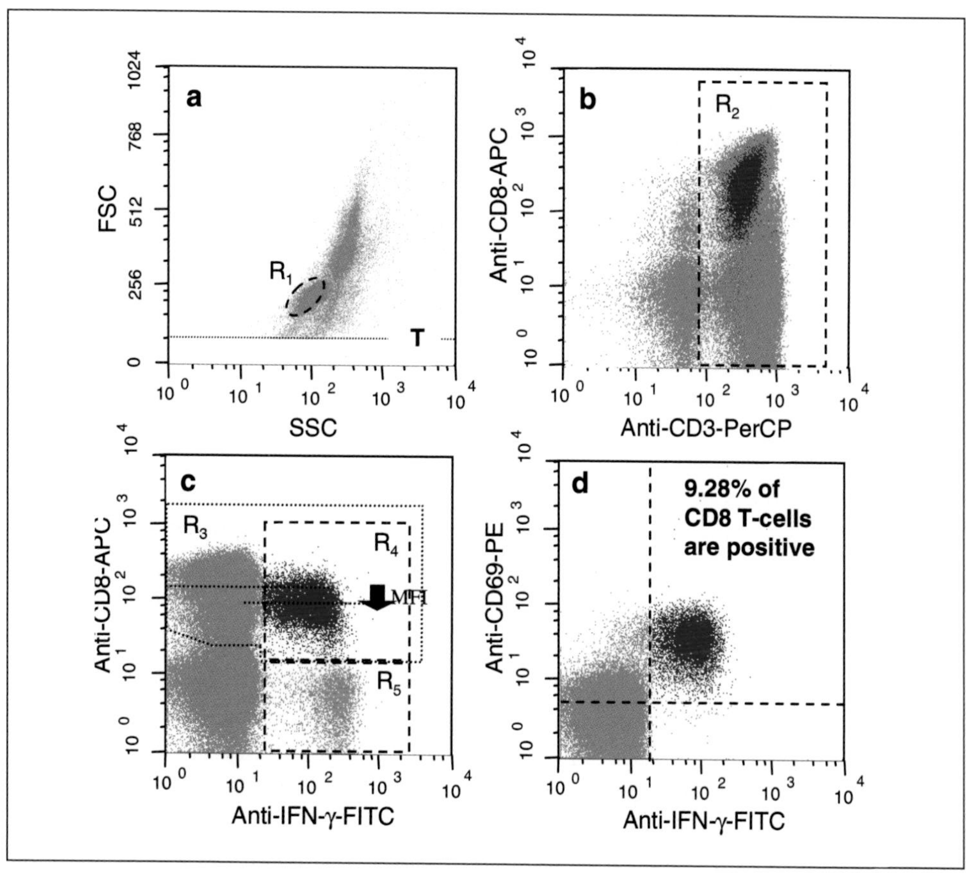

Fig. 2. The flow-cytometric analysis of IFN-γ+ T-cell responses should follow the same principles each time. PBMCs were stimulated for 6 h with 15 amino acid peptides (138 overlapping peptides covering the complete sequence of pp65). 250,000 cells were acquired. **a** Region 1 (R1) was set to include small, vital lymphocytes. **b** CD3+ T-cells were defined by region 2 (R2). Staining of surface parameters following fixation/permeabilization slightly increases background compared to staining prior to this procedure (see CD3− population). Downregulation of the TCR after activation includes both CD4 (red) and CD8 T-cells (blue). **c** In the example, this is indicated by the decreased geometric mean (CD8 staining) of responding compared to non-responding cells (see dotted horizontal lines/arrow). **d** Where responses occurred in both subpopulations (CD8+ und CD8−), the high staining intensity of anti-CD8-APC permits a clear distinction between responding and nonresponding populations even if downregulation is marked. For epitope mapping, the combination of anti-CD4 and anti-CD8 is recommended in order to allow direct identification of the responding population by its own relevant marker. CD69 is not necessary to delineate activated cells. In the example, it was used when testing individual peptides. It may be useful in combination with CD137 to replace intracellular staining at the expense of (some) specificity.

Use of Protein Lysates and Recombinant Proteins

The use of protein lysates or pathogen lysates or recombinant proteins mainly leads to the stimulation of CD4+ T-helper cells. However, the stimulation of cytotoxic/suppressor T-cells depends on whether suitable peptides are presented on class-I

MHC molecules. This only happens in approximately 30% of the population; however, even in this minority it is not particularly effective. A lack of CD8 T-cell stimulation when using pathogen lysates, for example, cannot be interpreted as inability of CD8 T-cells to be stimulated. This is simply because the stimulus is not is not appropriate. A very useful analysis of the stimulatory capacity of CMV lysate, whole proteins and corresponding peptides with regard to CD8 T-cells has been published [26].

Protocol for Antigen-Specific Proliferation

Principle

Conventional proliferation assays are based on the detection of DNA synthesis (^{3}H-thymidine incorporation) or the measurement of metabolic activity of cells [43]. Because these assays work on a population rather than a single-cell level, attribution of proliferation to a particular cell type is only possible if cell clones are used or purified populations of cells.

For the measurement of antigen-specific T-cell proliferation the majority of researchers use a dye dilution assay based on staining with carboxyfluorescein diacetate succinimidyl ester (CFDA-SE, or CFDA or CFSE) [44, 45]. CFDA-SE has the ability to penetrate the cell membrane and covalently bind to cellular proteins. The resulting fluorescent labeling is stable over weeks in nonproliferating cells; however, in proliferating cells it is halved with every division. As a result, up to 8 subsequent generations of proliferating cells can be detected as distinct populations by flow cytometry. CFDA-SE-labeled cells are cultured in vitro for 3–8 days and are subsequently analyzed by flow cytometry.

CFDA-SE labeling can be combined with conventional fluorescent antibody staining. As a result, the number of cells, the number of cell divisions as well as phenotypic and functional characteristics of antigen-specific cells can be evaluated simultaneously. The sensitivity of this approach is the same or higher than that of a classic ^{3}H-thymidine method [46].

Material

Reagents
- CFDA-SE (e.g. Molecular Probes, Eugene, OR, USA) and
- PI, stock solution at 0.1 mg/ml in H_2O (e.g. Sigma).

Cells
- Single-cell suspension, e.g. PBMCs, spleen or lymph node cells, but not whole blood.

Buffers

- PBS, PBS + 0.5% (w/v) BSA (PBS/BSA) and
- Culture medium: RPMI-1640 + FCS (10% end concentration, v/v) + glutamine (2 mmol/l) end concentration) + penicillin/streptomycin or other appropriate antibiotic;

Antigens

- Virus lysate, (e.g. HCMV AD169 purified viral lysate; adjust to 0.5 mg/ml protein in distilled water, store in aliquots at −80 °C);
- peptides (0.5 ml/ml of distilled water; store in aliquots at −80 °C);
- SEB (0.5 mg/ml in distilled water; store in aliquots at −80 °C).

Calibration

Unstained sample and single-stained samples as described above. Because of the high fluorescence intensity and variability of CDFA staining, single-stained samples for compensation are absolutely mandatory.

Attention: wait for a minimum of 24 h after staining cells with CFDA before sample acquisition (see also 'Comment: Critical Parameters', pp. 500). Acquisition on an instrument unable to address compensation between all channels (matrix) may prove difficult.

Quality Control

An unstimulated sample is run in parallel. A positive control is also run in parallel, e.g., stimulation with SEB. The quality of CFDA-SE staining by flow cytometry can be assessed 24 h after the staining at the earliest. CFDA-SE-stained cells should form a very small band in the fluorescence 1 channel. (FITC channel). Unstained and stained cells should be well separated; however, they should be able to be seen within the same dot plot or histogram using the same instrument settings. This corresponds to a 100- to 1,000-fold difference in intensity.

Protocol

CFDA-SE Staining Solution

CFDA-SE must be kept away from humidity and should be kept as a stock solution (5 mmol/l in water-free DMSO) in small aliquots at −20 °C. Aliquots should be thawed immediately before use and should not be refrozen for later use.

CFDA-SE Staining

- Cells are washed once with a protein-free buffer (PBS) and resuspended in PBS and adjusted to 1×10^7 cells/ml. There should be no cell aggregates in the solution. If necessary, the cell suspension should be filtered to remove aggregates (e.g. preseparation filter, Miltenyi Biotec).
- Additional CFDA stock solution. The end concentration of CFDA-SE should be between 0.5 and 5 μmol/l, the optimal concentration for each cell type must be determined by prior titration (see 'Comment: Critical Parameters', pp. 500).
- Immediately after adding CFDA-SE, samples should be thoroughly mixed by pipetting them up and down in order to avoid a gradient of concentration.
- Incubate for 4 min at room temperature in the dark, gently agitate from time to time to prevent cell sedimentation (the optimum incubation time for each cell type must be determined, the 4 min given here are only a rough guideline (see 'Comment: Critical Parameters', pp. 500).
- Following the incubation time, the reaction is stopped by addition of a 10-fold volume of PBS/BSA. Cells are washed at least twice with PBS/BSA and are subsequently resuspended in culture media.
- Flow-cytometric analysis should not be performed until at least 24 h later because until then cells will release unbound dye or dyes bound to cellular proteins with a short half life (see 'Comment: Critical Parameters', pp. 500).

In vitro Activation

- The so stained cells can now be activated in vitro. Details of this procedure are described under 'Antigen-Specific T-Cell stimulation' (pp. 481). The protocol given here is for the stimulation of human PBMCs.
- Cells are seeded at a concentration of 2×10^6/ml and 200 μl of cell suspension/well into a 96-well round-bottom plate, experiments should be run at least in triplicate. Antigen is added to the respective wells at its optimum stimulating concentration, which ought to be determined by titration. The same solvent/buffer used to dissolve the antigen added to the antigen-stimulated wells should be added to the control wells, using exactly the same volume. SEB is added to the positive control wells at an end concentration of 1 μg/ml.
- Cells are incubated for 4–5 days at 37 °C in a standard CO_2 incubator (mouse or murine cells should only be incubated for 3 days).
- After the incubation, cells are removed from the 96-well plate. Each well should be carefully flushed with PBS to ensure that activated adherent T-cells are recovered. Check plate under the microscope for remaining cells. Cells are washed with PBS/BSA and can then be stained with monoclonal antibodies for phenotype markers of interest (see example below). In order to obtain meaningful data, dead cells and cells binding antibody unspecifically, e.g., monocytes or B-cells, must be excluded from the analysis. This can be achieved by using phenotype markers for B-cells and monocytes in the same channel along with a live/

dead stain sample, e.g. PI. The assigned channel can then be used as a dump channel serving to exclude the set cell times and dead cells.

- An example for the simultaneous assessment of the proliferation of CD4 and CD8 T-cells using a flow cytometer equipped with a 488 nm and 633 nm laser:
  - FL1: CFDA-SE;
  - FL2: CD4 PE or CD8 PE;
  - FL4: CD3 Cy5 or APC;
  - FL3: dump channel – dead cells PI plus CD14 (monocytes) plus CD19 (B-cells), fluorochromes: PerCp, PerCP-Cy5 or PE-Cy5.

Data Acquisition

- Calibration of flow cytometer using the unstained sample and the single-stained samples (see above). Unstained cells should be visible as populations in all fluorescence channels. If CFDA-SE staining is too bright it will often be difficult to adjust compensation (this is particularly difficult if data acquisition is not digitalized).
- When acquiring the sample, it is important to ensure that a sufficient number of cells is analyzed because antigen-specific cells can be rare.
- No 'live gates' should be used to select cells for storage during acquisition because valuable information can be lost. If this is necessary because of the low concentration of target cells and limited storage capacity, only safe parameters should be used for exclusion. One such parameter is viability (PI). Do not make your gates too small, antigen-specific cells are usually rare and can easily be overlooked.

Data Analysis

Setting Windows and Regions for Analysis (fig. 3):
FSC/SSC: Exclusion of small particles and cell aggregates. The region should not be too small in order not to exclude activated proliferating cells (fig. 3A);

FL2/FL3: FL3 serves as a dump channel for dead cells, monocytes and B-cells;

FL2/FL4: CD3+ CD4+ and CD4– (= CD8+) T-cells.

Determination of proliferation in a dot plot or FL1 histogram: Statistical markers are set based on the negative control (no antigen stimulation) in order to be able to distinguish undivided from proliferating cells. Thus, the frequency of dividing cells can be determined (fig. 3B). In addition, the number of passed divisions can be determined. If individual generations of cells can be discriminated, the corresponding statistic markers can be used to directly determine a number of measured cells per generation.

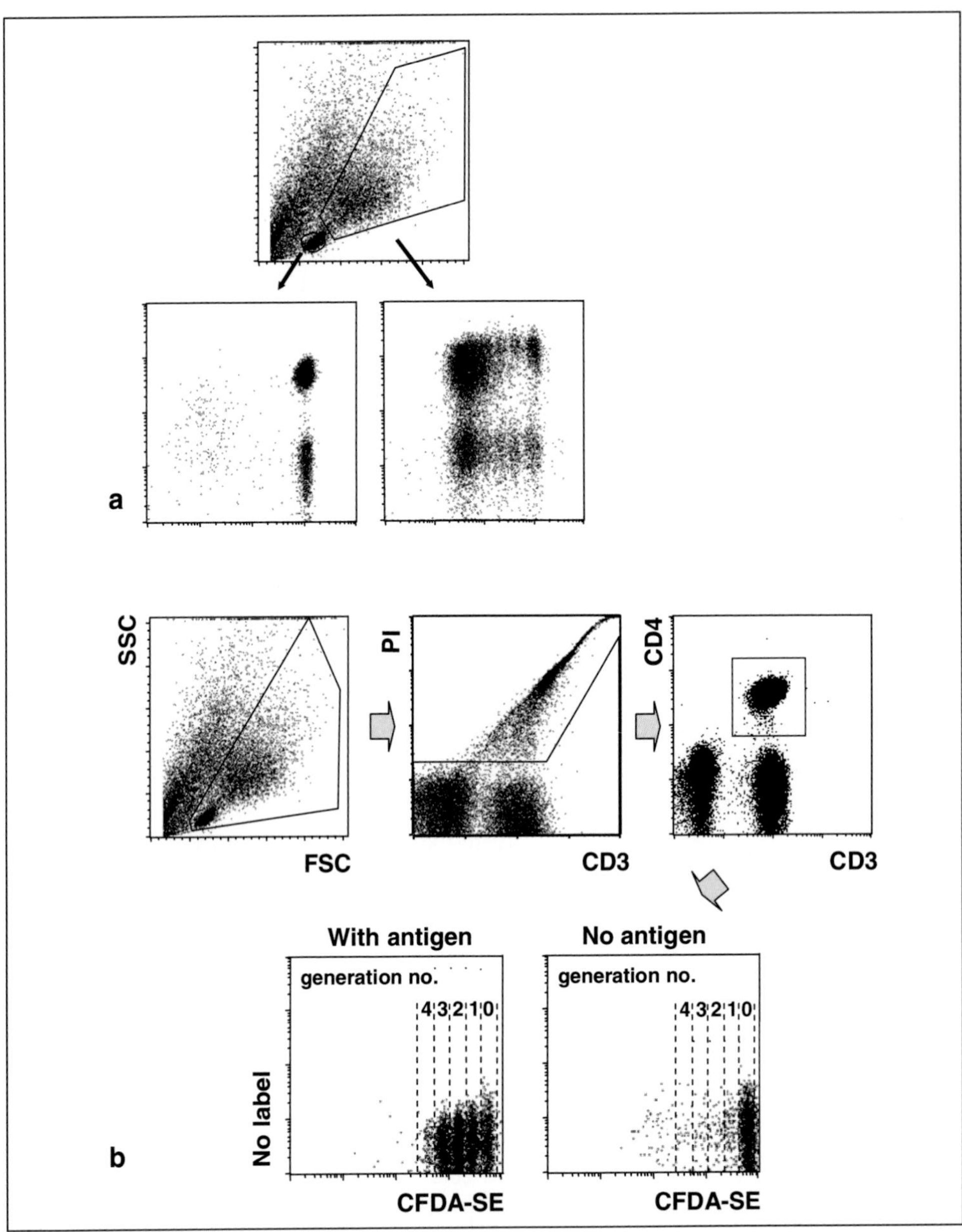

Fig. 3. Analysis of the CFDA dye dilution proliferation assay. **a** Proliferating cells exhibit increased scatter parameters (FSC and SSC). **b** Possible gating strategy und positioning of statistical markers (details and explanations of calculations are provided in the text).

Quantification of Proliferation Activity

Flow cytometry can only determine the frequency of proliferating cells. In order to be able to make a statement about the proliferation in different samples, the absolute number of cells must be considered. An acceptable simplification assumes that if the composition of the samples was identical at the beginning of the experiment and the culture conditions and survival rates were comparable, the frequency of dividing cells can be correlated to the absolute number of cells. If, however, the culture conditions and survival rates must be expected to have varied, the absolute number of cells per sample must be determined, for example, by counting in a Neubauer chamber or the use of quantification beads during the measurements, e.g., Trucount-Beads™ (BD Biosciences).

The simplest analysis is that of the frequency of proliferating T-cells. However, this does not take into account variations in the number of cell divisions. In order to be able to quantify proliferation in different samples which are comparable to those of, for example, the ^{3}H-thymidine incorporation assay, the number of cell divisions per the number of cells at the beginning of the experiment must be calculated.

Calculation of the Number of Divisions per Cells at the Outset

For this purpose, the number of precursor cells, p_i is calculated based on the absolute number of cells measured in one division generation x_i.

$$p_i = \frac{x_i}{2^i} \tag{1}$$

p_i = number of precursor cells of one generation; x_i = absolute number of cells in one generation; i = zero, 1, . . . n; number of generation, corresponds to the number of cell divisions.

Subsequently, the sum of all precursor cells P_s from which the measured sample results can be determined:

$$P_S = \sum_{i=0}^{n} p_i = \sum_{i=0}^{n} \frac{x_i}{2^i} \tag{2}$$

Based on the number of precursor cells, the total number of cell divisions C_{total} can now be calculated by adding up the numbers of cell divisions per precursor cells. P_s is used here to normalize for one single precursor cell.

$$C_{total} = \sum_{i=1}^{n} c_i, \quad \text{with} \quad c_i = \frac{x_i}{2P_S} \tag{3}$$

Attention: i in this case begins with 1 because only cells that have divided can be considered.

The number of total cell divisions C_{total} can now be multiplied with the number of cells at the beginning of the experiment in order to calculate the absolute number of cell divisions per seeded cell.

Alternatively, the geometric mean (not the arithmetic mean) can be used as a relative measure M for the strength of proliferation in different samples:

$$M = M_0 - M_P \tag{4}$$

M_0 = geometric mean of nondivided cells (Start value);
M_p = geometric mean of the sample in question (across all cells);

Please note that M_0 must be the same for all samples.

Determination of the Frequency of Antigen-Specific Cells in the Initial Sample

Based on the frequency of cells per cell division generation (F_i), the minimum frequency of precursor cells (P), i.e. the frequency of antigen-specific cells in the initial sample, can be calculated. For simplification, it is assumed that P is small (usually <1%) and as a result, the number of nondividing cells can be assumed to be constant. If this is true, then:

$$P = \sum_{i=1}^{n} \frac{x_i}{2^i} \tag{5}$$

i = Cell division generation 1, 2, ... n; F_i = frequency of cells in one generation; P = frequency of precursor cells.

This is only a rough estimate because the number of dead cells per generation is normally not known. A more detailed approach to this problem was described by Givan et al. [47].

Comment: Critical Parameters

CFDA-SE Labeling

- The brightness of the staining will be determined by the incubation time (use a stopwatch), the CFDA-SE concentration, the CFDA-SE quality (for example, the age of the stock solution, single or multiple use of aliquots), cell density, and temperature. In order to achieve reproducible staining intensity, these parameters must be kept constant in as much as possible.
- Cell size and/or status of activation will have an influence on the distribution of fluorescence. Small resting lymphocytes are labeled uniformly (distinct peaks), large activated cells, however, will exhibit an inhomogeneous distribution of size and as a result an inhomogeneous distribution of fluorescence will result. As a consequence, cell generation can sometimes not be distinguished.
- Flow-cytometric analysis should be done no sooner than 24 h after labeling cells with CFDA because the cells may still contain large amounts of non-stably bound CFDA. This may lead to high fluorescent signals, however would rapidly be released. Not until 24 h after labeling is the labeling stable for days to weeks.
- If the concentration of CFDA-SE is too high, cell viability may be affected negatively. This is another reason why optimum conditions have to be determined for each cell type.

T-Cell Activation

- The efficiency of T-cell activation depends on the antigen concentration, the cell concentration (cells per area), the geometry of the culture dish (round or flat bottom). These parameters may have to be optimized prior to running the experiment.

- The relation between antigen-specific cells and T-cells will also have an influence on activation. If unseparated cells (PBMCs, cells from lymph nodes) are used, the numeric relationship between antigen-presenting cells and T-cells is fixed. However, if antigen-presenting cells and T-cells are prepared separately, the ratio of antigen-presenting cells to T-cells can be as high as 4:1. This should also be optimized according to the cell types used.

Troubleshooting

Staining too Bright
- Measurement performed too early after CFDA-SE staining.
- Labeling time too long or CFDA-SE concentration too high.

Compensation Impossible, PE-Counterstain too Weak
- CFDA-SE staining too bright (see above).
- CFDA-SE emission strongly overlaps into the PE channel, therefore PE labeling should be used for very highly expressed antigens.

No Homogeneous Labeling, No Generations Distinguishable
- Cell aggregates were present during labeling, or cells were not sufficiently mixed during the labeling.
- Cell sizes variable, e.g. inactivated cells or cell lines.

Labeling too Weak
- CFDA-SE is too old: make new aliquots, use aliquots only once.
- Higher CFDA-SE concentration required or longer labeling time.

Many Dead Cells No Proliferation in Positive Control
- Check analysis windows/regions. Were proliferating cells excluded accidentally (FSC/SSC region may be too small, live/dead exclusion region may be too small or in the wrong place)?
- CFDA-SE labeling is too strong (see above).
- Cell density is too low or too high.
- Wrong culture plates were used for the given cell type (round bottom instead of flat bottom).
- Measurement was performed too early.

No Proliferating Cells Found in the Antigen Activated Sample
- Antigen concentration was not optimal.
- Longer stimulation time is required.
- The frequency of specific cells is below detection limit.
- Donor may not have a response to the antigen.

Big Variation between Triplicates

- Check positioning of your samples in the plate; wells near the edge of the plate are subject to stronger evaporation, which may affect cell viability. This may be avoided if all wells along the edge of the plate are filled with media or buffer to avoid evaporation of sample.

Time Required

- Total time required including activation: 4–5 days.
- CFDA-SE labelling: approximately 1 h (not including cell preparation).
- Labeling and cytometric analysis: approximately 2–3 h, depending on experience.

References

1 Frentsch M, Arbach O, Kirchhoff D, Moewes B, Worm M, Rothe M, Scheffold A, Thiel A: Direct access to CD4+ T cells specific for defined antigens according to CD154 expression. Nat Med 2005;11:1118–1124.

2 Betts MR, Brenchley JM, Price DA, De Rosa SC, Douek DC, Roederer M, Koup RA: Sensitive and viable identification of antigen-specific CD8+ T cells by a flow cytometric assay for degranulation. J Immunol Methods 2003;281:65–78.

3 Kern F, LiPira G, Gratama JW, Manca F, Roederer M: Measuring Ag-specific immune responses: Understanding immunopathogenesis and improving diagnostics in infectious disease, autoimmunity and cancer. Trends Immunol 2005;26:477–484.

4 Brosterhus H, Brings S, Leyendeckers H, Manz RA, Miltenyi S, Radbruch A, Assenmacher M, Schmitz J: Enrichment and detection of live antigen-specific CD4(+) and CD8(+) T cells based on cytokine secretion. Eur J Immunol 1999;29:4053–4059.

5 Day CL, Seth NP, Lucas M, Appel H, Gauthier L, Lauer GM, Robbins GK, Szczepiorkowski ZM, Casson DR, Chung RT, Bell S, Harcourt G, Walker BD, Klenerman P, Wucherpfennig KW: Ex vivo analysis of human memory CD4 T cells specific for hepatitis C virus using MHC class II tetramers. J Clin Invest 2003;112:831–842.

6 Leyendeckers H, Odendahl M, Lohndorf A, Irsch J, Spangfort M, Miltenyi S, Hunzelmann N, Assenmacher M, Radbruch A, Schmitz J: Correlation analysis between frequencies of circulating antigen-specific IgG-bearing memory B cells and serum titers of antigen-specific IgG. Eur J Immunol 1999;29:1406–1417.

7 Meyer AL, Trollmo C, Crawford F, Marrack P, Steere AC, Huber BT, Kappler J, Hafler DA: Direct enumeration of Borrelia-reactive CD4 T cells ex vivo by using MHC class II tetramers. Proc Natl Acad Sci U S A 2000;97:11433–11438.

8 Hayakawa K, Ishii R, Yamasaki K, Kishimoto T, Hardy RR: Isolation of high-affinity memory B cells: Phycoerythrin as a probe for antigen-binding cells. Proc Natl Acad Sci U S A 1987;84:1379–1383.

9 Altman JD, Moss PAH, Goulder PJR, Barouch DH, McHeyzer-Williams MG, Bell JI, McMichael AJ, Davis MM: Phenotypic analysis of antigen-specific T lymphocytes. Science 1996;274:94–96. Published erratum appeared in Science 1998;280:1821.

10 McMichael A, O'Callaghan C: A new look at T cells. J Exp Med 1998;187:1367–1371.

11 Busch DH, Pilip IM, Vijh S, Pamer EG: Coordinate regulation of complex T cell populations responding to bacterial infection. Immunity 1998;8:353–362.

12 Selin LK, Lin MY, Kraemer KA, Pardoll DM, Schneck JP, Varga SM, Santolucito PA, Pinto AK, Welsh RM: Attrition of T cell memory: selective loss of LCMV epitope-specific memory CD8 T cells following infections with heterologous viruses. Immunity 1999;11:733–742.

13 Toebes M, Coccoris M, Bins A, Rodenko B, Gomez R, Nieuwkoop NJ, van de Kasteele W, Rimmelzwaan GF, Haanen JB, Ovaa H, Schumacher TN: Design and use of conditional MHC class I ligands. Nat Med 2006;12:246–251.

14 Grotenbreg GM, Roan NR, Guillen E, Meijers R, Wang JH, Bell GW, Starnbach MN, Ploegh HL: Discovery of CD8+ T cell epitopes in *Chlamydia trachomatis* infection through use of caged class I MHC tetramers. Proc Natl Acad Sci U S A 2008;105:3831–3836.

15 Khanna KM, McNamara JT, Lefrançois L: In situ imaging of the endogenous CD8 T cell response to infection. Science 2007;318:116–120.

16 Kwok WW, Ptacek NA, Liu AW, Buckner JH: Use of class II tetramers for identification of CD4+ T cells. J Immunol Methods 2002;268:71–81.

17 McMichael A, Kelleher A: The arrival of HLA class II tetramers. J Clin Invest 1999;104:1669–1670.

18 Xu XN, Purbhoo MA, Chen N, Mongkolsapaya J, Cox JH, Meier UC, Tafuro S, Dunbar PR, Sewell AK, Hourigan CS, Appay V, Cerundolo V, Burrows SR, McMichael AJ, Screaton GR: A novel approach to antigen-specific deletion of CTL with minimal cellular activation using alpha3 domain mutants of MHC class I/peptide complex. Immunity 2001;14,:591–602.

19 Whelan JA, Dunbar PR, Price DA, Purbhoo MA, Lechner F, Ogg GS, Griffiths G, Phillips RE, Cerundolo V, Sewell AK: Specificity of CTL interactions with peptide-MHC class I tetrameric complexes is temperature dependent. J Immunol 1999;163:4342–4348.

20 Daniels MA, Jameson SC: Critical role for CD8 in T cell receptor binding and activation by peptide/major histocompatibility complex multimers. J Exp Med 2000,191:335–346.

21 Knabel M, Franz TJ, Schiemann M, Wulf A, Villmow B, Schmidt B, Bernhard H, Wagner W, Busch DH: Reversible MHC multimer staining for functional isolation of T-cell populations and effective adoptive transfer. Nat Med 2002;8:631–637.

22 Casazza JP, Betts MR, Price DA, Precopio ML, Ruff LE, Brenchley JM, Hill BJ, Roederer M, Douek DC, Koup RA: Acquisition of direct antiviral effector functions by CMV-specific CD4+ T lymphocytes with cellular maturation. J Exp Med 2006;203:2865–2877.

23 Manz R, Assenmacher M, Pfluger E, Miltenyi S, Radbruch A: Analysis and sorting of live cells according to secreted molecules, relocated to a cell-surface affinity matrix. Proc Natl Acad Sci U S A 1995;92:1921–1925.

24 Krensky AM: The HLA system, antigen processing and presentation. Kidney Int Suppl 1997;58:S2–S7.

25 Waldrop SL, Pitcher CJ, Peterson DM, Maino VC, Picker LJ: Determination of antigen-specific memory/effector CD4+ T cell frequencies by flow-cytometry: Evidence for a novel, antigen-specific homeostatic mechanism in HIV-associated immunodeficiency. J Clin Invest 1997;99:1739–1750.

26 Maecker HT, Ghanekar SA, Suni MA, He XS, Picker LJ, Maino VC: Factors affecting the efficiency of CD8+ T cell cross-priming with exogenous antigens. J Immunol 2001;166:7268–7275.

27 Kern F, Surel IP, Brock C, Freistedt B, Radtke H, Scheffold A, Blasczyk R, Reinke P, Schneider-Mergener J, Radbruch A, Walden P, Volk HD: T-cell epitope mapping by flow-cytometry. Nat Med 1998;4:975–978.

28 Waldrop SL, Davis KA, Maino VC, Picker LJ: Normal human CD4+ memory T cells display broad heterogeneity in their activation threshold for cytokine synthesis. J Immunol 1998;161:5284–5295.

29 Appay V, Nixon DF, Donahoe SM, Gillespie GM, Dong T, King A, Ogg GS, Spiegel HM, Conlon C, Spina CA, Havlir DV, Richman DD, Waters A, Easterbrook P, McMichael AJ, Rowland-Jones SL: HIV-specific CD8(+) T cells produce antiviral cytokines but are impaired in cytolytic function. J Exp Med 2000;192:63–75.

30 Kiecker F, Streitz M, Ay B, Cherepnev G, Volk HD, Volkmer-Engert R, Kern F: Analysis of antigen-specific T-cell responses with synthetic peptides – what kind of peptide for which purpose? Hum Immunol 2004;65:523–536.

31 Kostense S, Ogg GS, Manting EH, Gillespie G, Joling J, Vandenberghe K, Veenhof EZ, van Baarle D, Jurriaans S, Klein MR, Miedema F: High viral burden in the presence of major HIV-specific CD8(+) T cell expansions: Evidence for impaired CTL effector function. Eur J Immunol 2001;31:677–686.

32 Kern F, Faulhaber N, Frömmel C, Khatamzas E, Prösch S, Schönemann C, Kretzschmar I, Volkmer-Engert R, Volk H-D, Reinke P: Analysis of CD8 T cell reactivity to cytomegalovirus using protein-spanning pools of overlapping pentadecapeptides. Eur J Immunol 2000;30:1676–1682.

33 Maecker HT, Dunn HS, Suni MA, Khatamzas E, Pitcher CJ, Bunde T, Persaud N, Trigona W, Fu TM, Sinclair E, Bredt BM, McCune JM, Maino VC, Kern F, Picker LJ: Use of overlapping peptide mixtures as antigens for cytokine flow-cytometry. J Immunol Methods 2001;255:27–40.

34 Rammensee HG, Bachmann J, Stefanovic S: MHC Ligands and Binding Motifs. Georgetown, Landes Bioscience, 1997.

35 Fournier P, Ammerlaan W, Ziegler D, Giminez C, Rabourdin-Combe C, Fleckenstein BT, Wiesmuller KH, Jung G, Schneider F, Muller CP: Differential activation of T cells by antibody-modulated processing of the flanking sequences of class II-restricted peptides. Int Immunol 1996;8:1441–1451.

36 Muller CP, Ammerlaan W, Fleckenstein B, Krauss S, Kalbacher H, Schneider F, Jung G, Wiesmuller KH: Activation of T cells by the ragged tail of MHC class II-presented peptides of the measles virus fusion protein. Int Immunol 1996;8:445–456.

37 Arend SM, Geluk A, van Meijgaarden KE, van Dissel JT, Theisen M, Andersen P, Ottenhoff TH: Antigenic equivalence of human T-cell responses to *Mycobacterium tuberculosis*-specific RD1-encoded protein antigens ESAT-6 and culture filtrate protein 10 and to mixtures of synthetic peptides. Infect Immun 2000;68:3314–3321.

38 Accapezzato D, Nisini R, Paroli M, Bruno G, Bonino F, Houghton M, Barnaba V: Generation of an MHC class II-restricted T cell epitope by extracellular processing of hepatitis delta antigen. J Immunol 1998;160:5262–5266.

39 Eberl G, Renggli J, Men Y, Roggero MA, Lopez JA, Corradin G: Extracellular processing and presentation of a 69-mer synthetic polypetide to MHC class I-restricted T cells. Mol Immunol 1999;36:103–112.

40 Sherman LA, Burke TA, Biggs JA: Extracellular processing of peptide antigens that bind class I major histocompatibility molecules. J Exp Med 1992;175:1221–1226.

41 Suni MA, Picker LJ, Maino VC: Detection of antigen-specific T cell cytokine expression in whole blood by flow-cytometry. J Immunol Methods 1998;212:89–98.

42 Hoffmeister B, Bunde T, Rudawsky IM, Volk HD, Kern F: Detection of antigen-specific T cells by cytokine flow-cytometry: The use of whole blood may underestimate frequencies. Eur J Immunol 2003;33:3484–3492.

43 Pechhold K, Kabelitz D: Measurement of cellular proliferation (^{3}H-thymidine uptake, MTT, absolute cell numbers by SCDA); in Kaufmann SHE, Kabelitz D (eds): Methods in Microbiology: Immunology of Infection. San Diego, Academic Press, 1998, vol 25, pp 59–78.

44 Lyons AB: Divided we stand: Tracking cell proliferation with carboxyfluorescein diacetate succinimidyl ester. Immunol Cell Biol 1999;77:509–515.

45 Lyons AB, Parish CR: Determination of lymphocyte division by flow-cytometry. J Immunol Methods 1994;171:131–137.

46 Angulo R, Fulcher DA: Measurement of Candida-specific blastogenesis: Comparison of carboxyfluorescein succinimidyl ester labeling of T cells, thymidine incorporation, and CD69 expression. Cytometry 1998;34:143–151.

47 Givan AL, Fisher JL, Waugh M, Ernstoff MS, Wallace PK: A flow cytometric method to estimate the precursor frequencies of cells proliferating in response to specific antigens. J Immunol Methods 1999;230:99–112.

Sack U, Tárnok A, Rothe G (eds): Cellular Diagnostics. Basics, Methods and Clinical Applications of Flow Cytometry. Basel, Karger, 2009, pp 503–523

Immunophenotyping CD4 T-Cells to Monitor HIV Disease

Francis F. Mandy

International Centre for Infectious Diseases, Winnipeg, Manitoba, Canada

Introduction

At the beginning of the 1980s, the appearance of the acquired immunodeficiency syndrome (AIDS) provided an almost instantaneous opportunity in industrialized countries to introduce and establish flow cytometry as a standard clinical instrument to serve infectious-immunology laboratories. The exceptional conditions leading up to this technological breakthrough in diagnostic medicine were most remarkable. By 1980, the commercialization of a variety of monoclonal antibodies (mAbs) thanks to the rapid acclimatization to the wide range of applications of the hybridoma technology. The production of relatively reliable argon-ion lasers and personal computers at modest prices were reaching the market place in resource-rich countries. The synchronous appearance of these three components on the market place was the essential element that made it feasible to start commercial production of flow cytometers. The final critical component, the economic incentive, was provided by the spontaneous and voluminous clinical need created by the AIDS pandemic. There was a sudden large demand for routine counting of T-lymphocyte subpopulations; the CD4 T-cell set. With such novel yet practical application of biotechnology, it was possible to monitor an unusual infectious immunodeficiency syndrome that was appearing among young homosexual men in several major cosmopolitan centers in North America. It was a new infectious syndrome without a known causative agent. Because of the rapidly increasing demand for CD4 T-cell enumeration in diagnostic immunology laboratories, flow cytometry quickly replaced a large segment of the routine utilization of epifluorescent microscopy. This sequence of events precipitated a unique success story in medical diagnostic history. By the mid 1980s, a retrovirus was identified as the cause of this devastating lethal disease: the human immunodeficiency virus (HIV). Once the etiology of AIDS was known, scientists had to isolate the virus, produce viral-based antigenic material and begin to manufacture screening tests to identify the HIV antibody in blood of infected individuals. The initial screening tests were developed to protect blood bank supplies throughout the industrialized world from this deadly virus. In several Western countries this critical and essential harm reduction process did not go without complications. For example, in France and Canada, there were 'tainted blood' scandals that worked their way through the legal systems for years. Such unprecedented circumstances forced lethargic bureaucratic processes to uncover some of the flaws in their respective systems to provide faster federal response to protect national blood supplies and factor VIII extract production (required to treat certain forms of hemophilia). These unusual public trials exposed incidences where blood and blood products were not always effectively monitored for the presence of lethal contaminating agents such as HIV and later hepatitis C. Once more progressive legislations took effect, and the various roots of disease transmission became known, blood supplies in industrialized countries in general received a remarkable level of protection. With the epidemic under control in most resource-rich countries, the demand for massive-scale immunophenotyping of T-cell subsets began to diminish. However, by the end of the 1980s, flow cytometry was moving into other areas of cellular diagnostics. Flow cytometers turned out to be valuable tools with certain types of lymphoma and leukemia diagnostics. Hematopathologists began to take advantage of the available cell analysis capacity with a variety of newly available cell lineage- and function-identifying mAbs as the arsenal for treatment in the field of oncology began to expand. Many such methodologies are covered in other chapters in this book. A growing variety of cell surface markers became available with directly conjugated fluorescent tags. The role of flow cytometry as part of routine cellular-based diagnostic medical technology became secure. Before the end of the 20th century, flow cytometry became an irreplaceable clinical diagnostic tool and it was recognized as a unique 20th century biomedical diagnostic product. The whole remarkable evolutionary process was recognized as a 'new biological platform' by a pair of medical historians, Keating and Cambrosio [1].

The enumeration of CD4 T-cell numbers with epifluorescent microscopy from peripheral blood was considered to be a test to monitor clinical immune suppression even before the discovery of the HIV [2, 3]. Thirty years later, counting absolute CD4 T-lymphocytes is still considered to be the best indicator of HIV disease progression. In industrialized countries it is rarely used alone to monitor HIV disease. After a patient's HIV status is determined and confirmed, T-cell subset enumeration and HIV viral load are the two complementary assays used for regular monitoring during the asymptomatic phase of the disease. While viral load is the measure of drug efficacy during antiretroviral therapy (ART) [4, 5], absolute CD4 T-cell counts remain the overall best predictor of clinical outcome [6, 7], including patients who are receiving ART [8]. The CD4 T-cell count is the marker to monitor for HIV disease progression, level of immune suppression and overall assessment of immune restoration of patients receiving ART. CD4 T-cell counts are essential for decision making as to when to start or alter ART [9] and later with advanced disease progression, when to consider prophylactic interventions to prevent opportunistic infections [10]. CD4 T-cell numbers can also be used for epidemiological studies to forecast the magnitude and corresponding cost burden a regional HIV outbreak may have on the public health care system [11].

The AIDS pandemic continued to be out of control in many African and Asian countries. In 2003, the World Health Organization (WHO) responded to the African situation where about 70% of world HIV infected people live. WHO introduced a strategic attack on the pandemic called '3 by 5'. The goal was to treat 3 million HIV-infected individuals by the end of 2005. While the objective was not achieved, nevertheless a global momentum was established. It had significant and unprecedented impact on mobilizing mass treatment for a lethal infectious disease. The accelerated implementation of ART combined with persistent global political pressure has resulted in a two-log cost reduction of ART. In this chapter, the measurement of CD4 T-cell is reviewed for the two different economical realities: for resource-rich and poor nations.

The Beginning of Multiparametric Immunophenotyping

The first generation of commercial flow cytometers introduced for clinical applications was equipped with three parameters:

i) forward scatter (FSC),
ii) side scatter (SSC),
iii) a color or fluorescent signal (FL1).

The two light scatters are intrinsic parameters that define in two dimensions some of the morphological features of leukocytes. The FSC and SSC measure size and granularity, respectively. The single fluorescence signal measures extrinsic cell parameters that require the addition of a reagent tagged with a fluorescent dye for detection [12]. The fluorescent signal that is captured by a detector is referred as the FL1 signal. In the early days of monochromatic flow cytometry, it was a green fluorescence light emitted by the dye fluorescein isothiocyanate. This dye was conjugated directly to a mAb specifically binding to a receptor on the surface of a leukocyte. With these early immunophenotyping instruments only one antigen could be identified at a time on the cell's surface. In the mid 1980s clinical flow cytometers switched to deliver simultaneous dual fluorescent signal detection capacity. Thus from that time on dual-color immunophenotyping as part of routine clinical immunophenotyping was available for CD4 and CD8 T-cell enumeration. By the late 1980s, additional fluorochromes were developed to allow the simultaneous detection of three fluorochromes. The three tags could be detected from the surface of one cell subtype or from as many as three different cells based on lineage- or function-specific characteristics. From here on it was possible to detect simultaneously from a single specimen tube all the T-cell markers that were considered essential for HIV immunophenotyping (CD3, CD4 and CD8).

Immunopathology of HIV

The CD4 T-cell is the primary but not the only leukocyte that may be infected with HIV. The CD4 receptor is not present on all T-cells, only on the functionally characterized T-helper cells, whereas, albeit with lower density, CD4 receptors are present on all monocytes. The direct viral entry root is on the surface of a CD4 T-cell in the presence of a coreceptor CCR5. A fusion process takes place that permits viral penetration through the cell membrane [13, 14]. The T-helper lymphocytes have a critical role in the cellular immune response. This whole cellular response is compromised with the destruction of the CD4 T-cell. Some of the CD4 T-cell reaction that is altered is the normal interaction with both B-cells and cytotoxic T-cells (CTLs). CTLs are a subset of CD8 T-cells which have the capacity of mediating lysis of infected autologous cells. There are some CD8 T-cells that suppress or regulate ongoing immune responses, thus providing a mechanism to assure a dynamic cellular immune response. In HIV infection, it is believed that both CTLs and neutralizing antibodies play a role in controlling the virus for years after the onset of the initial infection. However, as immune competence fails, the host eventually succumbs to various opportunistic infections. CD4 T-cell counts are ordered by physicians to get a way to assess the overall damage to the immune system which is brought about indirectly by the HIV infection [6, 7]. Normal absolute CD4 T-cells counts are about 1000 cells/ml of blood with a range of approximately from 400 to 1500. A decrease in T-helper cell levels indicates that the virus has either directly or indirectly caused the death or redistribution in peripheral blood of these important lymphocyte subsets. CD4 T-cell levels less than 400 cells/ml usually indicate that the virus reached the point where the immune system will be functionally compromised [15, 16], and levels less than 200 cells/ml have a prognostic significance: the official onset of AIDS [17]. There is a strong correlation between CD4 T-cell count and clinically significant immunosuppression. The classic AIDS definition is based on a CD4 T-cell count less than 200 cells/ml with or without the presence of any clinical symptom [10]. The rate at which the CD4 T-cells decline can vary considerably. Some individuals have a very rapid disease course and the CD4 T-cells drop precipitously in 1–2 years. Others may have a lengthy steady reduction in CD4 cell count yet still others maintain relatively high levels for 10–15 years, with a rapid CD4 drop late in infection just before death [7, 18]. In the 21st century, most physicians use absolute CD4 T-cell counts to monitor disease progression in adults. Going back to the beginning of the use of flow cytometers, there was only one way to quantify CD4 T-cells. Initially, CD4 T-cells were calculated as the relative percentage of all the lymphocytes.

The original rough calculations were based on adding CD4, CD8 and CD19 cells all together to get the total lymphocyte count. With density gradient separation of whole blood, most of the NK cells were lost and the calculation was often based on the false assumption that anti-CD4 and anti-CD8 mAbs were lineage-, not function-specific markers. Yet, this ratio method was a remarkably reliable in the early days of flow cytometry. As the absolute count values were inconsistent when the lymphocytes were collected with density gradient separation technology that included some washing procedures. Once the red cell lysing whole blood sample processing replaced centrifugation the absolute count values were still acceptable only in theory. They remained relatively unreliable because the absolute count value relied on results from two different instruments. This method for generating absolute counts became known as the double-platform technology (DPT). The hematology instrument generated the total white cell and lymphocyte count. The flow cytometer furnished the CD4 T-cell count as a percentage of the lymphocytes (fig. 1). The counts generated by the hematology instrument were very sensitive to the age of the blood specimen. Over 6 h between phlebotomy and cell counting were problematic. The morphological degradation of granulocytes and monocytes had a deleterious effect on the calculation of CD4 T-cell absolute counts. In general, the percentage levels of 28 and 14% that are approximately equivalent to 500 and 200 cells/ml, expressed as absolute counts, respectively, remained the more frequently used option by most physicians [10]. As flow cytometers were initially designed for research application, the goal was to collect a finite number of cells from a specific subset of leukocytes from a whole blood specimen. Therefore, the total volume of blood required for such a protocol was variable and irrelevant. However, there was a persistent demand for absolute CD4 counts. The Center for Disease Control publication defined AIDS based on a CD4 T-cell count of 200 or less [10]. The flow cytometry industry was slow to respond. As the demand continued, there were two options:

i) Introduce a volumetric sample-handling system that will provide intrinsic volumetric counts or
ii) Modify the current protocol by adding an extrinsic volumetric strategy that requires additional microfluorospheres of known volume and concentration to each specimen to count (cells/ml).

This way it is possible to calculate the volume of the blood sample in order to express the CD4 T-cell count. It was not until the mid 1990s that the extrinsic single-platform volumetric immunophenotyping protocol for absolute counts was introduced for general clinical application. In figure 2 the various options are listed for determining absolute CD4 counts with the single-platform technology (SPT). The alternative options for SPT will be discussed later in this chapter. It is important to remember that absolute counts reported in the literature before the introduction of SPT are in general unreliable. The need to report CD4 T-cell count as percentage of lymphocytes still exists. Currently they are the required method for CD4 T-cell monitoring of HIV-positive children under the age of 5 years (table 1).

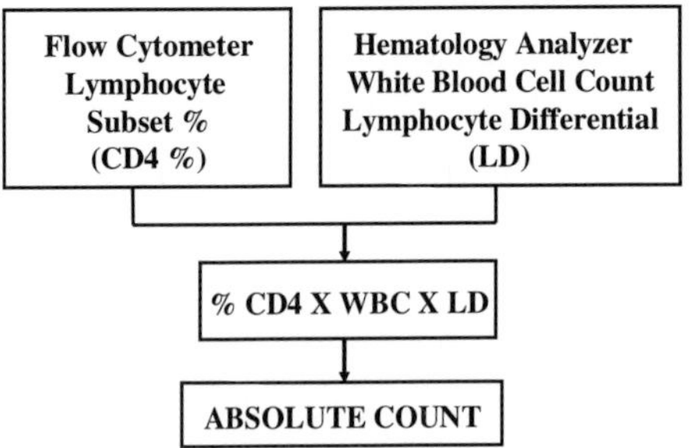

Fig. 1. Dual platform absolute CD4 T-cell count (DPT).
To obtain an absolute CD4 T-cell count with flow cytometry, the traditional approach required the implementation of a DPT solution. Specimen-specific data from two different instruments are collected. The top left side box represents the required information from the flow cytometer and the top right side box indicates the information that is required from a hematology instrument. Once the CD4 percentage is multiplied with the lymphocyte differential, the absolute CD4 T-cell count is obtained.

Since the introduction of ART, the measurement of CD4 and CD8 T-cells during the administration of experimental ART cocktails became an important part of selecting effective protocols for managing HIV infection. It is well known that CD4 T-cells rise following aggressive ART [8]. However, this rise does not return to previral exposure or to normal levels. Current therapies can significantly extend life expectancy of an infected patient. For various reasons viral resistance during ART can occur. The most common cause of drug resistance is noncompliance with the antiretroviral medication regimen. Even with the most rigorous drug compliance, HIV is never completely eliminated from an individual, thus taking antiretroviral medication is a life time commitment. Noncompliance usually accelerates the shift to drug-resistant virus and death often occurs as a result of complications with opportunistic infections. Thus, evaluating the effectiveness of therapies is critical and monitoring viral load and the immune status are the two most important measures. A number of studies based on flow cytometry have evaluated subsets of T-cells to determine whether there is depletion or expansion of specific subsets that may have prognostic value. Many of these studies have found increases or decreases in subsets that correspond to functional leukocyte subtypes, with various levels of activation marker expression. These clinical trials improve our understanding of how HIV impacts on the immune system. Additional uses of flow cytometry in HIV infection include evaluating cell function, cell activation, proliferation in vitro, CTL responses, detection of cytokine-producing cells, and measurement of HIV antigens in situ. None of these research applications found their way into the clinical service laboratories. The prognostic values of these measurements have yet to be established in large clinical studies. As the nature and complexity of the interaction between

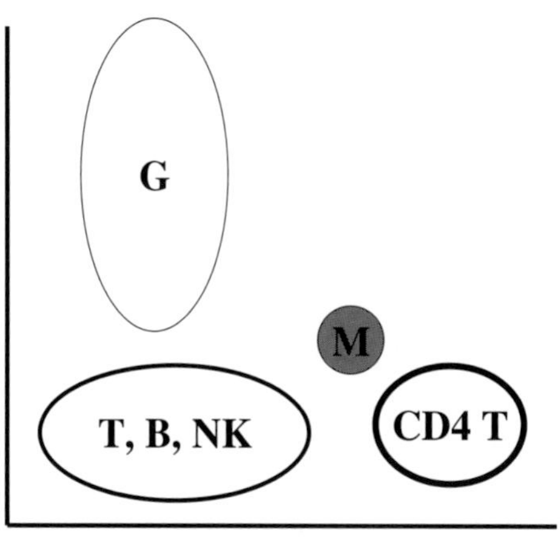

Fig. 2. Primary CD4 gating.
This immunophenotyping protocol requires only 1 mAb to perform a CD4 T-cell measurement. It utilizes a heterogeneous dual-parameter histogram with light scatter versus a CD4-associated fluorescent or a colloidal-gold-conjugated signal. It is possible to discriminate at least four subsets of leukocytes. The cluster with G, T/B/NK, M and CD4 T designations represent granulocytes, a combination of T suppressor cells (CD8), natural killer cells (CD16/CD56), monocytes (CD4) and CD4 T-cells, respectively. With such a protocol it is critical to have reliable spatial resolution between the monocyte cluster and the CD4 T-cell cluster as they are situated relatively close to each other and both have CD4 surface receptors. Currently, there are three alternative flow cytometers that use this protocol for CD4 T-cell enumeration. Two use a fluorescent tag on the anti-CD4 mAb, the other uses a colloidal gold tag attached to the anti-CD4 mAb.

Table 1. The WHO classification of HIV-associated immunodeficiency in children under 5 years of age[a]

Classification of HIV-associated immunodeficiency	Age-related CD4 values			
	<11 months, %	12–35 months, %	36–59 months, %	>5 years, cells/mm^3
Not significant	>35	>30	>25	>500
Mild	30–35	25–30	20–25	350–499
Advanced	25–29	20–24	15–19	200–349
Severe[b]	<25	<20	<15	<200 or <15%

[a] Healthy newborn babies have CD4 T-cell counts that can be four or five times higher than in adults. The higher CD4 counts remain significantly above adult values for the next few years. Accordingly, infants who are living with HIV also have relatively high CD4 counts. A count that is above normal adult value may already indicate immunodeficiency in a very young child. At the same time, while the percentage of lymphocytes that are CD4 T-positive is also higher in infants, it is considerably less subject to temporal fluctuations. Pediatricians found that the CD4 T-cell percent is a much reliable measure to use to classify the severity of immunodeficiency in children. The current WHO classification scheme is outlined.

[b] Level at which patient has >10% chance of progressing to AIDS or >5% to death.

antiretroviral medication and T-lymphocytes became better understood, the implementation of ART moved slowly from monotherapy to include eventually four distinct groups of drugs. Each type of drug category blocks a unique processing pathway that is critical for viral reproduction inside the CD4 T-cell. The categories include drugs that inhibit fusion that takes place on the surface of the CD4 T-cell. There are the reverse transcriptase inhibitors: they block RNA transcription to DNA strands that can be integrated into the host's DNA. The reverse transcriptase inhibitors represent the oldest class of antiretroviral drugs. The third type of drugs are the integrase inhibitors, they block the reintegration of the viral genetic material expressed as DNA from reversing back to RNA. Finally, there are the protease inhibitors that prevent the assembly of newly synthesized viral components into the next generation of HIV to be discharged into the circulating blood of the already infected host. In the past decade HIV viral load assay sensitivity has increased dramatically, thus it is becoming possible to measure small incremental change in viral load.

Over the years there has been a lot of interest to study alterations in CD4 T-cell subset redistribution during treatment trials. The goal was to determine if a particular T-cell subset is preferentially lost in HIV infection or if response to ART can be predicted by alterations in CD4 T-cell subset distribution. Early studies indicated that the CD4 T-memory cells (CD29 or CD45RO) were preferentially infected [19, 20] with HIV. There is evidence that both memory and naive cells return with therapy [21]. The overall conclusion is that counting CD4 naive and memory T-cells is not essential for routine monitoring of patient progress during ART. The clinical importance of monitoring the CD8 T-cell subset during HIV infection. CD8 T-cells increase in numbers and percent during the progression of HIV disease. Most of these cells possess antigens similar to those found in response to almost any viral infection. This includes increases in cells that express activation antigens such as HLA-DR and CD38 [22]. The utility of CD38 expression on CD8 T-cells has been controversial ever since Giorgi et al. [22] published their findings that indicated a direct relationship between numbers of CD38+ CD8 cells and viral load. It has been suggested by Janossy [23] that the CD38 expression can be monitored with flow cytometry and interpreted as a 'poor man's viral load' assay. CD8 T-cells function primarily as CTLs. They recognize viral particles presented on antigen-presenting cells. HIV can be taken up by these cells to be processed and presented to the specific T-cells. This CTL process is completed by the signal delivery to kill the infected cell. There is a vigorous in vitro CTL response which is active from the early days of the infection and is maintained for a considerable time throughout the disease [24]. During the late stages of HIV infection this response disappears. CTL activity can be identified by various activation antigens on the cell surface. However, they are not considered practical markers for routine clinical disease monitoring. CD8 cells that are CD28 negative can be apoptotic [25]. There is also a shift from naive cells to memory cells; naive cells are identified with a marker combination expressing both

CD45RA and CD62L [26]. In situations where first- and second-line ARTs have failed, some early activation markers can be helpful to predict which salvage combination therapy may be the most effective [27].

The Response to HIV in the Northern Hemisphere

This section offers a brief passage though the technical evolution of CD4 T-cell immunophenotyping. For obvious reasons, most of the biotechnological response to the HIV pandemic for the first 20 years came from the industrialized Northern Hemisphere. HIV-related laboratory technologies improved steadily over time. During the first decade, CD4 T-cell counting moved from manual epifluorescent microscopy, to polychromatic flow cytometry with automated software and some capacity for batch analysis. In the 1980s and early 1990s, there was an assumption that polychromatic marker application with multiparameters will be the most robust way to define the various T-cell subset populations. The number of mAbs conjugated to fluorochromes for CD4 T-cell immunophenotyping and the corresponding assay costs began to escalate dramatically. Traditional multiparametric analysis that always began with dual SS was becoming problematic as complex color compensation protocols were required for polychromatic data interpretation. Traditionally, clinical histograms depict two-parameter analysis. However, the listmode data files often contained five-parameter data sets. Synchronously with the emergence of these immunophenotyping options, immunologist gained considerable understanding of the difference between lineage, function and activation markers. This was a great opportunity to devise methods for leukocyte subset identification with fewer markers. The initial hurdle that had to be confronted was the chronic dependence on the homogeneous dual light scatter (FSC versus SSC) gating parameters for lymphocyte population identification. Once the concept to switch to a heterogeneous pair as initial parameter for gate selection was accepted, new options opened for immunophenotyping of CD4 T-cells. The options for significantly more robust, reliable and more cost-effective assays were availed. The original heterogeneous parameter-gating approach for simpler CD4 T-cell determinations used two tubes, CD3/CD4 and CD3/CD8. Once there were flow cytometers with tricolor capacity, a single tube containing CD45, CD3, and CD4 was quite effective [28]. With this triple color combination the heterogeneous gating combines SSC and CD45 fluorescence. CD45 expression varies between different leukocyte populations, and its combination with SSC allows for the identification of lymphocytes. Exploiting this phenomenon of heterogeneous (combination of intrinsic and extrinsic parameters) gates [29] immunophenotyping is simplified. Once the two families of tandem dyes, PerCP/ECD and Cy-Chrome, with good spectral separation were developed, the introduction of instruments with four-color detection was possible. The polychromatic strategy can be cumbersome for some applications but for CD4 T-cell enumeration with daily

Table 2. A summary of currently recommended four-color CD4 T-cell enumeration methods for adults[a]

Current clinical instruments	Automated sample processing	Required reagents				
		gating markers	lysing solution	counting beads	sheath fluid	total number of reagents
BD FACSCalibur	an option	CD45/CD3	yes	yes	yes	7
BD FACSCanto	an option	CD45/CD3	yes	yes	yes	7
BD FACSArray	an option	CD45/CD3	yes	optional	yes	6
BC Epics XL	an option	CD45/CD3	yes	yes	yes	7
BC FC 500	an option	CD45/CD3	yes	yes	yes	7

[a] All the frequently used flow cytometers that are in use in resource- rich countries for CD4 T-cell immunophenotyping are considered. The majority of laboratories use either three or four-color FDA-approved protocols. The reagents used for the four-color protocols depend on whether the instrument has one or two lasers and on the corresponding filter configuration available on the specific version of the instrument. Therefore, only certain combinations of fluorochromes work on instruments with four or more color combinations. The reagent list contains three additional reagents that are not always required with the alternative methods; they are: the reference beads, sheath fluid and lysing solution.

quality control, the flow cytometry laboratory operation was significantly simplified. This was accomplished by either adding a fourth photomultiplier tube for detecting a far red tandem dye (Beckman Coulter Corporation, Miami, FL, USA) or adding an additional laser as well as a fourth PMT for the excitation and detection of a dye allophycocyanin that is excited at a longer wavelength (red diode at 635 nm) than the commonly used dyes (488 nm) (Becton Dickinson Biological Systems, San Jose, CA, USA) (see 'Selection and Combination of Fluorescent Dyes', pp. 107). It is clear that the mAb combinations, such as the panel that is most appropriate for HIV immunophenotyping (CD45/CD3/CD4/CD8), work well with both single- and dual-laser instruments [29]. With simultaneous four-color analysis, it is possible to gate first on the bright CD45 population then identifying T-cells by using CD3, or to do the opposite gate on CD3 first and with a back-gating technique confirm that all those cells are also CD45 bright. Once the heterogeneous gating was accepted, immunophenotyping technology kept moving on to five- and six-color capacity including clinical instruments. The four-color protocol is still the current state-of-the art method for CD4 T-cell measurement [30, 31]. Table 2 illustrates the number of reagents each flow cytometer utilizes for CD4 T-cell measurement assays approved by the Federal Drug Administration (FDA). All of them have automated software for execution of gate selection for immunophenotyping, but automated sample processing is an additional option for all of them. In a situation when an HIV patient is just enrolled for long-term care, it is customary to run an additional four-color panel of markers that includes CD19, CD14, CD16/CD56 and CD45. The second four-color panel is usually not repeated again unless some unusual complications arise with that patient.

The management of a CD4 immunophenotyping laboratory requires a quality control strategy similar to the one that would apply to most diagnostic laboratories. The first level of quality control is the within laboratory daily quality control that covers the entire assay process [32]. The second type is the interlaboratory quality assessment program. By adhering to these two regimens, the CD4 assay most likely will yield less intra- and interlaboratory variation. Over the past two decades, some very efficient external quality assurance programs have been developed for HIV drug and vaccine trials. Performance evaluation programs for percent CD4 T-cells have indicated that with frequent participation laboratories can produce similar results and the error margin will diminish with time [33]. The daily quality control protocol can be further subdivided. The effective quality management of both, the specimen processing and result processing are critical. The quality of the specimen at the time of arrival in the laboratory can be an issue. Excess heat or freezing, of specimen during transport can cause significant problems. Reagents and methods for labeling white cells and lysing red cells also need to be quality controlled. The manufacturers of reagents have strict criteria for their use.

Some Problems and Pitfalls Related to CD4 Immunophenotyping

As with all cell-based diagnostic technologies, there are some occasional complications. The most fundamental problem with this type of cell-based immunoassay is that there is no stable reference or standard preparation available that can sit on a shelf for years to verify instantly instrument performance the way it is possible with an instrument that is to measure glucose levels, as glucose controls can be stored for years. In general, the currently recommended protocols are quite robust, and procedural complications are infrequent. The most likely problem is when there are missing or heat-damaged reagents or an abnormal cell cluster distribution occurs with some of the autogating methods. Proper controls will resolve the first problem. With the second situation, the automated software from time to time is unable to define the correct leukocyte subpopulation, and therefore will not provide an acceptable result. Often this situation can be salvaged by switching to a manual gating protocol. Unfortunately, there are many immunophenotyping facilities where not all the staff is trained to run an assay with a manual over-ride protocol. Most external quality assessment programs for HIV immunophenotyping offer to evaluate both the percent of lymphocytes and absolute CD4 T-cell counts. When a laboratory participates in an external quality assessment program and they are using DPT for absolute count technology, they will most likely not perform well compared to laboratories using SPT. The logical solution is to switch to SPT absolute-count technology or report only the percent CD4 values.

The global fight to bring the AIDS pandemic under control and the struggle to develop a truly affordable CD4 T-cell-counting technology; both have eluded us thus far. The implementation of ART is in progress in many resource-limited regions. Without laboratory infrastructure, the scaling-up ART effort in rural locations is an irresponsible act that needs to be improved immediately. The need for CD4-counting instrumentation that is affordable and sustainable is critical now. CD4 T-cell counting is essential for finding HIV-positive individuals who qualify for treatment based on their low CD4 T-cell count. The challenge is to design a CD4 T-cell counter that is simple to operate. It must be robust to survive in the tropical environment that is known to be hostile to sensitive instrumentation. The system must be reproducible, yet offer low cost per test. In the past few years the price for ART in many parts of resource-limited regions underwent a 50- to 100-fold reduction. If a similar cost reduction model is applied to immunophenotyping, the price of CD4 testing would drop from the current minimum of USD 4.– to 25.– to under a USD 1.– per test. The current goal is to make assays available that cost under USD 2.– per test. The emphasis has to be on a sustainable technology that is deliverable without compromise in assay quality. Howard Shapiro leads a small group of scientists who consider flow cytometry an overkill when the requirement is an accurate count of dual-labeled cells (CD4/CD3) from a unit volume of whole blood [34]. However, until his type of cytometry with digital imaging is ready, flow cytometry is the viable option. There are considerable varieties of alternative methods that have reached the market recently. Some of the technical issues that govern affordability having been addressed, at least in part, the challenge now facing the cellbased diagnostics industry is to create a CD4 T-cell counter that is simple to operate and also effectively reduces test costs per patient. Based on recent developments, there are three possible directions for future alternatives. First, is the extreme minimalist approach try to reduce the cost of a dedicated CD4 T-cell counting machine to about USD 5,000.– in the approach that Cytometry For Life (C4L) is developing [35]. The second option is the middle-of-the-road position where the instrument has some flexibility in terms of assay options and the USD 30,000.– price of such flow cytometry platform could be amortized with low-cost steady reagent kit purchases such as what Guava Technologies is currently offering. The third option is the other extreme where the idea is to reconcile the fact that a small clinic has to deal with several frequently encountered infectious diseases all at once. It is important to recognize that HIV cannot be isolated from the other major infectious diseases such as malaria, tuberculosis and various coinfections, especially in remote rural clinics. These same clinics must also deal with the various opportunistic infections that are the hallmark manifestations of chronic immune deficiency. Therefore, a robust diagnostic platform that cost USD 40,000.– that is a flow-cytometry platform, adopted to include suspension array technology, would deal with most immunoassays. Even without these three scenarios there are already an interesting variety of alternative flow cytometers available for CD4 T-cell counting. New reagents and novel gating strategies have been implemented. Various heterogeneous gating strategies have led to cost-effective gating strategies. There are three gating protocols that were adopted for alternative CD4 T-cell counting instrumentation. Some require dedicated new instruments; others work with existing flow cytometers. They include:

i) Primary CD4 gating,
ii) T-cell gating,
iii) CD45-fluorescence-based panleukogating.

Primary CD4 Gating

In 2000, the first, the primary CD4 gating protocol (using a single reagent), CD4 mAb, was developed [23, 36]. It is a heterogeneous combination of a function-specific marker anti-CD4 with SSC as the second parameter. Both the Partec CyFlow Counter and Apogee A40 use this protocol for absolute CD4 T-cell counting. These two instruments rely on clear identification and separation of CD4 T-cells from the monocyte population which in some situations can be problematic unless careful attention is given to identifying events generated with the SSC. See figure 2 for the heterogeneous primary gating histogram for a typical leukocyte distribution.

T-Cell Gating

About 15 years ago, anti CD3 mAbs were already recognized as a T-cell-lineage-specific marker. A duo mAb was introduced as a combination with an anti CD4, a function specific marker with a T-cell-lineage marker on the FACSCount instrument [37]. With this novel homogeneous automated gating strategy, fluorescence detection is triggered to identify dual-labeled cells (fig. 3). Therefore, in this case there is no need to lyse the red cells as only fluorescent labeled cells are acknowledged by the optical system. The FACSCount does require reference beads with each specimen to generate an accurate absolute count. This T-cell count strategy requires the total of three reagents, two mAbs (CD3 and CD4 mAb) and counting beads. A few years ago, Guava Technologies released an instrument that also uses similar gating strategy, but requires lysing reagent. The current configuration has automated gating software. This instrument does not require sheath fluid as it uses a μ-capillary system for sample delivery.

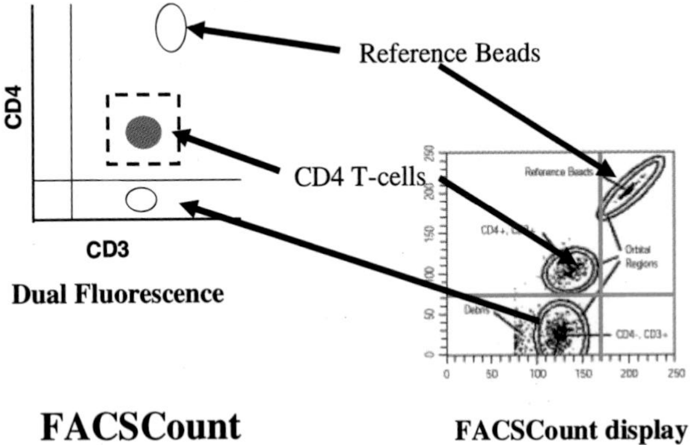

Fig. 3. Variations on the T-cell gating method.
The first commercially successful alternative flow cytometry based CD4 T-cell counting protocol uses 2 mAbs. The identification of the CD4 T-cell population is from a homogeneous dual-parameter histogram. Cell clusters are identified with the simultaneous detection of CD3 and CD4 fluorescent signals. This approach eliminates the potential problem with monocyte contamination as the protocol incorporates a lineage- and a function-specific marker CD3 and CD4, respectively. The left panel illustrates a theoretical dual-parameter histogram, the right panel is the actual graphic display that the instrument generates. This protocol uses an extrinsic SPT for calculating absolute counts. The reference beads are displayed on both panels. Currently, two alternative flow cytometers use this gating principle.

CD45 Fluorescence-Based Panleukogating

The panleukogating strategy was developed for conventional high-throughput hospital clinical flow cytometers where there is access to a hematology analyzer. This innovative approach requires only two mAbs [38, 39]. By obtaining the total white blood cell count from a hematology instrument, the need for one additional reagent for absolute count was eliminated. The omission of the counting beads is achieved without compromising the quality of the CD4 T-cell count [39, 40]. Figure 4 illustrates the simplicity of the panleukogating three-step strategy:

i) First, a large manual gate is drawn around all CD45 labeled cells called 'G' gate.
ii) A smaller second gate is constructed inside the large gate that is called 'g' gate. It includes only the lymphocytes, the brightly labeled CD45 cells. A heterogeneous histogram 'G' (SSC/CD45) provides the lymphocyte differential and CD4 T-lymphocyte as percentage.
iii) A second heterogeneous histogram (SSC/CD4) provides the location of all the CD4-positive lymphocytes.

The absolute CD4 T-cell count is generated by multiplying the two values from the flow cytometer with the one from the hematology analyzer (fig. 1). The original panleukogate protocol was further developed by Glencross et al. [41]. It is used in

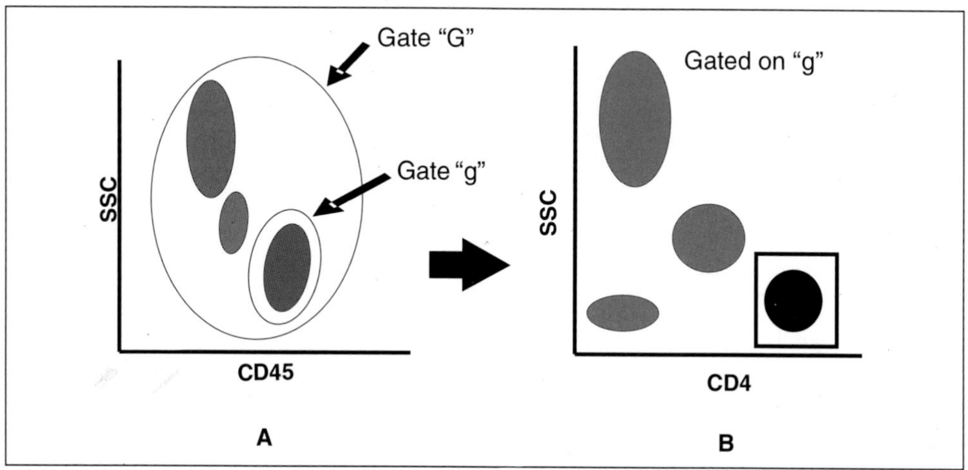

Fig. 4. The Panleukogating protocol.
This is another alternative protocol that requires only 2 mAbs; they are CD45 and CD4. This assay can be performed on any clinical flow cytometer. The first heterogeneous dual-parameter histogram has CD45 and SSC as the two parameters. They are illustrated on panel **A**. The G-gate includes all the leukocytes while the g-gate is restricted to all the lymphocytes. On panel **B** is the second heterogeneous dual-parameter histogram that uses CD4 and SSC as the two parameters. Because CD4+ cells are considered only from the bright CD45 population (from lymphocytes) there is no obvious issue with monocyte contamination. This protocol requires a total white blood cell count from a hematology instrument to generate an absolute CD4 T-cell count, therefore it is a DPT. There is another version of this protocol that requires reference beads. With that modification, the panleukogating protocol becomes an extrinsic SPT which does add to the test cost.

South Africa as the nationwide approved SPT. Because both these protocols use a heterogeneous gating strategy, they work well on samples that have been drawn up to 5 days earlier [41, 42].

All the above alternative protocols and instrument options reviewed use fewer numbers of reagents compared with currently approved methods in use in resource-rich countries. In general the cost reductions associated with the alternative methods do not represent a compromise in assay quality. In fact, by most standards they represent a significant improvement over the accepted Western immunophenotyping technology. The alternative methods presented here are still too complex and do not address the enormous economic disparities that are prevalent in Africa and in many other parts of the world. None of the options presented incorporated all the features and conditions that are essential features when selecting sustainable CD4-counting technologies for resource-poor regions. For example, methods requiring cold chain for reagent transportation, manual gate settings, manual pipetting, methods requiring shipped sheath fluid have not seriously considered the economic and geostrategic realities facing the less fortunate yet vast segment of humanity with profound and debilitating disease burden. On the bright side, there are two systems that work

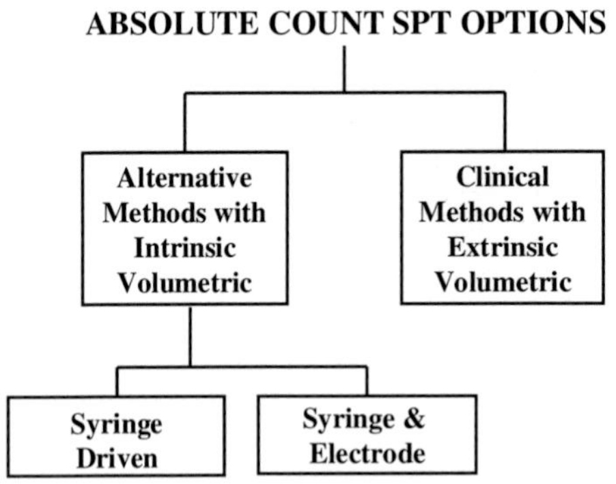

ABSOLUTE COUNT SPT OPTIONS

Alternative Methods with Intrinsic Volumetric

Clinical Methods with Extrinsic Volumetric

Syringe Driven

Syringe & Electrode

Fig. 5. Absolute count with single-platform technology (SPT).
Each manufacturer has the option of installing a volumetric delivery device on the flow cytometer, an intrinsic cell metric device or use reference beads to monitor flow rate to calculate unit volume. This latter option is referred to as the 'extrinsic volumetric method'. Most alternative manufacturers have decided to go with the intrinsic absolute count option. All but one use a volumetric delivery system driven with a precision syringe. This approach is depicted on the far left box on the illustration. In the second box is the other intrinsic volumetric solution, a system that has a feedback control to the syringe from electrode- driven systems. The electrodes detect air-liquid and liquid-air phase changes at the beginning and at the end of the pumping cycle of the specimen through the detection zone.

Table 3. Alternative CD4 T-cell percentage methodologies for pediatrics[a]

Alternative instruments	Method complexity	Total lymphocyte	All required reagents				
			MAbs markers	lysing solution	counting beads	sheath fluid	total number
Apogee A40	semi-automatic autogating	CD45	CD3/CD45 CD4	yes	no	yes	5
BD FACSCount	semi-automatic autogating	nuclear stain cocktail-2	CD3/CD4 CD14/CD15	no	yes	yes	6
Guava PC	semi-automatic autogating	cocktail-2	CD3/CD4 CD2/CD19	yes	no	no	5
Partec CyFlow Counter	manual gating	CD45	CD4	no	no	yes	3
PointCare NOW	fully automatic	lymph count	CD4	yes	no	yes	3

[a] There are five instruments that belong to the alternative category. The classification-related automation is defined as manual, semi-automated and as fully automated. A manual method is where both sample preparation and data acquisition require manual intervention. Semi-automatic means that sample processing is manual, but acquisition gates are set automatically. A fully automatic system is where both sample processing and data acquisition are a single seamless automated process. The total lymphocyte count is derived using four different strategies. One instrument has a four-part hematology differential capacity built in. Two instruments use the full range CD45 fluorescent signals to obtain the lymphocyte ratio. There are two systems that use a mAb cocktail to calculate the total lymphocyte count. One of them includes a cell nuclear stain. The total reagent requirement includes not just the mAbs but reference beads, lysing and sheath solutions.

Table 4. Alternative CD4 T-cell absolute-count methodologies for adults[a]

Alternative instrument	Method complexity	All required reagents				
		mAb markers	lysing solution	counting beads	sheath fluid	total number
Apogee A40	semi-automatic autogating	CD4	yes	no	yes	3
BD FACSCount	semi-automatic autogating	CD3/CD4	no	yes	yes	4
Guava PC	semi automatic autogating	CD3/CD4	yes	no	no	3
Partec CyFlow Counter	manual gating	CD4	no	no	yes	2
PointCare NOW	Fully automatic	CD4	yes	no	yes	3

[a] This is a most frequently requested alternative technology. Three out of five instruments listed here use the primary CD4 gating protocol for CD4 T-cell absolute count. Two instruments use the T-gating strategy (CD3/CD4). Three instruments use three reagents. The other two go with either one less or one more reagent.

with dry reagents, one technology operates without sheath fluid. Four alternative systems use intrinsic volumetric system for absolute cell counting (fig. 5). The Cy-Flow Counter has the most sophisticated volumetric solution where electrodes detect the air-liquid interphase to control the syringe assembly. These are major breakthroughs related to cost-effective management of low-volume rural clinics. Tables 3 and 4 offer a summary of various features these alternative technologies offer.

Problems and Pitfalls Related to Alternative Technologies for CD4 Immunophenotyping

The issue of quality control has not been emphasized adequately for these technologies that must work most of the time under extremely harsh environmental conditions with limited supervisory or qualified staff. For example, most alternative instruments do not have any onboard built-in automatic record-keeping capacity of daily quality control performance data. Very few of these alternative systems have undergone acceptable evaluation. They all require a multicenter study in the field where they will be performing. Such a validation studies must be conducted independent of the manufacturer and should be published in a reputable peer-reviewed international journals. Almost all the alternative technologies require rigorous cleaning/flushing cycles in order to provide dependable service over time, yet very few offer proper training and record-keeping procedure. They also often neglect to calculate the additional reagent and other required consumable costs that are related to daily routine maintenance. In some situations bio-hazardous waist disposal can be a very difficult and costly issue that is frequently neglected. Yet, biosafety is part of the cost of delivering diagnostic laboratory services. It is important to choose an alternative method that is compatible with one of the major international external quality assessment schemes and or the instrument's performance can be monitored remotely.

Conclusion

The section dealing with conventional CD4 T-cell counting technologies as they are used in the Northern Hemisphere included a few historical details. It was to help the reader to follow the not always logical evolutionary path that this relatively contemporary cell-based diagnostic technology pursued. It is quite remarkable that the acronym 'FACS', which stands for 'fluorescence-activated cell sorting', has been in use since the late 1970s, yet it was not until the mid 1990s before the homogeneous gating combination (the dual light scatter) was replaced with a heterogeneous gating strategy (using CD45 fluorescence and one scatter) to perform immunophenotyping on T-cell subsets. This reluctance to discard the well-characterized but flawed dual light scatter strategy was probably a way to keep the cell cluster pattern recognition concept that provided a direct link to traditional pathological analysis that holds deep affiliation with epifluorescent microscopy. In the late 1980s, the standard CD4-counting method required six tubes with dual-labeled specimens in each to perform a CD4 T-cell assay. Therefore a single test included 12 units of mAbs (table 5). This misguided and excessive choice was economically devastating and was inhuman to use of testing infants. Most current guidelines suggest a four-color single-tube assay [30, 31]. Clearly, 4 units of mAbs compared to 12 is a move in a more reasonable and cost-effective direction. Also having all the leukocytes processed simultaneously in the same test tube makes the assay far more robust compared to the earlier times when 6 different preparations processed, analyzed, interpreted individually and then collated to produce an integrated and unified result set (table 5). The section covering the alternative methods provides some hope that entrepreneurial spirit is alive in cellular diagnostics in general and more specifically as it relates to immunophenotyping with newly designed flow cytometers. There are hundreds of instruments running in Africa without hydrodynamic focusing and sheath fluid, and still others without fluorescent tagged mAbs. There are FDA-approved instruments out there with only 1 mAb to measure both absolute and percent CD4 T-cells. Unfortunately, there also some which have never been adequately validated. Some systems use dry reagents that never require refrigeration. While none of the available instruments have all the desired specifications, each company has made some commendable progress. It is a good time to reflect on a wish-list to assess what is really required

Table 5. Evolution of CD4 T-cell enumeration as revealed by the shift of number of mAbs required per assay over the years[a]

5-year intervals	Number of tubes	CD marker combination(s)	n	Absolute count	Lymphocytes %
1980	2	4[b] and 8[b]	1+1	–	Ratio
1985	1	4/8	2	–	+
1990	6	45/14, 3/4 8/4, 3/19 3/NK, isotype	12	–	+
1990	2	3/4, 3/8	4	?	+
1995	1	45/3/4	3	+	+
2000	1	3/4	2	+	+
2005	1	4	1	+	+

[a] The essential objectives are to provide quality-robust results at low cost. Initially, the emphasis was on finding methods that were reliable. More recently, since the improved understanding of cellular immunology, immunophenotyping is better understood. There are significant efforts ongoing to reduce the number of reagents for this assay. It seems that the industry has gone a full circle with the number of mAbs that are essential for CD4 T-cell counting.
[b] The instruments could only handle one color at a time so two separate assays were performed to get a CD4/CD8 ratio.

for a sustainable CD4-counting system in a resource poor rural regional laboratory. At the same time as the critical validation studies from these alternative technologies are published, it would also be a good time to reconsider which instruments are the required for CD4 T-cell counting for the industrialized world. It is an opportunity to revisit what are the essential requirements:

i) Must eliminate the need for cold-chain transport and local refrigeration of reagents. This is a key step to secure cost-effective inventory management in remote rural locations that include locations such as first-nation reserves in North America.
ii) An instrument must have a significantly reduced carbon footprint. This means either low-energy laser-or light-emitting diode as the light source with built-in effective protection against electric supply irregularities.
iii) The new technology should use a robust CD4 T-cell-counting protocol requiring minimal numbers and quantity of reagents. It must also be compatible with external quality assessment schemes.

Finally, a system that meets all the above requirements will most likely be also appropriate for serving laboratories in resource-rich countries. Cost-effective health care delivery is a universal concern. The need for dramatic reduction of the escalating health care costs in the US was elegantly articulated by Orszag and Ellis [43]. The sooner these new cellular diagnostic devices reach the market, the better it is for all of humanity regardless on which hemisphere you call as your home.

References

1 Keating P, Cambrosio A: Biomedical Platforms-Realining the Normal and the Pathological in Late-Twentieth-Century Medicine. Cambridge, MIT Press, 2003.

2 Ammann AJ, Abrams D, Conant M, Chudwin D, Cowan M, Volberding P: Acquired immune dysfunction in homosexual men: immunologic profiles. Clin Immunol Immunopathol 1983;27:315–325.

3 Schroff RW, Gottlieb MS, Prince HE, Chai LL, Fahey JL: Immunological studies of homosexual men with immunodeficiency and Kaposi's sarcoma. Clin Immunol Immunopathol 1983;27:300–314.

4 Ho DD, Moudgil T, Alam M: Quantitation of human immunodeficiency virus type 1 in the blood of infected persons. N Engl J Med 1989;321:1621–1625.

5 Nicholson JKA, Spira TJ, Aloisio CH, Jones BM, Kennedy S, Holman RC: Serial determinations of HIV-1 titers in HIV-infected homosexual men: association of rising riters with CD4 T cell cepletion and crogression to AIDS AIDS Res Hum Retrovir 1989;5:205–215.

6 Turner BJ, Hecht FM, Ismail RB: CD4+ T-lymphocyte measures in the treatment of individuals infected with human immunodeficiency virus type I. Arch Intern Med 1994;154:1561–1572.

7 Stein DS, Korvick JA, Vermund SH: CD4+ lymphocyte cell enumeration for prediction of clinical course of human immunodeficiency virus disease: a review. J Infect Dis 1992;165:352–363.

8 Koot M, Schellekens PTA, Mulder JW, Lange JMA, Roos MTL, Coutinho RA, Tersmette M, Miedema F: Viral phenotype and T cell reactivity in human immunodeficiency virus type 1-infected asymptomatic men treated with zidovudine. J Infect Dis 1993;168:733–736.

9 O'Shea S, Rostron T, Hamblin AS, Palmer SJ, Banatvala JE: Quantitation of HIV: correlation with clinical, virological, and immunological status. J Med Virol 1991;35:65–69.

10 Centers for Disease Control and Prevention: 1997 revised guidelines for performing CD4+ T-cell determinations in persons infected with human immunodeficiency virus (HIV). Morbid Mortal Wkly Rep 1997;46:1–29.

11 McAnulty JM, Modesitt SK, Yusem SH, Hyer CJ, Fleming DW: Improved AIDS surveillance through laboratory-initiated CD4 cell count reporting. J Acquir Immune Defic Syndr Hum Retrovirol 1997;16:362–366.

12 Shapiro HN: Parameters and probes; in Shapiro HN (ed): A Practical Flow Cytometry. New York, Wiley Liss, 1994.

13 McDougal JS, Kennedy MS, Sligh JM, Cort SP, Mawle A, Nicholson JKA: Binding of HTLV-III/LAV to CD4+ T cells by a complex of the 100K viral protein and the T4 molecule. Science 1986;231:382–385.

14 Jones NM: The role of receptors in the HIV-1 entry process. Eur J Med Res 2007;12:391–6.

15 Fahey JL, Taylor JMG, Detels R, Hofmann B, Melmed R, Nishanian P, Giorgi JV: The prognostic value of cellular and serologic markers in infection with human immunodeficiency virus type 1. N Engl J Med 1990;322:166–172.

16 Bottinger B, Morfeldt-Manson L, Putkonen P, Nilsson B, Julander I, Biberfeld G: Predictive markers of AIDS: A follow-up of lymphocyte subsets and HIV serology in a cohort of patients with lymphadenopathy. Scand J Infect Dis 1989;21:507–514.

17 Yarchoan R, Venzon DJ, Pluda JM, Lietzau J, Wyvill KM, Tsiatis AA, Steinberg SM, Broder S: CD4 count and the risk for death in patients infected with HIV receiving antiretroviral therapy. Ann Intern Med 1991;115: 184–189.

18 Giorgi JV, Detels R: T-Cell subset alterations in HIV-infected homosexual men: NIAID multicenter AIDS cohort study. Clin Immunol Immunopathol 1989;52:10–18.

19 Schnittman SM, Greenhouse JJ, Psallidopoulos MC, Baseler M, Salzman NP, Fauci AS, Lane HC: Increasing viral burden in CD4+ T cells from patients with human immunodeficiency virus (HIV) infection reflects rapidly progressive immunosuppression and clinical disease. Ann Intern Med 1990;113:438–443.

20 Cayota A, Vuillier F, Scott-Algara D, Dighiero G: Preferential replication of HIV-1 memory CD4+ subpopulation. Lancet 1990;336:941.

21 Dolan MJ, Lucey DR, Hendrix CW, Melcher GP, Spencer GA, Boswell RN: Early markers of HIV infection and subclinical disease progression. Vaccine 1993;11:548–551.

22 Giorgi JV, Liu Z, Hultin LE, Cumberland WG, Hennessey K, Detels R: Elevated levels of CD38+CD8+ T cells in HIV infection add to the prognostic value of low CD4+ T cell levels: Results of 6 years of follow-up. J AIDS 1993;12:904–912.

23 Janossy G: The changing pattern of 'smart' flow cytometry (S-FC) to assist the cost-effective diagnosis of HIV, tuberculosis, and leukemias in resource-restricted conditions. Biotechnol J 2008;3:32–42.

24 Walker BD, Plata F: Cytotoxic T lymphocytes against HIV. AIDS 1990;4: 177–184.

25 Lewis DE, Ng Tang DS, Adu-Oppong A, Schober W, Rodgers JR: Anergy and apoptosis in CD8+ T cells from HIV-infected persons. J Immunol 1994;153:412–420.

26 Roederer M, Dubs JG, Anderson MT, Raju PA, Herzenberg LA: CD8 naive T cell counts decrease progressively in HIV-infected adults. J Clin Invest 1995;95:2061–2066.

27 Shepard BD, Loutfy MR, Raboud J, Mandy F, Kovacs CM, Diong C, Bergeron M, Govan V, Rizza SA, Angel JB, Badley AD. Early changes in T-cell activation predict antiretroviral success in salvage therapy of HIV infection. J Acquir Immune Defic Syndr 2008;20.

28 Nicholson JKA, Jones BM, Hubbard M: CD4 T-lymphocyte determinations on whole blood specimens using a single-tube three-color assay. Cytometry 1993;14:685–689.

29 Mandy FF, Bergeron M, Minkus T: Principles of flow cytometry. Transf Sci 1995;16:303–314.

30 Mandy F, Nicholson JK, McDougal S: Guidelines for performing single-platform absolute CD4+ T-cell determinations with CD45 gating for persons infected with human immunodeficiency virus. Morbid Mortal Wkly Rep 2003;52/RR-2:243–260.

31 Schnizlein-Bick CT, Mandy F, O'Gorman MRG, Paxton H, Nicholson JKA, Hultin LE, Gelma RS, Wilkening CL, Livnat D. Use of CD45 gating in three and four-color flow cytometric immunophenotyping: guideline from the National Institute of Allergy and Infectious Diseases, Division of AIDS. Clin Cytometry 2002;50:46–52.

32 Mandy FF, Bergeron M, Minkus T: Evolution of leukocyte immunophenotyping as influenced by the HIV/AIDS pandemic: a short history of the development of gating strategies for CD4 T-cell enumeration. Cytometry 1997;30:157–165.

33 Bergeron M, Faucher S, Minkus T, Lacroix F, Ding T, Phaneuf S, Mandy F, Somorjai R, Summers R, and the Participating flow cytometry laboratories of the Canadian Clinical Trials Network for HIV/AIDS Therapies: The impact of unified procedures as implemented in the Canadian Quality Assurance Program for T-lymphocyte subset enumeration. Cytometry 1998;33:146–155.

34 Shapiro HM, Perlmutter NG: 'Killer' applications: towards affordable rapid cell-based diagnostics for malaria and tuberculosis. Cytometry Part B (Clin Cytometry) 2008;S152–S164.

35 *www.cytometryforlife.org*

36 Janossy G, Jani I, Gohde W. Affordable CD4(+) T-cell counts on 'single-platform' flow cytometers I. Primary CD4 gating. Br J Haematol 2000;111:1198–1208.

37 Mandy F, Bergeron M, Recktenwald D, Izaguirre CA: A simultaneous three-color T cell subsets analysis with single laser flow cytometers using T cell gating protocol. Comparison with conventional two-color immunophenotyping method. J Immunol Methods 1992;156:151–162.

38 Loken MR, Brosnan JM, Bach BA, Ault KA: Establishing optimal lymphocyte gates for immunophenotyping by flow cytometry. Cytometry 1990;11:453-459.

39 Glencross D, Scott LE, Jani IV, Barnett D, Janossy G: CD45-assisted PanLeucogating for accurate, cost-effective dual-platform CD4+ T-cell enumeration. Cytometry 2002;50:69–77.

40 Janossy G, Jani IV, Bradley NJ, Bikoue A, Pitfield T, Glencross DK: Affordable CD4(+)-T-cell counting by flow cytometry: CD45 gating for volumetric analysis. Clin Diagn Lab Immunol 2002;9:1085–1094.

41 Glencross DK, Aggett HM, Stevens WS, Bergeron M, Mandy F: African regional external quality assessment (AFREQAS) for CD4 T-cell enumeration: development, outcomes, and performance of laboratories. Cytometry Part B (Clin Cytometry) 2008;S69–S79.

42 Nicholson JK, Hubbard M, Jones BM. Use of CD45 fluorescence and side-scatter characteristics for gating lymphocytes when using the whole blood lysis procedure and flow cytometry. Cytometry 1996;26:16–21.

43 Orszag PR, Ellis P. Addressing Rising Health Care Costs- A view from the Congressional Budget Office N Engl. J Med 2007;357:1793-5,1885–1797.

Sack U, Tárnok A, Rothe G (eds): Cellular Diagnostics. Basics, Methods and Clinical Applications of Flow Cytometry. Basel, Karger, 2009, pp 524–539

Primary Immunodeficiency Diseases

Ilka Schulze[a] · Volker Wahn[b]

[a] Center for Pediatric and Adolescent Medicine, University Hospital Freiburg, Freiburg i. Br.,
[b] Charité Immunodeficiency Center, Berlin, Germany

Introduction

Primary immunodeficiency diseases (PID) are genetic disorders which mostly cause susceptibility to infections and are sometimes associated with autoimmune and malignant diseases. Most patients suffer from recurrent severe infections already from early childhood on. More than 120 distinct genes have been identified, whose mutations account for more than 150 different forms of PID [1]. The incidence of these diseases varies between 1 : 500 for selective IgA deficiency to 1 : 70,000 for severe combined immunodeficiency [2, 4]. Knowledge of the pathogen which causes infection can point to the specific defect in the patient's immune system, e.g. severe viral or fungal infections in patients with severe combined immunodeficiencies. According to the International Union of Immunological Societies Primary Immunodeficiency Diseases Classification Committee of the World Health Organization, PID can be classified into 8 groups: 1) combined T-cell and B-cell immunodeficiencies; 2) predominantly antibody deficiencies; 3) other well-defined immunodeficiency syndromes; 4). diseases of immune dysregulation (e.g. syndromes with defects of apoptosis); 5) congenital defects of phagocyte number, function or both; 6). defects in innate immunity; 7) autoinflammatory disorders (e.g. familial Mediterranean fever), and 8) complement deficiencies [1]. In addition to the investigation of the immunoglobulins, specific antibody production, blood count with leukocyte differentiation and microbial diagnostics, flow cytometry presents a valuable technology for investigating the immune system if a PID is suspected. Flow cytometry enables the detection of a relative or absolute decrease in a specific subset of leukocytes and the loss or abnormal expression of a specific cell-associated marker. Thus, defects in cell maturation, differentiation and function can be identified, often enabling the classification of PID [3]. This article gives an overview of the application of flow cytometry in the diagnosis and monitoring of patients with PID.

Application of Flow Cytometry in Primary Immunodeficiencies

Specific Immune System

All cells of the immune system express proteins on their surface and in the cytoplasm that are characteristic for the cell line and their developmental stage. Whenever an immunodeficiency is suspected, parallel to the differential blood count, the lymphocyte subpopulation should be evaluated by detection of the different surface markers. This applies to both cellular and humoral immunodeficiencies. The number of T-lymphocytes (CD3+), T-helper cells (CD3+ CD4+), cytotoxic T-cells (CD3+ CD8+), B-cells (CD3− CD19+ CD20+) and NK cells (CD3− CD56+) is counted. CD2 as the sum of T-cells and NK cells is also useful. When evaluating the subpopulation, the normal values for age should be taken into consideration [5]. If, for example, this analysis shows a marked reduction of T-cells in the absence of a secondary immunodeficiency, the evaluation of B-cells and NK cells can already provide an approximate classification of the cellular immunodeficiency (table 1). Irrespective of the absolute lymphocyte count, increased or decreased CD4/CD8 ratios calculated from the distribution of the subpopulations can be an indicator for a defect in the immune system, which may be useful during the follow-up of PID patients.

Lymphocyte subpopulations and their function should also be examined in patients suffering from antibody deficiency. Foremost, the number of peripheral B-cells provides information on a possible defect in B-cell maturation and differentiation, e.g. for X-chromosomal agammaglobulinemia (XLA, Bruton's disease) or variants of autosomal recessive agammaglobulinemia. Furthermore, T-cell abnormalities were shown in about half of the patients with antibody deficiency syndrome. About 20% of patients with common variable immunodeficiency (CVID) showed a decrease in CD4+ helper cells with a reduced CD4/CD8 ratio, and 40% a reduced lymphocyte proliferation response after stimulation with mitogens in the lymphocyte proliferation assay [6]. These perturbations of the cellular immune system often emerge in the course of the disease.

If the indicative immunophenotype points to a PID or if a defect of other gene products is suspected, further markers can be evaluated by flow cytometry [3, 7]. The detection of additional surface markers serves to assess stages of differentiation, activation, and maturation of the various subpopulations. For instance, naive T-cells express CD45RA, while memory T-cells show CD45RO on their surface. The α-chain of the IL-2 receptor (CD25) is not only an important marker for activated T-cells, but also for certain regulatory T-cells. In case of the X-linked severe combined immunodeficiency (X-SCID), the γc-chain (CD132) of the receptor for IL-2, IL-4, IL-7, IL-9, IL-15, IL-21 is missing on the surface of mononuclear cells [8]. Primary immunodeficiencies with faulty interaction between T- and B-cells can be diagnosed, e.g., by a lack of HLA-DR expression for the MHC class II defect, or

Table 1. Cellular phenotype of combined T-cell and B-cell immunodeficiencies (without maternal engraftment)

Immune phenotype	Possible combined immunodeficiencies
T− B+ NK−	common γ-chain of receptors for IL-2/-4/-7/-9/-15/-21, JAK3, CD45, PNP
T− B+ NK+	IL-7Rα, CD3δ/CD3ε/CD3ζ deficiency, IL-2R α-chain
T− B− NK+	RAG 1/2, Artemis, DNA ligase IV
T− B− NK−	ADA, reticular dysgenesis
T+ B− NK+	Omenn syndrome
CD4− B+ NK+	MHC class II
CD8− B+ NK+	ZAP-70, TAP-1/2/MHC class I, CD8 α-chain

nondetection of the CD40 ligand (CD145) for the X-chromosomal hyper IgM syndrome.

In addition to sequencing and blotting techniques, intracellular proteins can also be analyzed by flow cytometry. Thus, Bruton's tyrosine kinase (BTK) for instance, which causes XLA if defective, or WASP, the lacking protein responsible for Wiskott-Aldrich-syndrome (WAS), and also adenosine desaminase (ADA), which, if faulty, leads to a combined immunodeficiency, can all be detected by monoclonal antibodies [8–10]. In case of more specific problems (e.g. IL-12 defect), flow-cytometric methods can also be applied to determine and quantify intracellular cytokines after in vitro stimulation of mononuclear cells. With suitable equipment, flow cytometry further enables isolation of specific cell populations and subsequent testing of sorted cells.

Phagocytes

The function of phagocytosing cells can now be analyzed by flow cytometry. By testing the phagocyte system in suspected leukocyte adhesion deficiency, the surface adhesion proteins CD11a, b, c/CD18 and CD15s can be detected. Furthermore, phagocytosis and synthesis of reactive oxygen metabolites can be tested by flow cytometry [11]. Chronic granulomatous disease (CGD), the most significant disorder involving phagocytes, is caused by a defective NADPH oxidase and no oxygen radicals are synthesized for the intracellular 'killing' of bacteria. For flow-cytometric analysis, granulocytes and monocytes are incubated with a dye (e.g. dihydrorhodamine) and stimulated in vitro; oxidation of the dye by oxygen metabolites that appear in the fluorescent dichlorofluorescein is then measured. In two thirds of the cases, CGD is genetically transmitted through the X-chromosome. For this disorder, as well as for the X-chromosomal variants of hyper-IgM syndrome [12] and WAS, carrier status can be tested on maternal cells by flow-cytometric analysis.

Thus, flow-cytometric analyses are useful for diagnosing primary immunodeficiencies, for follow-ups during the course of the disease and after therapy (e.g. stem cell transplantation), and finally for testing the carrier status of family members. De Vries et al. [7] introduced a sound basic and extended protocol for immunophenotyping of pediatric PID patients.

Primary Immunodeficiencies – Clinical Presentations

This chapter addresses the central clinical presentations of various groups of congenital immunodeficiencies and the role of flow-cytometric analysis as a relevant diagnostic tool.

Combined Immunodeficiencies

This severe type of congenital immune deficiency is caused by a defective T-cell system, and usually coincides with distinct lymphopenia. Since normal function of B-cells is largely dependent on T-cells, a secondary B-cell deficiency with hypogammaglobulinemia is also commonly observed, resulting in a combined immune deficiency. From infancy, children with SCID often suffer from recurring and sometimes life-threatening bacterial, viral or fungal infections. Many of these pediatric patients show general failure to thrive, persistent thrush on skin and mucous membranes and chronic diarrhea. The classification of 2007 [1] distinguishes 24 different types of combined immunodeficiencies (table 2) and a selection of some well-defined types is described below.

X-Chromosomal and Autosomal Recessive SCID

With a proportion of approximately 45%, the X-chromosomally inherited form of SCID is the most frequently occurring disorder among the combined immunodeficiencies. It is caused by a mutation in the common γ-chain (γc-chain, CD132) of the interleukins-2, -4, -7, -9, -15 und -21 receptors [14, 15]. These cytokine receptor complexes are required for normal development of T- and NK cells. A defect in the γ-chain consequently leads to a considerable reduction of these subpopulations. Although B-lymphocytes are present, they show a naive phenotype and are unable to differentiate without the help of T-cells. To date, 95 mutations have been discovered, causing defective γ-chains in two thirds of the patients, whereas in one third the protein is not expressed at all [14].

About 6% of combined immunodeficiencies are acquired by autosomal recessive inheritance. This type of SCID cannot be distinguished clinically or immunologically from the X-chromosomal type and shows the same T− B+ NK− phenotype. This defect is caused by a mutation in the Janus kinase 3 (JAK3). This tyrosine kinase is associated with the γc-chain of the interleukin receptors described above and is activated in the first step of intracellular signal transduction upon binding of the relevant cytokines [16].

Table 2. Combined T-cell and B-cell immunodeficiencies [modified according to Geha et al. [1]

Disease	Inheri-tance	Pathogenesis/gene defect	Circulating T-cells	Circulating B-cells
γc deficiency	XL	defect in γ-chain of receptors for IL-2, -4, -7, -9, -15, -21	↓	N or ↑
JAK3 deficiency	AR	Defect in JAK3 signaling kinase	↓	N or↑
IL7Rα deficiency	AR	Defect in IL-7 receptor α-chain	↓	N or ↑
CD45 deficiency	AR	Defect in CD45	↓	N
CD3δ/CD3ε/CD3ζ deficiency	AR	defect in CD3δ, CD3ε or CD3ζ chains of T-cell antigen receptor	↓	N
RAG 1/2 deficiency	AR	defective VDJ recombination, complete defect of RAG 1 or 2	↓	↓
Omenn syndrome	AR	missense mutations allowing residual activity, usually in RAG 1 or 2 genes	N, restricted heterogeneity	N or ↓
Artemis deficiency	AR	defective VDJ recombination, defect in Artemis DNA recombinase-repair protein	↓	↓
ADA deficiency	AR	absent adenosine deaminase, elevated lymphotoxic metabolites	↓	↓
PNP defekt	AR	Absent purine nucleoside phosphorylase	↓	N
DNA ligase IV	AR	Defect in DNA ligase IV, impaired NHEJ	↓	↓
Cernunnos deficiency	AR	Defect in Cernunnos, impaired NHEJ	↓	↓
CD40 ligand deficiency	X	defects in CD40 ligand, defective B-cell and dendritic cell signaling	N	N, only IgM+ and IgD+ B-cells present
CD40 deficiency	AR	defects in CD40, defective B-cell and dendritic cell signaling	N	N, only IgM+ and IgD+ B-cells present
CD3γ deficiency	AR	Defect in CD3 γ-chain	N, reduced TCR expression	N
MHC class I deficiency	AR	mutations in TAP1, TAP2 or TAPBP genes	CD8 ↓, normal CD4	N

Table 2 continued on next page

Table 2. Continued

Disease	Inheritance	Pathogenesis/gene defect	Circulating T-cells	Circulating B-cells
MHC class II deficiency	AR	mutation in transcription factors for MHC class II proteins (CIITA, RFX5, RFXAP, RFXANK genes)	CD4 ↓	N
CD8 deficiency	AR	defects of CD8α chain	CD8 ↓, normal CD4	N
ZAP-70 deficiency	AR	Defect in ZAP-70 signaling kinase	CD8 ↓, normal CD4	N
Reticular dysgenesis	AR	defective maturation of T, B and myeloid cells	↓	N or ↓
Winged helix nude deficiency (WHN, human and mice)	AR	defects in forkhead box N1 transcription factor encoded by FOXN1, gene mutated in nude mice	↓	N
Ca^{2+} channel deficiency	AR	defect in Orai-1, a Ca^{2+} channel component	N, defective TCR-mediated activation	N
CD25 deficiency	AR	Defect in IL-2Rα chain	N to modestly decreased	N
STAT5b	AR	defects of STAT5B gene, impaired development and function of γδ-T-cells, T-regulatory and NK cells, impaired T-cell proliferation	modestly decreased	N

↓ = Decreased counts; ↑ = increased counts; N = normal counts; XL = X-linked inheritance; AR = autosomal recessive inheritance; JAK3 = Janus kinase 3; NHEJ = nonhomologous end joining; TAP = transporter associated with antigen processing; TAPBP = TAP binding protein; TCR = T-cell receptor.

IL-7 Receptor Defect

The IL-7 receptor consists of one α- and one γ-chain. About 1% of patients with CID show mutations in the gene for the α-chain (CD127). This missing function of IL-7 receptors leads to faulty T-cell maturation with a T− B+ NK+ phenotype. flow-cytometric analysis can measure missing or reduced expression of the receptor α-chain [8].

Metabolic Defects

Approximately 20% of combined immunodeficiencies are caused by a defect in purine metabolism. The more common defect in ADA leads to accumulation of the toxic metabolic metabolites deoxyadenosin and deoxy-ATP, while the rarely encountered purine nucleoside phosphorylase (PNP) defect mostly generates deoxy-GTP. Compared to other cells, lymphocytes exhibit highly sensitive reactions to these toxic metabolites and are therefore preferentially damaged. Besides a variably pronounced immunodeficiency, ADA-deficient patients also show skeletal abnormalities. PNP deficiency is characterized by a relatively normal B-cell immunity while T-cell immunity is obviously disrupted. Additionally and over time, these patients may develop neurological problems and autoimmune diseases such as autoimmune hemolytic anemia, autoimmune thyroiditis and immune thrombocytopenic purpura [17]. Besides RNA /DNA analysis or enzyme activity testing, intracellular staining by fluorescent antibodies for ADA may also be performed [10].

Defects of V(D)J Recombination

The recombination-activating genes (RAG)-1 and RAG-2 start the process of V(D)J-recombination of T-cell receptor and immunoglobulin molecules. If the V(D)J recombination is impaired, T- as well as B-cells are unable to differentiate. Assorted mutations in RAG-1 and RAG-2 can lead to various clinical symptoms. AT− B− NK+ SCID phenotype, for instance, shows a null mutation in RAG-1 or -2, while point mutations with frameshift can produce partially functioning truncated proteins and the immunodeficiency may present as the Omenn syndrome, where T-cells have a highly restricted T-cell receptor (TCR) repertoire. For guidance, the Vβ repertoire can also be measured by flow-cytometric analysis. Along with life-threatening infections, these patients also show generalized erythroderma, lymphadenopathy and hepatosplenomegaly.

Patients with mutations in the genes for Artemis, DNA ligase IV and Cernunnos show the same immunological phenotype of a severe combined immunodeficiency. These three genes exert vital functions for DNA repair during V(D)J recombination. Furthermore, these immunodeficiencies are associated with radiation sensitivity (RS-SCID, radiosensitive SCID). DNA ligase IV and Cernunnos defects are also associated with microcephaly [19].

Defects in Class I and Class II Molecules of the Major Histocompatibility Complex and Bare Lymphocyte Syndrome

T-lymphocytes can only recognize antigen peptides in association with major histocompatibility (MHC) class I or class II molecules. MHC class I molecules are expressed on nearly all nucleated cells, while class II molecules can only be detected on specialized antigen-presenting cells (e.g. monocytes, B-cells). Intracellularly processed viral peptides are delivered by a transporter associated with antigen processing (TAP) to the endoplasmatic reticulum (ER), where they bind to MHC class I molecules. This endogenous peptide/MHC class I molecule complex is then expressed on the cell surface and presented to the CD8+ cytotoxic T-lymphocytes. The heterodimer TAP is formed of two subunits, TAP1 and TAP2. If deletions or mutations occur in one of the subunits, the peptide translocation to the ER is abolished and patients develop MHC class I deficiency (bare lymphocyte syndrome type

1; BLS type 1). Clinically and immunologically, MHC class I deficiencies can be divided into three groups [20]. Flow-cytometric analysis can detect reduced MHC class I expression in all three patient groups. The first group shows the most severe symptoms with recurring bacterial, mycotic and parasitic infections beginning in infancy. Despite evidence of MHC class I molecule transcription, the complexes are barely expressed on the cell surface. This group also suffers from hypogammaglobulinemia. Two asymptomatic siblings with typical group 2 MCH class I deficiency also showed impaired transcription of the MHC class I molecules in addition to diminished cell surface expression. In addition to recurring bacterial infections, group 3 patients also display necrotizing granulomatous skin lesions. This group shows evidence of a defective TAP complex. Some cases of adult bronchiectasis with MHC class I deficiency have also been described [22].

MHC class II molecules are expressed on B-lymphocytes, activated T-cells, dendritic cells, monocytes, macrophages and thymic epithelial cells. This expression is in part constitutional, and to a certain extent regulated by IFN-γ. Some proteins (RFX5, RFXANK) bind directly to the promoter, while the class II transactivator CIITA – not a DNA-binding molecule – docks to DNA-binding proteins. Peptides from uptake of exogenous antigen are then presented by MHC class II molecules to CD4+ helper cells, which are activated and initiate humoral as well as cellular immune responses. Underlying MHC class II deficiency is characterized by mutations in proteins regulating MHC class II expression (CIITA, RFX5, RFXANK) [21, 23]. This immunodeficiency almost always leads to severe opportunistic infections, and, without stem cell transplantation, patients die in early childhood. Due to defects in the transcription factors, no MHC class II expression can be detected immunologically (double staining) on the surface of, e.g. B-cells or monocytes with very low numbers of CD4+ cells and normal levels of T- and B-lymphocytes. These patients also have moderate to severe hypogammaglobulinemia.

ZAP-70 Defect/CD8 Deficiency

ζ-associated protein 70 (ZAP-70) binds to the ζ-chain of the CD3 complex. ZAP-70 is a tyrosine kinase that transmits an activation signal from the receptor to the T-cell. This protein plays a pivotal role for the maturation of T-cells in the thymus, and especially for the development of CD8+ T-cells. Symptoms of ZAP-70 mutations are variable: they range from mild disorders to severe combined immunodeficiency. Patients present with lymphocytosis and CD8 lymphopenia; the lymphocyte proliferation assay shows reduced proliferation in circulating CD4+ cells [14]. Other CD8+ deficiency disorders (MHC I defect, CD8α defect) should be excluded by differential diagnosis.

CD40L and CD40 Defect (Hyper-IgM Syndromes, HIGM Syndromes)

The X-chromosomal variant of the hyper-IgM syndrome exhibits mutations in the gene for the CD40 ligand (CD40L, CD154), while one type of autosomal recessive hyper-IgM syndrome shows mutations in the gene for CD40. CD154 is mainly expressed on activated T-lymphocytes and binds to the surface marker CD40 on antigen-presenting cells. Binding of the CD40 ligand is essential for B-cell isotype switching. Thus, patients with this immunodeficiency display normal to elevated plasma concentrations of IgM and IgD, while the levels of isotypes IgG, IgE and IgA are very low or undetectable. In addition to recurrent infections including, among others, *Pneumocystis jirovecii* infections, many patients suffer from immune diseases and have a higher risk for malignancies. Flow-cytometric analysis can measure CD40L expression on CD4+ cells activated in vitro. About 90% of normal activated CD4+ cells express CD40L as compared to only 5% in patients with hyper-IgM syndrome. Also, expression of CD40 on B-cells should be at least 90% [3]. In the case of X-chromosomal hyper-IgM syndrome, carrier status of patients' mothers can be determined by flow-cytometric analysis as well [12].

Problems in the Diagnosis of SCID

All variants of SCID can present with chronic graft-versus-host disease when maternal T-cells are transferred to the newborn [18]. Analyzed cells thus stem from the mother and not from the child, which can be confirmed on isolated T-cells by HLA typing, enzyme polymorphisms or DNA fingerprinting.

Antibody Deficiencies

Antibodies have several functions e.g., neutralization of bacterial toxins, opsonization of microorganisms and activation of complement, to name a few. Patients with antibody deficiency suffer mainly from recurring bacterial infections. Antibody protection plays a decisive role, especially in the defense against encapsulated bacteria such as *Streptococcus* and *Haemophilus influenzae*, as well as enteroviruses. Additionally, antibody deficiency can be associated with autoimmune diseases and a higher propensity for malignant tumors [24]. The main diagnosis focuses on the evaluation of the humoral immune system with immunoglobulin isotypes, IgG subclasses, isohemagglutinins and specific antibody formation. For these patients, determination of lymphocyte subpopulations and in vitro testing of lymphocyte function by lymphocyte proliferation assay is also indicated. X-chromosomal and autosomal-recessive agammaglobulinemia exhibit a significantly lower level of peripheral mature B-cells (CD19, CD20) in flow-cytometric analysis, while most other antibody deficiency or antibody production disorders present with a normal B-cell count. Approximately 50% of patients with common variable immunodeficiency (CVID) have a defective T-cell system [6]. Below, we describe how flow-cytometric analysis can be used as a valuable diagnostic tool for some antibody deficiency syndromes.

X-Chromosomal Agammaglobulinemia (XLA, Bruton's Disease)

Ogden C. Bruton was the first to describe the clinical presentation of agammaglobulinemia and its therapy with immunoglobulins in 1952 [24]. Approximately 85% of patients with agammaglobulinemia are male and exhibit the X-linked disease. This humoral immunodeficiency occurs in 1 : 200,000 live births and is caused by a defect in BTK. More than 380 different mutations have been described [26]. There is no correlation between genotype and phenotype, and the severity of the disorder can vary even within one family [27]. This tyrosine kinase defect prevents signal transduction in B-lymphocytes, thus blocking subsequent differentiation from the pre-B-cell stage. Flow-cytometric analysis consequently detects very few mature peripheral B-cells (CD19, CD20) or plasma cells (CD38) and immunoglobulin isotype levels are significantly reduced or nondetectable. Recurring neutropenia in some of these patients is explained by expression of BTK in myeloid cell lines [28]. Patients and carriers can be identified by flow cytometry using intracellular staining with anti-BTK antibodies [9]. XLA patients are characterized by hypoplasia of the secondary lymphatic organs and suffer mainly from recurrent bacterial respiratory infections.

Autosomal-Recessive Agammaglobulinemia

Patients with autosomal recessive agammaglobulinemia represent a genetically heterogeneous group. Clinically and immunologically, these patients cannot be distinguished from XLA patients, and here, too, B-cell differentiation is blocked by faulty signal transduction. Defects causing autosomal recessive agammaglobulinemia are found in the pre-B-cell receptor complex or in the B-cell linker protein (BLNK). The pre-B-cell receptor complex is made up of a heavy μ-chain, a surrogate light chain composed of the two proteins VpreB and λ5, and two accessory chains Igα and Igβ. This complex is transiently expressed on the pre-B-cell and generates the signals for further B-cell development [29]. In short, the signal is transmitted first by the tyrosine kinase Syk and is then directed via BLNK and BTK to the cell nucleus. For patients with autosomal recessive agammaglobulinemia, mutations in the genes for μ-, VpreB-, λ5-, Igα- and Igβ-chain, for BLNK and tyrosine kinase Syk could be identified [24].

CVID

CVID represents a heterogeneous patient group and occurs at a rate of about 1 : 25,000. Patients usually suffer from hypogammaglobulinemia for IgG and IgA, in some cases, the isotype IgM may also be reduced. CVID clinically presents predominantly with recurring infections of the upper and lower respiratory tract. Approximately half of the patients additionally exhibit disorders of the gastrointestinal tract and a higher incidence of autoimmune disease and malignancies. Occasionally, the number of peripheral B-cells is reduced, and abormalities in the T-cell system can be detected in approximately half of the patients at the time of diagnosis or in the course of the disease [6, 24, 31] (see 'Diagnostics for Common Variable Immunodeficiency Syndrome', pp. 542)

Defects in the Phagocytic System

Patients with clinically relevant granulocytopenias or defective phagocyte functioning suffer from recurrent bacterial infections and mycoses, often accompanied by skin infections and sometimes with abscesses, ulcera and necrosis. In principle, neutrophil granulocytes can be tested for mobilization of calcium, cytoskeletal actin, adhesion molecules, phagocytosis, degranulation and generation of reactive oxygen metabolites by flow-cytometric analysis [11]. Immunophenotyping by flow cytometry is mainly relevant for the immunodeficiencies described below (see 'Oxidative Burst, Phagocytosis and Expression of Adhesion Molecules', pp. 343).

Chronic Granulomatous Disease

In septic or chronic granulomatous (CGD) disease, the intracellular 'killing' of phagocytosed microorganisms is affected. This defect is caused by mutations in components of the NADPH oxidase, and the most common mutation is found in the X-chromosomally inherited membrane-bound subunit gp91^{phox} (cytochrome b558, β-subunit). Three other autosomal recessive types are caused by defects in p22^{phox} (cytochrome b558, α-subunit) or the cytoplasmic subunits p47^{phox} and p67^{phox}. As a consequence of this defective NADPH oxidase, synthesis of reactive oxygen metabolites, and particularly the 'killing' of catalase-positive microorganisms, is prevented. Thus, the disorder clinically presents with occasionally life-threatening bacterial (e.g. *Staphylococcus aureus, Burkholderia cepacia*, mycobacteria) or fungal (e.g. *Aspergillus* spp.) infections as well as granulomatous inflammatory reactions [32, 33]. Flow-cytometric analysis can detect production of reactive oxygen metabolites using the dihydrorhodamine (DHR) test (see 'Application of Flow Cytometry in Primary Immunodeficiencies', pp. 527). For the X-chromosomal form, two granulocyte populations are detected in female carriers in this test. In addition to Western blot analysis of the NADPH oxidase subunits, the α- and β-subunits of cytochrome b558 can also be detected by monoclonal antibodies [34, 35].

Leukocyte Adhesion Deficiency

Phagocyte function is dependent on adhesion to cells or extracellular matrix and migration to the inflammation site. Adhesion to cells or tissue occurs mainly via sialyl-Lewis X (sLe-X, CD15s) and integrins expressed on the surface of neutrophil granulocytes and macrophages. β_2-integrins are heterodimers, each of which is made up of one α-chain CD11a (LFA-1), CD11b (Mac-1) or CD11c (CR4) and one common β_2-chain CD18. CD18 defects lead to leukocyte adhesion deficiency type 1 (LAD-1). For most patients, no expression of CD11/CD18 can be detected by flow-cytometric analysis, but patients have been described who express 3–10% of this glycoprotein, exhibiting milder clinical symptoms [36, 37]. Some variants have full surface expression of a nonfunctional protein [38]. Differential blood count shows leukocytosis with increased neutrophil granulocytes. Patients suffer from recurring life-threatening bacterial infections without pus formation or delayed wound healing. Recurrent pneumonia, sinusitis and otitis are often accompanied by necrotizing infections of soft tissues, especially skin, mucosa and gastrointestinal tract. Newborns with LAD-1 often have a delayed umbilical cord separation.

The rare autosomal-recessive inherited leukocyte adhesion deficiency type 2 (LAD-2) is caused by mutations in the GPD-fucose transporter which affects fucosylation of glycoproteins. One of these glycoproteins is sialyl-Lewis X (CD15s), the ligand for the selectin family of adhesion moleculues on leukocytes and endothelial cells. This defective ligand blocks the adhesion cascade after the first steps of initial adhesion and rolling of leukocytes [39]. Flow-cytometric analysis shows a markedly reduced CD15s expression, e.g. on neutrophil granulocytes. Compared to LAD-1, the immunodeficiency is less pronounced in LAD-2 patients, but due to defective fucosylation of further proteins, they additionally develop neurologic impairment, mental and growth retardation and microcephaly.

Other Immunodeficiency Syndromes

The international classification of 2007 [1] describes 15 other well-defined immunodeficiency syndromes. Flow-cytometric analysis is of limited use for these disorders and diagnostically only relevant for the immunodeficiencies described below.

Wiskott-Aldrich Syndrome

WAS and X-chromosomal thrombocytopenia, a milder variation, are caused by mutations in the WAS protein (WASP). WASP is involved in the organization of the actin cytoskeleton and guides formation of immunological synapses, e.g. between T-lymphocytes and antigen-presenting cells [40]. It influences T-cell function, cytotoxic activity of NK cells, motiliy of monocytes and formation of podosomes in dendritic cells. Patients present with the typical triad of congenital thrombocytopenia with too small thrombocytes, eczema, and combined immunodeficiency. Additionally, autoimmune diseases and lymphomas can occur. Examination of the humoral immune system shows low levels of IgM and defective specific antibody formation against polysaccharides and isohemagglutinins. With advancing age, a progressive reduction of T-lymphocytes can be observed by flow-cytometric analysis [41]. It can also detect a diminished expression of intracellular WASP with monoclonal antibodies in patients and female carriers [13].

Ataxia telangiectasia (Louis-Bar Syndrome)

Autosomal recessive inherited ataxia telangiectasia (AT), Nijmegen breakage syndrome (NBS) and Fanconi anemia all belong to the group of chromosome instability syndromes. The mutant ATM gene which causes AT is a tumor suppressor gene and plays an important role in recognizing and repairing double-strand breaks of DNA. Mutations in the ATM gene lead to ionization radiation sensitivity and a higher number of chromosomal breaks. Chromosomal breaks and translocations are found mostly in chromosomes 7 and 14, thereby causing alterations in the genes for the TCR (14q11.2, 7q35) or the heavy immunoglobulin chain (14q32). Clinical presentations are cerebellar ataxia, oculo-cutaneous telangiectasia (dilated blood vessels in the eyes and skin), predisposition to primarily lymphoid malignancies and variable combined immunodeficiency. Examinations of the humoral immune system usually show reduced levels of IgA, IgE, and IgG2/IgG4 subclasses as well as impaired antibody formation against polysaccharides. The full-blown syndrome exhibits decreased T-cell levels in the peripheral blood, mainly affecting αβ-TCR-carrying T-cells and naive T-helper cells. AT patients also show lymphocyte malfunctioning relating to proliferation and cytokine production, and reduced activity of cytotoxic T-cells against viral antigens. Along with molecular and cytogenetic methods, AT can also be diagnosed by flow cytometry cell cycle analysis, where lymphocytes are stained with Hoechst 33258 and ethidium bromide before in vitro stimulation. Flow-cytometric analysis showed increased levels of cells nonstimulatable by PHA and higher levels of cells in the G2 phase as compared to those in normal controls [42, 43]. A simple, but diagnostically highly significant measure is the almost obligatory increase of alpha-1-fetoprotein in plasma.

DiGeorge Anomaly (CATCH22)

DiGeorge anomaly, or CATCH22 (cardiac, abnormal facies, thymic hypoplasia, cleft palate, hypocalcemia, 22nd chromosome) is caused by a microdeletion in chromosome 22 (22q11.2) and occurs at a rate of 1 : 4,000–6,000. The disorder involves an embryologic defect, usually in the third and fourth pharyngeal pouch development and clinical presentation, even with identical deletions, can be very varied [44]. Affected children with fully formed sequence show congenital heart defect, abnormal facies, early-onset hypocalcemia through hyper- or hypoparathyroidism and thymic hypoplasia or thymic aplasia. The degree of thymic hypoplasia determines the extent of the cellular immunodeficiency. A total lymphopenia may be present. One study comparing 195 CATCH22 patients to age-matched controls found decreased levels of CD3, CD4, CD8 and γδ-positive T-lymphocytes while NK and B-cell count was elevated. Immature T-cell levels in peripheral blood (CD1/CD38), however, were not increased in these patients. In vitro lymphocyte proliferation response after stimulation with mitogens and antigens was not different from that of healthy controls [44]. As a rule, humoral immunity is rarely affected. Patients suffer from recurring viral, mycotic and also bacterial infections and autoimmune diseases. About 40% of patients are immunologically healthy.

Advanced Diagnostic Procedures in Primary Immunodeficiency Diseases

Flow-cytometric analysis can be employed to determine cell surface markers, cytokines and also for some functional tests, e.g. generation of reactive oxygen metabolites in phagocytosing cells, NK cell activity, rate of apoptosis and cell cycle analysis can all be determined with special tests by flow cytometry. For many primary immunodeficiency diseases, flow-cytometric analysis is an important tool for characterizing the immunodeficiency and as a basis for further diagnostic evaluation. Generally, a definitive classification of the defect can only be achieved through analysis by molecular genetics. Different genotypes can have the same immunological phenotype. Genetic testing is also essential for determining carrier status and for prenatal diagnostics. Cytogenetic and molecular cytogenetic diagnostics (chromosome analysis and fluorescent in situ hybridization) play an important role in chromosome instability syndromes and the microdeletion syndrome CATCH22. For examining primary immunodeficiencies, the lymphocyte proliferation assay is the method of choice to determine lymphocyte function by culture and stimulation with mitogens or antigens. ^{3}H-thymidine is then incorporated into the newly synthezised DNA and proliferation response is measured in a scintillation spectrophotometer. Assessment of immunoglobulins, IgG subclasses, isohemagglutinins and vaccine antibody levels is essential for diagnosing humoral immunodeficiencies. Complement deficiencies are evaluated by measuring CH50 and AP50 values for the classical and alternative activation pathway and determining the relevant complement factors. When employing flow-cytometric technology, it should be noted that some markers, although fully expressed, may not necessarily be functionally intact molecules.

Summary

Flow-cytometric methods hold an important position in the diagnostics and follow-up of primary immunodeficiency diseases. Especially in combined immunodeficiencies, the patient's immunological phenotype can be determined by flow-cytometric analysis. Detection of decreased expression of a surface marker (e.g., CD19 or CD8) can lead to diagnosis directly or indirectly. In addition to assessment of lymphocyte subpopulations, flow-cytometric analysis is also used for functional testing such as oxygen radical formation in phagocytes. For further testing, intracellular proteins (e.g. cytokines) can be measured and cell populations isolated. By combining clinical symptoms, laboratory and clinical parameters, results of flow-cytometric analysis and functional and molecular genetic analyses, it is possible to identify well-defined and new primary immunodeficiencies.

References

1 Geha RS, Notarangelo LD, Casanova JL, Chapel H, Conley ME, Fischer A, Hammarström L, Nonoyama S, Ochs HD, Puck JM, Roifman C, Seger R, Wedgwood J: Primary immunodeficiency diseases: an update from the International Union of Immunological Societies Primary Immunodeficiency Diseases Classification Committee. J Allergy Clin Immunol 2007;120:776–794.

2 IUIS Scientific Committee, report on primary immunodeficiency disease. Clin Exp Immunol 1999;118(suppl 1): 1–28.

3 O'Gorman MRG: Role of flow cytometry in the diagnosis and monitoring of primary immunodeficiency disease. Clin Lab Med 2007;27:591–626.

4 Ryser O, Morell A, Hitzig WH: Primary immunodeficiencies in Switzerland: first report of the national registry in adults and children. J Clin Immunol 1988;8:479–488.

5 Shearer WT, Rosenblatt HM, Gelman RS, Oyomopito R, Plaeger S, Stiehm ER, Wara DW, Douglas SD, Luzuriaga K, McFarland EJ, Yogev R, Rathore MH, Levy W, Graham BL, Spector SA: Lymphocyte subsets in healthy children from birth through 18 years of age: The Pediatric AIDS Clinical Trials Group P1009 study. J Allergy Clin Immunol 2003;112:973–980.

6 Cunningham-Rundles C: Common variable immunodeficiency. Curr Allergy Asthma Rep 2001;1:421–429.

7 de Vries E, Noordzij JG, Kuijpers TW, van Dongen JJM: Flow cytometric immunophenotyping in the diagnosis and follow-up of immunodeficient children. Eur J Pediatr 2001;160:583–591.

8 Illoh OC: Current applications of flow cytometry in the diagnosis of primary immunodeficiency diseases. Arch Pathol Lab Med 2004;128:23–31.

9 Futatani T, Miyawaki T, Tsukada S, Hashimoto S, Kunikata T, Arai S, Kurimoto M, Niida Y, Matsuoka H, Sakiyama Y, Iwata T, Tsuchiya S, Tatsuzawa O, Yoshizaki K, Kishimoto T: Deficient expression of Bruton's tyrosine kinase in monocytes from X-linked agammaglobulinemia as evaluated by a flow cytometric analysis and its clinical application to carrier detection. Blood 1998;91:595–602.

10 Otsu M, Hershfield MS, Tuschong LM, Muul LM, Onodera M, Ariga T, Sakiyama Y, Candotti F: Flow cytometry analysis of adenosine deaminase (ADA) expression: a simple and reliable tool for the assessment of ADA-deficient patients before and after gene therapy. Hum Gene Ther 2002;13:425–432.

11 van Eeden SF, Klut ME, Walker BAM, Hogg JC: The use of flow cytometry to measure neutrophil function. J Immunol Methods 1999;232:23–43.

12 O'Gorman MRG, Zaas D, Paniagua M, Corrochano V, Scholl PR, Pachman LM: Development of a rapid whole blood flow cytometry procedure for the diagnosis of X-linked hyper-IgM syndrome patients and carriers. Clin Immunol Immunopathol 1997;85:172–181.

13 Yamada M, Ariga T, Kawamura N, Yamaguchi K, Ohtsu M, Nelson DL, Kondoh T, Kobayashi I, Okano M, Kobayashi K, Skiyama Y: Determination of carrier status for the Wiskott-Aldrich syndrome by flow cytometric analysis of Wiskott-Aldrich syndrome protein expression in peripheral blood mononuclear cells. J Immunol 2000;165:1119–1122.

14 Buckley RH: Primary immunodeficiency diseases due to defects in lymphocytes. N Engl J Med 2000;343:1313–1324.

15 Ariga T, Yamaguchi k, Yoshida J, Miyanoshita A, Watanabe T, Date T, Miura JI, Kumaki S, Ishii N, Skiyama Y: The role of common gamma chain in human monocytes in vivo; evaluation from the studies of X-linked severe combined immunodeficiency (X-SCID) carriers and X-SCID patients who underwent cord blood stem cell transplantation. Br J Haematol 2002;118:858–863.

16 Leonard WJ: Cytokines and immunodeficiency diseases. Nat Rev Immunol 2001;1:200–208.

17 Hong R: Disorders of the T-cell system; in Stiehm ER (ed): Immunologic Disorders in Infants and Children. Philadelphia, Saunders, 1996, pp 356–360.

18 Villa A, Sobacchi C, Notarangelo LD, Bozzi F, Abinun M, Abrahamsen TG, Arkwright PD, Baniyash M, Brooks EG, Conley ME, Cortes P, Duse m, Fasth A, Filipovich AM, Infant AJ, Jones A, Mazzolari E, Muller SM, Pasic S, Rechavi G, Sacco MG, Santagata S, Vogler LB, Ochs H, Vezzoni P, Friedrich W, Schwarz K: V(D)J recombination defects in lymphocytes due to RAG mutations: severe immunodeficiency with a spectrum of clinical presentations. Blood 2001;97:81–88.

19 Fischer A: Human primary immunodeficiency diseases. Immunity 2007;27:835–845.

20 Gadola SD, Moins-Teisserenc HT, Trowsdale J, Gross WL, Cerundolo V: TAP deficiency syndrome. Clin Exp Immunol 2000;121:173–178.

21 Reith W, Mach B: The bare lymphocyte syndrome and the regulation of MHC expression. Annu Rev Immunol 2001;19:331–373.

22 Touraine JL: The bare lymphocyte syndrome: report on the registry. Lancet 1981;i:319–321.

23 Prod'homme T, Dekel B, Barbieri G, Lisowska-Grospierre B, Katz R, Charron D, Alcaide-Loridan C, Pollack S: Splicing defect in RFXANK results in a moderate combined immunodeficiency and long-duration clinical course. Immunogenetics 2003;55:530–539.

24 Ballow M: Primary immunodeficiency disorders: antibody deficiency. J Allergy Clin Immunol 2002;109: 581–591.

25 Bruton OC: Agammaglobulinemia. Pediatrics 1952;9:722–728.
26 Vihinen M, Mattsson PT, Smith CI: Bruton tyrosine kinase (BTK) in X-linked agammaglobulinemia (XLA). Front Biosci 2000;1:D917–D928.
27 Vihinen M, Kwan SP, Lester T, Ochs HD, Resnick I, Väliaho J, Conley ME, Smith CIE: Mutations of the human BTK gene coding for Bruton Tyrosine Kinase in X-linked agammaglobulinemia. Hum Mutat 1999;13: 280–285.
28 Farrar JE, Rohrer J, Conley ME: Neutropenia in X-linked agammaglobulinemia. Clin Immunol Immunopathol 1996;81:271–276.
29 Rudensky AY, Sprent J, Berg LJ, Hayday AC, Owen M, Swanborg RH, Jameson SC: The development and survival of lymphocytes; in Janeway CA, Travers P, Walport M, Shlomchik M (eds): Immunobiology. New York, Garland, 2001, pp 237–241.
30 Conley ME, Rohrer J, Rapalus L, Boylin EC, Minegishi Y: Defects in early B-cell development: comparing the consequences of abnormalities in pre-BCR signalling in the human and the mouse. Immunol Rev 2000;178:75–90.
31 Cunningham-Rundles C, Knight AK: Common variable immune deficiency: reviews, continued puzzles, and a new registry. Immunol Res 2007;38:78–86.
32 Winkelstein JA, Marino MC, Johnston RB, Boyle J, Curnutte J, Gallin JI, Malech HL, Holland SM, Ochs H, Quie P, Buckley RH, Foster CB, Chanock SJ, Dickler H: Chronic granulomatous disease. Report on a national registry of 368 patients. Medicine 2000;79:155–169.
33 Segal BH, Leto TL, Gallin JI, Malech Hl, Holland SM: Genetic, biochemical, and clinical features of chronic granulomatous disease. Medicine 2000;79:170–200.
34 Liese J, Kloos S, Jendrossek V, Petropoulou T, Wintergerst U, Notheis G, Gahr M, Belohradsky BH: Long-term follow-up and outcome of 39 patients with chronic granulomatous disease. J Pediatr 2000;137:687–693.
35 Batot G, Martel C, Capdeville N, Wientjes F, Morel F: Characterization of neutrophil NADPH oxidase activity reconstituted in a cell-free assay using specific monoclonal antibodies raised against cytochrome b558. Eur J Biochem 1995;234:208–215.
36 Rosenzweig SD, Uzel G, Holland SM: Phagocyte disorders; in Stiehm ER, Ochs HD, Winkelstein JA (eds): Immunologic Disorders in Infants und Children. Philadelphia, Saunders, 2004, pp 631–633.
37 Kuijpers TW, van Lier RAW, Hamann D, de Boer M, Thung LY, Weening RS, Verhoeven AJ, Roos D: Leukocyte adhesion deficiency type 1 (LAD-1)/variant. J Clin Invest 1997;100:1725–1733.
38 Hogg N, Stewart MP, Scarth SL, Newton R, Shaw JM, Law SK, Klein N: A novel leukocyte adhesion deficiency caused by expressed but non-functional beta2 integrins Mac-1 and LFA-1. J Clin Invest 1999;103:97–106.
39 Wild MK, Lühn K, Marquardt T, Vestweber D: Leukocyte adhesion deficiency. II. Therapy and genetic defect. Cells Tissues Organs 2002;172:161–173.
40 Notarangelo LD, Ochs HD: Wiskott-Aldrich Syndrome: a model for defective actin reorganization, cell trafficking and synapse formation. Curr Opin Immunol 2003;15:585–591.
41 Park JY, Kob M, Prodeus AP, Rosen FS, Shcherbina A, Remold-O'Donnell E: Early deficit of lymphocytes in Wiskott-Aldrich syndrome: possible role of WASP in human lymphocyte maturation. Clin Exp Immunol 2004;136:104–110.
42 Stuhrmann M, Dörk T, Karstens JH: Ataxia teleangiectatica. München, Verlag Medizinische Genetik, 1999.
43 Habermehl P, Zepp F: Störungen der zellulären Immunfunktion; in Wahn U, Seger R, Wahn V, Holländer GA (Hrsg): Pädiatrische Allergologie und Immunologie. München, Urban & Fischer, 2005, pp 580–583.
44 Jawad AF, McDonald-McGinn DM, Zackai E, Sullivan KE: Immunologic features of chromosome 22q11,2 deletion syndrome (DiGeorge syndrome/velocardiofacial syndrome). J Pediatr 2001;139:715–723.

Sack U, Tárnok A, Rothe G (eds): Cellular Diagnostics. Basics, Methods and Clinical Applications of Flow Cytometry. Basel, Karger, 2009, pp 540–557

Diagnostics for Common Variable Immunodeficiency Syndrome

Klaus Warnatz · Michael Schlesier

Department of Rheumatology and Clinical Immunology, Medical Clinic, University Hospital Freiburg, Freiburg i. Br., Germany

Introduction

Common variable immunodeficiency (CVID) is the most common, clinically relevant immunodeficiency disorder. In the western world, the incidence of this disease ranges between 1:20,000 and 1:200,000 [1]. CVID is an antibody deficiency syndrome with decreased serum concentrations of at least two of the three antibody classes IgG, IgM and IgA of at least two standard deviations below the age-appropriate mean value (*www.esid.org*). According to the diagnostic criteria of the European Society of Immunodeficiency (ESID), the immunodeficiency begins after the 2nd year of life, and immunization has been unsuccessful in most patients. Since CVID is a diagnosis by exclusion, well-known genetic, infectious and toxic causes of hypogammaglobulinemia have to be excluded (*www.esid.org*). Approximately 20% of the cases [2, 3] are familial and are assumed to have a genetic origin. There is a genetic association between CVID and the more frequent, but typically clinically inconspicuous selective IgA deficiency [4]. Clinically, CVID patients are frequently diagnosed as adolescents or young adults, particularly due to recurrent infections of the upper and lower respiratory tracts. Untreated, these lead to progressive destruction of the airways and lungs. In contrast to patients with severe combined immunodeficiency, CVID patients are particularly susceptible to infections with encapsulated bacteria, while atypical infections are rare. Additionally, many of the patients suffer from gastrointestinal infections during the course of their illness. Further symptoms include autoimmune phenomena, lymphoproliferation (splenomegaly, lymphadenopathy), lymphomas and granulomatous disease [1]. To a large extent, the pathogenesis of this syndrome is unclear and certainly heterogeneous in origin. Both disturbed

T-cell and B-cell differentiation and function have been described in the literature [5–7]. In addition to a thorough medical history and physical examination of the patient, the diagnosis of CVID is based on the determination of total and specific serum antibody titers, the exclusion of paraproteins, protein-loosing enteropathy and renal disease, bone marrow biopsy as well as functional and imaging procedures for the analysis of organ damage. Flow cytometry supplements the diagnostic work-up by assessing peripheral blood lymphocyte populations and investigating the presence of defined genetic defects. A detailed analysis of the lymphocyte population allows the exclusion of severe immune disturbances, the identification of large monoclonal populations and disturbances in early B-cell differentiation which result in very low numbers of peripheral blood B-cells as in X-chromosomal agammaglobulinemia (Bruton's disease) caused by a defect in the Bruton tyrosine kinase (BTK) [8].

Flow-cytometric analysis can point to specific primary immunodeficiency syndromes such as class switch recombination defects due to activation-induced cytidine deaminase (AID) or uracil-DNA glycosylase (UNG) deficiency which presents with absent IgM− IgD− CD27+ B-cells and concomitant expansion of IgM-only CD27+ B-cells [9]. Additional tests allow the exclusion of genetic defects abrogating the expression of CD40 ligand (CD40L) and CD40 [9]. So far, four genetic defects have been associated with CVID: ICOS, TACI, CD19, and BAFF-R deficiency [10–12]. Since 2003, 9 patients have been identified with a homozygous deletion in the gene encoding for the inducible costimulator (ICOS) [10]. ICOS belongs to the CD28 family and is exclusively expressed on activated T-cells [13]. This defect is detected by flow-cytometric analysis of ICOS expression on activated T-cells. TACI deficiency represents the most common monogenetic defect in CVID [14]. The flow-cytometric analysis of TACI, however, is unreliable, and genetic testing is required. Nevertheless, the few patients with CD19 deficiency and BAFF-R deficiency were detected by flow cytometry (see below).

The differential diagnosis of CVID includes some subforms of X-linked lymphoproliferative syndrome (XLP) [15]. Flow-cytometric analysis of peripheral blood mononuclear cells (PBMCs) of XLP patients show reduced class-switched memory B-cells, expanded transitional B-cells and low intracellular SAP expression [16].

In recent years, flow cytometry has been introduced as a diagnostic tool for the classification of CVID. Three schemes (the Freiburg, Paris and EUROClass Classifications) are based on the phenotyping of circulating B-cell subpopulations [17–19] and two on T-cell phenotypes [20, 21]. The classification schemes were introduced to permit a clinically and pathogenetically relevant discrimination of distinct patient groups.

Table 1. Antibody mixture for the basic lymphocyte panel[a]

	FITC 100 μl	PE 100 μl	PerCP 50 μl	APC 25 μl
K1	CD8	CD4	CD45	CD3
K2	CD19	CD56/16[b]	CD45	CD3

[a] 15 μl/50 μl whole blood.
[b] Per 50 μl.

Protocol

Principles

These protocols are appropriate for four-color flow cytometry. Both surface-only staining and combinations of stainings for various surface molecules and intracellular proteins can be used. All protocols are compiled on the basis of analyses using the FACSCalibur (BD Biosciences, Heidelberg, Germany) with 488- and 635-nm diode lasers and CellQuest software (BD Biosciences). Appropriate adaptation to other cytometers must be performed if necessary. The reader is expected to be familiar with the prerequisites for the operation, instrument settings, compensation, dot plot analyses and the setting of evaluation windows (region, gates).

Basic Lymphocyte Status

In whole-blood staining, the respective proportions of CD4+, CD8+ T-cells, B-cells and NK cells are determined (table 1). This serves as preliminary information as well as for the determination of absolute counts by simultaneous differential blood cell count.

Exclusion of Confounding Diagnoses
It is essential to understand that descriptive flow-cytometric analysis can only detect quantitative expression defects, rather than functional or minimum structural changes (such as point mutations). To rule out a hyper-IgM syndrome with defects in CD40 (former type 3), B-cells are stained for CD40. Both CD40L and ICOS defects can be proven only by analyzing activated T-cells. While ICOS induction can be examined in PBMNCs after stimulation, CD4+ T-cells should be isolated beforehand to optimize the determination of surface expression of CD40L. Further functional analyses are performed in specialized laboratories and are not dealt with in this chapter.

Classification of CVID

In the last 5 years, three classification schemes based on the quantification of B-cell subpopulations have been introduced: the Freiburg Classification [18], the Paris Classification [17] and a European consensus classification EUROclass [19], which overlap in several aspects (fig. 1). Patients can be classified according to all three schemes combining two stainings (table 2). The first staining, B1, allows differences to be determined between naïve (CD19+ CD27– IgM+ IgD+), IgM IgD double-positive memory (CD19+ CD27+ IgM+ IgD+) and class-switched memory B-cells (CD19+ CD27+ IgM– IgD–). The often very low levels of 'IgM-only' memory B-cells (CD19+ CD27+ IgM+ IgD–) can also be calculated from the analysis.

In a second staining, B2, the proportion of transitional (CD19+ $CD21^{int}$ CD38++ IgM++) B-cells, plasmablasts ($CD19^{low}$ $CD21^{low}$ CD38+++ IgM– or IgM^{low}) and an activated $CD21^{low}$ population ($CD19^{high}$ $CD21^{low}$ $CD38^{low}$ IgM+) is determined.

In addition, two groups have suggested a classification of CVID patients according to T-cell subpopulations [20, 21] based on staining lymphocytes for CD3, CD4, CD45R0 and CD45RA. The latter distinguishes three groups of patients according to the percentage of circulating naive CD3+ CD4+ CD45RA+ T-cells (group I <15%, group II 16–30% and group III >30%). The group with the strongest reduction of naïve peripheral CD4 T-cells was associated with an increased incidence of splenomegaly and inflammatory changes [21]. There was a significant correlation to the Freiburg B-cell-based classification scheme.

Material

Sample Material

8 ml EDTA blood (not older than 24 h); for the determination of ICOS and CD40L on activated T-cells, an additional 8 ml EDTA blood is required.

Reagents

- Ficoll separation medium (1,077 g/ml; Biochrom, Berlin, Germany)
- RPMI-1640 (Biochrom)
- FCS (PAN Biotech, Aidenbach, Germany)
- FACSFlow (BD Biosciences)
- Optilyse B (Beckman-Coulter, Krefeld, Germany)
- IntraPrep (Beckman-Coulter)
- Deionized water.

Equipment

- FACSCalibur with a 488-nm laser and a 635-nm diode laser
- standard pipettes

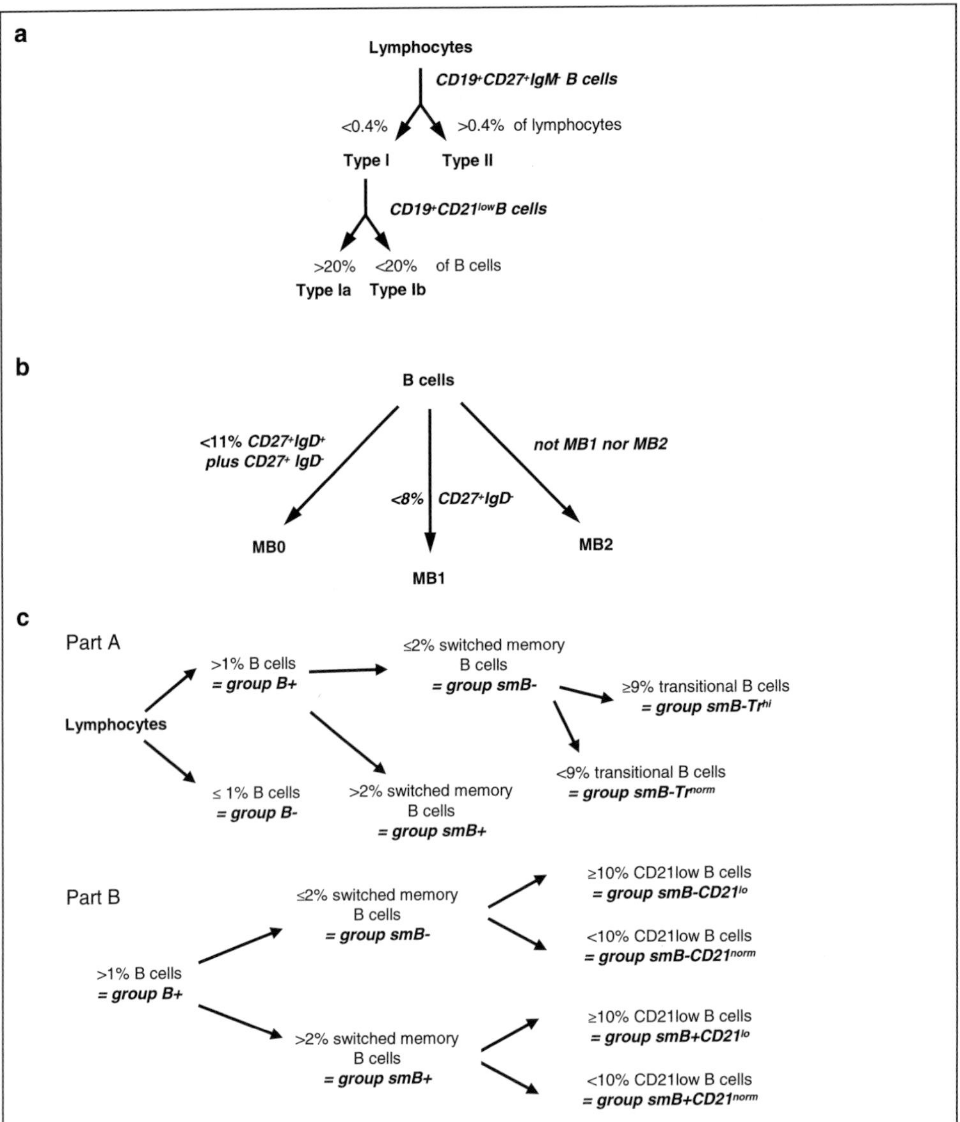

Fig. 1. Flow charts for the classification of patients with CVID. The classification is based on the analysis of different B-cell subpopulations. **a** Freiburg Classification: Type I is characterized by a decrease in class-switched B-cells to below 0.4% of lymphocytes while type II patients have a nearly normal number. Type I can be differentiated into a type Ia and a type Ib based on the expansion of the CD21^{low} B-cell population to over 20% of the B-cells in type Ia and less than 20% in type Ib [18]. **b** The Paris Classification distinguishes a group MB0, with a decrease in all CD27+ B-cells, from a group MB1 with a rather selective decrease in class-switched memory B-cells and a group of patients with nearly normal numbers of non-class-switched as well as class-switched B-cells (MB2) [17]. **c** The most recent classification, EUROClass, first of all distinguishes patients with severe B-cell deficiency (B–) from patients with B-cells above 1% (B+). The latter group is divided into patients with severely decreased switched memory B-cells (smB–) and patients with switched memory B-cells above 2% (smB+). Further subgroups can be differentiated according to the expansion of transitional B-cells or CD21^{low} B-cells.

Table 2. Antibody mixture for the Freiburg Classification[a]

	FITC 100 μl	PE 100 μl	PC7 25 μl	FL4 25 μl
B1	CD27 (1:5)	anti-IgD (1:40)	CD19	anti-IgM Cy5 (1:40)
B2	CD38	CD21	CD19	anti-IgM Cy5 (1:40)
B3	anti-κ (1:2)	anti-λ (1:2)	CD19	anti-IgM Cy5 (1:40)
B4[b]	CD40	??	CD19	anti-IgM Cy5 (1:40)

[a] 10 μl/50 μl cell suspension.
[b] Only if hyper-IgM is suspected.

- centrifuge
- counting chamber
- vortex
- 50-ml Falcon polypropylene test tubes with caps (BD Biosciences)
- 15-ml Falcon polypropylene test tubes with caps (BD Biosciences)
- 12- to 75-mm Falcon polystyrene test tubes (BD Biosciences)
- 0.5-ml sample test tubes with caps (Biozym, Hessisch Oldendorf, Germany).

Antibodies
The antibodies utilized are listed in table 3.

Protocol

Whole-Blood Staining for 'Basic-Lymphocyte Status'
The major lymphocyte population is analyzed by whole-blood staining from EDTA blood samples in lyse-no-wash assays, which are available from various manufacturers. One possible antibody panel is listed in table 1.

Staining of Peripheral Blood Mononuclear Cells (Classification, CD40 Expression)
The staining to exclude a CD40 defect or monoclonal populations and to classify CVID is carried out on PBMNCs following purification over Ficoll density gradients. The technical procedure for staining PBMNCs which have been isolated by Ficoll gradients is described in 'Characterization of B-Lymphocytes; 'Procedures' (pp. 214). The antibody mixtures of the Freiburg classification are summarized in table 2. It should be mentioned that B-cell phenotyping can also be performed in whole-blood assays and that the results are comparable. For a protocol, see Ferry et al. [22].

Determination of CD40L and ICOS Expression on Activated T-Cells/Activation of T-Cells
For the determination of ICOS expression, 1×10^6 Ficoll-isolated PBMNCs are plated into 24-well plates in RPMI-1640 with 10% (v/v) FCS and incubated for 16–20 h

Table 3. List of antibodies[a]

Antibody	Clone	Fluorochrome	Order No.	Provider
CD3	SK7	APC	345767	BD
CD4	13B8.2	PE	A07751	BC
CD8	B9.11	FITC	A07756	BC
CD16	3G8	PE	A07766	BC
CD19	J4,119	FITC	A07768	BC
CD19	J4,119	PC7	IM3628	BC
CD21	B-ly4	PE	555422	BD
CD27	M-T271	FITC	F7178	Dako
CD38	HIT2	FITC	555459	BD
CD40	5C3	FITC	555588	BD
CD40L	TRAP1	PE	555700	BD
CD45	2D1	PerCP	345809	BD
CD56	NKH-1	PE	A07788	BC
CD69	FN50	FITC	555530	BD
Anti-BAFF-R	goat IgG	purified	AF1162	R and D
Anti-goat-IgG	swine IgG	PE	G50004	Caltag
Anti-ICOS	ISA-3	PE	12–9948–71	eBioscience
Anti-IgM	goat F(ab)2	Cy5	109–176–129	JIR
Anti-IgD	goat F(ab)2	PE	2032–09	SB
Anti-κ	G20–193	FITC	555791	BD
Anti-λ	JDC-12	PE	555797	BD
Isotype control	mouse-IgG1	PE	IM0670	BC

BD = BD Biosciences; BC = Beckman Coulter; SB = SouthernBiotech; JIR = Jackson ImmunoResearch Laboratories; R and D = R and D Systems; Caltag = Caltag/Invitrogen.

[a] The antibodies listed are proven and recommended by the authors, but can be replaced by suitable alternative ones.

Table 4. Antibody panel to ascertain activation of T-cells[a]

	FITC	PE	PerCP	APC
A1	CD69	isotype G1	CD3	CD4
A2	CD69	ICOS	CD3	CD4
A3	CD69	CD40L (CD154)	CD3	CD4

[a] 5 μl of each of the pretitrated antibodies is administered to every $0.5–1.0 \times 10^6$ cells.

at 37°C and 5% CO_2 with and/or without 1 μg/ml PHA. The cells are subsequently transferred into a FACS tube, washed once in FACS medium (5 min, $300 \times g$) and stained with the antibody mixture according to the protocol for PBMNCs. To confirm that the T-cells are activated, they are additionally stained for CD69 (table 4, A1 + A2). The methodology to exclude a CD40L defect is described in 'Primary Immunodeficiency Diseases; CD40L and CD40 Defect (Hyper-IgM Syndromes, HIGM Syndromes)' (pp. 531) (table 4, A1 + A3).

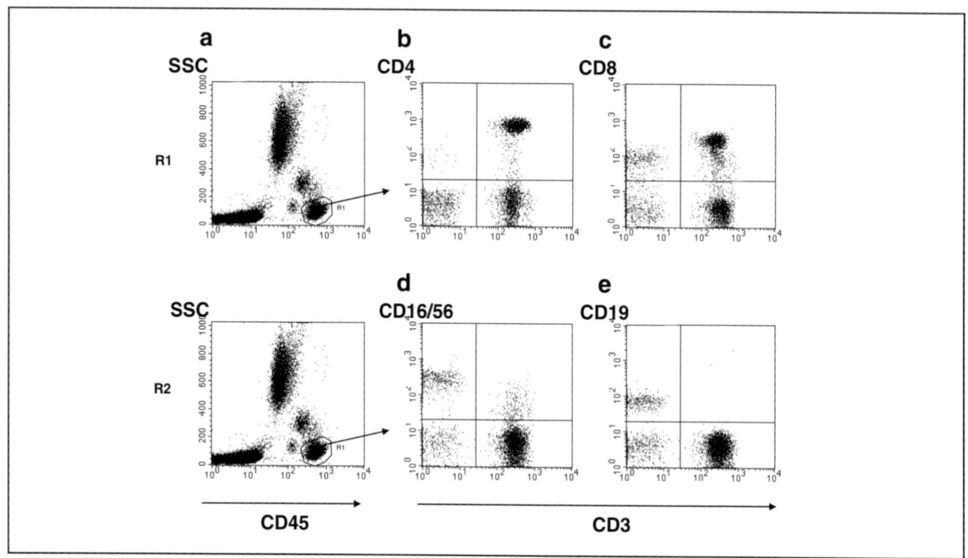

Fig. 2. Basic lymphocyte panel. **a** The lymphocyte region (R1) is depicted in window FL3 (CD45)/SSC. The different lymphocyte populations are characterized by the expression of CD3. T-cells are divided according to CD4+ (**b**), CD8+ (**c**), CD16+ CD56+ CD3− (NK cells) (**d**) and CD19+ (B-cells) (**e**).

Measurements

Accurate instrument settings for photomultipliers and compensation are the preconditions for B-cell determination. For well-titrated antibodies, the compensation is uniform within the panel. Usually, the instrument settings are very stable over time, and a control with beads of any color is sufficient. It is not necessary to prepare each measurement with single-color stainings.

Measurement of 'Basic Lymphocyte Status' (fig. 2)

The acquisition protocol contains a two-dimensional acquisition window with the FL3 (CD45) and side scatter (SSC) axes. The R1 analysis region defines the lymphocyte population through the expression of CD45 and low scatter. As further windows, the lymphocytes in R1 are represented as FL1 vs. FL4 and FL2 vs. FL4 plots. Data of at least 5,000 lymphocytes should be collected.

Measurements for the Classification of CVID (fig. 3)

The analysis of the B-cell subpopulation is described in detail in 'Characterization of B-lymphocytes' (pp. 211). Table 2 shows the recommended minimum staining supplemented by a staining for CD40 (see below). Staining B1 (fig. 3a–c, panel 1) enables a distinction between naïve B-cells (UL, CD27− IgD+ IgM+), non-class-switched (UR, CD27+ IgD+ IgM+) and class-switched (LR, CD27+ IgD− IgM−) memory B-cells [23]. The second staining, B2 (fig. 3, panel 2), shows the markers CD21 and CD38 for further B-cell differentiation levels. CD38 is strongly expressed

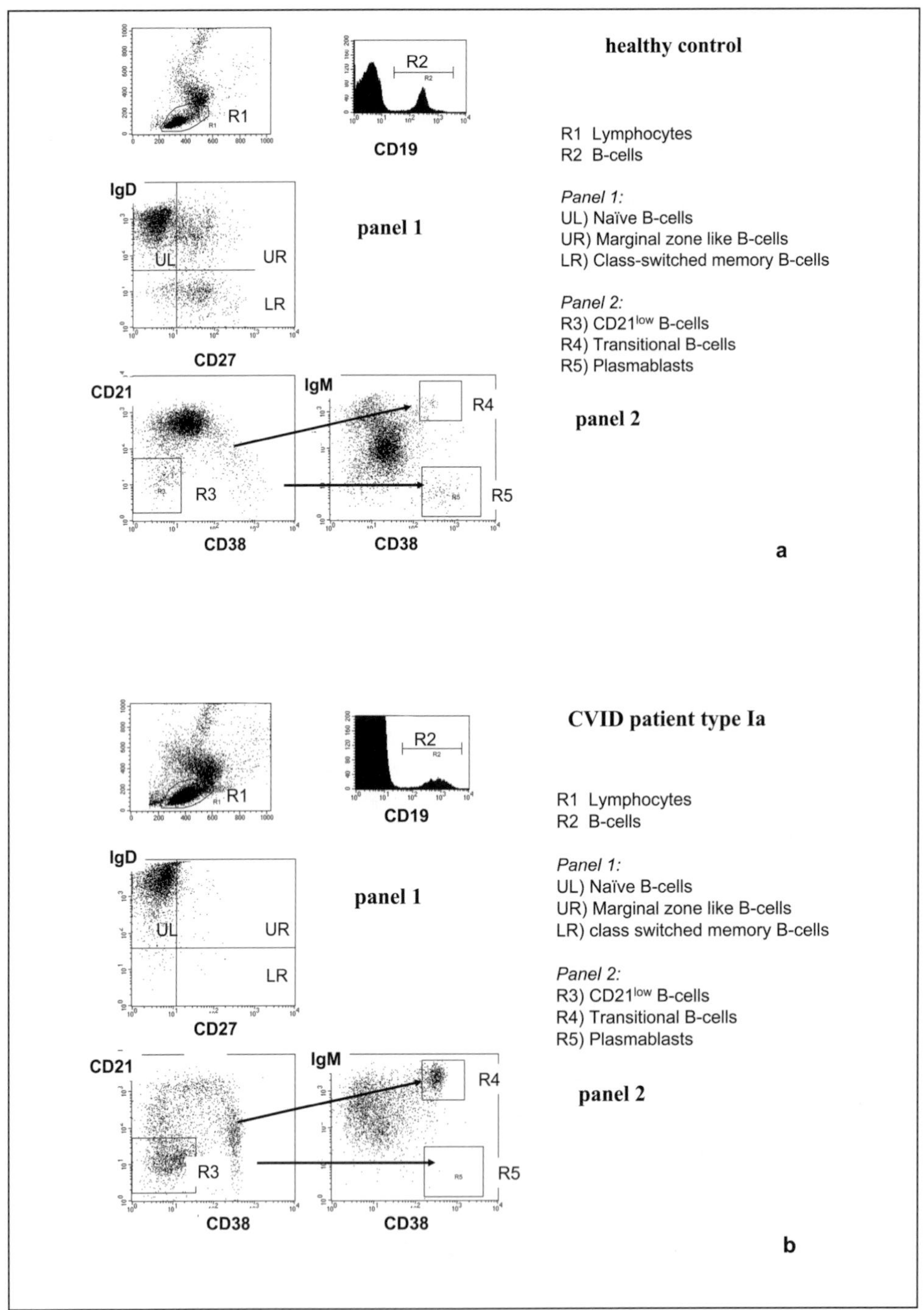
healthy control
R1
R2
CD19
R1 Lymphocytes
R2 B-cells
IgD
UL
UR
LR
CD27
panel 1
Panel 1:
UL) Naïve B-cells
UR) Marginal zone like B-cells
LR) Class-switched memory B-cells
Panel 2:
R3) CD21low B-cells
R4) Transitional B-cells
R5) Plasmablasts
CD21
IgM
R3
R4
R5
CD38
CD38
panel 2
a
CVID patient type Ia
R1
R2
CD19
R1 Lymphocytes
R2 B-cells
IgD
UL
UR
LR
CD27
panel 1
Panel 1:
UL) Naïve B-cells
UR) Marginal zone like B-cells
LR) class switched memory B-cells
Panel 2:
R3) CD21low B-cells
R4) Transitional B-cells
R5) Plasmablasts
CD21
IgM
R3
R4
R5
CD38
CD38
panel 2
b

Fig. 3 a, b

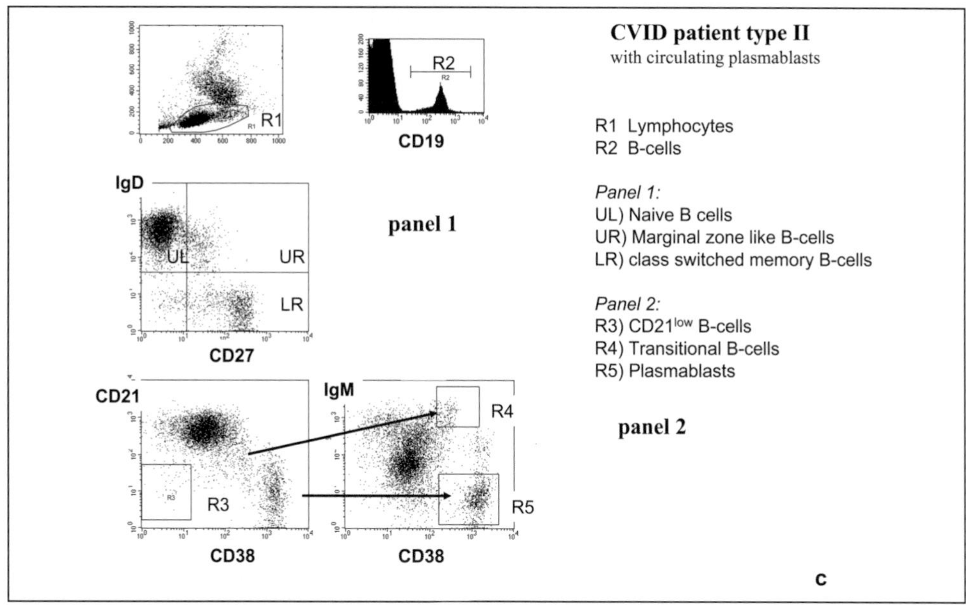

Fig. 3. Analysis of peripheral B-cell subpopulations for the classification of CVID. A lymphocyte region (R1) and a CD19+ region (R2) are combined to a B-cell gate, in which B1 and B2 stainings are analyzed. Panel 1 permits the distinction between naïve (UL), IgM memory (UR) and class-switched B-cells. In panel 2 the CD21^{low} B-cells (R3) can be defined in the first window (CD38/CD21). To identify transitional B-cells (R4) and the plasmablasts (R5), the representation CD38/IgM is preferred. The evaluation is partially done by means of quadrant statistics and partially by means of regional statistics. **a** Results in a normal person. **b** Typical findings in a CVID patient with expansion of CD21^{low} B-cells. **c** The uniqueness of the third example is the expansion of plasmablasts in the peripheral blood of a patient during an infection.

on all B-cell precursors (also see 'Characterization of B-Lymphocytes', pp. 211). Naïve B-cells show an intermediate CD38 surface expression, while memory B-cells are mostly negative for CD38. The low levels of expression of CD21 and CD38 (R3) defines the CD21^{low} B-cell population, which is included in the Freiburg and EUROClass Classification schemes. In addition, the high expression of IgM and CD38 defines transitional B-cells (R4) [24] whereas missing or low IgM expression and the highest CD38 expression are found on plasmablasts (R5) [25]. In case of a detectable plasmablast population, the gate for CD27+ IgM and memory B-cells in B1 needs to be adapted to exclude CD27++ plasmablasts.

Measurements to Detect ICOS and CD40L Expression on Activated T-Cells

The cells are defined first by a two-dimensional data collection window, which deliberately includes larger cells in the FSC/SSC display and subsequently by gating on CD4+ cells (*cave* after stimulation by CD3: CD3 is only weakly detectable on the surface of activated cells). To confirm activation, the expression of CD69 (x-axis) and

CD40L and/or ICOS (y-axis) on CD4+ T-cells is represented in a second window (not shown). Data of at least 5,000 CD4+ cells should be collected.

Data Analysis

Peripheral Lymphocyte Population

To assess the basic lymphocyte status, the lymphocyte region (R1) is drawn according to figure 2a with the FL3 (CD45) and SSC axes. The proportions of CD4+ and CD8+ T-cells (UR, fig. 2b, c), NK cells (UL, fig. 2d) and B-cells (UL, fig. 2e) are determined from the respective quadrants. The absolute values are calculated by multiplication of total lymphocyte counts from the differential blood cell count and the relative proportion of the individual populations of lymphocytes. The CD45 staining used in whole blood is used to define a lymphocyte window which excludes erythrocytes and basophils, from which the proportion of lymphocyte subpopulations can be accurately calculated.

Results

The patient's clinical symptoms and medication should be taken into account when evaluating the analytical data. Reference values for the different B-lymphocyte subpopulations for adults are specified in 'Characterization of B-Lymphocytes'; table 7 (pp. 222). The reference values for other lymphocyte populations in adults have been reported by Bisset et al. [26]. A clear reduction in B-cell number (<1%) gives an indication of the different forms of agammaglobulinemia. Clear inversions of the CD4+/CD8+ relationship can be a sign of viral infection. In particular in the presence of low CD4+ cell levels, an HIV infection should be ruled out. The diagnosis of an EBV infection should also be kept in mind, in particular in the case of male patients with positive family histories (XLP). Such a shift in lymphocyte profile without a clear cause is found more frequently in CVID patients. Lymphomas are an important possible diagnosis with regard to abnormal (e.g. immunoglobulin negative) B-cell populations or even monoclonal populations (characterized by a clearly shifted κ/λ relationship) (see 'Flow Cytometry in the Diagnosis of Non-Hodgkin's Lymphomas', pp. 642).

Classification

All classification schemes based on B-cell phenotype group patients according to the reduction in class-switched memory B-cells. The Paris classification includes non-class-switched memory B-cells, the Freiburg Classification includes CD21^{low} B-cells and the EUROClass includes CD21^{low} B-cells and transitional B-cells as additional discriminating parameters (fig. 1).

Results

For details on B-cell phenotyping, also see 'Characterization of B-Lymphocytes' (pp. 211).

The evaluation and reference values for the B-cell subpopulations are described in 'Characterization of B-Lymphocytes' (pp. 211). In general, it is meaningful to indicate the proportion of the entire B-cell population as percent of the lymphocytes and the subpopulation in percentage of B-cells. From these data, lymphocyte-specific or absolute values can be calculated if necessary.

Freiburg Classification (fig. 1a): Following the Freiburg Classification, the proportion of class-switched memory B-cells of the total lymphocyte population is determined first (fig. 3, panel 1). If the proportion is below 0.4%, the patient is diagnosed as type I (about 75% of the patients). Patients with more than 0.4% belong to type II (about 25% of the patients). Type I patients are subdivided according to the proportion of CD21^{low} B-cells into type Ia with an expansion of CD21^{low} B-cells of greater than 20% and type Ib with less of an expansion (fig. 3, panel 2). Freiburg type Ia patients present significantly more often with splenomegaly and granulomatous disease [18].

Paris Classification (fig. 1b): The division of CVID patients according to the Paris Classification [17] is based on the same staining procedure as the B1 staining of the Freiburg Classification. The groups are differentiated into MB2 with a normal percentage of memory B-cells, MB1 with decreased class-switched and normal non-class-switched memory B-cells and MB0 with a decrease in both CD27+ B-cell populations. 'Decreased' values are defined as 2 standard deviations below the mean. In the cohorts examined, the proportion of CD27+ B-cells for the MB0 group was <11%; the proportion of class-switched B-cells was <8% for the MB1 group. With this classification scheme, significant levels of splenomegaly and granulomatous inflammation were found in the MB0 group [17].

EUROClass (fig. 1c): The EUROClass classification [19] uses the same staining procedure and assessment as the Freiburg Classification and was the fruit of common efforts of several European immunodeficiency centers. It discriminates a B– group with equal or less than 1% B-cells and B+ group with more than 1% of B-cells. The latter group is divided into smB+ (>2% IgM– IgD– CD27+ B-cells) and smB– (< = 2% IgM– IgD– CD27+ B-cells) patients. In addition, patients with expansion of CD21^{low} B-cells $\geq$10% are classified as 'smB– or smB+ CD21^{low}' compared to 'CD21^{norm}'; patients with an expansion of transitional B-cells $\geq$9% are designated 'smB-Tr^{hi}' compared to 'Tr^{norm}'.

The evaluation of more than 300 patients confirmed the association of severely reduced class-switched memory and CD21^{low} B-cells with splenomegaly and granulomatous disease. In addition, it demonstrated an association between expanded transitional B-cells and lymphadenopathy.

T-Cell-Based Classification: A recent suggestion for the classification of CVID is based on the relative reduction of naive CD4 T-cells [21].

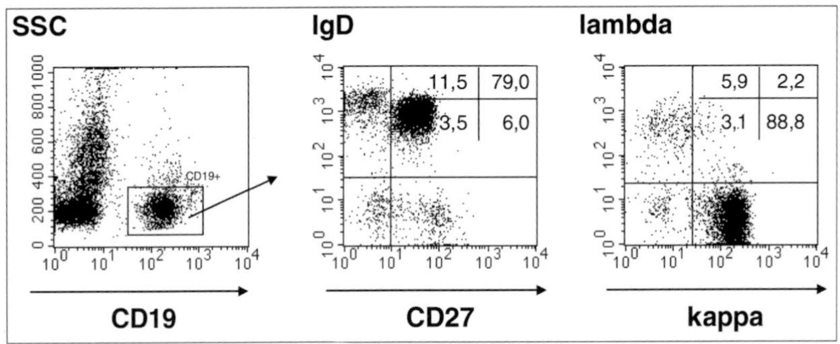

Fig. 4. Confirmation of a monoclonal population. In the representation of B-cells, a clear expansion of the CD27+ non-class-switched memory B-cells (UR quadrant: 79%) in the IgD/CD27 window is already noticeable. In the second window, a substantial expansion of κ+ B-cells (κ/λ: 15:1) suggests a monoclonal population. Further analyses in the case of this patient resulted in the diagnosis of a mantle cell lymphoma.

Diagnosing Common Variable Immunodeficiency

Exclusion of Monoclonal B-Cell Populations

The B3 staining gives insight into monoclonal B-cell population distributions (fig. 4). After gating on lymphocytes and B-cells, the κ versus λ light chain expression is represented. This relationship should come to approximately 50:50. Clear deviations or the presence of a substantial proportion of light-chain-negative B-cells can be considered ambiguous clues for atypical B-cell populations after the exclusion of plasmablasts (stain B2, R5). The presence of such abnormal populations should be clarified more precisely in cooperation with the department of hematology.

Exclusion of Class Switch Recombination Defect due to CD40 Deficiency

For this analysis, CD40 expression is examined after gating on B-cells. Normally, almost 100% of B-cells are positive for CD40. As previously mentioned, this analysis only allows the confirmation of severe structural defects in this molecule. The hyper-IgM syndrome, which is due to a defect in the CD40 receptor, is extremely rare (<1% the patients with hyper-IgM syndrome) and so far only early-childhood onset has been described. The investigation is easily accomplished and can be carried out without excessive expenditures. Thus far, none of the CVID patients we examined manifested abnormal CD40 expression. Thus, even with a positive result, further genetic and functional analysis would be necessary to confirm the diagnosis.

Exclusion of Class Switch Recombination Defect due to CD40 Ligand Deficiency

The exclusion of CD40L defects is complex. Therefore, this procedure should be used only if the suspicion is strong, i.e. in male patients, especially with a positive

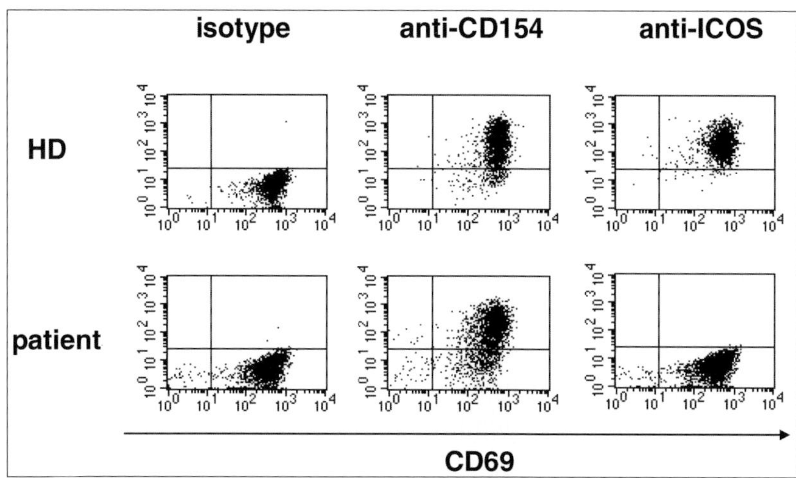

Fig. 5. Confirmation of an ICOS defect. After the activation of T-cells over 16 h in vitro (see text regarding details), surface expression of both CD40L and ICOS could be clearly proven after gating on CD4+ T-cells for healthy controls (HD). The increased expression of CD69 is used as control for activation. In the comparison, the patient shows an almost normal CD40L expression, but no ICOS expression. One of the 9 thus far identified CVID patients with a genetic ICOS defect [10] is shown. Additionally, a nonstimulated control should always be run in parallel (not shown here).

family history, whose serum IgM value is in the upper range of normal or is increased. After stimulation, activated CD4+ T-cells and nonactivated control cells should be evaluated according to figure 5. As an internal control for successful activation, activated CD4+ T-cells are evaluated for increased surface expression of CD69. Nonactivated T-cells should be CD40L–. Isolated CD4+ T-cells of a healthy control must be stimulated in parallel as a positive control.

Exclusion of AID and UNG Deficiency

AID and UNG deficiency cannot be excluded by flow cytometry, but the absence of class-switched B-cells combined with increased IgM-only memory B-cells (IgM+ IgD– CD27+) strongly suggests intrinsic B-cell defects of class switch recombination.

Exclusion of ICOS Deficiency

In CVID patients with autosomal recessive transmission of the illness and severely reduced class-switched memory B-cells, a defect in ICOS expression should be investigated. ICOS is expressed on activated T-cells within a few hours and, in contrast to nonstimulated cells, is readily detectable after 16 h (fig. 5). Due to its T-cell-specific expression, other lymphocyte populations can serve as negative controls. As positive control, PBMNCs from a healthy person must be tested in parallel. We carry out this evaluation after gating CD4+ T-cells as for the analysis of CD40L. The exclusive analysis of CD4+ T-cells is necessary since the expression of the ICOS receptor is higher on CD4+ T-cells than on CD8+ T-cells.

Exclusion of TACI Deficiency
TACI deficiency is the most common genetic defect in CVID patients [14]. It is not associated with a specific B- or T-cell phenotype [11]. TACI itself is expressed on memory B-cells and can readily be detected by flow cytometry with several monoclonal antibodies, but only very few TACI-deficient patients present with absent or severely reduced TACI surface expression, rendering the flow-cytometric evaluation unreliable.

Exclusion of CD19 Deficiency
CD19 deficiency [12] was readily detectable by flow cytometry in all five cases reported up to the present date since most laboratories use CD19 for the identification of B-cells. Therefore the discrepancy between the absence of CD19+ cells and the simultaneous detection of other B-cell markers (for example immunoglobulin or CD20) strongly suggests CD19 deficiency. In the standard lymphocyte panel, CD19-deficient B-cells may present as an undefined CD3– CD19– CD16/CD56– lymphocyte population, resulting in a summation of T-, B- and NK cells below 100%.

Exclusion of BAFF-R Deficiency
So far, only 2 patients have been identified with BAFF-R deficiency, thus making it a very rare defect. B-cell phenotyping showed a reduction in total B-cells and a significant accumulation of transitional B-cells in both affected patients (Warnatz et al., in preparation). In healthy donors, BAFF-R is highly expressed on all peripheral B-cells. It should be noted that some monoclonal antibodies such as 11C2 and 8A7 show weaker signals in some patients while this difference is not detectable by the polyclonal antiserum (M. Rizzi and H. Eibel, personal communication).

Comments

We recommend the standard evaluation of patients with hypogammaglobulinemia by the assessment of the basic lymphocyte panel and B-cell phenotyping for classification as well as exclusion of atypical lymphocyte populations. The evaluation of differential diagnoses specified under 'Diagnosing Common Variable Immunodeficiency' (see above) remains reserved for special cases. In our laboratory, B-cell staining is performed on Ficoll-separated PBMNCs. Several laboratories carry out B-cell staining with whole-blood assays. This procedure represents an alternative which requires the appropriate controls.

Quality Control

Apart from the staining with appropriate isotype controls, each analysis should be run together with healthy control samples in order to exclude technical errors. The stimulation experiments should likewise contain both a medium control (nonstimulated) and a healthy control sample. Highly pathological findings should be confirmed by an independent second analysis. All results should be interpreted considering the patient's clinical history.

Troubleshooting

From our experience, evaluations of B-cell subpopulations in patients with less than 1% B-cells are difficult and must be interpreted with caution. Different specific antibodies and/or conjugates (e.g. CD38-PE vs. CD38-APC) can lead to different results, which must be considered in the conclusions. In particular, the highly variable quality of the anti-ICOS antibodies must be taken into account. The problems of in vitro stimulation of T-cells are numerous (also see 'Detection of Antigen-Specific T-Cells using MHC Multimers or Functional Parameters', pp. 476).

Expected Results

The expected results for B-cell subpopulations in peripheral blood and bone marrow are specified in the respective sections and illustrations.

Required Time

The preparation of PBMNCs takes about 1.5 h. The same amount of time must be scheduled for surface staining. T-cell stimulation to confirm CD40L requires 0.5 h on the 1st day for the preparation of the plates, 2 h on the 2nd day for the cleaning and incubation of the PBMNCs, which can be further processed after 16–20 h of overnight culture. The data acquisition of 2,000–5,000 cells requires up to 5 min per tube.

Summary

Common variable immunodeficiency syndrome (CVID) is a heterogeneous antibody deficiency syndrome that manifests clinically by recurrent respiratory and gastrointestinal infections either in childhood (after the 2nd year of life) or in the 2nd–3rd decade. Since CVID represents a diagnosis by exclusion, well-known genetic, infectious and toxic causes must be excluded. The diagnostic criteria were established by the ESID (*www.esid.org*). Flow cytometry allows the detection of patients with class switch recombination deficiency syndromes and agammaglobulinemia (B-cell numbers <1%). Increasingly, flow cytometry also serves to classify this heterogeneous syndrome and, to a limited extent, to exclude genetic defects associated with CVID. Further flow-cytometric analyses are currently reserved for research.

References

1 Cunningham-Rundles C, Bodian C: Common variable immunodeficiency: clinical and immunological features of 248 patients. Clin Immunol 1999;92:34–48.

2 Spickett GP: Current perspectives on common variable immunodeficiency (CVID). Clin Exp Allergy 2001;31:536–542.

3 Schaffer AA, Salzer U, Hammarstrom L, Grimbacher B: Deconstructing common variable immunodeficiency by genetic analysis. Curr Opin Genet Dev 2007;17:201–212.

4 Vorechovsky I, Zetterquist H, Paganelli R, Koskinen S, Webster AD, Bjorkander J, Smith CI, Hammarstrom L: Family and linkage study of selective IgA deficiency and common variable immunodeficiency. Clin Immunol Immunopathol 1995;77:185–192.

5 Thon V, Wolf HM, Sasgary M, Litzman J, Samstag A, Hauber I, Lokaj J, Eibl MM: Defective integration of activating signals derived from the T cell receptor (TCR) and costimulatory molecules in both CD4+ and CD8+ T lymphocytes of common variable immunodeficiency (CVID) patients. Clin Exp Immunol 1997;110:174–181.

6 Di Renzo M, Zhou Z, George I, Becker K, Cunningham-Rundles C: Enhanced apoptosis of T cells in common variable immunodeficiency (CVID): role of defective CD28 co-stimulation. Clin Exp Immunol 2000;120:503–511.

7 Groth C, Drager R, Warnatz K, Wolff-Vorbeck G, Schmidt S, Eibel H, Schlesier M, Peter HH: Impaired up-regulation of CD70 and CD86 in naive (CD27–) B cells from patients with common variable immunodeficiency (CVID). Clin Exp Immunol 2002;129:133–139.

8 Kanegane H, Tsukada S, Iwata T, Futatani T, Nomura K, Yamamoto J, Yoshida T, Agematsu K, Komiyama A, Miyawaki T: Detection of Bruton's tyrosine kinase mutations in hypogammaglobulinaemic males registered as common variable immunodeficiency (CVID) in the Japanese Immunodeficiency Registry. Clin Exp Immunol 2000;120:512–517.

9 Durandy A, Revy P, Fischer A: Human models of inherited immunoglobulin class switch recombination and somatic hypermutation defects (hyper-IgM syndromes). Adv Immunol 2004;82:295–330.

10 Grimbacher B, Hutloff A, Schlesier M, Glocker E, Warnatz K, Drager R, Eibel H, Fischer B, Schaffer AA, Mages HW, Kroczek RA, Peter HH: Homozygous loss of ICOS is associated with adult-onset common variable immunodeficiency. Nat Immunol 2003;4:261–268.

11 Salzer U, Chapel HM, Webster AD, Pan-Hammarstrom Q, Schmitt-Graeff A, Schlesier M, Peter HH, Rockstroh JK, Schneider P, Schaffer AA, Hammarstrom L, Grimbacher B: Mutations in TNFRSF13B encoding TACI are associated with common variable immunodeficiency in humans. Nat Genet 2005;37:820–828.

12 van Zelm MC, Reisli I, van der BM, Castano D, van Noesel CJ, van Tol MJ, Woellner C, Grimbacher B, Patino PJ, van Dongen JJ, Franco JL: An antibody-deficiency syndrome due to mutations in the CD19 gene. N Engl J Med 2006;354:1901–1912.

13 Hutloff A, Dittrich AM, Beier KC, Eljaschewitsch B, Kraft R, Anagnostopoulos I, Kroczek RA: ICOS is an inducible T-cell co-stimulator structurally and functionally related to CD28. Nature 1999;397:263–266.

14 Pan-Hammarstrom Q, Salzer U, Du L, Bjorkander J, Cunningham-Rundles C, Nelson DL, Bacchelli C, Gaspar HB, Offer S, Behrens TW, Grimbacher B, Hammarstrom L: Reexamining the role of TACI coding variants in common variable immunodeficiency and selective IgA deficiency. Nat Genet 2007;39:429–430.

15 Engel P, Eck MJ, Terhorst C: The SAP and SLAM families in immune responses and X-linked lymphoproliferative disease. Nat Rev Immunol 2003;3:813–821.

16 Hare NJ, Ma CS, Alvaro F, Nichols KE, Tangye SG: Missense mutations in SH2D1A identified in patients with X-linked lymphoproliferative disease differentially affect the expression and function of SAP. Int Immunol 2006;18:1055–1065.

17 Piqueras B, Lavenu-Bombled C, Galicier L, Bergeron-van der Cruyssen F, Mouthon L, Chevret S, Debre P, Schmitt C, Oksenhendler E: Common variable immunodeficiency patient classification based on impaired B cell memory differentiation correlates with clinical aspects. J Clin Immunol 2003;23:385–400.

18 Warnatz K, Denz A, Drager R, Braun M, Groth C, Wolff-Vorbeck G, Eibel H, Schlesier M, Peter HH: Severe deficiency of switched memory B cells ($CD27^+IgM^-IgD^-$) in subgroups of patients with common variable immunodeficiency: a new approach to classify a heterogeneous disease. Blood 2002;99:1544–1551.

19 Wehr C, Kivioja T, Schmitt C, Ferry B, Witte T, Eren E, Vlkova M, Hernandez M, Detkova D, Bos PR, Poerksen G, von BH, Baumann U, Goldacker S, Gutenberger S, Schlesier M, Bergeron-van der CF, Le GM, Debre P, Jacobs R, Jones J, Bateman E, Litzman J, van Hagen PM, Plebani A, Schmidt RE, Thon V, Quinti I, Espanol T, Webster AD, Chapel H, Vihinen M, Oksenhendler E, Peter HH, Warnatz K: The EUROclass trial: defining subgroups in common variable immunodeficiency. Blood 2008;111:77–85.

20 Aukrust P, Lien E, Kristoffersen AK, Muller F, Haug CJ, Espevik T, Froland SS: Persistent activation of the tumor necrosis factor system in a subgroup of patients with common variable immunodeficiency – possible immunologic and clinical consequences. Blood 1996;87:674–681.

21 Giovannetti A, Pierdominici M, Mazzetta F, Marziali M, Renzi C, Mileo AM, De FM, Mora B, Esposito A, Carello R, Pizzuti A, Paggi MG, Paganelli R, Malorni W, Aiuti F: Unravelling the complexity of T cell abnormalities in common variable immunodeficiency. J Immunol 2007;178:3932–3943.

22 Ferry BL, Jones J, Bateman EA, Woodham N, Warnatz K, Schlesier M, Misbah SA, Peter HH, Chapel HM: Measurement of peripheral B cell subpopulations in common variable immunodeficiency (CVID) using a whole blood method. Clin Exp Immunol 2005;140:532–539.

23 Klein U, Rajewsky K, Kuppers R: Human immunoglobulin (Ig)M^+IgD^+ peripheral blood B cells expressing the CD27 cell surface antigen carry somatically mutated variable region genes: CD27 as a general marker for somatically mutated (memory) B cells. J Exp Med 1998;188:1679–1689.

24 Carsetti R, Rosado MM, Wardmann H: Peripheral development of B cells in mouse and man. Immunol Rev 2004;197:179–191.

25 Odendahl M, Jacobi A, Hansen A, Feist E, Hiepe F, Burmester GR, Lipsky PE, Radbruch A, Dorner T: Disturbed peripheral B lymphocyte homeostasis in systemic lupus erythematosus. J Immunol 2000;165:5970–5979.

26 Bisset LR, Lung TL, Kaelin M, Ludwig E, Dubs RW: Reference values for peripheral blood lymphocyte phenotypes applicable to the healthy adult population in Switzerland. Eur J Haematol 2004;72:203–212.

Sack U, Tárnok A, Rothe G (eds): Cellular Diagnostics. Basics, Methods and Clinical Applications of Flow Cytometry. Basel, Karger, 2009, pp 558–579

Monitoring of Organ-Transplanted Patients

Andreas Lun[a] · Ulrich Sack[b]

[a] Institute for Laboratory Medicine and Pathobiochemistry, Charité University Hospital, Berlin,
[b] Institute for Clinical Immunology and Transfusion Medicine, Medical Faculty of the University of Leipzig, Germany

Introduction

Solid-organ transplantation and has been used for many years in clinical practice as a therapeutic option in functional loss of the kidney, liver, heart, lung, pancreas, and intestine. Despite considerable progress in surgery and immunosuppression, transplant and patient survival is still limited (table 1) [1].

Multiple factors such as conservation damage by hypoxia and reactive oxygen radicals, surgical problems, bleeding, toxic side effects of immunosuppressive therapy, infection as well as reinfection and chronic rejection can lead to transplant loss and/or death. Moreover, episodes of acute rejection, mainly steroid-resistant rejection, reduce long-term survival of the transplant [2–4]. So, in order to detect and treat any complications in time, recipients of transplants must be closely monitored, especially during the critical early postoperative period.

Rejection of the transplant occurs because the recipient's immune system recognizes cellular structures of the donor as foreign and starts an immune response directed at the engrafted organ [5–7]. This response must be suppressed.

Immunosuppressive therapy must be precisely adapted to the transplanted organ and the individual patient's needs. Insufficient immunosuppression increases the risk of rejection. Aggressive immunosuppressive therapy can prevent early acute rejections almost completely, but may cause iatrogenic immunodeficiency [8]. This in turn may lead to repeated infections, for example with cytomegalovirus (CMV), or even septicemia (see 'Pathophysiology and Immune Monitoring of Sepsis', pp. 600), and severe delayed complications such as tumors or infections with polyoma virus which may cause transplant loss [9, 10].

Table 1. Organ transplants and patients' survival in 2006 [1]

Organ	Number of transplantations		1-year survival rate, %
	Eurotransplant	Germany	
Kidney	4,412	2,776	95
Liver	1,487	979	80[a]
Heart	587	412	80
Lung	496	253	70
Pancreas	234	141	<40–80[b]

[a] 1-year survival rate is 80% in the most frequent indications for liver transplantation. These indications are end-stage chronic hepatitis and biliary cirrhosis in adults, biliary atresia and inborn metabolic deficiency.
[b] Depending on recovery of a normal glycemia.

Immunopathology of Transplant Rejection

With the help of major histocompatibility complex (MHC) molecules, antigen-presenting cells (APCs) present donor-derived allogeneic peptides to CD4+ and CD8+ T lymphocytes via two pathways [11]:

- indirect pathway: peptides cleaved from donor MHC molecules are presented to T-cells by host APCs similarly to bacterial antigens.
- direct pathway: donor APCs present alloantigens to host T-cells.

The direct pathway seems to be responsible for acute rejections, while the indirect pathway is generally implicated in chronic rejections [12, 13]. CD4+ and CD8+ T-cells are crucial in transplant rejection [14]. More detailed information can be found in reviews published in the literature [e.g. 5, 7].

T-cells specifically recognize MHC antigen peptides with their T-cell receptor (first signal). CD8+ and CD4+ T-cells can be activated by presenting MHC class I and II molecules. Activated T-cells express molecules such as CD25 (α-chain of the IL-2 receptor), HLA-DR (a group of MHC class II molecules), CD71 (transferrin receptor), CD26 (dipeptidylpeptidase IV = EC 3.4.14.5) or CD11a, b, c (adhesion molecules for lymphocyte homing) that can be used for diagnostic purposes [15–20]. Further accessory molecules such as CD28/B7-1 (CD80), CD28/B7-2 (CD86), CTLA4/B7, LFA-1/ICAM-1, and CD40/CD40L strengthen T-cell and APC binding. They influence intracellular signaling, resulting in T-cell activation or induction of tolerance (second signal) [21–24].

Furthermore, antigen-specific T-cell activation induces the production and secretion of cytokines and chemokines. These molecules can also be used for the laboratory diagnosis of rejections. Interactions between innate and adaptive immunity and induction of alloantibody production and cellular cytotoxicity are regulated by these mediators [7].

Special attention must be paid to the IL-2/IL-2 receptor system [25]. While the T-cell receptor determines the specifity of the immune response, the IL-2/IL-2 receptor system regulates its type, duration, and strength. By interacting with other cytokines, IL-2 induces proliferation in B- and T-cells, stimulates antibody formation and activates macrophages, monocytes and natural killer (NK) cells. The inhibition of the IL-2/IL-2 receptor system by anti-CD25 antibodies does not completely suppress T-cell function: IL-7 and IL-15 can at least in part replace IL-2 functions [26].

Regulatory T-cells (Tregs) are also characterized by high expression of IL-2 receptor [27]. Although primarily characterized by the ability to suppress T-cell responses in an antigen-specific way, foxP3, a functionally relevant molecule, and, later on, low-grade expression of CD127 were found to facilitate diagnosis [28, 29]. Normally, 6–7% CD4+ T-cells are Tregs.

NK cells infiltrate the graft already within 12 h after transplantation [30]. They produce cytokines and mediators such as granzyme B. NK cells express CD158b, a receptor for MHC class I molecules, that inhibits cytotoxicity [31]. In acute rejections, NK cell (CD3– CD16/56+) counts in peripheral blood are enhanced [32]. The percentage of CD158+ NK cells (CD3– CD16/55+ CD158+) in peripheral blood moderately drops following kidney transplantation, but shows a very strong reduction in acute rejection [32]. Through this missing inhibitory molecule, NK cells can attack the transplant.

Immunosuppression

In contrast to animal models, long-term tolerance of allotransplants cannot be obtained without immunosuppression in humans. The prevention and therapy of acute rejection are based on the suppression of immune response by drugs, which are administered according to established therapeutic protocols depending on the transplanted organ.

The immunosuppressive effects of corticosteroids are based on the inhibition of the local synthesis of arachidonic acid by blocking phospholipase A2, and on their immunosuppressive and anti-inflammatory effect on T-cells and macrophages. Corticosteroids bind to membrane-bound and cytoplasmic receptors. The steroid-receptor complex enters the nucleus, binds to glucocorticoid response elements and inhibits the transcription of several cytokine genes. Furthermore, corticosteroids prevent the translocation of NFκB into the nucleus. This also inhibits the transcription of cytokine genes. Downregulation of proinflammatory cytokines, growth factors, receptors, chemokines, adhesion molecules, and enzymes decreases T-cell activation [33].

Polyclonal antithymocyte globulins (ATG) or antilymphocyte globulins (ALGs) and monoclonal pan T-cell antibodies (Orthoclone OKT3®, Ortho Biotech, Neuss,

Germany) lead to a fast elimination of lymphocytes from the peripheral blood and effective immunosuppression [34, 35].

Cyclosporin A (CsA; Sandimmun®; Novartis Pharma AG, Basel, Switzerland) and tacrolimus (Prograf, Astellas Pharma US, Inc., Deerfield, IL, USA) inhibit calcineurin and influence the regulation of IL-2 production. They inhibit calcium-dependent intracellular processes following antigen-specific T-cell activation. CsA binds to cyclophyllin which forms a complex with the phosphatase calcineurin. Normally, calcineurin dephosphorylates the nuclear factor of activated T-cells (NF-AT) and allows it to enter the nucleus and to bind the IL-2 promotor [34, 35]. The effect of tacrolimus is similar except that tacrolimus binds to another intracellular molecule, FKBP12. The FKBP12-tacrolimus complex binds to calcineurin and so inhibits the phosphatase activity of calcineurin. This prevents NF-AT dephosphorylation; as a consequence, it can neither penetrate into the nucleus nor bind the IL-2 promotor. This results in inhibited synthesis of IL-2, IL-3, IL-4, TNF-α and IFN-γ.

Chimeric antibodies such as basiliximab (Simulect®; Novartis Pharma AG) or humanized monoclonal antibodies like daclizumab (Zenapax®; F. Hoffmann-La Roche AG, Basel, Switzerland) bind the IL-2 receptor α-chain and thus block IL-2-dependent T-cell activation [34–38].

Similar to tacrolimus, rapamycin (Sirolimus®; Rapamune, Wyeth Pharmaceuticals, Inc., Philadelphia, PA, USA) and Everolimus (Certican®; Novartis Pharma AG) bind the immunophyllin FKBP12. This complex then binds mTOR and inhibits the proliferation of activated immunocompetent cells. The rapamycin-FKBP12 complex inhibits the costimulatory signaling cascade CD28/B7-2, and the rapamycin-FKBP12-mTOR complex suppresses the activation of kinases and thus the phosphorylation of ribosomes. Similarly, ckd2-cyclin complexes do not become activated. This results in a cell cycle arrest from the G1 to S phase [39].

Azathioprin (Imuran®; GlaxoSmithKline, London, UK), a purine analogue, strongly inhibits cell proliferation. It can be used to reduce T-cell proliferation induced by transplant antigens. During the mitotic S phase, azathioprin a precursor of the purines adenosine and guanosine, inhibits inosin acid synthesis. Azathioprin reduces the proliferation of B- and T-cells [34, 35].

Mycophenolate mofetil (Cellcept®; F. Hoffmann-La Roche AG) competitively inhibits inosine monophosphate dehydrogenase. It prevents the conversion of inosine to guanosine nucleotides. Due to the reduced concentration of guanosine nucleotides, DNA synthesis is inhibited. Cells stay in G1 arrest and cannot proliferate. Furthermore, the formation of adhesion molecules in response to antigens and penetration of mononuclear cells into the transplant are inhibited as well [34, 35, 40].

Methotrexate® (Methotrexate Lederle; Wyeth Pharmaceuticals, Inc.) represents another inhibitor of mitosis. It antagonizes folic acid and inhibits the biosynthesis of purines [34, 35].

All inhibitors of mitosis do not specifically influence the immune system, but impede the proliferation of all dividing cells.

Aims of Monitoring

The monitoring of transplanted patients is aimed at reducing complications such as rejection, infection, or other damage by toxic or ischemic influences. Clinical symptoms and biochemical parameters do not enable a reliable discrimination between these complications. Therefore, diagnosis is based on biopsy, which will mostly be performed in acute states.

Besides rejection, oversuppression of the immune system causing immune paralysis represents a serious risk. Immune paralysis can be detected by measuring HLA-DR expression on monocytes (see 'Pathophysiology and Immune Monitoring of Sepsis', pp. 600). At least for calcineurin inhibitors, detection of individualized immunosuppressive response enables an optimized dosage [41].

Infections constitute the third source of complications. Impaired cellular control of viral replication due to immunosuppression causes reactivation of viral diseases. This is often the case when engrafting liver transplants from donors who had suffered from viral hepatitis or CMV+ transplants into CMV− recipients. Preparation and preoperative administration of CMV-specific donor CD8+ T-cells could be a therapeutic option [42]. Detection of lymphocyte activation and seroconversion for CMV or hepatitis indicates acute infections also in transplanted, immunosuppressed patients.

Flow-Cytometric Parameters for Monitoring Transplanted Patients

T-cells and their activation are pivotal in the pathogenesis of acute rejections. Therefore, T-cell subpopulations (CD4+ or CD8+ T-cells; CD4+/CD8+ ratio; CD4− CD8− CD3+ T-cells; CD3+ CD4+ $CD25^{high}$ $CD127^{low}$ Tregs), activity markers (CD4+ CD25+, CD8+ CD25+, CD4+ CD71+, CD8+ HLA-DR+ T-cells), and co-stimulatory receptors (CD28 CD154, CD40 CD40L) have been investigated in peripheral blood as markers of acute rejection [15, 16, 18, 43–49]. It could be demonstrated that peripheral blood T-cells and transplant-infiltrating cells are much alike [50, 51].

It is important to remember that immunocompetent cells can be activated not only by the transplant but also by infections. The activation patterns can be very similar depending on disease stage and can be misinterpreted. Furthermore, as disease courses vary from one individual to the other, it would be necessary to monitor patients very frequently, which may represent a heavy economic burden. HLA-DR+ T-cells are a good example: their increase in kidney-transplanted patients rather indicates a CMV infection than a rejection [51–53].

It is difficult to draw conclusions on the criteria for diagnosing an acute rejection from the existing publications on T-cell markers, subpopulations, and activation markers. Most papers are based on small sample sizes, different transplanted organs and various immunosuppressive medications. Calculations of diagnostic specificity

Table 2. Diagnostic value of lymphocytes and clinical chemistry findings in the detection of acute rejection episodes

Parameter	Diagnostic sensitivity	Diagnostic specificity	Positive predictive value	Youden index	References
AP	68–83	55–80	32–70	0.33–0.48	[18, 42, 54–58]
ALT	50–97	17–84	29–38	0.14–0.34	[57–60]
Bilirubin	66–86	53–81	32–65	0.23–0.47	[18, 57, 58]
α-GST	63–100	39–75	20–31	0.02–0.73	[57–60]
γ-GT	91–94	23–58	60–71	0.17–0.49	[18, 60]
Eosinophils	49–87	62–89	29–82	0.18–0.57	[61–65]
Thromboplastin time	40	58	20	0.01	[58]
sIL-2R	58	96	83	0.54	[18]
ΔCD3	49	92	71	0.41	[18]
ΔCD4	55	92	74	0.47	[18]
ΔCD4/CD25	63	91	75	0.54	[18]
CD4/CD8 ratio and CD3 count					[16]
NTx	90	83			
LTx	92	41			
HTx	96	42			

Δ = Difference to previous value; NTx/LTx/HTX: renal/liver/heart transplantation.

and sensitivity or of predictive values are scarce [54]. Therefore, the following summary of findings, which is mainly based on liver, kidney, pancreas and intestine transplants, should be considered cautiously:

- The diagnostic sensitivity of most immunological parameters and biochemical parameters is sufficient to detect an acute rejection.
- In contrast, diagnostic specificity is mostly too low because other events such as toxic damage to the graft, infection or even cryopreservation of the transplant induced changes in clinical chemistry, cytokines, or lymphocyte activation [55]. Due to a more differentiated pattern of specific parameters, changes in lymphocyte subsets and activation markers can be superior to biochemical findings (table 2).
- Besides the parameters presented in table 2, additional parameters exhibited significant relationships to acute rejection. Since their diagnostic benchmarks were not stated, they were not included in the table. Some selected examples are presented here. Upregulation of double-negative T-lymphocytes (CD4− CD8− CD3+) and expression of CD26, CD28 and CD40L on CD4+ T-cells and CD71 on CD3+ cells were mentioned above [66–69].
- Different immunosuppressive therapy schemes impede the comparability of studies. Immunosuppression affects immune parameters and diagnostic markers. In renal transplantation, initial immunosuppression with ATG results in a long-lasting decrease in CD3+, CD25+ and CD45RO+ cells in peripheral

blood [70, 71]. Tacrolimus inhibits the secretion of IL-2 and the expression of CD25 and CD69 [72]. Following anti-IL-2 receptor therapy, CD25 is not detectable on CD4+ and CD8+ T-cells. Soluble IL-2R in serum is diminished and thus cannot be used for the diagnosis of acute rejection.

- Assessment of the time course of several parameters improves their diagnostic value (table 2). This is mainly true for T-cell subsets and their activation, which changes dependent on rejection/infection, immunosuppression, and transplanted organ [53, 73–76].
- Compared with lymphocyte subsets and cellular activation markers, intracellular parameters show higher diagnostic specificity. Intracellular cytokines in CD4+ T- cells of transplant recipients normally display a Th2 pattern (IL-10 increased), whereas they display a Th1 pattern (IFN-γ increased) during acute episodes of rejection [77].
- Specific T-cell response to donor MHC antigens can be a diagnostic alternative [13, 78].
- Animal experiments indicate that induction of tolerance could be a therapeutic option [72, 79–83]. Preconditions are well-prepared organs of healthy animals, a short cold ischemia time to avoid preservation damage, and recipients in a good health state. In humans, only living donor transplantation can meet these conditions. In the future, tolerance-inducing protocols are expected. As a consequence, demands for immune monitoring will change, but induction of tolerance will nonetheless modify cytokine production, lymphocyte subsets and activation markers [42, 83].

Protocols for Transplant Monitoring

Peripheral blood leukocyte, eosinophil, basophil and neutrophil granulocyte, lymphocyte and monocyte counts are performed with a hematology analyzer. Cells are further assessed on a FACScan together with monoclonal antibodies for lymphocyte subpopulations and lymphocyte activation characterized by the expression of special surface molecules. The expression of surface molecules on T-cells is suitable for the diagnosis of acute rejection [18, 53].

Protocol 1: Measurement of Lymphocyte Subpopulations and Lymphocyte Activation using the Three-Color Test (Protocol 1a) and the Four-Color Test (Protocol 1b)

Principle

Blood cells are incubated together with monoclonal antibodies labeled with fluorescent dyes against surface molecules which characterize lymphocyte subpopulations or lymphocyte activation. Erythrocytes are eliminated by cell lysis and platelets removed by washing the cell suspension. The stained leukocytes are stabilized and measured on a FACScan. Signals of forward scatter (FSC) (cell size), side scatter (SSC) (granularity) and the fluorescence signals according to the labeled antibodies are registered. Fluorescent dyes are chosen with the aim of examining three surface molecules simultaneously. Using a Multi Carousel Loader reduces pipetting.

Material

- Flow cytometer, e.g., FACScan (Becton-Dickinson), EPICS XL-MCL (Beckman-Coulter)
- Excitation at 488 nm (argon laser, FSC, SSC)
- Fluorescence detection 1 (FITC, 515–545 nm), (PE, 546–602 nm), 3 (PerCP, >650 nm (Becton-Dickinson) or 635 nm (Beckman-Coulter), 4 (PE/Cy5 = PC5, 670 nm (Beckman-Coulter)

  A four-color flow cytometer measurement is also possible with Becton-Dickinson systems, e.g. using a FACSCalibur. In this case a helium neon laser is needed for secondary excitation light (wavelength 633 nm) so labeling should be done with fluorescent dye Cy5 (661 nm)
- Sample preparation by TQ-Prep™ (Beckman-Coulter)
- Software for data acquisition, storage and analysis, e.g., XL System II (Beckman-Coulter) or Cell Quest (Becton Dickinson) or other
- 12 × 75 mm PS test tubes (Falcon) and compatible test tube racks
- Pipette tips (Sarstedt)
- Roller mixer and Vortex mixer
- Variable pipette tips for 1–10 µl, 10–100 µl, 100–1000 µl and compatible tips
- Storage room 4 °C for the test kits
- Centrifuge and adapter for 12 × 75 mm test tubes with cups
- Dispenser and tips.

Reagents for Three-Color Transplant Monitoring (Protocol 1a)

- Calibrite beads™ (Becton Dickinson), catalogue No. 340486 (polymethylmethacrylate microparticles about 6 µm, unlabeled, and FITC-, PE- or PerCP-labeled) for the preliminary calibration of the flow cytometer and to check instrument settings.
- BD Simulset control$_{\gamma1/\gamma2}$ (FITC/PE) used as isotype control, BD catalogue No. 342409 (1 : 10 diluted with PBS/acid).
- PBS/acid: dissolve 0.2 g NaN_3 and 0.037 g EDTA in 1 l PBS.
- Lysis stock solution (dissolve 82.9 g NH_4Cl, 10.2 g $KHCO_3$ and 292 mg EDTA in 1 l distilled water, prepare a fresh working solution daily by dilution of the stock solution 1 : 10 using distilled water.

Antibody mixtures were prepared according to tables 3 and 4, and are stable for at least 3 weeks when stored at 4 °C.

Table 3. Antibody panel for three-color transplant monitoring

Antibody	Labeling	Catalogue No. BD
CD45	PerCP	345809
CD14	PE	345785
CD3	FITC	345764
CD4	FITC/PE	345768/345769
CD8	FITC/PE	345772/345773
CD19	PE	345777
CD16/56	PE	337279/345812
CD25	PE	341011
HLA-DR	FITC/PE	347400/347401

Table 4. Antibody mixtures for three-color transplant monitoring

Tube	FITC		PE		Per-CP		PBS/acid, μl
	combined antibody	μl	combined antibody	μl	combined antibody	μl	
1	CD3	100	CD14	50	CD45	100	750
2	mouse IgG1/2a	FITC/PE mixture 100 μl			CD45	100	800
3	CD3	100	CD19	50	CD45	100	750
4	CD3	150	CD16/ CD56	75 75	CD45	100	600
5	CD3	100	CD4	50	CD45	100	750
6	CD3	100	CD8	50	CD45	100	750
7	CD4	100	CD25	50	CD45	100	750
8	CD8	100	HLA-DR	50	CD45	100	750
9	HLA-DR	150	CD14	50	CD45	100	700

Reagents for Transplant Monitoring using a Four-Color Panel (Protocol 1b)

- ImmunoPrep™ ReagentSystem (Beckman-Coulter)
- The antibodies used are listed in table 5.

Procedure

A venous blood sample is measured with a hematology analyzer to determine leukocyte differential count and especially lymphocyte count.

- Antibody mixtures are filled into marked tubes (12 × 75 mm) according to table 4 or 5 (protocol 1a or 1b).
- For protocol 1a, add 50 μl EDTA blood into the tubes. In case of a lymphocyte count below 200/μl, add 100 μl blood. For protocol 1b, always add 100 μl blood. Avoid contamination of the upper part of the tube wall with blood.

Further steps using automated technology:

- Insert the prepared tubes into a Coulter TQ-Prep and set incubation time to 10 min.

Table 5. Antibodies

Tubes	Antibodies	Producer	Volume, μl
1	CD45-FITC/CD4-PE/CD8-ECD/CD3-PC5	Beckman-Coulter	2.5
2	CD45-FITC/CD56-PE/CD19-ECD/CD3-PC5	Beckman-Coulter	2.5
	CD16-PE	Immunotech	2.5
3	CD45-FITC	Immunotech	5
	IgG2a-PE	Immunotech	2.5
	CD3-PC5	Immunotech	2.5
4	CD45-FITC	Immunotech	5
	HLA-DR-PE	BD Biosciences	5
	CD3-PC5	Immunotech	2.5
5	CD45-FITC	Immunotech	5
	CD25-(RD or PE)	Immunotech	5
	CD3-PC5	Beckman-Coulter	2.5
7	CD45-FITC	Immunotech	5
	CD57-PE	Immunotech	2.5
	CD8-PC5	Immunotech	2.5

- Lysis occurs automatically after 10 min.
- Until measurement, store the cell suspension in a refrigerator at 4 °C
- Measure the prepared cell suspension in a flow cytometer within 48 h.

Further steps using manually operated equipment:

- Shortly vortex and incubate for 10 min at room temperature in the dark.
- Add 3 ml lysis reagent, vortex shortly and incubate for 10 min at room temperature in the dark.
- Spin down (5 min 300 × *g*), discard supernatant and resuspend with 3 ml PBS/acid, vortex briefly
- Spin down (5 min 300 × *g*), discard supernatant, disperse cell pellet and measure the cell suspension in a flow cytometer within 48 h. Store the cell suspension in a refrigerator at 4 °C until measurement.

Flow-Cytometric Measurement

The flow cytometer has to be calibrated before measuring. As a general rule, the system settings are considered to be stable over time so a control with beads of every color is sufficient. Compensation within the panel is consistent for well-titrated antibodies. Protocols for data acquisition and analysis instrument settings are saved and loaded prior to measurement.

Manual Three-Color Measurement (Protocol 1a)

Cells are analyzed by flow cytometry using blue-green excitation light (488 nm argon ion laser), for the gating of lymphocytes (fluorescence 3, >650 nm; CD45 log, SSC lin and FSC lin; (fig. 1). Surface molecules are measured by FITC fluorescence 1

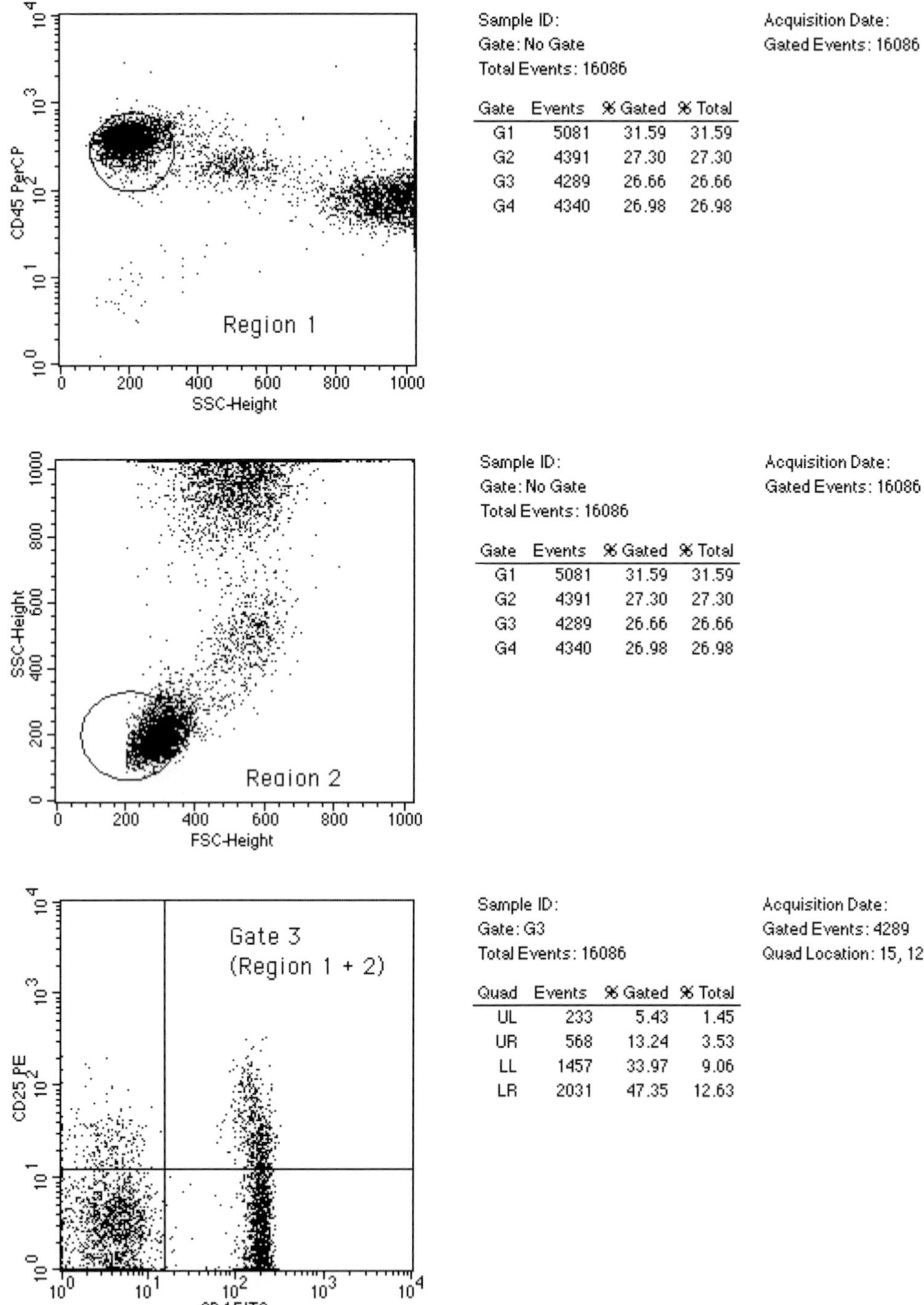

Sample ID:
Gate: No Gate
Total Events: 16086

Acquisition Date:
Gated Events: 16086

Gate	Events	% Gated	% Total
G1	5081	31.59	31.59
G2	4391	27.30	27.30
G3	4289	26.66	26.66
G4	4340	26.98	26.98

Sample ID:
Gate: No Gate
Total Events: 16086

Acquisition Date:
Gated Events: 16086

Gate	Events	% Gated	% Total
G1	5081	31.59	31.59
G2	4391	27.30	27.30
G3	4289	26.66	26.66
G4	4340	26.98	26.98

Sample ID:
Gate: G3
Total Events: 16086

Acquisition Date:
Gated Events: 4289
Quad Location: 15, 12

Quad	Events	% Gated	% Total
UL	233	5.43	1.45
UR	568	13.24	3.53
LL	1457	33.97	9.06
LR	2031	47.35	12.63

Fig. 1. Measurement of lymphocyte subpopulations and activation marker with the three-color technique and manual gating.

(515–545 nm) and PE fluorescence 2 (550–600 nm). 10,000–15,000 cells are measured and results are stored in one file per sample.

Automatic Four-Color Measurement (Protocol 1b)

Cells are analyzed by flow cytometry using blue-green excitation light (488 nm argon ion laser). Lymphocytes are gated and fluorescence is measured as follows: fluorescence 1 (525 nm; FITC); fluorescence 2 (575 nm; PE or RD); fluorescence 3 (620 nm; ECD) and fluorescence 4 (675 nm LP, PC5). 10,000–15,000 cells are measured and results are stored in one file per sample.

- Remove the sample carousel from the TQ-Prep sample preparation system and put it into the EPICS XL-MCL flow cytometer.
- Open the worklist, add Lab-No., patient identification data and blood count results; if not done by lab information system, select measuring panel.
- Start measuring procedure; a report is printed automatically when it is finished.

Data Analysis

Lymphocytes or other leukocytes are gated in an FSC/SSC and an SSC/CD45 scattergram and the gating is checked with a monocyte surface marker CD14–^{dim} (fig. 1). In the case of an insufficient discrimination of lymphocytes and monocytes, the gate has to be adjusted in such a way that the cell population contains more than 90% lymphocytes.

The threshold for the expression of the surface markers (i.e. the fluorescence value that 95% of patient cells labeled with Simultest fail to reach) is fixed using an isotype control (control$_{\gamma 1/\gamma 2a}$). The results are expressed as the percentage of cells with a surface fluorescence above this threshold.

Each data sheet showing the measured data and the histogram is printed. In addition, the correct selection of markers is checked. In the automated mode, this is done automatically by the software and the settings outlined in the protocol. Sometimes, the measurement range has to be set manually. The results are checked and commented online, then printed and saved in a database.

The diagnosis is established via appropriate database software and issued in a standardized fashion. The evaluation is added as a free comment. The following aspects are evaluated:

- Leukocyte and lymphocyte counts are compared to reference and previous values.
- One or more activation markers are assessed as needed.
- If previous findings are available, these are taken into account.

Quality Control

The manufacturer of the test kits guarantees that the kits are validated and authorized for in vitro diagnostics. Internal quality control of the automaton includes a weekly calibration of the flow cytometer by means of a flow-check. For each patient

Table 6. Troubleshooting

Problem	Cause	Corrective
High fluorescence of all cells	measurement chamber contaminated by fluorescence dye, unlabeled cells fluorescence	rinse measurement chamber
Monocytes cannot be delineated	old blood sample, monocytes depleted	CD14 labeling of monocytes
Lymphocyte acquisition too low	lymphopenia	application of blood according to lymphocytes in total blood count
Bad discrimination between monocytes and lymphocytes	too long storage of blood sample	fresh sample
	binding of fluorescent drugs to blood cells	washing of sample with PBS prior to labeling with antibody mixtures

sample, patient-matched negative controls (isotype controls, and lymphocyte subpopulations that are negative for a special marker) are used.

The external quality control includes participation in interlaboratory surveys. Further quality tests are listed below:

- The sum of lymphocyte populations CD3, CD19 and CD16/56 cells adds up to 100%.
- The sum of lymphocyte populations CD4 and CD8 adds up to the CD3 cell count.
- The count of a cell population like CD3 or CD8 cells has to be measured as well using different antibody mixtures.
- In addition, a quality control sample is measured to check if the CD3, CD4, CD8, CD19 and CD16/56 lymphocyte subpopulations are correctly measured.

Troubleshooting

Protocol 1: problems, causes and solutions are listed in table 6.

Expected Results, Interpretation and Pitfalls

The lymphocyte subpopulations are calculated in percent and per microliter based on the lymphocyte count performed using the hematology analyzer. For the evaluation, new and previous data should be compared, for example an increase in the CD4/CD8 ratio or the activation markers (CD25 or HLA-DR). However, the increased ratio has to be evaluated taking into account the absolute cell count of the previous measurement. Changes under 20% should be carefully interpreted. A considerable increase in the activation markers CD25 or HLA-DR or an increased CD4/CD8 ratio should be interpreted as evidence for an acute rejection, if there is no other explanation (fig. 2).

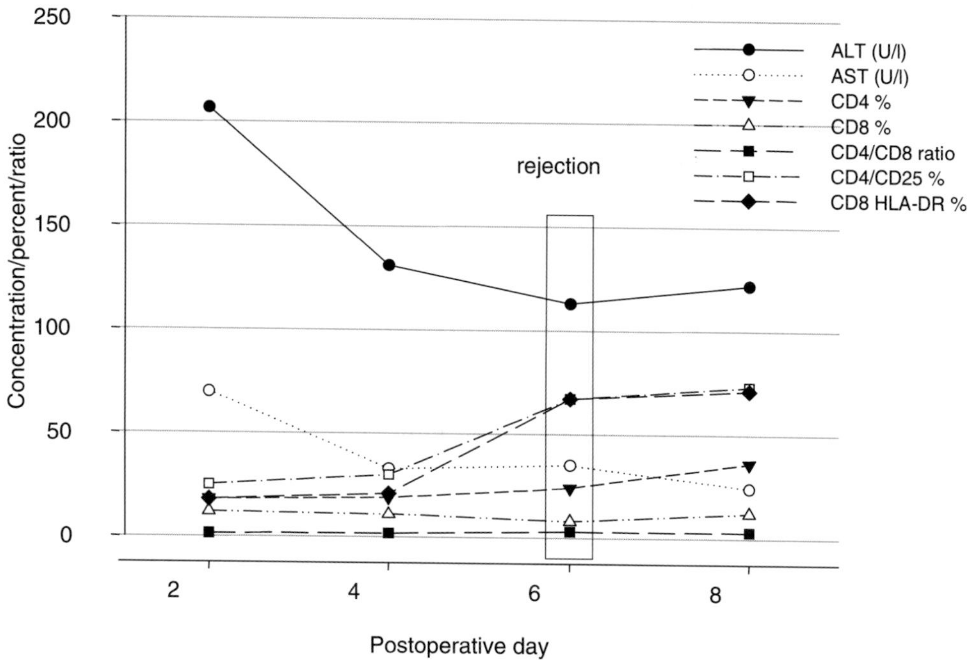

Fig. 2. Time course of various selected parameters following liver transplantation and acute second-degree rejection on day 6 after transplantation. During the first days after transplantation AST and ALT activities decreased, recovery or stagnation of enzyme activity in blood plasma is ambivalent evidence for acute rejection. A considerable increase in CD4+ CD25+ and CD8+ HLA-DR+ subpopulations indicates an acute rejection. In the case of a simultaneous mild increase in CRP and a mild decrease in leukocyte and lymphocyte count, acute rejection is more likely than a bacterial or viral infection [87]. An increased CD4 subpopulation and an increased CD4/CD8 ratio indicate acute rejection.

Increased lymphocyte markers can be the result of infections. Bacterial infections cause a clear increase in C reactive protein or procalcitonin and can therefore be distinguished from an acute rejection. Immunosuppressive therapy relying on antibodies like OKT3 or basiliximab can mask CD25 or CD3 on the cell surface. CD25+ cells cannot be measured. Nor can T-cells be measured in an OKT3 therapy, but the sum of CD4 and CD8 cells enables an exploratory assessment.

Time Required

The various steps of the protocol have to be performed up to the measurement. The procedure may be interrupted for 48 h at most and only if fixed cells are being measured. Measuring of a patient and a control sample with data acquisition takes approximately 90–120 min. The preparation of 9 additional patient samples takes about another 60 min. The data analysis of the first sample (patient or control) takes approximately 10 min and an additional 5 min per each extra patient sample. Automatic pipetting reduces the workload.

Protocol 2: Monitoring of Lymphocyte Count in Anti-Thymocyte Globulin Therapy

In the case of severe lymphopenia (10–150 lymphocytes/μl), determination of the lymphocytes fails with most hematology analyzers. However, to decide whether the immunosuppressive therapy should be prolonged or stopped, a correct lymphocyte count is needed [84]. Therefore the dual-platform technique is applied to assess the total leukocyte count with the hematology analyzer; the number of CD3+ lymphocytes will be measured in an SSC/CD3 plot after gating of all leukocytes (FSC/SSC plot for elimination of debris).

Material and Reagents
Analogous to the conditions in protocol 1, with the exception that only one tube with the antibody mixture CD45/PerCP, CD14/PE, CD3/FITC and a tube for isotype control are needed.

Procedure
A venous EDTA blood sample is used to measure total blood count with a hematology analyzer. The leukocytes are differentiated. The leukocyte differentiation will trigger an alarm due to the depleted lymphocyte count; a microscopic evaluation has to confirm the low numbers of lymphocytes.

Preparation: for each patient two Falcon tubes are labeled (patient 1, 1/1 and 1/2), and 25 μl antibody mixture are pipetted to the bottom of the Falcon tubes. Thereafter, about 50,000 cells are measured with the flow cytometer according to protocol 1. Quality control and troubleshooting are as in protocol 1.

Protocol 3: Measurement of Intracellular Cytokines
The diagnostic benefit of intracellular cytokines in transplant monitoring is not yet completely evaluated, but their use seems to be necessary in innovative immunosuppressive therapy.

Furthermore, intracellular cytokines may assign donor-specific reagibility in case of previous stimulations of donor cells, which can be important in tolerance-inducing therapies.

Host lymphocytes are specifically stimulated by host spleen cells or nonspecifically with phorbol-12,13-dibutyrate (PDB) (positive control) and ionomycin. Monensin prevents efflux of cytokines from lymphocytes. An unstimulated probe with ionomycin and monensin is used as a negative control. The cells are fixed with paraformaldehyde (PFA), the erythrocytes are lysed, and the intracellular cytokines and the lymphocyte surface antigens labeled with monoclonal antibodies are measured by flow cytometry [85, 86].

Material
The materials used are the same as stated in protocol 1.

Reagents

- AIM-V incubation medium (add 2.5 ml fungizone to 500 ml AIM-V; Gibco BRL, catalogue No. 12055-091).
- PDB (molecular weight, MW = 504.6, Sigma No. P 1269; stock solution 10^{-3} mol/l = 1 mg PDB in 1.98 ml DMSO; dilute PDB stock solution 1 : 100 with AIM-V.
- Ionomycin stock solution (500 μg ionomycin calcium salt, Calbiochem, No. 407952), dissolved in 1 ml DMSO and diluted with 9 ml AIM-V. Ionomycin solution (50 μg/ml) is stored frozen, ready for use and portioned.
- Monensin stock solution, 20 mmol/l used as an inhibitor of protein transport (MW = 692.9; Sigma No. M 5273); monensin solution: dissolve 0.1 g monensin in 7.2 ml ethanol and dilute monensin stock solution for use 1 : 100 with AIM-V.
- PFA solution 2% (prepare fresh daily); dilute 2 g PFA with 10 ml distilled water in 50-ml tubes and incubate at 70 °C in a water bath for about 2 h, add 1 drop of NaOH 16% and 0.5 g glucose; PFA solution: 90 ml phosphate buffer and 10 ml PFA glucose solution; prepare phosphate buffer by mixing with 83 ml buffer solution A (NaH_2PO_4 2.2 g ad 100 ml distilled water) and 13 ml buffer solution B (NaOH 2.5 g ad 100 ml distilled water); the pH of the phosphate buffer solution is between 7.4 and 7.6).
- Saponin solution (prepare saponin solution by mixing 5 ml saponin stock solution and 5 ml HEPES buffer solution and 490 ml PBS; saponin stock solution 10%, dissolve 2 g saponin with 20 ml PBS; stable for 1 week in a refrigerator at 4–8 °C)
- Antibodies, see table 7.

Procedure

Preparation: in 12 × 75 mm polystyrene tubes with cups. For further processing, see table 8.

Flow Cytometric Analysis

The cells are analyzed by flow cytometry using blue-green excitation light (488-nm argon ion laser), as in protocol 1. Quality control and troubleshooting are as in protocol 1.

To verify the measurement, one has to check for correct allocation and quantification of cell populations in FSC lin versus SSC lin and compare the size of the leukocyte populations with the total count obtained with the hematology analyzer.

Expected Values, Interpretation and Pitfalls

- The intracellular antibodies have to be diluted with saponin solution; without saponin the staining will fail.
- The results cannot be evaluated if the negative control exhibits high amounts of intracellular cytokines.

Table 7. Antibody-panel-specific T-cell response

Antibodies	Labeling	Catalogue-No. BD
CD45	PerCP	345809
IL-2	PE	Ph 18955A
CD14	PE	345785
CD3	PerCP	345766
CD8	FITC	345772
IFN-γ	PE	Ph 18905A
IL-4	PE	Ph 18505A
IL-10	PE	Ph 18555A
Isotype control γ1/γ2	FITC/PE	342409

Table 8. Intracellular cytokine staining[a]

	Negative control	Positive control	Specific stimulation
(1) Preparation of the mixtures			
Heparinized blood (host), μl	100	100	100
AIM-V, μl	950	920	920
PDB 10^{-5} mol/l, μl		10	
Ionomycin 50 μg/ml		10	10
Monensin 20 μmol/l, μl		10	10
Spleen cells (donor), μl			10
(2) Subsequent processing			
Incubator 37 °C	Vortex, subsequently 4 h incubation		
Washing	3 ml PBS, vortex, spin down (5 min, 300 × *g*), discard supernatant		
Fixation	add 0.5 ml PFA 2%, incubate for 10 min at room temperature		
Washing	3 ml PBS, vortex, spin down (5 min, 300 × *g*), discard supernatant		
Resuspend pellet	in 1 ml PBS		
Storage overnight	refrigerator at 4 °C spin down (5 min, 300 × *g*), discard supernatant		
Permeabilization	0.5 ml 0.1% saponin buffer, incubate for 10 min at room temperature; spin down (5 min, 300 × *g*), discard supernatant		
Staining intracellular cytokines	Add intracellular antibodies, 10 μl, diluted 1:8 with saponin buffer		
Incubation	20 min at room temperature		
Washing	3 ml saponin buffer		
Washing	3 ml PBS		
Surface molecule labeling	50 μl, incubation 15 min at room temperature		
Antibody dilution (1:10 with PBS/acid)			
Washing	3 ml PBS/acid, vortex, ; spin down (5 min, 300 ×*g*), discharge supernatant		

[a] Alternative protocol, see 'Flow-Cytometric Analysis of Intracellularly Stained Cytokines, Phosphoproteins, and Microbial Antigens' (pp. 459).

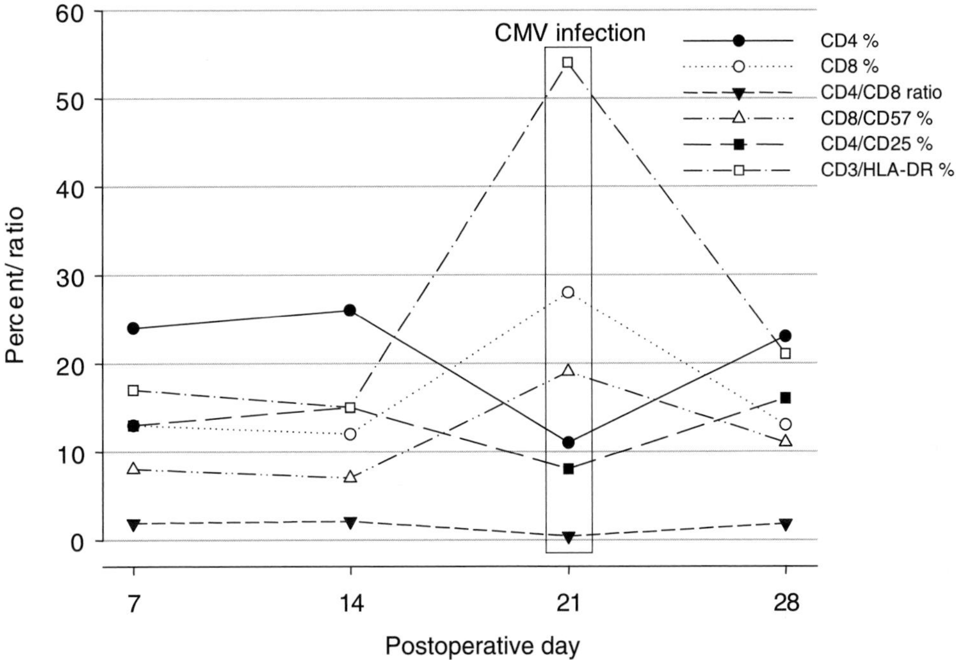

Fig. 3. Time course of various selected parameters following kidney transplantation and acute CMV reactivation during the 3rd week. A simultaneous decrease in CD4+ CD25+ and an increase in HLA-DR+ CD3− cells after renal transplantation suggest a CMV infection [16, 52, 53]. A simultaneous increase in CD8+ CD57+ lymphocytes suggests reactivation of the CMV infection [16, 52]. An increase in CD8+ and decrease in the CD4/CD8 ratio suggest a CMV infection [16, 52, 53].

- EDTA or citrate blood is unsuitable because Ca^{2+} undergoes complex binding to these anticoagulants, but is needed in the intracellular signal cascade.
- As CD4 antibodies are unsuitable, one avoids the problem by defining CD3+ CD8− lymphocytes as CD4 T-cells.
- Ammonium heparin probes are better suited than lithium heparin probes. The supernatant has to be carefully removed; decanting may result in cell loss.
- PFA solution has to be prepared fresh daily as old solutions may be cloudy and cell fixation may be insufficient.
- Pellets must be carefully resuspended to remove cell clots.

References

1 Eurotransplant. *http://www.eurotransplant.nl/index.php?id = statistics.* Electronic citation, 2003.

2 Quiroga J, Colina I, Demetris AJ, Starzl TE, Van Thiel DH: Cause and timing of first allograft failure in orthotopic liver transplantation: a study of 177 consecutive patients. Hepatology 1991;14:1054–1062.

3 Lindholm A, Ohlman S, Albrechtsen D, Tufveson G, Persson H, Persson NH: The impact of acute rejection episodes on long-term graft function and outcome in 1347 primary renal transplants treated by 3 cyclosporine regimens. Transplantation 1993;56:307–315.

4 Tullius SG, Nieminen M, Bechstein WO, Jonas S, Qun Y, Rayes N, et al: Prompt treatment of initial acute rejection episodes may improve long-term graft outcome. Transpl Int 1998;11(suppl 1):S3-S4.

5 Janeway CA, Travers P, Walport M, Shlomchik M: Autoimmunität und Transplantation; in Janeway CA, Travers P, Walport M, Shlomchik M (Hrsg): Immunologie. Heidelberg, Spektrum Akademischer Verlag, 2004, pp 537–591.

6 Sayegh MH, Vella JP: Acute renal allograft rejection: Diagnosis. *www.uptodateonline.com*, 2004.

7 Vella JP, Sayegh MH: Transplantation immunobiology. *www.uptodateonline.com*, 2004.

8 Machold KP, Stamm TA, Nell VP, Pflugbeil S, Aletaha D, Steiner G, et al: Very recent onset rheumatoid arthritis: clinical and serological patient characteristics associated with radiographic progression over the first years of disease. Rheumatology (Oxford) 2007;46:342–349.

9 Cecka JM: The UNOS Renal Transplant Registry. Clin Transpl 2002;1–20.

10 Nalesnik MA: Clinicopathologic characteristics of post-transplant lymphoproliferative disorders. Recent Results Cancer Res 2002;159:9–18.

11 Sayegh MH, Turka LA: The role of T-cell costimulatory activation pathways in transplant rejection. N Engl J Med 1998;338:1813–1821.

12 Ciubotariu R, Liu Z, Colovai AI, Ho E, Itescu S, Ravalli S, et al: Persistent allopeptide reactivity and epitope spreading in chronic rejection of organ allografts. J Clin Invest 1998;101:398–405.

13 Vella JP, Spadafora-Ferreira M, Murphy B, Alexander SI, Harmon W, Carpenter CB, et al: Indirect allorecognition of major histocompatibility complex allopeptides in human renal transplant recipients with chronic graft dysfunction. Transplantation 1997;64:795–800.

14 Watson CJ, Cobbold SP, Davies HF, Rebello PR, Waldmann H, Calne RY, et al: CD4 and CD8 monoclonal antibody therapy in canine renal allografts. Transpl Int 1994;7(suppl 1):S322–S324.

15 Jindal RM, Greer G, Popescu I, Sidner RA: Lymphocyte subset analysis for the diagnosis of rejection and infection in recipients of liver transplants. Am Surg 1999;65:77–80.

16 Döcke WD, Reinke P, Staffa G, Settmacher U, Höger T, Groth J, et al: An immune monitoring program for the management of immunosuppressive therapy in early phase after transplantation. Tex Med 1994;6:13–28.

17 Korom S, De M, I, Onodera K, Stadlbauer TH, Borloo M, Lambeir AM, et al: The effects of CD26/DPP IV-targeted therapy on acute allograft rejection. Transplant Proc 1997;29:1274–1275.

18 Lun A, Cho MY, Muller C, Staffa G, Bechstein WO, Radke C, et al: Diagnostic value of peripheral blood T-cell activation and soluble IL-2 receptor for acute rejection in liver transplantation. Clin Chim Acta 2002; 320:69–78.

19 Perkins JD, Nelson DL, Rakela J, Grambsch PM, Krom RA: Soluble interleukin-2 receptor level as an indicator of liver allograft rejection. Transplantation 1989;47:77–81.

20 Stockenhuber F, Gnant M, Gottsauner-Wolf MT, Apperl A, Steininger R, Sautner T, et al: Soluble interleukin-2 receptor in liver transplant recipients. Transplant Proc 1991;23:1417–1418.

21 June CH, Bluestone JA, Nadler LM, Thompson CB: The B7 and CD28 receptor families. Immunol Today 1994;15:321–331.

22 Linsley PS, Brady W, Urnes M, Grosmaire LS, Damle NK, Ledbetter JA: CTLA-4 is a second receptor for the B cell activation antigen B7. J Exp Med 1991;174:561–569.

23 Linsley PS, Ledbetter JA: The role of the CD28 receptor during T cell responses to antigen. Annu Rev Immunol 1993;11:191–212.

24 Walunas TL, Bakker CY, Bluestone JA: CTLA-4 ligation blocks CD28-dependent T cell activation. J Exp Med 1996;183:2541–2550.

25 Dai Z, Konieczny BT, Baddoura FK, Lakkis FG: Impaired alloantigen-mediated T cell apoptosis and failure to induce long-term allograft survival in IL-2-deficient mice. J Immunol 1998;161:1659–1663.

26 Baan CC, Boelaars-van Haperen MJ, van Riemsdijk IC, van der Plas AJ, Weimar W: IL-7 and IL-15 bypass the immunosuppressive action of anti-CD25 monoclonal antibodies. Transplant Proc 2001;33:2244–2246.

27 Waldmann H, Graca L, Cobbold S, Adams E, Tone M, Tone Y: Regulatory T cells and organ transplantation. Semin Immunol 2004;16:119–126.

28 Banham AH: Cell-surface IL-7 receptor expression facilitates the purification of FOXP3(+) regulatory T cells. Trends Immunol 2006;27:541–544.

29 Liu W, Putnam AL, Xu-Yu Z, Szot GL, Lee MR, Zhu S, et al: CD127 expression inversely correlates with FoxP3 and suppressive function of human CD4+ T reg cells. J Exp Med 2006;203:1701–1711.

30 Hsieh CL, Obara H, Ogura Y, Martinez OM, Krams SM: NK cells and transplantation. Transpl Immunol 2002;9:111–114.

31 Guerra N, Guillard M, Angevin E, Echchakir H, Escudier B, Moretta A, et al: Killer inhibitory receptor (CD158b) modulates the lytic activity of tumor-specific T lymphocytes infiltrating renal cell carcinomas. Blood 2000;95:2883–2889.

32 Kang N, Guan D, Xing N, Xia C: Expression of CD158b on peripheral blood lymphocytic cell after kidney transplantation. Transplant Proc 2005;37:782–784.

33 Gärtner R, Haen E: Endokrinpharmakologie. In: Forth W, Henschler D, Rummel W, Förstermann U, Starke K(Hrsg): Allgemeine und spezielle Pharmakologie und Toxikologie. Jena, Urban & Fischer, 2001, pp 671–737.

34 Breidenbach T: Immunsystem/Immunsuppressiva; in Novartis (Hrsg): Handbuch Transplantation – der schnelle Ratgeber. Basel, Novartis, 2004, pp155–175.

35 Kaever V, Resch K: Antiphlogistika und Immuntherapeutika; in Forth W, Henschler D, Rummel W, Förstermann U, Starke K (Hrsg): Allgemeine und spezielle Pharmakologie und Toxikologie. Jena, Urban & Fischer, 2001, pp 393–427.

36 Waldmann TA, O'Shea J: The use of antibodies against the IL-2 receptor in transplantation. Curr Opin Immunol 1998;10:507–512.

37 Lemmens HP, Langrehr JM, Bechstein WO, Blumhardt G, Keck H, Lusebrink R, et al: Interleukin-2 receptor antibody vs ATG for induction immunosuppression after liver transplantation: initial results of a prospective randomized trial. Transplant Proc 1995;27:1140–1141.

38 Lohmann R, Langrehr JM, Klupp J, Neumann U, Guckelberger O, Muller AR, et al: Quadruple induction therapy including antithymocyte globulin or interleukin-2 receptor antibody or FK 506-based induction therapy after liver transplantation. Transplant Proc 1999;31:380.

39 Sehgal SN: Sirolimus: its discovery, biological properties, and mechanism of action. Transplant Proc 2003;35(suppl):7S–14S.

40 Pescovitz MD, Conti D, Dunn J, Gonwa T, Halloran P, Sollinger H, et al: Intravenous mycophenolate mofetil: safety, tolerability, and pharmacokinetics. Clin Transplant 2000;14:179–188.

41 Giese T, Zeier M, Schemmer P, Uhl W, Schoels M, Dengler T, et al: Monitoring of NFAT-regulated gene expression in the peripheral blood of allograft recipients: a novel perspective toward individually optimized drug doses of cyclosporine A. Transplantation 2004;7:339–344.

42 Keenan RD, Ainsworth J, Khan N, Bruton R, Cobbold M, Assenmacher M, et al: Purification of cytomegalovirus-specific CD8 T cells from peripheral blood using HLA-peptide tetramers. Br J Haematol 2001;115:428–434.

43 Crosbie OM, Norris S, Hegarty JE, O'Farrelly C: T lymphocyte subsets and activation status in patients following liver transplantation. Immunol Invest 1998;27:237- 241.

44 Herrod HG, Williams J, Dean PJ: Alterations in immunologic measurements in patients experiencing early hepatic allograft rejection. Transplantation 1988;45:923–925.

45 Jindal RM, Popescu I, Emre S, Schwartz ME, Boccagni P, Meneses P, et al: Serum lipid changes in liver transplant recipients in a prospective trial of cyclosporine versus FK506. Transplantation 1994;57:1395–1398.

46 Ninova DI, Wiesner RH, Gores GJ, Harrison JM, Krom RA, Homburger HA: Soluble T lymphocyte markers in the diagnosis of cellular rejection and cytomegalovirus hepatitis in liver transplant recipients. J Hepatol 1994;21:1080–1085.

47 Pross M, Manger T, Weiss G, Lippert H, König w, Kunz D: Die diagnostische Wertigkeit von Procalcitonin und IL-6 in Kombination mit dem zellulären Immunstatus für das therapeutische Management nach Lebertransplantation. Tex Med 1998;10:195–201.

48 Shanahan T: Application of flow cytometry in transplantation medicine. Immunol Invest 1997;26:91–101.

49 Demirkiran A, Kok A, Kwekkeboom J, Kusters JG, Metselaar HJ, Tilanus HW, et al: Low circulating regulatory T-cell levels after acute rejection in liver transplantation. Liver Transpl 2006;12:277–284.

50 Bestard O, Cruzado JM, Mestre M, Caldes A, Bas J, Carrera M, et al: Achieving donor-specific hyporesponsiveness is associated with FOXP3+ regulatory T cell recruitment in human renal allograft infiltrates. J Immunol 2007;179:4901–4909.

51 Reinke P, Fietze E, Docke WD, Kern F, Ewert R, Volk HD: Late acute rejection in long-term renal allograft recipients. Diagnostic and predictive value of circulating activated T cells. Transplantation 1994;58:35–41.

52 Lehmann I, Borte M, Sack U. Diagnosis of immunodeficiencies. J Lab Med 2001;25:495–511.

53 Oertel M, Sack U, Kohlhaw K, Lehmann I, Emmrich F, Berr F, et al: Induction therapy including antithymocyte globulin induces marked alterations in T lymphocyte subpopulations after liver transplantation: results of a long-term study. Transpl Int 2002;15:463–471.

54 Abraham SC, Furth EE: Receiver operating characteristic analysis of serum chemical parameters as tests of liver transplant rejection and correlation with histology. Transplantation 1995;59:740–746.

55 Tilg H, Vogel W, Aulitzky WE, Herold M , Konigsrainer A, Margreiter R, et al: Evaluation of cytokines and cytokine-induced secondary messages in sera of patients after liver transplantation. Transplantation 1990;49:1074–1080.

56 Hickman PE, Lynch SV, Potter JM, Walker NI, Strong RW, Clouston AD: Gamma glutamyl transferase as a marker of liver transplant rejection. Transplantation 1994;57:1278–1280.

57 Hughes VF, Melvin DG, Niranjan M, Alexander GA, Trull AK: Clinical validation of an artificial neural network trained to identify acute allograft rejection in liver transplant recipients. Liver Transpl 2001;7:496–503.
58 Trull AK, Facey SP, Rees GW, Wight DG, Noble-Jamieson G, Joughin C, et al: Serum alpha-glutathione S-transferase – a sensitive marker of hepatocellular damage associated with acute liver allograft rejection. Transplantation 1994;58:1345–1351.
59 Azoulay D, Lemoine A, Salvucci M, Adam R, Bismuth H: Prospective evaluation of serum GST a in a liver transplantation (abstract). Gut 1996;39(suppl 3):A26.
60 Nagral A, Butler P, Sabin CA, Rolles K, Burroughs AK: Alpha-glutathione-S-transferase in acute rejection of liver transplant recipients. Transplantation 1998;65:401–405.
61 Trull AK, Akhlaghi F, Charman SC, Endenberg S, Majid O, Cornelissen J, et al: Immunosuppression, eotaxin and the diagnostic changes in eosinophils that precede early acute heart allograft rejection. Transpl Immunol 2004;12:159–166.
62 Trull AK, Steel LA, Sharples LD, Akhlaghi F, Parameshwar J, Cary N, et al: Randomized trial of blood eosinophil count monitoring as a guide to corticosteroid dosage adjustment after heart transplantation. Transplantation 2000;70:802–809.
63 Nagral A, Ben Ari Z, Dhillon AP, Burroughs AK: Eosinophils in acute cellular rejection in liver allografts. Liver Transpl Surg 1998;4:355–362.
64 Barnes EJ, Abdel-Rehim MM, Goulis Y, Ragab MA, Davies S, Dhillon A, et al: Applications and limitations of blood eosinophilia for the diagnosis of acute cellular rejection in liver transplantation. Am J Transplant 2003;3:432–438.
65 Foster PF, Sankary H, Williams JW: Study of eosinophilia and hepatic dysfunction as a predictor of rejection in human liver transplantation. Transplant Proc 1988;20(suppl 1):676–677.
66 Crosbie OM, Costello PJ, O'Farrelly C, Hegarty JE: Changes in peripheral blood double-negative T-lymphocyte (CD3+ CD4– CD8–) populations associated with acute cellular rejection after liver transplantation. Liver Transpl Surg 1998;4(2):141–145.
67 Lederer SR, Friedrich N, Gruber R, Landgraf R, Toepfer M, Sitter T: Reduced CD40L Expression on ex vivo Activated CD4+ T-Lymphocytes from patients with excellent renal allograft function measured with a rapid whole blood flow cytometry procedure. Int Arch Allergy Immunol 2004;133:276–284.
68 May WS Jr, Cuatrecasas P: Transferrin receptor: its biological significance. J Membr Biol 1985;88:205–215.
69 Minguela A, Garcia-Alonso AM, Marin L, Torio A, Sanchez-Bueno F, Bermejo J, et al: Evidence of CD28 upregulation in peripheral T cells before liver transplant acute rejection. Transplant Proc 1997;29:499–500.
70 Lange H, Muller TF, Ebel H, Kuhlmann U, Grebe SO, Heymanns J, et al: Immediate and long-term results of ATG induction therapy for delayed graft function compared to conventional therapy for immediate graft function. Transpl Int 1999;12:2–9.
71 Martins L, Ventura A, Dias L, Henriques A , Sarmento A, Guimaraes S: Long-term effects of ATG therapy on lymphocyte subsets. Transplant Proc 2001;33:2186–2187.
72 Hartel C, Schumacher N, Fricke L, Ebel B, Kirchner H, Muller-Steinhardt M: Sensitivity of whole-blood T lymphocytes in individual patients to tacrolimus (FK 506): impact of interleukin-2 mRNA expression as surrogate measure of immunosuppressive effect. Clin Chem 2004;50:141–151.
73 Deng MC, Breithardt G, Scheld HH: The Interdisciplinary Heart Failure and Transplant Program Munster: a 5-year experience. Int J Cardiol 1995;50:7–17.
74 Kootte AM, Henny FC, Tanke HJ, Slats J, van Es LA, Paul LC: Enumeration of Leu2a+, Leu2a+-DR+, and Leu2a+-Leu15+ cells in peripheral blood of renal transplant patients. Transplantation 1988;45:132–138.
75 Rayes N, Bechstein WO, Volk HP, Tullius SG, Nussler N, Naumann U, et al: Distribution of lymphocyte subtypes in liver transplant recipients. Transplant Proc 1997;29:501–502.
76 Rayes N, Bechstein WO, Tullius SG, Nussler NC, Naumann U, Jonas S, et al: Distribution of lymphocyte subtypes in liver transplant recipients with viral reinfection or de novo malignancy. Transplant Proc 1998;30:1846–1847.
77 Ganschow R, Broering DC, Nolkemper D, Albani J, Kemper MJ, Rogiers X, et al: Th2 cytokine profile in infants predisposes to improved graft acceptance after liver transplantation. Transplantation 2001;72:929–934.
78 Molajoni ER, Cinti P, Orlandini A, Molajoni J, Tugulea S, Ho E, et al: Mechanism of liver allograft rejection: the indirect recognition pathway. Hum Immunol 1997;53:57–63.
79 Karim M, Kingsley CI, Bushell AR, Sawitzki BS, Wood KJ: Alloantigen-induced CD25+CD4+ regulatory T cells can develop in vivo from CD25-CD4+ precursors in a thymus-independent process. J Immunol 2004;172:923–928.
80 Khoury S, Sayegh MH, Turka LA: Blocking costimulatory signals to induce transplantation tolerance and prevent autoimmune disease. Int Rev Immunol 1999;18:185–199.

81 Lee RS, Yamada K, Womer KL, Pillsbury EP, Allison KS, Marolewski AE, et al: Blockade of CD28-B7, but not CD40-CD154, prevents costimulation of allogeneic porcine and xenogeneic human anti-porcine T cell responses. J Immunol 2000;164:3434–3444.
82 Li XC, Strom TB. Blocking T-cell costimulation in transplantation: opportunities and challenges. Transfusion 2000;40:139–142.
83 Sawitzki B, Lehmann M, Ritter T, Graser E , Kupiec-Weglinski JW, Volk HD: Regulatory tolerance-mediating T cells in transplantation tolerance. Transplant Proc 2001;33:2092–2093.
84 Reckzeh B, Grebe SO, Rhiele I, Neubauer A , Muller TF: Monitoring of ATG therapy by flow cytometry: comparison of one single-platform and two different dual-platform methods. Transplant Proc 2001;33:2234–2236.
85 Buttner C, Lun A, Splettstoesser T, Kunkel G, Renz H: Monoclonal anti-interleukin-5 treatment suppresses eosinophil but not T-cell functions. Eur Respir J 2003;21:799–803.
86 Meissner N, Kussebi F, Jung T, Ratti H, Baumgarten C, Werfel T et al: A subset of CD8+ T cells from allergic patients produce IL-4 and stimulate IgE production in vitro. Clin Exp Allergy 1997;27:1402–1411.
87 Kaden J, Schutze B, May G: A critical analysis of soluble interleukin-2 receptor levels in kidney allograft recipients. Transpl Int 1996;9(suppl 1):S63–S67.

Sack U, Tárnok A, Rothe G (eds): Cellular Diagnostics. Basics, Methods and Clinical Applications of Flow Cytometry. Basel, Karger, 2009, pp 580–599

Flow-Cytometric Analysis of Cells from the Respiratory Tract: Bronchoalveolar Lavage Fluid and Induced Sputum

Detlef Loppow

Laboratory Dr. Kramer & Colleagues, LADR GmbH, Geesthacht, Germany

Introduction/Background

Alterations of the respiratory tract are reflected, among others, in the composition of respiratory secretions, the analysis of which – especially by flow cytometry – can be a valuable contributor to the diagnosis of respiratory tract and lung diseases [1, 2]. During the last decade, research on respiratory tract diseases has mainly focused on the cellular, biochemical and molecular-biological processes underlying these diseases, which are well characterized from a clinical point of view. Inflammatory processes in the lungs and airways involve a multitude of cells. Lymphocytes, in particular, play a central role in some of the most widespread respiratory and lung diseases. Beyond the purely descriptive analysis of cell populations by their surface markers, the analysis of cell function is becoming significantly more important.

Bronchoalveolar lavage (BAL) is an established procedure in the pneumological differential diagnosis of interstitial lung disease. In combination with medical history, general condition, X-ray results and histology, assessment of bronchoalveolar lavage fluid (BALF) contributes to the classification of numerous inflammatory, granulomatous and tumorous lung diseases. Additionally, the evaluation of cellular and noncellular components in the BALF allows the assessment of the activity and course (progression) of generalized lung disease [3–9].

To measure changes on the cellular or mediator level, biological samples are repeatedly needed and – if possible – after different therapeutic interventions. This requirement appears problematic, especially for patients with severe pulmonary diseases, since the necessary invasive techniques (BAL, transbronchial biopsy) could involve an undue burden or insufficiently justifiable risk for the patient. For this reason, new analytical methods were established to allow measurements in noninvasive material such as induced sputum (iSP). Compared to bronchoscopy, inhalation of ultrasonically vaporized saline solution was found to be less of a strain on the patient.

Bronchoscopic examinations, including BAL, are carried out in hospitals as well as in pneumologists' private practices. Analysis of the aspirated fluid should follow relatively quickly in a diagnostic laboratory. To date, sputum induction is performed only in specialized centers for scientific and research purposes, and the samples are analyzed only there. The BALF and sputum samples should be transported to the laboratory without delay; if necessary, they should be stored in a cool place (2–8 °C).

Compared with peripheral blood, cell count and cell quality are decreased in BALF and iSP; as a consequence, new techniques had to be established to process these small amounts without loss in analytical accuracy. This requirement is met by flow cytometry, especially when analyzing lymphocytes and their subpopulations. Flow cytometry is superior to analysis by light microscopy with stained cytospin preparations due to its greater statistical precision and the ability to cover multiple cell parameters simultaneously. The flow-cytometric analysis of BALF was optimized in the 1990s [10–12]. Only when three and more fluorescence colors had become available were the discrete identification of lymphocyte populations by CD45 expression (fig. 1) and the precise identification of lymphocyte subpopulations by two further markers possible [13]. As for blood, certain requirements can be formulated for the analysis of BALF, such as how *gates* should be defined and which plausibility criteria have to be met (checksums: $T_4 + T_8 \approx T$ und $T + B + NK \approx 100\%$) [14]. A consensus protocol on the diagnosis of BALF through flow-cytometric immunophenotyping has been elaborated, but has not been published yet [5, 15]. Analogous to the BAL procedure, flow-cytometric analysis of lymphocyte subpopulations in iSP has already been employed by several research groups, and first tests have yielded representative results [16–23].

A flow cytometer is indispensable for visualizing intracellular cytokine production and to characterize certain cell types such as Th1 (IFN-γ, IL-2) or Th2 (IL-4, IL-5). The concentration of cytokines in lymphocytes is very low since the proteins are released immediately after synthesis. Flow-cytometric measurement of fluorescent-marked intracellular lymphocytes of peripheral heparinized blood, BALF or iSP is therefore preceded by 4 h stimulation with phorbol-12-myristate-13-acetate (PMA) and ionomycin [24–27]. At the same time, secretion of the produced cytokines is suppressed by monensin or brefeldin A; thus, cytokines are able to accumulate in the cell.

In case of the very rare occurrence of histiocytosis X, it is possible to detect Langerhans cells in the BALF by CD1a expression; however, the level of CD1a+ alveolar macrophages is also increased in asthmatic patients [28, 29]. Far more frequently, the question arises whether the BALF or iSP shows eosinophilia because it is known since 1889 that bronchial asthma and sputum eosinophilia are closely associated [30]. Various protocols for the detection of eosinophil granulocytes in the blood, which permit discrete separation of eosinophils and neutrophils, e.g. through the selective binding of

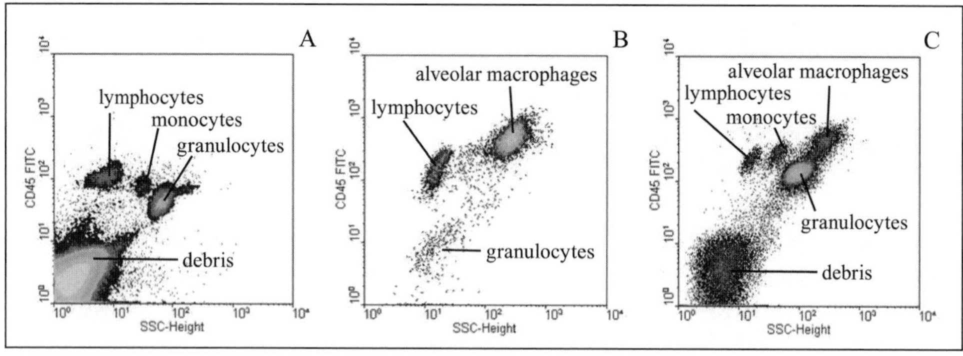

Fig. 1. Comparison of lysed whole blood, BALF and iSP. Leukocyte differentiation according to SSC and CD45 expression in lysed whole blood (**A**), BALF (**B**) and iSP (**C**).

unconjugated FITC or CD49d expression, respectively, have been published, [31, 32]. For analyzing eosinophils in BALF or iSP, detection with a depolarized side scatter is superior to other methods, but requires the installation of a polarization filter [33–37].

Protocols

Performing a Bronchoalveolar Lavage

BAL is conducted according to the recommendations of the German Society of Pneumology [38]. Following premedication (atropine, hydrocodone), a flexible bronchoscope is introduced under local anesthesia (lidocaine). Since lidocaine influences cell vitality and cell function, it should be suctioned off before the BAL [39]. After reaching the 'wedge position' and closing of the middle lobe bronchus, 100 ml of warm (37 °C) physiological saline solution is instilled and aspirated in aliquots of 20 ml each. Ideally, between 50 and 70% of BAL fluid should be recovered. The aspirate should be placed in sterile polypropylene tubes since alveolar macrophages can stick to glass and polystyrene.

Morphological Analysis of the Bronchoalveolar Lavage

The cell suspension is visually evaluated and the aspiration volume measured. Solid components (mucus, flakes) are removed by filtration through two layers of gauze and processed separately (smear preparation). BALF is centrifuged for 5 min at 500 × *g*; the supernatant carefully removed, and the cell pellet is resuspended in a defined volume of PBS. If the supernatant is obtained under sterile conditions, further microbiological testing is possible.

Table 1. Reagents

Lymphocyte subtyping	Item No.	Manufacturer
Flow count	7547053	Beckmann Coulter, Krefeld
CD45 FITC	0782	Beckmann Coulter, Krefeld
Propidium iodide		Beckmann Coulter, Krefeld
Glycophorin A PE	340947	BD Biosciences, Heidelberg
MultiTest CD3/CD8/CD45/CD4	342417	BD Biosciences, Heidelberg
MultiTest CD3/CD16+56/CD45/CD19	342416	BD Biosciences, Heidelberg
Simultest CD3/HLA-DR	340048	BD Biosciences, Heidelberg
Sputum processing PBS, 10× concentrate		Gibco BRL, Berlin
Sputolysin® (contains 6.5 mmol/l DTT in PBS)		Calbiochem, Bad Soden
Intracellular Cytokine Detection RPMI-1640 medium		Gibco BRL, Berlin
PMA	P 1585	Sigma-Aldrich, Taufkirchen
Ionomycin	I 0634	Sigma-Aldrich, Taufkirchen
Monensin	M 5273	Sigma-Aldrich, Taufkirchen
Fix & Perm	GAS004	Caltag, Hamburg
CD3 PerCP-Cy5.5	340949	BD Biosciences, Heidelberg
CD8 APC	345775	BD Biosciences, Heidelberg
IFN-γ FITC	340449	BD Biosciences, Heidelberg
IL-2 FITC	340448	BD Biosciences, Heidelberg
IL-4 PE	340451	BD Biosciences, Heidelberg
IL-5 PE	559332	BD Biosciences, Heidelberg

First, one aliquot of the concentrated cell suspension is analyzed by an experienced cytologist using conventional cytological methods. The primary goal of the morphological testing of BALF is the assessment of cellular and acellular components (leukocyte differentiation, tumor cell screening). Special dyes can be applied (iron, lipid, Grocott, or auramin staining) in order to detect exposure to inhaled pollutants or for direct pathogen identification.

Flow-Cytometric Analysis of Bronchoalveolar Lavage Fluid

100 μl each of concentrated BALF cell suspension is incubated with the appropriate antibodies for 15 min in the dark (table 1), and mixed with 500 μl PBS containing 100 μl Coulter Flow-Count Beads with a previously defined quantity of particles to determine the absolute cell count (table 2). This is followed by flow-cytometric analysis.

In all protocols, the lymphocytes are identified through side scatter (SSC) and CD45 expression. Initially, the viability of lymphocytes is determined in the first protocol with 20 μl CD45-FITC and 5 μl propidium iodide (fig. 2A–D). Immunologically intact lymphocytes in figure 2A are depicted in the light scatter diagram; the proportion of PI– 'viable' lymphocytes is 56.2% (fig. 2C, lower right quadrant). All other protocols, however, are based on immunologically and morphologically intact

Table 2. Mean values for 149 lavages processed with the described protocol

Parameter	
Instillation volume, ml	106
Aspiration volume, ml	55
Recovered fluid, %	52
Total cell count $\times 10^6$	16.4
Cell concentration $\times 10^5$/ml	3.2
Alveolar macrophages, %	76.7
Neutrophils, %	7.0
Eosinophils, %	2.0
Lymphocytes, %	13.8
Lymphocyte viability, %	66.1
Mature T-lymphocytes, %	93.7
CD4+ T-cells, %	52.0
CD8+ T-cells, %	38.2
CD4/CD8-ratio	1.4
Activated T-cells, %	38.6
Mature B-lymphocytes, %	0.5
NK cells, %	4.4
Checksums	
T + B + NK, %	98.6
T-(TH + TS), %	3.5

lymphocytes (fig. 2 B, oval gate). Here, their viability is 91.9% (fig. 2D, lower right quadrant).

In the second protocol, erythrocytes are identified among CD45– cells (fig. 3A, B, histogram) using 20 μl CD45-FITC and 5 μl glycophorin A-PE, and erythrocytic contamination of the lymphocyte gate is examined (not shown). The result is given as the percentage of erythrocytes compared to the sum of erythrocytes and CD45+ events.

The last three protocols are structured similarly. Mature T-lymphocytes as well as T_4- and T_8-cells are determined with 20 μl CD3-FITC, CD8-PE, CD45-PerCP, CD4-APC, B-lymphocytes and NK cells with 20 μl CD3-FITC, CD16-PE, CD56-PE, CD45-PerCP, CD19-APC, and activated T-lymphocytes and pan-B-cells with 20 μl CD3-FITC, HLA-DR-PE and 20 μl CD45-PerCP/Cy5.5. Immunologically and morphologically intact cells with high viability (fig. 2D, 4B) are depicted in diagrams comparing CD3 expression against the respective target parameters (fig. 4C–G).

Analysis and interpretation of flow-cytometric testing of BALF require a high degree of experience. Figures 5 and 6 show CellQuest analysis protocols of BAL T_4- and T_8-lymphocytes, where only the threshold is modified whereas numbers of events and percentages remain unchanged as long as the gates remain the same. The identified number (#) and percentage (%) of events from the analysis protocols are then transferred to a self-programmed and validated Excel workbook. Results are

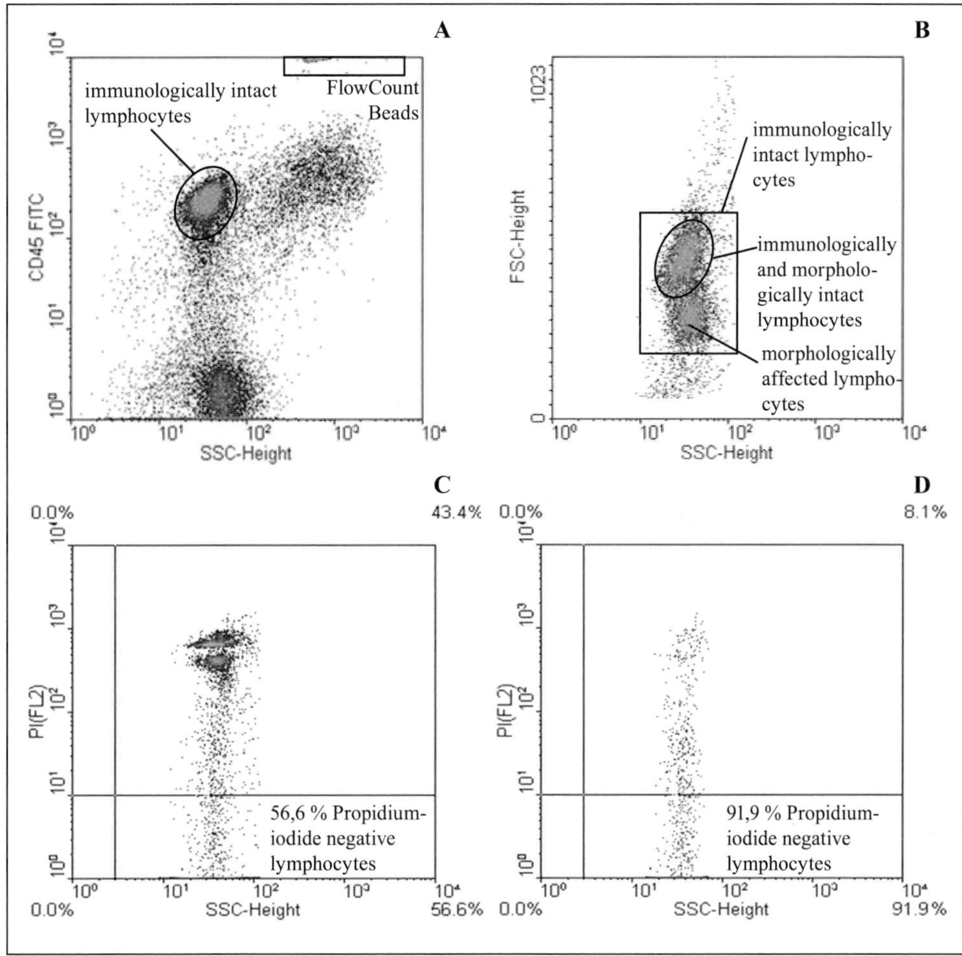

Fig. 2. BAL lymphocyte viability. **A** Immunologically intact lymphocytes were identified according to SSC and CD45 expression. **B** These cells were further differentiated as immunologically and morphologically intact or morphologically affected cells due to their FSC. **C** Lymphocyte viability was assessed by propidium iodide exclusion for all immunologically intact lymphocytes. **D** Immunologically and morphologically intact lymphocytes. Viability of all immunologically intact lymphocytes was 56.6%. Only immunologically and morphologically intact lymphocytes showing a viability of 91.9% were used for the flow-cytometric analyses of lymphocyte subpopulations.

recorded on a worksheet (fig. 7), whose nomenclature for gates and regions corresponds to that of the analysis protocols. On another worksheet (fig. 8), the results are depicted as absolute cell count, percentages, median and coefficient of variation (CV) for multiple parameter measurements, checksums, plausibility controls and required volumes for cytospin preparations. The data may be exported to a laboratory information system for diagnostic purposes.

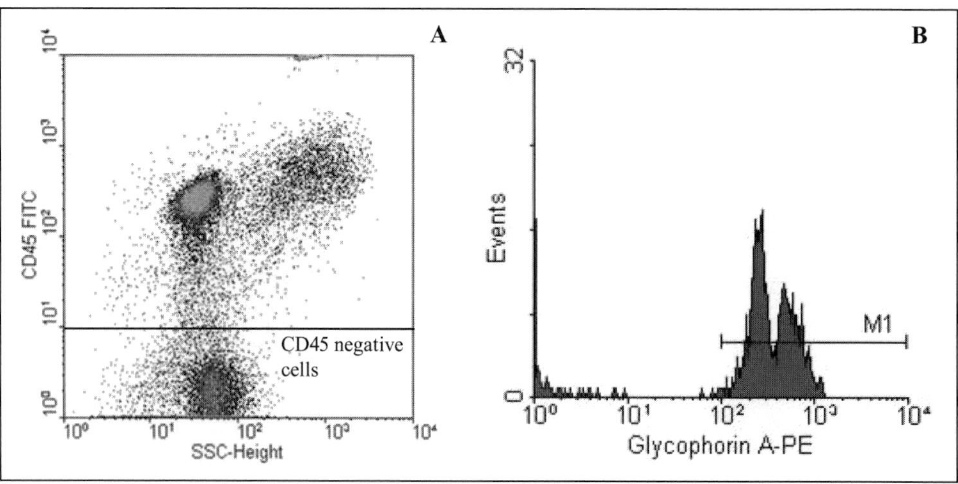

Fig. 3. Erythrocytes contaminating BALF. **A** First, nonleucokyte cells were identified according to SSC and CD45 expression. **B** Then glycophorin-A-positive erythrocytes were determined. Finally, the percentage of erythrocytes based on all events was calculated for the physician's report.

Acquisition of Induced Sputum and Sample Processing

Sputum is induced by inhaling ultrasonically vaporized saline solution so that sputum flakes can be coughed up from the upper and middle bronchi. After 15 min incubation in sputolysin solution at 37 °C in a water bath, sputum flakes are washed with PBS and resuspended in PBS/BSA (1%) [40].

Flow-Cytometric Analysis of Induced Sputum

As done for BALF, 100 µl each of sputum cell suspension is incubated and measured with the appropriate number of antibodies for lymphocyte subtyping (fig. 9).

Flow Cytometric Analysis of Intracellular Cytokines in Bronchoalveolar Lavage Fluid and Induced Sputum

Two solutions containing 100–500 µl each of BALF or iSP cell suspension are mixed with 500–900 µl RPMI-1640 medium. One preparation is incubated for 4 h with 20 ng/ml PMA and 1 µg/ml ionomycin at 37 °C and 5% CO_2, while protein secretion is suppressed by 5 mmol/l monensin in both the stimulated and unstimulated solution. After fixation with 100 µl Fix & Perm reagent A for 15 min in the dark at room temperature, cells are washed with 4 ml PBS for 5 min at 300 × *g* and mixed with

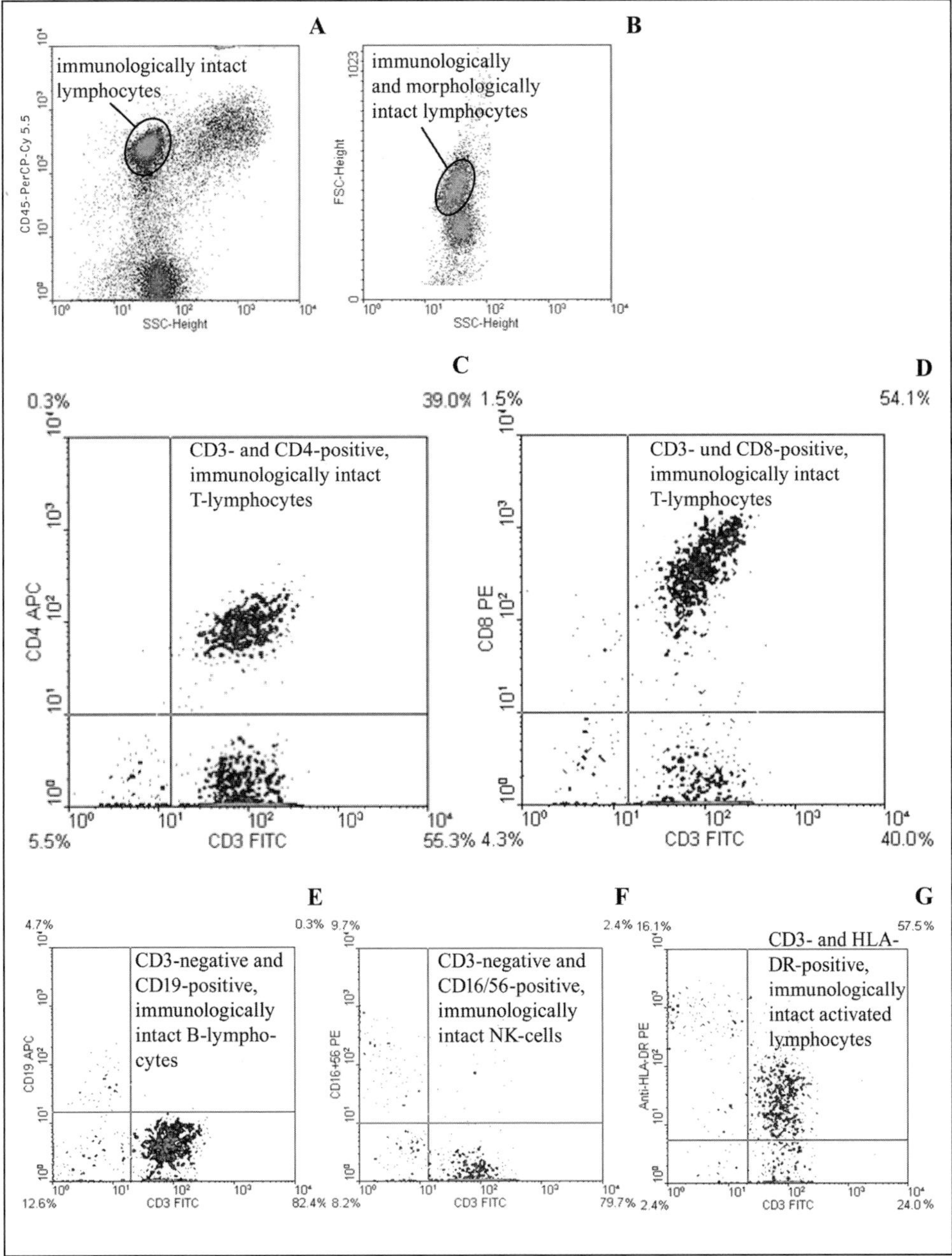

Fig. 4. BALF lymphocyte subpopulations. **A** Immunologically intact lymphocytes were identified according to SSC and CD45 expression. **B** Immunologically and morphologically intact lymphocytes were then determined based on FSC. Thereafter, CD3+ and CD4+ T-lymphocytes (**C**), CD3+ and CD8+ T-lymphocytes (**D**), CD3− and CD19+ B-lymphocytes (**E**), CD3− and CD16+ or CD56+ NK lymphocytes (**F**) and CD3− and HLA-DR+ pan-B-lymphocytes or CD3+ and HLA-DR+ activated T-lymphocytes (**G**) were determined.

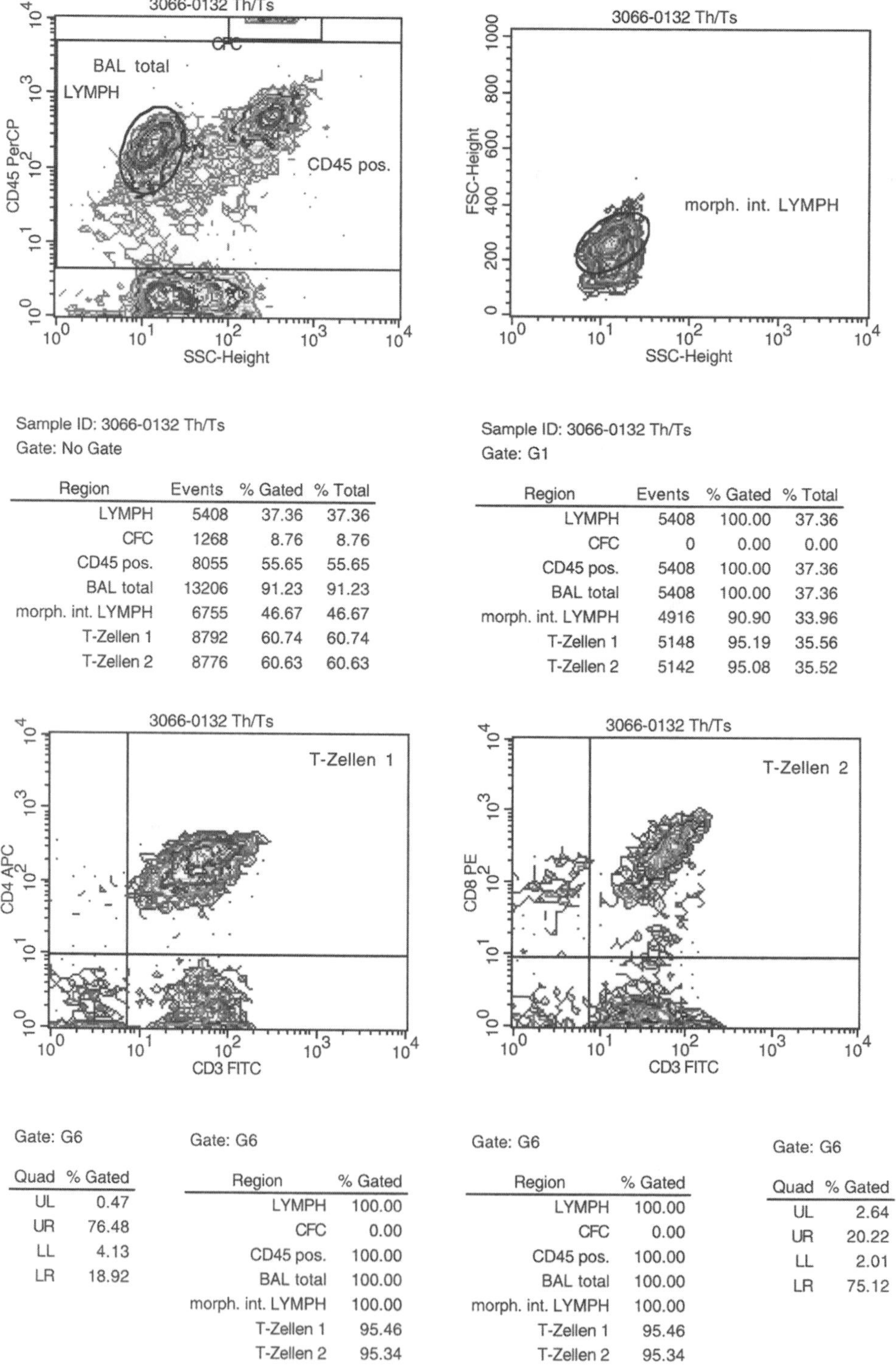

Sample ID: 3066-0132 Th/Ts
Gate: No Gate

Region	Events	% Gated	% Total
LYMPH	5408	37.36	37.36
CFC	1268	8.76	8.76
CD45 pos.	8055	55.65	55.65
BAL total	13206	91.23	91.23
morph. int. LYMPH	6755	46.67	46.67
T-Zellen 1	8792	60.74	60.74
T-Zellen 2	8776	60.63	60.63

Sample ID: 3066-0132 Th/Ts
Gate: G1

Region	Events	% Gated	% Total
LYMPH	5408	100.00	37.36
CFC	0	0.00	0.00
CD45 pos.	5408	100.00	37.36
BAL total	5408	100.00	37.36
morph. int. LYMPH	4916	90.90	33.96
T-Zellen 1	5148	95.19	35.56
T-Zellen 2	5142	95.08	35.52

Gate: G6

Quad	% Gated
UL	0.47
UR	76.48
LL	4.13
LR	18.92

Gate: G6

Region	% Gated
LYMPH	100.00
CFC	0.00
CD45 pos.	100.00
BAL total	100.00
morph. int. LYMPH	100.00
T-Zellen 1	95.46
T-Zellen 2	95.34

Gate: G6

Region	% Gated
LYMPH	100.00
CFC	0.00
CD45 pos.	100.00
BAL total	100.00
morph. int. LYMPH	100.00
T-Zellen 1	95.46
T-Zellen 2	95.34

Gate: G6

Quad	% Gated
UL	2.64
UR	20.22
LL	2.01
LR	75.12

Fig. 5. BAL protocol. CellQuest BAL T_4- and T_8-lymphocyte protocol showing the right threshold. T-Zellen = T-cells; UL = bottom left; UR = bottom right; OL = top left; OR = top right.

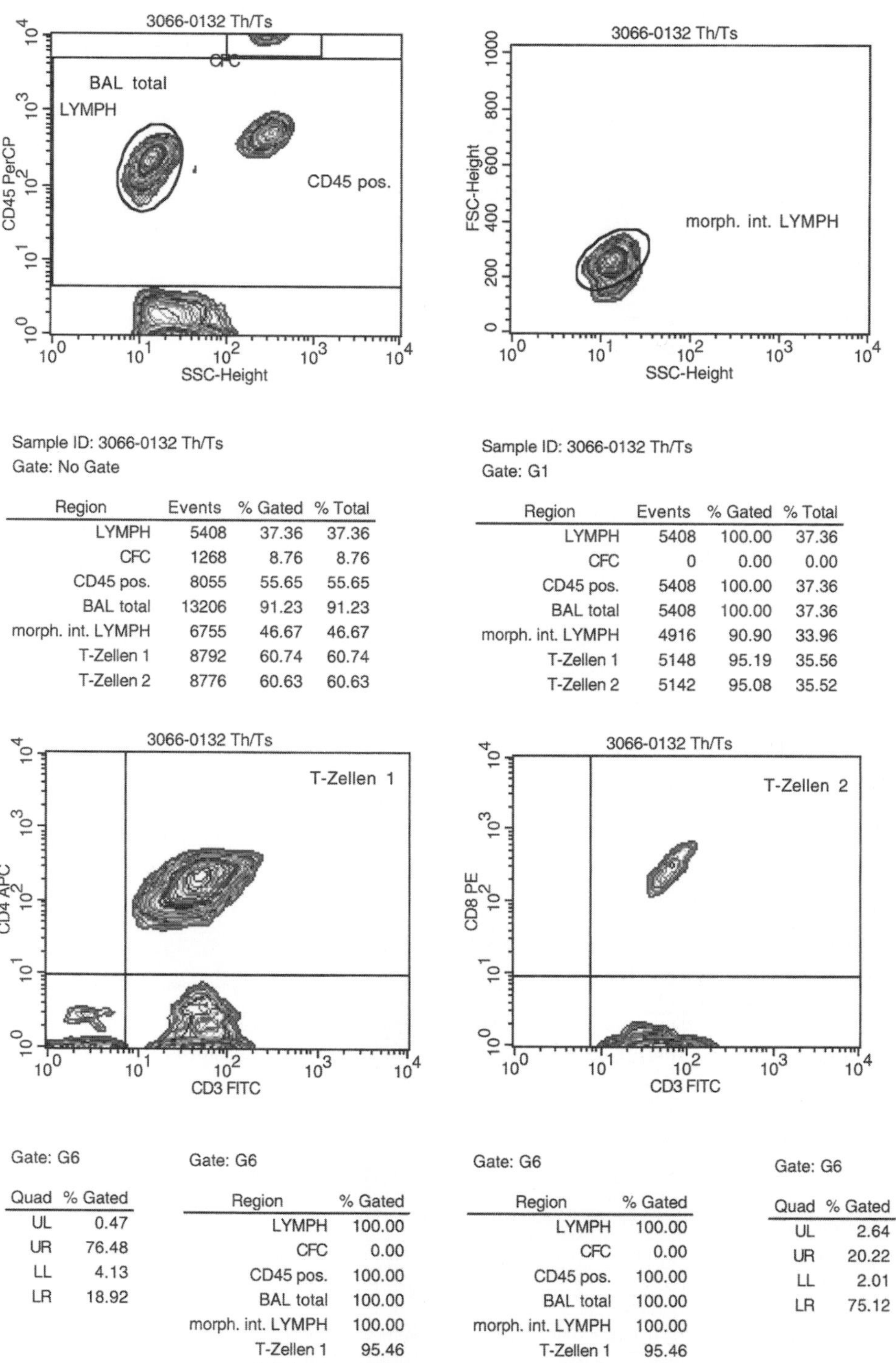

Sample ID: 3066-0132 Th/Ts
Gate: No Gate

Region	Events	% Gated	% Total
LYMPH	5408	37.36	37.36
CFC	1268	8.76	8.76
CD45 pos.	8055	55.65	55.65
BAL total	13206	91.23	91.23
morph. int. LYMPH	6755	46.67	46.67
T-Zellen 1	8792	60.74	60.74
T-Zellen 2	8776	60.63	60.63

Sample ID: 3066-0132 Th/Ts
Gate: G1

Region	Events	% Gated	% Total
LYMPH	5408	100.00	37.36
CFC	0	0.00	0.00
CD45 pos.	5408	100.00	37.36
BAL total	5408	100.00	37.36
morph. int. LYMPH	4916	90.90	33.96
T-Zellen 1	5148	95.19	35.56
T-Zellen 2	5142	95.08	35.52

Gate: G6

Quad	% Gated
UL	0.47
UR	76.48
LL	4.13
LR	18.92

Gate: G6

Region	% Gated
LYMPH	100.00
CFC	0.00
CD45 pos.	100.00
BAL total	100.00
morph. int. LYMPH	100.00
T-Zellen 1	95.46
T-Zellen 2	95.34

Gate: G6

Region	% Gated
LYMPH	100.00
CFC	0.00
CD45 pos.	100.00
BAL total	100.00
morph. int. LYMPH	100.00
T-Zellen 1	95.46
T-Zellen 2	95.34

Gate: G6

Quad	% Gated
UL	2.64
UR	20.22
LL	2.01
LR	75.12

Fig. 6. BAL protocol. CellQuest BAL T_4- and T_8-lymphocyte protocol showing an increased threshold. Compared to figure 5, events and percentages will not change as long as the gates remain unchanged! T-Zellen = T-cells; UL = bottom left; UR = bottom right; OL = top left; OR = top right.

Allgemein	CFC	1020	[#/µL]
Name, first name	Name, first name		
Date of birth	11.11.02		
Nr.	99 1234 5678		
Instilliated volume		100	
Aspirated volumen	(received volume)	60	[mL]
Suspension volume	(reduced to)	5500	[µL]
BAL Erythrocytes			
Gate: No Gate; Region:	CFC	2047	[#]
Gate: No Gate; Region:	CD45 pos	2054	[#]
Gate: No Gate; Region:	Lymph	184	[#]
Gate: No Gate; Region:	BAL total	2255	[#]
Gate: G5; Marker:	Ery/CD45 neg	79	[#]
Gate: G3; Region:	morph. int. Lymph	97,3	[%]
BAL Vitality			
Gate: No Gate; Region:	Lymph	237	[#]
Gate: No Gate; Region:	CFC	2412	[#]
Gate: No Gate; Region:	CD45 pos. Z.	2935	[#]
Gate: No Gate; Region:	BAL total	3256	[#]
Gate: G1; Region:	PI neg. alle	78,9	[%]
Gate: G6; Region:	PI neg. int.LY	79,0	[%]
Gate: G1; Region:	morph.int.LYMPH	80,2	[%]
BAL 3/8/45/4			
Gate: No Gate; Region:	LYMPH	470	[#]
Gate: No Gate; Region:	CFC	5786	[#]
Gate: No Gate; Region:	CD45 pos.	5256	[#]
Gate: No Gate; Region:	BAL total	7143	[#]
Gate: G6; Region:	T-Zellen 1	91,9	[%]
Gate: G6; Region:	T-Zellen 2	91,9	[%]
Gate: G6; Quad CD3/CD4	UR	72,3	[%]
Gate: G6; Quad CD3/CD8	UR	17,8	[%]
Gate: G1; Region:	morph. int. LYMPH	99,2	[%]
BAL 3/16+56/45/19			
Gate: No Gate; Region:	LYMPH	477	[#]
Gate: No Gate; Region:	CFC	6056	[#]
Gate: No Gate; Region:	CD45 pos.	5384	[#]
Gate: No Gate; Region:	BAL total	7008	[#]
Gate: G6; Region:	T-Zellen 1	90,5	[#]
Gate: G6; Region:	T-Zellen 2	90,5	[%]
Gate: G6; Quad CD19	UL	1,3	[%]
Gate: G6; Quad CD16/56	UL	6,9	[%]
Gate: G1; Region:	morph. int. LYMPH	97,1	[%]
BAL 3/DR/45			
Gate: No Gate; Region:	Lymph	343	[#]
Gate: No Gate; Region:	CFC	3975	[#]
Gate: No Gate; Region:	CD45+	3522	[#]
Gate: No Gate; Region:	BAL total	5084	[#]
Gate: G6; Region:	T-Zellen 1	88,9	[%]
Gate: G6; Region:	T-Zellen 2	88,9	[%]
Gate: G6; Quad CD3-/DR	UL (pan-B)	7,7	[%]
Gate: G6; Quad CD3+/DR	UR (akt. T)	26,9	[%]
Gate: G1; Region:	morph. Int. Lymph	94,5	[%]

Fig. 7. BAL standard report. Data were collected on an Excel worksheet. T-Zellen = T-cells; UL = bottom left; UR = bottom right.

Name, first name		Name, first name		
Date of birth		11.11.2002		
Nr.		99 1234 5678		
Instillated volume [mL]	100	-	-	-
Aspirated volume [mL]	60	-	-	-
Recovery [%]	60	-	-	-
Suspension volume [µL]	5500	-	-	-
	MW	**VK**	**mean**	**VK**
	[%]	**[%]**	**[Cells/µL]**	**[%]**
Erythrocytes in BAL	**4**	-	-	-
CD45 pos. cells *in BAL* *(n=5 / Prot. 1- 5)*	**80**	**12,3**	**1000**	**14,3**
CD45 pos. cells *in BAL (n=3 / Prot. 3-5)*	73	5,2	912	1,4
CD45 pos. cells *in BAL (n=2 / Prot. 1+2)*	91	0,7	1132	13,6
Lymphocytes in *CD45-pos. BAL-cells* *(n=5 / Prot.1- 5)*	**9**	**6,6**	**89**	-
thereof morphologically intact cells (n=5 / Prot.1-5)	94	8,2	-	-
Lymphocytes in *CD45-pos. BAL-cells (n=3 / Prot.3- 5)*	9	5,3	84	-
thereof morphologically intact cells (n=3 / Prot.3-5)	97	2,4	-	-
Lymphocytes in *CD45-pos. BAL-cells (n=2 / Prot.1+2)*	9	7,3	96	-
thereof morphologically intact cells (n=2 / Prot. 1+2)	89	13,6	-	-
PI-negative lymphocytes	**79**	-	-	-
T-lymphocytes	**90**	**1,7**	-	-
CD4-positive T-lymphocytes (T_4)	**72**	-	-	-
CD8-positive T-lymphocytes (T_8)	**18**	-	-	-
CD4/CD8-Quotient	**4,1**	-	-	-
Checksum: ***$T - (T_4 + T_8)$***	*0,3*	-	-	-
aktivated T-cells	**27**	-	-	-
NK-cells	**7**	-	-	-
B-cells	**1**	-	-	-
pan-B-cells	8	-	-	-
Checksum: ***100 - (T + B + NK-cells)***	*1,4*	-	-	-
CD45 pos. cells in whole *BAL* ***(n = 5 / Prot. 1 - 5)***	**5,50**	**[10^6 cells]**	-	-
CD45 pos. cells *per mL Aspirat* ***(n = 5 / Prot. 1 - 5)***	**0,92**	**[10^5 cells/mL]**	-	-
Zytospin: cellsuspension [µL]	***60***	***[µL]***	-	-
Zytospin: CellWash [µL]	***140***	***[µL]***	-	-
CD45 pos. cells in whole BAL *(n = 3 / Prot. 3 - 5)*	5,02	[10^6 cells]	-	-
CD45 pos. cells per *ml Aspirat* *(n = 3 / Prot. 3 - 5)*	0,84	[10^5 cells/mL]	-	-
CD45 pos. cells in whole *BAL* *(n = 2 / Prot. 1 & 2)*	6,23	[10^6 cells]	-	-

Fig. 8. BAL report. The results are shown as an Excel worksheet with means (MW) and CVs (VK) for multiple measured parameters and the computed volumes for the preparation of cytospins.

100 µl Fix & Perm reagent B for permeabilization. Fluorochrome-marked antibodies are added against the cell surface antigens CD3 and CD8 as well as against the intracellular cytokines IFN-γ und IL-4, or IL-2 and IL-5, respectively. Cells are then incubated for 20 min in the dark at room temperature, washed again in 4 ml PBS for 5 min at 300 × *g*, mixed with 500 µl PBS and sorted by flow cytometry [41].

iSP lymphocytes are first identified according to their light scatter characteristics (fig. 10A). CD3+ T-lymphocytes (fig. 10B) are then depicted in diagrams showing

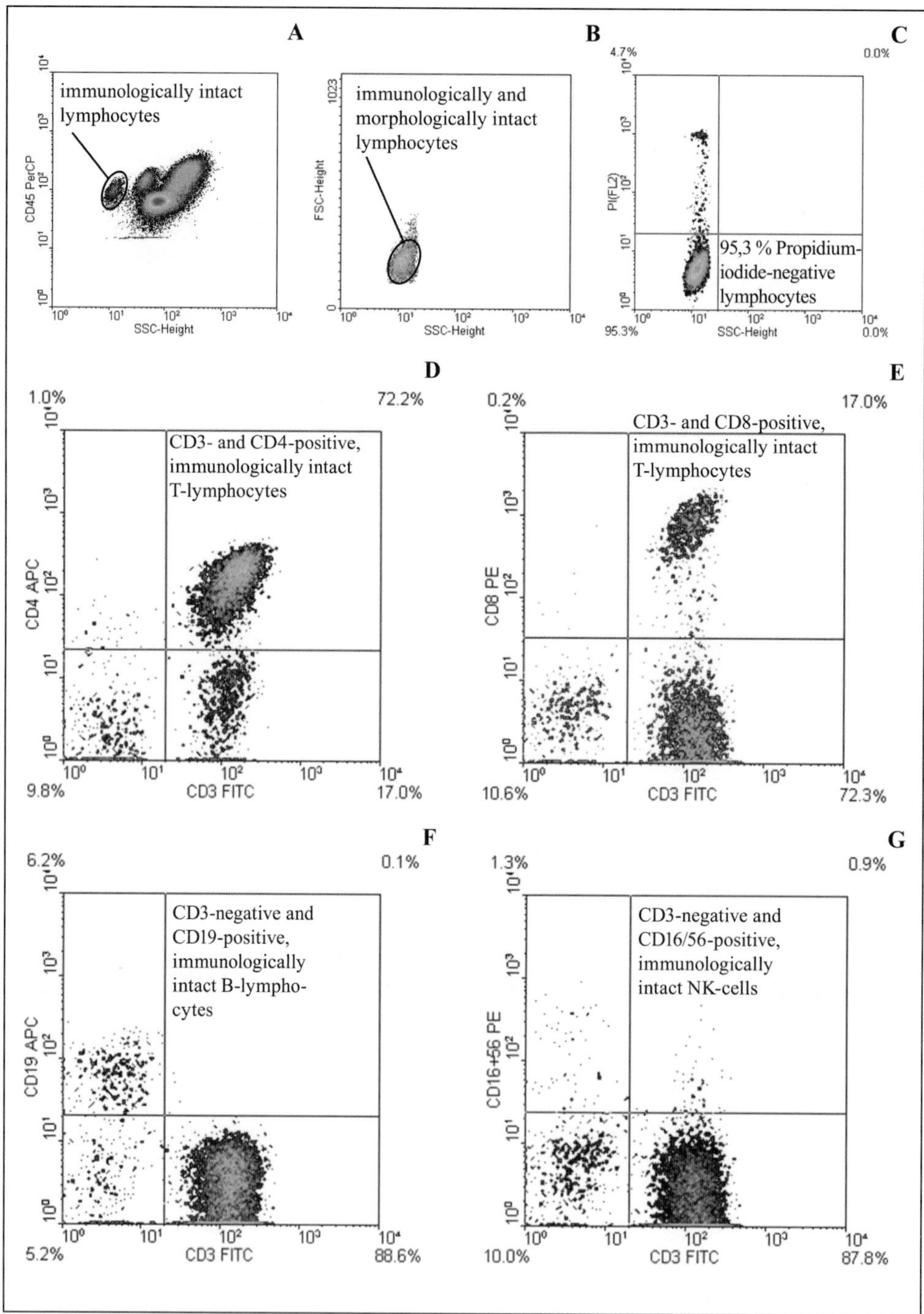

Fig. 9. Sputum lymphocyte subpopulations. **A** Immunologically intact lymphocytes were identified according to SSC and CD45 expression. **B** Immunologically and morphologically intact lymphocytes were then determined based on FSC. **C** Lymphocyte viability was assessed by propidium iodide exclusion. CD3+ and CD4+ T-lymphocytes (**D**), CD3− and CD8+ T-lymphocytes (**E**), CD3− and CD19+ B-lymphocytes (**F**) and CD3− and CD16+ or CD56+ NK-lymphocytes (**G**) were then determined.

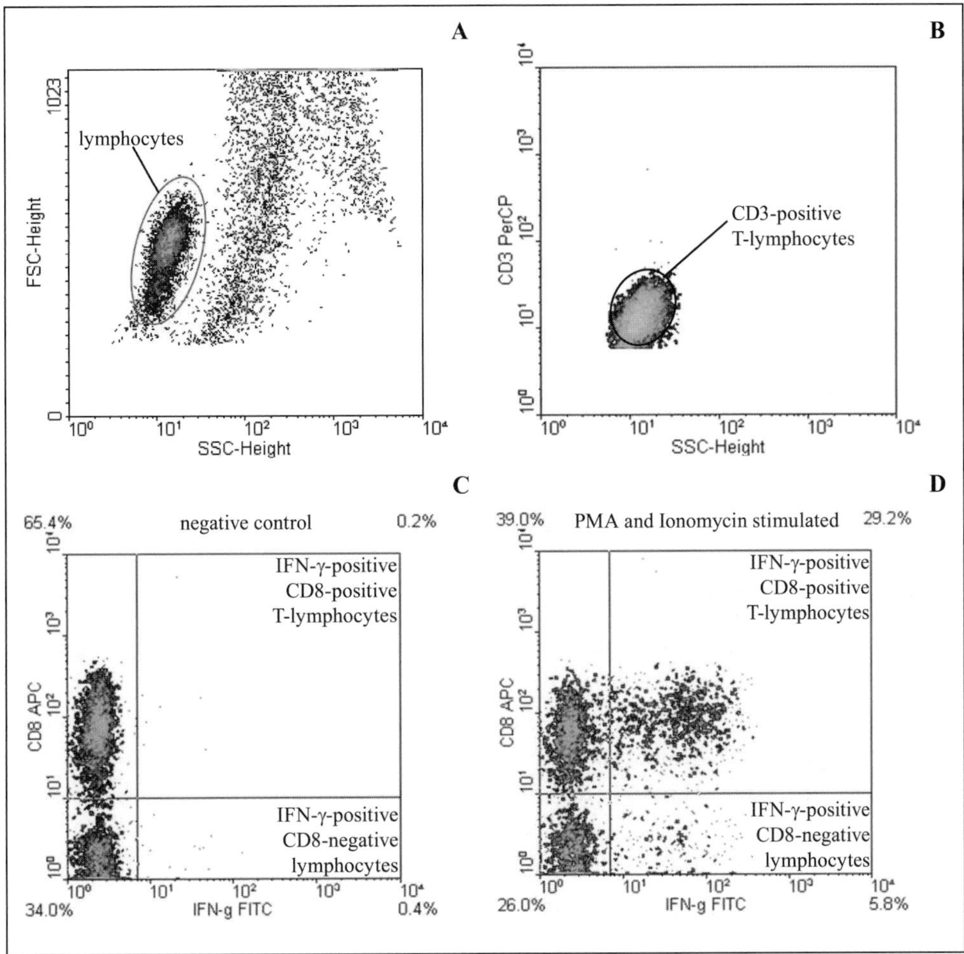

Fig. 10. Intracellular cytokine production of sputum lymphocytes. **A** Lymphocytes were identified morphologically based on FSC and SSC. **B** Immunologically intact T-lymphocytes were determined according to CD3 expression. For the negative control (C) and the PMA/ionomycin-stimulated sample, IFN-γ production was plotted versus CD8 expression to determine cytokine production of CD8+ and CD8− T-lymphocytes.

cytokine production versus CD8 expression for both unstimulated (fig 10C) and stimulated preparations (fig. 10D). The percentage of cytokine-producing T_4-lymphocytes is displayed in the lower right quadrant while the upper right quadrant shows the percentage of T_8-lymphocytes. Defining T_4- or Th-lymphocytes as CD8− mature T-lymphocytes seems rational since PMA stimulation may lead to a transient loss of CD4 expression [42].

All reagents used are listed in table 1.

Comment

Cell count and cell quality are decreased in BALF and iSP compared with peripheral blood (fig. 1). As a consequence, analysis and interpretation of results are increasingly difficult. Therefore, it is important to recognize and consider limiting factors before interpretation of the final result.

Critical Parameters

Good collaboration between the analyst, the cytologist and the physician performing the BAL, who should forward all relevant details to the hospital or external laboratory, is an important prerequisite for a valid BAL interpretation. Information on the patient's medical history (nicotine abuse, allergen- or dust exposition), clinical diagnoses, and details on the BAL procedure itself (Instillation volume, aspiration volume, artificially added blood) are particularly important. If, for example, there are clinical signs of lung disease, the a posteriori likelihood of sarcoidosis according to Bayes is doubled (with moderate lymphocytosis and normal granulocyte count) as compared to the a priori likelihood when no clinical diagnosis is available [43].

Besides the small cell count, the use of dithiothreitol during iSP processing represents a critical part of the analysis. For the antibodies used for lymphocyte subtyping it could be shown that this reagent, which cleaves disulfide bridges, has no influence on the epitopes needed for the immunological detection reaction [44].

Duration of stimulation is the critical point when measuring intracellular cytokine production in BALF or iSP lymphocytes: it should not exceed 4 h.

Advantages of the Described Protocol

If carried out in an experienced laboratory, flow-cytometric analysis of BALF enables pneumologists and hospitals that do not have their own flow cytometer to obtain valid and statistically safer differential-diagnostic results than with other immunocytochemical procedures. Moreover, whereas immunocytochemical staining of older material is not practicable since cell viability decreases over time and nonviable cells do not stick to microscope slides, flow cytometry still yields good results since the important CD4/CD8 ratio remains stable even in nonviable lymphocytes.

Quality Control

The plausibility of flow-cytometric measurements of the lymphocyte subpopulations in BALF and iSP must be verified by the following checksums: 'T-cells + B-

Table 3. Mean values and SEM for a total of 77 sputum samples (Epics XL, 27 samples; FACSCalibur, 50 samples) measured on an Epics XL and a FACSCalibur flow cytometer

Lymphocytes	Viability, %	T-cells, %	T_4-cells, %	T_8-cells, %	NK cells, %	B-cells, %	T_4+T_8-cells, %	T+B+NK, %
Mean value	84.4	88.6	63.6	23.3	2.3	3.3	86.8	94.1
Epics XL[a]	86.2	85.4	60.5	22.7	2.0	2.6	83.2	89.9
FACSCalibur[b]	83.1	90.3	65.3	23.5	2.5	3.6	88.8	96.4
SEM	1.8	0.8	1.8	1.8	0.3	0.4	1.0	0.6
Epics XL	2.2	1.6	2.1	2.2	0.5	0.7	2.1	1.2
FACSCalibur	2.7	0.8	2.5	2.6	0.3	0.5	0.9	0.4

[a]n = 27. [b]n = 52.

cells + NK cells = 100% lymphocytes' and 'T_4-cells + T_8-cells = mature T-lymphocytes'. The deviation should always be below 5%. As shown, 149 lavages (table 2) and 77 sputum samples (table 3) analyzed with our methodology exhibited only small mean deviations. For repeatedly determined parameters like CD3, it is advisable to calculate the CV, which is a sensitive indicator of the quality of the tests. It should be below 5%, which is not always achievable with strongly heterogeneous material containing mucus and flakes.

Expected Results

Table 4 shows typical alterations of parameters and/or percentage of cell populations in some diseases. Classification of leukocyte percentages in the BALF is presented in table 5 whereas table 6 shows the laboratory reference regions for flow-cytometric lymphocyte subtyping, taking smoker status into account. The quantity of activated T_4-cells increases with advancing age [45]. Meaningful reference regions for T_4- and T_8-lymphocytes are not yet available.

Required Time

Preparation, filtration, centrifugation and incubation for BALF or sputum lymphocyte subtyping takes approximately 1 h. Sorting of the five preparations on a four-color flow cytometer takes between 10 and 60 min, depending on cell concentration. For analysis and interpretation, a further 20–60 min is required.

Preparation of stimulation for intracellular cytokine detection should take approximately 30 min if all stock solutions are sensibly aliquoted and frozen. After the 4-hour incubation period, the samples are ready for sorting after 60–80 min. Testing

Table 4. Diseases and typical alterations of cell populations

Disease	T_4/T_8	Ly	Neu	Eos	AM
Sarcoidosis	↑↑	↑↑	↔	↔	
Idiopathic pulmonary fibrosis	↔	(↑)	↑↑	↑	
Exogenous allergic alveolitis	↓↓	↑↑↑	(↑)	↑	
Eosinophilic pneumonia		(↑)	↔	↑↑[1]	
Idiopathic pulmonary hemosiderosis					iron+ inclusions
Asbestosis		↑	↑		
Alveolar proteinosis					PAS + inclusions,
Lamellar bodies					
Histiocytosis X				(↑)	CD1a+ > 4%
BOOP	↓	↑↑	↑	↑[2]	

T_4/T_8 = CD4/CD8-ratio; Ly = lymphocytes; Neu = neutrophils; Eos = eosinophils; AM = alveolar macrophages; BOOP = Bronchiolitis obliterans organizing pneumonia ↓↓ = decreased; ↓ = slightly decreased; ↔ = variable; (↑) = possibly increased; ↑ = slightly increased; ↑↑ = increased; ↑↑↑ = strongly increased.

[1] > 20%. [2] < 20%.

Table 5. Distribution of leukocytes in BALF and classification into degrees of severity

Cells	Degree of severity			
	normal	slight	medium	high
Lymphocytes	0–15	16–30	31–50	>50
Neutrophils	0–3	4–20	21–50	>50
Eosinophils	0–1	2–10	11–20	>20

Table 6. Reference range for BAL results in the author's laboratories

Parameter	Smoker	Nonsmoker
Total cell count × 10^6	4–10	11–35
Cell concentration × 10^5/ml		
Macrophages, %	88–96	93–99
Neutrophils, %		
Eosinophils, %		
Lymphocytes, %	4–10	1–5
Erythrocytes, %		
Viability of lymphocytes, %		
Mature T-lymphocytes, %	78–95	
Mature B-lymphocytes, %	<1	
CD4+ T-cells, %		
CD8+ T-cells, %		
T_4/T_8-ratio	1.1–3.5	0.5–1.5
Activated T-cells, %	<10	
NK lymphocytes, %	2–8	

of the four tubes varies between 10 and 45 min, depending on cell concentration. Analysis of the results requires 15 min.

Summary

Flow-cytometric analysis of cells from the respiratory tract, bronchoalveolar fluid or iSP combined with medical history, general condition, X-ray results and histology can provide valuable information for the diagnosis of pulmonary (respiratory) and lung disease. After macroscopic evaluation of the cell suspension and microscopic cell differentiation, including tumor cell screening, immunophenotyping of lymphocytes is carried out by flow cytometry with at least three fluorescence channels since lymphocytes should be identified according to their CD45 expression in each test. Next to the most frequently requested CD4/CD8 ratio of CD4+ T-cells (T_4, Th, CD3+ and CD4+) and CD8+ T-cells (T_8, CD3+ and CD8+), B-cells (B, CD3− and CD19+) and NK cells (NK, CD3− and CD16+ or CD56+) should also be determined to verify the plausibility of the measurements using checksums ($T_4 + T_8 \approx T$ and $T + B + NK \approx 100\%$).

In scientific research, special protocols are employed for the flow-cytometric analysis of, e.g. eosinophil granulocytes, plasma cells, alveolar macrophages and the intracellular cytokine production of lymphocytes from BALF and iSP

References

1 Keatings VM, Collins PD, Scott DM, Barnes PJ: Differences in interleukin-8 and tumor necrosis factor-α in induced sputum from patients with chronic obstructive pulmonary disease or asthma. Am J Respir Crit Care Med 1996;153:530–534.

2 Pizzichini E, Marcia MM, Efthimiadis A, Dolovich J, Hargreave FE: Measuring airway inflammation in asthma: eosinophils and eosinophilic cationic protein in induced sputum compared with peripheral blood. J Allergy Clin Immunol 1997;99:539–544.

3 Costabel U: Atlas der bronchoalveolären Lavage. Stuttgart, Thieme, 1994.

4 The BAL Cooperative Steering Committee: Bronchoalveolar lavage constituents in healthy individuals, idiopathic pulmonary fibrosis and selected comparison groups. Am Rev Respir Dis 1990;141:S169–S202.

5 Sack U, Rothe G, Barlage S, et al: Durchflußzytometrie in der klinischen Diagnostik. J Lab Med 2000;2:77–297.

6 Cobben NAM, Jacobs JA, van Dieijen-Visser MP, et al: Diagnostic value of BAL fluid cellular profile and enzymes in infectious pulmonary disorders. Eur Respir J 1999;14:496–502.

7 Bernhardt K: Flow cytometry for extensive thoracic diagnosis. Acta Med Austriaca 1997;24:5–7.

8 Klech H, Hutter C, Costabel U (eds): Clinical guidelines and indications for bronchoalveolar lavage (BAL): Report of the European Society of Pneumology Task Group on BAL. Eur Respir Rev 1992;2:47–127.

9 Klech H, Hutter C (eds): Clinical Guidelines and Indications for Bronchoalveolar Lavage (BAL): Report of the European Society of Pneumology Task Group on BAL. Eur Respir J 1990;3:937–974.

10 Dauber JH, Wagner M, Brunsvold S, et al: Flow cytometric analysis of lymphocyte phenotypes in bronchoalveolar lavage fluid: comparison of a two-color technique with a standard immunoperoxidase assay. Am J Respir Cell Mol Biol 1992;7:531–541.

11 Brandt B, Thmas M, von Eiff M, et al: Immunophenotyping of lymphocytes obtained by bronchoalveolar lavage: description of an all-purpose tricolor flow cytometric application. J Immunol Methods 1996;19:195–102.

12 Böttcher M, Abel G: Immunphänotypisierung von Lymphozyten in der bronchoalveolären Lavage (BAL). Krefeld, Coulter Electronics GmbH, 1994.

13 Nicholson JK, Hubbard M, Jones BM: Use of CD45 fluorescence and side-scatter characteristics for gating lymphocytes when using the whole blood lysis procedure and flow cytometry. Cytometry 1996;26:16–21.

14 Calvelli T, Denny TN, Paxton H, et al: Guideline for flow cytometric immunophenotyping: a report from the national institute of allergy and infectious diseases, division of AIDS. Cytometry 1993;14:702–715.

15 Konsensus-Protokoll für die Diagnostik der bronchoalveolären Lavage mittels durchflußzytometrischer Immunphänotypisierung (Draft, approximately 1995, cited in Sack et al [5].

16 Perna F, Vatrella A, Parrella R et al: Flow cytometric evaluation of lymphocyte subsets on induced sputum samples. Eur Respir J 1997;10:61s.

17 Vatrella A, Perna F, Parrella R, et al: Flow cytometric analysis of lymphocyte subsets in induced sputum of asthmatic subjects before and after eight weeks of treatment with inhaled budesonide. Eur Respir J 1998;12:226s.

18 Kidney JC, Wong AG, Efthimiadis A et al: Sputum lymphocyte subclasses and activation measured by flow cytometry. Am J Respir Crit Care Med 1994;149:A572.

19 Kidney JC, Efthimiadis A, Dolovich J, Hargreave FE: Three-color flow cytometry on sputum lymphocytes. Am J Respir Crit Care Med 1996;153:A291.

20 Grootendorst DC, Sont JK, Willems LNA, et al: Comparison of inflammatory cell counts in asthma: induced sputum vs bronchoalveolar lavage and bronchial biopsies. Clin Exp Allergy 1997;27:769–779.

21 Fireman E, Topilsky I, Greif J, et al: Induced sputum compared to bronchoalveolar lavage for evaluating patients with sarcoidosis and non-granulomatous interstitial lung disease. Respir Med 1999;93:827–834.

22 Pizzichini E, Pizzichini MMM, Kidney JC, et al: Induced sputum, bronchoalveolar lavage and blood from mild asthmatics: inflammatory cells, lymphocyte subsets and soluble markers compared. Eur Respir J 1988;11:828–834.

23 Tsiligianni J, Tzanakis N, Kyriakou D, et al: Comparison of sputum induction with bronchoalveolar lavage cell differential counts in patients with sarcoidosis. Sarcoidosis Vasc Diffuse Lung Dis 2002;19:205–210.

24 Jung T, Schauer U, Heusser C, et al: Detection of intracellular cytokines by flow cytometry. J Immunol Methods 1993;159:197–207.

25 Krouwels FH, Nocker RET, Snoek M, et al: Immunocytochemical and flow cytofluorometric detection of intracellular IL-4, IL-5 and IFN-γ: applications using blood- and airway-derived cells. J Immunol Methods 1997;203:89–101.

26 Krug N, Madden J, Redington AE, et al: T-cell cytokine profile evaluated at the single cell level in BAL and blood in allergic asthma. Am J Respir Cell Mol Biol 1996;14:319–326.

27 Loppow D, Böttcher M, Gercken G, et al: Flow cytometric assessment of sputum lymphocyte cytokine profiles in patients with asthma and COPD. Eur Respir J 2001;18:316s.

28 Müller-Quernheim J: Bronchoalveoläre Lavage; in Thomas L: Labor und Diagnose. Frankfurt/Main, TH-Books-Verl-Ges, 2005, pp 1853–1857.

29 Agea E, Forenza N, Piattoni S, et al: Expression of B7 co-stimulatory molecules and CD1a antigen by alveolar macrophages in allergic bronchial asthma. Clin Exp Allergy 1998;28:1359–1367.

30 Gollasch H: Zur Kenntnis des asthmatischen Sputums. Fortschr Med 1889;7:361–365.

31 Efthimiadis A, Hargreave FE, Dolovich J: Use of selective binding of fluorescein isothiocyanate to detect eosinophils by flow cytometry. Cytometry 1996;26:75–76.

32 Thurau AM, Schulz U, Wolf V, et al: Identification of eosinophils by flow cytometry. Cytometry 1996;23:150–158.

33 Shapiro HM: Practical Flow Cytometry. New York, Wiley-Liss, 1995.

34 de Grooth BG, Terstappen LW, Puppels GJ, Greve J: Light-scattering polarization measurements as a new parameter in flow cytometry. Cytometry 1987;3:139–544.

35 Lavigne S, Bossé M, Boulet LP, et al: Identification and analysis of eosinophils by flow cytometry using the depolarized side scatter-saponin method. Cytometry 1997;29:197–203.

36 Mengellers HJ, Maikoe T, Brinkmann L, et al: Immunophenotyping of eosinophils recovered from blood and BAL of allergic asthmatics. Am J Respir Crit Care Med 1994;149:345–351.

37 Loppow D, Böttcher M, Jörres RA, Gercken G, Magnussen H: Flow cytometric detection of induced sputum eosinophils due to their depolarized sideward scatter properties. Am J Respir Crit Care Med 2000;161:A179.

38 Deutsche Gesellschaft für Pneumologie: Empfehlungen zur diagnostischen bronchoalveolären Lavage. Pneumologie 1993;47:607–619.

39 Klech H, Pohl W (eds): Technical Recommendations and Guidelines for Bronchoalveolar Lavage (BAL). Report of the European Society of Pneumology Task Group on BAL. Eur Respir J 1989;2:561–585.

40 Holz O, Kips J, Magnussen H: Update on sputum methodology. Eur Respir J 2000;16:355–359.

41 Caltag Laboratories Inc: Applications guide to intracellular flow cytometry, ed 2. Burlingame, Caltag Laboratories Inc.

42 Picker LJ, Singh MK, Zdraveski Z et al: Direct demonstration of cytokine synthesis heterogenity among human memory/effector T cells by flow cytometry. Blood 1995;86:1408–1419.

43 Welker L, Jörres R, Costabel U, et al: Predictive value of BAL cell differentials in the diagnosis of interstitial lung diseases. Eur Respir J 2004;24:1000–1006.

44 Loppow D, Böttcher M, Gercken G, et al: Flow cytometric analysis of the effect of dithiothreitol on leukocyte surface markers. Eur Respir J 2000;16:324–329.

45 Meyer KC, Soegel P: Variation of bronchoalveolar lymphocyte phenotypes with age in the physiologically normal human lung. Thorax 1999;54:697–700.

Sack U, Tárnok A, Rothe G (eds): Cellular Diagnostics. Basics, Methods and Clinical Applications of Flow Cytometry. Basel, Karger, 2009, pp 600–611

Pathophysiology and Immune Monitoring of Sepsis

Christian Meisel · Conny Höflich · Hans-Dieter Volk

Institute for Medical Immunology, Charité University Hospital, Campus Mitte, Berlin, Germany

Introduction

Sepsis remains a major challenge in medicine. Despite considerable progress in critical-care medicine, the outcome of sepsis has improved only little over the last three decades, and mortality rates remain high. For many years, based on early studies in animal models, the prevailing theory of sepsis has been that mortality is largely a consequence of an overwhelming host inflammatory response to invasion by microorganisms and/or microbial products. However, the failure of virtually all clinical trials targeting inflammatory mediators to improve sepsis outcome as well as recent insights from clinically more relevant experimental models and from patient studies prompted to reconsider this concept. It has become evident that during sepsis the host immune response changes over time, resulting in excessive inflammation and immune suppression, which both may have fatal consequences. Thus, monitoring of the patient's immune function using appropriate diagnostic markers and standardized assays may not only provide information on the clinical course of the disease but is also crucial for devising effective immunomodulatory therapies in sepsis.

Role of Proinflammatory Cytokines in Local Antimicrobial Host Defense

The physiological response to infection is characterized by the induction of an acute inflammatory reaction leading to the local containment and killing of invading pathogens but also to the initiation of processes involved in tissue repair and regeneration. Innate immune cells (e.g. mast cells and macrophages) recognize conserved microbial pathogen-associated molecular patterns (PAMP) such as endotoxin (or lipopolysaccharide; LPS), lipoteichoic acid, and bacterial nonmethylated DNA via pattern recognition receptors (PRR), including the Toll-like receptor (TLR) family.

PRRs not only facilitate the uptake pathogens, but also trigger the activation of intracellular signaling cascades leading to the release of proinflammatory cytokines such as TNF-α. TNF-α amplifies inflammatory processes by inducing further cytokines, including IL-1β, IL-6 or high-mobility group B1 and other inflammatory mediators (e.g. prostaglandins, carbon monoxide and reactive oxygen species), which lead to local vasodilatation, increased vascular permeability, and activation of the coagulation cascade. In addition, proinflammatory mediators activate the endothelium to upregulate adhesion molecules and attract further cellular components of the inflammatory system (e.g., platelets, granulocytes, natural killer cells) from the circulation. The release of TNF-α by activated mast cells is not only crucial for the recruitment and activation of neutrophils [1] but also for the accumulation of T-cells in draining lymph nodes [2]. Local inflammation also activates tissue-resident professional antigen-presenting cells (e.g. Langerhans cells of the skin) by delivering 'danger' signals. Activated antigen-presenting cells pick up soluble and particulate antigenic material and move rapidly to local lymph nodes, where they present processed antigens to antigen-specific T-cells and initiate an adaptive immune response. Proinflammatory mediators are not only essential for the induction of an effective antimicrobial host response: they also initiate counterregulatory mechanisms to limit inflammatory processes. For example, TNF-α induces the release of the anti-inflammatory cytokine IL-10 by monocytes/macrophages in an autocrine/paracrine fashion [3]. In addition, unequivocal evidence now exists that immune responses are under reflex-like control of the central nervous system (CNS). Proinflammatory cytokines can engage with their cognate receptors on neurons in the central or peripheral autonomic nervous system (e.g. local afferent fibers of the vagal nerve in the inflamed tissue) from which signals are relayed to control centers of neural-immune interaction in the hypothalamus and the brain stem [4]. Input processing in these brain sites results in the activation of the hypothalamus-pituitary-adrenal axis, the sympathetic nervous system as well as of the efferent part of the vagal nerve and, ultimately, in the release of glucocorticoids, catecholamines, acetylcholine and other neuropeptides into the circulation and peripheral tissues. These neurotransmitters and neuroendocrine mediators can inhibit inflammatory responses directly or indirectly via the release of anti-inflammatory mediators such as IL-10.

Systemic Inflammatory Response

Local infections or inflammatory reactions are frequently associated with systemic changes collectively referred to as the 'acute-phase response'. Enhanced production and systemic release of proinflammatory cytokines such as TNF-α, IL-6, and granulocyte colony-stimulating factor (G-CSF) induce the production of acute-phase proteins in the liver, activate myelopoiesis and enhance the release of myeloid leukocytes (granulocytes, monocytes) from the bone marrow, induce fever, and mobilize energy stores by promoting catabolic energy-generating processes in the liver, adipose and muscle tissues. Strong systemic responses may manifest clinically as a systemic inflammatory response syndrome (SIRS), which marks the transition from a protective host response to a pathophysiological condition.

SIRS is a frequent clinical phenomenon in intensive-care units and may have infectious as well as noninfectious etiologies (e.g. extensive tissue damage after severe burn, major trauma or surgery). Infectious SIRS or sepsis is associated with high mortality, in particular in the presence of septic shock and multiple-organ failure. If the host fails to rapidly eliminate invading pathogens and to contain the infection locally, microbes and microbial products may be released into the circulation and induce a hyperinflammatory reaction caused by a systemic activation of immune cells (macrophages, mast cells, neutrophils) and of the complement cascade. The resulting systemic vasodilatation, increased vascular permeability and activation of the clotting cascade may lead to severe hypovolemia, disseminated intravascular coagulation and multiple-organ failure.

Immunopathogenesis of Sepsis – Too Much Inflammation?

Animal studies in the 1980s demonstrated that an excessive systemic release of proinflammatory cytokines in response to bacterial infections or after administration of bacterial toxins such as endotoxin plays a central role in the pathogenesis of septic shock and organ failure. Administration of neutralizing anti-TNF-α antibodies dramatically improved survival in preclinical models of endotoxemia and Gram-negative bacteremia [5, 6]. Additionally, high doses of recombinant TNF-α or IL-1β administered to animals in the absence of infection induces circulatory collapse and organ dysfunction similar to that observed in humans with septic shock [7]. Based on this new pathophysiological understanding of sepsis, several clinical trials on anti-inflammatory agents, including anti-TNF-α antibodies, IL-1 receptor antagonist (IL-1RA) and anti-endotoxin antibodies, have been conducted in the 1990s. However, the disappointing results of these trials to improve outcome in critically ill septic patients indicated that animal models with bolus injections of bacteria or high amounts of endotoxin do not adequately reflect the complex host-pathogen interactions during sepsis.

Emerging experimental and clinical observations rather challenged the pathophysiological concept of sepsis-related morbidity and mortality being solely a result of too much inflammation. For example, while TNF-α targeted therapy was demonstrated to be successful in reducing mortality following a bolus administration of endotoxin or bacteria in many animal species, the same therapeutic approach showed no benefit or even deleterious effects in clinically more relevant sepsis models such as in the cecal ligation and puncture peritonitis model [8–10]. Moreover, mice with a genetic defect in the p55 TNF-α receptor are resistant to endotoxic shock, yet they are unable to control infection with certain living bacteria [11]. Similar observations have been made in mice that are genetically deficient in the LPS-binding protein (LBP) [12], and in mice with a loss-of-function mutation in the LPS receptor TLR4 gene [13]. Functional TLR4 mutations have also been identified in humans and may increase their susceptibility to infection [14–16]. Thus, although endotoxin may have deleterious effects, total blockade of endotoxin recognition by immune cells may be detrimental, too. Furthermore, studies have shown that many sepsis patients do not show high plasma levels of TNF-α and other proinflammatory mediators [17, 18]. Instead, many intensive-care patients show signs of temporary depression of both innate and adaptive immune responses, as will be further discussed below. Importantly, persistent immunodepression has been shown to be associated with increased mortality in sepsis [19], indicating that a well-balanced immune response is important to control bacterial infections. Too much inflammation in response to infection resulting in hyperinflammation can be as lethal as too little.

Mechanisms of Immunodepression in Sepsis Patients

In recent years it became clear that, similar to local inflammatory processes that trigger a local counterregulatory anti-inflammatory response, systemic inflammatory reactions (both infectious and noninfectious SIRS) induce a systemic compensatory anti-inflammatory response (CARS) [20]. Although one or the other may dominate, frequently both responses can be observed simultaneously, resulting in a mixed antagonistic response syndrome (MARS). Strong systemic anti-inflammatory responses may not only severely impair antimicrobial host responses to the initial infectious pathogen, but may also render the organism highly susceptible to opportunistic infections. The extreme form of this loss of immune competence has been called 'immunoparalysis', as both innate and adaptive immune responses are strongly diminished [19].

Several groups, including ours, have described functional deactivation of monocytes in sepsis patients, in particular during the late stages of disease. Altered monocyte function is reflected by a reduced capacity to produce proinflammatory cytokines (part of the innate response), diminished expression of HLA-DR and costimulatory molecules (e.g. CD80/CD86), as well as severely impaired antigen-presenting capacity (part of the adaptive response). However, monocyte function is not completely turned off as their ability to produce anti-inflammatory mediators such as IL-1RA and, at least partially, also IL-10 is largely preserved [21, 22]. This monocyte reprogramming is in part due to enhanced levels of anti-inflammatory molecules, including IL-10 and prostaglandin E2, during systemic inflammatory responses that act on circulating monocytes even before they migrate into peripheral tissues to become macrophages. In addition, as discussed above, high levels of proinflammatory mediators (in particular TNF-α and IL-1β) activate control centers of brain-immune interaction within the CNS, resulting in the activation of stress pathways and release of stress mediators (glucocorticoids, catecholamines) that inhibit monocyte and other immune cell functions. In addition, during prolonged sepsis there is a lack of immunostimulatory cytokines such as IFN-γ produced by T-cells and natural killer (NK) cells which increase the proinflammatory potential of monocytes/macrophages and enhance their antigen-presenting capacity by upregulating MHC class II and CD80/CD86 expression. In sepsis patients, severely reduced numbers of lymphocytes, in particular of T-cells, in peripheral blood are frequently observed. Experimental as well as clinical studies have demonstrated that the massive loss of lymphocytes during sepsis is due to excessive apoptosis in primary and secondary lymphoid organs [23–25]. Prevention of lymphocyte apoptosis dramatically improved outcome in animal models of sepsis [26, 27]. Besides a loss of T-cells, many studies have shown disturbed T-cell function in intensive-care patients, including reduced delayed-type hypersensitivity skin test responses in vivo, as well as diminished proliferation and cytokine production ex vivo [28–33]. Altered ex vivo T-cell function, particularly a decrease in the IFN-γ/IL-4 cytokine ratio was found to be associated with a worse clinical outcome. Interestingly, this dysbalance in type-1/type-2 cytokine production was predominantly found in CD8+ T-cells [32]. Disturbed T-cell function is probably mediated through direct effects of inhibitory mediators (e.g. stress mediators) as well as indirectly through impaired stimulation by antigen-presentation cells. The uptake of apoptotic cells by professional phagocytes (e.g., macrophages and dendritic cells) after tissue injury not only blocks the synthesis of proinflammatory mediators and enhances the release of the anti-inflammatory cytokine IL-10 but may also lead to reduced presentation of antigenic peptides by antigen-presenting cells.

Although it is widely held that sepsis pathology is characterized by an initial hyperinflammation that is followed by a counterregulatory or compensatory anti-inflammation, several groups proposed that an ineffective activation of host defenses, leading to uncontrolled and disseminated infections, is primary rather than secondary to the systemic inflammatory response [34–36]. Indeed, infectious complications are more often observed in older patients and patients with advanced tumors, most of whom show diminished immune responsiveness such as impaired T-cell proliferation and reduced endotoxin-induced monocytic cytokine production already preoperatively [31, 37].

Immune Monitoring in Sepsis Patients

Inflammatory and Infection Parameters

The classical laboratory parameters of inflammation and infection include acute-phase proteins such as C reactive protein as well as changes in total and differential white blood cell count (e.g. granulocytosis and elevated immature forms). Although highly sensitive, these parameters are unspecific markers of inflammatory processes and of limited diagnostic value to discriminate between infectious and noninfectious causes of SIRS. More specific markers of infection include LBP and procalcitonin (PCT) [38, 39]. High plasma levels of PCT strongly indicate the presence of systemic bacterial or fungal infection although PCT levels may also temporarily rise after translocation of bacterial products such as endotoxin, for example in patients with abdominal surgery or cardiogenic shock [40, 41]. Because the half-life of PCT of about 24 h is shorter than that of acute-phase proteins but longer than that of plasma cytokines, and since PCT levels promptly decrease in response to effective antimicrobial treatment, procalcitonin is now widely used for the monitoring of intensive care patients.

Elevated plasma levels of proinflammatory cytokines indicating the presence of systemic inflammatory responses can be reliably and rapidly determined using automated ELISA systems (e.g. Immulite®, Siemens Healthcare Diagnostics). New systems using flow-cytometric bead arrays (e.g. Cytometric Bead Array, BD Biosciences, Franklin Lakes, NJ, USA) facilitate the simultaneous measurement of several analytes (multiplex assays) in a small sample volume (<100 μl). However, the half-life of most cytokines in plasma is measured in minutes, and they are not exclusively produced by activated immune cells. For example, IL-6 is released by endothelial cells and fibroblasts after tissue hypoxia or injury, which complicates the interpretation of elevated plasma cytokine levels with regard to an infectious or noninfectious cause of inflammation. However, concurrent increase in plasma TNF-α and IL-6 may indicate the activation of monocytes/macrophages during acute systemic inflammatory processes in response to infections.

Markers of Immune Competence

The monitoring of a patient's immune competence is an important recent diagnostic development in intensive care medicine. Because temporary immunodepression by itself does not present with clinical symptoms, paraclinical parameters are required for its early diagnosis. Several markers, including soluble plasma mediators (e.g. the ratio of proinflammatory to anti-inflammatory cytokines), functional tests (e.g. endotoxin-induced cytokine production in whole blood, mitogen-induced T-cell cytokine production and proliferation), and cell surface marker (e.g. HLA-DR expression on monocytes), have been evaluated for monitoring temporary immunodepression in critically ill patients. In particular, monocytic HLA-DR expression has been found to be a promising diagnostic tool to assess the magnitude and persistence of immunodepression [36, 42]. HLA-DR belongs to the family of MHC class II molecules that are constitutively expressed on antigen-presenting cells, including monocytes. The level of HLA-DR expression on monocytes is influenced by the balance of immunostimulatory and inhibitory mediators in vivo. For example, it is increased by the immunostimulatory cytokine IFN-γ and downregulated by anti-inflammatory mediators such as IL-10 and the stress hormones glucocorticoids and catecholamines. Diminished monocytic HLA-DR expression has been shown to correlate with impaired cellular functions (particularly proinflammatory cytokine secretory capacity and induction of antigen-specific Th cell responses), indicating that monocytic HLA-DR expression may serve as a global marker for immunodepression/immunoparalysis [22].

About two decades ago, decreased expression of HLA-DR on monocytes was first described as a prognostic marker for the development of infections in trauma and transplant patients [43–45]. Subsequent studies confirmed the association between low monocytic HLA-DR expression and an increased risk for infectious complications also in patients after major surgery [46–50], severe burn injury [51, 52], acute CNS injury [53, 54], patients with acute pancreatitis [55] and transplant patients [56–58]. However, other studies failed to show an association between low monocytic HLA-DR levels and the occurrence of infectious complications after major surgery or clinical outcome in septic patients [59, 60]. Several reasons may explain the conflicting results regarding the predictive value of monocytic HLA-DR in critically ill patients. First, the consequences of immunodepression for the clinical outcome of these patients appear to depend on the magnitude of immunodepression, its time course and the underlying trigger. For example, at early time points following cardiopulmonary bypass surgery, all patients presented with low monocytic HLA-DR expression, indicating severe immunodepression [49]. Whereas most patients recovered within 2–3 days after surgery without infectious complications, some patients showed persistent signs of severe immunodepression, leading to the development of infection. Similarly, many sepsis patients show severely reduced monocytic HLA-DR expression as a sign of immunoparalysis, but this only predicts poor outcome if persistent for more than 3 days [61–63]. Secondly, all above-mentioned studies were single-center studies. In fact, no multicenter trials have been performed so far to ascertain the diagnostic value of monocytic HLA-DR expression because of the lack of a standardized flow-cytometric assay. Differences in the preanalytical handling of samples (blood coagulant, sample storage, time between blood draw and sample processing), the use of different HLA-DR antibodies, different flow cytometers and instrument settings, and different quantification strategies (HLA-DR levels expressed either as percentage of positive monocytes or as mean fluorescence intensity) have made it difficult to compare the results of the various studies. In collaboration with BD Biosciences our laboratory has recently developed a standardized assay for the quantification of monocytic HLA-DR expression (BD QuantiBRITE® Anti-HLA-DR/Anti-Monocyte), which is independent of the flow cytometer and instrument settings [64]. A recent multicenter comparison demonstrated excellent interassay and interlaboratory coefficients of variance of less than 10 and 25%, respectively [64]. The assay requires less than 45 min if the 'lyse-no wash' method is applied. The use of EDTA blood, sample storage on ice and sample processing within 4 h after blood draw have proved best to minimize preanalytical errors due to the upregulation of surface HLA-DR levels on monocytes ex vivo during sample storage. In comparison to the old assay (percentage of HLA-DR+ monocytes), monocytic HLA-DR expression determined with the new quantitative assay is >15,000 HLA-DR molecules/cell (>85% HLA-DR+ CD14+ cells) in healthy subjects, while HLA-DR levels of 5,000–10,000 (30–45%) and <5,000 (<30%) molecules/cell indicate severe immunodepression and immunoparalysis, respectively. In a recent study in patients after cardiopulmonary bypass surgery, a cut-off value of <5,000 HLA-DR molecules/cell at day 1 after surgery was demonstrated as highly predictive for patients who developed postoperative infectious complications [49].

A second assay for the assessment of immune competence is the ex vivo LPS-induced monocytic TNF-α secretion in whole blood cultures. The intra- and inter-assay variance of a commercially available standardized test assay (Milenia® Ex vivo Whole Blood Stimulation) was less than 5 and 20%, respectively, in healthy volunteers over a period of more than 1 year. TNF-α levels of <300 pg/ml (Immulite® TNF-α assay) after 4 h stimulation of heparinized blood (50 μl, 1:10 dilution with RPMI-1640 medium) with 500 pg/ml LPS indicate immunoparalysis (normal range 500–2,500 pg/ml). The overall assay time is about 5.5 h.

T-cell dysfunction, as indicated for example by reduced proliferative responses and a diminished IFN-γ/IL-4 ratio, is another feature of immunodepression, besides deactivation of monocytes, which is often observed in critically ill patients [28, 31, 32, 36]. A simple assay to determine the cytokine secretion pattern of T-cells is the stimulation of whole blood with the mitogen concanavalin A. In our lab, we stimulate 200 μl of heparinized blood (1:5 dilution with RPMI-1640 medium) with 100 μg/ml of concanavalin A for 24 h. Cytokine production in supernatant is determined by multiplex flow-cytometric assays (cytometric bead array; CBA, BD Biosciences). The interassay variance of this assay is less than 20%. Other applications of flow cytometry to assess T-cell function include quantification of intracellular cytokine production and proliferation of fluorescently labeled cells (e.g. using carboxyfluorescein diacetate succinimidyl ester; CFDA-SE) after polyclonal or antigen-specific stimulation. However, standardized assays for the assessment of T-cell reactivity in patients within the framework of multicenter studies are lacking so far.

Conclusion

Our understanding of the pathophysiological mechanisms of sepsis has considerably advanced in recent years. We have learned that sepsis is not only characterized by an overwhelming hyperinflammation in response to infections but also that failure of the organ immune system to eliminate invading pathogens is a major determinant of the clinical outcome in affected patients. The assessment of the functional status of the immune system is therefore as important as the monitoring of other organ systems (kidney, liver, cardiovascular, lung, coagulation) in critically ill patients. The failure of early sepsis trials on immune interventions without immune monitoring indicates that a better immunological characterization of patients will be crucial to determine which and when patients may benefit from specific immunomodulatory therapies. Results from recent pilot studies (IFN-γ, granulocyte-monocyte colony-stimulating factor; GM-CSF therapy) suggest that immunostimulatory therapies in sepsis patients with diagnosed immunoparalysis may provide a novel approach [19, 65]. The availability of standardized assays to determine the status of the immune system will help to validate the efficacy of immunomodulatory therapies in sepsis patients in multicentre trials. Flow cytometric analysis of monocytic HLA-DR expression as well as ex vivo LPS-induced TNF-α secretion in whole blood have been demonstrated as promising diagnostic tools to assess immune competence, whereas the magnitude of systemic inflammation and tissue damage may be evaluated by measuring soluble plasma mediators including TNF-α and IL-6 using automated ELISA systems. Furthermore, measurement of PCT or LBP plasma levels may be helpful for the early detection of systemic bacterial/fungal infections.

Finally, in our view, preventative therapy of high-risk patients would be a much better approach than treatment of established sepsis. To improve the cost/benefit ratio of immunomodulatory interventions, measurement of immune competence may help to identify patients at high risk for developing infections after major surgery or trauma.

References

1 Malaviya R, Ikeda T, Ross E, Abraham SN: Mast cell modulation of neutrophil influx and bacterial clearance at sites of infection through TNF-alpha. Nature 1996;381:77–80.

2 McLachlan JB, Hart JP, Pizzo SV, Shelburne CP, Staats HF, Gunn MD, Abraham SN: Mast cell-derived tumor necrosis factor induces hypertrophy of draining lymph nodes during infection. Nat Immunol 2003;4:1199–1205.

3 Platzer C, Meisel C, Vogt K, Platzer M, Volk HD: Up-regulation of monocytic IL-10 by tumor necrosis factor-alpha and cAMP elevating drugs. Int Immunol 1995;7:517–523.

4 Tracey KJ: The inflammatory reflex. Nature 2002;420:853–859.

5 Beutler B, Milsark IW, Cerami AC, Tracey KJ, Fong Y, Hesse DG, Manogue KR, Lee AT, Kuo GC, Lowry SF, Cerami A: Passive immunization against cachectin/tumor necrosis factor protects mice from lethal effect of endotoxin. Science 1985;229:869–871.

6 Tracey KJ, Fong Y, Hesse DG, Manogue KR, Lee AT, Kuo GC, Lowry SF, Cerami A: Anti-cachectin/TNF monoclonal antibodies prevent septic shock during lethal bacteraemia. Nature 1987;330:662–664.

7 Okusawa S, Gelfand JA, Ikejima T, Connolly RJ, Dinarello CA: Interleukin 1 induces a shock-like state in rabbits. Synergism with tumor necrosis factor and the effect of cyclooxygenase inhibition. J Clin Invest 1988;81:1162–1172.

8 Echtenacher B, Falk W, Mannel DN, Krammer PH: Requirement of endogenous tumor necrosis factor/cachectin for recovery from experimental peritonitis. J Immunol 1990;145:3762–3766.

9 Eskandari MK, Bolgos G, Miller C, Nguyen DT, DeForge LE, Remick DG: Anti-tumor necrosis factor antibody therapy fails to prevent lethality after cecal ligation and puncture or endotoxemia. J Immunol 1992;148:2724–2730.

10 Echtenacher B, Weigl K, Lehn N, Mannel DN: Tumor necrosis factor-dependent adhesions as a major protective mechanism early in septic peritonitis in mice. Infect Immun 2001;69:3550–3555.

11 Pfeffer K, Matsuyama T, Kundig TM, Wakeham A, Kishihara K, Shahinian A, Wiegmann K, Ohashi PS, Kronke M, Mak TW: Mice deficient for the 55 kd tumor necrosis factor receptor are resistant to endotoxic shock, yet succumb to *L. monocytogenes* infection. Cell 1993;73:457–467.

12 Jack RS, Fan X, Bernheiden M, Rune G, Ehlers M, Weber A, Kirsch G, Mentel R, Furll B, Freudenberg M, Schmitz G, Stelter F, Schutt C: Lipopolysaccharide-binding protein is required to combat a murine gram-negative bacterial infection. Nature 1997;389:742–745.

13 Hagberg L, Briles DE, Eden CS: Evidence for separate genetic defects in C3H/HeJ and C3HeB/FeJ mice that affect susceptibility to gram-negative infections. J Immunol 1985;134:4118–4122.

14 Arbour NC, Lorenz E, Schutte BC, Zabner J, Kline JN, Jones M, Frees K, Watt JL, Schwartz DA: TLR4 mutations are associated with endotoxin hyporesponsiveness in humans. Nat Genet 2000;25:187–191.

15 Agnese DM, Calvano JE, Hahm SJ, Coyle SM, Corbett SA, Calvano SE, Lowry SF: Human toll-like receptor 4 mutations but not CD14 polymorphisms are associated with an increased risk of gram-negative infections. J Infect Dis 2002;186:1522–1525.

16 Lorenz E, Mira JP, Frees KL, Schwartz DA: Relevance of mutations in the TLR4 receptor in patients with gram-negative septic shock. Arch Intern Med 2002;162:1028–1032.

17 Debets JM, Kampmeijer R, van der Linden MP, Buurman WA, van der Linden CJ: Plasma tumor necrosis factor and mortality in critically ill septic patients. Crit Care Med 1989;17:489–494.

18 Pruitt JH, Welborn MB, Edwards PD, Harward TR, Seeger JW, Martin TD, Smith C, Kenney JA, Wesdorp RI, Meijer S, Cuesta MA, Abouhanze A, Copeland EM, 3rd, Giri J, Sims JE, Moldawer LL, Oldenburg HS: Increased soluble interleukin-1 type II receptor concentrations in postoperative patients and in patients with sepsis syndrome. Blood 1996;87:3282–3288.

19 Docke WD, Randow F, Syrbe U, Krausch D, Asadullah K, Reinke P, Volk HD, Kox W: Monocyte deactivation in septic patients: restoration by IFN-gamma treatment. Nat Med 1997;3:678–681.

20 Kox WJ, Volk T, Kox SN, Volk HD: Immunomodulatory therapies in sepsis. Intensive Care Med 2000;26(suppl 1):S124–128.

21 Randow F, Syrbe U, Meisel C, Krausch D, Zuckermann H, Platzer C, Volk HD: Mechanism of endotoxin desensitization: involvement of interleukin 10 and transforming growth factor beta. J Exp Med 1995;181:1887–1892.

22 Wolk K, Docke WD, von Baehr V, Volk HD, Sabat R: Impaired antigen presentation by human monocytes during endotoxin tolerance. Blood 2000;96:218–223.

23 Hotchkiss RS, Swanson PE, Cobb JP, Jacobson A, Buchman TG, Karl IE: Apoptosis in lymphoid and parenchymal cells during sepsis: findings in normal and T- and B-cell-deficient mice. Crit Care Med 1997;25:1298–1307.

24 Hotchkiss RS, Swanson PE, Freeman BD, Tinsley KW, Cobb JP, Matuschak GM, Buchman TG, Karl IE: Apoptotic cell death in patients with sepsis, shock, and multiple organ dysfunction. Crit Care Med 1999;27:1230–1251.

25 Ayala A: Lymphoid apoptosis during sepsis: now that we've found it, what do we do with it? Crit Care Med 1997;25:1261–1262.

26 Hotchkiss RS, Tinsley KW, Swanson PE, Chang KC, Cobb JP, Buchman TG, Korsmeyer SJ, Karl IE: Prevention of lymphocyte cell death in sepsis improves survival in mice. Proc Natl Acad Sci U S A 1999;96:14541–14546.

27 Prass K, Meisel C, Hoflich C, Braun J, Halle E, Wolf T, Ruscher K, Victorov IV, Priller J, Dirnagl U, Volk HD, Meisel A: Stroke-induced immunodeficiency promotes spontaneous bacterial infections and is mediated by sympathetic activation reversal by poststroke T helper cell type 1-like immunostimulation. J Exp Med 2003;198:725–736.

28 Meakins JL, Pietsch JB, Bubenick O, Kelly R, Rode H, Gordon J, MacLean LD: Delayed hypersensitivity: indicator of acquired failure of host defenses in sepsis and trauma. Ann Surg 1977;186:241–250.

29 Teodorczyk-Injeyan JA, Sparkes BG, Mills GB, Peters WJ, Falk RE: Impairment of T cell activation in burn patients: a possible mechanism of thermal injury-induced immunosuppression. Clin Exp Immunol 1986;65:570–581.

30 O'Sullivan ST, Lederer JA, Horgan AF, Chin DH, Mannick JA, Rodrick ML: Major injury leads to predominance of the T helper-2 lymphocyte phenotype and diminished interleukin-12 production associated with decreased resistance to infection. Ann Surg 1995;222:482–490; discussion 490–482.

31 Heidecke CD, Hensler T, Weighardt H, Zantl N, Wagner H, Siewert JR, Holzmann B: Selective defects of T lymphocyte function in patients with lethal intraabdominal infection. Am J Surg 1999;178:288–292.

32 Zedler S, Bone RC, Baue AE, von Donnersmarck GH, Faist E: T-cell reactivity and its predictive role in immunosuppression after burns. Crit Care Med 1999;27:66–72.

33 Spolarics Z, Siddiqi M, Siegel JH, Garcia ZC, Stein DS, Denny T, Deitch EA: Depressed interleukin-12-producing activity by monocytes correlates with adverse clinical course and a shift toward Th2-type lymphocyte pattern in severely injured male trauma patients. Crit Care Med 2003;31:1722–1729.

34 Munford RS, Pugin J: Normal responses to injury prevent systemic inflammation and can be immunosuppressive. Am J Respir Crit Care Med 2001;163:316–321.
35 Netea MG, van der Meer JW, van Deuren M, Kullberg BJ: Proinflammatory cytokines and sepsis syndrome: not enough, or too much of a good thing? Trends Immunol 2003;24:254–258.
36 Volk HD, Reinke P, Docke WD: Clinical aspects: from systemic inflammation to 'immunoparalysis'. Chem Immunol 2000;74:162–177.
37 Weighardt H, Heidecke CD, Westerholt A, Emmanuilidis K, Maier S, Veit M, Gerauer K, Matevossian E, Ulm K, Siewert JR, Holzmann B, Bartels H, Hensler T, Hecker H, Heeg K, Zantl N, Wagner H, Barthlen W: Impaired monocyte IL-12 production before surgery as a predictive factor for the lethal outcome of postoperative sepsis. Ann Surg 2002;235:560–567.
38 Schumann RR, Zweigner J: A novel acute-phase marker: lipopolysaccharide binding protein (LBP). Clin Chem Lab Med 1999;37:271–274.
39 Steinbach G, Grunert A: Procalcitonin – a new indicator for bacterial infections. Exp Clin Endocrinol Diabetes 1998;106:164–167.
40 Assicot M, Gendrel D, Carsin H, Raymond J, Guilbaud J, Bohuon C: High serum procalcitonin concentrations in patients with sepsis and infection. Lancet 1993;341:515–518.
41 Brunkhorst FM, Clark AL, Forycki ZF, Anker SD: Pyrexia, procalcitonin, immune activation and survival in cardiogenic shock: the potential importance of bacterial translocation. Int J Cardiol 1999;72:3–10.
42 Monneret G, Finck ME, Venet F, Debard AL, Bohe J, Bienvenu J, Lepape A: The anti-inflammatory response dominates after septic shock: association of low monocyte HLA-DR expression and high interleukin-10 concentration. Immunol Lett 2004;95:193–198.
43 Livingston DH, Appel SH, Wellhausen SR, Sonnenfeld G, Polk HC Jr: Depressed interferon gamma production and monocyte HLA-DR expression after severe injury. Arch Surg 1988;123:1309–1312.
44 Baehr RV, Volk HD, Reinke P, Falck P, Wolff H: An immune monitoring program for controlling immunosuppressive therapy. Transplant Proc 1989;21:1189–1191.
45 Hershman MJ, Cheadle WG, Wellhausen SR, Davidson PF, Polk HC, Jr.: Monocyte HLA-DR antigen expression characterizes clinical outcome in the trauma patient. Br J Surg 1990;77:204–207.
46 Wakefield CH, Carey PD, Foulds S, Monson JR, Guillou PJ: Changes in major histocompatibility complex class II expression in monocytes and T cells of patients developing infection after surgery. Br J Surg 1993;80:205–209.
47 Schinkel C, Sendtner R, Zimmer S, Faist E: Functional analysis of monocyte subsets in surgical sepsis. J Trauma 1998;44:743–748; discussion 748–749.
48 Allen ML, Peters MJ, Goldman A, Elliott M, James I, Callard R, Klein NJ: Early postoperative monocyte deactivation predicts systemic inflammation and prolonged stay in pediatric cardiac intensive care. Crit Care Med 2002;30:1140–1145.
49 Strohmeyer JC, Blume C, Meisel C, Doecke WD, Hummel M, Hoeflich C, Thiele K, Unbehaun A, Hetzer R, Volk HD: Standardized immune monitoring for the prediction of infections after cardiopulmonary bypass surgery in risk patients. Cytometry 2003;53B:54–62.
50 Asadullah K, Woiciechowsky C, Docke WD, Liebenthal C, Wauer H, Kox W, Volk HD, Vogel S, Von Baehr R: Immunodepression following neurosurgical procedures. Crit Care Med 1995;23:1976–1983.
51 Venet F, Tissot S, Debard AL, Faudot C, Crampe C, Pachot A, Ayala A, Monneret G: Decreased monocyte human leukocyte antigen-DR expression after severe burn injury: Correlation with severity and secondary septic shock. Crit Care Med 2007;35:1910–1917.
52 Sachse C, Prigge M, Cramer G, Pallua N, Henkel E: Association between reduced human leukocyte antigen (HLA)-DR expression on blood monocytes and increased plasma level of interleukin-10 in patients with severe burns. Clin Chem Lab Med 1999;37:193–198.
53 Haeusler KG, Schmidt WU, Fohring F, Meisel C, Helms T, Jungehulsing GJ, Nolte CH, Schmolke K, Wegner B, Meisel A, Dirnagl U, Villringer A, Volk HD: Cellular immunodepression preceding infectious complications after acute ischemic stroke in humans. Cerebrovasc Dis 2007;25:50–58.
54 Woiciechowsky C, Asadullah K, Nestler D, Eberhardt B, Platzer C, Schoning B, Glockner F, Lanksch WR, Volk HD, Docke WD: Sympathetic activation triggers systemic interleukin-10 release in immunodepression induced by brain injury. Nat Med 1998;4:808–813.
55 Satoh A, Miura T, Satoh K, Masamune A, Yamagiwa T, Sakai Y, Shibuya K, Takeda K, Kaku M, Shimosegawa T: Human leukocyte antigen-DR expression on peripheral monocytes as a predictive marker of sepsis during acute pancreatitis. Pancreas 2002;25:245–250.
56 Reinke P, Volk HD: Diagnostic and predictive value of an immune monitoring program for complications after kidney transplantation. Urol Int 1992;49:69–75.

57 van den Berk JM, Oldenburger RH, van den Berg AP, Klompmaker IJ, Mesander G, van Son WJ, van der Bij W, Sloof MJ, The TH: Low HLA-DR expression on monocytes as a prognostic marker for bacterial sepsis after liver transplantation. Transplantation 1997;63:1846–1848.
58 Haveman JW, van den Berg AP, van den Berk JM, Mesander G, Slooff MJ, de Leij LH, The TH: Low HLA-DR expression on peripheral blood monocytes predicts bacterial sepsis after liver transplantation: relation with prednisolone intake. Transpl Infect Dis 1999;1:146–152.
59 Oczenski W, Krenn H, Jilch R, Watzka H, Waldenberger F, Koller U, Schwarz S, Fitzgerald RD: HLA-DR as a marker for increased risk for systemic inflammation and septic complications after cardiac surgery. Intensive Care Med 2003;29:1253–1257.
60 Perry SE, Mostafa SM, Wenstone R, Shenkin A, McLaughlin PJ: Is low monocyte HLA-DR expression helpful to predict outcome in severe sepsis? Intensive Care Med 2003;29:1245–1252.
61 Döcke WD, Syrbe U, Meinecke A, Platzer C, Makki A, Asadullah K, Klug C, Zuckermann H, Reinke P, Brunner H, Baehr RV, Volk HD: Improvement of monocyte function – a new therapeutic approach?; in Reinhart K, Eyrich K, Sprung C (eds): Sepsis-current perspectives in pathophysiology and therapy. Berlin, Springer, 1994, pp 473–500.
62 Monneret G, Elmenkouri N, Bohe J, Debard AL, Gutowski MC, Bienvenu J, Lepape A: Analytical requirements for measuring monocytic human lymphocyte antigen DR by flow cytometry: application to the monitoring of patients with septic shock. Clin Chem 2002;48:1589–1592.
63 Monneret G, Lepape A, Voirin N, Bohe J, Venet F, Debard AL, Thizy H, Bienvenu J, Gueyffier F, Vanhems P: Persisting low monocyte human leukocyte antigen-DR expression predicts mortality in septic shock. Intensive Care Med 2006;32:1175–1183.
64 Docke WD, Hoflich C, Davis KA, Rottgers K, Meisel C, Kiefer P, Weber SU, Hedwig-Geissing M, Kreuzfelder E, Tschentscher P, Nebe T, Engel A, Monneret G, Spittler A, Schmolke K, Reinke P, Volk HD, Kunz D: Monitoring temporary immunodepression by flow cytometric measurement of monocytic HLA-DR expression: a multicenter standardized study. Clin Chem 2005;51:2341–2347.
65 Nierhaus A, Montag B, Timmler N, Frings DP, Gutensohn K, Jung R, Schneider CG, Pothmann W, Brassel AK, Schulte Am Esch J: Reversal of immunoparalysis by recombinant human granulocyte-macrophage colony-stimulating factor in patients with severe sepsis. Intensive Care Med 2003;29:646–651.

Sack U, Tárnok A, Rothe G (eds): Cellular Diagnostics. Basics, Methods and Clinical Applications of Flow Cytometry. Basel, Karger, 2009, pp 612–641

Flow-Cytometric Immunophenotyping of Acute Leukemias

Richard Ratei[a] · Thomas Nebe[b] · Richard Schabath[a] · Hans-Dieter Kleine[c] · Leonid Karawajew[a] · Wolf-Dieter Ludwig[a]

[a] HELIOS Klinikum Berlin-Buch, Charité University Hospital, Campus Buch, Berlin,
[b] University of Heidelberg, Medical Faculty of Mannheim,
[c] Ansomed AG, Rostock, Germany

Introduction

For many years, the diagnosis of acute leukemia has exclusively relied on the morphological enumeration of blast cells in cytological smears of bone marrow. Acute leukemia is diagnosed if at least 20% of the total nucleated cells in the bone marrow are blasts, or if the bone marrow shows erythroid predominance (erythroblasts comprising ≥50% of nucleated cells), at least 20% of nonerythroid cells are blasts (lymphocytes, plasma cells and macrophages also being excluded from the differential count of nonerythroid cells), or if the characteristic morphological features of hypergranular promyelocytic leukemia are present [1–3]. As such, morphology remains the mainstay for the diagnosis of acute leukemia and the differentiation of nonhematological neoplasias like histiocytosis, neuroblastoma, rhabdomyosarcoma, carcinoma or melanoma, which can be overlooked with flow-cytometric immunophenotyping (FCI) only. But morphology is insufficient to assess the biological heterogeneity of acute leukemias, which is of critical importance for treatment decisions; prognosis and risk assignment can only be realized with modern techniques like FCI, cytogenetics and molecular studies [4–8]. The evolution of leukemia nomenclatures started in 1976 with the FAB classification based on morphology and cytochemistry. It was further influenced by rapidly advancing new techniques which were able to incorporate immunological, cytogenetic and molecular characteristics of the leukemic cell clone and eventually led to the MIC and EGIL classifications and to the WHO classification of acute leukemias [2, 3, 5, 9]. Within these orchestra of new methodologies, FCI has evolved as a diagnostic platform with a central role for the diagnosis, classification and monitoring of therapy response of acute leukemias [10, 11].

FCI has been available for almost three decades. In clinical medicine it was initially utilized to assess lymphocyte subpopulations in patients with acquired immunodeficiency in the course of the AIDS epidemic. Recent improvements in modern technologies such as fluidics, lasers, optics, analog and digital electronics, computers and software, paralleled by the introduction of new fluorochromes and antibodies, have contributed immensely to the development of flow cytometry from a one-color, three-parameter technique with only single-antibody staining to a multicolor approach with simultaneous cellular staining with 6–9 monoclonal antibodies (mAbs) [12–14]. The intriguing possibility of detecting phenotypic features at a single-cell level in large cell populations forms the rational basis for flow cytometry to be used in the evaluation of minimal residual disease (MRD) [15]. Despite the ideal conditions of this versatile and robust technique for the diagnosis, classification and monitoring of acute leukemias, its widespread and broad application has suffered from a lack of standardization and the dependence of data interpretation on individual experience.

This need for standardization has been taken note of in several consensus statements generated by a group of US, Canadian and European hematopathologists, hematologists and laboratory scientists at first in 1995 and recently with an update in 2006 [16–26]. These recommendations addressed a number of topics including the standardization and validation of laboratory procedures, the selection of antibody combinations, data analysis, interpretation and data reporting.

In this chapter we will focus on the laboratory procedure for the diagnosis and monitoring of MRD of acute leukemias by FCI with regard to the pre-, intra- and postanalytical stages of the diagnostic procedure with special consideration of the clinical implications.

Preanalysis

Medical Indications

Clinical suspicion, morphological evidence or the exclusion of an acute leukemic cell population in any body fluid such as blood, bone marrow, pleural or pericardial effusion, cerebrospinal fluid or ascites justifies the validation of the diagnostic process with FCI beyond any cost-effectiveness consideration. Lineage determination of morphologically undifferentiated blast cells is the prime task of FCI followed by subclassification according to the immunological profile. In certain cases of acute leukemia, characteristic leukemia-associated immunophenotypes (LAIPs) have been defined and correlated with morphological and cytogenetic features for risk estimation. Furthermore, the immunological counting of leukemic cells in a sample is much more sensitive than the morphological estimation of residual blast cells during the course of the disease [27]. In acute myeloid leukemia (AML), FCI is indispensable for the diagnosis of morphologically undifferentiated subtypes like FAB-M0 or subtypes with megakaryocytic differentiation, FAB-M7, and sometimes it is helpful to distinguish between the variant form of acute promyelocytic leukemia (FAB-M3v) and acute monocytic leukemias (FAB-M5) [28–30].

In acute lymphoblastic leukemia (ALL), FCI is absolutely necessary for the immunological subclassification of precursor B-cell ALL and precursor T-cell ALL. Prognostically important subtypes in the B-lineage like mature B-cell ALL and pro-B-ALL, as well as in the T-lineage, like pro-T-ALL and intermediate (thymic) T-ALL, can only be defined by their immunophenotype.

In the era of targeted therapies, these molecules of interest are to be detected in the blast cells first, which is easily accomplished with FCI in the majority of cases. These targets mainly include surface membrane molecules like CD20 (rituximab), CD52 (Campath®) or CD33 (Mylotarg®), but can also include cytoplasmic tyrosine kinases such as FLT3 (HerbimycinA) [31–35].

Choice of mAbs

There have been numerous consensus recommendations and guidelines for reagents to be used in the diagnosis of hematopoietic neoplasia and especially acute leukemia, the latest being those from the 2006 Bethesda Consensus Conference [19]. Participants of the conference conducted a survey to facilitate the recognition of common reagents that laboratories would be likely to use in the initial evaluation of specimens for a number of different medical indications including 'blasts in bone marrow' and for secondary assignment and subclassification (table 1). Another approach is given by the consensus panel of the German competence network for acute and chronic leukemias, which facilitates both lineage determination according to the EGIL score and recognition of prognostically relevant subtypes (e.g., CD1a for cortical-T-ALL and mAb 7.1 for leukemias with MLL rearrangements) (table 2).

Choice of Antibody Panel

The requirements for the construction of a panel for acute leukemias are predetermined by several conditions. Most important, the clinical necessities for lineage determination, subclassification and quantification of the leukemic cell population have to be satisfied because each of these issues can guide clinical decisions for therapy stratification, prognosis and risk assessment. The composition of the panel has to respect the laboratory facilities and the flow-cytometric equipment, i.e., number of lasers and photomultipliers, filter combinations and lastly the availability of mAbs and their respective fluorochrome conjugates according to the expected expression intensity of the antigen. Highly expressed antigens should be detected with mAbs conjugated with dim fluorochromes and vice versa dimly expressed antigens should

Table 1. Primary and secondary[a] reagents recommended by the 2006 Bethesda Consensus Conference survey for 'blasts in bone marrow'

Lineage	Primary reagents
B-cells	CD10, CD5, CD19, CD20, CD45, Kappa, Lambda
T-cells and NK cells	CD2, CD3, CD4, CD5, CD7, CD8, CD45, CD56
Myelomonocytic cells	CD7, CD11b, CD13, CD14, CD15, CD16, CD33, CD34, CD38, CD45, CD56, CD117, HLA-DR
	Secondary reagents
B-cells	CD9, CD11c, CD15, CD22, cyCD22, CD23, CD25, CD13, CD33, CD34, CD38, CD43, CD58, cCD79a, CD79b, CD103, FMC7, Bcl-2, cykappa, cylambda, TdT, ZAP-70, cyIgM
T-cells and NK cells	CD1a, cyCD3, CD10, CD16, CD25, CD26, CD30, CD34, CD45RA, CD45RO, CD57, αβ-TCR, γδ-TCR, cyTIA-1, Tâ chain isoforms, TdT
Myelomonocytic cells	CD2, CD4, CD25, CD36, CD38, CD41, CD61, cCD61, CD64, CD71, MPO, CD123, CD163, CD235a

[a] Secondary reagents were not included in the survey. They are listed with the anticipation that only a limited number would be used for any particular specimen and indication.

Table 2. German competence network consensus panel recommended for the diagnosis of acute leukemia

Lineage	Primary reagents
B-cells	cyCD79a, CD19, κ λ
T-cells and NK cells	cy/sCD3, CD2, CD7, CD1a, CD56
Myelomonocytic cells	MPO, CD13, CD33, CD65, CD61, CD14
Progenitor/miscellaneous	HLA-DR, CD45, mAb 7.1, TdT, CD34, CD117
	Secondary reagents
B-cells	cy/sCD22, CD20, CD5, cyIgM, sIg
T-cells and NK cells	TCR-αβ, TCR-γδ, CD4, CD8, CD5
Myelomonocytic cells	cyCD61, CD64, CD235a, CD41, CD15, LF

be stained with bright fluorochromes. In order to track populations between tubes, a 'common denominator' like scatter characteristics, immunological characteristics, i.e., mostly lineage-specific antigens, or both features as in CD45 versus side scatter (SSC) have to be considered.

The logistic and personnel structure of the laboratory may also guide the choice of the panel. In a hospital-based hematopathology laboratory with morphological and clinical information from experienced personnel at hand a sophisticated prediction can guide the choice of a restricted and targeted panel to confirm the preconceived diagnosis. On the contrary, laboratories devoid of clinical information and with morphologically unexperienced personnel will have to apply a much broader panel to come to a correct diagnosis.

Gating Strategy

According to the aforementioned issues on the choice of the panel, different gating strategies will come into use. In childhood ALL, a clinically and morphologically guided panel decision allows for an immunological gating strategy using either CD19 or cCD3 as a common reagent in the majority of specimens. In most cases, the blast populations in acute leukemia are discernible according to light scatter characteristics, which allows an easy tracking between tubes. This strategy permits the assignment of specific populations in the scatter gate according to the composition of the panel, but does not allow the exact allocation of all antigens in the panel on the blast cell population due to the lack of a lineage-restricted common reagent. The combination of an immunological parameter like CD45 with side scatter (SSC) is a powerful tool for the identification of basic hematopoietic populations, e.g. lymphocytes, monocytes, neutrophils, myeloid or lymphoid blast cells, and is used in many laboratories for tracking populations between tubes [36].

Sampling and Shipment of the Specimen

Taking a bone marrow sample has to be carefully planned because many specifications are to be served simultaneously and sample preparation and storage requirements of the individual laboratory can be different since assay procedures may vary. In case of a 'dry tab' due to concomitant bone marrow fibrosis, as in some patients with acute megakaryocytic leukemia, or a 'packed marrow', FCI can only be performed from peripheral blood. In either case, the specimen has to be of a large enough volume in order to supply sufficient cells for cytomorphology, cytochemistry, immunohistochemistry, FCI, cytogenetics, molecular genetics and cryopreservation. In order to perform a meaningful four-color FCI, a total of at least 2×10^6 mononuclear cells with an input of 10^5 to 2×10^5 cells into each of 5–10 tubes is required. Most often, not all laboratory facilities are at the bedside or even in the same hospital, so vials and containers for shipment have to be orderly prepared. The sample should be taken in the morning hours and express shipment should guarantee a delivery in the laboratory on the same day or within the next 24 h. For FCI, 0.5–2 ml bone marrow aspirate or 5–10 ml peripheral blood may be collected in a 10-ml syringe containing either EDTA or heparin. Acid citrate dextrose is not recommended for bone marrow anticoagulation since relatively large volumes of ACD are sometimes needed which can lead to pH shifts with reduced cell viability. Sterile and properly anticoagulated samples are stable at room temperature for 24 h prior to staining, but according to our own experience samples can even give sufficient results after 48–72 h of transportation if there are no extreme temperature shifts below 4 °C and above 30 °C, but these long transportation times are not recommended [37]. For safety reasons, the container has to be prepared in accordance with the national or international regulations for the transportation of biological materials [38]. As the correlation of morphological findings is often helpful in the interpretation of cytometry data, samples for the flow cytometry laboratory should always be accompanied by an unstained smear of the same material as the specimen.

Sample Preparation and Viability [25, 39]

Once the specimen has reached the laboratory and has been checked for adequate identity, a thorough and gentle mixing without shaking is necessary to ensure homogeneous aliquoting. If there is no unstained smear with the sample, it has to be prepared on site and viewed before any further processing of the sample. Viability testing of the specimen can either be done with trypan blue staining using conventional light microscopy or thiazole orange/propidium iodide staining using flow cytometry. Propidium iodide or 7-aminoactinomycin-D will give a red stain only to the DNA in the nucleus of dead cells with a permeable membrane whereas thiazole orange will give a green stain to all RNA-containing leukocytes. The amount of red (dead) cells within all green leukocytes in 1 ml of diluted blood gives an estimate of the viability of the sample. Instead of propidium iodide, 7-aminoactinomycin-D can be used which is also compatible with simultaneous staining with FITC, phycoerythrin (PE) and allophycocyanin. Furthermore, light scatter characteristics can give a clue on the condition of the probe because debris and nonviable cells will form a cluster with low forward scatter (FSC) and SSC compared to vital cells.

Preparation of Cells for Staining

Preparation and isolation of cells can either be accomplished with whole blood lysis or Ficoll gradient centrifugation. Historically, gradient centrifugation has been used first for this purpose. It offers the advantage of enriching the mononuclear blast population in most instances and at the same time losing debris and nonviable cells. Disadvantages of the procedure are the loss of cells, especially of granulocytic and erythroid origin, which in general does not hamper the diagnosis and classification of acute leukemia, but which can be a drawback for the analysis and interpretation of samples from patients with myelodysplastic syndrome. In situations with suspected erythroid leukemia or mature B-ALL, gradient centrifugation should be used with caution because the sometimes highly vacuolated blasts of mature B-ALL or the large proerythroblasts of erythroid leukemia within these specimens can be diminished or lost by the procedure. Due to multiple centrifugations and washings with a consecutive loss of cells, gradient centrifugation is not the appropriate method for specimens with low cell numbers. After centrifugation, the monocyte layer is separated and washed. Most often, some erythroid cells still remain in the sample and have to be lysed in a further ammonium-chloride-based lysis step.

Most laboratories involved in the immunophenotyping of hematolymphoid neoplasms use the whole blood lysis method for the preparation of cells. Besides the low demand of sample volume, it offers the advantage of being much faster than Ficoll gradient centrifugation, but incomplete lysis can lead to interferences, especially with remaining erythroid progenitor cells in bone marrow specimens. There are many different lysis reagents with varying ingredients on the market, and results can vary substantially between manufacturers. Not commercially prepared reagents for hypotonic lysis and non-pharmacy grade ammonium-chloride-based lysis often have the disadvantage of decreased stability and worsening of light scatter characteristics. It is beyond the scope of this chapter to assess and compare all the commercially available products.

Staining

Staining is a thermodynamic process with antibody and antigen as the critical components in the equation. Antibodies have to be used in saturating concentration in order to be sure that the fluorescence intensity of a stained cell is directly and linearly proportional to the protein or antigen that is measured. Theoretically, this would require the titration of each of the antibodies used in the panel with the same cells of the specimen to be tested because the antigen density of a given leukemic sample is not known before. This of course is impracticable. Commercially available mAbs are usually tested for linearity by the manufacturer using cell lines strongly expressing the respective antigen. Therefore, recommendations by the manufacturer usually include a high enough concentration of antibody so that an effective equilibrium between bound and unbound antibody will be achieved in a reasonable staining time, usually 15–20 min at room temperature protected from light. In case of uncertainties concerning staining conditions for a given mAb, they have to be reassured by antibody titration and if necessary readjusted. For the reduction or blocking of unspecific bindings, it is often useful to saturate the sample with serum from the same species used for the detection antibody.

With either of the above-mentioned methods of cell preparation, cell counts are adjusted, so that 10^5–10^6 cells in a volume of 50–100 μl are aliquoted into each tube of the panel for staining with the mAbs according to the manufacturer's instructions.

Intracellular stainings are of great importance for lineage determination of hematolymphoid neoplasias, especially acute leukemias [40, 41]. There are a lot of different methods and formulas commercially available and many of them have been tested in comparison. According to two independent publications and unpublished experience from other laboratories, the method developed by W. Knapp in Vienna yields superior results for a broad spectrum of intracellular antigens [42, 43]. Before addition of the detecting mAbs, it is very important to block unspecific binding because this is much stronger for cytoplasmic than for surface staining. Combined staining for surface and intracellular antigens is possible and useful for the detection of surface progenitor markers and cytoplasmic lineage antigens like myeloperoxidase in AML.

For the detection of immunoglobulins and light chains on the cell surface or in the cytoplasm, the cells have to be washed 2–3 times in order to get rid of soluble immunoglobulins in the sample. Additional blocking with serum from the same species as the mAb used for the detection of the target antigen is necessary to avoid unspecific binding.

State of the art in many laboratories involved in immunophenotyping of acute leukemias is a direct multicolor staining with fluorochrome-conjugated mAbs, usually with a three- to six-color approach depending on the available flow cytometer. Only in rare situations where there is no conjugated mAb available is indirect staining necessary. Here, the unconjugated-antigen-specific primary mAb requires a staining procedure with a second mAb conjugated to a fluorochrome and directed against the Fc domain of the primary antibody.

The one- or two-color stainings commonly applied in the past used isotype control antibodies in the same fluorochrome conjugation as the detecting antibody for negative control in an extra control tube. With modern multicolor staining panels, many different isotypes and fluorochromes are applied, and negative controls with isotype controls are not useful anymore. In multicolor tubes, the antibody reaction can almost always be controlled on cells in the same tube that either do not stain or on residual normal cells that stain with the applied panel, thus offering the possibility of a negative and positive control within the test sample and with the same staining.

Analysis

Instrument Settings

The instrument setting for the immunophenotyping of acute leukemias is the same as for normal lymphocytes or of lymphoproliferative diseases [44]. Amplification of the linear light scatter parameters should ensure detection of all cellular components in the bone marrow or peripheral blood, including small normoblasts, myeloid progenitors and eosinophils with strong SSC. A logarithmic amplification of SSC is possible and can enlarge the area for blast cells, thus enabling a more accurate gating on these populations. Fluorescence intensities are displayed on a logarithmic scale; photomultipliers (PMTs) have to be adjusted such that unstained lymphocytes are positioned in the first log decade – notable, all compensation settings have to be at zero while adjusting PMTs. For the subsequent compensation procedure, PMT settings are not to be changed.

Due to spectral overlapping of fluorochrome emissions, it is necessary to subtract the spillover of a given fluorochrome into other detection channels. The spectral emissions and spillover in other channels are best detected with beads coated with mAbs that react with the Fc portion of the fluorochrome-conjugated specific antibodies applied in the staining. Compared to cells, these beads have the advantage of a homogeneous staining independent of antigen density variation on lymphocytes or blast cells. In contrast to the two-color stainings of the past, spectral overlapping cannot be compensated manually when using multicolor stainings; the complex compensation matrix for these stainings has to be calculated by the software controlling modern flow cytometers [45]. Some of these software packages will even respect the resulting negative values of the compensation in their depiction of the plot points.

Acquisition

At the time of diagnosis of acute leukemia, bone marrow or peripheral blood specimens usually contain a high number of blast cells, so the acquisition of 10,000 cells per tube will suffice for an FCI diagnosis. In situations with low blast counts in the specimen either due to dilution of the bone marrow aspirate with peripheral blood or a case of aleukemic leukemia with no blasts on the peripheral smear, it is advisable to acquire at least 50,000–100,000 events. During acquisition, the light scatter features and the preliminary aspects of the stainings of the specimen will often enable early adjustment of the panel in order to complete the applied panel.

Quality Control

The quality control for an FCI analysis has to comprise the following internal plausibility controls [46]:

- comparison of the flow-cytometric results with morphology either in the peripheral blood or bone marrow smears produced in the own or an external laboratory;
- comparison of the leukemic cell population with remaining normal cells, i.e. lymphocytes, granulocytes, in the probe;
- comparison of the expected reaction patterns in the staining panel with certain cell populations in the sample, which includes no myeloperoxidase staining in lymphocytes and no lactoferrin in monocytes, and
- comparison of blood and bone marrow specimens of the same patient.

Instrument stability is mandatory to ensure robust results over a period of years, especially in clinical trials. There are several approaches and recommendations to monitor inter- and intralaboratory longitudinal instrument performance [47, 48]. Most important is to check for optical alignment, fluorescence resolution and intensity. This can be accomplished with fluorescence beads of different types [49]. Once the window of analysis is defined by the PMT settings and the compensation matrix, stability of the optical alignment can be checked with the daily measurement of type IIa fluorescent beads giving a single emission peak in each channel. The linearity and longitudinal stability of the PMTs is best controlled with repeated measurements of type III fluorescent beads giving several peaks in each channel distributed over the entire log scale, which also offers the possibility to calculate MESF [47, 50]. The graphical presentation of these measurements in Levey-Jennings plots easily demonstrates variations in the performance of the instrument. Additionally, the appearance of normal residual cell populations in the leukemic sample can be used for internal control purposes [51]. For interlaboratory quality control, participation in external quality assessment schemes is needed [52]. These can be designed to control the whole analytical procedure with sending out a real patient sample or a prepared cell suspension, which of course requires the stabilization of the cells. The control of the postanalytical capabilities of the FCI practitioner to interpret flow cytometry data is best accomplished with sending out listmode data in the scheme.

Postanalytical Methodology

Data Analysis

Many efforts to standardize the interpretation of flow cytometry data in clinical flow cytometry undertaken in the past have been hampered by the complexity and variety of pre- and postanalytical methodology. One of the cornerstones of a standardized postanalytical concept is the CD45/SSC gating strategy [36]. Nonetheless, there still is an enormous diversity of different approaches, which become even more difficult to interpret with the introduction of three- or four-color or even multiparameter flow cytometry. Basically, there are three different approaches for flow cytometry data analysis: i) simplistic two-dimensional manual gating with Boolean combinations on flow-cytometric data and sample classification based on individual knowledge and experience; ii) sample classification using a computer-based automated multivariate learning algorithm generated by a two-dimensional gating decision procedure, and iii) supervised classification with support vector machines and automated in silico gating of multivariate flow cytometry data [53–55]. Common practice in most laboratories involved in leukemia immunophenotyping is a procedure with manual gating of the blast population either by light scatter or immunological features or in the case of CD45 vs. SSC by using both parameters [23, 36]. The expression of antigens on the blast population should be reported both quantitatively as percentage of positive cells compared to a negative control and with a qualitative remark considering the fluorescence intensity compared to a normal reference population or a negative control.

Data Interpretation and Diagnosis

Acute Myeloid Leukemia

For the diagnosis of undifferentiated or megakaryocytic AML, FCI is indispensable, especially if there is coexpression of lymphoid markers. FCI is also quite helpful for the differentiation of myelomonocytic and minimally myeloid-differentiated cases of AML. Subtypes with granulocytic differentiation, e.g. with AML-M2 or AML-M3 morphology, can be confirmed with FCI. The diagnostic significance and sensitivity of antibody panels that comprise lineage-specific antibodies as well as progenitor markers has been shown for AML in adults and children.

None of the myeloid-lineage-specific mAbs like CD13, CD33, CD65 or MPO alone can detect all leukemic blasts in all patients, but with the combination of these markers nearly all patients with AML can be identified. In contrast to the mAb which detects both the proenzymatic and enzymatic forms of MPO, cytochemical stainings will only react with the active enzyme. Numerous efforts to correlate immunophenotypic features with the FAB classification have failed in the past. Although there are some AML subtypes which display characteristic expression patterns like AML-M3, only a few of these yield strong and consistent correlations. Therefore, cases with identical antigen expression patterns can belong to different FAB subtype and vice versa.

The interpretation of FCI data in AML can be difficult due to the mixture of normal hematopoietic with leukemic cells and the mostly heterogeneous expression patterns of the blast population. CD45 versus SSC gating and the introduction of multicolor stainings have been very helpful to detect the leukemia-associated immunophenotype, which enhances the discrimination of leukemic cells from the remaining normal cell populations. Asynchronous or aberrant antigen expression patterns of leukemic blast populations have also been useful for the monitoring of therapy response and MRD detection. In the following, the immunophenotypic features of the different FAB subtypes of AML will be described in correlation to their cytogenetic and molecular characteristics.

Undifferentiated or Minimally Differentiated AML (FAB-M0)

The AML-M0 subtype is found in 3–6% of childhood AML and in about 10% of adult AML. Diagnosis is not possible based on morphology and cytochemistry alone, but compulsorily needs FCI [30]. The criteria listed by the FAB group in 1991 were revised in 1999 and 2001 and include less than 3% positivity of myeloperoxidase/sudan black, lack of lymphatic differentiation with negativity for CD3, CD22, CD79a and TCR-α/β and lack of megakaroycytic differentiation with negativity of CD61, CD41 as detected by FCI [56–58]. The detection of myeloid-lineage-specific antigens as MPO, CD13 or CD33 by either FCI or electron microscopy is of course mandatory. Some cases also express CD65 and often progenitor-associated markers like CD34, HLA-DR and CD117 are positive. The granulomonocytic differentiation markers CD15, CD14 and CD64 are expressed only rarely. More than 80% of AML-M0 exhibit a complex immunophenotype with expression of the non-lineage-specific lymphoid markers CD2, CD4, CD7, CD19 and TdT, which sometimes hampers a lineage-specific classification; so the diagnosis of a biphenotypic acute leukemia must be given. The demonstration of a multitude of chromosomal aberrations including chromosome 5, trisomy 8 or 13 and numerous others reflects the biological heterogeneity of AML-M0.

AML-M2/t(8;21)

The characteristic features of AML-M2 have been described in many studies. Sometimes, the panmyeloid antigens CD13 and CD33 can be found in very low expression although CD15, CD65, HLA-DR and MPO are clearly positive, and asynchronous detection of MPO by FCI can occur in contrast to cytochemistry [59]. Coexpression of CD19 and CD56 is correlated with the t(8;21), but CD19 expression has to be interpreted carefully because variations can occur depending on the clone and fluorochrome conjugation [60, 61]. Especially CD19 PE-Cy5 conjugates have shown inconsistent positive results which were not confirmed by other fluorochrome conjugates of the same clone, as PE-Cy5 in general tends to unspecifically bind to myeloid cells.

Acute Promyelocytic Leukemia with t(15;17)

The immunophenotypic hallmarks of acute promyelocytic leukemia (APL) with t(15;17) are not always obvious due to their weak or absent expression of HLA-DR, CD7, CD14, CD15 and CD34 as well as their variable expression of CD11b, CD65 and CD117 [62, 63]. Most often, there is a low expression for CD64 combined with a strong positivity for the early panmyeloid antigens CD13, CD33 as well as for CD9, CD68 and MPO. Sensitivity and specificity for the FCI diagnosis of APL, especially for the AML-M3 variant and for rare t(15;17)-positive cases, which morphologically resemble AML-M1 or AML-M2, have increased due to the combination of three immunophenotypic criteria: singular population of blast cells, heterogeneous expression pattern for CD13 and characteristic distribution of the simultaneous CD34/CD15 staining [64]. Moreover, mAbs against the PML protein are available [65]. FCI is very helpful in situation where there is no clear morphologic differentiation between a monocytic AML and a microgranular variant form of APL which, in contrast to AML-M4 or AML-M5, does not show expression for CD14, CD4, CD36 and HLA-DR. The biological and morphological heterogeneity of this leukemia is partially reflected in the immunophenotype because the coexpression of CD2 is associated with the microgranular variant form, and the coexpression of CD56 has been found to be a possible indicator for poor treatment outcome [66, 67].

AML-M4Eo

FCI can distinguish between blast populations expressing the panmyeloid lineage markers CD13, CD33 and the markers correlating to granulocytic differentiation (CD65, CD15) or monocytic differentiation (CD14, CD64, CD4). As in AML-M2 with t(8;21), there is often strong positivity for CD34 without expression of CD7. In AML-M4Eo with inv(16) or t(16;16), aberrant expression of CD2 is occasionally detectable [68]. With the identification and characterization of the chimeric protein product of CBFβ/MYH11 gene fusion in inv(16)-positive AML, mAbs against this protein are available and can be used to identify this type of AML with FCI [69].

AML-M5 with 11q23 Aberrations

In myelomonocytic AML with rearrangement of the MLL gene, especially with t(9;11) there is strong expression of HLA-DR, CD33, CD65, CD4 and coexpression of CD56 whereas other myeloid markers like CD13 or CD14 and the progenitor marker CD34 are much less expressed [70]. The mAb 7.1 which detects the human analogue of the rat NG2 molecule is found to be strongly expressed in blast populations with 11q23 rearrangements is implicated as well [71]. Some studies have even provided clear evidence for a correlation between NG2 expression and 11q23 rearrangements not only in AML with monocytic differentiation but also in precursor B-cell ALL, so mAb 7.1 is a highly sensitive, but not totally specific marker for AML with MLL rearrangement. For the detection of a monocytic differentiation of the blast population by FCI, it can be helpful to visualize SSC on a logarithmic scale against CD45, CD14 or CD64 expression [72].

AML with Megakaryocytic Differentiation

Morphology and cytochemistry are not sufficient to distinguish AML with megakaryocytic differentiation from ALL, undifferentiated AML-M0, round-cell tumors like sarcomas or transient myeloproliferative disease in childhood [73, 74]. FCI, immunohistology or electron microscopy are mandatory to safely diagnose megakaryocytic differentiation of the leukemic population. Megakaryocytic features are demonstrated by the expression of CD61, CD41 or less often CD42b on the cell membrane. Sometimes, with low or absent CD61 on the cell membrane, it is helpful to do an intracytoplasmic staining for CD61 because the molecule appears early in the cytoplasm, before its integration into the cell membrane [75]. Panmyeloid markers like CD13, CD33, MPO and the progenitor antigens CD34, CD4, CD7, CD117 and coexpression of the lymphatic antigen CD2 can be found as well.

Prognostic Implications of FCI in AML

The prognostic value of FCI in AML, especially for the prediction of overall survival and complete remission rates, has been discussed in many studies [62, 76–79]. The expression of CD7, CD9, CD11b, CD13, CD14, HLA-DR, CD34 and TdT has been shown to have a prognostically unfavorable effect whereas the detection of the granulocytic differentiation antigens CD15 and CD65 and the coexpression of CD2 (mostly found in AML-M3) are correlated with a favorable outcome. Contradictory results have been reported for the expression of CD34, CD2, CD7 and TdT [80–82]. The comparability of these studies is hampered by methodical differences including the application of different mAb clones, fluorochromes, different staining condition; thus, the results have to be interpreted with great caution. In view of these contradictory results it is obvious that it is not the expression of a single antigen but rather the definition of complex leukemia-associated immunophenotypes that can be correlated with prognostically relevant biological entities of the disease. Large series of untreated children and adults with de novo AML have not shown any influence of the expression of individual myeloid-, lymphoid-, and progenitor-cell-associated antigens on prognosis and thus do not indicate that FCI alone can be applied for risk stratification in AML at the time of diagnosis.

Acute Lymphoblastic Leukemia

FCI is an indispensable tool for the diagnosis and classification of ALL, not only to distinguish precursor B- from precursor T-ALL, but also to identify prognostically relevant subtypes; this has substantially contributed to a more precise and biologically orientated classification of the disease. The expression patterns of normal B- and T-cell ontogeny are at least partially reflected in ALL phenotypes and can therefore be used for lineage determination and definition of the maturational stage. This has been considered by the EGIL group in their 1995 proposal to classify and define the biological subtypes of ALL (table 2) [9]. More recently, FCI in conjunction with cytogenetic and molecular studies, has identified associations between immunophenotypic features and numerical and/or structural chromosomal abnormalities that have contributed to a refined ALL classification, especially in precursor B-cell ALL [3, 83]. Moreover, the differences between normal and neoplastic lymphopoiesis have been clearly described and the definition of LAIPs are used for disease monitoring or the detection of MRD with FCI [84, 85].

Precursor B-Cell Acute Lymphoblastic Leukemia/Lymphoma

The disease is mainly present in the bone marrow or peripheral blood, but can also show nodal or extranodal manifestations, e.g. testis or central nervous system. Morphology can only distinguish between small and large blast cells of the FAB-L1 and FAB-L2 subtypes and the densely vacuolated FAB-L3 blasts with a dark blue cytoplasm of mature B-ALL or Burkitt lymphoma [86]. Cytochemistry will show negativity for MPO, and PAS staining will give a positive reaction with a small rim around the nucleus [1].

FCI has shown that all precursor B-cell ALLs are positive for TdT, HLA-DR, cytoplasmic CD79a and CD19 [86, 87]. The expression of other B-lymphoid markers like CD20, CD22 or of CD10 and CD45 is variable [88–90]. The progenitor marker CD34 is found to be positive in 60–70% of all precursor B-cell ALLs and early myeloid antigens are coexpressed in about 30% of ALLs [91, 92]. Surface immunoglobulins are not expressed although intracytoplasmic light chains and IgM can be detected. In contrast to the immunological EGIL classification, the WHO nomenclature also comprises entities of ALL with distinct cytogenetic aberrations [3, 9].

Early Precursor B-Cell ALL/Pro-B ALL

This is the most common ALL of infancy and is most often diagnosed with a high leukocyte count [93]. Girls are more frequently affected than boys and often with organomegaly and manifestation within the CNS. The immunophenotype is characterized by a lack of CD10 expression and the coexpression of the myeloid markers CD65 and CD15. In 2–6% of childhood and adult ALL, the MLL rearrangement with t(4;11) can be found and is clearly associated with the immunophenotype of pro-B ALL [94, 95]. The sensitivity and specificity of FCI in predicting the molecular and cytogenetic detection of the t(4;11) translocation has been very much enhanced by mAb 7.1 which detects the overexpression of the human homologue of the rat chondroitin sulfate proteoglycan, NG2, in blast cells with MLL gene rearrangements [96].

Intermediate Precursor B-Cell ALL/'Common ALL'

In contrast to pro-B ALL, the leukemic cells of this intermediate stage of early B-cell maturation express the common acute lymphoblastic leukemia antigen ('CALLA' antigen), a surface enzyme with neutral metalloendopeptidase activity, clustered as CD10. The mature B-cell markers like CD20, CD24 and the leukocyte common antigen CD45 are variably expressed, with 10–15% of all common ALLs being negative for CD45 [89, 97]. Complete immunoglobulins are not yet detectable either on the membrane or in the cytoplasm, but light chains are already present. Coexpression of one or more myeloid markers can be found in about 30–40% of common ALLs, but has no correlation to cytogenetic features or any prognostic implication, except for the t(12;21) translocation [92, 98]. The Philadelphia chromosome with t(9;22)(q34;q11) is found in 15–30% of adult and in 3–5% of childhood ALL, mostly associated with the phenotype of a common ALL or pre-B ALL [99, 100]. Although several studies describe an association of the KOR-SA354 antigen expression or the coexpression of CD34 and CD10 or even CD25, no definite correlation between the Philadelphia chromosome and any immunophenotype has been found [101–103]. The most common chromosomal aberration in childhood ALL is the t(12;21)(p13;q22) translocation with the fusion transcript TEL/AML1 [104]. It occurs in about 20–25% of childhood ALL and only in 3% of adult ALL and is associated with a phenotype of common ALL with coexpression of the myeloid antigens CD13 and CD33 [105, 106]. Moreover, the lack of expression for CD9, CD66c and CD20 is highly predictive of a TEL/AML1 rearrangement.

Mature Precursor B-Cell ALL/Pre-B ALL

Except for the presence of cytoplasmic IgM, the immunophenotype of pre-B ALL resembles that of the CD10+ common ALL and is only rarely associated with chromosomal aberrations. In about 5–6% of all childhood (25% of all pre-B ALL) and in less than 5% of adult ALL, a t(1;19)(q23;p13.3) translocation with a chimeric PBX/E2A fusion protein occurs which is significantly correlated with an immunophenotypic pattern of a cyIgM+, CD19+, CD22+, CD20±, CD34− and CD45^{high} pre-B

ALL [107]. Moreover, specific mAbs are available for the intranuclear detection of the PBX/E2A fusion protein within leukemic blast cells [108].

Mature B-ALL and Burkitt Lymphoma

In contrast to AML, there is no correlation between morphological features and specific chromosomal aberrations in ALL, except for the FAB-L3 subtype. Immunologically defined by the expression of surface immunoglobulin, the t(14;18)(q24;q11) translocation is detected in 75–85% of patients with mature B-ALL. Although there is a very strong correlation between morphology, immunology and cytogenetics, there are variations with a typical L3 appearance and detection of t(14;18): instead of the mature B-ALL phenotype, the immunological profile of a more immature precursor B-cell ALL is expressed on the blast cells. This emphasizes the importance of simultaneously applying all the available diagnostic methods for accurate risk assessment and therapy stratification [109–111].

Precursor T-Lymphoblastic Leukemia/Lymphoma

About 15% of childhood and 25% of adult ALLs are precursor T-ALLs. Neither morphology, which resembles the L1 L2 subtypes of the FAB classification nor cytochemistry, with negativity for MPO and a focal positivity for PAS, can distinguish between precursor T-ALL blast cells and precursor B-ALL. Only FCI is able to differentiate and classify subtypes according to the maturational stage in analogy to normal T-cell ontogenesis. In general, the immunophenotype is characterized by the expression of TdT, CD7 and CD3 on the membrane or within the cytoplasm. The more mature T-cell markers CD2, CD5, CD4, CD8 and TCR are variably expressed. Often positivity for CD10, CD34 and coexpression of myeloid markers like CD13, CD33 and CD117 is noticed. The EGIL group defined the following four subtypes according to the degree of thymic differentiation [9]:

- the most immature subtype, pro-T-ALL (T-I) is characterized by the expression of CD7 and cyCD3 alone;
- the acquisition of CD2 and/or CD5 and/or CD8 is still consistent with an early phenotype denotes as pre-T-ALL (T-II);
- with the progression of development in the thymus, CD1a appears on the cell surface; expression of CD1a, regardless of other markers defines cortical T-ALL (T-III);
- mature-T-ALL express membrane CD3 and are distinguished according to the expression of TCR-α/β or TCR-γ/δ. They are negative for CD1a.

The German ALL study group (GMALL) merged the subtypes pro- and pre-T as early T-ALL and in the WHO classification, even pro-, pre- and cortical T-ALL are placed under one label, despite their strikingly different biological and clinical behavior. These immunological and clinically defined subentities have additionally been characterized according to their chromosomal aberrations and molecular features only recently [112, 113]. The most common translocations detected involve the α- and δ-TCR locus at 14q11.2, the β-locus at 7q35 and the γ-locus at 7p14–15 with a variety of partner genes. These include the transcription factor genes MYC, TAL1, RBTN1, RBTN2, HOX11 and the cytoplasmic tyrosine kinase LCK [114]. The frequency of these chromosomal abnormalities varies between adult and childhood ALL [115]. More than 50% of human T-ALLs exhibit activating mutations that involve the extracellular heterodimerization domain and/or the C-terminal PEST domain of NOTCH1 implicating NOTCH1 as a major player in the etiology of T-ALL [116]. Significant correlations with clinical impact between the immunophenotype and chromosomal aberrations in T-ALL have only been described for childhood T-ALL with t(11;14) and expression of a CD4+, CD8+, CD3+ phenotype [117].

Acute Leukemias of Ambiguous Lineage

In almost all cases of acute leukemia, a clear lineage assignment according to the expression of lineage-specific markers, e.g., CD19, cyCD22, cyCD79a or cyIgM for B-cell lineage, s/cyCD3, anti-TCR-α/β or anti-TCR-γ/δ for the T-cell lineage and MPO, CD13 and CD33 for the myeloid lineage is possible. Myeloid coexpression (My+ ALL) is found in about 30% of childhood ALL, and roughly 10–20% of AML coexpress lymphoid markers (Ly+ AML). Whether this is only considered as an aberrant expression of cross-lineage markers or as a biphenotypic acute leukemia depends on lineage specificity and on the number of markers that are coexpressed. The EGIL group has proposed a scoring system for the definition of biphenotypic acute leukemia which has been integrated into the WHO classification (table 3) [3, 9, 118]. This scoring system defines biphenotypic acute leukemia if there are concomitant features of lymphoid and myeloid lineage with a score of >2 present on the blast population. Specific chromosomal aberrations have not been significantly correlated with biphenotypic acute leukemia, but the Philadelphia chromosome with t(9;22) is detected quite often and in one third of cases an 11q23 rearrangement is present [119].

Acute Dendritic Cell Leukemia

Acute leukemias of dendritic cell origin are very uncommon and present with a clinically distinct picture showing erythematous nodular skin lesions, generalized lymphadenopathy and pancytopenia [121–123]. Blast cells are defined according to their strong expression of CD4 and CD56, while lacking specific markers for the T-lymphoid, B-lymphoid or myeloid lineage, displaying the following immunophenotypic pattern: CD2–, sCD3–, CD4+, CD7±, CD8–, CD16–, CD34–, CD36+,

Table 3. Scoring system for markers proposed by the European Group for the Immunological Classification of Leukemia (EGIL) [9]

Score	B-lymphoid	T-lymphoid	Myeloid
2	cyCD79a* cyIgM cyCD22	s/cyCD3 anti-TCR	MPO
1	CD19 CD20 CD10	CD2 CD5 CD8 CD10	CD117 CD13 CD33 CD65
0.5	TdT CD24	TdT CD7 CD1a	CD14 CD15 CD64

* CD79a may also be expressed in some cases of precursor T lymphoblastic leukemia [120].

CD38±, CD45+(weak), CD56+, CD57−, CD68+, HLA-DR− [124]. They have to be distinguished from other CD56+ hematologic neoplasias, which can be accomplished with FCI, but complementary consideration of clinical symptoms and cytomorphology is helpful.

Precursor NK Cell Leukemia/Lymphoma

Immature NK cell neoplasias are very rare and comprise the aggressive NK cell progenitor leukemia, sometimes with a myeloid/NK cell phenotype and the blastic NK cell lymphoma [125–127]. Progenitor NK cell leukemias often present with extramedullary manifestations like lymphadenopathy and dermal infiltrations. Blast cells show an FAB-L2 morphology and display an immunophenotype with: CD2+, sCD3−, CD4±, CD5±, CD7+, CD11b+, CD11c+, CD13+ and/or CD33+, MPO±, CD56+, CD16−, CD57− and CD34+. Patients with blastic NK cell lymphoma are sometimes difficult to distinguish from patients with dendritic cell leukemia or monocytic leukemia because they often have dermal infiltrations and involvement of the bone marrow as well and the immunophenotype is only slightly different than in dendritic-cell or monocytic leukemia: CD2±, sCD3−, CD4+, CD5±, CD7±, CD33±(weak), CD56+, CD57−, TdT± and CD34±.

Prognostic Significance of Flow-Cytometric Immunophenotyping in Acute Lymphoblastic Leukemia

The assessment of the prognostic impact of FCI in ALL has been hampered in the past by the lack of standardized criteria for the classification of immunophenotypic subgroups, the scarcity of controlled prospective studies on the treatment outcome of ALL subsets and the different treatment strategies pursued in these studies. In addition, the value of FCI as an independent prognostic predictor of outcome has to be interpreted with caution in view of the strong correlations between certain immunophenotypic subgroups and cytogenetic or clinical features. Moreover, the improved efficacy of modern therapeutic regimens has diminished the prognostic impact of both, immunophenotypic subgroups and chromosomal abnormalities; thus, any prognostic statement has to be interpreted carefully within the context of the delivered therapy [128].

In precursor B-cell ALLs no significant differences in the remission induction rates between immunophenotypic subgroups were recorded, but several trials have demonstrated an association between maturational stage and duration of remission. The most immature CD10–, pro-B ALL phenotype has been reported in both childhood and adult studies to have a worse prognosis, which was, however, frequently associated with adverse biological (e.g. 11q23 rearrangements) and clinical features (e.g. high tumor burden and infant age) [129–131].

Cytogenetic and molecular studies have provided conclusive evidence that there is a significant difference between children and adults with common and pre-B ALL with respect to their incidence of known favorable and unfavorable chromosomal translocations. For instance, the unfavorable t(9;22) translocation accounts for up to 55% of adult and only 5% of childhood CD10+ precursor B-cell ALLs whereas the favorable t(12;21) translocation is present in 12–36% of childhood ALL and only rarely occurs in adult patients. These findings at least partially explain the striking differences in treatment outcome between children and adults with common or pre-B ALL [100, 132, 133]. Earlier studies from the Pediatric Oncology Group have indicated a poorer treatment outcome for a subgroup of pre-B ALL within the group of precursor B-cell ALLs, which was subsequently shown to be associated with the occurrence of the t(1;19)(q23;p13) chromosomal translocation, but later trials from the German ALL-BFM group and from the Medical Research Council in Britain did not reveal any differences in the duration of remission between childhood common ALL and pre-B ALL [134].

In precursor T-cell ALL, various immunophenotypic features seem to be associated with an increased risk of treatment failure, including an immature pro/pre-T ALL phenotype, membrane expression of CD3 or MHC class II antigen, and negativity for CD2, CD5, CD1a, or CD10 [135–137]. The prognostic impact of these factors, however, has differed according to the treatment strategies used, and immunophenotypic features are still not routinely used for risk classification or assignment to novel treatment strategies in high-risk precursor T-cell ALL patients. Several multicenter trials in childhood ALL and data of adult patients treated in a CALGB trial have lent strong support to evidence that patients with a cortical (CD1a+) precursor T-cell ALL have a better early response to treatment and a significant longer duration of event-free survival than patients with an immature or mature precursor

T-cell ALL phenotype, which can be partially explained by a higher rate of spontaneous apoptosis [131, 138–140].

Minimal Residual Disease Detection

Recent studies have indicated a significant role for the detection of MRD in adult and childhood acute leukemia [141, 142]. Blast reduction kinetics during induction therapy has proven to be of prognostic importance and is used for risk-adapted therapy stratification in many studies. Today, three highly specific and sensitive methods are available for MRD detection: multiparametric FCI, real-time quantitative PCR (RQ-PCR)-based detection of fusion gene transcripts or breakpoints, and RQ-PCR-based detection of clonal immunoglobulin and T-cell receptor gene rearrangements. Multiparametric FCI relies on the detection of the LAIP, which separates leukemic blast populations from normal regenerating myelo- or lymphopoiesis. LAIPs generally involve cross-lineage antigen expression (i.e. expression of lymphoid-associated markers in myeloid blast cells or vice versa), asynchronous antigen expression (coexpression of sequentially expressed antigens that are not present at once in the same maturational stage in normal hematopoiesis), antigen over- or underexpression (presence of an antigen in leukemic cells at abnormally high or low quantities) or ectopic expression of markers that are normally confined to specific tissues (the phenotype of T-ALL reflects that of normal thymocytes, which are never found outside the thymus). LAIPs can be found in about 80% of AML and nearly 95% of all ALL cases.

Sensitivity of Minimal Residual Disease Detection

The ability of flow cytometry to detect rare events in clinical samples has been proven and validated not only in dilution experiments but also in numerous clinical trials [15, 143, 144]. Leukemic cells displaying an aberrant immunophenotype can be reliably detected when present at concentrations of 10^{-4}–10^{-5} (1 leukemic cell in 10,000–100,000 normal cells). However, the level of sensitivity depends on the quality and cellularity of the sample, the type of phenotypic aberration, the combination of mAb reagents used and the number of cells/events acquired for the FCI analysis. In an FCI analysis of 100,000 acquired events, one single aberrant event is not enough to define MRD; thus, by definition, a cluster of 10 aberrant events is needed. Hence, a homogeneous sample carrying 1 leukemic cell/100,000 normal cells will not be considered positive for MRD if only 300,000 cells are acquired but with 1–2 × 10^6 cells a cluster with 10–20 aberrant cells will definitely allow the detection of MRD. The recent developments of modern flow cytometers with enhanced technologies for fluidics and with multiple lasers and fluorochrome channels have facilitated rare-event detection and help to increase the sensitivity and reliability of the method. Concordance rates of RQ-PCR results for clonal immunoglobulin gene rearrangements with FCI MRD results are variable depending on the method used for FCI MRD (lysis or Ficoll gradient centrifugation, number of cells analyzed) and the sensitivity level compared, but can reach more than 90% with a sensitivity $\geq 10^{-3}$ [145–147].

Clinical Impact of FCI Detection of Minimal Residual Disease in Acute Lymphoblastic Leukemia

Pioneering studies by Campana and Coustan-Smith [27] and Coustan-Smith et al. [143] assessing the potential advantage of measuring treatment response in childhood ALL with FCI showed that patients with high levels of MRD ($= 10^{-2}$) at the end of the induction phase (6 weeks) or moderate to high levels of MRD ($\geq 10^{-3}$) at week 14 of continuation therapy had a particularly poor outcome. The predictive power of FCI MRD remained significant even after adjusting for adverse features. These results were supported by an Austrian study on children treated by the ALL-BFM protocol [148]. In that study, sequential monitoring at day 33 and week 12 of treatment proved to be particularly useful, because patients with persistent disease (≥1 blast/μl) had a 100% probability of relapse, compared to only 6% in all others. It was also shown that flow-cytometric detection of MRD was associated with a significantly higher relapse rate even in adult ALL patients [149]. Although there are slight methodological differences between these studies, they emphasize the importance and clinical significance of FCI MRD detection independent of age, methodology and therapeutic regimen applied.

Clinical Impact of FCI Detection of Minimal Residual Disease in Acute Myeloid Leukemia

Quantification of MRD in AML is mainly accomplished with real-time quantitative polymerase chain reaction (RQ-PCR) relying on the detection of fusion gene transcripts or breakpoints which occur in almost 50% of all patients with AML. However, in one half of the patients, the blast cells do not carry a leukemia-specific genetic alteration, rendering them undetectable by molecular techniques [7, 150]. In contrast, flow cytometry is applicable to nearly all patients with AML when pursued in a comprehensive approach in order to define the LAIP [151, 152]. In a recent study by Kern et al. [152], at least one LAIP was defined for each patient and the degree of reduction in LAIP-positive cells between diagnosis and complete remission was significantly related to relapse-free survival both after induction and consolidation therapy. In a multivariate analysis, the reduction rate remained an independent prognostic factor for relapse-free survival at both time points. In a similar study performed at the University of Salamanca four risk groups could be defined according to the level of MRD found in the bone marrow samples after induction and in morphological remission [153]. Relapse-free survival of 15, 55, 85 and 100% at 3 years was found to be correlated with very low ($<10^{-4}$), low ($>10^{-4}$ and $<10^{-3}$) intermediate ($>10^{-3}$ and $<10^{-2}$) and high ($>10^{-2}$) MRD levels, respectively. Moreover, high levels of MRD were also associated with adverse cytogenetic features; in a multivariate analysis, FCI MRD detection was confirmed to be the most powerful independent prognostic factor for the prediction of disease-free and overall survival followed by cytogenetics. These results show that FCI MRD detection is a useful tool in the management of AML patients and provide a strong rationale to support its implementation in future trials of new treatment modalities in AML.

Summary

Multiparametric flow cytometry is most important for the characterization of immunological subtypes of acute leukemia and for the immunological detection of MRD. Flow-cytometric immunophenotyping (FCI) is indispensable, especially for the diagnosis and classification of acute lymphoblastic leukemia; in correlation with cytogenetics and molecular biology, it can provide valuable prognostic information. Recent technological advances have accelerated the development of a new generation of flow cytometers with enhanced fluidics, computer hardware and software, and an increased signal resolution that allows the complex detection of 6–9 fluorochromes in a multiparameter approach on a single-cell level. With the appropriate clinical and preanalytical prerequisites, FCI can be utilized far beyond lineage assignment and classification of acute leukemia, notably for the detection of minimal residual disease and the definition of rare leukemia entities like NK- and dendritic cell leukemia and biphenotypic or bilineage acute leukemias. Moreover, future developments of antibodies, fluorochromes and fluorescent labeling techniques will likely facilitate the functional characterization of cellular compartments in correlation with the immunophenotype, which can give an even more complex flow-cytometric picture of the blast cell population.

References

1 Bain BJ: Leukemia Diagnosis. A Guide to the FAB Classification. London, Mosby Year Book Europe, 1993.

2 Bennett JM, Catovsky D, Daniel MT, Flandrin G, Galton DA, Gralnick HR, Sultan C: Proposals for the classification of the acute leukaemias. French-American-British (FAB) co-operative group. Br J Haematol 1976;33:451–458.

3 Jaffe ES, Harris NL, Stein H, Vardimann JW: Tumors of Haematopoietic and Lymphoid Tissues. Lyon, IARC Press, 2001.

4 Bain BJ: Classification of acute leukaemia: the need to incorporate cytogenetic and molecular genetic information. J Clin Pathol 1998;51:420–423.

5 FMCSG: Morphologic, immunologic, and cytogenetic (MIC) working classification of acute lymphoblastic leukemias. Report of the workshop held in Leuven, Belgium, April 22–23, 1985. First MIC Cooperative Study Group. Cancer Genet Cytogenet 1986;23:189–197.

6 Loffler H: Morphologic basis for the MIC classification in acute myeloid leukemia. Recent Results Cancer Res 1993;131:339–343.

7 Haferlach T, Schnittger S, Kern W, Hiddemann W, Schoch C: Genetic classification of acute myeloid leukemia (AML). Ann Hematol 2004;83(suppl 1):S97–S100.

8 Haferlach T, Schoch C, Löffler H, Gassmann W, Schnittger S, Fonatsch C, Haase D, Lengfelder E, Staib P, Ludwig W-D, Wörmann B, Sauerland MC, Büchner T, Hiddemann W: Cytomorphology and cytogenetics in de novo AML: importance for the definition of biological entities. Blood 1999;94:291a.

9 Bene MC, Castoldi G, Knapp W, Ludwig WD, Matutes E, Orfao A, van't Veer MB: Proposals for the immunological classification of acute leukemias. European Group for the Immunological Characterization of Leukemias (EGIL). Leukemia 1995;9:1783–1786.

10 Janossy G. Clinical flow cytometry, a hypothesis-driven discipline of modern cytomics. Cytometry A 2004;58:87–97.

11 Keating P, Cambrosio A Biomedical Platforms. Realigning the Normal and the Pathological in Late-Twentieth-Century Medicine. Cambridge, Mass, MIT Press, 2003.

12 Jennings CD, Foon KA: Recent advances in flow cytometry: application to the diagnosis of hematologic malignancy. Blood 1997;90:2863–2892.

13 Roederer M, De Rosa S, Gerstein R, Anderson M, Bigos M, Stovel R, Nozaki T, Parks D, Herzenberg L, Herzenberg L: 8-color, 10-parameter flow cytometry to elucidate complex leukocyte heterogeneity. Cytometry 1997;29:328–339.

14 Wood B: 9-color and 10-color flow cytometry in the clinical laboratory. Arch Pathol Lab Med 2006;130:680–690.

15 Gross HJ, Verwer B, Houck D, Recktenwald D: Detection of rare cells at a frequency of one per million by flow cytometry. Cytometry 1993;14:519–526.

16 Davis BH, Holden JT, Bene MC, Borowitz MJ, Braylan RC, Cornfield D, Gorczyca W, Lee R, Maiese R, Orfao A, Wells D, Wood BL, Stetler-Stevenson M: 2006 Bethesda International Consensus recommendations on the flow cytometric immunophenotypic analysis of hematolymphoid neoplasia: medical indications. Cytometry B Clin Cytom 2007;72(suppl 1):S5–S13.

17 Greig B, Oldaker T, Warzynski M, Wood B: 2006 Bethesda International Consensus recommendations on the immunophenotypic analysis of hematolymphoid neoplasia by flow cytometry: recommendations for training and education to perform clinical flow cytometry. Cytometry B Clin Cytom 2007;72(suppl 1):S23–S33.

18 Stetler-Stevenson M, Davis B, Wood B, Braylan R: 2006 Bethesda International Consensus Conference on Flow Cytometric Immunophenotyping of Hematolymphoid Neoplasia. Cytometry B Clin Cytom. 2007;72(suppl 1):S3.

19 Wood BL, Arroz M, Barnett D, DiGiuseppe J, Greig B, Kussick SJ, Oldaker T, Shenkin M, Stone E, Wallace P: 2006 Bethesda International Consensus recommendations on the immunophenotypic analysis of hematolymphoid neoplasia by flow cytometry: optimal reagents and reporting for the flow cytometric diagnosis of hematopoietic neoplasia. Cytometry B Clin Cytom 2007;72(suppl 1):S14–S22.

20 Rothe G, Schmitz G: Consensus protocol for the flow cytometric immunophenotyping of hematopoietic malignancies. Working Group on Flow Cytometry and Image Analysis. Leukemia 1996;10:877–895.

21 Davis BH, Foucar K, Szczarkowski W, Ball E, Witzig T, Foon KA, Wells D, Kotylo P, Johnson R, Hanson C, Bessman D: US-Canadian consensus recommendations on the immunophenotypic analysis of hematologic neoplasia by flow cytometry: medical indications. Cytometry 1997;30:249–263.

22 Braylan RC, Atwater SK, Diamond L, Hassett JM, Johnson M, Kidd PG, Leith C, Nguyen D: US-Canadian consensus recommendations on the immunophenotypic analysis of hematologic neoplasia by flow cytometry: data reporting. Cytometry 1997;30:245–248.

23 Borowitz MJ, Bray R, Gascoyne R, Melnick S, Parker JW, Picker L, Stetler-Stevenson M. U.S.-Canadian Consensus recommendations on the immunophenotypic analysis of hematologic neoplasia by flow cytometry: data analysis and interpretation. Cytometry. 1997;30:236–244.

24 Stewart CC, Behm FG, Carey JL, Cornbleet J, Duque RE, Hudnall SD, Hurtubise PE, Loken M, Tubbs RR, Wormsley S. U.S.-Canadian Consensus recommendations on the immunophenotypic analysis of hematologic neoplasia by flow cytometry: selection of antibody combinations. Cytometry 1997;30:231–235.

25 Stelzer GT, Marti G, Hurley A, McCoy P Jr, Lovett EJ, Schwartz A: U.S.-Canadian Consensus recommendations on the immunophenotypic analysis of hematologic neoplasia by flow cytometry: standardization and validation of laboratory procedures. Cytometry 1997;30:214–230.

26 US-Canadian consensus recommendations on the immunophenotypic analysis of hematologic neoplasia by flow cytometry. Bethesda, MD, 1995. Cytometry 1997;30:213–274.

27 Campana D, Coustan-Smith E: Detection of minimal residual disease in acute leukemia by flow cytometry. Cytometry 1999;38:139–152.

28 Bennett JM, Catovsky D, Daniel MT, Flandrin G, Galton DA, Gralnick HR, Sultan C: A variant form of hypergranular promyelocytic leukaemia (M3). Br J Haematol 1980;44:169–170.

29 Bennett JM, Catovsky D, Daniel MT, Flandrin G, Galton DA, Gralnick HR, Sultan C: Criteria for the diagnosis of acute leukemia of megakaryocyte lineage (M7). A report of the French-American-British Cooperative Group. Ann Intern Med 1985;103:460–462.

30 Bennett JM, Catovsky D, Daniel MT, Flandrin G, Galton DA, Gralnick HR, Sultan C: Proposal for the recognition of minimally differentiated acute myeloid leukaemia (AML-M0). Br J Haematol 1991;78:325–329.

31 Lo-Coco F, Cimino G, Breccia M, Noguera NI, Diverio D, Finolezzi E, Pogliani EM, Di Bona E, Micalizzi C, Kropp M, Venditti A, Tafuri A, Mandelli F: Gemtuzumab ozogamicin (Mylotarg) as a single agent for molecularly relapsed acute promyelocytic leukemia. Blood 2004;104:1995–1999.

32 van Der Velden VH, te Marvelde JG, Hoogeveen PG, Bernstein ID, Houtsmuller AB, Berger MS, van Dongen JJ: Targeting of the CD33-calicheamicin immunoconjugate Mylotarg (CMA-676) in acute myeloid leukemia: in vivo and in vitro saturation and internalization by leukemic and normal myeloid cells. Blood 2001;97:3197–3204.

33 Arceci RJ, Sande J, Lange B, Shannon K, Franklin J, Hutchinson R, Vik TA, Flowers D, Aplenc R, Berger MS, Sherman ML, Smith FO, Bernstein I, Sievers EL: Safety and efficacy of gemtuzumab ozogamicin in pediatric patients with advanced CD33+ acute myeloid leukemia. Blood 2005;106:1183–1188.

34 Golay J, Cortiana C, Manganini M, Cazzaniga G, Salvi A, Spinelli O, Bassan R, Barbui T, Biondi A, Rambaldi A, Introna M: The sensitivity of acute lymphoblastic leukemia cells carrying the t(12;21) translocation to campath-1H-mediated cell lysis. Haematologica 2006;91:322–330.

35 van der Velden VH: Cytotoxicity of Campath-1H for acute lymphoblastic leukemia cells carrying the t(12;21) translocation. Haematologica 2006;91:291A.

36 Borowitz MJ, Guenther KL, Shults KE, Stelzer GT: Immunophenotyping of acute leukemia by flow cytometric analysis. Use of CD45 and right-angle light scatter to gate on leukemic blasts in three-color analysis. Am J Clin Pathol 1993;100:534–540.

37 Paxton H, Bendele T: Effect of time, temperature, and anticoagulant on flow cytometry and hematological values. Ann NY Acad Sci 1993;677:440–443.

38 United Nations: Recommendations on the Transport of Dangerous Goods – Model Regulations. Geneva, United Nations Publications, 2005, vol 14.

39 Paietta E: How to optimize multiparameter flow cytometry for leukaemia/lymphoma diagnosis. Best Pract Res Clin Haematol 2003;16:671–683.

40 Knapp W, Strobl H, Majdic O: Flow cytometric analysis of cell-surface and intracellular antigens in leukemia diagnosis. Cytometry 1994;18:187–198.

41 Knapp W, Majdic O, Strobl H: Flow cytometric analysis of intracellular myeloperoxidase and lactoferrin in leukemia diagnosis. Recent Results Cancer Res 1993;131:31–40.

42 Van Lochem EG, Groeneveld K, Te Marvelde JG, Van den Beemd MW, Hooijkaas H, Van Dongen JJ: Flow cytometric detection of intracellular antigens for immunophenotyping of normal and malignant leukocytes: testing of a new fixation-permeabilization solution. Leukemia 1997;11:2208–2210.

43 Groeneveld K, te Marvelde JG, van den Beemd MW, Hooijkaas H, van Dongen JJ: Flow cytometric detection of intracellular antigens for immunophenotyping of normal and malignant leukocytes. Leukemia 1996;10:1383–1389.

44 Kraan J, Gratama JW, Keeney M, D'Hautcourt JL: Setting-up and calibration of a flow cytometer for multicolor immunophenotyping. J Biol Regul Homeost Agents 2003;17:223–233.

45 Roederer M: Spectral compensation for flow cytometry: visualization artifacts, limitations, and caveats. Cytometry 2001;45:194–205.

46 Oldaker TA: Quality control in clinical flow cytometry. Clin Lab Med 2007;27:671–685.

47 Gratama JW, Bolhuis RL, Van 't Veer MB: Quality control of flow cytometric immunophenotyping of haematological malignancies. Clin Lab Haematol 1999;21:155–160.

48 Owens MA, Vall HG, Hurley AA, Wormsley SB: Validation and quality control of immunophenotyping in clinical flow cytometry. J Immunol Methods 2000;243:33–50.

49 Schwartz A, Marti GE, Poon R, Gratama JW, Fernandez-Repollet E: Standardizing flow cytometry: a classification system of fluorescence standards used for flow cytometry. Cytometry 1998;33:106–114.

50 Lenkei R, Gratama JW, Rothe G, Schmitz G, D'Hautcourt J L, Arekrans A, Mandy F, Marti G: Performance of calibration standards for antigen quantitation with flow cytometry. Cytometry 1998;33:188–196.

51 Ratei R, Karawajew L, Lacombe F, Jagoda K, Del Poeta G, Kraan J, De Santiago M, Kappelmayer J, Bjorklund E, Ludwig WD, Gratama J, Orfao A: Normal lymphocytes from leukemic samples as an internal quality control for fluorescence intensity in immunophenotyping of acute leukemias. Cytometry B Clin Cytom 2006;70:1–9.

52 Kluin-Nelemans J, Van Wering E, Van Der Schoot C, Adriaansen H, Van TVM, Van Dongen J, Gratama J: SIHONSCORE: a scoring system for external quality control of leukaemia/lymphoma immunophenotyping measuring all analytical phases of laboratory performance. Br J Haematol 2001;112:337–343.

53 De Zen L, Bicciato S, te Kronnie G, Basso G: Computational analysis of flow-cytometry antigen expression profiles in childhood acute lymphoblastic leukemia: an MLL/AF4 identification. Leukemia 2003;17:1557–1565.

54 Ratei R, Karawajew L, Lacombe F, Jagoda K, Del Poeta G, Kraan J, De Santiago M, Kappelmayer J, Bjorklund E, Ludwig WD, Gratama JW, Orfao A: Discriminant function analysis as decision support system for the diagnosis of acute leukemia with a minimal four color screening panel and multiparameter flow cytometry immunophenotyping. Leukemia 2007;21:1204–1211.

55 Toedling J, Rhein P, Ratei R, Karawajew L, Spang R: Automated in-silico detection of cell populations in flow cytometry readouts and its application to leukemia disease monitoring. BMC Bioinformatics 2006;7:282.

56 Bennett JM, Catovsky D, Daniel MT, Flandrin G, Galton DAG, Gralnick HR, Sultan C: Proposal for the recognition of minimally differentiated acute myeloid leukemia (AML-M0). Br J Haematol 1991;78:325–329.

57 Venditti A, Del Poeta G, Buccisano F, Tamburini A, Cox MC, Stasi R, Bruno A, Aronica G, Maffei L, Suppo G, Simone MD, Forte L, Cordero V, Postorino M, Tufilli V, Isacchi G, Masi M, Papa G, Amadori S: Minimally differentiated acute myeloid leukemia (AML-M0): comparison of 25 cases with other French-American-British subtypes. Blood 1997;89:621–629.

58 Bene MC, Bernier M, Casasnovas RO, Castoldi G, Doekharan D, van der Holt B, Knapp W, Lemez P, Ludwig WD, Matutes E, Orfao A, Schoch C, Sperling C, van't Veer MB: Acute myeloid leukaemia M0: haematological, immunophenotypic and cytogenetic characteristics and their prognostic significance: an analysis in 241 patients. Br J Haematol 2001;113:737–745.

59 Hurwitz CA, Raimondi SC, Head D, Krance R, Mirro J, Jr., Kalwinsky DK, Ayers GD, Behm FG: Distinctive immunophenotypic features of t(8;21)(q22;q22) acute myeloblastic leukemia in children. Blood 1992;80:3182–3188.

60 Andrieu V, Radford Weiss I, Troussard X, Chane C, Valensi F, Guesnu M, Haddad E, Viguier F, Dreyfus F, Varet B, Flandrin G, Macintyre E: Molecular detection of t(8;21)/AML1-ETO in AML M1/M2: correlation with cytogenetics, morphology and immunophenotype. Br J Haematol 1996;92:855–865.

61 Kita K, Nakase K, Miwa H, Masuya M, Nishii K, Morita N, Takakura N, Otsuji A, Shirakawa S, Ueda T, et al: Phenotypical characteristics of acute myelocytic leukemia associated with the t(8;21)(q22;q22) chromosomal abnormality: frequent expression of immature B-cell antigen CD19 together with stem cell antigen CD34. Blood 1992;80:470–477.

62 Creutzig U, Harbott J, Sperling C, et al: Clinical significance of surface antigen expression in children with acute myeloid leukemia: results of study AML-BFM-87. Blood 1995;86:3097–3108.

63 Paietta E, Andersen J, Gallagher R, Bennett J, Yunis J, Cassileth P, Rowe J, Wiernik PH: The immunophenotype of acute promyelocytic leukemia (APL): an ECOG study. Leukemia 1994;8:1108–1112.

64 Orfao A, Chillon MC, Bortoluci AM, Lopez Berges MC, Garcia Sanz R, Gonzalez M, Tabernero MD, Garcia Marcos MA, Rasillo AI, Hernandez Rivas J, San Miguel JF: The flow cytometric pattern of CD34, CD15 and CD13 expression in acute myeloblastic leukemia is highly characteristic of the presence of PML-RARα gene rearrangements. Haematologica 1999;84:405–412.

65 Falini B, Flenghi L, Fagioli M, Coco FL, Cordone I, Diverio D, Pasqualucci L, Biondi A, Riganelli D, Orleth A, Liso A, Martelli MF, Pelicci PG, Pileri S: Immunocytochemical diagnosis of acute promyelocytic leukemia (M3) with the monoclonal antibody PG-M3 (anti-PML). Blood 1997;90:4046–4053.

66 Biondi A, Luciano A, Bassan R, Mininni D, Specchia G, Lanzi E, Castagna S, Cantu-Rajnoldi A, Liso V, Masera G, et al: CD2 expression in acute promyelocytic leukemia is associated with microgranular morphology (FAB M3v) but not with any PML gene breakpoint. Leukemia 1995;9:1461–1466.

67 Claxton DF, Reading CL, Nagarajan L, Tsujimoto Y, Andersson BS, Estey E, Cork A, Huh YO, Trujillo J, Deisseroth AB: Correlation of CD2 expression with PML gene breakpoints in patients with acute promyelocytic leukemia. Blood 1992;80:582–586.

68 Adriaansen HJ, te Boekhorst PA, Hagemeijer AM, van der Schoot CE, Delwel HR, van Dongen JJ: Acute myeloid leukemia M4 with bone marrow eosinophilia (M4Eo) and inv(16)(p13q22) exhibits a specific immunophenotype with CD2 expression. Blood 1993;81:3043–3051.

69 Liu PP, Wijmenga C, Hajra A, Blake TB, Kelley CA, Adelstein RS, Bagg A, Rector J, Cotelingam J, Willman CL, Collins FS: Identification of the chimeric protein product of the CBFB-MYH11 fusion gene in inv(16) leukemia cells. Genes Chromosomes Cancer 1996;16:77–87.

70 Baer MR, Stewart CC, Lawrence D, Arthur DC, Mrozek K, Strout MP, Davey FR, Schiffer CA, Bloomfield CD: Acute myeloid leukemia with 11q23 translocations: myelomonocytic immunophenotype by multiparameter flow cytometry. Leukemia 1998;12:317–325.

71 Wuchter C, Karawajew L, Ruppert V, Büchner T, Schoch C, Haferlach T, Ratei R, Dörken B, Ludwig WD: Clinical significance of CD95, Bcl-2 and Bax expression and CD95 function in adult de novo acute myeloid leukemia in context of P-glycoprotein function, maturation stage, and cytogenetics. Leukemia 1999;13:1943–1953.

72 Krasinskas AM, Wasik MA, Kamoun M, Schretzenmair R, Moore J, Salhany KE. The usefulness of CD64, other monocyte-associated antigens, and CD45 gating in the subclassification of acute myeloid leukemias with monocytic differentiation. Am J Clin Pathol 1998;110:797–805.

73 Langebrake C, Creutzig U, Reinhardt D: Immunophenotype of Down syndrome acute myeloid leukemia and transient myeloproliferative disease differs significantly from other diseases with morphologically identical or similar blasts. Klin Pädiatr 2005;217:126–134.

74 Bennett JM, Catovsky D, Daniel MT, Flandrin G, Galton DA, Gralnick HR, Sultan C: Criteria for the diagnosis of acute leukemia of megakaryocyte lineage (M7). A report of the French-American-British Cooperative Group. Ann Intern Med 1985;103:460–462.

75 Kafer G, Willer A, Ludwig W, Kramer A, Hehlmann R, Hastka J: Intracellular expression of CD61 precedes surface expression. Ann Hematol 1999;78:472–474.

76 Creutzig U, Zimmermann M, Ritter J, Henze G, Graf N, Löffler H, Schellong G: Definition of a standard-risk group in children with AML. Br J Haematol 1999;104:630–639.

77 Del Poeta G, Stasi R, Venditti A, Suppo G, Aronica G, Bruno A, Masi M, Tabilio A, Papa G: Prognostic value of cell marker analysis in de novo acute myeloid leukemia. Leukemia 1994;8:388–394.

78 Schabath R, Ratei R, Ludwig WD: The prognostic significance of antigen expression in leukaemia. Best Pract Res Clin Haematol 2003;16:613–628.

79 Smith FO, Lampkin BC, Versteeg C, Flowers DA, Dinndorf PA, Buckley JD, Woods WG, Hammond GD, Bernstein ID: Expression of lymphoid-associated cell surface antigens by childhood acute myeloid leukemia cells lacks prognostic significance. Blood 1992;79:2415–2422.

80 Legrand O, Perrot JY, Baudard M, Cordier A, Lautier R, Simonin G, Zittoun R, Casadevall N, Marie JP: The immunophenotype of 177 adults with acute myeloid leukemia: proposal of a prognostic score. Blood 2000;96:870–877.

81 Sperling C, Büchner T, Creutzig U, Ritter J, Harbott J, Fonatsch C, Sauerland C, Mielcarek M, Maschmeyer G, Löffler H, Ludwig WD: Clinical, morphologic, cytogenetic and prognostic implications of CD34 expression in childhood and adult de novo AML. Leuk Lymphoma 1995;17:417–426.

82 Sperling C, Büchner T, Sauerland C, Fonatsch C, Thiel E, Ludwig WD: CD34 expression in de novo acute myeloid leukaemia. Br J Haematol 1993;85:635–637.

83 Armstrong SA, Look AT: Molecular genetics of acute lymphoblastic leukemia. J Clin Oncol 2005;23:6306–6315.

84 Lucio P, Gaipa G, van Lochem EG, van Wering ER, Porwit-MacDonald A, Faria T, Bjorklund E, Biondi A, van den Beemd MW, Baars E, Vidriales B, Parreira A, van Dongen JJ, San Miguel JF, Orfao A: BIOMED-I concerted action report: flow cytometric immunophenotyping of precursor B-ALL with standardized triple-stainings. BIOMED-1 Concerted Action Investigation of Minimal Residual Disease in Acute Leukemia: International Standardization and Clinical Evaluation. Leukemia 2001;15:1185–1192.

85 Porwit-MacDonald A, Bjorklund E, Lucio P, van Lochem EG, Mazur J, Parreira A, van den Beemd MW, van Wering ER, Baars E, Gaipa G, Biondi A, Ciudad J, van Dongen JJ, San Miguel JF, Orfao A: BIOMED-1 concerted action report: flow cytometric characterization of CD7+ cell subsets in normal bone marrow as a basis for the diagnosis and follow-up of T cell acute lymphoblastic leukemia (T-ALL). Leukemia 2000;14:816–825.

86 Pui CH: Childhood leukemias. N Engl J Med 1995;332:1618–1630.

87 Pui CH, Behm FG, Crist WM: Clinical and biologic relevance of immunologic marker studies in childhood acute lymphoblastic leukemia. Blood 1993;82:343–362.

88 Pui CH, Rivera GK, Hancock ML, Raimondi SC, Sandlund JT, Mahmoud HH, Ribeiro RC, Furman WL, Hurwitz CA, Crist WM, et al: Clinical significance of CD10 expression in childhood acute lymphoblastic leukemia. Leukemia 1993;7:35–40.

89 Ratei R, Sperling C, Karawajew L, Schott G, Schrappe M, Harbott J, Riehm H, Ludwig WD: Immunophenotype and clinical characteristics of CD45-negative and CD45-positive childhood acute lymphoblastic leukemia. Ann Hematol 1998;77:107–114.
90 Rego EM, Tone LG, Garcia AB, Falcao RP: CD10 and CD19 fluorescence intensity of B-cell precursors in normal and leukemic bone marrow. Clinical characterization of CD10(+strong) and CD10(+weak) common acute lymphoblastic leukemia. Leuk Res 1999;23:441–450.
91 Ludwig WD, Harbott J, Rieder H, et al: Incidence, biologic features and treatment outcome of myeloid-antigen-positive acute lymphoblastic leukemia (My + ALL); in Büchner T, et al (eds): Acute Leukemias. IV. Prognostic Factors. Berlin, Springer-Verlag, 1994, p 24.
92 Pui CH, Behm FG, Singh B, Rivera GK, Schell MJ, Roberts WM, Crist WM, Mirro J Jr: Myeloid-associated antigen expression lacks prognostic value in childhood acute lymphoblastic leukemia treated with intensive multiagent chemotherapy. Blood 1990;75:198–202.
93 Biondi A, Camino G, Pieters R, Pui CH: Biological and therapeutic aspects of infant leukemia. Blood 2000;96:24–33.
94 Rubnitz JE, Behm FG, Downing JR: 11q23 rearrangements in acute leukemia. Leukemia 1996;10:74–82.
95 Ludwig WD, Rieder H, Bartram CR, Heinze B, Schwartz S, Gassmann W, Löffler H, Hossfeld D, Heil G, Handt S, Heyll A, Diedrich H, Fischer K, Weiss A, Volkers B, Aydemir U, Fonatsch C, Gökbuget N, Thiel E, Hoelzer D: Immunophenotypic and genotypic features, clinical characteristics, and treatment outcome of adult pro-B acute lymphoblastic leukemia: results of the German multicenter trials GMALL 03/87 and 04/89. Blood 1998;92:1898–1909.
96 Wuchter C, Harbott J, Schoch C, et al: Detection of acute leukemia cells with mixed lineage leukemia (MLL) gene rearrangements by flow cytometry using monoclonal antibody 7.1. Leukemia 2000;14:1232–1238.
97 Behm FG, Raimondi SC, Schell MJ, Look AT, Rivera GK, Pui CH: Lack of CD45 antigen on blast cells in childhood acute lymphoblastic leukemia is associated with chromosomal hyperdiploidy and other favorable prognostic features. Blood 1992;79:1011–1016.
98 Drexler HG, Ludwig WD: Incidence and clinical relevance of myeloid antigen-positive acute lymphoblastic leukemia. Recent Results Cancer Res 1993;131:53–66.
99 Secker-Walker LM, Craig JM, Hawkins JM, Hoffbrand AV: Philadelphia positive acute lymphoblastic leukemia in adults: age distribution, BCR breakpoint and prognostic significance. Leukemia 1991;5:196–199.
100 Maurer J, Janssen JWG, Thiel E, van Denderen J, Ludwig WD, Aydemir Ü, Heinze B, Fonatsch C, Harbott J, Reiter A, Riehm H, Hoelzer D, Bartram CR: Detection of chimeric BCR-ABL genes in acute lymphoblastic leukaemia by the polymerase chain reaction. Lancet 1991;337:1055–1058.
101 Hanenberg H, Baumann M, Quentin I, Nagel G, Grosse Wilde H, von Kleist S, Gobel U, Burdach S, Grunert F: Expression of the CEA gene family members NCA-50/90 and NCA-160 (CD66) in childhood acute lymphoblastic leukemias (ALLs) and in cell lines of B-cell origin. Leukemia 1994;8:2127–2133.
102 Hrusak O, Porwit MacDonald A: Antigen expression patterns reflecting genotype of acute leukemias. Leukemia 2002;16:1233–1258.
103 Hrusak O, Trka J, Zuna J, Houskova J, Bartunkova J, Stary J: Aberrant expression of KOR-SA3544 antigen in childhood acute lymphoblastic leukemia predicts TEL-AML1 negativity. The Pediatric Hematology Working Group in the Czech Republic. Leukemia 1998;12:1064–1070.
104 Baruchel A, Cayuela JM, Ballerini P, Landman Parker J, Cezard V, Firat H, Haddad E, Auclerc MF, Valensi F, Cayre YE, Macintyre EA, Sigaux F: The majority of myeloid-antigen-positive (My+) childhood B-cell precursor acute lymphoblastic leukaemias express TEL-AML1 fusion transcripts. Br J Haematol 1997;99:101–106.
105 Borowitz MJ, Rubnitz J, Nash M, Pullen DJ, Camitta B: Surface antigen phenotype can predict TEL-AML1 rearrangement in childhood B-precursor ALL: a Pediatric Oncology Group study. Leukemia 1998;12:1764–1770.
106 Lanza C, Volpe G, Basso G, Gottardi E, Barisone E, Spinelli M, Ricotti E, Cilli V, Perfetto F, Madon E, Saglio G: Outcome and lineage involvement in t(12;21) childhood acute lymphoblastic leukaemia. Br J Haematol 1997;97:460–462.
107 Secker-Walker LM, Berger R, Fenaux P, Lai JL, Nelken B, Garson M, Michael PM, Hagemeijer A, Harrison CJ, Kaneko Y, et al: Prognostic significance of the balanced t(1;19) and unbalanced der(19)t(1;19) translocations in acute lymphoblastic leukemia. Leukemia 1992;6:363–369.
108 Sang BC, Shi L, Dias P, Liu L, Wei J, Wang ZX, Monell CR, Behm F, Gruenwald S: Monoclonal antibodies specific to the acute lymphoblastic leukemia t(1;19)-associated E2A/pbx1 chimeric protein: characterization and diagnostic utility. Blood 1997;89:2909–2914.
109 Kaplinsky C, Rechavi G: Acute lymphoblastic leukemia of Burkitt type (L3 ALL) with t(8;14) lacking surface and cytoplasmic immunoglobulins. Med Pediatr Oncol 1998;31:36–38.

110 Mitelman F, Heim S: Quantitative acute leukemia cytogenetics. Genes Chromosomes Cancer 1992;5:57–66.

111 Vasef MA, Brynes RK, Murata Collins JL, Arber DA, Medeiros LJ: Surface immunoglobulin light chain-positive acute lymphoblastic leukemia of FAB L1 or L2 type: a report of 6 cases in adults. Am J Clin Pathol 1998;110:143–149.

112 Graux C, Cools J, Michaux L, Vandenberghe P, Hagemeijer A: Cytogenetics and molecular genetics of T-cell acute lymphoblastic leukemia: from thymocyte to lymphoblast. Leukemia 2006;20:1496–1510.

113 Pui CH, Relling MV, Downing JR: Acute lymphoblastic leukemia. N Engl J Med 2004;350:1535–1548.

114 Ballerini P, Blaise A, Busson Le Coniat M, Su XY, Zucman Rossi J, Adam M, van den Akker J, Perot C, Pellegrino B, Landman Parker J, Douay L, Berger R, Bernard OA: HOX11L2 expression defines a clinical subtype of pediatric T-ALL associated with poor prognosis. Blood 2002;100:991–997.

115 Garand R, Vannier JP, Béné MC, Faure G, Favre M, Bernard A: Comparison of outcome, clinical, laboratory, and immunological features in 164 children and adults with T-ALL. Groupe d'Etude Immunologique des Leucémies. Leukemia 1990;4:739–744.

116 Weng AP, Ferrando AA, Lee W, Morris JP 4th, Silverman LB, Sanchez-Irizarry C, Blacklow SC, Look AT, Aster JC: Activating mutations of NOTCH1 in human T cell acute lymphoblastic leukemia. Science 2004;306:269–271.

117 Ribeiro RC, Raimondi SC, Behm FG, Cherrie J, Crist WM, Pui CH: Clinical and biologic features of childhood T-cell leukemia with the t(11;14). Blood 1991;78:466–470.

118 Matutes E, Morilla R, Farahat N, Carbonell F, Swansbury J, Dyer M, Catovsky D: Definition of acute biphenotypic leukemia. Haematologica 1997;82:64–66.

119 Carbonell F, Swansbury J, Min T, Matutes E, Farahat N, Buccheri V, Morilla R, Secker-Walker L, Catovsky D: Cytogenetic findings in acute biphenotypic leukaemia. Leukemia 1996;10:1283–1287.

120 Kappelmayer J, Gratama JW, Karaszi E, Menendez P, Ciudad J, Rivas R, Orfao A: Flow cytometric detection of intracellular myeloperoxidase, CD3 and CD79a. Interaction between monoclonal antibody clones, fluorochromes and sample preparation protocols. J Immunol Methods 2000;242:53–65.

121 Shortman K, Naik SH: Steady-state and inflammatory dendritic-cell development. Nat Rev Immunol 2007;7:19–30.

122 Jacob MC, Chaperot L, Mossuz P, Feuillard J, Valensi F, Leroux D, Bene MC, Bensa JC, Briere F, Plumas J: CD4+ CD56+ lineage negative malignancies: a new entity developed from malignant early plasmacytoid dendritic cells. Haematologica 2003;88:941–955.

123 Bene MC, Feuillard J, Jacob MC: Plasmacytoid dendritic cells: from the plasmacytoid T-cell to type 2 dendritic cells CD4+CD56+ malignancies. Semin Hematol 2003;40:257–266.

124 Trimoreau F, Donnard M, Turlure P, Gachard N, Bordessoule D, Feuillard J: The CD4+ CD56+ CD116– CD123+ CD45RA+ CD45RO– profile is specific of DC2 malignancies. Haematologica 2003;88:ELT10.

125 Scott AA, Head DR, Kopecky KJ, et al: HLA-DR-, CD33+, CD56+, CD16– myeloid/natural killer cell acute leukemia: a previously unrecognized form of acute leukemia potentially misdiagnosed as French-American-British acute myeloid leukemia-M3. Blood 1994;84:244–255.

126 Suzuki R, Nakamura S: Malignancies of natural killer (NK) cell precursor: myeloid/NK cell precursor acute leukemia and blastic NK cell lymphoma/leukemia. Leuk Res 1999;23:615–624.

127 Vidriales MB, Orfao A, Gonzalez M, Hernandez JM, Lopez-Berges MC, Garcia MA, Canizo MC, Caballero MD, Macedo A, Landolfi C, et al: Expression of NK and lymphoid-associated antigens in blast cells of acute myeloblastic leukemia. Leukemia 1993;7:2026–2029.

128 Ludwig WD, Schoch C, Haferlach T: Classification of acute leukemias: perspective 1; in Pui CH(ed): Treatment of Acute Leukemias: New Directions for Clinical Research. Totowa, Humana Press, 2003.

129 Consolini R, Legitimo A, Rondelli R, Guguelmi C, Barisone E, Lippi A, Cantu-Rajnoldi A, Arico M, Conter V, Cocito MG, Putti MC, Pession A, Masera G, Biondi A, Basso G: Clinical relevance of CD10 expression in childhood ALL. The Italian Association for Pediatric Hematology and Oncology (AIEOP). Haematologica 1998;83:967–973.

130 Lenormand B, Béné MC, Lesesve JF, et al: PreB1 (CD10-) acute lymphoblastic leukemia: immunophenotypic and genomic characteristics, clinical features and outcome in 38 adults and 26 children. Groupe d'Etude Immunologique des Leucémies. Leuk Lymphoma 1998;28:329–342.

131 Ludwig WD, Harbott J, Bartram CR, Komischke B, Sperling C, Teichmann JV, Seibt Jung H, Notter M, Odenwald E, Nehmer A, et al: Incidence and prognostic significance of immunophenotypic subgroups in childhood acute lymphoblastic leukemia: experience of the BFM study 86. Recent Results Cancer Res 1993;131:269–282.

132 Harbott J: Cytogenetics in childhood acute lymphoblastic leukemia. Rev Clin Exp Hematol 1998;5:25–43.

133 Westbrook CA, Hooberman AL, Spino C, Dodge RK, Larson RA, Davey F, Wurster Hill DH, Sobol RE, Schiffer C, Bloomfield CD: Clinical significance of the BCR-ABL fusion gene in adult acute lymphoblastic leukemia: a Cancer and Leukemia Group B Study (8762). Blood 1992;80:2983–2990.

134 Crist WM, Carroll AJ, Shuster JJ, Behm FG, Whitehead M, Vietti TJ, Look AT, Mahoney D, Ragab A, Pullen DJ: Poor prognosis of children with pre-B acute lymphoblastic leukemia is associated with the t(1;19)(q23;p13): a Pediatric Oncology Group study. Blood 1990;76:117–122.

135 Garand R, Voisin S, Papin S, Praloran V, Lenormand B, Favre M, Philip P, Bernier M, Vanhaecke D, Falkenrodt A, et al: Characteristics of pro-T ALL subgroups: comparison with late T-ALL. Groupe d'Etude Immunologique des Leucémies. Leukemia 1993;7:161–167.

136 Ludwig WD, Thiel E, Bartram CR, Kranz BR, Raghavachar A, Löffler H, Ganser A, Büchner T, Hiddemann W, Heil G, et al: Clinical importance of T-ALL subclassification according to thymic or prethymic maturation stage. Hamatol Bluttransfus 1990;33:419–427.

137 Uckun FM, Sensel MG, Sun L, Steinherz PG, Trigg ME, Heerema NA, Sather HN, Reaman GH, Gaynon PS: Biology and treatment of childhood T-lineage acute lymphoblastic leukemia. Blood 1998;91:735–746.

138 Karawajew L, Ruppert V, Wuchter C, et al: Inhibition of in vitro spontaneous apoptosis by IL-7 correlates with upregulation of Bcl-2, cortical/mature immunphenotype, and better early cytoreduction in childhood T-ALL. Blood 2000;96:297–306.

139 Ludwig WD, Reiter A, Löffler H, Gokbuget, Hoelzer D, Riehm H, Thiel E: Immunophenotypic features of childhood and adult acute lymphoblastic leukemia (ALL): experience of the German Multicentre Trials ALL-BFM and GMALL. Leuk Lymphoma. 1994;13(suppl 1):71–76.

140 Niehues T, Kapaun P, Harms DO, Burdach S, Kramm C, Korholz D, Janka Schaub G, Gobel U: A classification based on T cell selection-related phenotypes identifies a subgroup of childhood T-ALL with favorable outcome in the COALL studies. Leukemia 1999;13:614–617.

141 Campana D, Pui C-H: Detection of minimal residual disease in acute leukemia: methodologic advances and clinical significance. Blood 1995;85:1416–1434.

142 Szczepanski T, Orfao A, van der Velden VH, San Miguel JF, van Dongen JJ: Minimal residual disease in leukaemia patients. Lancet Oncol 2001;2:409–417.

143 Coustan-Smith E, Behm FG, Sanchez J, Boyett JM, Hancock ML, Raimondi SC, Rubnitz JE, Rivera GK, Sandlund JT, Pui C-H, Campana D: Immunological detection of minimal residual disease in children with acute lymphoblastic leukaemia. Lancet 1998;351:550–554.

144 Orfao A, Ciudad J, Lopez-Berges MC, Lopez A, Vidriales B, Caballero MD, Valverde B, Gonzalez M, San Miguel JF: Acute lymphoblastic leukemia (ALL): detection of minimal residual disease (MRD) at flow cytometry. Leuk Lymphoma 1994;13(suppl 1):87–90.

145 Kerst G, Kreyenberg H, Roth C, Well C, Dietz K, Coustan-Smith E, Campana D, Koscielniak E, Niemeyer C, Schlegel PG, Muller I, Niethammer D, Bader P: Concurrent detection of minimal residual disease (MRD) in childhood acute lymphoblastic leukaemia by flow cytometry and real-time PCR. Br J Haematol 2005;128:774–782.

146 Malec M, van der Velden VH, Bjorklund E, Wijkhuijs JM, Soderhall S, Mazur J, Bjorkholm M, Porwit-MacDonald A: Analysis of minimal residual disease in childhood acute lymphoblastic leukemia: comparison between RQ-PCR analysis of Ig/TcR gene rearrangements and multicolor flow cytometric immunophenotyping. Leukemia 2004;18:1630–1636.

147 Neale GA, Coustan-Smith E, Stow P, Pan Q, Chen X, Pui CH, Campana D: Comparative analysis of flow cytometry and polymerase chain reaction for the detection of minimal residual disease in childhood acute lymphoblastic leukemia. Leukemia 2004;18:934–938.

148 Dworzak MN, Froschl G, Printz D, Mann G, Potschger U, Muhlegger N, Fritsch G, Gadner H: Prognostic significance and modalities of flow cytometric minimal residual disease detection in childhood acute lymphoblastic leukemia. Blood 2002;99:1952–1958.

149 Vidriales MB, Perez JJ, Lopez-Berges MC, Gutierrez N, Ciudad J, Lucio P, Vazquez L, Garcia-Sanz R, del Canizo MC, Fernandez-Calvo J, Ramos F, Rodriguez MJ, Calmuntia MJ, Porwith A, Orfao A, San-Miguel JF: Minimal residual disease in adolescent (older than 14 years) and adult acute lymphoblastic leukemias: early immunophenotypic evaluation has high clinical value. Blood 2003;101:4695–4700.

150 Kern W, Schoch C, Haferlach T, Braess J, Unterhalt M, Wörmann B, Büchner T, Hiddemann W: Multivariate analysis of prognostic factors in patients with refractory and relapsed acute myeloid leukemia undergoing sequential high-dose cytosine arabinoside and mitoxantrone (S-HAM) salvage therapy: relevance of cytogenetic abnormalities. Leukemia 2000;14:226–231.

151 Kern W, Haferlach T, Schoch C, Loffler H, Gassmann W, Heinecke A, Sauerland MC, Berdel W, Büchner T, Hiddemann W: Early blast clearance by remission induction therapy is a major independent prognostic factor

for both achievement of complete remission and long-term outcome in acute myeloid leukemia: data from the German AML Cooperative Group (AMLCG) 1992 Trial. Blood 2003;101:64–70.

152 Kern W, Voskova D, Schoch C, Hiddemann W, Schnittger S, Haferlach T: Determination of relapse risk based on assessment of minimal residual disease during complete remission by multiparameter flow cytometry in unselected patients with acute myeloid leukemia. Blood 2004;104:3078–3085.

153 San Miguel JF, Martinez A, Macedo A, Vidriales MB, Lopez-Berges C, Gonzalez M, Caballero D, Garcia-Marcos MA, Ramos F, Fernandez-Calvo J, Calmuntia MJ, Diaz-Mediavilla J, Orfao A: Immunophenotyping investigation of minimal residual disease is a useful approach for predicting relapse in acute myeloid leukemia patients. Blood 1997;90:2465–2470.

Sack U, Tárnok A, Rothe G (eds): Cellular Diagnostics. Basics, Methods and Clinical Applications of Flow Cytometry. Basel, Karger, 2009, pp 642–667

Flow Cytometry in the Diagnosis of Non-Hodgkin's Lymphomas

Mohammed Wattad

Kliniken Essen Süd, Evangelisches Krankenhaus Essen-Werden, Germany

Introduction/Background

In 2003, the incidence of non-Hodgkin's lymphoma (NHL) per 100,000 inhabitants was 14.9 in men and 15.7 in women [1], with a rising tendency. Until 1994, the histopathological assessment of NHL was based on various classification systems. The Kiel Classification was established in Germany and Europe. Its fundamental principle is the grouping of the lymphatic cell neoplasias according to cytomorphological, cytochemical (cytic/blastic) and immunologic (T-cell or B-cell origin) features. The consensus worked out by a project of the National Cancer Institute, the Working Formulation, was adopted in the USA in 1982 with the objective of harmonizing the recognized classifications (e.g. Rappaport and Lukes). This classification focuses on a subdivision in nodular and diffuse patterns of growth, the cytological findings being only of secondary importance. In the early 1990s, several European and American hemapathologists joined together and presented a list of lymphoma entities which is now called the REAL Classification (Revised European American Lymphoma). The REAL classification achieved a high percentage of reproducibility as documented by the International Lymphoma Classification Project. With the introduction of the WHO Classification (WHO Classification of Neoplastic Diseases of the Hematopoietic and Lymphoid Tissues) in November 1997 [2, 3], an international standard language for classifying lymphomas became available, making diagnostic and therapeutic measures comparable. In addition to the (B- or T-cell) lineage affiliation, the forms of differentiation and maturation were described, and the terms 'precursor cell lymphoma' and 'peripheral lymphoma' were introduced. The most relevant lymphoma entities for flow cytometric diagnosis are listed in table 1.

There are several complementary methods for diagnosing lymphomas which provide important elements for the diagnosis and prognosis of lymphoproliferative dis-

Table 1. Lymphoproliferative diseases according to WHO [3, 4]

B-cell lymphomas
B-cell chronic lymphoid leukemia
Prolymphocytic leukemia
Immunocytoma/lymphoplasmacytic lymphoma
Splenic marginal cell lymphoma extranodal/splenic lymphoma with villous lymphocytes
Hair cell leukemia
Multiple myeloma
Nonsecretory myeloma
Indolent myeloma
Smoldering myeloma
Plasma cell myeloma
Plasmocytoma
Solitary plasmocytoma of bone
Extramedullary plasmocytoma
Follicular lymphoma
Stage 1: 0–5% of the blasts
Stage 2: 6–15% of the blasts
Mantle cell lymphoma
Classical variant (mantle cell lymphoma)
Pleomorphic variant
Blastoid variant
T-cell lymphomas
T-cell chronic lymphocytic leukemia
Prolymphocytic leukemia
Large granular lymphocyte leukemia
Adult T-cell leukemia (HTLV-1+)
NK cell leukemia
Mycosis fungoides/Sézary syndrome

eases [5]. The most important methods are histopathological examinations (cytology/cytochemistry or histology/immunohistochemistry), cytogenetics or molecular biology, flow cytometry [6] and clinical examination. Flow cytometry significantly contributes in phenotyping lymphoma with a leukemic course by enabling a differentiation between the B- or T-cell lineage affiliation and the stage of maturation of the neoplasia. Due to an easy and rapid demonstration of the risk factors (e.g. Zap 70 for chronic lymphatic leukemia; CLL) and of minimal residual disease (MRD), it permits a quick implementation of appropriate treatment strategies.

Figure 1 summarizes the differentiation and antigen expression profiles during lymphopoiesis.

NHLs are characterized by the following immunological criteria:

- B- or T-cell proliferation;
- monoclonality;
- coexpression

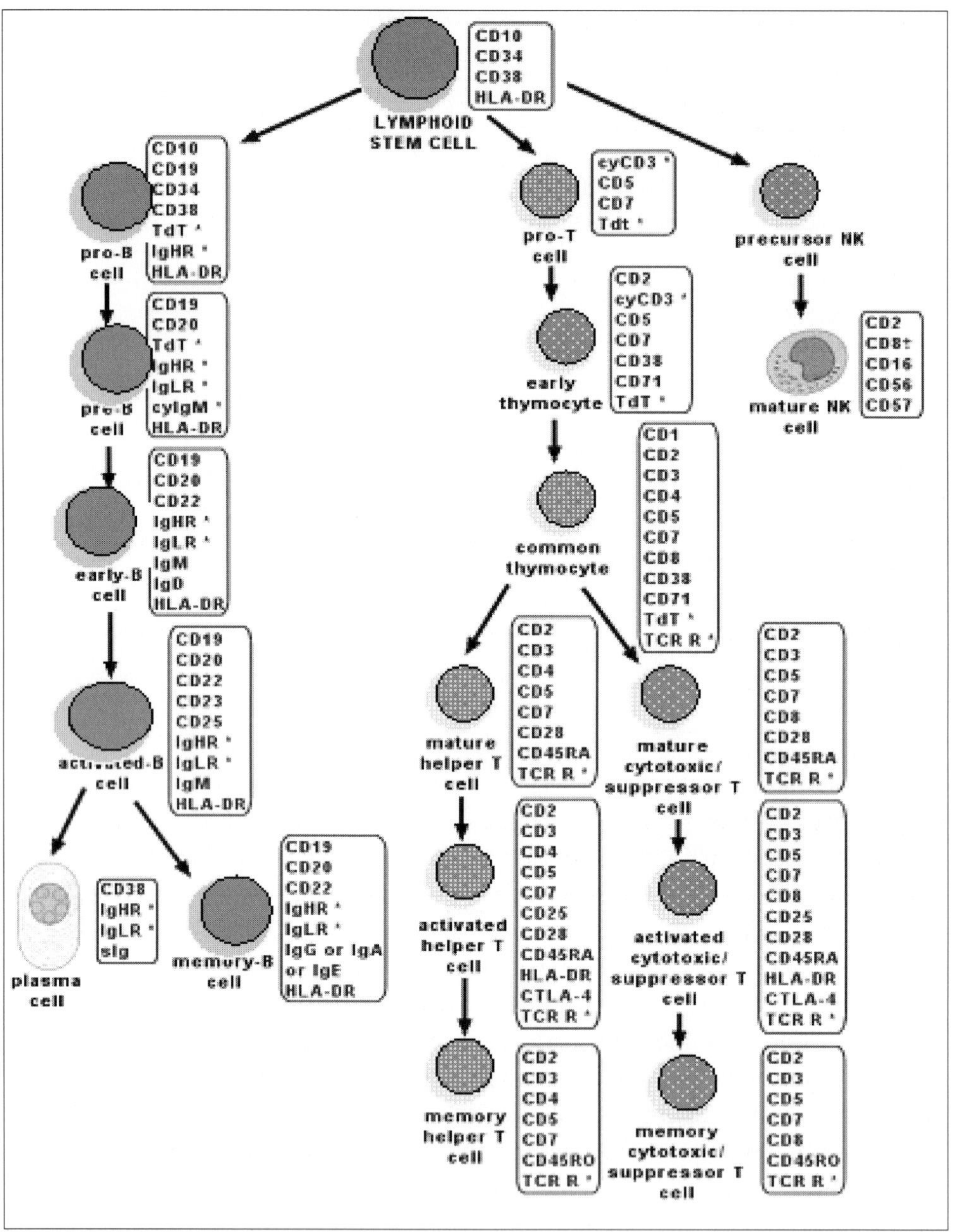

Fig. 1. Differentiation of B- and T-cell lymphocytes.

- of antigens of immature B- or T-cells and
- activation features;

– loss of lineage-affiliated antigens (frequently in T-lymphomas).

B-cell lymphomas are essentially characterized by the expression of pan-B-cell markers (CD19, CD20, CD22, CD23) [7, 8]. B-cell markers are expressed in different antigen densities. Depending on their state of maturity, B-cell lymphomas contain

Table 2. Antigen expression of lymphoproliferative B-cell diseases [modified according to 4, 9, 10]

	B-CLL	B-PLL	LPL/IC	HCL	FL	MCL	MZL
SIg	+ low	+ strong	+ strong	+ strong	+ strong	+	+
CD5	+	−/+ low	+/−	−	−	+	−
CD10	−	−	−	−	+	−	−
CD11c		−/+	−/+	+ strong	−	−	+
CD19	+	+	+	+	+	+	+
CD20	+ (low)	+	+	+	+	+	+
CD23	+	−	−/+	−	−/+	−	−
CD25				+		+	
CD38	−/+	−	+	−/+ low	−/+ low	−	+
CD79b	−/+ low	+	+	+	+	+	+
CD103	−	−	−	+ strong	−	−	−
FMC7	−/+ low	+	−/+	+	+	+	+

FL = Follicular lymphoma; HCL = hair cell lymphoma; IC = immunocytoma; MCL = mantle cell lymphoma; MZL = marginal zone lymphoma; PLL = prolymphocytic leukemia; + = always positive; − always negative; +/− = expression > 50% of the patients; −/+: expression <50% of the patients; low = low antigen density; strong = high antigen density; empty slots = are not relevant for this entity.

membrane-bound monoclonal immunoglobulins (surface immunoglobulins; sIg – predominantly IgM or IgM + IgD) and a clonal light chain restriction of the κ- or λ-type. Intracytoplasmic immunoglobulins are detected as well. However, for diagnosing the lymphoma entities, the expression of B-cell markers and/or activation markers is decisive. Besides leukocytosis, T-cell lymphomas show various expression patterns or a variable loss of T-cell antigens. Examination of the TCR-αβ/γδ is only of little relevance. The molecular examination is crucial.

Table 2 depicts the marker expressions of the individual entities of B-cell lymphoma. The markers of T-cell lymphomas are shown in table 3.

In general, differentiating B-cell chronic lymphatic leukemia (B-CLL) does not cause major problems. The other entities often have similar antigen patterns, thus their accurate assignment requires experience. Figure 2 shows how to simply distinguish B-cell lymphomas with several antigens. In addition, cell morphology, the results of the clinical examination and, if necessary, cytogenetics will influence the assessment as well.

For the routine diagnosis of leukemia, NHL and the so-called immune state, two-color markings with fluorescein isothiocyanate (FITC)- and phycoerythrin (PE)-conjugated antibodies were generally used up to now. The technological development of flow cytometers and their software, which makes it possible now to detect up to five different fluorescences at the same time and to compensate them completely as well as the further development of fluorochromes which are excitable at 488 nm (PE tandem conjugates such as ECD (trade name PE-TexasRed Conjugate), PE-

Table 3. Antigen expression of lymphoproliferative T-cell diseases [modified according to 3, 9, 11]

	T-CLL/PLL	T-LGL	NK-LGL	ATL	Sézary syndrome CTCL
CD2	+	+	+	+	+
CD3	+	+	−	+	+
CD4	+	−	−	+	+
CD5	+	−	−	+/−	+
CD7	+	−	+/−	−	−
CD8	−/+	+	+/−	−	−
CD16	−	+	+	−	−
CD25	−	−	−	+	−
CD45 RA				+	
CD56	−	−	+	−	−
CD4/CD8+	possible			Rare	

ATL = Adult T-cell leukemia/lymphoma; PLL = prolymphocytic leukemia; + = always positive; − = always negative; +/− = expression > 50% of the patients; −/+ = expression <50% of the patients.

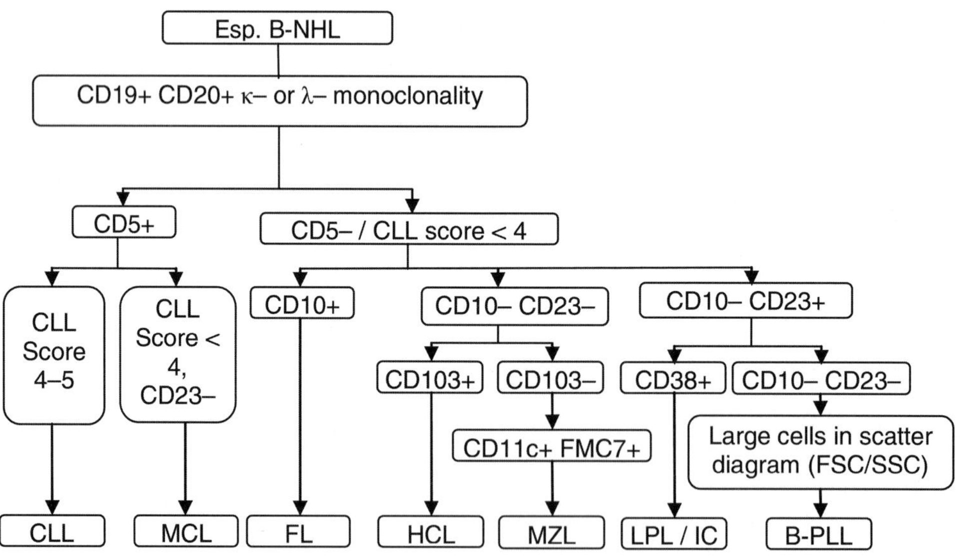

Fig. 2. Diagram of the diagnosis of B-NHL.

cyanin-5 (PE-Cy5) and PE-Cy7) now permit user-friendly standardized staining with three-, four- and five-color stains. Thus these innovative multiparameter analyses give the opportunity to obtain important additional diagnostic and prognostic information. Thus, only protocols with five-color stains are dealt with in this paper. The properties of the most important currently used fluorochromes for routine diagnosis are portrayed in 'Selection and Combination of Fluorescent Dyes', pp. 107.

Protocol

Principle

The cellular components of the samples are analyzed in the flow cytometer by examining the light scatter and emission of fluorescence light with the help of laser light. The lymphocytes can thus be distinguished from other leukocyte fractions due to their forward scatter (FSC) and side scatter (SSC) signals (granularity). Lymphocyte subpopulations are differentiated with the help of directly fluorochrome-conjugated antibodies by marking surface antigens with different four- or fivefold combinations (see 'Technical and Methological Basics of the Flow Cytometry', pp. 53).

The choice of the examination panel depends on the clinical problem. When a B- or T-NHL is suspected, performance of the whole panel is generally favored to save time and antibodies (e.g. isotype controls). When there is an increase in B- or T-cells, expansion of the examination for subtyping the B-cells as well as for proving their monoclonality is recommended. When there is clinical suspicion of multiple myeloma (MM), the measurements must be done with a special MM panel immediately upon arrival of the sample.

The basis for preparing the samples is the filtration of the material (40- to 70-μm filters) and determination of the number of leukocytes and thrombocytes with a blood count analyzer and determination of the hemoglobin concentration. Thereafter, the cell count is done with appropriate media to 10,000/μl in samples above 15,000/μl. This cell count results in standardization and a favorable antigen-to-antibody ratio.

These preparative steps are carried out using full-blood lysis and the 'wash'-method. The full-blood lysis method has the following advantages:

- low material consumption (50–100 μl);
- low cell, in particular low B-cell loss;
- maintenance of all cell fractions so that relevant additional information can be gathered;
- saving of time and material.

A cell-friendly lysis is a condition for choosing lysis reagents. Ammonium chloride lysis and the VersaLysis are suitable in this context. For the diagnosis of lymphoma, 'staining-lysing-washing (twice)' is recommended except for the preparation of the samples for κ- and λ-differentiation because such samples must be washed several times (lysing-washing (twice)-staining-washing (once)) to remove soluble immunglobulins.

Material

Material for Examination

The following material is suitable for the examination:

- peripheral blood, 5–10 ml (anticoagulant: ammonium-heparin or EDTA);
- bone marrow, 2–5 ml (anticoagulant: ammonium-heparin or EDTA);

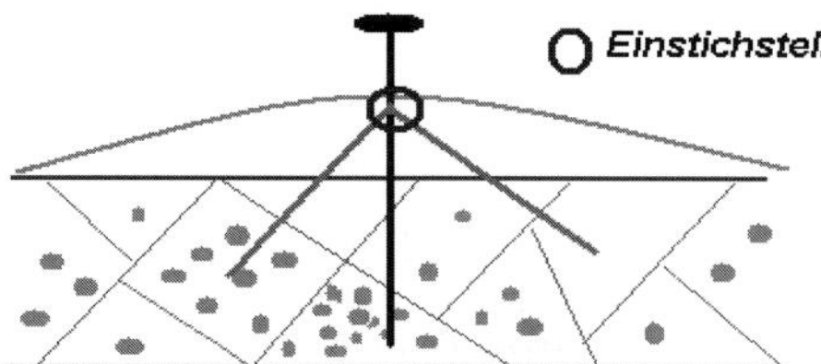

Fig. 3. Change of the aspiration areas when taking a bone marrow sample. 'Einstichstelle' = site of puncture.

- ascites aspirate/pleural effusion, 50–100 ml (anticoagulant: ammonium-heparin or EDTA);
- cerebrospinal fluid (CSF), 2–5 ml (native);
- bronchoalveolar lavage fluid (BALF), 20–50 ml (native);
- material for aspiration (e.g., lymph nodes, liver foci), 1–2 aspirations in 10 ml medium.

Reagents
We refrained from listing the products commonly used in flow cytometry and the laboratory.

The cell count can be carried out with different media. We cannot give advice on choosing companies for purchasing the antibodies here. However, it is reasonable to use combined antibodies to avoid mistakes by pipetting.

PreAnalysis Phase
The preanalysis phase has a considerable influence on the results; therefore, standardization is advisable to achieve reproducible results.

Sampling Technique
When taking a bone marrow sample, the first sample has to be used for flow cytometry. Otherwise, the puncture needle has to be placed in a new bone marrow compartment (fig. 3).

Anticoagulants
The following anticoagulants are generally available:

- Ammonium-heparin (S-Monovette®, 7.5 ml AH, Sarstedt, Nürnbrecht, Germany). The cells are stable for at least 24–48 h.
  Cave: In our experience, low-molecular heparins and lithium-heparin are not suitable.
- EDTA: the cells are stable for up to 6 h.
  In our experience, citrate is not optimal for flow cytometry.

When choosing the anticoagulants, the lapse until the sample cana be further processed should be taken into account.

Storage and Transport
Basically, the material should be used as quickly as possible. If temporary storage cannot be avoided, the material is to be stored at room temperature. Figure 4 shows

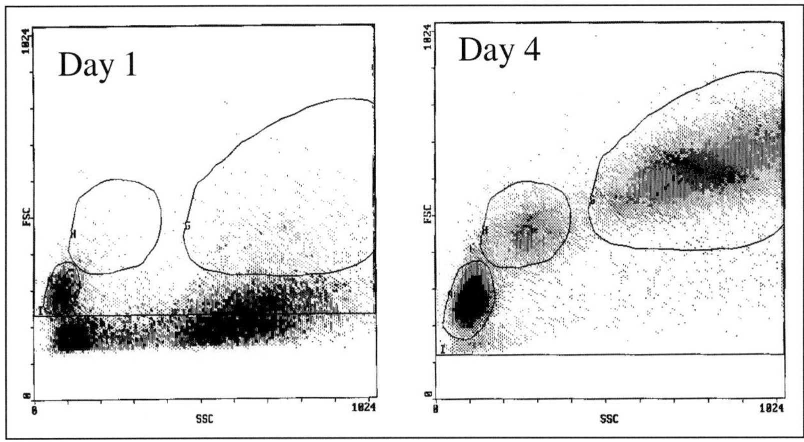

Fig. 4. Influence of time on measurement results.

test results from material that was used on the first and fourth day after sampling.

Bone marrow and peripheral blood should be used at once, or else the sample should be stored at room temperature (18–22 °C) in the dark. Too high temperatures will destroy the fragile cells. Low temperatures can lead to selective loss of cells and antigens, and adsorption of immunoglobulins to malignant cells. Heparin-stabilized samples may be stored for a longer period of time (48 h) than EDTA specimens (6–8 h). Body fluids (ascites aspirates/pleural effusion, CSF or BALF) have to be used immediately.

Requirement and Problem

- Details on material (bone marrow, peripheral blood, BALF) and date of sampling; if applicable, time,
- (suspected) diagnosis
- therapies (e.g. anti-CD20 monoclonal antibody rituximab),
- requirement profile, e.g. initial examination, suspected MRD, CD20 expression, CD52 expression.

The details mentioned above lead to an efficient, economical and valid gathering of results. The appropriate reagents can be prepared (dilution, concentration, filtration) based on details concerning the type of material and the date of sampling. Details concerning previous therapies can notably explain the lack of several antigens (e.g. CD20 on rituximab therapy) during measurement and can avoid unnecessary repetitions of examinations. The requirement specification is the basis for accounting the medical services, and working with standardized panels can save resources.

Lysing Methods

Lysis means destruction of erythrocytes. For that, several reagents are suitable. In the following, some lysis reagents are described as examples without claiming completeness.

- Q-Prep (Beckman-Coulter, Krefeld, Germany) consists of the Immuno-Prep-Reagenzien-System A/B/C: the A system lyses, the B system stabilizes and the C system fixes. This lysis is very aggressive (formic acid) and is not suitable for lymphoma diagnosis.
- Ammonium chloride lysis: pH-induced gentle lysis that causes less destruction of epitopes. Suitable for lymphoma diagnosis.
- VersaLyse (Beckman-Coulter): non-pH-induced, ready for use, stable scatter features. The main active substance is a ring-like amine that turns into a highly lysing active substance for the cells when coming into contact with the carbonic acid hydratases that are contained in the erythrocytes. Gentle lysis that is suitable for lymphoma diagnosis.

Instruments
- Vortex mixer
- Wash centrifuge or table centrifuge
- Flow cytometer.

Working Steps

Preparation of the Sample

Preparation of the sample for demonstrating κ/λ is fundamentally different from the remaining antigen evidence within the framework of NHL diagnosis. Thus both methods are portrayed separately.

During preanalysis phase, before processing the material, the following details have to be documented.
- Quality of the material
  - color/transparency
  - viscosity (glutinous/liquid)
  - large tissue pieces
  - formation of blood clots.
- date and time of arrival of the material.
- name of the medical-technical assistant.

By filtering the material with a 40- to 70-μm filter, protein chunks and blood clots are removed. DNase can be used in the case of bone marrow clots.

B-Cell Non-Hodgkin's Lymphoma without Demonstration of κ/λ
The method is based on the following principle: staining-lysing-washing (once or twice). Only material that contains erythrocytes (peripheral blood, bone marrow) is lysed. When working with CSF, BALF and acites aspirates/pleural effusion, lysing is not necessary. The material is used completely and concentrated in Falcon tubes.

(1) Mix the patient material thoroughly.
(2) Draw up a blood count.
(3) Adjust the material:
  - If the number of leukocytes is >15/nl, the cells have to be set at 10/nl.
  - If the hemoglobin concentration is >17g/dl, the material has to be diluted 1 : 2.
(4) If a cell count was done, a new blood count has to be done. (If not, this step is dropped.)
(5) Put up the test tubes and label them with the patients' data.
(6) Pipette 10 μl each of the appropriate antibodies.
(7) Add 50 μl of the patient material.
(8) Vortex the mixture.
(9) Incubate for 15 min (protected from daylight, at room temperature!).
(10) Add 2 ml VersaLyse and vortex immediately.
(11) Incubate for 15 min (protected from daylight, at room temperature!).
(12) Check lysis.
(13) Add 2 ml Isoton II, stir and then centrifuge at 300 × *g*.
(14) Cautiously pipette the surplus and pick up the pellet with 2 ml Isoton II, stir and centrifuge again at 300 × *g*.
  If a wash centrifuge (Dade Serocent) is available, it can be used for steps (13) and (14).
(15) Cautiously pipette the surplus and resuspend the pellet in 1 ml Isoton II.
(16) Store the mixture in the refrigerator pending measurement (2 h at most).

B-Cell Non-Hodgkin's Lymphoma only Demonstration of κ/λ

The method is based on the following principle: lysing-washing (twice)-staining-washing.

When working with CSF, BALF and ascites aspirates/pleural effusion, lysing is not necessary. The material is used completely and concentrated in Falcon tubes. Steps (1)–(5) are as in the preceding method (B-NHL without demonstration of κ/λ).

(6) Extract 50 μl of the adjusted sample and mix with 20 μl of rabbit serum.
(7) Add 1 ml VersaLyse and vortex immediately.
(8) Incubate for 15 min at room temperature and protected from daylight.
(9) Check lysis.
(10) Add 2 ml Isoton II, stir and then centrifuge at 300 × *g*.
(11) Cautiously pipette the surplus and pick up the pellet with 2 ml Isoton II, stir and centrifuge again at 300 × *g*.
  If a wash centrifuge (Dade Serocent) is available, it can be used for steps (10) and (11).
(12) Cautiously pipette the surplus up to 200 μl liquid, resupend the cell pellet and mix the cell suspension with 20 μl κ-FITC/λ-PE/CD19-ECD and 10 μl CD20-PE-Cy5.
(13) Incubate the cells for 15 min at room temperature, protected from daylight

(14) Add 2 ml Isoton II, stir and then centrifuge at 300 × *g*.
(15) Discard the surplus and add 2 ml Isoton II.
(16) Store the mixture in the refrigerator pending measurement (2 h at most).

Choice of Antibodies

In May 2001 we switched our lymphoma antibody panel from two-color protocols (FITC- and PE-conjugated antibodies) to four-color protocols (FITC-, PE-, ECD- and PE-Cy5-conjugated antibodies) and since spring 2005 we have been using five-color protocols (FITC-, PE-, ECD-, PE-Cy5- and PE-Cy7-conjugated antibodies). We employ preassembled three-color reagents with the fluorochromes FITC, PE and ECD (IOTest3-Reagenzlinie). The fourth and fifth color is provided by antibodies that can be chosen flexibly ('drop in').

This panel combination enables the demonstration of NHL and its subclassification according to the WHO Classification and provides reliable results for MRD diagnosis. Additionally, using panel combinations saves time and antibodies.

Protocols for the First Diagnosis of B-Cell Non-Hodgkin's Lymphoma

- Four-color B-NHL panel without demonstration of κ/λ (table 4).
- Supplement for four-color B-NHL panel for demonstration of κ/λ (table 5).
- Five-color B-NHL panel (table 6).

Protocols for the Diagnosis of Minimal Residual Disease of B-Cell Non-Hodgkin's Lymphoma

The diagnosis of MRD becomes increasingly important due to modern treatment methods such as chemotherapy or antibody therapy, stem cell transplantation and radio-immunotherapy. The diagnosis of MRD contributes to the determination of the tumor load and provides useful information on the kinetics of the response to treatment. The following points should be observed when choosing the panels and designing the protocol:

- careful choice of antibodies and conjugates;
- efficient instruments (multicolor analyses, multidimensional data analyses, two lasers if applicable);
- choice between a 'single-platform' or a 'dual-platform' method; furthermore, one has to select the precision (table 7), and
- definition of the 'gating' strategy (FSC/SCC or pan-leukocyte marker CD45); we consider CD45 advantageous, especially for bone marrow diagnosis.

A protocol for the diagnosis of MRD is given in table 8.

Protocol for T-Cell Non-Hodgkin's Lymphoma

A protocol for T-NHL is given in table 9.

Table 4. Four-color B-NHL panel without demonstration of κ/λ[a]

Tube No.	Fluorochrome-marked antibody				Volume of antibodies, μl	
	FITC	PE	ECD	PE-Cy5 'drop in'	FITC/ PE/ECD	PE-Cy5
1	isotype	isotype	Isotype	isotype	10	10
2[b]	CD8	CD4	CD3	CD19	10	10
3	CD5	CD23	CD19	CD38	10	10
4	FMC7	CD23	CD19	CD10	10	10
5[c]	CD 1003	CD11c	CD19	CD25	10	10

[a] Lysing method VersaLyse; quantity of material: 50 μl; examination protected from sunlight.
[b] Tube 2 shows the distribution of the T- and B-cells in the lymphocyte slot. In this tube, CD19 is conjugated with PE-Cy5 and serves as the quality control for B-cell evidence in the further tubes (CD19-ECD).
[c] Especially in the case of hair cell leukemia, addition of tube 5.

Table 5. Addition to the four-color B-NHL panel for demonstration of κ/λ[a]

Tube No.	Fluorochrome-marked antibody				Volume of antibodies, μl	
	FITC	PE	ECD	PE-Cy5 'drop in'	FITC/ PE/ECD	PE-Cy5
1[b]	isotype	isotype	CD19	Isotype	10	10
2	κ	λ	CD19	CD20	10	10

[a] Lysing method VersaLyse; quantity of material: 50 μl; examination protected from sunlight.
[b] Tube 1 is only prepared if κ/λ is to be examined separately. If that is not the case, tube 2 can be measured in the lymphoma panel as well.

Table 6. Five-color B-NHL panel[a]

Tube No.	Fluorochrome-marked antibody					Volume of antibodies, μl		
	FITC	PE	ECD	PE-Cy5 'drop in'	PE-Cy5 'drop in'	FITC/PE/ ECD	PE-Cy5	PE-Cy7
1	CD8	CD4	CD3	CD38	CD19	10	10	10
2[b]	FMC7	CD23	CD3	CD10	CD5	10	10	10
3	κ	λ	CD19	CD22	CD20	10	10	10
4[b]	CD103	CD11c	CD19	CD25		10	10	

[a] Lysing method VersaLyse; quantity of material: 50 μl; examination protected from sunlight.
[b] Especially in the case of hair cell leukemia, addition of tube 5.

Table 7. Calculation of the number of measurements necessary to achieve the required precision [12]

Frequency of the positive results		Total number of results that have to be measured to record a certain number of positive results with a mandatory coefficient of variation				
%	1:n	10,000 for CV 1.0%	1,600 for CV 2.5%	400 for CV 5.0%	100 for CV 10%	25 for CV 20%
10	10	100,000	16,000	4,000	1,000	250
1	100	10^6	160,000	40,000	10,000	2,500
0.1	1,000	10^7	1.6×10^6	400,000	100,000	25,000
0.01	10,000	10^8	1.6×10^7	4×10^6	10^6	250,000
0.001	100,000	10^9	1.6×10^8	4×10^7	10^7	

Table 8. Four-color B-NHL panel for evidence of MRD[a]

Tube No,[b]	Fluorochrome marked antibody				Volume of antibodies, µl	
	FITC	PE	ECD	PE-Cy5 'drop in'	FITC/PE/ECD	PE-Cy5
1	isotype	isotype	CD19	CD45	10	10
2	CD5	CD23	CD19	CD45	10	10
3	CD20	CD10	CD19	CD45	10	10
4	κ	λ	CD19	CD45	10	10
5	CD 1003	CD11c	CD19	CD45	10	10

[a] Lysing method VersaLyse; quantity of material: 50 µl; examination protected from sunlight.
[b] For CLL, follicular lymphoma, mantle cell lymphoma and marginal zone lymphoma tubes 1–4, for hair cell lymphoma tubes 1 + 5.

Table 9. Four-color T-NHL panel[a]

Tube No.	Fluorochrome-marked antibody				Volume of antibodies, µl	
	FITC	PE	ECD	PE-Cy5 'drop in'	FITC/PE/ECD	PE-Cy5
1	isotype	isotype	CD45	isotype	10	10
2	CD8	CD4	CD45	CD3	10	10
3	HLA-DR	CD7	CD45	CD2	10	10
4	CD5	CD1a	CD45	CD56	10	10
5	CD45RA	CD25	CD45	CD4	10	10
5	TCR γδ	TCR αβ	CD45	CD3	10	10

[a] Lysing method VersaLyse; quantity of material: 50 µl; examination protected from sunlight.

Table 10. Five-color panel for multiple myeloma[a]

Tube number	Fluorochrome-marked antibody					Volume of antibodies, µl		
	FITC	PE	ECD	PE-Cy5 'drop in'	PE-Cy7 'drop in'	FITC/PE/EDC	PE-Cy5	PE-Cy7
1	CD14	CD138	CD45	CD38	CD156	10	10	10
2	κ	λ	CD19	CD38	CD45 (optional)	10	10	10

[a] Lysing method VersaLyse; quantity of material: 50 µl; examination protected from sunlight.

Protocol for Multiple Myeloma

A protocol for MM is given in table 10.

Measuring with the Flow Cytometer

Instrument Adjustment

The adjustment, maintenance and inspection of the instruments are essential parts of the quality assurance in the laboratory and should be conducted according to the manufacturer's guidelines. The data have to be documented in a logbook. In our opinion, the following tests should be part of the quality control:

- daily optical inspection of the instruments;
- daily flow check;
- color compensation once a week.

Data Analysis

When interpreting the measured data, not only malignant cells but also normal or nonmalignant cells have to be considered. This information should be taken from the nongated diagram. As the malignant cell population is marked by an amorphous gate, the following measurements of the individual parameters only refer to this gate. The percentage of individual cell populations can be determined by quadrant statistics. For the individual antibody, the percentage is obtained from the sum of two quadrants, either 1 + 2 or 2 + 4. The double expression of the cell population can be read in quadrant four. For the weakly expressed antigens such as CD23, CD38, C25 and sIg, the 'multigraph option' (fig. 5) is very useful for the re-correction due to the fact that the calculated percentages in the quadrant statistics are too low. Quadrant statistics must often be corrected again manually for certain antigens. To prove the monoclonality of B-cells, a κ/λ-light chain factor of ?5 is applied. Evi-

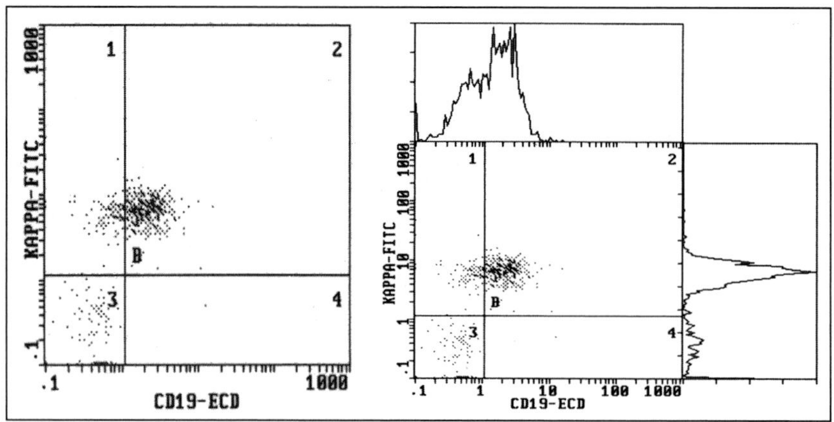

Fig. 5. sIg depicted with the 'multigraph option'.

dence of an 'abnormal' phenotype without proof of the monoclonality is not sufficient for the diagnosis of peripheral B-cell neoplasia. After chemotherapy, nonneoplastic B-cells can express up to 50% CD5 or CD10, reflecting a regeneration of the B-cell system.

The demonstration of T-cell diseases is more difficult due to 'contamination' with normal T-cells. Therefore, multicolor analysis plays an important role in identifying aberrant antigen patterns of lymphocytes. The assessment is limited to the distribution of subpopulations and the number of T-cells with unusual antigen expression (loss of an antigen or co-expression of CD8 or CD4).

Interpretation

In the following, some peripheral B- and T-lymphoproliferative diseases are introduced in which flow-cytometric diagnosis plays an important role.

Characteristics of Peripheral B-Lymphoproliferative Diseases

B-Cell Chronic Lymphoid Leukemia

B-CLL represents 6.7% of NHLs [4, 13–19] (fig. 6).

Clinical examination:

- mostly inconspicuous;
- splenomegaly;
- lymphoma.

Morphology:

- small lymphocytes;
- crude chromatin;

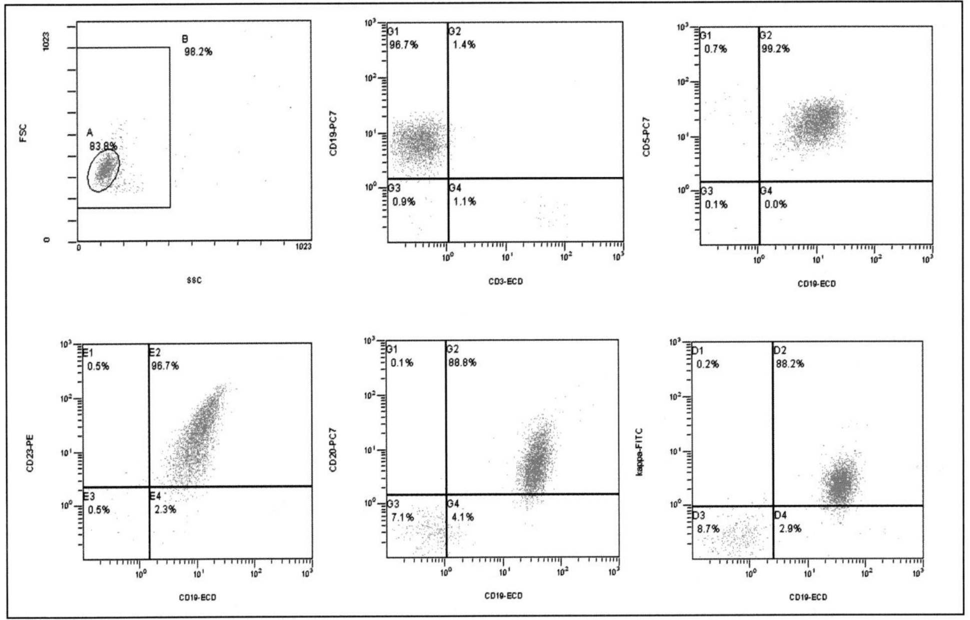

Fig. 6. Typical flow-cytometric result in case of B-CLL.

– narrow cytoplasm;
  nucleus shadow.

Immunocytology:

- CD19+ CD5+ CD23+;
- sIg+, low antigen density;
- CD20+, partly low antigen density.

Genetics and prognosis:

- del(13q): >15 years;
- del(11q): 6.6 years;
- del(17q): 3.6 years;
- trisomy 3.8: 12 years.

Prognosis markers:

- CD38 (CD38+ >30% = bad prognosis)
- Zap 70 (Zap 70+ >20% = bad prognosis)
- IgVH-gene (nonmutated IgVH gene = bad prognosis).

The diagnosis of CLL is precisely defined in the WHO classification [4].

B-Prolymphocyte Leukemia

Prolymphocyte leukemia is a rare disease [4, 20].

Clinical examination:

- massive splenomegaly;
- highly leukemic;
- quickly resistant to therapy.

Morphology:
- large leukocytes;
- moderately crude chromatin;
- enlarged cytoplasm;
- clear, round nucleus.

Immunocytology:
- CD19+ CD5± CD23−;
- sIg+, very high antigen density;
- CD20+, very high antigen density;
- FMC7+.

Genetics:
- t(11;14);
- del(11q23), del(13q14);
- p53 mutations.

Lymphoplasmocytic Lymphoma/Immunocytoma

Lymphoplasmocytic lymphoma/immunocytoma represent 1.5% of NHLs [4].

Clinical examination:
- splenomegaly;
- lymphoma;
- IgM paraprotein.

Morphology:
- larger lymphocytes;
- crude chromatin;
- enlarged cytoplasm, nucleus partly eccentric;
- partly plasma-cellular differentiation.

Immunocytology:
- CD19+ CD5− CD38+;
- immunocytoma, generally CD23−;
- sIg+, high antigen density;
- CD20+, high antigen density.

Genetics:
- t(9;14) (p13;q23);
- PAX-5 gene rearrangement.

Hair Cell Leukemia

Hair cell leukemias represent 2% of NHLs [4, 21].

Clinical examination:
- normal form:
  - massive splenomegaly;
  - pancytopenia;
  - opportunistic infections;

- variant:
  - massive splenomegaly,
  - leukemic, and
  - no cytopenia.

Morphology:
- medium-sized lymphocyte;
- relatively homogeneous chromatin;
- wide cytoplasm, fine extensions;
- tartrate-resistant acid phosphatase.

Immunocytology:
- normal form:
  - CD19+ CD5−,
  - CD25+, high antigen density,
  - CD103+,
  - CD11c, high antigen density,
  - sIg+, high antigen density,
  - CD20+, very high antigen density,
  - FMC7+;
- variant:
  - CD19+ CD5− CD25− CD103±,
  - CD11c, high antigen density,
  - sIg+, very high antigen density,
  - CD20+, very high antigen density, and
  - FMC7+.

Genetics:
- cyclin-D1 overexpression.

Mantle Cell Lymphoma

Mantle cell lymphomas represent 6% of NHLs [3, 4, 22].

Clinical examination:
- lymphadenopathy;
- hepato- and splenomegaly;
- often affection of the bone marrow.

Morphology:
- small- to medium-sized lymphocytes;
- moderately condensed chromatin;
- indentations of the nucleus:
- few prominent nucleoli;
- narrow, hardly definable cytoplasm.

Immunocytology:
- CD19+ CD5+ CD23− CD10−;
- sIg+ (more often λ than κ);

- CD20+;
- FMC7+.

Genetics:

- t(11;14), (q13,q23);
- cyclin-D1 overexpression;
- del 17p, del 13(q14), trisomy +12.

Follicular Lymphoma

Follicular lymphomas represent 22% of NHLs [3, 4, 23].

Clinical examination:

- lymphadenopathy;
- splenomegaly;
- often affection of the bone marrow.

Morphology:

- polymorph centrocytes.

Immunocytology:

- CD19+ CD5− CD10+ CD23± CD11c−;
- sIg+, very high antigen density;
- CD20+.

Genetics:

- t(14;18) 70–95% of the cases;
- bcl-2 overexpression.

Splenic Marginal Zone Lymphoma/Splenic Lymphoma with Villous Lymphocytes

Splenic marginal zone lymphomas represent 1.8% of NHL [4, 24, 25].

Clinical examination:

- lymphadenopathy;
- splenomegaly;
- often affection of the bone marrow.

Morphology:

- cellular heterogeneity, centrocyte-like;
- voluminous cytoplasm;
- monocytoid B-cells, small lymphocytes and plasma cells.

Immunocytology:

- CD19+ CD5− CD10− CD23− CD25− CD103− CD11c+;
- sIg+, moderately high antigen density;
- CD20+.

Genetics:

- t(11;18) 70–95% of the cases;
- del 7(q21–32), trisomy +3 (SLVL).

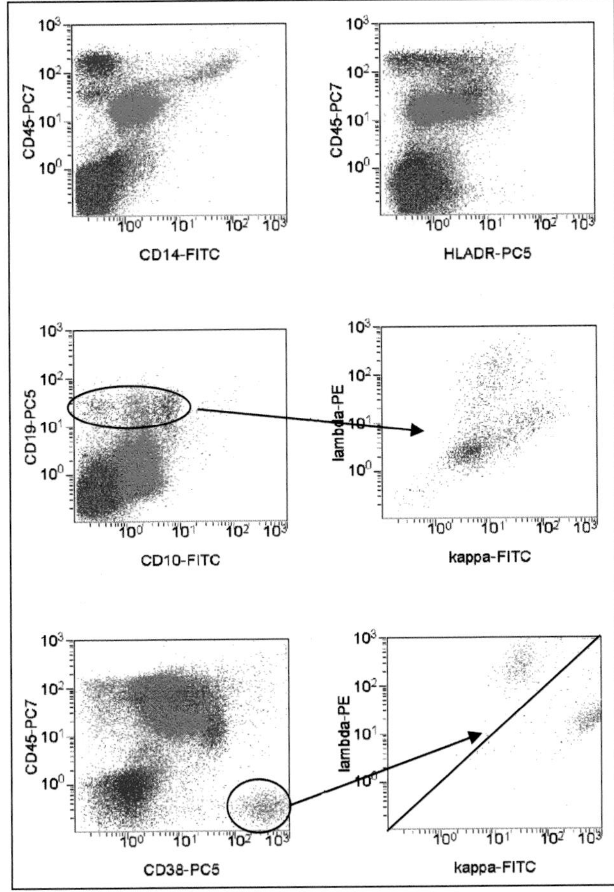

Fig. 7. Confirmation of partially nondifferentiated, but polyclonal plasma cells (figure was kindly provided by Gregor Rothe, Bremen).

Multiple Myeloma [4, 26–28]

Immunophenotyping is not a standard method for the diagnosis of MM. Immunophenotyping often enables distinction between normal and pathological plasma cells (fig. 7). The demonstration of aberrant expressions and/or cytoplasmic light chain expression is of vital importance to establish the diagnosis and predict the course of MRD. The cytomorphological degree of infiltration of multiple myeloma does not correlate with the results of flow cytometry.

Clinical examination:

- backache;
- signs of hematopoietic insufficiency.

Morphology:

- multiplication of plasma cells;
- mature cells, polymorph, anaplastic.

Immunocytology:

- CD38+ CD138+ CD56± CD45− cyIg+.

Genetics:
- structural/numeric 1; 11; 14;
- deletions (8, 13, 14, X), del13(q14), del7q, del17(p13), t(11;14) (q13;q32).

Characteristics of the Peripheral T-Lymphoproliferative Diseases [3, 4, 29–34]

T-Cell Chronic Lymphocytic Leukemia/T-Prolymphocytic Leukemia
Clinical examination:
- leukocytosis > 100,000/μl;
- lymphadenopathy;
- splenomegaly.

Morphology:
- barb- or cloverleaf-shaped nucleus;
- nucleoli mostly missing.

Immunocytology:
- CD2+ CD3+ CD5+ CD4 CD8±;
- CD7–.

Genetics and Prognosis:
- abnormalities on chromosome 14 and 8, e.g., t14(q11) t14(q32), t(14;14) (q11;q32) TCR-αβ/TCL1 t(8;8) (p11–12;q12); trisomy 8q; del 12(p13); del 11(q23) ATM, t(X;14) (q28;q11) TCR αβ/MTC;
- aggressive disease;
- survival <1 year.

Large Granular Lymphocyte Leukemia
Clinical examination:
- moderate lymphocytosis;
- often neutropenia.

Morphology:
- large lymphocytes;
- basophil cytoplasm with azurophil granules;
- variants: small lymphocytes, T-CLL-like (if indolent).

Immunocytology:
- T-LGL type:
  - CD2+ CD3+ CD8+ TRC αβ+ CD16+
  - CD4– CD5– CD56–, and
  - CD7–;
- NK-LGL type:
  - CD2+ CD16+ CD56+,
  - CD3– CD4– CD5– TRC αβ–,

- variants:
  - CD8+ CD56− CD57− and
  - CD8± CD56− CD57+.

Genetics and Prognosis:
- clonal rearrangement of the TCR-β-gene;
- T-LGL type mostly indolent and not aggressive;
- NK-LGL aggressive disease:
  - del6(q21q25).

Adult T-Cell Leukemia/Lymphoma

Clinical examination:
- HTLV-1 associated;
- lymphadenopathy;
- splenomegaly;
- leukocytosis, increased levels of LDH.

Morphology:
- variable appearance;
- pleomorph, hyperlobular nuclei;
- giant cells with multiple nuclei.

Immunocytology:
- CD2+ CD3+ CD4+ CD5+;
- CD7− CD8−.

Genetics:
- HTLV-1, clonal rearrangement of the TCR gene.

Clinical examination:
- acute adult T-cell leukemia/lymphoma;
- lymphomatous variant;
- 'smoldering' variant.

Mycosis Fungoides / Sézary Syndrome

Mycosis fungoides and Sézary syndrome represent 0.5–1% of NHLs.

Clinical examination:
- primary skin manifestation;
- lymphadenopathy;
- affection of the bone marrow.

Morphology:
- Sézary cells: cerebriform appearance;
- small nucleoli in general.

Immunocytology:
- CD2+ CD3+ CD5+ CD4+;
- CD7− (75% of the cases), CD8−.

Genetics and Prognosis:
- clonal rearrangement of the TCR-β gene;
- percent-year actuarial 5-year survival rate 10–20%.

Interpretation of the Results

The interpretation of the results starts with the description of the preanalyses, the clinical details and the question. Multi-attribute analysis enables a varied evaluation of the measured data:
- correlation of the scatter features;
- correlation of the fluorescence intensity;
- correlation of the scatter features and the fluorescence intensity.

The results have to be summarized in a written interpretation. The report comprises the following parameters:
- patients' data;
- preanalytical results;
- clinical details and question;
- cell count;
- multicolor analysis (three-color, four-color analysis);
- type of cytometer;
- gating strategy and relevant gates;
- expressions and coexpressions of the cells in the gate as percentage;
- description of the diagrams, especially antigen density;
- assessment and recommendation.

Quality Assurance/Quality Control

The standard operating procedures are the basis of quality assurance for all the processes in the laboratory. In addition, the adjustment of the instruments and regular employee training are important parts of the quality assurance. Internal quality controls (phenotyping) should be carried out on normal cell samples and reference material. It is useful as well to participate in an interlaboratory test. The results of immunophenotyping have to be compared with other results such as morphology/cytochemistry, fluorescence microscopy or PCR.

The examination of normal lymphocytes with monoclonal antibodies and knowledge of physiological antigen patterns permit the assessment of the staining behavior of individual antibodies which will enable the verification of the sensitivity and specificity of antibodies and their clones. CD19, CD20 and CD22 for mature B-cells and CD2, CD3, CD5, CD7 for peripheral T-cells should be distributed more or less equally. Moreover, logical sum controls should be carried out; and at this point, the following should be mentioned:

- CD3+ CD4+ cells + CD3+ CD8+ cells = CD3+ cells ;
- κ+ cells + λ+ cells = entire B-cells (CD19+);
- T-cells + NK cells + B-cells (in the lymphocyte gate) = 100%.

Troubleshooting

Problems with Lysis

- Coagulation during lysis:
- • reason: among others, cold agglutinins,
  • troubleshooting: transport, storage and preparation should be carried out at 37 °C (incubator);
- Partial lysis:
  • reason: among others, cold agglutinins; hemoglobin is high,
  • troubleshooting: incubate lysis at 37 °C (incubator) or new preparation with further dilution of the sample (HBSS/FCS);
- Absent lysis:
  • reason: resistance of erythrocytes due to concomitant diseases and/or medicaments,
  • troubleshooting: incubate lysis at 37°C (incubator).

Necessary Time

For the complete analysis, 2 h are necessary.

Summary

With the introduction of the WHO classification, a classification accepted worldwide accepted classification is available for the first time; it notably takes into consideration insights from the immunologically determined lineage affiliation. According to this classification, NHLs are classified as B- or T-cell lymphomas.

Flow cytometry permits an easy and quick classification of such lymphomas. The monoclonality of B-cells is demonstrated by light-chain restriction. By examining the co-expressed antigens with multiattribute analyses, the stage of maturation can be determined, especiallyin diseases such as chronic lymphatic leukemia that exhibit a leukemic courses. An antigen shift, which serves as a sign or rather as evidence of a transformation can be demonstrated early. Flow cytometry also offers an important prospective area of activity in the diagnosis of minimal residual disease, especially in the frame of curative treatments. In this chapter, the main emphasis was put on B-cell lymphomas.

Malignant lymphomas of the T/NK type are rare diseases that represent 10% of malignant lymphomas at most. The sensitivity and specificity of flow cytometry are clearly lower in the diagnosis of T/NK than in that of B-NHL due to the fact that a proof of monoclonality comparable to that in the diagnosis of B-NHL diagnosis (κ/λ-ratio) is lacking. Because of similar antigen expression patterns of normal and pathological T-cells (abnormal distribution and/or an abnormal antigen expression), false-negative results have to be taken into account. In the diagnosis of MRD of T-NHL, flow cytometry is as useful as in the diagnosis of B-NHL if there is an abnormal phenotype.

The working instructions and protocols result from experience and about ten years of continuous work. The mentioned antibodies and instrument equipment are merely examples.

References

1 Krebs in Deutschland, 4. überarbeitete, aktualisierte Ausgabe. Saarbrücken, Arbeitsgemeinschaft Bevölkerungsbezogener Krebsregister in Deutschland, 2004.

2 Jaffe ES, Harris NL, Diebold J, Müller-Hermelink HK: World Health Organization classification of neoplastic diseases of the hematopoietic and lymphoid tissues. Am J Clin Pathol 1999;111(suppl 1):8–12.

3 Harris NL, Jaffe ES, Diebold J, Flandrin G, Müller-Hermelink HK, Vardiman J, Lister TA, Bloomfield CD: World Health Organization classification of neoplastic diseases of the hematopoietic and lymphoid tissues: report of the clinical advisory committee meeting – Airlie House, Va., Nov 1997. J Clin Oncol 1999;17:3835–3849.

4 Jaffe ES, Harris NL, Stein H, Vardiman JW (eds): World Health Organization Classification of Tumours. Pathology and Genetics of Tumours of Haematopoietic and Lymphoid Tissues. Lyon, IARC Press, 2001.

5 Huber H, Fasching B, Pohl P, et al: Non-Hodgkin-Lymphome; in Huber H, Löffler H, Pastner D (Hrsg): Diagnostische Hämatologie. Berlin, Springer-Verlag, 1992, pp 440–553.

6 Tbakhi A, Edinger M, Myles J, Pohlman B, Tubbs RR: Flow cytometric immunophenotyping of non-Hodgkin's lymphomas and related disorders. Cytometry 1996;25:113–124.

7 Ludwig WD, Komischke B, Böttcher S, Martin M: Immunphänotypisierung akuter Leukämien und leukämisch verlaufender niedrig-maligner Non-Hodgkin-Lymphome (Methoden, relevante Antigene, Interpretation); in Schmitz G, Rothe G (Hrsg): Durchflußzytometrie in der klinischen Zelldiagnostik. Stuttgart, Schattauer, 1994, pp 77–104.

8 Bettelheim P: Immunphänotypisierung von Non-Hodgkin-Lymphomen; in Thomas L (Hrsg): Labor und Diagnose. Frankfurt, TH-Books Verlagsgesellschaft mbH, 1998, pp 556–558.

9 Harris NL, Jaffe ES, Diebold J, Flandrin G, Muller-Hermelink HK, Vardiman J, Lister TA, Bloomfield CD: World Health Organization classification of neoplastic diseases of the hematopoietic and lymphoid tissues: report of the Clinical Advisory Committee meeting, Airlie House, Va., Nov 1997. J Clin Oncol 1999;12:3835–3849.

10 Fruehauf S, Topaly J, Wilmes A, Ho AD: Durchflusszytometrische Diagnostik maligner hämatologischer Erkrankungen in Onkologie. Landsberg, ecomed, 2001.

11 Royston I, Majda JA, Baird SM, Meserve BL, Griffiths JC: Human T-cell antigens defined by monoclonal antibodies. J Immunol 1980;125:725–731.

12 Hoy T: Analysis and isolation of minor cell populations; in McCarthy DA, Macey MG (eds): Cytometric Analysis of Cell Phenotype and Function. Cambridge, Cambridge University Press, 2001, pp 165–181.

13 O'Brien S, del Giglio A, Keating M: Advances in the biology and treatment of B-cell chronic lymphocytic leukemia. Blood 1995;85:307–318.

14 Baldini L, Cro L, Calori R, Nobili L, Silvestris I, Maiolo AT: Differential expression of very late activation antigen-3 (VLA-3)/VLA4 in B-cell non-Hodgkin lymphoma and B-cell chronic lymphocytic leukemia. Blood 1992;79:2688–2693.
15 Baldini L, Cro L, Cortelezzi A, Calori R, Nobili L, Maiolo AT: Immunophenotypes in 'classical' B-cell lymphocytic leukemia: correlation with normal cellular counterpart and clinical findings. Cancer 1990;66:1738–1742.
16 Cro L, Guffanti A, Colombi M, Cesana B, Grimoldi MG, Patriarca C, Goldaniga M, Neri A, Intini D, Cortelezzi A, Maiolo AT, Baldini L: Diagnostic role and prognostic significance of a simplified immunophenotypic classification of mature B cell chronic lymphoid leukemias. Leukemia 2003;17:125–132.
17 Hubl W, Iturraspe J, Braylan RC: FMC7 antigen expression on normal and malignant B-cells can be predicted by expression of CD20. Cytometry 1998;34:71–74.
18 18 Durig J, Naschar M, Schmucker U, Renzing-Kohler K, Holter T, Huttmann A, Duhrsen U: CD38 expression is an important prognostic marker in chronic lymphocytic leukaemia. Leukemia 2002;16:30–35.
19 Crespo M, Bosch F, Villamor N, Bellosillo B, Colomer D, Rozman M, Marce S, Lopez-Guillermo A, Campo E, Montserrat E: ZAP-70 expression as a surrogate for immunoglobulin-variable-region mutations in chronic lymphocytic leukemia. N Engl J Med 2003;348:1764–1775.
20 Salomon-Nguyen F, Valensi F, Merle-Beral H, Flandrin G: A scoring system for the classification of CD5− B-CLL versus CD5+ B-CLL and B-PLL. Leuk Lymphoma 1995;16:445–450.
21 Matutes E, Morilla R, Owusu-Ankomah K, Houliham A, Meeus P, Catovsky D: The immunophenotype of hairy cell leukemia (HCL). Proposal for a scoring system to distinguish HCL from B-cell disorders with hairy or villous lymphocytes. Leuk Lymphoma 1994;14(suppl 1):57–61.
22 Finn WG, Thangavelu M, Yelavarthi KK, Goolsby CL, Tallman MS, Traynor A, Peterson LC: Karyotype correlates with peripheral blood morphology and immunophenotype in chronic lymphocytic leukemia. Am J Clin Pathol 1996;105:458–467.
23 Cornfield DB, Mitchell DM, Almasri NM, Anderson JB, Ahrens KP, Dooley EO, Braylan R: Follicular lymphoma can be distinguished from benign follicular hyperplasia by flow cytometry using simultaneous staining of cytoplasmic bcl-2 and cell surface CD20. Am J Clin Pathol 2000;114:258–263.
24 Matutes E, Morilla R, Owusu-Ankomah K, Houlihan A, Catovsky D: The immunophenotype of splenic lymphoma with villous lymphocytes and its relevance to the differential diagnosis with other B cell disorders. Blood 1994;83:1558–1562.
25 Baldini L, Fracchiolla NS, Cro LM, Trecca D, Romitti L, Polli E, Maiolo AT, Neri A: Frequent p53 gene involvement in splenic B-cell leukemia/lymphomas of possible marginal zone origin. Blood 1994;84:270–278.
26 Ocqueteau M, Orfao A, Almeida J, Blade J, Gonzalez M, Garcia-Sanz R, Lopez-Berges C, Moro MJ, Hernandez J, Escribano L, Caballero D, Rozman M, San Miguel JF: Immunophenotypic characterization of plasma cells from monoclonal gammopathy of undetermined significance patients. Implications for the differential diagnosis between MGUS and multiple myeloma. Am J Pathol 1998;152:1655–1665.
27 Lima M, Teixeira MA, Fonseca S, Goncalves C, Guerra M, Queiros ML, Santos AH, Coutinho A, Pinho L, Marques L, Cunha M, Ribeiro P, Xavier L, Vieira H, Pinto P, Justica B: Immunophenotypic aberrations, DNA content, and cell cycle analysis of plasma cells in patients with myeloma and monoclonal gammopathies. Blood Cells Mol Dis 2000;26:634–645.
28 Kyle RA: Multiple myeloma. Diagnostic challenges and standard therapy. Semin Hematol 2001;38(suppl 3):11–14.
29 Bennet JM, Catovsky D, Daniel M-T, Flandrin G, Galton DAG, Gralnick HR, Sultan C, the French-American-British (FAB) Cooperative Group: Proposals for the classification of chronic (mature) B and T lymphoid leukemias. J Clin Pathol 1989;42:567–584.
30 Uppenkamp M, Feller AC: Classification of malignant lymphoma. Onkologie 2002;25:563–570.
31 Neri A, Fracchiolla NS, Roscetti E, Garatti S, Trecca D, Boletini A, Perletti L, Baldini L, Maiolo AT, Berti E: Molecular analysis of cutaneous B- and T-cell lymphomas. Blood 1995;86:3160–3172.
32 Pirruccello SJ, Lang MS: Differential expression of CD24-related epitopes in mycosis fungoides/Sézary syndrome: a potential marker for circulating Sezary cells. Blood 1990;76:2343–2347.
33 Puel A, Ziegler SF, Buckley RH, Leonard WJ: Defective IL7R expression in T− B+ NK+ severe combined immunodeficiency. Nat Genet 1998;20:394–397.
34 Rothe G, Schmitz G: Consensus protocol for the flow cytometric immunophenotyping of hematopoietic malignancies. Working Group on Flow Cytometry and Image Analysis. Leukemia 1996;10:877–895.

Sack U, Tárnok A, Rothe G (eds): Cellular Diagnostics. Basics, Methods and Clinical Applications of Flow Cytometry. Basel, Karger, 2009, pp 668–679

Paroxysmal Nocturnal Hemoglobinuria

Martin Grünewald

Medizinische Klinik I, Klinikum Heidenheim, Germany

Introduction

Paroxysmal nocturnal hemoglobinuria (PNH) belongs to a group of corpuscular hemolytic anemias. Within this group, PNH is the only acquired form, and is caused by a somatic mutation arising in a hematopoietic stem cell. The mutation occurs in the gene that encodes the synthesis of the glycosylphosphatidylinositol (GPI) anchor protein (PIG-A gene); the gene is localized on the X-chromosome. The disturbed membrane function characteristic for PNH is caused by reduced or missing expression of cell surface proteins, anchored there by a GPI molecule (table 1) [1–3].

The denotation of the disease derives from the observation that hemolysis often sets off at night during sleep, resulting in hemoglobinuria in the morning urine which may be discolored. Most likely, hemolysis is triggered by the resorption of lipopolysaccharides from the gut during nighttime.

It is assumed that the causal event for the development of PNH is damage to a hematopoietic stem cell. Depending on the mode and intensity of the damage, the ensuing disease may manifest as aplastic anemia (AA), myelodysplasia, or PNH. In accordance with this hypothesis, the clinical presentation of these diseases may overlap; GPI-deficient cells are not only found in PNH, but also in AA and myelodysplasia. The phenotype of the resulting disease, however, is not only determined by the characteristics of the causal damage: predisposing factors of the bone marrow environment most probably play a major role as well [4].

Clinically, PNH is characterized by the triad of intravascular, corpuscular hemolysis (chronic or episodic), thromboembolic events, frequently of atypic localization (e.g. Budd-Chiari syndrome, portal vein or mesenteric vein thrombosis) and cytopenia (commonly as tricytopenia).

Table 1. GPI-anchored proteins on hematopoietic cells[a]

	Expression on all hematopoietic cell lines	Expression on few hematopoietic cell lines	Expression on a single hematopoietic cell line
Erythropoiesis	CD55, CD59	CD58	Acetylcholine esterase
Granulocytes	CD55, CD59	CD24, CD87, CD157	CD16, CD66b, alkaline phosphatase
Monocytes	CD55, CD59	CD48, CD52, CD87, CD157	CD14
Lymphocytes	CD55[b], CD59	CD24, CD48, CD52, CD58	CD73
Platelets	CD55, CD59	(CD87[c])	GP 500, GP 175

[a] Modified from [5].
[b] Heterogeneous expression [17].
[c] Contradictory data.

Classification, Quantification and Clinical Relevance

The increasing ability to discern different types of PNH, together with growing knowledge of the natural disease course, in particular the association of specific complications with defined subtypes, enable increasingly differentiated treatment of the various PNH types. To avoid both overtreatment and undertreatment, precise knowledge of PNH type and individual disease course is mandatory. The basis for a precise categorization of PNH types is the exact classification and quantification of GPI-deficient cells.

Different Types of PNH

The detection of a GPI-deficient cell population (called PNH clone) is not sufficient to diagnose classic hemolytic PNH. PNH clones are also commonly found in AA and less frequently in myelodysplasia. Of note, an overlap is notable in particular between AA and PNH; starting as primary AA, PNH may evolve, and, vice versa, during PNH, aplastic crises may occur. The AA overlap type of PNH is characterized by a relatively small PNH clone (the GPI-deficient granulocyte clone is typically <20%), in contrast to severe (tri-)cytopenia. Bone marrow examination will demonstrate the hypoplastic nature of the cytopenia. In contrast to this form, classic hemolytic PNH is characterized by large PNH clones (the GPI-deficient granulocyte clone is typically >50%) and hyperplastic bone marrow. Peripheral cytopenia, a common feature in classic PNH as well, is caused by increased turnover [5]. The relevance of the (rare) detection of PNH clones in myelodysplasia remains to be determined. Typical features of myelodysplasia are advanced age, peripheral cytopenia, bone marrow dysplasia, and cytogenetic aberrations in an increasing number of patients.

Together with the flow-cytometric quantification of PNH clones, the patient's history, laboratory findings, and, if necessary, bone marrow examination or cytogenetics, enable the important discrimination between classic hemolytic PNH and hypoplastic PNH, which is often associated with AA.

Clinical Relevance of PNH Classification

Hemolytic Anemia

Hemolytic anemia of variable severity is a constant finding in both types of PNH. The intensity of hemolysis is dependent on the degree of complement susceptibility of erythrocytes. Type III PNH cells are characterized by complete GPI deficiency, while type II PNH cells only display partial GPI deficiency [5]. The median life span of type III PNH erythrocytes is reduced to 20 days, that of type II PNH erythrocytes, to 45 days. Overall, the intensity of hemolysis is dependent on two major factors: the degree of GPI deficiency of individual cells and the total size of the PNH clone. In patients with PNH clones of <20%, almost invariably some degree of hemolysis and a certain hemosiderinuria are detectable; yet hemoglobinuria rarely occurs. In contrast, patients with more than 60% PNH clones suffer frequent to daily hemolytic episodes, resulting in hemoglobinuria in many cases. Viral (in particular of gastrointestinal origin) and bacterial infections lead to complement activation and may trigger intense hemolytic episodes, called hemolytic crises [5, 6].

In cases with highly active, chronic hemolysis, symptomatic anemia and high transfusion demands, bone marrow transplantation should be considered early in the disease course to avoid potentially life-threatening complications [7]. If bone marrow transplantation is not feasible, treatment with the complement factor 5 (C5) antibody eculizumab (Alexion®) represents an approved alternative. Eculizumab binds to C5, thus inhibiting the terminal complement-activating pathway [8]. Furthermore, short-time administration of corticosteroids should be considered for the treatment of hemolytic episodes or hemolytic crises [9].

Thromboembolism, Abdominal Crises

Thromboembolism is the most severe complication of PNH. It occurs in 50% of patients suffering from hemolytic PNH, and is the cause of death in 30% [10, 11]. In hypoplastic PNH, the thromboembolic risk is much lower and ranges only around 5%, resulting in distinctly lower mortality [12]. It was shown, however, that the excessively increased thromboembolic risk of patients with large PNH clones could be reduced with a primary prophylaxis using vitamin K antagonists. A potential indication to introduce primary prophylaxis with vitamin K antagonists must therefore be considered in all patients with classic hemolytic PNH and granulocyte GPI deficiency >50%. In addition to all general contraindications to vitamin K antagonists, low and variable platelet counts are of great concern and represent relative contraindications to vitamin K antagonist treatment in this setting [13]. Along with the primary thromboembolic risk, the risk of a recurrence is elevated as well. Recurrences frequently manifest themselves in the same location as the primary event, which constitutes a high risk of cirrhosis, especially in the case of hepatic vein thromboses.

Due to the dismal prognosis, bone marrow transplantation must therefore be considered in all patients with severe thromboembolic complications, chiefly in patients with favorable preconditions (age <50 years, HLA-identical family donor).

In PNH, thromboses preferentially affect intra-abdominal veins, often hepatic veins. Portal and splenic veins are frequently affected as well, resulting in increased risk of splenomegaly and hypersplenism. Microvascular thromboses in the splanchnic vasculature can cause attacks of violent abdominal pain, termed 'abdominal crises'. An extensive diagnostic workup in such cases will typically exclude acute hepatic vein, portal vein or splenic vein thrombosis, cholecystitis, cholelithiasis and pancreatitis as causes of the abdominal discomfort. As direct evidence of the causative microvascular thrombosis cannot be obtained, abdominal crises in PNH remain a diagnosis by exclusion.

Cytopenia, Myelodysplasia

Cytopenia is common in both types of PNH. Mortality due to cytopenia-associated complications, such as infections or bleedings, is estimated at about 10%. The results of immunosuppressive treatment for peripheral cytopenia in hypoplastic PNH are similar to those observed in true AA. If immunosuppressive treatment fails, bone marrow transplantation must be considered early during the disease course in suitable patients.

In most instances, anemia in PNH is caused by hemolysis; an alternative cause may be iron deficiency, mostly due to renal iron loss mediated by hemosiderinuria. Iron substitution must be initiated cautiously as the administration of iron could trigger a hemolytic episode. In patients with high transfusion demands, iron deficiency would be an extremely uncommon finding, as iron is substituted regularly via erythrocyte transfusion. In active hemolysis, folic acid should be substituted in addition as severe hemolysis is closely associated with folic acid depletion. A therapeutic trial of erythropoietin is feasible in patients in whom anemia is caused by a reduced erythropoietic reproduction capacity.

The life span of leukocytes and platelets is not shortened as a result of cellular lysis in PNH. Nevertheless, leukopenia and reduced platelet counts are common findings in PNH, usually caused by a reduced reproduction capacity combined with an increased turnover, e.g. caused by hypersplenism. If significant granulocytopenia develops, a therapeutic trial of granulocyte growth factor (G-CSF) may be judicious.

There are only few and contradictory data on the risk of acute myelogenous leukemia in patients with PNH. The overall risk is estimated to be low, in the order of <1%, which is the same as that of patients with true AA [10]. In some of the cases, leukemic cells also showed a GPI deficiency; it can therefore be concluded that the malignant cell clone in fact evolved from an (abnormal) PNH clone. Besides, a PNH clone was found in 5–9% of patients with myelodysplasia or a myeloproliferative syndrome [14]. The significance of this observation, with regard to the disease course remains unclear, however. In established myelodysplasia or myeloproliferative syndromes, treatment strategies should follow accepted guidelines, regardless of the presence of a PNH clone.

Uncomplicated Disease Course

Apart from the patient groups with an increased morbidity and mortality risk, there is a group of PNH patients showing a median overall survival of more than 25 years, free from relevant PNH complications, and comprising up to 28% of all PNH patients [10]. Even spontaneous complete remissions are observed in up to 15% of PNH patients. However, so far no predictor has been found either for an uncomplicated disease course or for spontaneous remissions.

Diagnosis

Introduction

The differential diagnostic workup of PNH aims at distinguishing PNH from other forms of hemolytic corpuscular anemia (table 2) [15]. Corpuscular hemolytic anemias are almost invariably due to hereditary defects of the erythrocyte cell membrane, hemoglobin molecule, or erythrocyte enzymes. PNH, which is caused by an acquired stem cell defect, is the only exception. Extracorpuscular hemolytic anemias are always acquired and are due to immune mechanisms, metabolic disorders or mechanic causes [16].

PNH is characterized by a normochromic hemolytic anemia, accompanied by hemoglobinuria and sometimes hemosiderinuria. Hemolysis may be chronic or episodic (e.g. triggered by infections). As the causal event in PNH is a defect in a myelogenous stem cell, not only erythrocytes are affected, but platelets, granulocytes, monocytes, and lymphocytes are similarly affected, resulting in concomitant leukopenia and thrombopenia.

The putative diagnosis PNH was formerly established by the acid test and Ham's (sucrose) test. Both tests work by complement activation via the alternative pathway, either by means of milieu acidification or by reduced ion density. As erythrocytes in PNH suffer from increased vulnerability to complement attack, the resulting hemolysis of PNH erythrocytes is distinctly increased in both tests.

Once the causal defect of PNH and its sequelae had been unraveled, a much faster and much more specific laboratory diagnosis became available with the advent of flow cytometry. A multitude of proteins are tethered to the cell surface via the GPI

Table 2. Clinical and laboratory characteristics of PNH /indications for PNH analysis

- Coombs-negative, intravascular hemolysis (chronic or episodic)
- Discoloration of the (morning) urine of unknown cause
- Hemolysis/anemia with pancytopenia
- Hemolysis with iron deficiency
- (recurrent) Thromboembolism of atypical localization
- Attacks of violent abdominal pain of unknown cause
- Aplastic anemia (primary diagnosis and serial monitoring)

anchor. The complement-controlling proteins DAF (decay-accelerating factor; CD55) and MIRL (membrane inhibitor of reactive lysis; CD59) are of utmost importance for the pathogenesis of PNH. PNH cells synthesize only little or no GPI; consequently, the examination of GPI-anchored proteins on the cell surface reveals either strongly reduced or a totally absent expression of those proteins. The entity of cells with strongly reduced or absent surface expression of GPI-anchored proteins is termed 'PNH clone'.

The complement-controlling proteins DAF and MIRL are expressed on all hematopoietic cells; other GPI-anchored proteins, such as enzymes (alkaline neutrophil phosphatase), adhesion molecules (CD58, CD66b), specific receptors (CD14, UPAR (CD87), Fcγ receptor), and others (Campath-1 (CD52), CD24, CEA) are only found on single or sparse cell types (table 1) [5]. Intimate knowledge of the normal expression pattern of GPI-anchored proteins is the basis for the detection of reduced or missing expression of GPI-anchored proteins, and thus for the diagnosis of a PNH clone. Separate analysis of cellular subpopulations (erythrocytes, reticulocytes, granulocytes, monocytes, lymphocytes and platelets) allows conclusions to be drawn on the clone size of all cell lines involved. Surface expression of GPI-anchored proteins on lymphocytes and within lymphocyte subgroups is heterogeneous; no normal values for lymphocyte subsets have been defined so far. Analysis of lymphocytes is therefore not suitable for the routine laboratory diagnosis of PNH: its use is investigational [17].

As the synthesis of GPI-anchored proteins in itself is not affected in PNH, increased amounts of proteins which cannot be anchored correctly to the cell surface enter the blood stream. The plasma concentration of these proteins can be determined quantitatively by specific tests. Congruously, a positive correlation could be demonstrated e.g. for the plasma concentration of the soluble urokinase-receptor (UPAR; CD87) with the size of GPI-deficient cell clones determined by flow cytometry [18].

Principles of the Flow-Cytometric Diagnosis of PNH

To perform a precise and reproducible flow-cytometric analysis, a target antigen must be expressed with adequate power. This criterion is ideally fulfilled by CD59.

As CD55 expression is weaker, its analysis is more prone to interference, and therefore the application of a bright fluorochrome such as phycoerythrin (PE), is advisable for CD55 analysis. Moreover, the expression of the target antigen should not be altered during cell maturation to avoid false-negative results. Since CD55 and CD59 are already expressed on early progenitors, they are exquisitely suited for the laboratory diagnosis of PNH, as opposed e.g. to CD14, which is expressed on mature cells only. Nor would antigens expressing polymorphisms of the binding region, which would influence antibody binding, be suitable for the laboratory diagnosis of PNH. Diagnostic antibodies should therefore always be directed against constant regions of the target antigen.

The life span of GPI-deficient erythrocytes in PNH may be strongly reduced due to intravascular hemolysis. As a consequence, the analysis of erythrocytes is not suited for precise determination of the PNH clone size during active hemolysis. In addition, transfusion of erythrocytes would result in a further relative reduction in the proportion of GPI-deficient erythrocytes. Both variations would lead to an underestimation of the actual size of the GPI-deficient cell clone. If a reliable estimation of the red cell clone size is essential e.g. for serial monitoring, analysis of the GPI-deficient reticulocyte clone size can be performed alternatively. The life span of granulocytes is not altered by hemolysis. For this reason, analysis of granulocytes offers the most reliable approach for the determination of the size of a GPI-deficient cell clone. Conversely, owing to the long life cycle of lymphocytes, GPI-deficient lymphocyte clones are generally small and suited neither for the assessment of the size of a GPI-deficient cell clone, nor for serial monitoring of disease course. The analysis of platelets is still poorly standardized and is reserved for investigational purposes [19].

Flow-Cytometric Test Chart

Anticoagulated full blood drawn from a peripheral vein is used for all flow-cytometric analyses. Containers used for automated blood counts and containing EDTA or heparin as anticoagulant are also suited as special containers for flow-cytometric analyses. The exception is the analysis of platelets: citrate-anticoagulated blood is the preferred material in this instance [19]. As the artifact-prone analysis of granulocytes is of crucial importance in the laboratory diagnosis of PNH, the material should not be used more than 24 h after sampling (48 h must be the exception) [20].

CD55 (DAF) and CD59 (MIRL) are the best-characterized surface markers for the laboratory diagnosis of PNH. Both are normally expressed on all hematopoietic cells. In PNH, their expression is reduced or absent. In addition to these ubiquitously present markers of all hematopoietic cells, cell-line-specific GPI-anchored markers are co-analyzed, as CD14 and CD48 for monocytes, CD16 and CD66b for granulocytes, and CD48 and CD52 for lymphocytes. To allow for a secure discrimination of individual leukocyte populations, a double-gating strategy is recommended, combining forward scatter (FSC)/side scatter (SSC) with SSC/CD45 (with SSC/CD41 for platelets). This procedure permits both a comparison of clone sizes of different cell lines and reliable determination of the clone size of individual cell lines.

Analysis employing two separate preparations is the preferred approach in the laboratory diagnosis of PNH. Unmanipulated full blood, anticoagulated with EDTA or heparin is used for the analysis of reticulocytes and, if desired, of erythrocytes. For the analysis of granulocytes, monocytes and lymphocytes, EDTA- or heparin-anticoagulated full blood is used as well but only after erythrocyte lysis (fig.1). Erythrocyte lysis aims at a reduction of antibody binding on erythrocyte surfaces, which is undesirable during leukocyte analysis as it would be associated with reduced availability of antibodies for binding on leukocyte surfaces. As the proportion of cells in a given sample would be extremely disparate, erythrocytes present in a sample for leukocyte analysis would create an important background signal that would hamper the ability to discriminate small leukocyte populations. Commercially available reagents can be used for erythrocyte lysis; good results are obtained employing ammonium chloride lysis (e.g. IO test 3 lysing solution®, Beckman Coulter). As ammonium chloride is not stable in solution, the lysis reagent must be freshly prepared daily.

'Nonlyse-nonwash' protocols are described in the literature [21]; in our experience, however, the 'lyse-wash-and-then-stain' process (fig.1) allows the most sensitive distinction between deficient, partially deficient, and normal populations.

Owing to the weak binding capacity of diagnostic antibodies to CD55 and to CD59, a washing strategy avoiding serum (e.g. with PBS) is of utmost importance, specifically for the analysis of reticulocytes and erythrocytes. Conversely, addition of 0.1% bovine serum albumin (BSA) to 0.1% buffered saline (PBS) is recommended for the analysis of leukocytes, notably granulocytes (fig.1). BSA reduces the nonspecific binding of antibodies to leukocyte surfaces and decreases the degranulation and agglutination of granulocytes. Whilst the analytic process up to the addition of the antibodies is little time sensitive, the analysis should be completed within 30 min after adding the staining antibodies. After resuspending the pellet in 0.5–1 ml PBS (or PBS-0.1% BSA, no additional washing is allowed.

Analysis of platelets is still only poorly standardized. Citrate-anticoagulated blood is typically used; erythrocyte lysis is neither required nor recommended to avoid in vitro activation and alteration of platelets.

Notably in AA, the duration of the flow-cytometric analysis may pose a problem, due to the very low cell counts; even so, flow velocity should not be more than 1500 cells/s. To avoid artifacts due to segregation of the sample, a maximum duration of analysis of 5 min should be defined as a stop criterion. Further criteria could be a total cell count of 100,000 for the analysis of reticulocytes, of 10,000 for granulocytes, or of 70,000 for leukocytes.

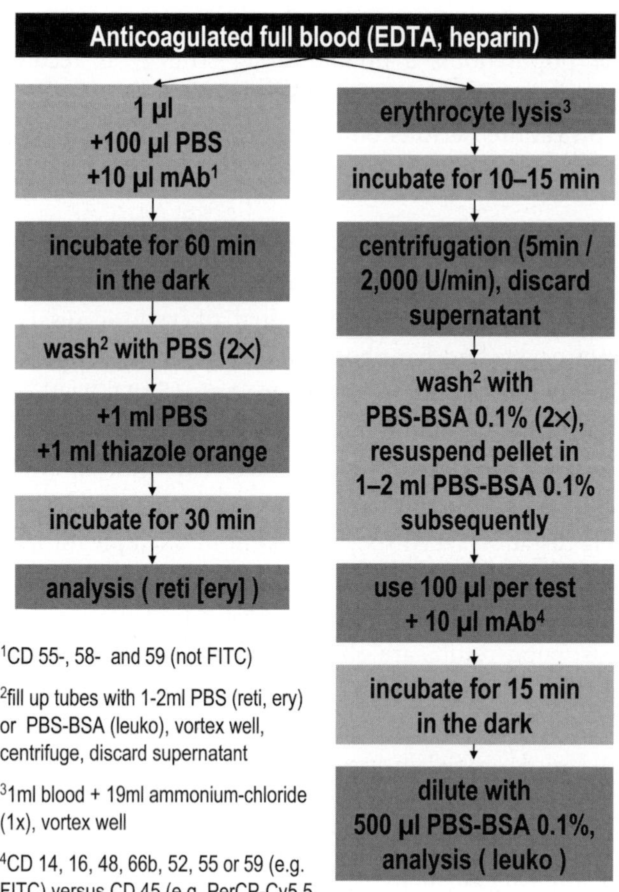
Anticoagulated full blood (EDTA, heparin)
1 µl
+100 µl PBS
+10 µl mAb[1]
incubate for 60 min
in the dark
wash[2] with PBS (2×)
+1 ml PBS
+1 ml thiazole orange
incubate for 30 min
analysis (reti [ery])
erythrocyte lysis[3]
incubate for 10–15 min
centrifugation (5min / 2,000 U/min), discard supernatant
wash[2] with PBS-BSA 0.1% (2×), resuspend pellet in 1–2 ml PBS-BSA 0.1% subsequently
use 100 µl per test
+ 10 µl mAb[4]
incubate for 15 min
in the dark
dilute with
500 µl PBS-BSA 0.1%,
analysis (leuko)
[1]CD 55-, 58- and 59 (not FITC)
[2]fill up tubes with 1-2ml PBS (reti, ery) or PBS-BSA (leuko), vortex well, centrifuge, discard supernatant
[3]1ml blood + 19ml ammonium-chloride (1x), vortex well
[4]CD 14, 16, 48, 66b, 52, 55 or 59 (e.g. FITC) versus CD 45 (e.g. PerCP-Cy5.5, PE or PC 5)

Fig. 1. Flow sheet for the flow-cytometric diagnosis of PNH following the 'lyse-wash-and-then-stain' procedure

Flow-Cytometric Classification and Quantification

Flow-cytometric analysis in the differential diagnosis of PNH should enable the differentiation of cell populations with complete expression of GPI-anchored proteins, from populations with reduced or absent expression. Cells with normal expression of GPI-anchored proteins are referred to as PNH type I cells; cells with partial expression of GPI-anchored proteins are referred to as PNH type II cells, and those with absence of GPI-anchored proteins are referred to as PNH type III cells (figs. 2, 3) [5, 20]. GPI-deficient clones with a minimal size of 1% can be detected with sophisticated assays. For the clinical routine, however, a cut-off of 5% has proved of value for the definition of a pathological result. The quantitative analysis of GPI-deficient cell populations is indicated for the primary diagnosis of anemias and the serial monitoring of disease activity in proven PNH or AA.

In clinical routine, poor sample quality frequently poses a problem. A reliable differentiation of cell populations with a type II deficiency from normal cells is not always possible (fig. 3).

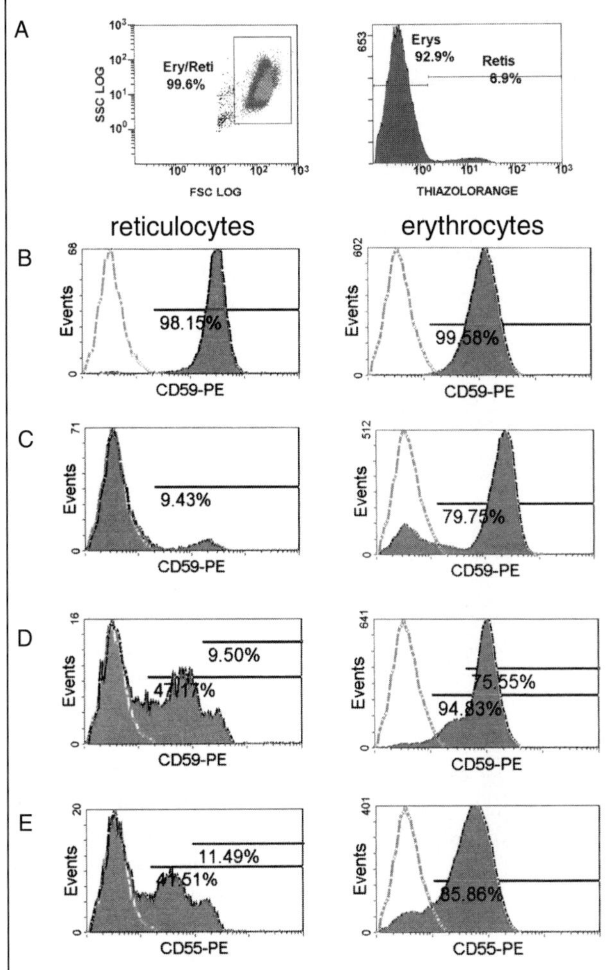

Fig. 2. Results chart of a flow-cytometric PNH analysis on reticulocytes and erythrocytes.
A Results of the FSC/SSC gating (dot plot) and of the thiazole-orange staining (histogram). The thiazole histogram defines the analytical regions for analyses **B–E**. **B** Reticulocyte (left) and erythrocyte analysis (right) of a normal person. **C** Analysis of a PNH patient with type III deficiency (reticulocyte clone size 90.57%). **D** Analysis of a PNH patient with mixed type II and III deficiency (reticulocyte type II clone size 37.67%; type III clone size 52.83%). **E** Analysis of the same PNH patient as in **D** with CD55 analysis (Reticulocyte type II clone size 30.02%; type III clone size 58.49%).
A comparison of the results of reticulocyte and erythrocyte analyses illustrates the marked underestimation of the clone size in erythrocyte analysis; the actual clone size ranges around 90%, but seems to account for only 15–25% in erythrocyte analysis.

The detection of a GPI-deficient cell clone is not sufficient to establish a diagnosis of PNH. A reliable diagnosis of PNH requires intimate knowledge of anamnestic, clinical and laboratory data, and consideration of potential alternative diagnoses (table 2). With respect to differing prognoses and treatments, differentiation between classic hemolytic, and hypoplastic PNH, frequently associated with AA, should be sought once the diagnosis PNH is established [5, 13].

The sizes of GPI-deficient clones of distinct cell lineages do not correlate well with each other. Varying susceptibility to lytic attacks, differing life spans and reproduction rates of the diverse cell systems are the most likely reasons for this observation. Variations in the alteration of surface protein expression intensity in response to activating stimuli could also be relevant [22]. The size of the CD59-deficient granulocyte clone is generally accepted as the single parameter with the highest diagnostic and prognostic impact in the laboratory diagnosis of PNH to date.

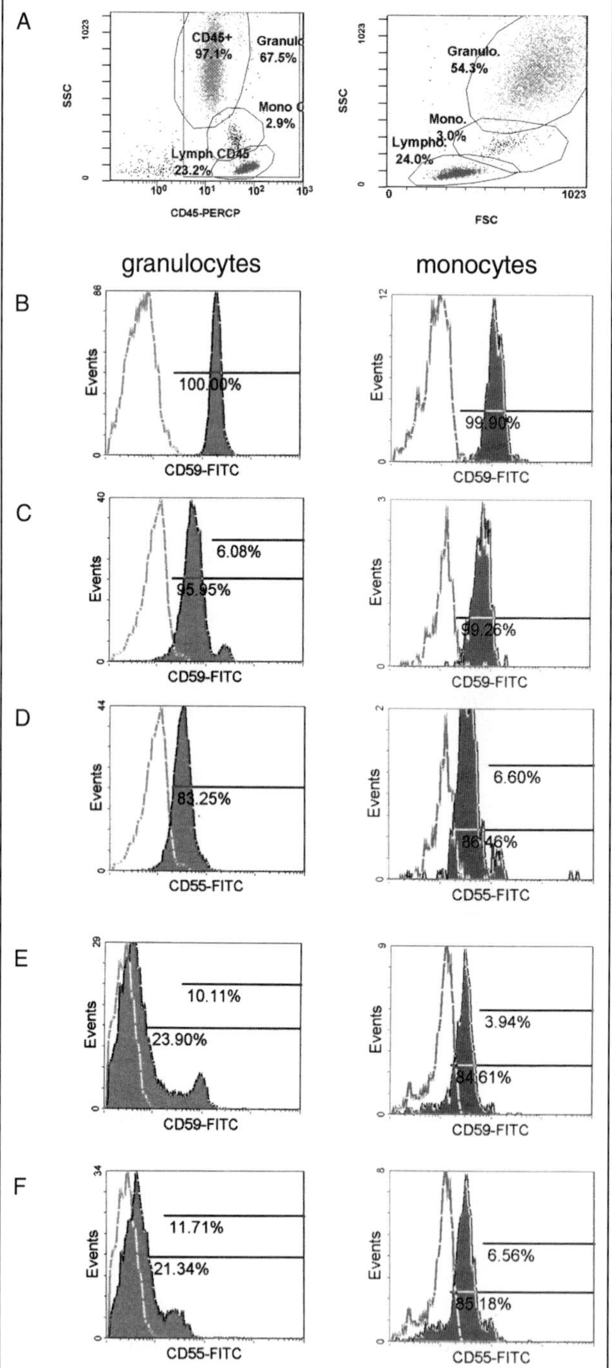

Fig. 3. Results chart of a flow-cytometric PNH analysis on granulocytes and monocytes. **A** Results of the FSC/SSC and SSC/CD45 double gating. **B** Analysis of the CD59 expression on granulocytes (left) and monocytes (right) of a normal person. **C** Analysis of a PNH patient with a large type II deficient clone (around 90%) and a small normal cell population (around 6%). **D** Analysis of the CD55 expression of the same PNH patient as in **C**; the small fraction of normal cells (around 6%) is not demonstrable in all analyses, a reason for the necessity to analyze the expression of different GPI-anchored proteins. **E** Analysis of a PNH patient with a large type III deficient granulocyte clone (76.1%), and a small normal population (10.11%) and presumably a small type II deficient population (13.79%); monocytes demonstrate a type II deficiency. **F** Analysis of the CD55 expression of the same PNH patient as in **E**; a large clone with type III deficiency (78.66%), and also difficulties in the differentiation of the putative type II deficient population (9.63%); again, monocytes alone demonstrate a type II deficiency.

Comparison of the results confirms that a reliable PNH diagnosis requires the analysis of various cell types, employing different markers.

References

1 Rotoli B, Bessler M, Alfinito F, del Vecchio L: Membrane proteins in paroxysmal nocturnal hemoglobinuria. Blood Rev 1993;7:75–86.

2 Yeh ETH, Rosse WF: Paroxysmal nocturnal hemoglobinuria and the glycosylphosphatidylinositol anchor. J Clin Invest 1994;93:2305–2310.

3 Hall C, Richards SJ, Hillmen P: The glycosylphosphatidylinositol anchor and paroxysmal nocturnal hemoglobinuria/aplasia model. Acta Haematol 2002;108:219–230.

4 Johnson RJ, Hillmen P: Paroxysmal nocturnal hemoglobinuria: nature's gene therapy? Mol Pathol 2002;55:145–152.

5 Hillmen P, Richards SJ: Implications of recent insights into the pathophysiology of paroxysmal nocturnal hemoglobinuria. Br J Haematol 2000;108:470–479.

6 Rosse WF: The life-span of complement-sensitive and -insensitive red cells in paroxysmal nocturnal hemoglobinuria. Blood 1971;37:5565–62.

7 Saso R, Marsh J, Cevreska L, Szer J, Gale RP, Rowlings PA, et al: Bone marrow transplants for paroxysmal nocturnal hemoglobinuria. Br J Haematol 1999;104:392–396.

8 Hillmen P, Hall C, Marsh JC, Elebute M, Bombara MP, Petro BE, et al: Effect of eculizumab on hemolysis and transfusion requirements in patients with paroxysmal nocturnal hemoglobinuria. N Engl J Med 2004;350:552–559.

9 Rosse WF: Treatment of paroxysmal nocturnal hemoglobinuria. Blood 1982;60:20–23.

10 Hillmen P, Lewis SM, Bessler M, Luzzatto L, Dacie JV: Natural history of paroxysmal nocturnal hemoglobinuria. N Engl J Med 1995;333:1253–1258.

11 Socié G, Mary J, de Gramont A, Rio B, Leporrier M, Rose C, et al: Paroxysmal nocturnal hemoglobinuria: long-term follow-up and prognostic factors. Lancet 1996;348:573–577.

12 Ray JG, Burows RF, Ginsberg JS, Burrows EA: Paroxysmal nocturnal hemoglobinuria and the risk of venous thrombosis: review and recommendations for management of the pregnant and nonpregnant patient. Hemostasis 2000;30:103–117.

13 Hall C, Richards S, Hillmen P: Primary prophylaxis with warfarin prevents thrombosis in paroxysmal nocturnal hemoglobinuria (PNH). Blood 2003;102:358733–3591.

14 Graham DL, Gastineau DA: Paroxysmal nocturnal hemoglobinuria as a marker for clonal myelopathy. Am J Med 1992;93:671–674.

15 Berger D, Engelhardt R. Hämolytische Anämien; in: Berger D, Engelhardt R, Mertelsmann R (eds): Das Rote Buch. Landsberg/Lech, ecomed Verlag, 2002, pp 486–496.

16 Heimpel H, Hoelzer D, Lohrmann H-P, Seifried E, Kleihauer E: Hämatologie in der Praxis. Jena, Fischer, 1996.

17 Cui W, Fan Y, Yang M, Zhang Z: Expression of CD59 on lymphocyte and the subsets and its potential clinical application for paroxysmal nocturnal hemoglobinuria diagnosis. Clin Lab Hematol 2004;26:95–100.

18 Grünewald M, Siegemund A, Grünewald A, Schmid A, Koksch M, Schöpflin C, et al: Plasmatic coagulation and fibrinolytic system alterations in PNH: relation to clone size. Blood Coagul Fibrinolysis 2003;14:685–695.

19 Grünewald M, Grünewald A, Schmid A, Schöpflin C, Schauer S, Griesshammer M, et al: The platelet function defect of paroxysmal nocturnal hemoglobinuria. Platelets 2004;15:145–154.

20 Hall SE, Rosse WF: The use of monoclonal antibodies and flow cytometry in the diagnosis of paroxysmal nocturnal hemoglobinuria. Blood 1996;87:5332–5340.

21 Hernandez-Campo PM, Martin-Ayuso M, Almeida J, Lopez A, Orfao A: Comparative analysis of different flow cytometry-based immunophenotypic methods for the analysis of CD59 and CD55 expression on major peripheral blood cell subsets. Cytometry 2002;50:191–201.

22 Holada K, Simak J, Risitano AM, Maciejewski J, Young NS, Vostal JG: Activated platelets of patients with paroxysmal nocturnal hemoglobinuria express cellular prion protein. Blood 2002;100:341–343.

Sack U, Tárnok A, Rothe G (eds): Cellular Diagnostics. Basics, Methods and Clinical Applications of Flow Cytometry. Basel, Karger, 2009, pp 680–686

Fetal Erythrocytes

Peter Sedlmayr

Institute of Cell Biology, Histology and Embryology, Center for Molecular Medicine, Medical University of Graz, Austria

Introduction/Background

Fetal erythrocytes in maternal blood are analyzed for the purpose of measuring the extent of fetal blood transfer to the mother's circulation (fetomaternal hemorrhage). This phenomenon can be triggered during delivery, in the course of pregnancy or by medical procedures such as amniocentesis or external version of a fetus in breech presentation. This transfer plays a role in rhesus D incompatibility (mother Rh(D)–, child Rh(D)+), a situation in which it is necessary to administer an appropriate dose of anti-Rh(D) immunoglobulin to suppress sensitization of the maternal immune system, preventing the risks of subsequent pregnancies with another Rh(D)+ fetus. 10 μg Rh(D) immunoglobulin can suppress immunization with 1 ml fetal blood (approximately 0.5 ml fetal erythrocytes). In German-speaking countries, a standard dose of 250–300 μg/ml is administered within 72 h postpartum; this dose should be increased if a substantial fetomaternal blood transfer is suspected.

The long-established Kleihauer-Betke acid elution test is based on the different solubility of HbF and HbA and allows the detection and quantification of fetal erythrocytes in maternal blood smears [1]. However, due to its low sensitivity and poor reproducibility, this method is being replaced by flow-cytometric techniques.

The appearance of HbF+ erythrocytes (F-cells) is not limited to the fetal period: this test is also indicated in the following disorders: hereditary persistence of HbF [2], monitoring of hydroxybutyrate and hydroxyurea therapy for sickle cell anemia [3] and prognostic evaluation of myelodysplasia [4, 5]. During pregnancy, maternal hematopoiesis also generates low levels of HbF+ erythrocytes, and low concentrations of HbF+ cells can be found in healthy adults [6].

Two protocols are presented for measuring fetal erythrocytes: one stains for HbF in permeabilized erythrocytes [see also 7–10], the other – a double-fluorescence method – additionally stains for the Rh(D) antigen on the erythrocytes. The second method is more elaborate but more precise as well, particularly in the case of rhesus incompatibility since it enables better discrimination of fetal erythrocytes and maternal HbF+ cells.

Protocol for Measuring HbF+ Erythrocytes

Principle

Erythrocytes from maternal blood drawn postpartum are fixed with glutaraldehyde, permeabilized with Triton X-100 and marked with an FITC-labeled antibody against HbF. Erythrocytes of fetal origin are HbF++.

Material

- Phosphate-buffered saline solution (PBS, pH 7.4)
- PBS-BSA (0.1% BSA in PBS)
- 0.05% glutaraldehyde (Sigma) in PBS (freshly prepared daily, storage until use at 4 °C)
- 0.1% Triton X-100 (Sigma) in PBS-BSA, storage at 4 °C
- 1% paraformaldehyde (Sigma) in PBS-BSA
- Anti-HbF-FITC (Millipore; catalogue-No. MAB3433F)
- Blood sample (anticoagulated with EDTA): maternal blood sample (taken after delivery, before administration of anti-Rh(D) immunoglobulin)
- Controls:
  - blood from a nonpregnant individual;
  - umbilical cord blood;
  - positive control blood: mixture of blood from a nonpregnant individual with approximately 3% umbilical cord blood; before mixing, both blood samples are washed 3 times (commercially available as FETALtrol, Trillium Diagnostics, Scarborough, ME, USA);
  - an assay without addition of antibodies serves as negative control.

Procedure

- 20 µl of whole blood is mixed for 15 s in 1 ml glutaraldehyde and incubated for 10 min at room temperature;
- 100 µl of the mixture is then agitated for 15 s with 400 µl Triton X-100 and incubated for 5 min at room temperature;
- 10 µl of this mixture is then incubated together with the antibody solution (concentration to be titrated, typically 5 µl in 100 µl PBS-BSA) for 15 min at room temperature in the dark;

postfixation by adding 0.5 ml paraformaldehyde.

Quantification

Parameters: forward scatter (FSC), side scatter (SSC), green (530 nm, FL1) and orange (585 nm, FL2) fluorescence are measured on a logarithmic scale. Threshold for

FSC should be placed so as to exclude the majority of events that are smaller than erythrocytes. Calibration of emission intensity (PMT voltage) should be such that the unmarked glutaraldehyde-fixed erythrocytes are approximately in the middle of the first decade of the respective logarithmic scale (note: glutaraldehyde fixation leads to enhanced autofluorescence). Correct fluorescence compensation (see 'Technical Background and Methodological Principles of Flow Cytometry', pp. 53) will display green and orange leukocyte autofluorescence at almost identical values. Plotting the events in a two-parameter display shows them arranged on a 45° line. A minimum of 50,000 events should be acquired. Data are saved in list mode. Prepared samples are stable over several days (storage at 4 °C in the dark).

Data Analysis

A gate is set for the erythrocyte population in FSC and SSC (fig. 1). Intensely autofluorescent leukocytes are excluded via a further gate in SSC/FL2 displaying the unlabeled control (fig. 2). Only data included in both gates are further processed and analyzed in a one-parameter histogram. Fetal erythrocytes will show markedly higher green fluorescence than autofluorescent erythrocytes, with a discrete peak in frequency distribution (fig. 3). Fluorescence intensity of HbF+ maternal cells is comparatively lower. The area of fluorescence intensity, which is significant for quantifying the HbF++ cells, is determined with the positive control. (Determination of the number of cells showing green fluorescence in more than 99.95% of the control is another definition of positive cells that will include a slightly higher number of maternal cells).

Result

The number of fetal erythrocytes is given as the percentage of total number of erythrocytes (accurate up to 2 digits after the decimal point). The fetal blood volume transferred to the maternal circulation can then be estimated based on a total maternal blood volume of 5 l.

Quality Control

Calibration of the flow cytometer is carried out according to manufacturer's instructions (e.g. with CaliBrite beads).

The Fetal Control Kit (Millipore; catalogue No. FT100) contains stabilized mixed blood at different concentrations.

Expected Results

A concentration of fetal erythrocytes of more than 0.5% indicates that the standard dose of 250 μl anti-Rh(DF) immunoglobulin is not sufficient.

Time Frame

60 min.

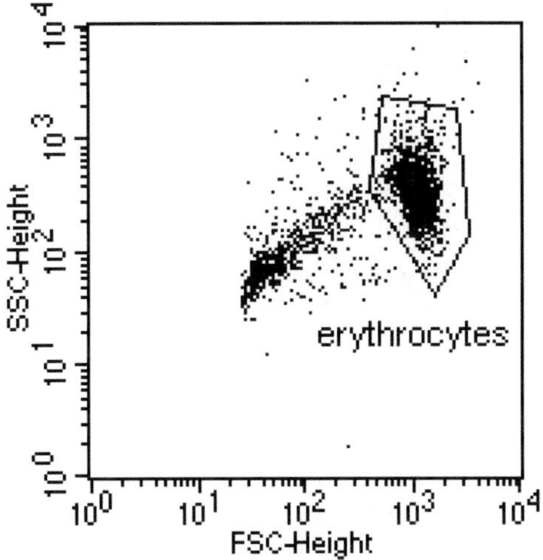

Fig. 1. Gating of erythrocytes. FSC and SSC on a logarithmic scale.

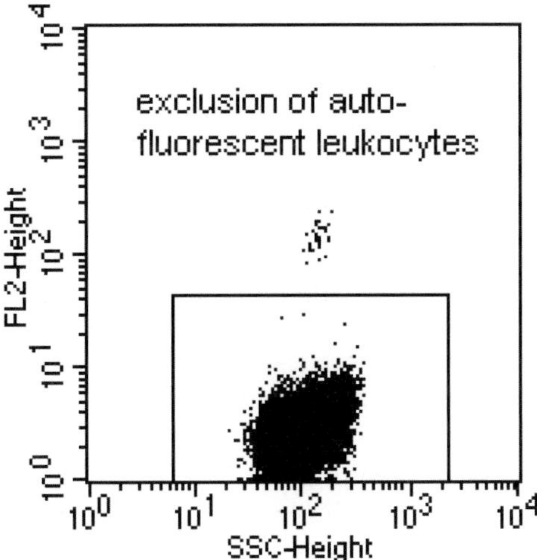

Fig. 2. Exclusion of autofluorescent leukocytes. The dot plot is gated on data as shown in figure 1. Leukocytes end erythroblasts show higher autofluorescence than erythrocytes in the FITC as well as in the PE channel. With correct fluorescence compensation, these cells can be excluded from further analysis.

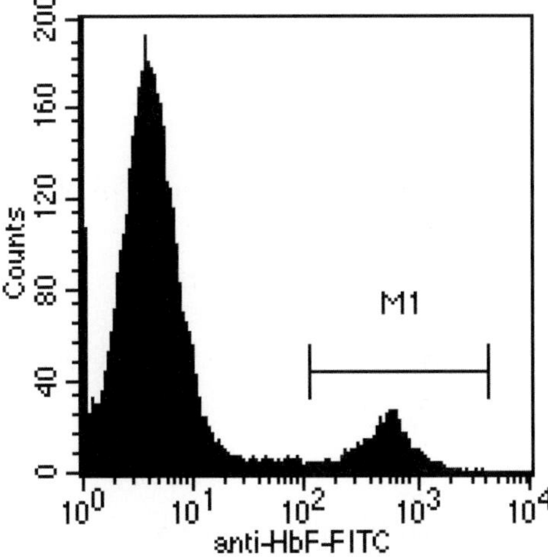

Fig. 3. Quantification of HbF+ erythrocytes. Data in the histogram result from gating the original data on the regions shown in figures 1 and 2.

Protocol for the Quantification of HbF+ Rh(D)+ Erythrocytes

Principle

Glutaraldehyde-fixed erythrocytes are permeabilized with Triton X-100 and stained with anti-HbF-FITC and anti-Rh(D)-RPE. In case of a rhesus incompatibility, maternal erythrocytes are negative, with a small number of HbF+ cells. In contrast, fetal erythrocytes are double positive and can be easily separated.

Material

- Washing buffer: PBS (pH 7.4) with 0.1% sodium azide and 0.2% BSA
- Glutaraldehyde solution: 0.05% glutaraldehyde (Sigma) in PBS (freshly prepared daily, storage until use at 4 °C)
- Permeabilizing solution: 0.1% Triton X-100 (Sigma) in PBS with 0.1% BSA
- 1% paraformaldehyde (Sigma) in PBS
- Anti-HbF-FITC (Millipore; catalogue No. MAB3433F)
- Anti-Rh(D)-PE (Millipore, Anti-Quanti-D Rh D Antigen; catalogue-No. MAB3434H)
  (both antibodies can be purchased from Millipore premixed under the name Anti-Com-DF (Rh-D PE/HbF-FITC) Reagent; catalogue-No. MAB3435F)
- Blood sample and controls, see 'Protocol for Measuring HbF+ Erythrocytes' (pp. 681).

Controls: see 'Protocol for Measuring HbF+ Erythrocytes' (pp. 681).

Procedure

- Obtain EDTA blood (before treatment with anti-h(D) immunoglobulin);
- dilution of blood in washing buffer 1 : 10;
- add 1 ml of glutaraldehyde solution to 20 μl diluted blood;
- vortex and incubate 10 ml at room temperature;
- centrifuge 3 times with 2 ml washing buffer;
- discard supernatant;
- carefully resuspend pellet in 0.5 ml permeabilization solution;
- incubate for 4 min at room temperature;
- centrifuge with 2 ml washing buffer;
- discard supernatant;
- add antibodies: 10 μl anti-HbF-FITC and 50 μl anti-Rh(D)-PE (or 50 μl of the combination reagent);
- vortex carefully, incubate for 15 min at room temperature in the dark;
- Wash at 600 × *g;*
- discard supernatant;
- resuspend pellet in 0.6 ml paraformaldehyde.

Measurement

See 'Protocol for Measuring HbF+ Erythrocytes', 'Quantification' (pp. 681).

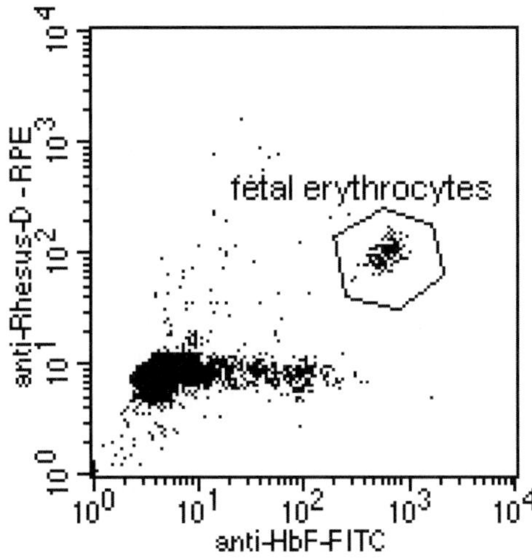

Fig. 4. Quantification of HbF+ Rh(D)+ erythrocytes. Data were gated as shown in figure 1.

Data Analysis

Set gate on erythrocyte population (fig. 1) and analyze green and orange fluorescence in a two-parameter histogram. In case of a rhesus constellation (fetal erythrocytes Rh(D)+, maternal Rh(D)−), orange fluorescence displays fetal erythrocytes distinctly different from maternal ones, thereby also permitting fetal and maternal HbF+ cell differentiation (fig. 4).

Result

See 'Protocol for Measuring HbF+ erythrocytes', 'Result' (p. 682).

Quality Control

See 'Protocol for Measuring HbF+ Erythrocytes', 'Quality Control' (p. 682).

Expected Results

See 'Protocol for Measuring HbF+ Erythrocytes', 'Expected Results' (p. 682).

Time Frame

3 h.

Comments

As is the standard for rare-cell analysis, care should be taken not to transfer cells from one tube to another during analysis. This can be achieved, e.g. by running a tube with fresh PBS solution for a few seconds in between measurements. It is not necessary to use isotype controls [9]. The protocol for measuring HbF+ erythrocytes described above can also be applied to monitor anti-Rh(D) prophylaxis.

Comparing Methods

It should be noted that HbF expression (albeit at low levels) is also induced in maternal erythropoiesis during pregnancy. For this reason, less intensely stained positive cells of nonfetal origin can also be found between the high peak of autofluorescence and the small HbF++ cell peak. An absolutely exact separation can be achieved by double-fluorescence analyses, which enable simultaneous detection of the Rh(D) antigen (as stated in the second protocol), allowing exact differentiation of cells in rhesus incompatibility. The fact that fetal erythrocytes contain less carboanhydrase than those of adults [11] offers an alternative to Rh(D) detection, as propagated by IQ Products Groningen, Netherlands *www.iqproducts.nl*). Irrespective of previous application of anti-Rh(D) immunoglobulin, the Rh(D) status of fetal erythrocytes can be determined by labeling with anti-Rh(D)-FITC as well as anti-IgG-FITC [12].

Sample Storage

Blood samples can be stored at 4 °C for 14 days [7]; however, if indicated, anti-Rh(D) prophylaxis must be administered not later than 72 h after delivery.

References

1 Kleihauer E, Braun H, Betke K: Demonstration von fetalem Hämoglobin in den Erythrocyten eines Blutausstriches. Klin Wochenschr 1957;35:637.

2 Hoyer JD, Katzmann JA, Fairbanks VF, Penz CS: A flow cytometric measurement of HbF in red cells; applications in hereditary persistence of high fetal hemoglobin (HPFH) and other conditions with elevated HbF levels. Blood 1998;92(suppl 1):40b.

3 Atweh GF, Sutton M, Nassif I, Boosalis V, Dover GJ, Wallenstein S, et al: Sustained induction of fetal hemoglobin by pulse butyrate therapy in sickle cell disease. Blood 1999;93:1790–1797.

4 Craig JE, Sampietro M, Oscier DG, Contreras M, Thein S. Myelodysplastic syndrome with karyotype abnormality is associated with elevated F-cell production. Br J Haematol 1996;93:601–605.

5 Reinhardt D, Haase D, Schoch C, Wollenweber S, Hinkelmann E, v Heyden W, et al: Hemoglobin F in myelodysplastic syndrome. Ann Hematol 1998;76:135–138.

6 Boyer SH, Belding TK, Margolte L, Noyes AN, Burke PJ, Bell WR: Variations in the frequency of fetal hemoglobin-bearing erythrocytes (F-cells) in well adults, pregnant women, and adult leukemics. Johns Hopkins Med J 1975;137:105–115.

7 Mundee Y, Bigelow NC, Davis BH, Porter JB: Simplified flow cytometric method for fetal hemoglobin containing red blood cells. Cytometry 2000;42:389–393.

8 Davis BH, Olsen S, Bigelow NC, Chen JC: Detection of fetal red cells in fetomaternal hemorrhage using a fetal hemoglobin monoclonal antibody by flow cytometry. Transfusion 1998;38:749–756.

9 Chen JC, Bigelow N, Davis BH: Proposed flow cytometric reference method for the determination of erythroid F-cell counts. Cytometry 2000;42:239–246.

10 Chen JC, Davis BH, Wood B, Warzynski MJ: Multicenter clinical experience with flow cytometric method for fetomaternal hemorrhage detection. Cytometry 2002;50:285–290.

11 Brady HJ, Edwards M, Linch DC, Knott L, Barlow JH, Butterworth PH: Expression of the human carbonic anhydrase I gene is activated late in fetal erythroid development and regulated by stage-specific trans-acting factors. Br J Haematol 1990;76:135–142.

12 Kumpel BM: Labeling D+ RBCs for flow cytometric quantification of fetomaternal hemorrhage after the RBCs have been coated with anti-D. Transfusion 2001;41:1059–1063.

Sack U, Tárnok A, Rothe G (eds): Cellular Diagnostics. Basics, Methods and Clinical Applications of Flow Cytometry. Basel, Karger, 2009, pp 687–692

Flow-Cytometric Cross-Match for the Determination of Platelet Antibodies

Rainer Lynen · Michael Köhler · Tobias J. Legler

Department of Transfusion Medicine, University Medical Center, Georg August University, Göttingen, Germany

Introduction

Refractoriness following platelet transfusion may be caused by nonimmune factors like fever, sepsis, or splenomegaly, but may also occur due to incompatibilities originating from genetic differences in the HLA (leukocyte), the HPA (platelet), and/or the ABO system. Alloantibodies are most frequently directed against HLA antigens; additionally, anti-HPA antibodies may also cause refractoriness after platelet transfusion. Usually, HLA-immunized patients develop antibodies against several HLA specificities. Monospecific HLA alloantibodies are rather rare. Irrespective of nonimmune causes of refractoriness, platelet antibodies should always be suspected in patients demonstrating an insufficient increase in platelets following transfusion.

Because of the extensive polymorphism of the HLA system, it is nearly impossible to select single-donor apheresis platelet concentrates that are completely HLA-identical from the beginning of transfusion therapy. Therefore, different cross-matching methods have been developed to identify compatible platelet donors and provide adequate platelet concentrates for immunized patients [1–9]. In our method, donor leukocytes and platelets are incubated with patient serum. Whether or not antibodies have bound and the extent of antibody binding to the respective cells is subsequently determined using fluorescence-labeled antihuman IgG antibodies. This method detects anti-HLA and anti-HPA antibodies on the respective test cells. Our protocol can also be used for the detection of HPA antibodies inducing neonatal alloimmune thrombocytopenia (NAIT) or posttransfusion purpura (PTP). The method is suitable as well for the detection of antibodies causing autoimmune thrombocytopenia (AITP).

Test Principle

When performing cross-matching analysis, the sera of patients are investigated for free antibodies using leukocytes and platelets from donors typed for their HLA class I and HPA antigens. Cells isolated from peripheral blood should be used within 1 day. However, we also successfully used aliquots from stored platelet concentrates during the whole storage period (5 days) prior to platelet transfusion. In a first step, antibody binding to the respective cells is accomplished by incubation of the serum with the cells. In control reactions, donor cells are incubated with autologous serum. After several washings, the cells are additionally incubated with fluorescein-labeled antihuman IgG. Bound antihuman IgG is analyzed with the flow cytometer. For the diagnosis of AITP, the patient's platelets are directly labeled with fluorescent antihuman IgG. In all experiments we use $F(ab)_2$ fragments in order to reduce nonspecific binding via the platelet's Fc receptor.

Material

- Patient and donor serum
- Heparin-anticoagulated blood or platelet concentrate aliquots for the preparation of test cells
- Lymphoprep (Nycomed Pharma, Unterschleissheim, Germany)
- Hanks balanced salt solution (HBSS; Gibco Ltd., Paisley, UK)
- Bovine serum albumin (BSA; Ortho Diagnostics, Neckargemünd, Germany)
- Phosphate-buffered saline (PBS), pH 7.22 + 0.1% BSA
- Cellwash Solution (BD Biosciences, Heidelberg, Germany)
- Antihuman IgG $F(ab)_2$ fragments, FITC-labeled (Beckman Coulter, Krefeld, Germany)
- Sodium citrate 3.8% in PBS
- U-bottom microplates 8 × 12 (Greiner, Frickenhausen, Germany)
- Flow cytometer.

Test Procedure

Isolation of donor leukocytes and platelets:
- 3 ml of heparin-anticoagulated donor blood are mixed with 3 ml of a balanced salt solution (e.g. HBBS).
- 6 ml of Lymphoprep are added to a centrifuge tube.
- The diluted blood (6 ml) is layered carefully over the Lymphoprep
- Centrifugation (15 min, 1,200 × g).

- The cellular interphase between plasma and Lymphoprep is transferred to a new tube using a Pasteur pipette (for cross-matching, 200 μl of the platelet concentrate may be used instead).
- The cells are resuspended in 1 ml of 3.8% sodium citrate and further diluted with 4 ml salt solution (e.g. HBBS).
- Centrifugation (5 min, 700 × *g*).
- Resuspension of the cells in 5 ml of balanced salt solution (e.g. HBBS), centrifugation (5 min, 700 × *g*).
- Platelets are adjusted to a concentration of 40,000/μl

Twenty-five microliters of patient serum are mixed with 25 μl of the donor cell suspension in one microplate well; 25 μl of donor serum are mixed with 25 μl of the donor cell suspension in a second microplate well:

- incubation for 30 min at room temperature,
- addition of 150 μl of PBS/BSA, centrifugation (3 min, 700 × *g*),
- discard supernatant, dry by inversely tapping on filter paper,
- this step is repeated once.

To each cavity, add 25 μl of FITC-labeled anti-IgG $F(ab)_2$:

- incubation (in the dark) for 20–60 min at room temperature,
- wash twice in 150 μl of PBS (centrifugation 3 min, 700 × *g*),
- resuspension of the cells in 150 μl PBS, transfer to measuring tube,
- count 10,000 cells in the flow cytometer.

Standardization

Two populations of cells from the density gradient can be distinguished by co-incubation with two antisera: platelet-specific phycoerythrin-labeled anti-CD41 serum, and T-lymphocyte-specific FITC-labeled anti-CD3-serum. After dot plot identification, i.e. plotting the forward scatter (FSC) against the sideward scatter (SSC) of the labeled cell clusters, two gates can be set (R1 = lymphocytes, R2 = platelets) for further evaluation (fig. 1). Each run contains one positive control (known widely reactive platelet antibodies) and donor serum as negative control.

Interpretation

With flow cytometry, the intensity of fluorescence is determined during the measurement of 10,000 events. Evaluation is performed according to the histogram statistics of the leukocyte and platelet gates. The median channel values of the histogram statistics are used as the criterion of fluorescence (fig. 2).

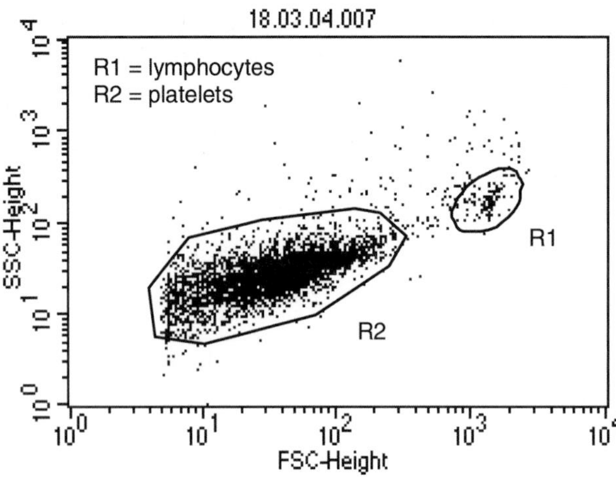

Fig. 1. Dot plot diagram of the lymphocyte (R1) and platelet preparation (R2).

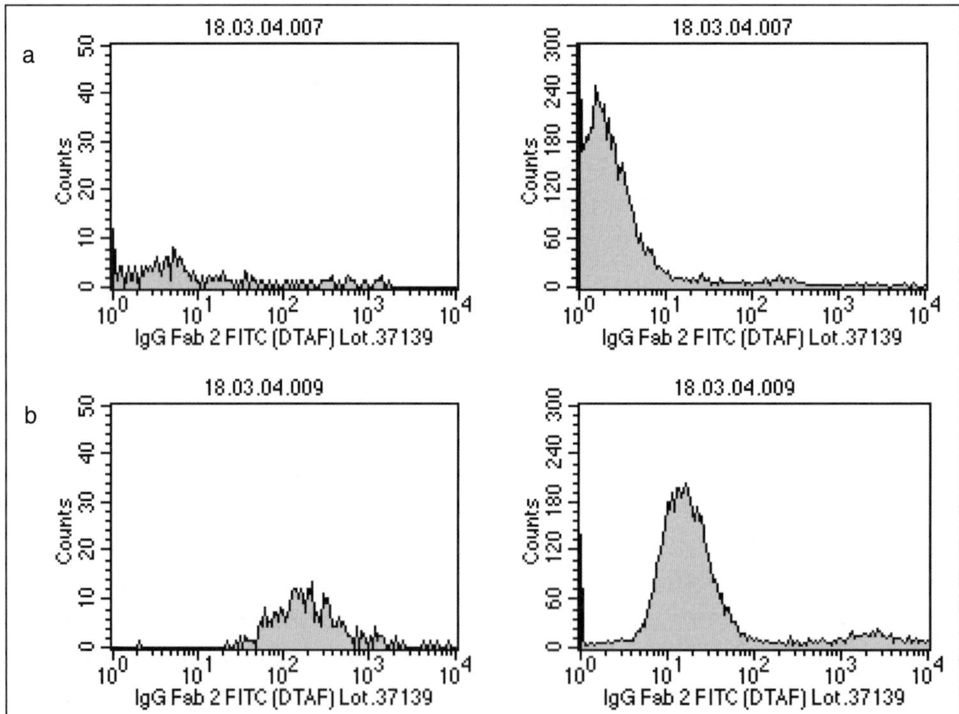

Fig. 2. Histogram of the fluorescence of lymphocytes (left) and platelets (right) using donor AB serum in **a**, and anti-HLA serum in **b**.

For determination of the cut-off values, we analyzed 48 pairs of donors with ABO major compatibility, and 17 pairs of donors with ABO major incompatibility. The ratio of median channel fluorescence of donor-donor crossmatches divided by median channel fluorescence of autologous controls was calculated. In the case of ABO compatibility of the donors, the cut-off for this ratio was 4.3 for lymphocytes

Table 1. Examples of flow-cytometric cross-match results – ratio of the allogeneic patient-derived median fluorescence divided by the autologous donor-derived median fluorescence in HLA- or HPA- alloimmunized patients, and in negative controls (donor-donor cross-match)

Serum	Test cells	Lymphocyte gate	Platelet gate
Case 1 (NAIT)			
Serum of the mother	of the father	8	62
Anti-HPA-1a	HPA-1aa donor	1	865
Anti-HPA + anti-HLA	HPA-1bb donor	2	8
Case 2			
Non-Hodgkin's lymphoma, anti-HLA	incompatible donor	23	2
Case 3			
Aplastic anemia, anti-HLA	incompatible donor	57	14
Control			
Donor	donor	1	1

and 3.0 for platelets. In the case of ABO incompatibility, the cut-off for this ratio was higher: 7.1 for lymphocytes and 10.0 for platelets.

Results

Typical examples for positive and negative results are shown in table 1. Strong exclusive fluorescence of platelets is found when only platelet-specific antibodies are present, as in the cases of NAIT or PTP. A negative result in the leukocyte gate excludes HLA antibodies. In the case of NAIT, the mother's serum gives positive results with the newborn's and the father's platelets, whereas the autologous control is negative (case 1, table 1). Most often, NAIT is caused by anti-HPA-1a antibodies. In this situation, the screening is positive against about 98% of platelet donors. To treat the newborn's thrombocytopenia, it is advisable to transfuse platelets from a compatible platelet donor (HPA-1bb) or maternal platelets after reducing the plasma volume.

In polytransfused patients, positive results are found much more frequently in the leukocyte gate, clearly indicating HLA-class I antibodies. The corresponding antigens are clearly visible on leukocytes, but less so on platelets. Low levels of anti-HLA antibodies cause a positive cross-match in the lymphocyte gate but a negative cross-match in the platelet gate. Sometimes, antibodies against some HLA antigens (e.g. B44, B57) cause the same flow-cytometric cross-match result even if they are present in higher concentrations because these antigens may be weakly expressed on platelets. A negative cross-match result in the platelet gate does not indicate compatibility and platelet transfusion success when the leukocyte result is positive (case 2, table 1). However, transfusion of such a platelet concentrate can be considered if no donors with a completely negative cross-match (leukocytes and platelets) are available.

In the case of positive results with both leukocytes and platelets, it is not possible to clearly distinguish HLA from HPA antibodies (case 3, table 1). One might be dealing with HLA antibodies only, with a concentration high enough to give positive results also with platelets, but on the other hand HPA antibodies might be present as well. In such cases, further glycoprotein-specific serological platelet antibody assays are required to allow a clear differentiation of the antibody specificity.

Summary

Prophylactic and/or therapeutic administration of single-donor platelet concentrates is essential in the treatment of thrombocytopenic patients. Incompatibilities involving the HLA system, the platelet antigen (HPA) system, and/or the ABO system can be uncovered by flow cytometry cross-match analysis, thus enabling us to circumvent refractoriness to platelet transfusions and to find compatible platelet donors also in cases of HLA antigen mismatches. In emergency situations, compatible platelet concentrates can be rapidly identified by taking aliquots from all single-donor platelet concentrates from the stock. The sensitivity and specificity of flow cytometry cross-match in predicting a low corrected count increment are higher than those achieved by lymphocytotoxicity tests.

References

1 Freedman J, Garvey MB, Salomon de Friedberg Z, Hornstein A, Blanchette V: Random donor platelet crossmatching: comparison of four platelet antibody detection methods. Am J Hematol 1988;28:1–7.

2 International Forum: Detection of platelet-reactive antibodies in patients who are refractory to platelet transfusions, and the selection of compatible donors. Vox Sang 2003;84:73–88.

3 Kakaiya R, Gudino MD, Miller WV, Sherman LA, Katz AJ, Wakem CJ, Krugman DJ, Klatsky AU, Kiraly TR: Four crossmatch methods to select platelet donors. Transfusion 1984;24:35–41.

4 Köhler M, Dittmann J, Legler TJ, Lynen R, Humpe A, Riggert J, Neumeyer H, Pies A, Panzer S, Mayr WR: Flow cytometric detection of platelet-reactive antibodies and application in platelet crossmatching. Transfusion 1996;36:250–255.

5 Legler TJ, Humpe A, Leib U, Riggert J, Köhler M: Flow-cytometric analysis of platelet-reactive allo- and autoantibodies. J Lab Med 1997;21:676–680.

6 McFarland JG, Aster RH: Evaluation of four methods for platelet compatibility testing. Blood 1987;69:1425–1430.

7 Myers TJ, Kim BK, Steiner M, Baldini MG: Selection of donor platelets for alloimmunized patients using a platelet-associated IgG assay. Blood 1981;58:444–450.

8 Tosato G, Appelbaum FR, Trapani RJ, Dowling R, Deisseroth AB: Use of in vitro assays in selection of compatible platelet donors. Transfusion 1980;20:47–54.

9 Wu KK, Hoak JC, Koepke JA, Thompson JS: Selection of compatible platelet donors: a prospective evaluation of three cross-matching techniques. Transfusion 1977;17:638–643.

Sack U, Tárnok A, Rothe G (eds): Cellular Diagnostics. Basics, Methods and Clinical Applications of Flow Cytometry. Basel, Karger, 2009, pp 693–713

Residual Cells in Blood Products

Michael Wötzel

Laboratory Dr. Reising-Ackermann and Partners, Leipzig, Germany

Introduction

For decades, blood components have been used as substitutes for transfusion of whole blood. For this purpose, whole blood units must be separated into their components; red blood cells (RBCs), platelets and plasma. White blood cells (WBCs) are separated as 'buffy coat' and are only seldom further used for clinical purposes.

Initially, separation of blood components was carried out for economic reasons: human blood represents a valuable resource which should only be used when absolutely necessary. Additionally, the separation of unnecessary blood components reduces unwanted side effects such as viral transmission, transfusion induced graft-versus-host reactions, HLA alloimmunization and immunomodulation/immunosuppression, caused in particular by contaminating leukocytes. Very helpful reviews on the immunological aspects of transfusion have been published by Dzik et al. [1] and Blajchman et al. [2].

In order to improve the safety of blood transfusion, the elimination of unwanted cellular subsets, especially leukocytes, by cell separation and/or filtration is regulated by law. The recommended standards for Europe are published in the *Guide to the Preparation, Use and Quality Assurance of Blood Components* [3] and for Germany in *Richtlinien zur Gewinnung von Blut und Blutbestandteilen und zur Anwendung von Blutprodukten (Hämotherapie)* [4]. Table 1 contains a short summary of the valid criteria for leukocyte-depleted blood products. These standards, however, will change over time and will have to be adapted to the most recent scientific and technical developments.

The levels of residual cells (WBCs, RBCs or platelets) in the blood product are regulated in particular by the definition of upper limits. For cellular blood products, the content of leukocytes per transfusion unit is limited to $<1 \times 10^6$ leukocytes or

Table 1. Some criteria for quality control of blood components (for a complete list see [3])

Parameter to be checked	Quality requirement	Frequency of control	Control executed by
Red cells, leukocyte-depleted			
Residual leukocytes[a]	$<1 \times 10^6$ per unit by count	1% of all units with a minimum of 10 units/month	QC lab
Hemoglobin	min 40 g/unit	1% of all units with a minimum of 4 units/month	QC lab
Hemolysis at the end of storage	<0.8% of red cell mass	4 units/month	QC lab
Whole-blood-derived platelets, leukocyte-depleted			
Volume	>40 ml	all units	processing lab
Platelet count	$>60 \times 10^9$/single unit equivalent	1% of all units with a minimum of 10 units/month	QC lab
Residual leukocytes before leukocyte depletion[b]			
(a) From platelet-rich plasma	$<0.2 \times 10^9$/single unit equivalent	1% of all units with a minimum of 10 units/month	QC lab
(b) Prepared from buffy coat	$<0.05 \times 10^9$/single unit equivalent	1% of all units with a minimum of 10 units/month	QC lab
Residual leukocytes after leukocyte depletion	$<0.2 \times 10^6$/single unit equivalent	1% of all units with a minimum of 10 units/month	QC lab
pH measured at the end of the recommended shelf life	6.8–7.4	1% of all units with a minimum of 4 units/month	QC lab
Apheresis-derived platelets, leukocyte-depleted			
Volume	>40 ml per 60×10^9 platelets	all units	processing lab
Platelet count	$>200 \times 10^9$/unit	1% of all units with a minimum of 10 units/month	QC lab
Residual leukocytes after leukocyte depletion[a]	$<1 \times 10^6$/unit	1% of all units with a minimum of 10 units/month	QC lab
pH measured at the end of the recommended shelf life	6.8–7.4	1% of all units with a minimum of 4 units/month	QC lab
Fresh frozen plasma			
Volume	stated volume ±10%	all units	processing lab
Factor VIIIc	≥ 70% of the average normal value	every two months a) pool of 6 units ... during first month of storage b) pool of 6 units ... during last month of storage	QC lab
Residual cells[c]	RBCs $<6.0 \times 10^9$/l WBCs $<0.1 \times 10^9$/l platelets $<50 \times 10^9$/l	1% of all units with a minimum of 4 units/month	QC lab
Leakage	no leakage at any part processing of container	all units	processing and receiving laboratory
Visual changes	no abnormal color or visible clots	all units	processing and receiving laboratory

[a] These requirements shall be deemed to have been met if 90% of the units tested fall within the values indicated.
[b] These requirements shall be deemed to have been met if 75% of the units tested fall within the values indicated.
[c] Cell counting performed before freezing.

<3.3 leukocytes/μl. For quality control purposes, 1% of all produced units must be tested, at a minimum of 10 units per month.

For fresh frozen plasma, elevated upper limits are valid, but the manufacturers of blood components must rule out virus transmission through residual leukocytes. Some manufacturers offer leukocyte-depleted plasma with less than 1×10^6 WBC per unit, obtained by plasmapheresis or produced by filtration of whole blood. These plasma products are frozen and stored for at least 4 months in quarantine to achieve maximum protection against virus transmissions.

Compliance with the levels of residual cells must be verified by appropriate laboratory tests. The manual counting of residual cells with the Nageotte chamber [5] is accepted as a classical process control and as a reference method. The related procedure for counting with this chamber is attached as an auxiliary protocol. However, in the range of 1–3 cells/μl, the results of the Nageotte chamber are nonlinear and inaccurate, and depend on the routine and the subjective assessment of the individual examiner. The inclusion of several examiners further increases this variability.

Meanwhile, more modern methods have been developed, which have been compared with the Nageotte chamber [6, 7]. The most widespread is the counting of residual cells by flow cytometry. These new techniques are independent of manual assessment under the microscope and are sensitive, and both more rapid and more precise [7–10].

Protocols

Principle

To count residual leukocytes in a unit of RBCs, a platelet concentrate or a plasma preparation using flow cytometry, a small aliquot of the unit with a defined volume is mixed with a reagent to permeabilize the cell membrane. This makes the cell nuclei accessible to staining with the nucleic acid dye propidium iodide (PI), which is excited by laser light at 488 nm and has an emission maximum at 620 nm. Thereafter, an accurate volume with a known number of fluorescent polymeric microspheres (beads) is added to this suspension. Alternatively, this test can be performed in tubes with a known quantity of lyophilized beads (TruCOUNT tubes). The beads and the leukocytes in this tube are simultaneously counted by flow cytometry. From the ratio of beads to leukocytes, the number of leukocytes per transfusion unit may be calculated. Instead of beads, which are preferred by commercial providers, fixed and stained chicken erythrocytes have also been successfully used as counting particles [11].

Residual RBCs in plasma or in platelet concentrates are identified with an antibody against glycophorin A, an antigen specifically expressed on the surface of

RBCs. This antibody is conjugated with phycoerythrin (PE), a fluorescent dye which is excited at 488 nm and reaches its emission maximum at 585 nm. Since PI has its emission maximum at 620 nm, both dyes can sufficiently be separated and WBCs and RBCs can be measured simultaneously.

Beckman Coulter (Krefeld, Germany) recommends using the LeukoSure Kit, however, it is necessary to measure each cell type separately because permeabilization of the leukocyte membrane may interfere with the detection of RBCs. BD Biosciences (Heidelberg, Germany) is currently developing an assay (Thrombo Count Kit) for the simultaneous measurement of residual WBCs and RBCs in platelet concentrates by stabilizing the cell membranes with a commercial buffer. This buffer allows PI and RNAse to diffuse across the cell membrane of WBCs whereas the glycophorin-A-staining of RBCs remains unaffected. Fixation of cells is not necessary.

To label residual platelets in plasma, Beckman Coulter recommends the PE conjugate of an antibody against CD41 (or CD61), which recognizes the fibrinogen receptor. This approach detects the platelets separately, through the adjustment of photomultipliers which are not compatible with erythrocytes and leukocytes.

The Plasma Count Kit (BD Biosciences) uses a different principle [12]: leukocytes are stained with thiazole orange, which is detectable in all three fluorescence channels. Erythrocytes are labeled with glycophorin A-FITC and platelets with CD41a-PerCP-Cy5.5. A separate step to permeabilize the membranes is not necessary and the test is done in a TruCOUNT tube with counting beads for volume measurement.

Test Samples

At least 100 μl of each unit of RBCs, platelets or plasma are necessary. The preparations should contain blood stabilizers such as CPD or ACD. According to a study by BD Biosciences, EDTA-anticoagulated blood is not suitable because the separation of leukocytes from background is difficult. For the same reason, some authors [11] warn against the use of crystalline Na_2EDTA and recommend stabilization of blood samples in a solution of K_3EDTA in water.

Beckman Coulter Procedure: Reagents, Materials, Equipment, Software

- LeukoSure™ Enumeration Kit, PN 175621 (Beckman Coulter) (contains membrane-permeabilizing reagent, staining reagent and counting particles);
- additional counting particles may be ordered separately as LeukoSure™ Fluorospheres (Beckman Coulter);
- Leuko-Trol™ RBC Control Cells, high and low (Beckman Coulter);

- Leuko-Trol™ Platelet Control Cells, high and low (Beckman Coulter);
- Coulter Clenz® (Beckman Coulter);
- Glycophorin A-PE (Beckman Coulter);
- CD41-PE (Beckman Coulter);
- Dulbecco's PBS (PAA Laboratories GmbH, Cölbe, Germany) or an equivalent product;
- Falcon vials (BD Biosciences) or an equivalent product;
- appropriate racks for these vials;
- pipettes (100 µl, 1 ml) and appropriate pipette tips;
- flow cytometer (Epics XL-MCL, Cytomics FC 500), equipped with all necessary materials and adjusted according to the manufacturer's recommendations;
- Protocol 'Cleanse';
- acquisition protocol for counting residual cells.

Designing the Acquisition Protocols for Epics XL or FC 500

Counting of Residual Leukocytes
- PI shows strongest emission in FL3 and lower light intensity in FL2 und FL4; (in this chapter, the terms 'FL3' and 'FL4' are used according to the practice of Beckman Coulter);
- the Flow-Count beads emit equally strong in all channels;
- for better separation of the leukocytes from background, the use of the quotient FL3/FL2 is recommended.

From these considerations, Beckman Coulter has developed an acquisition protocol with optimized regions, see figure 1. The following histograms are shown:

Leukocytes
- FL2 lin versus count;
- FL3 lin versus count;
- ratio FL3/FL2 versus FL3 lin;
- FL2 log versus FSC log.

LeukoSure Fluorospheres
- FL1 log versus count.

The cytometer can be set up with Flow-Set™ beads according to a procedure specified in the instructions included in the package insert which is included with the LeukoSure Kit. As an alternative, fresh whole blood can be diluted 1:10 with buffer; the cells should be stained according to the procedure described below. The mean of the fluorescence signal should be used to adjust the amplifiers when running this sample. Commercially available control blood samples cannot be used for this procedure! The following target values are recommended (table 2):

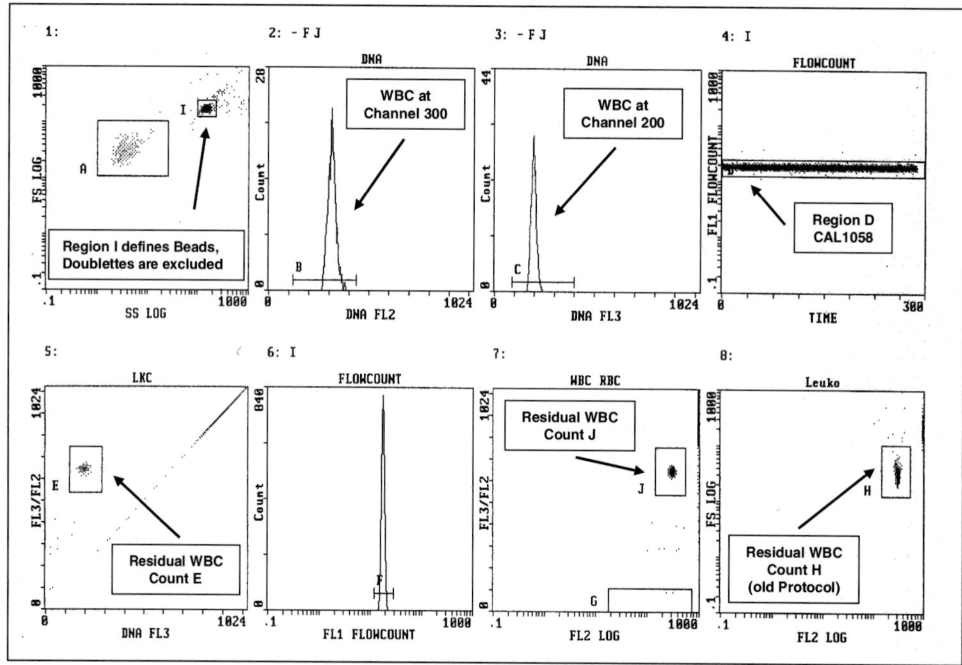

Fig. 1. Acquisition protocol provided by Beckman Coulter for the setup of the Epics XL and the residual cell counting with the LeukoSure® Kit. The regions I, D and F are necessary to count the beads (LeukoSure Fluorospheres). The amplifiers are set to move the peak of the leukocytes in FL2 to channel 300 and in FL3 to channel 200. For the counting of leukocytes, region E is preferred to region J or H in order to achieve the best possible separation from background. This protocol is also applicable to the counting of residual RBCs with glycophorin A (region G).

Table 2. Recommended instrument setup for counting residual WBC on an Epics XL

	Parameter	Amplifier	Mean channel
Bead signal	FL1	LIN	140
PI signal of the leukocytes	FL2	LIN	300
	FL3	LIN	200
Discriminator	set FL2 to 20		
Compensation	set all compensations to 60%		

Counting of Residual Erythrocytes (fig. 2)

- Labeling of the erythrocytes with glycophorin A-PE;
- the separate measurement of leukocytes and erythrocytes is recommended because of the effect of the permeabilizing reagent on erythrocytes;
- erythrocytes are shown in an additional histogram:
  - erythrocytes: FL2 log versus FL3 log; adjust discriminator in FL2 to 20.

Counting of Residual Platelets

- Labeling of the platelets with CD41-PE;
- count LeukoSure Fluorospheres® beads in FL1;

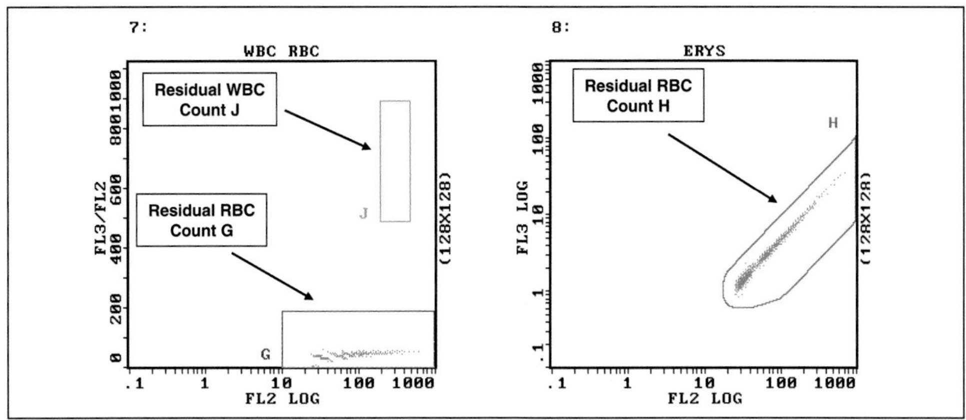

Fig. 2. Protocol for the counting of residual RBCs. Histograms 7 and 8 of the acquisition protocol (see fig. 1) are used, but histogram 8 must be modified to FL2 log vs. FL3 log and region H must be adapted as shown. The RBCs can be counted in regions G or H.

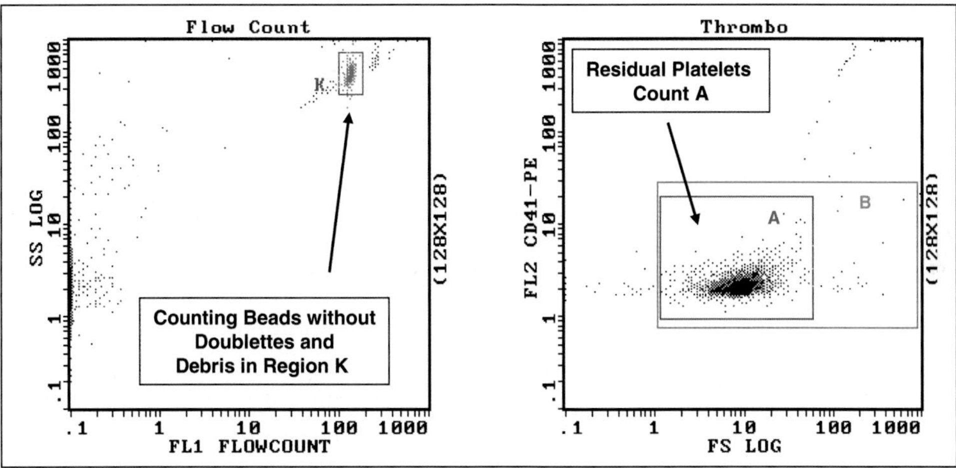

Fig. 3. Protocol for the counting of residual platelets. Region K contains the Fluorospheres. Region A yields the platelets and region B contains region A and the giant platelets.

– histograms in the protocol for counting residual platelets (fig. 3):
  • platelets: FL2 log versus FSC log, and
  • LeukoSure Fluorospheres: FL1 log versus SSC log;
  • adjust discriminator in FL2 to 1.

Stop Criterion

At least 10 μl of a blood sample should be analyzed to quantify residual leukocytes with sufficient accuracy. The blood sample and the Flow Count™ beads are pipetted with exactly the same volume into the test vial. If the concentration of the Flow

Count^(TM) beads is 1,058 beads/μl, for example (the exact concentration is printed on the package insert), acquisition should be stopped, when 10,580 beads are counted in the bead region. Thus, 10 μl of the original blood sample have been tested. The larger the volumes analyzed, the more accurate the test result.

CAL Factor

By using the 'CAL factor' function, the software calculates the result in cells/μl (Menu *Acquisition → Setup → Protocol → Statistics → CAL factor)*. The concentration specific for this batch of the Flow Count™ beads/μl has to be entered and the concentration in cells/μl is provided, without a decimal, however. Therefore, the bead amount in 10 μl should be entered to receive the cell number in 10 μl. When using a new batch of the Flow-Count™ beads, the stop criterion and the CAL factor must be adjusted.

Procedure for Staining of WBCs

(a) Add exactly 100 μl of blood to a cytometer vial by reverse pipetting (see below); after aspiration of the blood, the outer surface of the pipette tip must be wiped clean;
(b) add 100 μl LeukoSure permeabilizing reagent to this tube and vortex immediately;
(c) add 500 μl LeukoSure staining reagent and vortex immediately;
(d) repeat steps (b) and (c) for all samples to be tested;
(e) incubate for 15 min at 20–25 °C in the dark;
(f) mix the LeukoSure™ Fluorospheres thoroughly in the bottle; add exactly 100 μl of beads by reverse pipetting to each tube; after aspiration of the beads, the outer surface of the pipette tip must be wiped clean; mix test samples carefully;
(g) for best results, analyze the samples immediately, but not later than 2 h after addition of the fluorospheres; store samples at 4 °C in the dark.

Procedure for Staining of RBCs

(a) Add exactly 100 μl of blood to a cytometer vial by reverse pipetting; after aspiration of the blood, the outer surface of the pipette tip must be wiped clean;
(b) add 20 μl glycophorin A-PE; mix thoroughly; incubate for 10 min at 20–25 °C in the dark;
(c) add 2 ml of PBS buffer;
(d) mix the LeukoSure Fluorospheres thoroughly in the bottle; add exactly 100 μl of beads by reverse pipetting to each tube; wipe the pipette tip clean after aspiration; vortex all samples;
(e) analyze the samples immediately, but not later than 2 h after addition of the fluorospheres; store the vials at 4 °C in the dark.

Procedures for Staining of Platelets

(a) Add exactly 100 µl of blood to a cytometer vial by reverse pipetting;
after aspiration of the blood, the outer surface of the pipette tip must be wiped clean;
(b) add 20 µl CD41-PE; mix thoroughly; incubate 10 min at 20–25 °C in the dark;
(c) add 2 ml PBS buffer;
(d) mix the LeukoSure fluorospheres thoroughly in the bottle; add exactly 100 µl of beads by reverse pipetting to each tube; wipe the pipette tip clean after aspiration; vortex all samples;
(e) analyze the samples immediately, but not later than 2 h after addition of the Fluorospheres. In the meantime, store the vials at 4 °C in the dark.

Acquisition

- Run the panel 'Cleanse' with 1 tube Coulter Clenz and 3 tubes of bidistilled water;
- write up the worklist; the Multi Carousel Loader is recommended;
- analyze the tubes and print the results.

Calculation of the Residual Cells

The stop criterion as the x-fold number of beads/µl and the counted events in the leukocyte region are printed in the protocol sheet. The residual leukocytes per microliter of the undiluted blood sample are calculated using the following formula:

$$\text{Leukocytes}/\mu\text{l} = \frac{\text{counts in the leukocyte region}}{\text{counts in the beads region}} \times \text{LeukoSure Fluorospheres}/\mu\text{l} \qquad (1)$$

The CAL factor function performs this calculation in the background: if a tenfold amount of the beads/µl is used as CAL factor, the leukocyte count in 10 µl of the original blood sample appears in region E, no matter how much volume was really acquired by entering a stop criterion. To calculate the leukocytes/µl, the cell number in region E is divided by 10. Similar calculations have to be performed for counting residual erythrocytes and residual platelets.

Procedures of BD Biosciences

BD LeucoCOUNT™ Kit

Reagents, Material, Instruments, Software

- LeucoCOUNT™ Kit (BD Biosciences);
- RBC Control Kit (BD Biosciences);
- PLT Control Kit (BD Biosciences);
- FACS Clean (BD Biosciences);
- FACS Rinse (BD Biosciences);

Table 3. Recommended instrument setup for counting residual WBCs on a FACSCalibur

	Parameter	Amplifier	Mean channel
Bead signal	FSC	LIN	
	SSC	LOG	700 ± 20
	FL1	LOG	800 ± 20
	FL2	LOG	700 ± 20
Discriminator (threshold)	set FL2 to about 300		
Compensation	set all compensations to 0%		

- Dulbecco's PBS, cat. no. H31-002, (PAA Laboratories GmbH);
- Falcon tubes (BD Biosciences) with suitable racks;
- pipettes (100 µl, 1 ml), suitable pipette tips;
- FACSCalibur, adjusted with CaliBRITE Beads (BD Biosciences) and FACSComp (version 4.0 and higher generates automatically the settings file 'LeucoCOUNT CalibFile.LNW'), according to the manufacturer's recommendations;
- acquisition protocol 'LeucoCOUNT' (provided by BD Biosciences).

Manual Setup of the Cytometer
- Dilute 10 µl of any test blood with 1 ml PBS and mark with 1:100;
- mark 2 TruCOUNT tubes with '1' (bead control) and '2' (Leukogate control):
  - tube 1: add approximately 500 µl PBS, mix gently;
  - tube 2: add 100 µl of the 1:100 dilution + 400 µl LeucoCOUNT reagent, mix gently;
- incubate 5 min in the dark at 20–25 °C;
- obtain cytometer settings from a former list mode file or adjust the instrument manually;
- adjust the amplifiers with beads;
- use of the setup protocol provided by BD Biosciences is recommended;
- alternatively: click on *Menu* → *Plots* → *Log Data Units* → *Channel Values*;
  load histogram plots of side scatter (SSC), FL1 and FL2 with their statistics windows, place markers; load *Detector/Amps* window and set the PMTs as shown in table 3;
- leave setup mode, define storage folder; count 5,000 beads and store as reference file; store these instrument settings.

Acquisition of the Leukogate Control
- Load LeucoCOUNT Worksheet (fig. 4);
- run tube 2 in setup mode;
- readjust gates R1 (beads) and R2 (leukocyte) if necessary, and
- put 'threshold' in FL2 and choose it high enough to exclude debris and platelet aggregates.

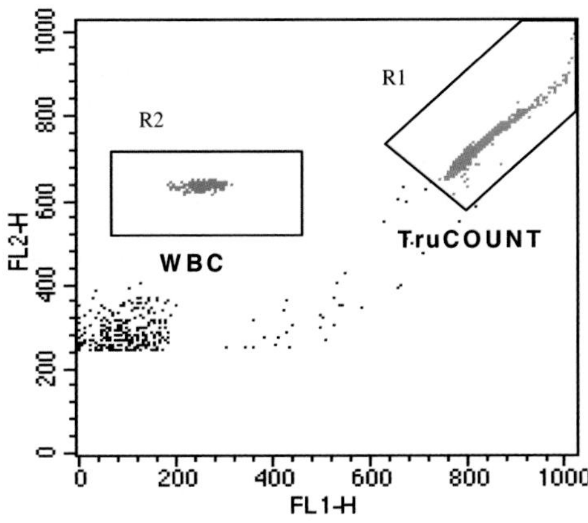

Fig. 4. LeucoCOUNT protocol. R1 contains the TruCOUNT-Beads. In R2 the leukocytes are shown. They are well separated from debris.

Cleaning/Rinsing of the Cytometer
Run FACSClean for 5 min; thereafter run bidistilled water for 5 min.

Staining of the Samples
- Mark one TruCOUNT tube per blood sample;
- add exactly 100 μl of blood to the TruCOUNT tube by reverse pipetting; after aspiration of the blood, the outer surface of the pipette tip must be wiped clean;
- add 400 μl LeucoCOUNT reagent, mix immediately and gently; incubate for 5 min at 20–25 °C in the dark.

Acquisition
- Open LeucoCOUNT Worksheet (fig. 4);
- menu *Acquire* → *Acquisition and Storage* → count at minimum 10,000 beads in R1, select *Store all*;
- go to menu *Acquire* → *Parameter Description* → determine storage folder, all samples may now be analyzed.

Calculation of the Residual Leukocytes

$$\text{Residual leukocytes}/\mu\text{l} = \frac{\text{leukocyte count } (= \text{R2})}{\text{bead count } (= \text{R1})} \times \frac{\text{beads per TruCOUNT tube}}{\text{volume of the blood sample } (\mu\text{l})} \tag{2}$$

BD^{TM} Plasma Count Kit
This test is used to count leukocytes, erythrocytes and platelets in cases where the sensitivity of an automatic cell analyzer is not sufficient, e.g. when residual cells are to be counted in plasma preparations.

The test combines known counting procedures for the different blood cells. Testing is performed in one tube [12]. For nuclear staining of viable leukocytes, thiazole orange is used as a fluorescence dye without permeabilization of the cell membrane (reagent A). Thiazole orange emits in all three fluorescence channels. In the two-color dot plots, the leukocytes are shown in the diagonal and they appear well separated from the other cells and the beads.

Reagent B contains the antibody mixture CD235a-FITC (glycophorin A) and CD41a-PerCP-Cy5.5. The measurement of the sample volume is performed with a TruCOUNT tube, which is provided with a known amount of 4.2 μm diameter lyophilized, fluorescent beads.

Reagents, Material, Instruments, Software

- Plasma Count Kit (BD Biosciences) with reagents A and B and TruCOUNT tubes
- FACS Clean (BD Biosciences)
- FACS Rinse (BD Biosciences)
- CellWASH (BD Biosciences) or PBS (filtered, 0.2 μm) from any manufacturer;
- Falcon tubes (BD Biosciences) and suitable tube racks
- Pipettes (25, 50 and 100 μl, 1 ml), suitable pipette tips
- FACSCalibur, adjusted with CaliBRITE beads, according to the manufacturer's recommendations and with FACSComp (version 4.0 and higher automatically generates an instrument setting file 'LeucoCOUNT CalibFile.LNW')
- Acquisition protocol 'Plasma Count Kit' for CellQUEST Pro and Attractors software (provided by BD Biosciences, see fig. 5).

Procedure

- Add to a TruCOUNT tube: 25 μl donor plasma (in case of cell-poor samples 50 μl), 100 μl reagent A, 20 μl reagent B;
- mix sample gently;
- incubate for 15 min at 20–25 °C;
- add 1 ml CellWASH (agglutination of erythrocytes must be avoided – at low concentrations of erythrocytes, less CellWASH may be added);
- analyze within 1 h after preparation; unfixed cells must always be tested fresh.

Setup of the Cytometer

Run FACSComp in 'lyse-no-wash' mode; then make the following changes to the instrument setting file 'CalibFile.LNW' according to the manufacturer's recommendations: FSC → LOG; SSC → LOG; 'Threshold' in FSC → 100–150 channels; optimization of the compensation (→ reduce compensation in FL2-% FL1 to about 14%); store the new file.

Sample preparation takes about 15–18 min; acquisition, approximately 2 min. BD recommends automatic testing with a loader; analysis is performed with Attractors software or with CellQUEST Pro.

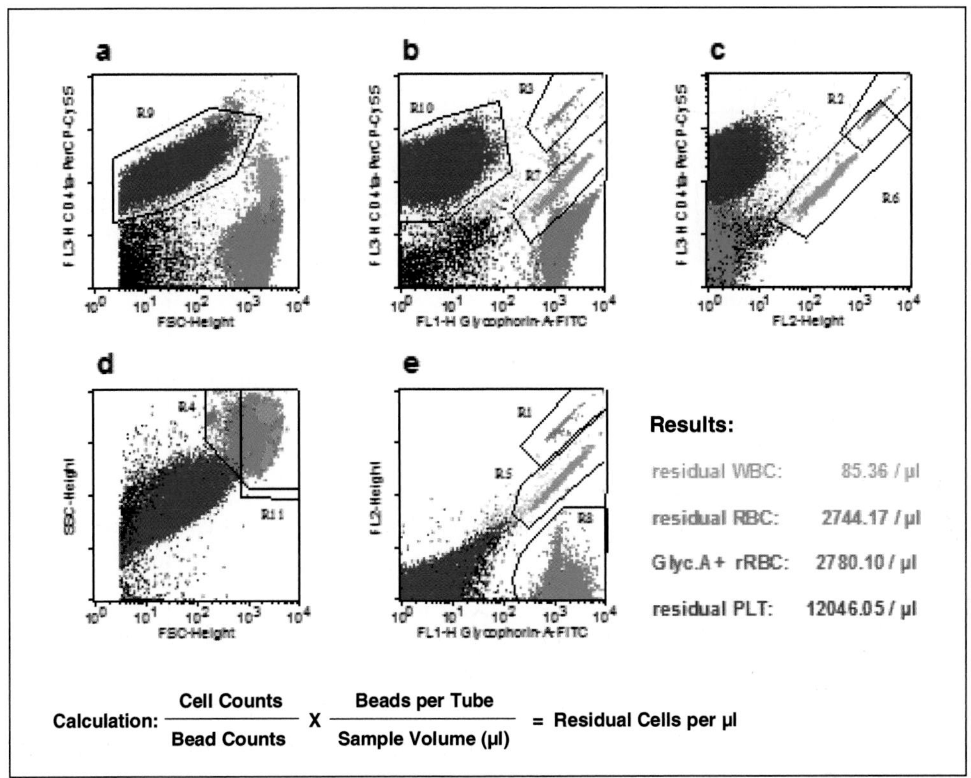

Fig. 5. Analysis protocol of the Plasma Count Kit. Regions R1–R3 contain the beads, regions R4–R7 and R11 the leukocytes, regions R4 and R8 the erythrocytes and R9–R10 the platelets. (Region R8 = 'Glyc. A + RBC' contains residual erythrocytes and glycophorin-A-positive microvesicles.)

BD Thrombo Count Kit

This test has been developed to enumerate residual WBCs and RBCs in platelet concentrates in combination with a TruCOUNT tube. WBCs are stained by a PI-containing buffer and RBCs are labeled with CD235a conjugated with PE, which emit strong fluorescence above 580 nm and may be detected in FL2. Therefore, a low fluorescence threshold set on FL2 (~ channel 360) reduces acquisition of unspecific dimly stained particles. The mode of analysis is shown in figure 6. For more details, consult the information provided by the manufacturer.

Auxiliary Protocols

Procedure for Counting Residual White Blood Cells in Leukocyte-Depleted Red Blood Cells Using a Nageotte Chamber (According to the Regulations of the American Association of Blood Banks)

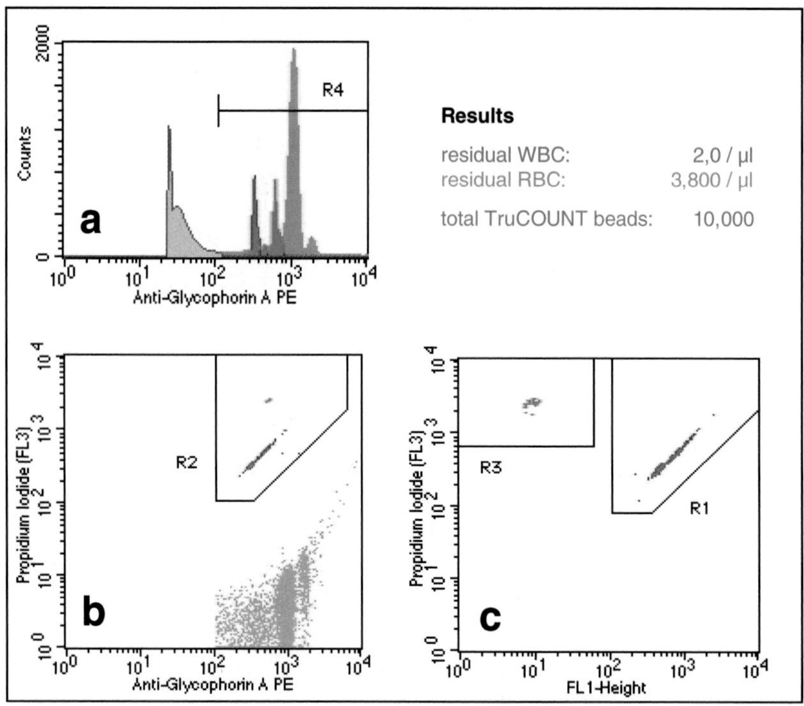

Fig. 6. Analysis protocol of the Thrombo Count Kit. Region 4 includes beads, WBCs and RBCs but not debris. Region R2 is set around beads and WBC. Region R1 calculates only the beads and region R3 only the WBCs. RBCs are calculated by gate definition excluding beads and WBCs. **a** Overlay histogram plot, all events are displayed. **b** Only 10% of events are shown. **c** All events are displayed.

Principle

The residual WBCs must be counted in a large-volume microscopic counting chamber such as the Nageotte chamber. This chamber has more than the 50-fold sample volume of a usual hemocytometer. Samples are prepared by RBC lysis and staining of leukocyte nuclei with crystal violet. Before microscopic inspection of the Nageotte chamber, sufficient sedimentation of WBCs in a humidified atmosphere is important. Counting accuracy is increased when evaluating a larger volume and a less diluted sample.

Material

- Hemocytometer with 50 µl of sample volume (Nageotte chamber, Karl Hecht KG, Sondheim, Germany)
- Crystal violet staining solution (0.01% w/v crystal violet in 1% v/v acetic acid; Türk's solution)
- Erythrocyte-lysing solution (e.g. Zapoglobin®, Beckman Coulter)
- Pipette and tips, 40 µl
- Powder-free gloves, dust free test tubes (perhaps with caps)

- Large Petri dishes and filter paper, to be used as a moist chamber to prevent evaporation;
- Light microscope with a 10× eyepiece and 20× objective.

Procedure

- The blood sample to be tested has a hematocrit ≤ 60%;
- add 40 μl of lysing reagent to a test tube;
- add 100 μl of the blood sample to this tube; vigorously mix with a pipette until all RBCs are lysed;
- add 360 μl crystal violet staining solution and mix well (now the blood is diluted 1:5);
- place a cover slide on a Nageotte chamber; transfer the stained sample to the chamber;
- place the hemocytometer into a humidified chamber (like a Petri dish); close this chamber, and allow the cells to settle undisturbed for approximately 10–15 min;
- count 50 μl of stained sample in the Nageotte chamber by light microscopy;
- the cell concentration in the blood sample (cells/μl) = counted cells in 50 μl/10 (50 μl of the 1:5 diluted sample were counted, corresponding to 10 μl of neat sample).

Remarks

- This procedure allows sufficiently precise counting without additional centrifugation down to 1 leukocyte/μl;
- The reliable recognition of crystal violet stained leukocytes must be trained with undiluted and serially diluted samples. Fluorescent DNA stains like SYTO-16 [13] are a useful alternative to crystal violet;
- for statistical evaluation, see comments on evaluation of flow-cytometric analysis.

Comment

In order to precisely and reproducibly count residual leukocytes by flow cytometry, the following problems need to be solved:

(1) Improvement of the signal-to-noise ratio by
   - using fluorescent dyes with stronger light intensity, which bind to nucleic acid (instead of labeling with CD45);
   - using stronger permeabilizing solutions and RNAse; to stain only the DNA of the leukocytes, all leukocytes should appear only in a narrow region; the RNA of the reticulocytes and platelets should be removed;
   - applying of suitable gating strategies (multiple gating); the background (RBCs and platelets) should remain beyond the analysis region.

(2) Analysis of larger volumes, so that enough leukocytes can be counted:
   - problem of sedimentation of cells and of counting particles during analysis;
   - concentration of the samples by lysis of RBCs and centrifugation can lead to cell loss.

(3) The statistics of rare events is to be used: the accuracy of test results in the lower test area increases with the number N of counted cells (Poisson distribution). The variation coefficient (CV) is calculated by:

$$CV(\%) = (1/\sqrt{N}) \times 100\% \tag{3}$$

Therefore the analyzed sample volume is correlated with the lower limit of detection. If the CV remains below 20%, then about 30 cells must be counted. This corresponds to a test volume of 9 µl if a sample contains 3.3 cells/µl. Further arithmetic examples are shown in the literature [14]. Acquiring larger volumes or a larger number of counted cells will decrease the CV.

Newer technical developments and the comparison of different cell separation technologies require improved residual cell counting. The aim is to detect leukocyte concentrations which lie 100 to 1,000-fold below the actually accepted threshold of 3.3 leukocytes/µl. At present, this can only be achieved with lysis of RBCs and reduction of larger volumes by centrifugation [15]. Another option would be a cytometer with a substantially increased acquisition volume per time unit [11].

Quality Control

The validation of a counting test should be performed with a series of calibrated leukocyte dilutions within the relevant analytic range of the assay. The following criteria must be checked: linearity and lower limit of detection, precision, accuracy, portability (comparability between different laboratories). A more detailed description is provided in the literature [14, 16].

Linearity and Lower Limit of Detection

Principle

At least 5 different leukocyte concentrations, which exceed the upper and the lower limit of the relevant test range, are produced through a stepwise dilution of whole blood and then tested.

Test range: 1–50 cells/µl, including the range of less than 3.3 leukocytes/µl (this corresponds to less than 1×10^6 WBCs per unit of RBCs).

Table 4. Dilution series

Dilution	Leukocytes/μl	
	calculated	counted
1:10	500	
1:100	50	53
1:200	25	23
1:400	12.5	11
1:800	6.25	7.5
1:1,600	3.1	2.5
1:3,200	1.6	1.8
1:6,400	0.8	1.4

Preparation of Several Dilutions of Leukocytes and Statistical Evaluation

- Prepare an extremely leukocyte-depleted sample of RBCs by two-fold filtration and use this as a diluent with a concentration of RBCs similar to the samples to be tested;
- dilute a blood sample with about 5,000 WBCs/μl using this diluent, and prepare a series of calibrated specimens, as shown in table 4;
- count these samples with a flow cytometer;
- calculate the linear correlation coefficient R (fig. 7), the target goal is $R^2 > 0.98$;
- perform multiple counting of the leukocyte-free diluent and a diluted sample with a WBC concentration near the theoretical lower limit of detection; calculate the 95% confidence interval for both series; the overlap of both confidence intervals can be used as the lower limit of detection [14, 16, 17].

Precision and Reproducibility

Assay precision or reproducibility is determined by repeated measurement of the same sample. Two diluted samples with 3.1 WBCs/μl and with 25 WBCs/μl are selected from the above-mentioned series of dilutions. They are aliquoted to 10 tubes for each concentration, processed and analyzed. Alternatively, commercially available control cells for residual cell counting can also be used. From these data, the mean, standard deviation (SD) and coefficient of variation (CV) are calculated.

Aim: Observed CV < expected CV ± 10%; expected CV = 20% ± 2%

Under the present conditions (Poisson statistics, less than 1×10^6 leukocytes per unit of 300 ml), an expected CV of 20% is a realistic value [17].

Examples

10 individual measurements of the sample with theoretically 3.1 leukocytes/μl: 2.8; 3.8; 3.3; 2.7; 3.2; 3.1; 2.6; 3.3; 2.6; 2.5; mean = 2.99; SD = 0.42; CV = 14%;

10 individual measurements of the sample with theoretically 25 leukocytes/μl: 23.2; 24.1; 26.5; 27.0; 25.2; 22.2; 22.9; 28.3; 24.5; 22.8; mean = 24.7; SD = 2.0; CV = 8.3%.

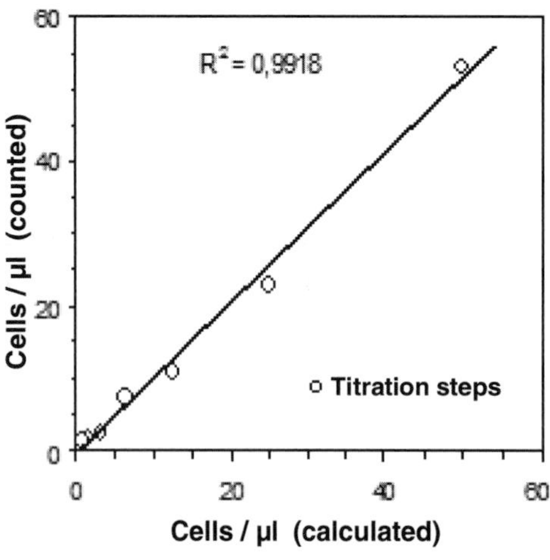

Fig. 7. Graph of a series of dilutions from table 2. The degree of linearity of the assay can be determined by calculating a regression line and the correlation coefficient as a measure of linearity.

Accuracy of Residual Cell Counting

Accuracy can be described as the percentage of data located in an expected range ±X%. The theoretically calculated cell concentration as a result of titration can be used as the expected test value, or the cell count of a commercially available control blood sample for residual cell counting can be used.

Aim: 80% of all test results are in the range of the expected value ±20% [16].

Examples

With the above-mentioned analysis of the sample repeated 10 times with 3.1 leukocytes/µl, 90% of the data fall within the target area of 3.1 ± 20% = 3.1 ± 0.6 cells = 2.5–3.7 leukocytes/µl.

With the above-mentioned analysis of the sample repeated 10 times with 25 leukocytes/µl, 100% of the data fall within the target area of 25 ± 20% = 25 ± 5 cells = 20–30 leukocytes/µl.

The loss of cells through sample processing, cell aggregates or the destruction of fragile leukocytes by the reagents used to lyse background RBCs or to make leukocytes permeable to stain can all cause an inherent bias which consistently underreports the true value (underestimation bias). The tendency to consistently overreport the true value (overestimation bias) may be a result of too high a background.

Comparability of the Results between Different Laboratories (Portability)

The ability to perform an assay at different test sites with equivalent results is an important feature of such a test and requires multicenter validation studies. To meet these requirements, the following alternatives should be considered:

- different laboratories are given identical samples from a certified institution and report their results to this institution for comparison and evaluation (not yet offered in many countries);
- commercially available and certified control blood samples with a known leukocyte count are processed and analyzed repeatedly;
- identical samples are analyzed by flow cytometry as well as with a Nageotte chamber. The results are compared and documented (comparison with the 'gold standard').

Troubleshooting

The precision of the assay can be improved by different measures:

- Run a cleansing program (e.g., 5 min FACS Clean and 5 min buffer) immediately before acquisition of the blood sample on the cytometer.
- Perform 'reverse pipetting' for optimal precision and reproducibility: aspirate a selected volume plus an excess of blood or bead suspension by pressing the control button of a manual pipette slightly past the first stop when aspirating. Dispense the liquid by pressing the control button gently and steadily to the first stop. The delivery should be done against the wall of the receiving tube. A small volume will remain in the tip and should be discarded prior to removal of the tip.
- The pipette tip is wiped clean without touching the opening to remove adhering liquid, cells or beads at the outer side of the tip.
- Cell loss during processing and counting can lead to an underestimation of the actual leukocyte content, e.g. because cells get stuck on the tube wall or are measured as doublets by the cytometer and are not recognized as cell aggregates.
- Analysis within 24 h following blood withdrawal provides results similar to fresh blood while counting 2–3 days after preparation of the blood unit provides less accurate results.
- The acquired sample volume can be increased from 10 to 20 μl (double acquisition time).
- The test tubes should be dust free (use tubes with caps, e.g., Falcon, cat. no. 352054). Avoid the use of powdered gloves.
- If serial dilutions of leukocytes for linearity measurements are prepared, RBCs free of residual leukocytes are to be used as diluent instead of saline buffer in order to assess the influence of the background correctly. For this purpose, a twofold filtered RBC sample assumed to be free of residual leukocytes can be used.
- The danger of a carry-over of cells from a tube to the following can be decreased if a tube with a buffer is measured between two samples.
- Stabilized blood samples such as control blood cannot be used to set the amplifiers and the regions: they show clearly altered signals.

Expected Results

If leukocyte depletion and residual WBC counting are carried out by well-trained staff, test results are expected to be below the permissible limits (table 1). Without additional volume reduction, the residual leukocyte count of most of the blood units should be below the limit of detection.

Required Time

The preparation of several tests, including controls, writing of the work list, setup of the cytometer, requires at least 30 min; running a sample takes approximately 2 min (as a function of the counted volume).

Summary

The efforts to make blood transfusion safer have led to the development of filters and cell separators for the removal of undesirable cell fractions, in particular for the depletion of leukocytes. This has been accompanied by the development of precise and reproducible methods for cell counting and by the development of legal rules for the safety of blood products. Modern separation techniques achieve a residual leukocyte count of far below 1×10^6 leukocytes per blood unit. Nevertheless, a continuous process control must be performed including the counting of residual leukocytes. For the reliable counting of residual cells, flow-cytometric methods and commercial IVD test kits are increasingly used. They replace the manual and laborious counting by means of a Nageotte chamber and simultaneously produce more precise data.

Acknowledgement

The author wishes to thank Dr. Spengler (BD Biosciences) and Dr. Kaymer (Beckman Coulter) for valuable comments and helpful discussions.

References

1 Dzik S, Blajchman MA, Blumberg N, Kirkley SA, Heal JM, Wood K: Current research on the immunomodulatory effect of allogeneic blood transfusion. Vox Sang 1996;70:187–194.

2 Blajchman MA, Dzik S, Vamvakas EC, Sweeney J, Snyder EL: Clinical and molecular basis of transfusion-induced immunomodulation: summary of the proceedings of a state-of-the-art conference. Transfus Med Rev 2001;15:108–135.

3 Guide to the Preparation, Use and Quality Assurance of Blood Components, ed 9. Strasbourg, Council of Europe Publishing, 2003.

4 ‚Richtlinien zur Gewinnung von Blut und Blutbestandteilen und zur Anwendung von Blutprodukten (Hämotherapie)', rev 2000. Bundesgesundheitsblatt 2000;7:555–589.

5 Lutz P, Dzik WH: Large-volume hemocytometer chamber for accurate counting of white cells (WBC) in WBC-reduced platelets: Validation and application for quality control of WBC-reduced platelets prepared by apheresis and filtration. Transfusion 1993;33: 409–412.

6 Andea A, Garritsen HSP, Cassens U, Kelsch R, Sibrowski W: A flow-cytometric method for assessing leukocyte contamination in white-cell-depleted blood products. Infusionsther Transfusionsmed 1999;26:53–56.

7 Dzik S, Moroff G, Dumont L for the Biomedical Excellence for Safer Transfusion (BEST) Working Party of the ISBT: A multicenter study evaluating three methods for counting residual WBCs in WBC-reduced blood components: nageotte hemocytometry, flow cytometry, and microfluorometry. Transfusion 2000;40:513–520.

8 Jilma-Stohlawetz P, Marsik C, Horvath M, Siegmeth H, Höcker P, Jilma B: A new flow cytometric method for simultaneous measurement of residual platelets and RBCs in plasma: validation and application for QC. Transfusion 2001;41:87–92.

9 Krailadsiri P, Seghatchian J: Residual red cell and platelet content in WBC-reduced plasma measured by a novel flow cytometric method. Transfus Apheresis Sci 2001;24:279–286.

10 Pichler J, Printz D, Scharner D, Trbojevic D, Siekmann J, Fritsch G: Improved flow cytometric method to enumerate residual cells: minimal linear detection limits for platelets, erythrocytes and leukocytes. Cytometry 2002;50:231–237.

11 Dumont LJ, Dumont DF: Enhanced flow cytometric method for counting very low numbers of white cells in platelet products. Cytometry 1996;26:311–316.

12 Lambrecht B, Spengler HP, Bauerfeind U, Mohr H: Simultaneous quantitation of contaminating leukocytes, erythrocytes and platelets in fresh frozen plasma in a single tube assay by flow cytometry (abstract). Infus Ther Transfus Med 2001;29(suppl 1):52–53.

13 Roehrig O, Endres W, Sugg U: Improvement of Nageotte chamber counting by the novel cell-permeant DNA-binding fluorescent dye Syto-16. (abstract P7.22). Infus Ther Transfus Med 2002;29(suppl 1):1–70.

14 Dzik S: Counting low numbers of leukocytes in leukoreduced blood components. Infusionsther Transfusionsmed 1999;26:62–65.

15 Szuflad P, Dzik WH: A general method for concentrating blood samples in preparation for counting very low numbers of white cells. Transfusion 1997;37:277–283.

16 Dzik WH: Leukocyte counting during process control of leukoreduced blood components. Vox Sang 2000;78(suppl 2):223–226.

Sack U, Tárnok A, Rothe G (eds): Cellular Diagnostics. Basics, Methods and Clinical Applications of Flow Cytometry. Basel, Karger, 2009, pp 714–723

Enrichment of Disseminated Tumor Cells

Jozsef Bocsi[a] · Bela Molnar[b] · Anja Mittag[c] · Dominik Lenz[a]

[a] Pediatric Cardiology, Heart Center, University of Leipzig, Germany
[b] Internal Medicine, Semmelweis University, Budapest, Hungary
[c] Translational Center for Regenerative Medicine, University of Leipzig, Germany

Introduction

Malignant primary tumors release tumor cells already at early stages. When these cells reach the circulation, they can form metastases locally but also at a distance from the primary tumor. This can affect the patient's prognosis.

State-of-the-art imaging systems are able to detect tumors already at early stages. However, if the tumors are small, the classical prognostic parameters, e.g. tumor size and the staging of sentinel lymph nodes, are not adequate for present-day diagnostic requirements and therapeutic possibilities. The latest results have shown that the prognostic grading of some tumors by determination of micrometastases can contribute to the improvement of therapy [1–3].

Certain characteristics of disseminated tumor cells (DTCs) enable their detection and enrichment. Cytokeratin 19 and 20 [4], for instance, are specific markers for colon carcinoma; melanoma can be detected by tyrosinase [5], and prostate carcinoma by prostate-specific antigen (PSA) [6]. Blood, bone marrow cells or cells of a close lymph node are suitable for analysis as well. Isolated DTCs can even be quantified by flow cytometry [7]. Once the target tumor cells are separated from healthy cells, their specific gene expression can be investigated by the polymerase chain reaction (PCR) or reverse transcription (RT) PCR [8].

Protocols for Disseminated Tumor Cell Enrichment

Two types of cell isolation may be distinguished: positive and negative selection. In positive selection, the target population is specifically labeled by antibodies and separated from other cells. Negative selection (or *depletion*) consists in labeling all unwanted cells (the majority of the cells) in the sample with antibodies. In the fol-

Table 1. Comparison of positive and negative selection

	Positive selection	Negative selection (depletion)
Viability	95–99%	99%
Typical rate of yield	60–99%	99%
Antibody or bead consumption	low	increased
Affected cell physiology due to antibody contact	yes	no
Flow-cytometric analysis of the target cell population	specific antigen epitope is covered by the magnetic antibody that was used for bead separation	no disturbance
Cell culture	possibly affected physiology	no disturbance

lowing preparation step, the labeled cells are removed, leaving only the target cells. Table 1 shows a short comparison of both approaches.

Both selection strategies concentrate the target cell population, which can be analyzed, e.g., by immunocytochemistry , cytometry or PCR. Basically, there are different methods for positive and negative selection. In this chapter, some typical examples will be described.

Positive Selection by Dynabeads

Principle

One method uses superparamagnetic particles of defined size (4.5 ± 0.2 µm) to concentrate tumor cells. These particles, called 'magnetobeads' or 'Dynabeads'(Dynal Biotech, Oslo, Norway), move to the magnet in a strong magnetic field. The beads are conjugated with a tumor-specific antibody; they thus bind specifically to tumor cells only. After removing the magnetic beads from cells, cells can be resuspended. Thereby, it is possible to concentrate tumor cells from the larger normal cell population.

Material und Reagents

- Biological sample (anticoagulated blood, cell suspension from blood, bone marrow or lymph nodes)
- Dynal magnetic particle concentrator (MPC-6) for six 10-ml tubes at most
- Dynal magnetic particle concentrator (MPC-M) for twelve Eppendorf tubes (1.5 ml) at most
- 10-ml glass tubes or not cone-shaped polystyrene tubes for Dynal MPC-6
- 1.5-ml Eppendorf tubes
- EDTA Monovettes (Sarstedt, Nümbrecht, Germany) for anticoagulation of blood

- Dynabeads with tumor-specific antibodies
- Rotary suspension mixer, rotation 15–20 min^{-1} (e.g. MX-2, Dynal Biotech; Stuart SB-2, Jencons Scientific Inc., Bridgeville, PA, USA)
- Washing buffer (PBS with 1% fetal calf serum and 0.6% sodium citrate)
- 50-µm filter (e.g. Partec CellTrics®, Partec Münster, Germany), pipette, tips
- Eppendorf tubes.

Preparative Steps

Washing of Dynabeads

1) Mix the original bead vial well.
2) Add the calculated quantity of beads (e.g. 100 µl) into an Eppendorf tube.
3) Pipette 1–2 ml washing buffer into the tube.
4) Place the tube into the MPC-M for 1 min.
5) Remove the supernatant.
6) Recover the original bead concentration from step 2 (e.g. 100 µl) with washing buffer.

Isolation of Tumor Cells

1) Cool 5 ml anticoagulated blood to 4–8 °C. As examination material bone marrow or lymph node cell suspensions can be used as well. Note: lymph nodes should be finely cut with scissors or a scalpel in 0.5% BSA-PBS. Suspend the cells and filter through a 50-µm mesh. After counting the cell number, the suspension is ready for magnetic separation.
2) Add 5×10^7 washed Dynabeads to the sample.
3) Incubate the suspension on the rotator at low speed (15–20 rpm) for 30 min at 2–8 °C. During incubation, the beads bind (for example BerEP4 for tumor cells of epithelial origin) to the cell surface via antigen to antibody contact. The tumor cells with the beads are present in rosettes (fig. 1).
4) Place the tubes into the magnetic particle concentrator (MPC-6). In the magnetic field, the tumor cells bound to beads move to the wall of the tubes. Non-bound cells remain in the suspension and can later easily be separated by removing the fluid.
5) Carefully remove the supernatant; the beads are attached to the tube wall.
6) Add 800 µl cold washing buffer and remove the tube from the MPC-6 to mix the bead-cell suspension. Resuspend the beads outside of the magnetic field. Transfer the suspension into an Eppendorf tube.
7) Place the tubes into an MPC-M. Wait for 2–3 min. Carefully aspirate the supernatant; the beads are attached to the tube wall.
8) Add 800 µl cold washing buffer to the tube and remove it from the MPC-M to mix the bead-cell suspension. Beads can be resuspended at a distance from the magnetic field.
9) Place the tubes into an MPC-M. Wait for 2–3 min. Aspirate the supernatant carefully; the beads are attached to the tube wall.

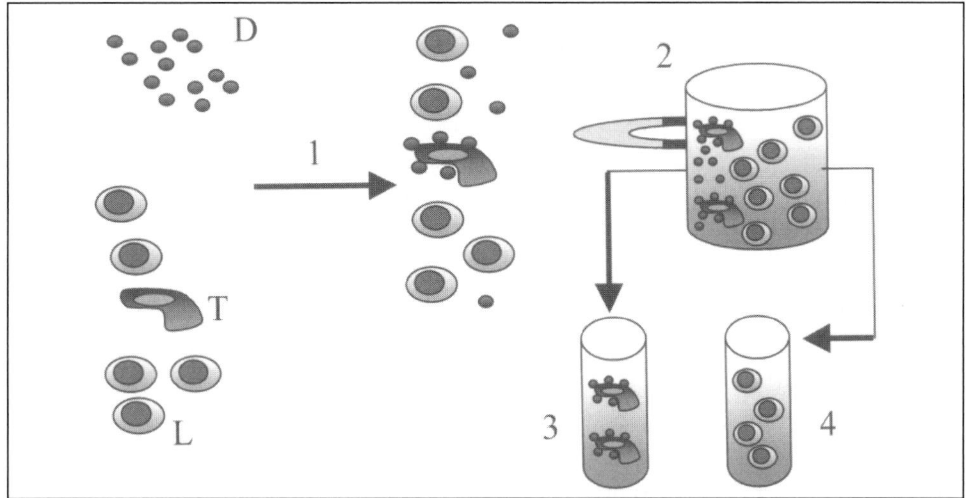

Fig. 1. Scheme of positive selection with magnetic beads.
D = Beads linked with antibodies against tumor marker; T = tumor cells, L = leukocytes.
1 = Incubation; 2 = magnetic field activation; 3 = positively isolated cells, 4 = supernatant.

10) Repeat washing steps 8–9 twice. The supernatant will be cleared of erythrocytes during the washing step.
11) After adding 800 μl cold washing buffer, the suspension should be transferred into a new tube. This suspension should be kept on ice until the following steps (for example: mRNA isolation and RT-PCR).

Positive Selection using Microbead-Labeled Antibodies

Principle
The tumor-specific antibodies are conjugated with magnetic microbeads (Miltenyi Biotech, Bergisch Gladbach, Germany). Tumor cells can be purified by magnetic separation (MACS – magnetic-activated cell sorting). The incidence of DTCs in the sample is increased by this method as well.

Material und Reagents
- Biological sample (anticoagulated blood, cell suspension from bone marrow or lymph nodes)
- EDTA-PBS (PBS with 2 mmol/l EDTA)
- Ficoll-Paque ($\varrho = 1.077$ g/ml)
- Cone-shaped tubes
- Centrifuge
- Pipette, tips
- PBS with 0.5% BSA

- FcR-blocking reagent
- MACS with selected tumor-specific antibody
- MiniMACS Separator, MACS columns
- Filter (with 30-μm pores)
- PBS.

Preparative Steps

1) Dilute blood sample 2–4 times with EDTA-PBS.
2) Add 15 ml Ficoll-Paque into each conical 50-ml tube.
3) Overlay with at most 35 ml of the diluted blood sample. Avoid mixing blood and Ficoll-Paque.
4) Centrifuge for 30–40 min at 400 × *g*.
5) Remove the upper cell-free layer.
6) Aspirate the cells and transfer them into a new conical tube (50 ml).
7) Fill the tube with EDTA-PBS and resuspend the pellet.
8) Centrifuge for 10 min at 300 × *g*, decant.
9) Repeat steps 7 and 8 twice.
10) Resuspend the cells in 300 μl PBS (0.5% BSA). BSA reduces nonspecific binding of antibodies to the cell surface. Instead of BSA, gelatine or FCS can be used as well.
11) For 5×10^7 cells (approximate leukocyte cell count in 10 ml blood) add 100 μl FcR-blocking reagent. Without blocking reagent, the proportion of nonspecific binding is increased.
12) For 5×10^7 cells add 100 μl MACS antibody (e.g. HEA-125 in case of tumor cells of epithelial origin). Vortex.
13) 30 min incubation at 6–12 °C. Longer incubation decreases the efficiency due to increased nonspecific binding. Dead cells bind antibodies nonspecifically. If the sample contains too many dead cells 'MACS dead-cell removal' should be applied.
14) Add 5–10 ml PBS to the sample.
15) Centrifuge (300 × *g*, 10 min) and remove the supernatant.
16) Place the MS+/RS+ columns into the separator. Add 500 μl PBS to one column. One column is usually enough for the cell amount in 50 ml blood. If the number of analyzed cells is higher than 2×10^8 or the number of bound cells is higher than 10^7, a bigger column is needed. Gas bubbles in the sample decrease the capacity of the column and the flow rate decelerates.
17) Wash the cell suspension in the column with PBS using a filter (30-μm pore size). During repeated centrifugation cell clumps can form, which later clog up the column. A dry filter could catch part of this small volume.
18) Washing steps (4 × 500 μl PBS) Nonbound cells flow through the column; positive cells will be collected by the magnetic field of the MACS Separator.
19) Take the column from the MACS Separator, add 1 ml buffer and transfer the cells into a new tube.

20) Repeat isolation steps 16–19 with a new column.
21) After this magnetic enrichment, the suspension should contain most of the antigen-expressing cells and a small part of negative cells from the blood sample. The suspension is now ready for further preparative steps (immunohistochemistry, RT-PCR or flow cytometry).

Negative Selection by Tetramer Antibody

Principle
In negative selection, the target cells are not labeled, but all undesirable cells are removed from the sample material. The physiological features of tumor cells will not be affected by the antibodies (see 'Comments', pp. 720). A tetramer-antibody complex (Celltech, Slough, UK) containing two subunits (two monoclonal antibodies and two binding units; fig 2) is used here. This complex has two different binding capacities: one against erythrocytes (anti-glycophorin-A antigen specificity) and another against leukocyte (CD45, CD36, CD66b antigen specificity). In the case of blood, the separation does not need application of magnetic tools.

Material und Reagents
- Biological sample (anticoagulated blood)
- EDTA tubes (Sarstedt)
- Cone-shaped polypropylene tubes (50 ml)
- Polystyrene tubes (5 ml)
- Tetramer antibody
- Ficoll-Paque ($\varrho = 1.077$ g/ml)
- Centrifuge
- Washing buffer
- Pipette, tips.

Preparative Steps
1) Cool down 5 ml of anticoagulated blood to 4–8 °C.
2) Add tetramer antibodies to blood sample.
3) 20 min of incubation at 4–8 °C; during this time, the antibodies bind specifically to erythrocytes and leukocytes. The leukocytes form rosettes. Tumor cells do not react with the tetramer antibodies (fig. 2).
4) Pipette the sample on the Ficoll solution.
5) Centrifuge the tubes (800 rpm for 15 min). In this step, the erythrocytes and leukocyte rosettes settle at the bottom of the tube. Tumor cells and other non-bound cells remain between the Ficoll solution and serum.
6) Remove the cells with an automatic pipette and transfer them into an Eppendorf tube.

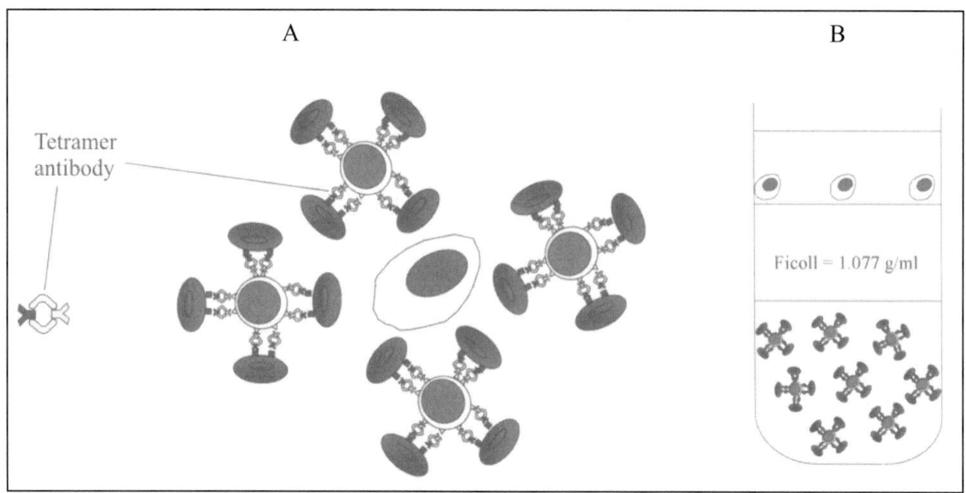

Fig. 2. Scheme of negative selection (depletion) with tetramer antibodies. **A** Antibodies are directed against erythrocytes and leukocytes. Cells and tetramers form rosettes. Tumor cells are not bound by antibodies. **B** Density gradient centrifugation (Ficoll) separates these single unbound cells.

7) Add 800 μl of cold washing buffer and resuspend the cells gently, centrifuge (300 × *g*, 10 min).
8) Remove the supernatant carefully with a pipette. The cells should remain at the bottom of the tube.
9) Repeat three times washing steps 7–8. The supernatant will be cleared from Ficoll during the washing.
10) Add 800 μl of washing buffer into the tube and transfer the suspension into a new tube. From this step on, the suspension should be stored in an ice bath for further preparative and analytic steps (cell culture, cytometry or isolation of mRNA, RT-PCR).

Comments

Background

Malignant primary tumors release cells into the circulation already at early stages. Thus metastases can be formed. DTCs can spread potentially in all directions via the bloodstream or the lymph.

Tumor cells can be detected by immunocytochemistry, PCR, RT-PCRand antibody-based enrichment in conjunction with flow cytometry or image cytometry (see 'Technical and Methodological Basics of Slide-Based Cytometry', pp. 89). Samples containing tumor cells can be taken from various sources. Bone marrow biopsies or samples from lymph nodes that supply the tumorous tissue are reported to be the best source of DTCs [9, 10]. Blood is an essential source as well, but a smaller number of tumor cells is expected in peripheral blood samples. However, advantages are evident: blood can be taken relatively often and examined frequently for the patient's follow-up. The detection of tumor cells in peripheral blood is diagnostically and prognostically more efficient than detection in other tissues.

Immunocytochemistry demonstrates the pathomorphologic criteria of tumor cells. However, this method is very laborious for routine investigations. The combination of immunocytochemistry and enrichment methods is the optimal way to discriminate tumor cells from healthy cells. As RT-PCR enables the use of more tissue- or tumor-specific markers, the detection limit of this system is relatively low (1 cell/ml). However, there is no morphological documentation and evaluation. But, due to the high sensitivity of this method, false-positive artifacts should be ruled out carefully to avoid wrong diagnoses.

In magnetic enrichment, tumor-cell-specific antibodies conjugated with magnetic particles for the purification of DTCs are used. However, the detection of DTCs with these methods is also possible using negative selection, i.e. removal of CD45+ (pan-leukocyte antigen) cells from the blood sample. The proportion of DTCs per volume will be significantly increased by separation. The technique of immune cytomagnetic enrichment for the separation of leukocyte subpopulations was further developed during the last decades. Basically, there are two different principles:

- antibody-coated supermagnetic particles (2–5 μm diameter) and
- supermagnetic microbeads (approximately 50 nm in size) bound to antibodies.

The results from first attempts at immunomagnetic separation by flow cytometry were promising. On the one hand, the lowest detection limit of these methods is 1 tumor cell per 5 ml of blood, which is lower than that of RT-PCR. In addition, with immunomagnetic enrichment further characterization of purified cells is possible. On the other hand, morphological analysis and detection of cell clusters [11] by flow-cytometric methods are not possible from a technical point of view. In clinical research, conflicting results have been obtained, but several studies suggest the vast potential of DTC detection in cancer prognosis and follow-up [12].

Advantages of the Methods

Positive Selection

Positive selection yields more homogeneous isolated cell populations than negative selection, i.e. the portion of non-target cells in the prepared suspension is very low (approximately 1%). Changes in the physiological cellular processes may be a possible drawback.

Positive selection by microbead-labeled antibodies , as shown in this chapter, can be performed automatically. Advantages of the automatic determination are: easier preparation and faster execution with strongly increased cell numbers, as well.

Table 2. Comparison of different selection methods

	Positive selection		Negative selection
	macrobead technique	microbead technique	
Visual control of preparation	yes	limited	limited
Use of columns	no	yes	no
Specimen			
Peripheral blood	yes	yes	yes rosette and magnetic
Bone marrow	yes	yes	yes magnetic
Lymph node	yes	yes	yes magnetic
Other		automation	

Negative Selection

In negative selection (e.g. with anti-CD45 antibody) the nonbound cells are inhomogeneous. But with this method, DTC separation is also possible in the absence of specific tumor cell markers. The advantage of this technique is that the physiology of target tumor cells is not changed by antigen-antibody contact (table 2).

Expected Results

The antibody-mediated separation of DTCs from the peripheral blood circulation is the most effective method after malignant primary tumor removal to quantify the probability of the existence of metastases before their manifestation and the possibility of a clinical diagnosis. Other advantages (easier to handle compared to other very laborious experiments and the specimens that can be repeatedly obtained) plead for this method as well.

The presented methods enable the concentration of DTCs in a biological sample e.g. blood (1 cell/ml) or bone marrow (1 cell in 100,000–1,000,000 bone marrow cells) as well as preprocessing for cytometric analysis (FACS, and laser scanning cytometry (LSC) slide-based cytometry (SBC) (see 'Technical and Methodological Basics of Slide-Based Cytometry', pp. 89) and morphologic examination.

Time Required

Protocol 2.1: Incubation of cells with beads takes 30 min. The following washing steps need a further 10 min.

Protocol 2.2: Ficoll separation takes approximately 40 min, the subsequent washing, 30 min and incubation with the MACS antibodies again 30 min. A further 40 min are required for the next repeated washing and separation steps.

Protocol 2.3: Incubation with antibodies takes 20 min; the Ficoll centrifugation takes 15 min, followed by 30 min washing.

Summary

Primary malignant tumors release tumor cells into the circulation already at an early stage. This may lead to metastases. Although these cells are present only at a very low frequency in peripheral blood (~1 cell per 1 ml blood), methods are available for the detection of such cells before the first metastasis is detected. Techniques such as immunocytochemistry and PCR or RT-PCR are sensitive enough to detect 1 tumor cell in 100,000–1,000,000 bone marrow cells or leukocytes.

For cytometric identification, an increased relative incidence of tumor cells in a sample volume is required. Thus preliminary separation of tumor cells is recommended, e.g. with immunomagnetic enrichment. Using these methods, the sample volume is reduced, but the loss of tumor cells is low.

In this chapter, several methods for the enrichment of DTCs from peripheral blood, bone marrow or lymph node were described and discussed.

References

1 Pantel K, Woelfle U: Micrometastasis in breast cancer and other solid tumors. J Biol Regul Homeost Agents 2004;18:120–125.

2 Müller V, Pantel K: Bone marrow micrometastases and circulating tumor cells: current aspects and future perspectives. Breast Cancer Res 2004;6:258–261.

3 Pantel K, Woelfle U: Detection and molecular characterisation of disseminated tumor cells: implications for anticancer therapy. Biochim Biophys Acta 2005;1756:53–64.

4 Bustin SA, Gyselman VG, Williams NS, Dorudi S: Detection of cytokeratins 19/20 and guanylyl cyclase in peripheral blood of colorectal cancer patients. Br J Cancer 1999;79:1813–1820.

5 Smith B, Selby P, Southgate J, Pittmann K, Bradly C, Blair GE: Detection of melanoma cells in peripheral blood by means of reverse transcriptase and polymerase chain reaction. Lancet 1991;338:1227–1236.

6 Bretton PR, Melamed MR, Fair WR, Cote RJ: Detection of occult micrometastases in the bone marrow of patients with prostate carcinoma. Prostate 1994;25:108–122.

7 Racila E, Euhus D, Wiess A, Rao C, McConnell J, Terstappen L, Uhr JW: Detection and characterization of carcinoma cells in the blood. Proc Natl Acad Sci U S A 1998;95:4589–4594.

8 Gudemann CJ, Weitz J, Kienle P, Lacroix J, Wiesel MJ, Soder M, Benner A, Staehler G, Doeberitz MV: Detection of hematogenous micrometastasis in patients with transitional cell carcinoma. J Urol 2000;164:532–536.

9 Kraeft SK, Sutherland R, Gravelin L, Hu GH, Ferland LH, Richardson P, Elias A, Chen LB: Detection and analysis of cancer cells in blood and bone marrow using a rare event imaging system. Clin Cancer Res 2000;6:434–442.

10 Kruger W, Datta C, Badbaran A, Togel F, Gutensohn K, Carrero I, Kröger N, Jänicke F, Zander AR: Immunomagnetic tumor cell selection-implications for the detection of disseminated cancer cells. Transfusion 2000;40:1489–1493.

11 Molnar B, Ladanyi A, Tanko L, Sreter L, Tulassay Z: Circulating tumor cell clusters in the peripheral blood of colorectal cancer patients. Clin Cancer Res 2001;7:4080–4085.

12 Paterlini-Brechot P, Benali NL: Circulating tumor cells (CTC) detection: clinical impact and future directions. Cancer Lett 2007;253:180–204.

Subject Index